OCEAN

NEW EDITION

American Museum of Natural History

OCEAN

THE DEFINITIVE VISUAL GUIDE

NEW EDITION

DK | Penguin Random House

THIRD EDITION
Senior Editor Miezan van Zyl
Managing Editor Angeles Gavira Guerrero
Managing Art Editor Michael Duffy
Production Controller Meskerem Berhane
Production Editor Jacqueline Street
Jacket Designers Surabhi Wadhwa-Gandhi, Juhi Sheth
Jacket Design Development Manager Sophia MTT
Art Director Karen Self
Associate Publishing Director Liz Wheeler
Publishing Director Jonathan Metcalf

Produced for DK by
Dynamo Limited
1 Cathedral Court, Southernhay East, Exeter, EX1 1AF

Publishing Manager Claire Lister
Senior Design Project Manager Judy Caley

FIRST EDITION
Senior Editors Peter Frances, Angeles Gavira Guerrero
Project Editor Rob Houston
Editors Rebecca Warren, Miezan van Zyl, Ruth O'Rourke, Amber Tokeley
Senior Art Editor Ina Stradins
Project Art Editors Peter Laws, Kenny Grant, Maxine Lea, Mark Lloyd
Designers Francis Wong, Matt Schofield, Steve Knowlden
Cartographers Roger Bullen, Paul Eames, David Roberts, Iowerth Watkins

SCHERMULY DESIGN COMPANY
Senior Editor Cathy Meeus
Art Editor Hugh Schermuly
Editors Gill Pitts, Paul Docherty
Designers Dave Ball, Lee Riches, Steve Woosnam-Savage
Cartographer Sally Geeve
Design Assistant Tom Callingham

Picture Researcher Louise Thomas
Illustrators Mick Posen (the Art Agency), John Woodcock, John Plumer, Barry Croucher (the Art Agency), Planetary Visions
Managing Editors Sarah Larter, Liz Wheeler
Managing Art Editor Philip Ormerod
Publishing Director Jonathan Metcalf
Art Director Bryn Walls

This American Edition, 2022
First American Edition, 2006
Published in the United States by DK Publishing
1450 Broadway, Suite 801, New York, NY 10018

Copyright © 2006, 2014, 2022 Dorling Kindersley Limited
DK, a Division of Penguin Random House LLC
22 23 24 25 26 10 9 8 7 6 5 4 3 2 1
001–326967–May/2022

A catalog record for this book
is available from the Library of Congress.
ISBN 978-0-7440-5170-4

DK books are available at special discounts when purchased in bulk for sales promotions, premiums, fund-raising, or educational use. For details, contact: DK Publishing Special Markets, 1450 Broadway, Suite 801, New York, NY 10018
SpecialSales@dk.com

Printed and bound in Dubai

For the curious
www.dk.com

MIX
Paper from responsible sources
FSC™ C018179

This book was made with Forest Stewardship Council ™ certified paper – one small step in DK's commitment to a sustainable future.

For more information go to www.dk.com/our-green-pledge

CONTENTS

OCEAN LIFE

ATLAS OF THE OCEANS

About this Book

THIS BOOK IS DIVIDED INTO four chapters. An overview of the physical and chemical features of the oceans is given in the INTRODUCTION, OCEAN ENVIRONMENTS looks at the main zones of the oceans, and OCEAN LIFE examines the life-forms that inhabit them. The ATLAS OF THE OCEANS contains detailed maps of the oceans. Most chapters are divided into smaller sections.

Introduction

This opening chapter is divided into four sections. In OCEAN WATER, the properties of water itself are examined. OCEAN GEOLOGY covers the materials of the ocean floor and the way that it changes over time. CIRCULATION AND CLIMATE is about the interaction between oceans and the atmosphere and the large-scale movement of water, while TIDES AND WAVES looks at movements and disturbances of water on a smaller scale.

◀ OCEAN WATER
This section covers the properties of the water molecule; the chemistry of seawater; and the way that attributes such as temperature, pressure, and light transmission change with depth in ocean water.

OCEAN GEOLOGY ▶
As well as describing the composition of the ocean floor, this section looks at the processes that shape it, tracing the origin of the oceans and their changing size and shape over geological time.

◀ CIRCULATION AND CLIMATE
This section describes the large-scale circulation of the oceans, both deep down and at the surface. It also looks at ocean climates and the many ways that the oceans and the atmosphere influence one another.

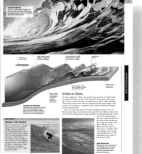

TIDES AND WAVES ▶
The regular movements of the tides are described here, as well as the way that disturbances spread out across the surface in the form of waves.

Ocean Environments

This chapter looks at specific parts of the oceans. It is divided into sections on different zones, starting with COASTS AND THE SEASHORE and then moving to progressively deeper waters, first with SHALLOW SEAS and then THE OPEN OCEAN AND OCEAN FLOOR. A final section, POLAR OCEANS, looks at the frozen waters around the North and South Poles. In each section, explanatory pages describe typical features and formative processes, while the succeeding pages contain profiles of actual features. The profiles are arranged by geographical location, starting with the Arctic Ocean and followed, in order, by the Atlantic, Pacific, Indian, and Southern Oceans.

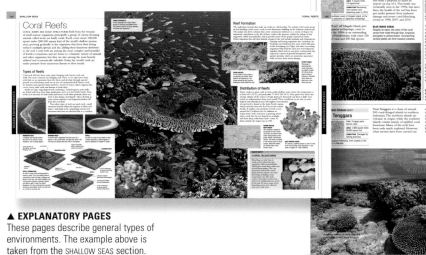

▲ EXPLANATORY PAGES
These pages describe general types of environments. The example above is taken from the SHALLOW SEAS section.

Ocean Life

This chapter contains two sections. The INTRODUCTION TO OCEAN LIFE covers the ecology and history of marine life and the way that marine organisms are classified. It is followed by a larger section, KINGDOMS OF OCEAN LIFE. This is divided into domains or kingdoms and, in the case of the plant and animal kingdoms, further divided into smaller groups. In each case, a general overview of the organisms that make up the group is followed by profiles of a selection of individual species. The section begins with the smallest forms of life, the bacteria and archaea, and ends with the animal kingdom.

DOMAIN	Eucarya
KINGDOM	Animalia
PHYLUM	Mollusca
CLASSES	8
SPECIES	73,683

color-coded panel shows position of group being described (indicated with white outline) in the classification hierarchy

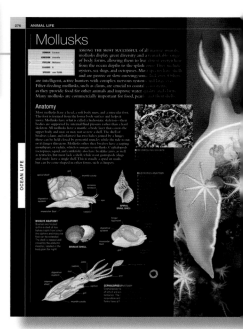

GROUP INTRODUCTION ▶
Pages such as the ones shown here describe groups of organisms in general. All introductions contain an account of the defining physical characteristics, usually followed by further information on behavior, habitats, and classification.

name of ocean in which feature is found

PACIFIC OCEAN SOUTHWEST

Great Barrier Reef

TYPE Barrier reef

AREA 14,300 square miles (37,000 square km)

CONDITION Damaged by Crown-of-thorns starfish; coral bleaching

LOCATION Parallel to Queensland coast, northeastern Australia

compass direction indicates position within ocean

map shows location and, in most profiles, geographical extent of feature

table of summary information (categories vary between sections)

all maps are accompanied by a written description of location

◀ **FEATURE PROFILES**
In most cases, explanatory pages are followed by profiles of actual features. For example, the profiles shown here describe coral reefs from around the world. Most profiles are illustrated with color photographs.

Atlas of the Oceans

The final chapter of the book is an atlas of the world's oceans. It includes maps of the five major oceans. The pages that immediately follow each whole-ocean map contain more detailed maps of selected regions of that ocean. All the maps have been produced using data collected from a combination of satellite- and ship-borne instruments. They are labeled to show the names of the seas, undersea features (such as ridges, trenches, and seamounts), and prominent coastal features. They also show ocean depths and the boundaries between tectonic plates. Features identified on the maps have been included in the index at the end of the book.

▼ **REGIONAL MAP**
As well as maps, these pages also include profiles of individual seas or undersea features. The example shown here is from the section on the Pacific Ocean.

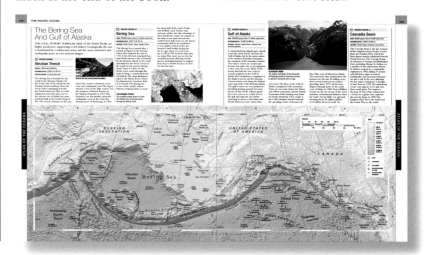

◀ **SPECIES PROFILES**
All species profiles contain a text description and, in most cases, a color photograph and distribution map.

CLASS GASTROPODA

Flamingo Tongue

Cyphoma gibbosum

LENGTH
1–1½ in (3–4 cm)

HABITAT
Coral reefs at about 50 ft (15 m)

DISTRIBUTION Western Atlantic, from North Carolina to Brazil; Gulf of Mexico, Caribbean Sea

name of group to which species belongs

common name of species is followed by scientific name

table of summary information (varies between categories)

all distribution maps are accompanied by a written summary of the range of the species

shaded area of map shows known natural range of species

CONSULTANTS

John Sparks is a curator in the Department of Ichthyology at the American Museum of Natural History and an adjunct professor at Columbia University in New York City.

CONTRIBUTORS

Richard Beatty Glossary

Kim Bryan Introduction to Ocean Life, Bacteria and Archaea, Protists, Fungi, Mollusks, Arthropods, Red Crab Migration

David Burnie Animal Life, Reptiles, Birds, Mammals

Robert Dinwiddie Ocean Water, Circulation and Climate, Tides and Waves, Coasts and the Seashore, Shallow Seas, Polar Oceans, Ocean Yacht Racing, Shutting Down the Atlantic Conveyor, Hurricane Katrina, Global Warming and Sea-level Rise, Coastal Defenses, The Titanic Disaster

Frances Dipper Introduction to Ocean Life, Sponges, Cnidarians, Segmented Worms, Flatworms, Ribbon Worms, Bryozoans, Echinoderms, Small Bottom-living Phyla, Planktonic Phyla, Tunicates and Lancelets, Jawless Fishes, Cartilaginous Fishes, Bony Fishes

Philip Eales Ocean Geology, Atlas of the Oceans, Oceanography from Space, The Indian Ocean Tsunami, Ice-shelf Breakup

Monty Halls Diving Tourism

Sue Scott Shallow Seas, Red and Brown Seaweeds, Plant Life, Green Seaweeds, Green Algae, Mosses, Flowering Plants, Fishing

Michael Scott The Open Ocean and Ocean Floor, Exploration with Submersibles, Cold Water Reefs, Biodiversity Hotspots, Whale Migration, Wind Farming in the Baltic

This revised edition was prepared with the help of Kim Bryan, Robert Dinwiddie, Frances Dipper, Philip Eales, Liz Grimwade, Robert Hume, Dean Lomax, and Michael Scott.

FOREWORD

We should call our planet Ocean. A small orb floating in the endless darkness of space, it is a beacon of life in the otherwise forbidding cold of the endless universe. Against all odds, it is also the Petri dish from which all life known to us springs.

Without water, our planet would be just one of billions of lifeless rocks floating endlessly in the vastness of the inky-black void. Even statisticians revel in the improbability that it exists at all, with such a rich abundance of life, much less that we as a species survive on its surface. Yet, despite the maze of improbability, we have somehow found our way to where we are today. Humans were enchanted by the sea even before the Greek poet Homer wrote his epic tale of ocean adventure, the *Odyssey*. It is this fascination that has driven us to delve into this foreign realm in search of answers, but the sea has always been reluctant to give up its secrets easily. Even with the monumental achievements of past explorers, scientists, and oceanographers, we have barely ventured through its surface.

It is estimated that over 90 percent of the world's biodiversity resides in its oceans. From the heartbeat-like pulsing of the jellyfish to the life-and-death battle between an octopus and a mantis shrimp, discoveries await us at every turn. And for every mystery solved, a dozen more present themselves. These are certainly exciting times as we dive into the planet's final frontier. Aided by new technology, we can now explore beyond the two percent or so of the oceans that previous generations observed. But even with the advent of modern technology, it will take several more generations to achieve a knowledge base similar to the one we have about the land.

No matter how remote we feel we are from the oceans, every act each one of us takes in our everyday lives affects our planet's water cycle and in return affects us. All the water that falls on land, from the highest peaks to the flattest plains, ends up draining into the oceans. And although this has happened for countless millions of years, the growing ecological footprint of our species in the last century has affected the cycle in profound ways. From fertilizer overuse in landlocked areas, which creates life-choking algal blooms thousand of miles away, to everyday plastic items washing up in even the most remote areas of the globe, our actions affect the health of this, our sole life-support system.

This statement is not here to make us feel that we are doomed by our actions, but rather to illustrate that through improved knowledge of the ocean system and its inhabitants, we can become impassioned to work toward curing our planet's faltering health. By taking simple steps, such as paying a little more attention to our daily routines, each one of us can have a significant positive impact on the future of our planet and on the world our children will inherit. In short, it would be much healthier for us to learn to dance nature's waltz than to try to change the music.

FABIEN COUSTEAU

MOVING EN MASSE
The fast, coordinated movement of a shoal of fish is one of the most spectacular sights in the oceans. These blackfin barracuda have formed a spiraling shoal in water around the Solomon Islands. Such shoals are often found in the same place several months, or even years, apart.

GOLDEN JELLYFISH
Drawn to the light, jellyfish in the genus *Mastigias* follow the sunlight around their brackish lake home in Palau. Stained golden-brown by algae in their tissues, they carry their photosynthetic passengers to the best-lit areas. In return, the algae provide them synthesized food. Isolated from the sea, these jellyfish are found nowhere else in the world.

PERFECT SPIRAL
A chain of salps—a group of barrel-shaped, mostly transparent, gelatinous marine animals—float near the sea surface off the Coronado Islands, northwestern Mexico. Salps belong to a subphylum of animals called tunicates, which also includes sea squirts. Salp colonies frequently form chains or spirals that can be several feet in length.

HUNGRY GANNETS
Three northern gannets dive to feed on discarded fish off Shetland, to the north of the Scottish mainland, while scores of fellow gannets hover overhead. Gannets are archetypal plunge-diving seabirds, with streamlined bodies and bills, forward-facing eyes, and wings that can fold back as they propel themselves underwater.

MUTUAL BENEFITS
A slow-moving, medium-sized shark, the oceanic whitetip occurs in warm waters worldwide—this one was spotted in the Red Sea, off Egypt. It is frequently seen, as here, accompanied by pilotfish, which feed on parasites on the shark's skin. The relationship between pilotfish and shark is mutually beneficial: the pilotfish are protected from predators, while the shark is relieved of parasites.

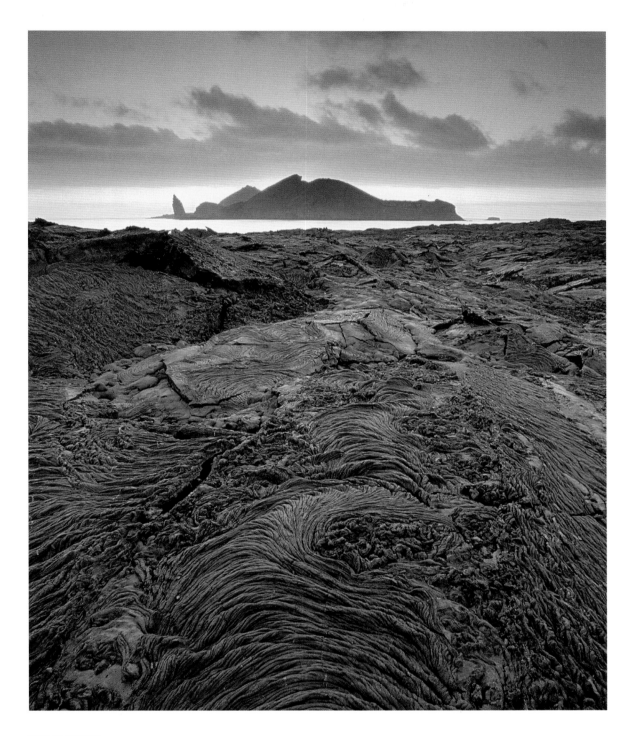

NEW COASTLINE
The shape of a coastline is determined by a balance of forces. The coastlines of the Galápagos Islands in the eastern Pacific are relatively new, having formed when the islands were created by volcanic eruptions. The lava seen here solidified about 100 years ago, but more recent eruptions have occurred on some of the group's younger islands.

THE FALL OF THE APOSTLES
On this part of the southern Australian coast, marine erosion is the dominant force. A line of limestone cliffs is slowly being worn back by the sea, leaving behind isolated stacks of rock. The stacks are collectively known as the Twelve Apostles, although when they were named, there were only nine of them and there are now just eight.

GAPING PREDATOR
A great hammerhead shark, mouth wide open, feeds in
shallow water, surrounded by fish, in the Bahamas. As with
all hammerhead sharks, its eyes are located at the ends of
its T-shaped head, called a cephalofoil. When hunting on the
seabed, it swings its head from side to side and uses special
sensors on the underside of the cephalofoil to detect the
electric signatures of buried prey, such as stingrays.

SUBAQUA BALLET
A California sea lion and her pup frolic in a kelp forest off Santa Barbara Island, California. Sea lions have especially flexible spines—individuals have been observed bending their heads backward to reach their flippers—and this allows these underwater acrobats to make particularly tight, agile turns when maneuvering to catch prey.

LEARNING TO SWIM
Being able to swim is a useful skill for polar bears, which for part of the year track their prey across shifting sea ice in the Arctic Ocean and are sometimes seen in the water many miles from land. While underwater, they keep their eyes open but close their nostrils. They can remain submerged for up to two minutes.

FILTERING FISH
A shoal of Indian mackerel moves through surface waters of the Egyptian Red Sea with mouths agape, filtering tiny zooplankton with their gill rakers. The gill rakers are featherlike bony or cartilaginous projections into the mouth cavity, bearing rows of minuscule hooks that trap any zooplankton passing by.

THE GREENLAND COAST
In this satellite image of part of Greenland's eastern coast, taken during a summer thaw, the brown areas are rocky land. Penetrating it are many long fiords, some partly filled by glaciers and large, flat icebergs. Other icebergs of varying size, formed from a disintegrated ice shelf, float offshore in vast numbers.

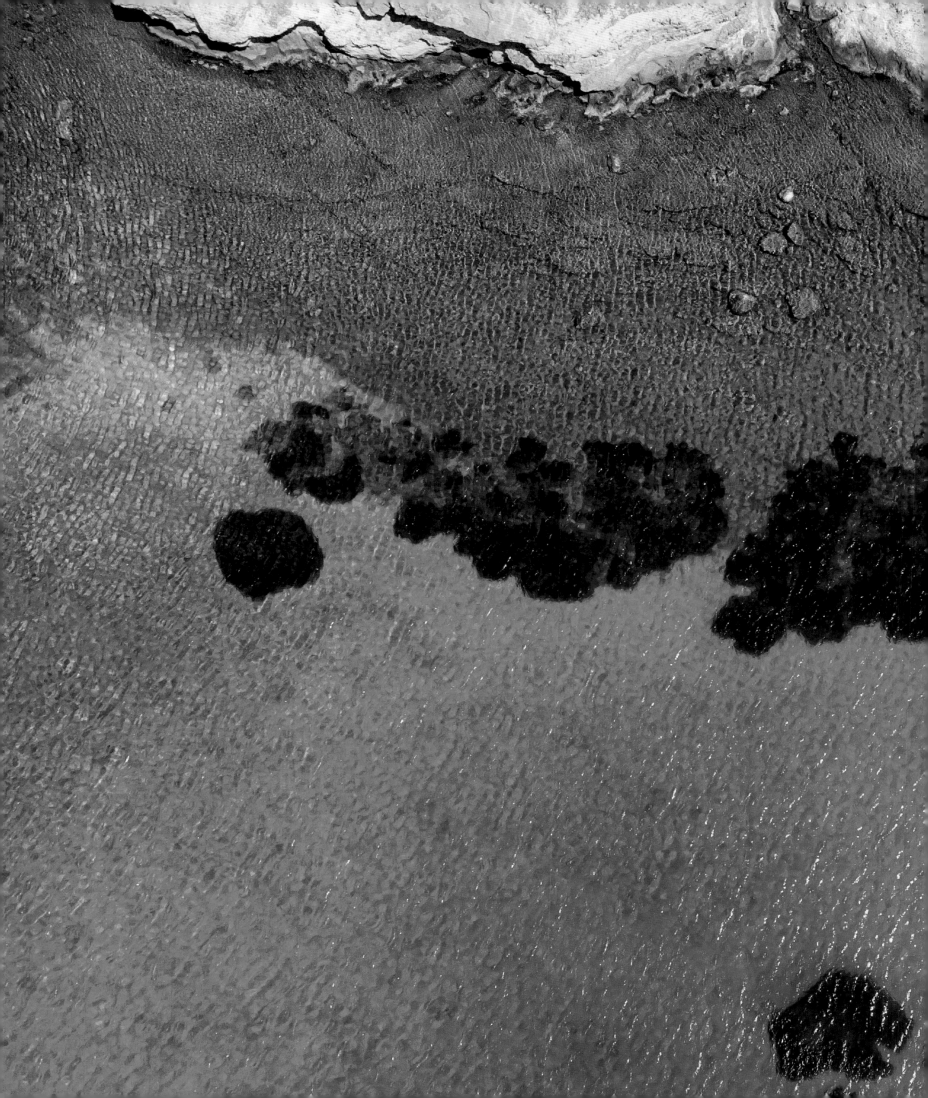

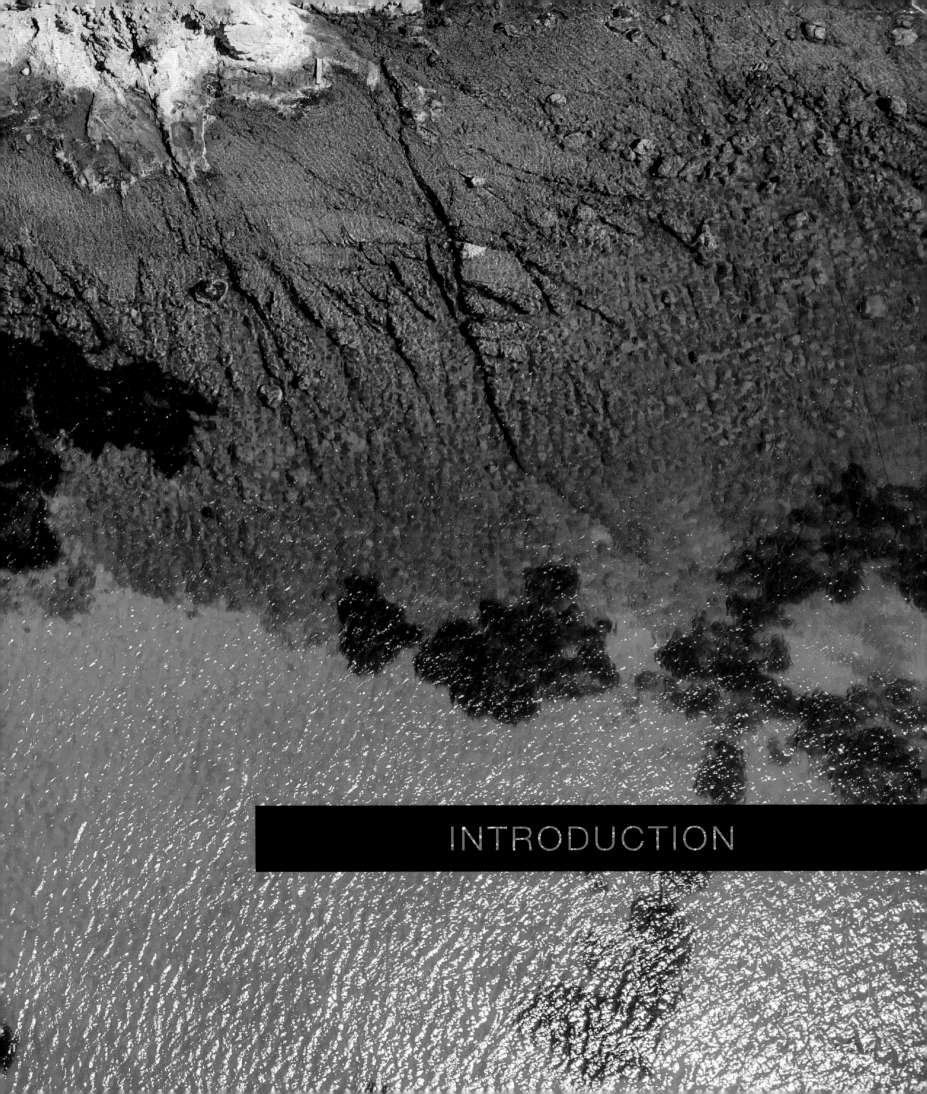

INTRODUCTION

EARTH'S OCEANS CONTAIN about 321 million cubic miles (1.34 billion cubic kilometers) of seawater. Dissolved in this are around 53 million billion tons (48 million billion metric tons) of salts, gases, and other substances. The base substance, water itself, has many unusual properties, such as its high surface tension and heat capacity. These are of tremendous significance to everything from the oceans' ability to support life, to their stabilizing effect on the world's climate, to their ability to transmit waves. Also of significance is the variability of ocean water—the sea is not uniform but varies spatially and sometimes seasonally in attributes such as its temperature, pressure, dissolved-oxygen content, and level and quality of light illumination. These attributes are important in numerous key respects.

OCEAN WATER

CRASHING WAVE
This "barrel" wave is crashing onto the north shore of the island of Oahu, Hawaii. Inspiring sights such as this are only possible because of some of the unusual properties of water.

The Properties of Water

THE MAIN CONSTITUENT OF THE OCEANS IS, of course, water. The presence of large amounts of liquid water on Earth's surface over much of its history has resulted from a fortunate combination of factors. Among them are water's unusually high freezing and boiling points for a molecule of its size, and its relative chemical stability. Water also has other remarkable properties that contribute to the characteristics of oceans—from their ability to support life to effects on climate. Underlying these properties is water's molecular structure.

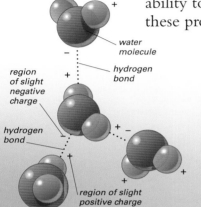

water molecule

hydrogen bond

region of slight negative charge

hydrogen bond

region of slight positive charge

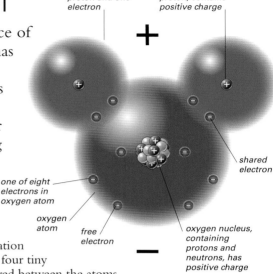

hydrogen atom consists of one proton and one electron

hydrogen nucleus, consisting of single proton, contains positive charge

shared electron

one of eight electrons in oxygen atom

oxygen atom

free electron

oxygen nucleus, containing protons and neutrons, has positive charge

CHARGE IMBALANCE
The distribution of negative charges (electrons) and regions of positive charge in an H_2O molecule causes one side to carry a slight positive charge and the other side a slight negative charge.

The Water Molecule

A molecule of water (H_2O) consists of two hydrogen (H) atoms bound to one atom of oxygen (O). Crucial to formation of the bonds between the oxygen and hydrogen atoms are four tiny negatively charged particles called electrons, which are shared between the atoms. In addition, six other electrons move around within different regions of the oxygen atom. This electron arrangement makes the H_2O molecule chemically stable but gives it an unusual shape. It also produces a small imbalance in the distribution of electrical charge within the molecule. An important result of this is that neighboring water molecules are drawn to each other by forces called hydrogen bonds.

HYDROGEN BONDS
A hydrogen bond is an attractive electrostatic force between regions of slight positive and negative charge on neighboring water molecules. Several bonds are visible here.

Surface Tension

One special property of liquid water that can be directly attributed to the attractive forces between its molecules is its high surface tension. In any aggregation of water molecules, the surface molecules tend to be drawn together and inward toward the center of the aggregation, forming a surface "skin" that is resistant to disruption. Surface tension can be thought of as the force that has to be exerted or countered to break through this skin. Water's high surface tension has various important effects. Perhaps the most crucial is that it is vital to certain processes within living organisms—for example, water transport in plants and blood transport in animals. Surface tension also allows small insects such as sea skaters to walk and feed on the ocean surface, and it even plays a part in the formation of ocean waves (see p.76).

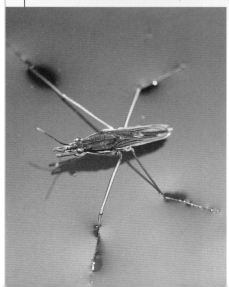

WALKING ON WATER
Certain insects, such as sea skaters and water striders (pictured below), exploit surface tension to walk, feed, and mate on the surface of the sea, lakes, or ponds.

water molecule at surface

hydrogen bonds

water molecule below surface

CAUSE OF SURFACE TENSION
In a drop of water, molecules are pulled in all directions by hydrogen bonding with their neighbors. But at the surface, the only forces act inward, or sideways, toward other surface molecules.

WATER DROPLETS
The shape of these droplets results from surface tension. The forces pulling their surface molecules together are stronger than the gravitational forces flattening them.

INTRODUCTION

LAND AND SEA
Water's high heat capacity means that the Sun warms the sea more slowly than land. In this satellite-generated temperature map of south California during a heat wave, much of the land (red) is 122°F (50°C) or more, but the sea (left) is cool, at 50°F (10°C).

SPECIFIC HEAT CAPACITY

Specific heat capacity (SHC) is the energy (in joules) needed to raise the temperature of 1 gram of a substance by 1°C. Listed below are the SHCs of 13 liquids, measured at room temperature unless otherwise stated.

SUBSTANCE	JOULES/GRAM °C
Liquid ammonia at −40°F (−40°C)	4.7
Fresh water	**4.18**
Seawater at 35°F (2°C)	**3.93**
Glycerine	2.43
Ethanol (ethyl alcohol)	2.4
Acetone	2.15
Kerosene	2.01
Olive oil	1.97
Benzene	1.8
Turpentine	1.72
Molten sodium chloride (salt)	1.56
Bromine	0.47
Mercury	0.14

Heat Capacity

A second property of liquid water that can be attributed to hydrogen bonding is its unusually high heat capacity, which exceeds that of nearly all other known liquids (see table, left). When heat is added to water, most of the heat is used to break hydrogen bonds linking the molecules. Only a fraction of the energy increases the vibrations of the water molecules, which are detected as a rise in temperature. This means that areas of ocean can absorb and release huge amounts of heat energy with little change in temperature. It also means that movements of water—ocean currents—transfer enormous amounts of heat energy around the planet. This role of ocean currents is vital to Earth's climate (see p.66).

HEAT ON THE MOVE
In this temperature map of part of the northwest Atlantic, the water surface ranges from about 41°F (5°C) (blue) to 77°F (25°C) (red). A warm current, the Gulf Stream is visible in red.

WATER TWISTER
The effects of surface tension can cause moving sheets, jets, and streams of water to assume or hold together in some surprising forms, as in this slightly spiral-shaped water jet.

Three States of Water

The temperatures at which water changes between its three states—melting point (ice to liquid water) and boiling point (liquid water to water vapor)—are both high compared with substances having similarly sized molecules. For ice to melt and water to vaporize, high levels of energy are needed to break all the hydrogen bonds. Water is also unusual in that its solid form is slightly less dense than its liquid form, so ice floats in liquid water. The reason for this is that the molecules in ice are loosely packed, whereas those in liquid water move around in snugly packed groups. The fact that ice floats on liquid water is important because it allows the existence of large areas of polar sea ice (see pp. 198–199). These affect heat flow between ocean and atmosphere and help stabilize ocean temperatures and Earth's climate.

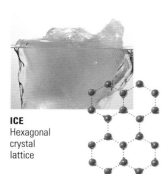

ICE
Hexagonal crystal lattice

LIQUID WATER
Small clumps of bonded molecules

WATER VAPOR
Widely spaced unbonded molecules

SOLID, LIQUID, AND GAS
In ice, hydrogen bonds hold the water molecules together in a rigid structure. In liquid water, the bonds hold the molecules in small, moving clumps. In water vapor, there are no hydrogen bonds.

SIDE BY SIDE
Water is the only natural substance found in all three states at Earth's surface. Sometimes, ice, liquid water, and condensing water vapor can be seen side by side, as here at the fjord in Spitsbergen.

The Chemistry of Seawater

THE OCEANS CONTAIN MILLIONS OF DISSOLVED chemical substances. Most of these are present in exceedingly small concentrations. Those present in significant concentrations include sea salt, which is not a single substance but a mixture of charged particles called ions. Other constituents include gases such as oxygen and carbon dioxide. One reason the oceans contain so many dissolved substances is that water is an excellent solvent.

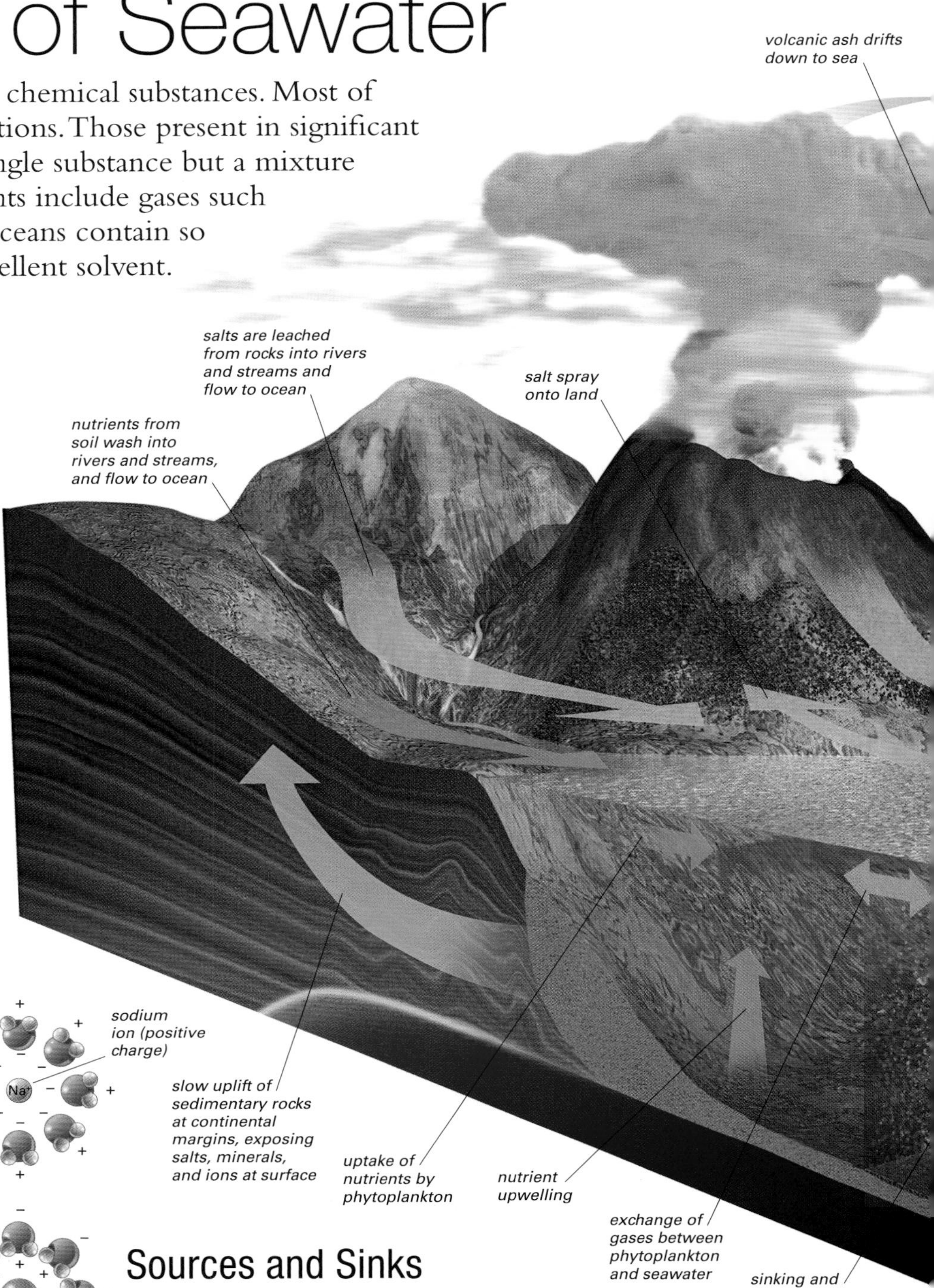

volcanic ash drifts down to sea

The Salty Sea

The salt in the oceans exists in the form of charged particles, called ions, some positively charged and some negatively charged. The most common of these are sodium and chloride ions, the components of ordinary table salt (sodium chloride). Together they make up about 85 percent by mass of all the salt in the sea. Nearly all the rest is made up of the next four most common ions, which are sulfate, magnesium, calcium, and potassium. All these ions, together with several others present in smaller quantities, exist throughout the oceans in fixed proportions. Each is distributed extremely uniformly—this is in contrast to some other dissolved substances in seawater, which are unevenly distributed.

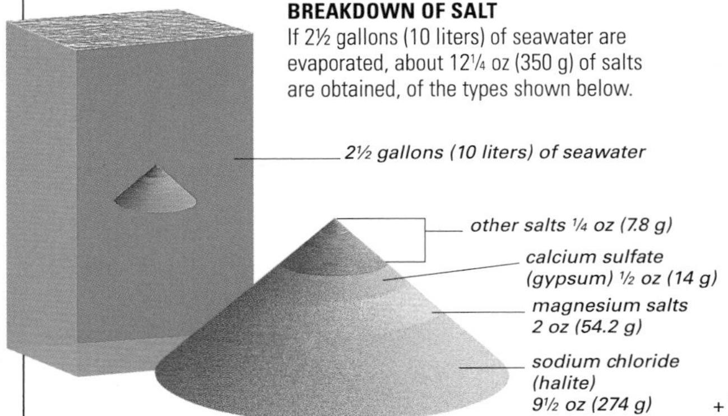

salts are leached from rocks into rivers and streams and flow to ocean

salt spray onto land

nutrients from soil wash into rivers and streams, and flow to ocean

BREAKDOWN OF SALT
If 2½ gallons (10 liters) of seawater are evaporated, about 12¼ oz (350 g) of salts are obtained, of the types shown below.

2½ gallons (10 liters) of seawater

other salts ¼ oz (7.8 g)

calcium sulfate (gypsum) ½ oz (14 g)

magnesium salts 2 oz (54.2 g)

sodium chloride (halite) 9½ oz (274 g)

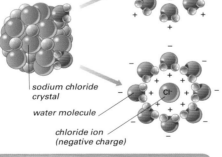

sodium ion (positive charge)

slow uplift of sedimentary rocks at continental margins, exposing salts, minerals, and ions at surface

uptake of nutrients by phytoplankton

nutrient upwelling

WATER AS A SOLVENT
The charge imbalance on its molecules makes water a good solvent. When dissolving and holding sodium chloride in solution, the positive ends of the molecules face the chloride ions and the negative ends face the sodium ions.

sodium chloride crystal

water molecule

chloride ion (negative charge)

exchange of gases between phytoplankton and seawater

sinking and decomposition of dead organisms

Sources and Sinks

The ions that make up the salt in the oceans have arrived there through various processes. Some were dissolved out of rocks on land by the action of rainwater and carried to the sea in rivers. Others entered the sea in the emanations of hydrothermal vents (see p.188), in dust blown off the land, or came from volcanic ash. There are also "sinks" for every type of ion—processes that remove them from seawater. These range from salt spray onto land to the precipitation of various ions onto the seafloor as mineral deposits. Each type of ion has a characteristic residence time. This is the time that an ion remains in seawater before it is removed. The common ions in seawater have long residence times, ranging from a few hundred years to hundreds of millions of years.

ALEXANDER MARCET

The Swiss chemist and doctor Alexander Marcet (1770–1822) carried out some of the earliest research in marine chemistry. He is best known for his discovery, in 1819, that all the main chemical ions in seawater (such as sodium, chloride, and magnesium ions) are present in exactly the same proportions throughout the world's oceans. The unchanging ratio between the ions holds true regardless of any variations in the salinity of water and is known today as the principle of constant proportions.

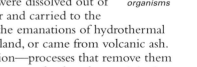

RIVER DISCHARGE
River discharge is a mechanism by which ions and nutrients enter the oceans. Here, the Noosa River empties into the sea on the coast of Queensland, Australia.

SOURCES, SINKS, AND EXCHANGES
Shown here are various sources, sinks, and exchange processes for the ions, salts, and minerals (yellow arrows), gases (pink arrows), and plant nutrients (turquoise arrows) in seawater.

KEY
- gases
- ions, salts, and minerals
- plant nutrients

Gases in Seawater

The main gases dissolved in seawater are nitrogen (N), oxygen (O_2), and carbon dioxide (CO_2). The levels of O_2 and CO_2 vary in response to the activities of photosynthesizing organisms (phytoplankton) and animals. The level of O_2 is generally highest near the surface, where the gas is absorbed from the air and also produced by photosynthesizers. Its concentration drops to a minimum in a zone between about 660 ft (200 m) and 3,300 ft (1,000 m), where oxygen is consumed by bacterial oxidation of dead organic matter and by animals feeding on this matter. Deeper down, the O_2 level increases again. CO_2 levels are highest at depth and lowest at the surface. Overall, CO_2 levels in the oceans are increasing, leading to ocean acidification (see Carbon in the Oceans, p.67).

CARBON SINK
Many marine animals, such as nautiluses (below), use carbonate (a compound of carbon and oxygen) in seawater to make their shells. After they die, the shells may form sediments and eventually rocks.

OXYGEN PRODUCER AND CONSUMER
Oxygen levels in the upper ocean depend on the balance between its production by photo-synthesizing organisms, such as kelp, and its consumption by animals, such as fish.

Nutrients

Numerous substances present in small amounts in seawater are essential for marine organisms to grow. At the base of the oceanic food chain are phytoplankton—microscopic floating life-forms that obtain energy by photosynthesis. Phytoplankton need nitrates, iron, and phosphates in order to grow and multiply. If the supply of these nutrients dries up, their growth stops; conversely, blooms (rapid growth phases) occur if it increases. Although the sea receives some input of nutrients from sources such as rivers, the main supply comes from a continuous cycle within the ocean. As organisms die, they sink to the ocean floor, where their tissues decompose and release nutrients. Upwelling of seawater from the ocean floor (see p.60) recharges the surface waters with vital substances, where they are taken up by the phytoplankton, refueling the chain.

PLANKTON BLOOM
This satellite image of the Skagerrak (a strait linking the North and Baltic seas) shows a bloom of phytoplankton, visible as a turquoise discoloration in the water.

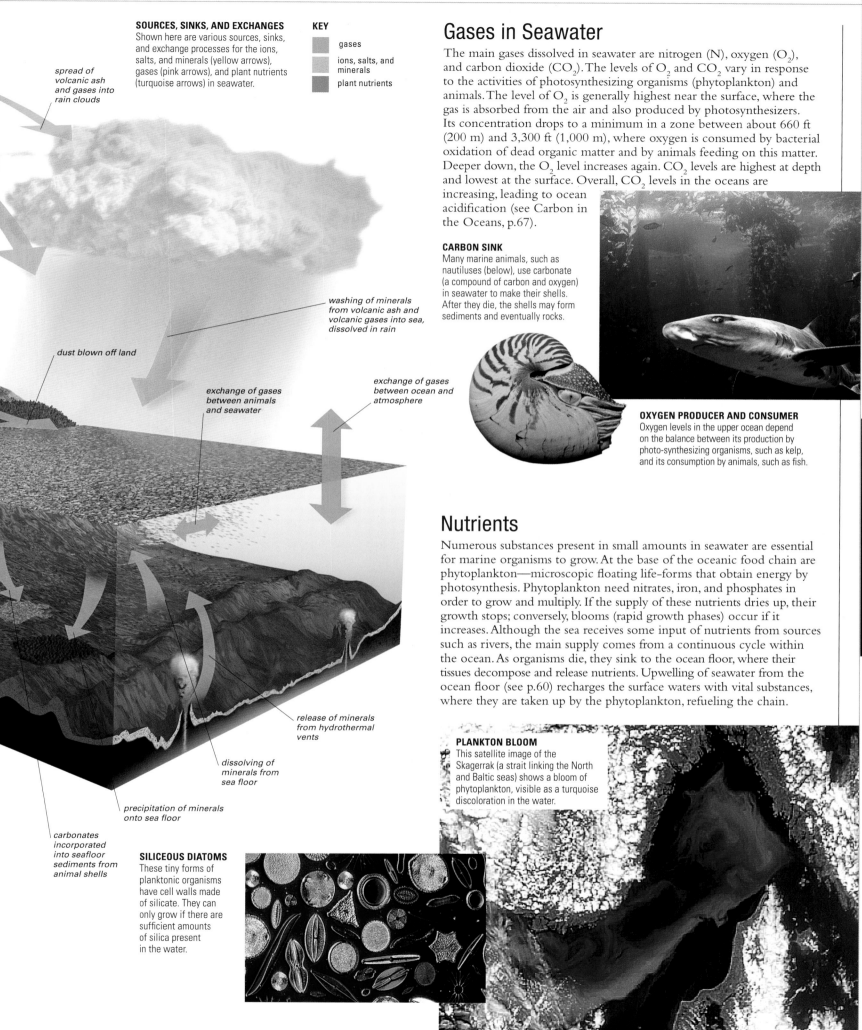

spread of volcanic ash and gases into rain clouds

dust blown off land

washing of minerals from volcanic ash and volcanic gases into sea, dissolved in rain

exchange of gases between animals and seawater

exchange of gases between ocean and atmosphere

release of minerals from hydrothermal vents

dissolving of minerals from sea floor

precipitation of minerals onto sea floor

carbonates incorporated into seafloor sediments from animal shells

SILICEOUS DIATOMS
These tiny forms of planktonic organisms have cell walls made of silicate. They can only grow if there are sufficient amounts of silica present in the water.

INTRODUCTION

INTRODUCTION

Temperature and Salinity

OCEAN WATER IS NOT UNIFORM BUT VARIES in several physical attributes, including temperature, salinity, pressure, and density. These vary vertically (dividing the oceans into layers), horizontally (between tropical and temperate regions, for example), and seasonally. The basic variables, temperature and salinity, in turn produce variations in density that help drive deep-water ocean circulation.

EL NIÑO TEMPERATURE ANOMALY IN PACIFIC
The Pacific experiences long-term fluctuations in the temperature patterns of its surface waters, which are linked to climatic disturbances known as El Niño and La Niña. This visualization, based on satellite data, shows a strong El Niño–type surface temperature pattern that developed in late 2015, with maroon-colored areas indicating higher than normal temperatures in the eastern Pacific.

Temperature

Temperature varies considerably over the upper areas of the oceans. In the tropics, solar heating keeps the ocean surface warm throughout the year. Below the surface, the temperature drops steeply to about 46–50°F (8–10°C) at a depth of 3,300 ft (1,000 m). This region of steep decline is called a thermocline. Deeper still, temperature decreases more gradually to a uniform, near-freezing value of about 36°F (2°C) on the sea floor. In midlatitudes, there is a much more marked seasonal variation in surface temperature. In high latitudes and polar oceans, the water is constantly cold, sometimes below 32°F (0°C). On average, the surface of the oceans has warmed by about 2.3°F (1.3°C) over the past century.

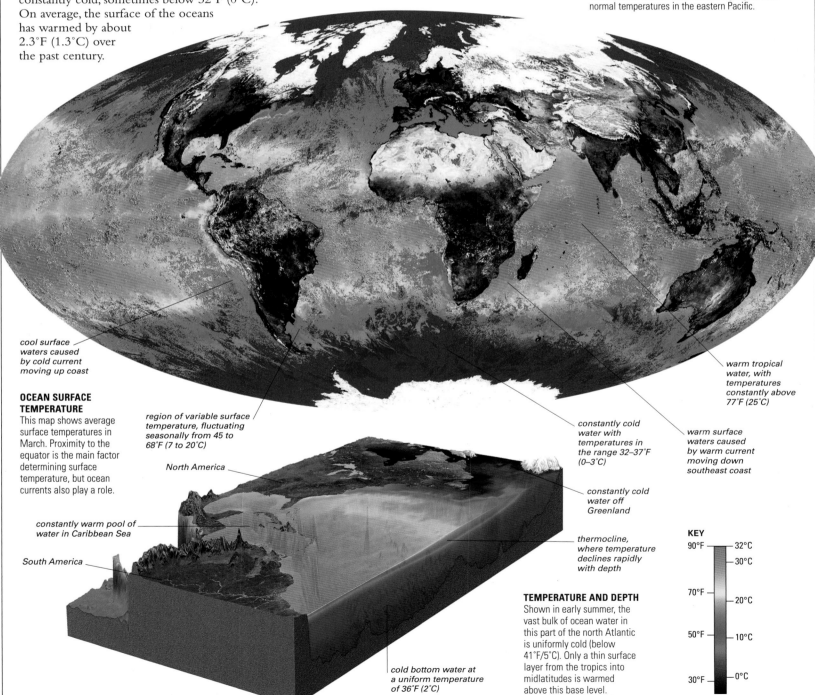

cool surface waters caused by cold current moving up coast

OCEAN SURFACE TEMPERATURE
This map shows average surface temperatures in March. Proximity to the equator is the main factor determining surface temperature, but ocean currents also play a role.

region of variable surface temperature, fluctuating seasonally from 45 to 68°F (7 to 20°C)

warm tropical water, with temperatures constantly above 77°F (25°C)

constantly cold water with temperatures in the range 32–37°F (0–3°C)

warm surface waters caused by warm current moving down southeast coast

constantly cold water off Greenland

North America

constantly warm pool of water in Caribbean Sea

South America

thermocline, where temperature declines rapidly with depth

TEMPERATURE AND DEPTH
Shown in early summer, the vast bulk of ocean water in this part of the north Atlantic is uniformly cold (below 41°F/5°C). Only a thin surface layer from the tropics into midlatitudes is warmed above this base level.

cold bottom water at a uniform temperature of 36°F (2°C)

KEY

90°F	32°C
	30°C
70°F	20°C
50°F	10°C
30°F	0°C

Salinity

Salinity is an expression of the amount of salt in a fixed mass of seawater. It is determined by measuring a seawater sample's electrical conductivity and averages about 35 grams of salt per kilogram of seawater. Salinity varies considerably over the surface of oceans—its value at any particular spot depends on what processes or factors are operating at that location that either add or remove water. Factors that add water, causing low salinity, include high rainfall, river input, or melting of sea ice. Processes that remove water, causing high salinity, include high evaporative losses and sea ice formation. At depth, salinity is near constant throughout the oceans. Between the surface and deep water is a region called a halocline, where salinity gradually increases or decreases with depth. Salinity affects the freezing point of seawater—the higher the salinity, the lower the freezing point.

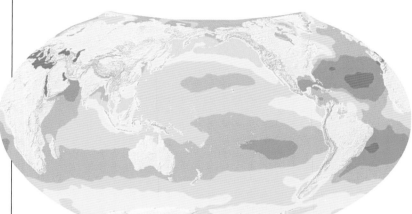

EASY FLOATING
In some enclosed seas where evaporative losses are high and there is little rainfall or river inflow, the seawater can become so saline and dense that floating becomes easy. This is the case here in the Dead Sea.

GLOBAL SALINITY
Surface salinity is highest in the subtropics, where evaporative losses of water are high, or in enclosed or semienclosed basins (such as the Mediterranean). It is lowest in colder regions or where there are large inflows of river water.

KEY

37	
36	
35	
34	
33	
32	
31	
30	
29	
under 29	

parts per thousand (‰)

Density

The density of any small portion of seawater depends primarily on its temperature and salinity. Any decrease in temperature or increase in salinity makes seawater denser—an exception being a temperature drop below 39°F (4°C), which actually makes it a little less dense. In any part of the ocean, the density of the water increases with depth because dense water always sinks if there is less dense water below it.

Processes that change the density of seawater cause it to either rise or sink and drive large-scale circulation in the oceans between the surface and deep water (see p.60). Most important is water carried toward Antarctica and the Arctic Ocean fringes. This becomes denser as it cools and through an increase in its salinity as a result of sea ice formation. In these regions, large quantities of cold, dense, salty water continually form and sink toward the ocean floor.

DENSITY LAYERS IN ATLANTIC
The oceans each contain distinct, named water masses that increase in density from the surface downward. The denser, cooler masses sink and move slowly toward the Equator. The cold, high-density deep and bottom waters comprise 80 percent of the total volume of the ocean.

Pressure

Scientists measure pressure in units called bars. At sea level, the weight of the atmosphere exerts a pressure of about one bar. Underwater, pressure increases at the rate of one bar for every 33 ft (10 m) increase in depth due to the weight of the overlying water. This means that at 230 ft (70 m), for example, the total pressure is eight bars or eight times the surface pressure. This pressure increase poses a challenge to human exploration of the oceans. To inflate their lungs underwater, divers have to breathe pressurized air or other gas mixtures, but doing so can cause additional problems (arising from the dissolution of excess gas in body tissues). These problems limit the depths attainable.

DECOMPRESSION STOP
To avoid a condition called "bends" that can arise from decompressing too quickly, on their way to the surface, scuba divers make one or more timed stops to release excess gas.

NATURAL ADAPTATION
Elephant seals can dive to depths of up to 5,100 ft (1,550 m). They have evolved various adaptations for coping with the high pressure, including collapsible ribcages.

DISCOVERY

DECOMPRESSION

After working underwater for hours at a time, professional divers routinely undergo controlled decompression in a purpose-built pressure chamber. These facilities are also used to treat pressure-related diving illnesses and for research into diving physiology.

PRESSURE CHAMBER
The person being decompressed may have to breathe a special gas mixture while the ambient pressure is slowly reduced.

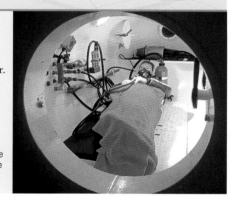

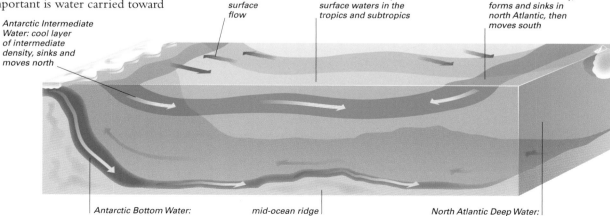

warm surface flow

Atlantic Central Water: warm, low-density surface waters in the tropics and subtropics

Atlantic Intermediate Water: cool layer of intermediate density, forms and sinks in north Atlantic, then moves south

Antarctic Intermediate Water: cool layer of intermediate density, sinks and moves north

Antarctic Bottom Water: coldest and densest layer, forms close to Antarctica, sinks then moves north

mid-ocean ridge

North Atlantic Deep Water: cold, dense water, forms and sinks in north Atlantic, then moves south

INTRODUCTION

Light and Sound

LIGHT AND SOUND BEHAVE VERY DIFFERENTLY in water than in air. Most light wavelengths are quickly absorbed by water, a fact that both explains why a calm sea appears blue and why ocean life is concentrated near its surface—almost the entire marine food chain relies on light energy driving plant growth. Sound, in contrast, travels better in water, a fact exploited by animals such as dolphins.

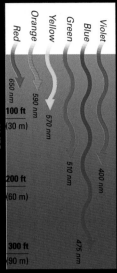

DEPTH

Violet
Blue
Green
Yellow
Orange
Red

650 nm
590 nm
570 nm
510 nm
475 nm
400 nm

100 ft
(30 m)

200 ft
(60 m)

300 ft
(90 m)

LIGHT PENETRATION
The red and orange components of sunlight are absorbed in the top 50 ft (15 m) of the ocean. Most other colors are absorbed in the next 130 ft (40 m). Wavelength is measured in nanometers (nm).

Light in the Ocean

White light, such as sunlight, contains a mixture of light wavelengths, ranging from long (red) to short (violet). Ocean water strongly absorbs red, orange, and yellow light, so only some blue and a little green and violet light reach beyond a depth of about 130 ft (40 m). At 300 ft (90 m), most of even the blue light (the most penetrating) has been absorbed, while below 650 ft (200 m), the only light comes from bioluminescent organisms, which produce their own light (see p.224). Because they rely on light to photosynthesize, phytoplankton are restricted to the upper layers of the ocean, and this in turn affects the distribution of other marine organisms. Intriguingly, many bright red animals live at depths that are devoid of red light: their color provides effective camouflage, since they appear black.

COLOR RESTORATION
At a depth of 65 ft (20 m), most animals and plants look blue-green under ambient light conditions (top). Lighting up the scene with a photographic flash or torch reveals the true colors of the marine life (bottom).

FIREFLY SQUID
This squid produces a pattern of glowing spots (photophores). When viewed by a predator swimming below, the spots help camouflage its outline against the moonlit waters above.

FISH VISION

Fish have excellent vision, which helps them find food and avoid predators. Many can see in color. The lens of a fish's eye is almost spherical and made of a material with a high refractive index. It can be moved backward and forward to focus light on the retina.

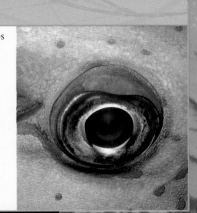

FISH EYE
The lens of a fish's eye bulges through the iris (the dark central part) almost touching the cornea (outer part). This helps to gather the maximum amount of light and gives a wide field of view.

Sea Colors

Seawater has no intrinsic color—a glass of seawater is transparent. But on a clear, sunny day, the sea usually looks blue or turquoise. In part, this is due to the sea surface reflecting the sky, but the main reason is that most of the light coming off the surface has already penetrated it and been reflected back by particles in the water or by the seabed. During its journey through the water, most of the light is absorbed, except for some blue and green light, which are the colors seen. Other factors can modify the sea's color. In windy weather, the surface becomes flecked with white, caused by trapped bubbles of air, which reflect most of the light that hits them. Rain interferes with seawater's light-transmitting properties, so rainy, overcast weather generally produces dark, gray-green seas. Occasionally, living organisms, such as "blooms" of plankton can turn patches of the sea vivid colors.

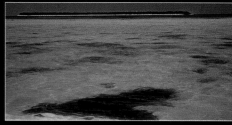

VIVID GREEN FROM ALGAL BLOOM

TROPICAL TURQUOISE

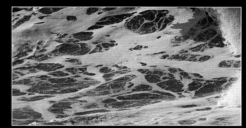

GRAY FOAMY TEMPERATE SEA

OCEAN SHADES
A green sea (top) is sometimes caused by the presence of algae. Turquoise is the usual shade in clear tropical waters, while gray water flecked with white foam is typical of windy, overcast days.

PEOPLE

WALTER MUNK

The Austrian–American scientist Walter Munk (1917–2019) pioneered the use of sound waves in oceanography. A professor at the Scripps Institute of Oceanography in San Diego, California, Munk demonstrated that by studying the patterns and speed of sound propagation underwater, information can be obtained about the large-scale structure of ocean basins.

Underwater Sounds

The oceans are noisier than might be imagined. Sources of sound include ships, submarines, earthquakes, underwater landslides, and the sounds of icebergs breaking off glaciers and ice shelves. In addition, by transmitting sound waves or bouncing them off underwater objects (echolocation) whales and dolphins use sound for navigation, hunting, and communication. Sound waves travel faster and further underwater than they do in air. Their speed underwater is about 5,000 ft (1,500 m) per second and is increased by a rise in the pressure (depth) of the water and decreased by a drop in temperature. Combining these two effects, in most ocean regions, there is a layer of minimum sound velocity at a depth of about 3,300 ft (1,000 m). This layer is called the SOFAR (Sound Fixing and Ranging) channel. The properties of the SOFAR channel are exploited by people using underwater listening devices and, it has been theorized, by animals such as whales and dolphins.

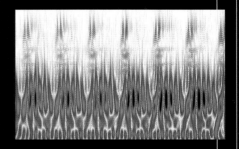

HUMPBACK WHALE SONG
The peaks and troughs in this spectrogram show the changes in frequency of a few seconds of repeated sound made by a Humpback Whale.

THE SOFAR CHANNEL
Low-frequency sounds generated in the SOFAR channel are "trapped" in it by inward refraction from the edges of the channel. As a result, sounds can travel very long distances in this ocean layer.

Sea level

sound travels slower within channel

3,300 ft
(1,000 m)

SOFAR channel

6,600 ft
(2,000 m)

9,800 ft
(3,000 m)

DEPTH

4,900 ft/s   5,000 ft/s   5,085 ft/s
(1,500 m/s)   (1,525 m/s)   (1,550 m/s)

SPEED OF SOUND UNDERWATER

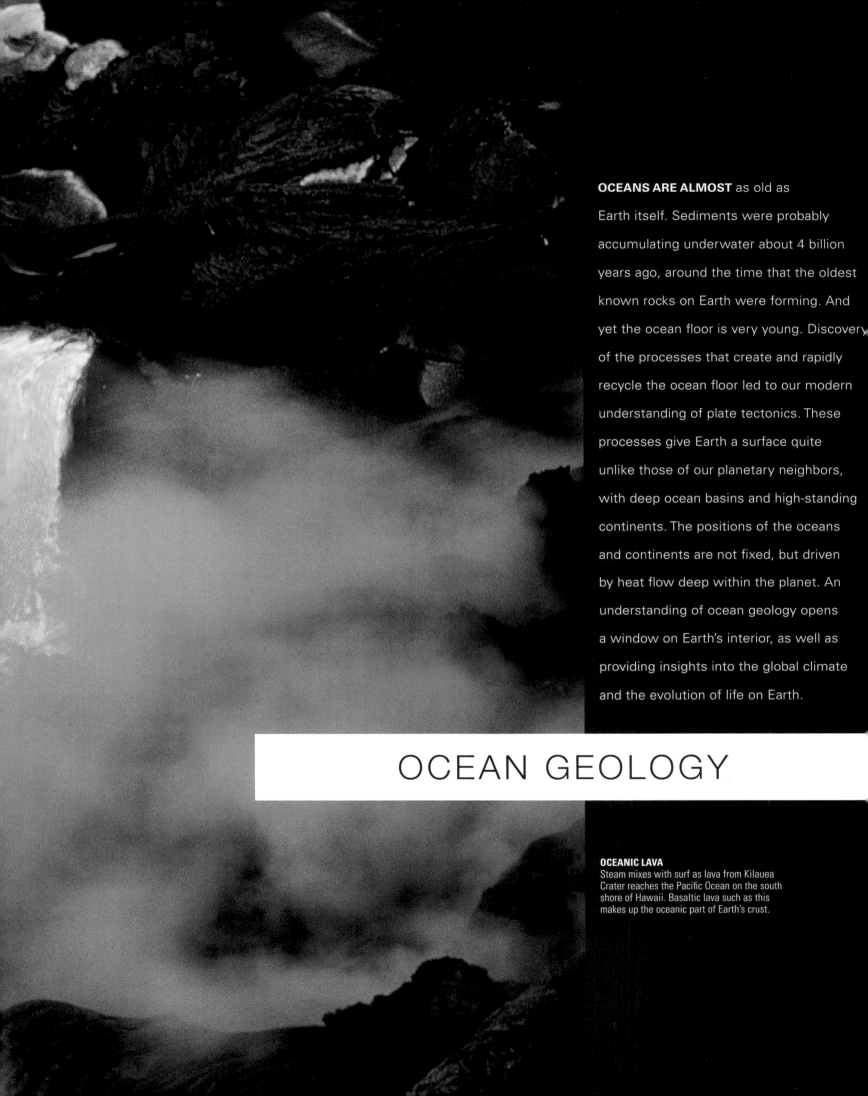

OCEANS ARE ALMOST as old as Earth itself. Sediments were probably accumulating underwater about 4 billion years ago, around the time that the oldest known rocks on Earth were forming. And yet the ocean floor is very young. Discovery of the processes that create and rapidly recycle the ocean floor led to our modern understanding of plate tectonics. These processes give Earth a surface quite unlike those of our planetary neighbors, with deep ocean basins and high-standing continents. The positions of the oceans and continents are not fixed, but driven by heat flow deep within the planet. An understanding of ocean geology opens a window on Earth's interior, as well as providing insights into the global climate and the evolution of life on Earth.

OCEAN GEOLOGY

OCEANIC LAVA
Steam mixes with surf as lava from Kilauea Crater reaches the Pacific Ocean on the south shore of Hawaii. Basaltic lava such as this makes up the oceanic part of Earth's crust.

The Formation of the Earth

THE EARTH STARTED TO FORM MORE THAN 4,500 million years ago in a disk of gas, dust, and ice around the early Sun. This protoplanetary disk, as it is known, was held in orbit by the gravitational field of the young star. Gravitational attraction between dust particles in the disk produced small rocks, and collisions concentrated the rocks into several rings. The most densely populated rings went on to form the planets of the Solar System.

EARLY SOLAR SYSTEM
The early Solar System contained a disk of dust, ice, and gas, from which the rocky inner planets and gaseous outer planets formed.

small pieces of rock and ice pulled together by gravitational attraction

planetesimals start to form in protoplanetary disk around Sun

1 COLD ACCRETION
Under gravity, pieces of rock and ice coalesced. Material sharing the same orbit around the Sun clumped together to form planetesimals.

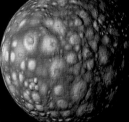

2 PROTOPLANET
By attracting more clumps of rock and ice, and through many collisions, planetesimals grew into protoplanets. As the size of these increased, gravity smoothed out their surfaces.

rocks accelerate toward primordial Earth

Birth of the Earth

Initially, the rocks within each ring drifted together, due to their mutual gravitational attraction, in a process known as cold accretion. The largest bodies in each ring attracted the most material and grew to form objects larger than 0.7 mile (1 km) across, called planetesimals. Planetesimals are loose collections of rock and ice, with a uniform structure. As the mass of a planetesimal grows larger, it exerts a stronger gravitational pull, becoming more tightly held together and attracting nearby rocks with greater force. Collisions between planetesimals broke them apart or grouped them together. In the inner Solar System, the planetesimals in each orbiting ring came together to form much larger objects, called protoplanets, and these later collided to form the rocky planets. The Earth was born in this way about 4.560 million years ago.

3 HEAVY BOMBARDMENT
Each growing protoplanet attracted more planetesimals, which impacted through more energetic, high-speed collisions. Finally the protoplanets themselves underwent a series of collisions to form the rocky planets, including Earth.

impacts generate surface heat and local melting

impact of Mars-sized body leads to total melting

Internal Heat

Although there was heat in the early Earth, it was mostly solid and had a fairly uniform internal composition. Today, it has layers of different compositions, including a dense, partially liquid core of iron and nickel. The transition between these two states may have its roots in several different energy sources. Localized surface melting would have occurred when the kinetic energy of incoming rocks was converted to heat during impacts. More significant heat sources would have been the decay of radioactive elements in the interior rocks and the heat released by the Earth's contraction under the force of its own gravity—a process that led to an event called the iron catastrophe (see below). Impact with a sufficiently large body might have released enough heat to melt the Earth's interior, and this may have happened more than once.

MOON FORMATION
It is thought that early in Earth's history, it was struck by a large protoplanet, creating the Moon, tilting the Earth's axis of rotation, and leaving it with a slightly eccentric orbit.

material ejected during collision later cooled and coalesced to form the Moon

THE IRON CATASTROPHE

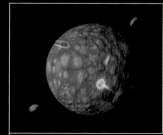

1 As the Earth grew larger, the strength of its gravitational field increased, which in turn attracted more material.

2 Eventually, the gravitational field was strong enough to cause the Earth to contract, converting gravitational potential energy into heat.

3 Enough heat was released to melt the iron contained in the Earth's rocks, allowing it to flow down to the center of the Earth.

4 The sinking of large amounts of iron released further heat, enough to melt the entire interior of the planet in the event called the iron catastrophe.

Convection and Differentiation

After the interior of the Earth melted, its heaviest constituents were able to sink to the center and the lighter ones to rise toward the surface. One-third of the planet's mass pooled at the center and formed a dense core consisting mainly of iron, the heaviest of the common elements making up the Earth. The core became the hottest part of the planet, up to 11,700°F (6,500°C), and a source of heat for the molten rocks above. Most materials expand as they are heated, becoming less dense and more buoyant. This is the basis of convection, which provided a mechanism for carrying heat and material from the interior of the Earth toward the surface. Vigorous convection cells carried hot, buoyant material upward, where it lost heat by conduction near the surface, before sinking again. Lighter materials such as aluminum were left behind at the surface, forming a thin crust. In this way, the Earth became differentiated into layers of different chemical composition: a metallic core, a rocky mantle, and a buoyant crust. This occurred as early as 4,500 million years ago.

The Earth Today

The Earth's interior is now split into three chemically distinct layers, which can be further differentiated by changes in their physical properties due to temperature and pressure variations with depth. The core consists of an iron-nickel alloy, with some impurities, at a temperature of 7,200–11,700°F (4,000–6,500°C). Iron in the inner part of the core has solidified under the immense pressure, but the outer part is still a free-flowing liquid. The mantle of silicate rock surrounding the core has also solidified, but a form of convection called "solid-state creep" still takes place, with material in the lower mantle moving a few inches per year. The upper mantle, within about 255 miles (410 km) of the surface, is a more easily deformed "plastic" region. Above it floats a thin crust enriched in lighter elements, with average thickness ranging from 5 miles (8 km) beneath the oceans to 28 miles (45 km) beneath the continents.

A LAYERED EARTH
The early Earth had a uniform composition but melting allowed chemical "zoning" to develop.

convection carries internal heat to surface

lighter materials rise up through semifluid mantle

heavy materials sink to form dense core

carbon dioxide

water vapor

nitrogen

ATMOSPHERE AND OCEAN
The lightest materials of all, gases and water, were expelled from the interior to form the outer atmospheric and ocean layers at an early stage in the Earth's history.

atmosphere

INSIDE THE EARTH
Earth has a layered internal structure, the main layers being the core, mantle, and crust. The density and temperature of the layers increases with depth. Heat from the core flows through the mantle, eventually reaching the cooler crust, where it escapes.

the transition zone is slightly denser than the upper mantle and forms a distinct layer between upper and lower mantle

reservoir of magma (hot, melted rock) under Yellowstone Park, on North American Plate

hotspot under Hawaii, probably caused by a plume of hot material rising from deep in the mantle

liquid outer core

solid inner core

lower mantle

upper mantle

consisting of the uppermost layer of the upper mantle together with overlying crust, the rigid lithosphere makes up tectonic plates

oceanic crust

continental crust

the Chile Rise is a ridge marking the divergence of two tectonic

The Origin of Oceans and Continents

EARTH'S OCEANS FORMED MORE THAN 4 billion years ago, mainly from water vapor that condensed from its primitive atmosphere but also from water brought from space by comets. Initially, after acquiring a layered internal structure, the Earth had a uniform crust that was enriched in lighter elements and floated on an upper mantle made of denser materials. Later, the crust became differentiated into two types as continents began to form, made from rocks that were chemically distinct from those underlying the oceans.

Continental Crust

The continents include a wide range of rock types, including granitic igneous rocks, sedimentary rocks, and the metamorphic rocks formed by the alteration of both. They contain a lot of quartz, a mineral absent in oceanic crust. The first continental rocks were the result of repeated melting, cooling, and remixing of oceanic crust, driven by volcanic activity above mantle convection cells, which were much more numerous and vigorous than today's. Each cycle left more of the heavier components in the upper mantle and concentrated more of the lighter components in the crust. The first microcontinents grew as lighter fragments of crust collided and fused. Thickening of the crust led to melting at its base and underplating with granitic igneous rocks. Weathering accelerated the process of continental rock formation, retaining the most resistant components, such as quartz, while washing solubles into the ocean.

THE OLDEST ROCKS
These sedimentary rocks on Baffin Island lie on the Canadian Shield. The stable continental shields contain the world's most ancient rocks, which are around 4 billion years old.

zircon crystals, among the earliest continental crust materials

ZIRCON

primitive continental crust thickens above sinking mantle flow, without mantle interference

sedimentary rocks

primitive oceanic crust

volcanic activity adds igneous rocks to surface above rising flows

Oceanic Crust

The oceanic crust has a higher density than the continental crust, making it less buoyant. Both types of crust can be thought of as floating on the "plastic" upper mantle, and the oceanic crust lies lower due to its lower buoyancy. It is relatively thin, with a depth of never more than 7 miles (11 km), compared with a thickness of 15–43 miles (25–70 km) for most continental crust. It consists mainly of basalt, an igneous rock that is low in silica compared with continental rocks, and richer in calcium than the mantle. Basalt lava is created when hot material in the upper mantle is decompressed, allowing it to melt and form liquid magma. The decompression occurs beneath rifts in the crust, such as those found at the mid-ocean ridges, and it is through these rifts that lava is extruded onto the surface to create new ocean crust.

basaltic lava · rift · basalt sheets (dikes) · sediment · ocean surface · gabbro · peridotite · ocean crust · lithosphere · Moho · asthenosphere · magma rises to surface · top layer of upper mantle

OCEAN-FLOOR STRUCTURE
Three layers of basalt in the crust (basaltic lava, dikes, and gabbro) are separated from the mantle by the Mohorovičić discontinuity (the Moho). The top layer of the upper mantle is fused to the base of the crust to form the rigid lithosphere, which makes up tectonic plates. The asthenosphere is the soft zone over which the plates of the lithosphere glide.

MANTLE ROCKS
Peridotite is the dominant rock type found in the mantle, consisting of silicates of magnesium, iron, and other metals. Sometimes it is brought to the surface when parts of the ocean floor are uplifted, as here in Newfoundland, Canada, or as fragments from volcanic activity.

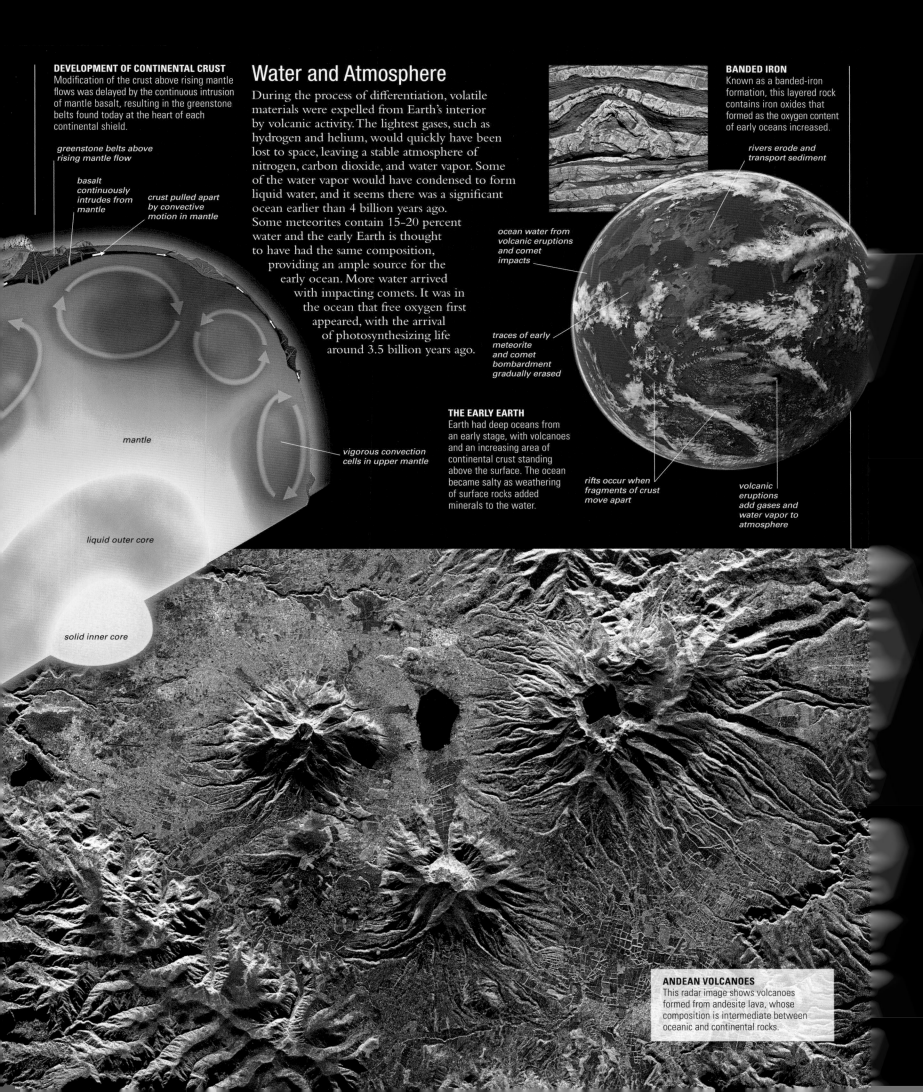

DEVELOPMENT OF CONTINENTAL CRUST
Modification of the crust above rising mantle flows was delayed by the continuous intrusion of mantle basalt, resulting in the greenstone belts found today at the heart of each continental shield.

greenstone belts above rising mantle flow

basalt continuously intrudes from mantle

crust pulled apart by convective motion in mantle

mantle

vigorous convection cells in upper mantle

liquid outer core

solid inner core

Water and Atmosphere

During the process of differentiation, volatile materials were expelled from Earth's interior by volcanic activity. The lightest gases, such as hydrogen and helium, would quickly have been lost to space, leaving a stable atmosphere of nitrogen, carbon dioxide, and water vapor. Some of the water vapor would have condensed to form liquid water, and it seems there was a significant ocean earlier than 4 billion years ago. Some meteorites contain 15–20 percent water and the early Earth is thought to have had the same composition, providing an ample source for the early ocean. More water arrived with impacting comets. It was in the ocean that free oxygen first appeared, with the arrival of photosynthesizing life around 3.5 billion years ago.

THE EARLY EARTH
Earth had deep oceans from an early stage, with volcanoes and an increasing area of continental crust standing above the surface. The ocean became salty as weathering of surface rocks added minerals to the water.

ocean water from volcanic eruptions and comet impacts

traces of early meteorite and comet bombardment gradually erased

rifts occur when fragments of crust move apart

volcanic eruptions add gases and water vapor to atmosphere

BANDED IRON
Known as a banded-iron formation, this layered rock contains iron oxides that formed as the oxygen content of early oceans increased.

rivers erode and transport sediment

ANDEAN VOLCANOES
This radar image shows volcanoes formed from andesite lava, whose composition is intermediate between oceanic and continental rocks.

The Evolution of the Oceans

EVER SINCE THE ATLANTIC COASTS OF SOUTH AMERICA and Africa were accurately charted, it has been apparent that they match like the pieces of a jigsaw puzzle. We now know that the continents move, that they were once joined together, and that today's oceans arose when the landmasses split apart. The evolving oceans have modified the global climate, and sea level has fluctuated in response to climate change and geological factors.

spreading ridge

continent carried on plates

subduction zone

convection cell drives plate motion

PLATE MOVEMENT
Crustal plates move around under the influence of convection cells, which probably reach deep down to the boundary between the outer core and the mantle.

Plate Tectonics

The numerous convection cells (see p.41) that gave rise to the first fragments of continental crust gradually gave way to fewer, larger-scale convection cells as the mantle cooled. The continental fragments became consolidated into larger areas, and rifts formed at the thinnest parts of the ocean crust, splitting it into large plates. When the density of the oceanic and continental plates became sufficiently different, the oceanic crust sank where it met the more buoyant continental crust, creating subduction zones. Since then, the evolution of the oceans and continents has been dominated by plate tectonics (see pp.48–49). As the plates move, they carry the continents with them, with oceans opening and closing in between.

1. CAMBRIAN (500 MYA)
The remains of the first supercontinent, Rodinia, were scattered, with the largest piece, Gondwana, lying in the south. The Iapetus Ocean separated Laurentia (North America) from Baltica (northern Europe). The Panthalassic Ocean occupied most of the Northern Hemisphere.

PANTHALASSIC OCEAN

LAURENTIA

SIBERIA

IAPETUS OCEAN

BALTICA

GONDWANA

"ancestral" North Atlantic lies between North America and Europe

scattered remnants of Rodinia

2. DEVONIAN (400 MYA)
The Rheic Ocean opened when a string of islands, which were to become western and southern Europe, broke away from Gondwana and moved toward Euramerica, closing the Iapetus Ocean in the process.

SIBERIA

PANTHALASSIC OCEAN

EURAMERICA

AUSTRALIA

RHEIC OCEAN

GONDWANA

southern Europe joins Euramerica (Laurentia and Baltica) as Iapetus Ocean closes

first plants on land form vegetated areas

shallow continental-shelf seas

Ural Mountains

Through the Ages

As Earth's plates have moved around, largely driven by the spreading ridges and subduction zones of the rapidly recycling oceanic crust (see p.48), continents have come together and moved apart—periodically grouping together to form "supercontinents." The German scientist Alfred Wegener proposed that 250 million years ago (mya), there was a supercontinent called Pangea, centered on the equator and surrounded by one great ocean. It seems there was another grouping about 1 billion years ago called Rodinia, and perhaps an earlier grouping before that. Each time the continental landmasses have come together, they have eventually been broken apart as deep rifts opened up in their interiors, as is happening today in the Red Sea and the East African Rift. Computer models of the crustal fragments and the locations of spreading and subduction have allowed fairly reliable reconstructions of the geography of earlier times back to 500 million years ago.

PANTHALASSIC OCEAN

SIBERIA

PALEO-TETHYS SEA

PANGEA

SOUTH AMERICA

AFRICA

AUSTRALIA

GONDWANA

extensive deserts

southern ice cap covers most of South America, Africa, and Australia

3. CARBONIFEROUS (300 MYA)
As the supercontinent Pangea came together, continental masses stretched from pole to pole, almost encircling the Paleo-Tethys Sea to the east. Today's coal seams were laid down in swampy forests along the shores of equatorial shelf seas. An extensive ice cap built up as Gondwana moved over the South Pole.

KEY

 subduction zone

spreading ridge

outline of modern landmass

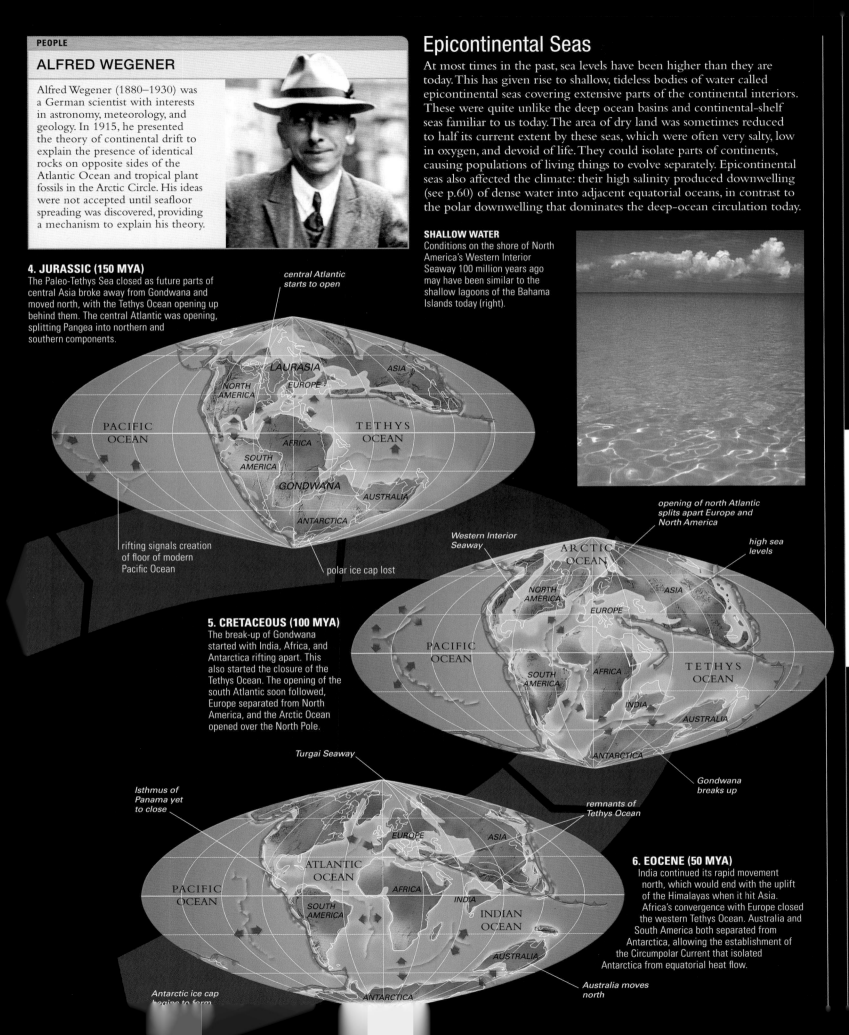

PEOPLE

ALFRED WEGENER

Alfred Wegener (1880–1930) was a German scientist with interests in astronomy, meteorology, and geology. In 1915, he presented the theory of continental drift to explain the presence of identical rocks on opposite sides of the Atlantic Ocean and tropical plant fossils in the Arctic Circle. His ideas were not accepted until seafloor spreading was discovered, providing a mechanism to explain his theory.

Epicontinental Seas

At most times in the past, sea levels have been higher than they are today. This has given rise to shallow, tideless bodies of water called epicontinental seas covering extensive parts of the continental interiors. These were quite unlike the deep ocean basins and continental-shelf seas familiar to us today. The area of dry land was sometimes reduced to half its current extent by these seas, which were often very salty, low in oxygen, and devoid of life. They could isolate parts of continents, causing populations of living things to evolve separately. Epicontinental seas also affected the climate: their high salinity produced downwelling (see p.60) of dense water into adjacent equatorial oceans, in contrast to the polar downwelling that dominates the deep-ocean circulation today.

SHALLOW WATER
Conditions on the shore of North America's Western Interior Seaway 100 million years ago may have been similar to the shallow lagoons of the Bahama Islands today (right).

4. JURASSIC (150 MYA)
The Paleo-Tethys Sea closed as future parts of central Asia broke away from Gondwana and moved north, with the Tethys Ocean opening up behind them. The central Atlantic was opening, splitting Pangea into northern and southern components.

central Atlantic starts to open

LAURASIA · ASIA · NORTH AMERICA · EUROPE · PACIFIC OCEAN · TETHYS OCEAN · AFRICA · SOUTH AMERICA · GONDWANA · AUSTRALIA · ANTARCTICA

rifting signals creation of floor of modern Pacific Ocean

polar ice cap lost

5. CRETACEOUS (100 MYA)
The break-up of Gondwana started with India, Africa, and Antarctica rifting apart. This also started the closure of the Tethys Ocean. The opening of the south Atlantic soon followed, Europe separated from North America, and the Arctic Ocean opened over the North Pole.

Western Interior Seaway

opening of north Atlantic splits apart Europe and North America

high sea levels

ARCTIC OCEAN · NORTH AMERICA · ASIA · EUROPE · PACIFIC OCEAN · SOUTH AMERICA · AFRICA · TETHYS OCEAN · INDIA · AUSTRALIA · ANTARCTICA

Gondwana breaks up

Turgai Seaway

Isthmus of Panama yet to close

remnants of Tethys Ocean

PACIFIC OCEAN · ATLANTIC OCEAN · EUROPE · ASIA · AFRICA · SOUTH AMERICA · INDIA · INDIAN OCEAN · AUSTRALIA · ANTARCTICA

Antarctic ice cap begins to form

Australia moves north

6. EOCENE (50 MYA)
India continued its rapid movement north, which would end with the uplift of the Himalayas when it hit Asia. Africa's convergence with Europe closed the western Tethys Ocean. Australia and South America both separated from Antarctica, allowing the establishment of the Circumpolar Current that isolated Antarctica from equatorial heat flow.

Currents, Continents, and Climate

Along with the atmosphere, the oceans are the means by which heat is redistributed around Earth. Most energy arriving from the Sun is absorbed as heat near the equator. It is then redistributed to colder regions. About 40 percent of the heat reaching the poles from the equator comes via ocean currents. The pattern of circulation in the oceans therefore has a large influence on Earth's climate (see pp.66–67). As continents, oceans, and currents have shifted through geological time, major climate changes have occurred. Conversely, warmer and colder periods affect sea level and the extent of seas. There is even speculation that the ocean froze to a depth of 6,500 ft (2,000 m) in places during a series of "snowball" events 775–635 million years ago, and possibly earlier, each event lasting up to 15 million years.

SNOWBALL EARTH
During snowball events, global glaciation would have left only the peaks of the highest mountains free of ice, as is the case today in Antarctica.

MESOZOIC CURRENTS
100 million years ago, ocean currents flowed through a continuous seaway from the Tethys Ocean in the east, through what is now the Mediterranean, the Central Atlantic between North and South America, and into the Pacific in the west.

TODAY'S CIRCULATION
Today, equatorial ocean currents are blocked by landmasses, and the South Circumpolar Current is the strongest current, blocking heat flow to the South Pole. The polar regions are colder.

Greenhouse to Icehouse

During the Mesozoic Era (252–65 million years ago), the climate was warmer than it is today, with a more even temperature distribution and no polar ice caps. Ocean currents freely flowed around the Equator, absorbing energy as they went, and carried heat to higher latitudes. The transition from this "greenhouse" climate to today's cooler "icehouse" is due to shifts in ocean currents following the breakup of Gondwana. When the other continents moved north, the Antarctic was surrounded by the Circum-polar Current, blocking heat flow from the equator. Equatorial flow between the oceans finally stopped when the Isthmus of Panama closed 5–3 million years ago. Antarctica now lies over the South Pole, allowing snow to accumulate into a thick ice cap, which reflects energy rather than absorbing it.

LAST GLACIAL (21,500 YEARS AGO)
Earth's climate swings between ice ages and warmer periods over cycles lasting 100,000 years or more. Within ice ages, there are colder periods called glacials and warmer periods called interglacials. During glacials (the last of which peaked 21,500 years ago), the world's ice sheets expand, lowering global sea levels and revealing land bridges.

Beringia land bridge
English Channel land bridge
Gulf of Persia dry
Yellow Sea dry
Greenland Ice Sheet
Siberian Ice Sheet
Scandinavian Ice Sheet
Cordilleran Ice Sheet
Laurentide Ice Sheet
Patagonian Ice Sheet
sea ice
Sunda land bridge
Sahul land bridge
Antarctic Ice Sheet

Sea-level Change

Sea level has constantly changed through history, being up to 1,300 ft (400 m) higher in the past. One of the factors controlling sea level is the global climate. Thermal expansion of ocean water increases global sea level by about 3 in (7.5 cm) for every 1.8°F (1°C) increase in temperature. The transfer of water between ice caps and the oceans during glacial cycles accounts for a global change of 330–655 ft (100–200 m) over a few tens of thousands of years. The rate of seafloor spreading also affects global sea levels and has outweighed climatic factors at some times. Faster-spreading ridges reduce the volume of the ocean basins as the younger, hotter crust rises higher, causing sea levels to rise (see p.88). Local changes also occur as a result of crustal movement.

TEMPERATURE AND SEA LEVEL
Over the last 100 million years, climate has controlled sea levels, and these (in blue on graph) have dropped as the climate has cooled (temperature in yellow). At other times, low sea levels were due to reduced rates of seafloor spreading.

MEDITERRANEAN BASIN HISTORY

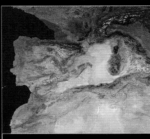

1 The Mediterranean was isolated from the Atlantic by the closure of the Strait of Gibraltar 5 million years ago and evaporated to a salty desert.

2 21,000 years ago, sea levels were 390 ft (120 m) lower than they are today due to water being locked up in ice caps at the height of the last glacial.

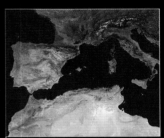

3 10,000 years ago, water from melting ice started to flood the continental shelves exposed during the glacial, leaving today's familiar shoreline.

Sedimentary Basins

Most of the world's sedimentary rocks were laid down in water over continental shelves or in inland seas. The movements of the continents and changes in sea level have determined where this deposition occurred at particular times, and many former marine sedimentary basins are now far inland.

Oil and gas deposits are found in marine sedimentary rocks, the result of animal and plant remains decomposing and then being buried and compressed. About 30 percent of the world's oil and gas production comes from offshore fields, but many offshore basins remain to be explored.

BASINS AND OILFIELDS
Sedimentary basins are found on the continental shelves and adjacent ocean floor, but also well inland where areas were once covered with water.

KEY
- onshore sedimentary deposits
- offshore sedimentary deposits
- △△△ Oil and gas deposits

GLACIAL COAST
During glacials, sea ice forms at lower latitudes than it does today. This scene may have been typical of the shores of western Europe 21,000 years ago.

Tectonics and the Ocean Floor

THE THEORY OF PLATE TECTONICS HAS REVOLUTIONIZED geology over the last half century, explaining many of the Earth's physical features. Tectonic plates are huge fragments of the Earth's lithosphere, which consists of the crust fused with the top layer of the upper mantle. They move over a more deformable layer of the mantle called the asthenosphere. Plate motion builds mountain ranges, but plate-tectonic processes are perhaps most clearly seen on the ocean floor, where most plate boundaries are found.

BASALT
The ocean floor is largely made of basalt, a fine-grained igneous rock derived from the upper mantle. It is a dense rock due to a high proportion of iron and magnesium.

Recycling Ocean Crust

The oldest rocks on the ocean floor are 180 million years old. This is young compared with the oldest continental rocks, which date from about 3.8 billion years ago. While the continental crust has been steadily accumulating throughout the Earth's history, it seems the oceanic crust is created and destroyed rather quickly. It is created at the mid-ocean ridges from hot material rising in the mantle, and then spreads away from the ridges, before eventually being recycled into the mantle at subduction zones. Continental crust is always less dense and more buoyant than oceanic crust, so where they meet, it is the oceanic crust that gives way, sinking (subducting) back into the mantle.

AGE OF THE OCEAN FLOOR
The age of the ocean floor increases away from the spreading ridges where new crust is forming. The map below shows the East Pacific Rise to be the fastest-spreading ridge, because it is flanked by the broadest spread of young rock (shaded red and orange).

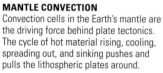

hotspot produces volcanic activity

mid-ocean ridge

lithospheric plate pushed away from ridge

rising mantle plume forms hotspot at surface

incipient mantle plume

convection cell

oceanic lithosphere descends at subduction zone

MANTLE CONVECTION
Convection cells in the Earth's mantle are the driving force behind plate tectonics. The cycle of hot material rising, cooling, spreading out, and sinking pushes and pulls the lithospheric plates around.

KEY age (millions of years)	144	89	54.8	24	1.8	
	154	127	65	33.5	5	0	undated

Plate Boundaries

The boundaries of a tectonic plate may be divergent, convergent, or transform. At divergent boundaries, the crust is extended, thinned, and fractured by the upwelling of hot mantle material. The crust buoys up, producing a mid-ocean ridge, and lava is extruded through a central rift valley to create new oceanic crust. Seamount volcanoes may also arise (see p.174).

Plates collide at convergent boundaries. Where oceanic lithosphere meets continental lithosphere, the crust on the continental side may be compressed and thickened, resulting in mountain-building. The oceanic lithosphere sinks beneath the lighter continental lithosphere, forming an ocean trench (see p.183), and volcanic activity occurs above the descending plate. Where slabs of oceanic lithosphere converge, the oldest, most dense is subducted and an arc of volcanic islands is formed parallel to the trench. Transform boundaries arise where plates are moving past each other. No plate is created or destroyed. They can occur where segments of a divergent boundary are offset, and extensive fracture zones can result.

ocean ridges at divergent plate boundaries

transform plate boundary

direction of plate movement

DIVERGENT AND TRANSFORM BOUNDARIES
At divergent boundaries, parallel ridges emerge as new ocean floor spreads out on either side of an ocean ridge. A transform boundary arises when sections of the ridge are offset from each other.

magma rises from mantle

movement of oceanic lithosphere

CONVERGENT BOUNDARIES
Ocean lithosphere is destroyed by subduction at convergent boundaries. The subducting plate carries water with it, which allows the surrounding mantle to melt, forming explosive volcanoes above.

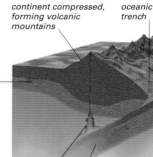

continent compressed, forming volcanic mountains

oceanic trench

movement of oceanic lithosphere

movement of continental lithosphere

magma forms as plate descends

oceanic lithosphere subducted beneath continental lithosphere

oceanic crust

Earthquakes and Tsunamis

Earthquakes are associated with all plate boundaries, but they are particularly frequent at convergent boundaries, such as subduction zones. Stress builds up at faults in the crust until it overcomes the strength of the rock and the fault slips. When this happens, a huge amount of energy can be released in a short time. The earthquake that produced the 2011 Tohoku Tsunami (see pp.462–463) in Japan released 45,000 times more energy than the Hiroshima atomic bomb.

A tsunami may be triggered if an earthquake results in the uplift or subsidence of part of the seafloor. The water above suddenly rises or sinks, then flows to regain equilibrium. Surface waves radiate out at 310–497 mph (500–800 kph) and can quickly cross an entire ocean basin.

TSUNAMI ALERTS

Tsunamis can be very destructive, so systems have been established to look out for their distinctive signs and give warning of their approach. These systems use networks of seismic stations to detect earthquakes, and automated deep-sea buoys with seafloor pressure sensors to confirm whether a tsunami has been generated. The prototype buoy pictured here is destined for seismic monitoring off the Caribbean coast of Grenada.

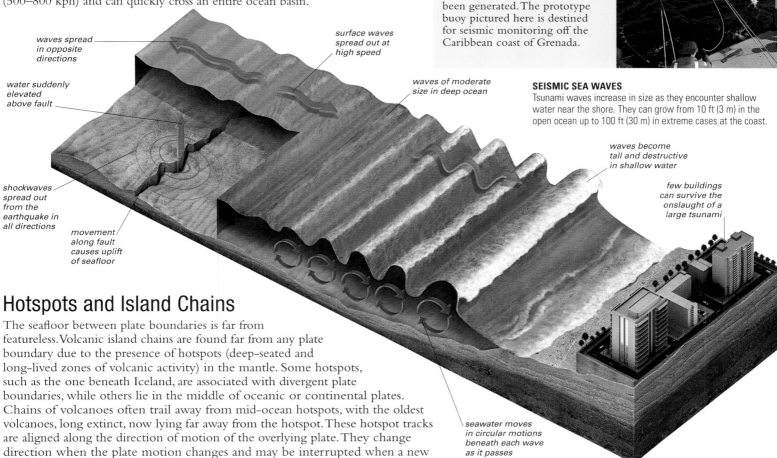

waves spread in opposite directions

surface waves spread out at high speed

water suddenly elevated above fault

waves of moderate size in deep ocean

SEISMIC SEA WAVES
Tsunami waves increase in size as they encounter shallow water near the shore. They can grow from 10 ft (3 m) in the open ocean up to 100 ft (30 m) in extreme cases at the coast.

shockwaves spread out from the earthquake in all directions

waves become tall and destructive in shallow water

movement along fault causes uplift of seafloor

few buildings can survive the onslaught of a large tsunami

Hotspots and Island Chains

The seafloor between plate boundaries is far from featureless. Volcanic island chains are found far from any plate boundary due to the presence of hotspots (deep-seated and long-lived zones of volcanic activity) in the mantle. Some hotspots, such as the one beneath Iceland, are associated with divergent plate boundaries, while others lie in the middle of oceanic or continental plates. Chains of volcanoes often trail away from mid-ocean hotspots, with the oldest volcanoes, long extinct, now lying far away from the hotspot. These hotspot tracks are aligned along the direction of motion of the overlying plate. They change direction when the plate motion changes and may be interrupted when a new spreading ridge opens up, as it has between India and the Réunion hotspot.

seawater moves in circular motions beneath each wave as it passes

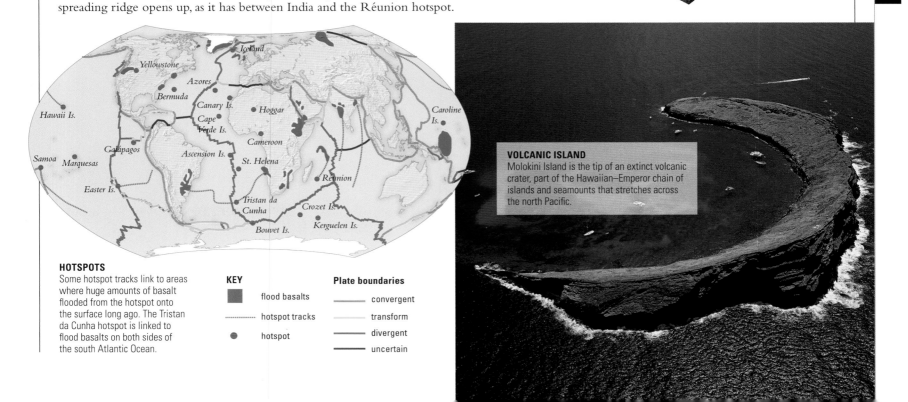

HOTSPOTS
Some hotspot tracks link to areas where huge amounts of basalt flooded from the hotspot onto the surface long ago. The Tristan da Cunha hotspot is linked to flood basalts on both sides of the south Atlantic Ocean.

KEY

flood basalts

hotspot tracks

hotspot

Plate boundaries

convergent

transform

divergent

uncertain

Iceland
Yellowstone
Azores
Bermuda
Canary Is.
Cape Verde Is.
Hoggar
Caroline Is.
Hawaii Is.
Cameroon
Galapagos
Ascension Is.
St. Helena
Samoa
Marquesas
Réunion
Easter Is.
Tristan da Cunha
Crozet Is.
Kerguelen Is.
Bouvet Is.

VOLCANIC ISLAND
Molokini Island is the tip of an extinct volcanic crater, part of the Hawaiian–Emperor chain of islands and seamounts that stretches across the north Pacific.

ERUPTION OF A SUBMARINE VOLCANO
Much of the world's volcanic activity occurs below the ocean
surface. Occasionally, an undersea volcano that has been growing
for millennia reaches the sea surface and produces a dramatic
eruption—as can be seen here, in an event that occurred near
Hunga Ha'apai, Tonga, in the southwest Pacific, in March 2009.

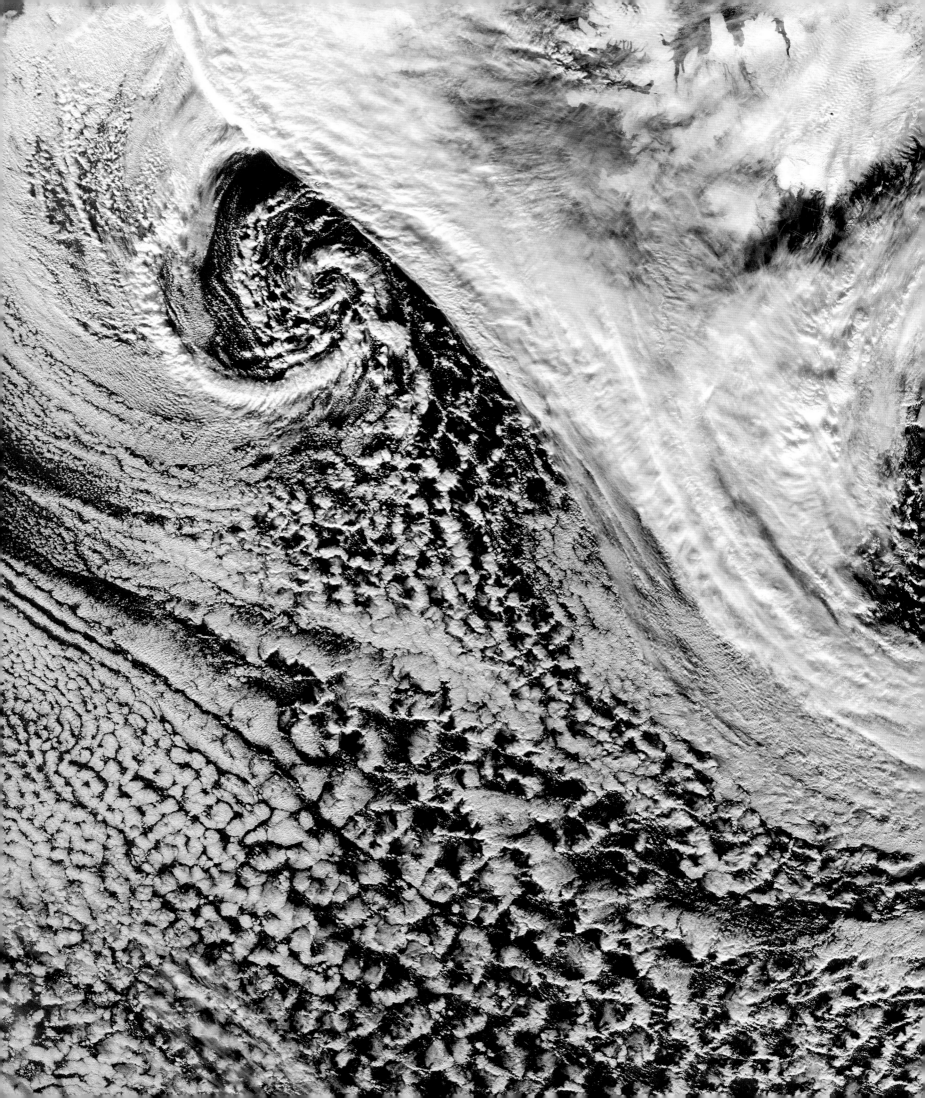

OCEAN WATER IS constantly in motion, and not simply in the form of waves. Throughout the oceans, there is a continuous circulation of seawater, both across the surface and more slowly deeper down. Several related processes play a part in causing and maintaining these ocean currents. They include solar heating of the atmosphere, prevailing winds, the effect of Earth's rotation, and processes that affect the temperature and salinity of surface waters. The various surface currents that are generated, some warm, some cold, have profound effects on climate in many parts of the world. Oceanic processes also play a part in the periodic climatic disturbances called El Niño and La Niña, and they help generate the extreme weather phenomena known as hurricanes and typhoons.

CIRCULATION AND CLIMATE

SPIRALING STORMS
Two cyclones—spiraling areas of low atmospheric pressure accompanied by cloud—are visible in this satellite image of part of the North Atlantic, taken in late 2006. The cyclones are moving eastward to the south of Iceland, which can be seen at top center.

Ocean Winds

THE PATTERN OF AIR MOVEMENT over the oceans results from solar heating of the atmosphere and Earth's rotation. This pattern of winds is modified by linked areas of low and high pressure (cyclones and anticyclones), which continually move over the oceans' surface. Near coasts, additional onshore and offshore breezes are common. These are caused by differences in the capacity of sea and land to absorb heat.

CIRCULATION CELLS
The atmospheric cells produce north–south airflows. These are modified by Earth's spin, producing winds that blow diagonally.

polar easterly — polar cell — air rises in subpolar latitudes
polar-front jet stream—narrow ribbon of strong wind at high altitude at top of front
Ferrel cell
southwesterly wind
air descends in subtropical latitudes
Hadley cell
air rises at equator
direction of Earth's spin
north-easterly trade wind
subtropical jet stream
southeasterly trade wind
polar-front jet stream
air descends at pole
trade winds meet at Intertropical Convergence Zone

Atmospheric Cells

Solar heating causes the air in Earth's atmosphere to cycle around the globe in three sets of giant loops, called atmospheric cells. Hadley cells are produced by warm air rising near the equator, cooling in the upper atmosphere, and descending to the surface around subtropical latitudes (30°N and S). Then the air moves back toward the equator. Ferrel cells are produced by air rising around subpolar latitudes (60°N and S), cooling and falling in the subtropics, and then moving toward the poles. Polar cells are caused by air descending at the poles and moving toward the equator.

The Coriolis Effect

initial direction of air movement
air deflected to right

The atmospheric cells cause north–south air movements. These are altered by the Coriolis effect. As the Earth spins, parcels of air at different latitudes in the atmosphere have different west-to-east velocities (air at the Equator moves fastest). When they change latitude by moving to the north or south, they retain these west-to-east velocities, which differ from those of air in the latitudes they move into. Hence, the air veers to the east (in the direction of Earth's spin) when moving away from the Equator and to the west when moving toward it.

AIR DEFLECTIONS
In the Northern Hemisphere, the Coriolis effect causes all air movements to be deflected to the right of their initial direction. In the Southern Hemisphere, they veer to the left.

Earth's rotation
air deflected to left
initial direction of air movement

SATELLITE IMAGING

Ocean winds are monitored by instruments called scatterometers, such as an instrument called ASCAT on the METOP-C satellite (right), launched in 2018. A scatterometer is a radar device that can measures wind speed and direction.

ASCAT antenna (one of three)

Prevailing Winds

The winds produced by pressure differences and modified by the Coriolis effect are called the prevailing winds. In the tropics and subtropics, the air movements toward the equator in Hadley cells are deflected to the west. These are known as trade winds. They comprise the northeasterly trades in the Northern Hemisphere, and southeasterly trades in the south. At higher latitudes, the surface winds in Ferrel cells deflect to the east, producing the westerlies. In the Southern Hemisphere, these winds blow from west to east without meeting land. Those around latitudes of 40°S are known as the Roaring Forties. In polar regions, winds deflect to the west as they move away from the poles. These are known as polar northeasterlies and southeasterlies.

westerlies
polar north-easterlies
westerlies
northeasterly monsoon (Nov–Mar)
northeasterly trade winds
Tropic of Cancer
Intertropical Convergence Zone
equator
Tropic of Capricorn
southeasterly trade winds
westerlies
southeasterlies
southeasterly trade winds
westerlies
southwesterly monsoon (Apr–Oct)
southeasterly trade winds

KEY

→	prevailing warm
→	prevailing cool
→	local warm
→	local cool

PATTERN OF WINDS
Year-round, the winds over most oceans are trades or westerlies. An exception is the northern Indian Ocean—this has a monsoon climate, in which a seasonal switch in wind direction occurs.

LONG-HAUL SAILING
Winds can blow with a consistent strength and direction over large areas of ocean. Consequently, on long-haul sailing trips, the same basic sail settings can often be used for days on end.

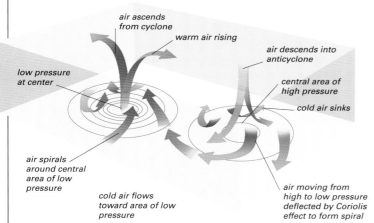

CYCLONES AND ANTICYCLONES
Air moves from an area of high pressure toward one of low pressure, but the Coriolis effect modifies this, producing circular winds.

Pressure-system Winds

In any area of ocean where air sinks—often at subtropical latitudes—a zone of high atmospheric pressure, or anticyclone, develops. Where warm air rises, areas of low pressure, called cyclones or depressions, occur. These often develop near the equator and subpolar latitudes. Cyclones and anticyclones create linked, circulating wind patterns, which continually move and change. In the Northern Hemisphere, there is a clockwise movement of air around an anticyclone, and a counterclockwise motion around a cyclone. This pattern is reversed in the Southern Hemisphere. Local pressure systems can affect the general pattern of prevailing winds. In particular, cyclones move swiftly over the ocean and can produce rapid changes in wind strength and direction.

Coastal Breezes

Local winds, called onshore and offshore breezes, are generated near coasts, especially in sunny climes. Onshore breezes—sometimes called sea breezes—develop during the day. These are caused by the land heating up more quickly than the sea, as both absorb solar radiation. This occurs because the sea absorbs large quantities of heat energy with only a small rise in temperature, whereas the same amount of heat energy is likely to cause the land temperature to rise sharply (see p.31).

As the land warms up, it heats the air above it, causing the air to rise. Cooler air then blows in from the sea to take its place. In the evening, and at night, the opposite effect occurs. At nightfall, the land quickly cools down, but the sea remains warm and continues to heat the air above it. As this warm air rises, it sucks the cooler air off the land, and so generates an offshore breeze. This is sometimes called a "land breeze."

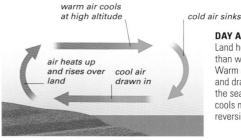

DAY AND NIGHT
Land heats up faster than water during the day. Warm air rises over the land and draws in cold air from the sea. At night, the land cools more quickly, reversing the airflow.

ONSHORE BREEZE

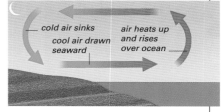

OFFSHORE BREEZE

BREEZY COAST
On warm coasts, there is often a noticeable drop in temperature from midday as a cool sea breeze blows in off the water. The breeze typically reverses in the evening and at night.

TRIMMING THE SAILS
A crew sets their sails as they set off on the Sydney-to-Hobart yacht race. The course crosses the often stormy Bass Strait between Australia and Tasmania.

Ocean yacht racing is the sport of competitive sailing, held over long distances and in open water. These races range from short but robust challenges lasting a few days, such as the annual Sydney-to-Hobart race, to long, multistage, around-the-world races, which can involve up to 6 months at sea. Some races are multihanders, with crews that can be as large as 20 or more per yacht; others are single-handers. The participants are typically highly experienced sailors. In multihanded races, there will be a skipper/tactician, a navigator, and general crew whose responsibilities include sail changing and trimming. Solo racers have to do everything themselves. One race, the Clipper Round the World Yacht Race, is unusual in that its crew consists of amateurs, some with little previous sailing experience, who have paid to take part under the leadership of a professional skipper.

To make races as equitable as possible, usually competing boats are identical or a handicapping system is used to adjust the times of different classes of boats. The use of computer technology is paramount in modern racing. Navigation is electronically assisted, and computers are employed to monitor and help optimize boat performance. During a race, vast amounts of weather data are downloaded using satellite communications hardware on board. An important skill is to be able to interpret this data, so as to know, for example, where the most wind is likely to be in the area ahead. Otherwise, doing well in a race is mainly down to tactics and seamanship—for example, knowing how to get the best out of a boat in both strong and light winds, or judging when best to tack (change course when sailing upwind).

Racing Around the Globe

Three of the most famous ocean yacht races are global circumnavigations. They go "round the right way" in the Southern Ocean (west to east, the same direction as the prevailing winds and currents). The Vendée Globe is a single-handed, nonstop race held once every four years. The Ocean Race (formerly called the Volvo Ocean Race) is a team event in stages, held once every three or four years. The Clipper Round the World Race is also held in stages, with some changes of crew between stages. It usually occurs biennially (once every two years), but the COVID-19 pandemic that started in 2020 severely disrupted the 2019–2020 race and the biennial schedule.

Seattle
New York
Newport
Liverpool
The Hague
Aarhus
Les Sables-d'Olonne
Genoa
Alicante
Shenzhen
Sanya
Qingdao
ATLANTIC OCEAN
PACIFIC OCEAN
Panama Canal
Cape Verde Islands
INDIAN OCEAN
PACIFIC OCEAN
PACIFIC OCEAN
Hajai
Punta del Este
Cape Town
Fremantle
Sydney
Hobart
Auckland
SOUTHERN OCEAN

→ Vendée Globe
→ Clipper Round the World Yacht Race 2017–2018
→ The Ocean Race 2022–2023

EQUIPPED TO WIN

YACHTMANSHIP

BEATING UPWIND *Competing in the 2017–2018 Volvo Ocean Race, Team Mapfre's crew strive to keep the boat balanced in a strong wind. To sail upwind (into the wind), a yacht must follow a zigzag course consisting of straight sailing sections, each at a 45° angle to the wind direction, punctuated by 90° turns. This is called "beating upwind" and is a major test of yachtmanship.*

YACHT CREWS

ALL HANDS ON DECK *The crews for some races are large— in the Clipper Round the World Yacht Race, there can be as many as 22 on board at any one time. For this race, the crew, who are recently trained amateurs, participate in all duties on the yacht, including trimming sails, navigating, helming, cooking, and so on.*

HIGH-TECH AIDS

CONTROL CENTER
Technology is all-important in modern racing, including the use of electronic charts and global positioning systems. Here, Frenchman Marc Thiercelin prepares for the Vendée Globe 2005–2006 in the control center on his yacht Pro-Form.

GROUNDED *During a 2014 race, Team Vestas Wind crashed into an Indian Ocean reef during the night, exposing its crew to great peril. They were rescued and are seen here retrieving equipment.*

HANGING ON *The crew of the yacht Astra (shown here) risk being swept overboard, the greatest disaster that can befall a sailor. To assist recovery in this eventuality, crew members usually wear radio beacons. It is also usual to be attached to the boat via a safety harness in anything other than calm daylight conditions.*

RISKY BUSINESS

Surface Currents

FLOWING FOR ENORMOUS DISTANCES within the upper regions of
the oceans are various wind-driven currents. Many join to produce
large circular fluxes of water, called gyres, around the surfaces of the
main ocean basins. Surface currents affect only about 10 percent of ocean water, but
they are important to the world's climate (see p.66), because their overall effect is to
transfer huge amounts of heat energy from the tropics to cooler parts of the globe.
They also impact shipping and the world's fishing industries.

Wind on Water

When wind blows over the sea, it causes the upper ocean to move, creating a
current. However, the water does not move in the same direction as the wind.
Instead, it moves off at an angle—to the right in the Northern Hemisphere and
to the left in the Southern Hemisphere. This phenomenon was first explained in
1902 by a Swedish scientist, Walfrid Ekman, using a model of the effect of wind
on water now called the Ekman spiral. The model assumes that the movement of
water in each layer of the upper ocean is produced by a combination of frictional
drag from the layer above (or, in the top layer, from wind drag) and the Coriolis
effect (see p.54). The model predicts that, overall, a mass of water will be pushed
at right angles to the wind direction, an effect known as Ekman transport.

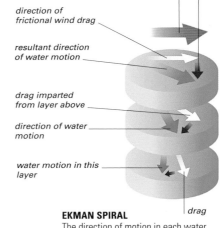

Coriolis deflection

wind

direction of frictional wind drag

resultant direction of water motion

drag imparted from layer above

direction of water motion

water motion in this layer

drag

EKMAN SPIRAL
The direction of motion in each water
layer results from a combination of the
drag from the layer above and a deflection
caused by the Coriolis effect. This diagram
shows the Ekman spiral in the Northern
Hemisphere. In the Southern Hemisphere,
deflection
is to the left of
wind direction.

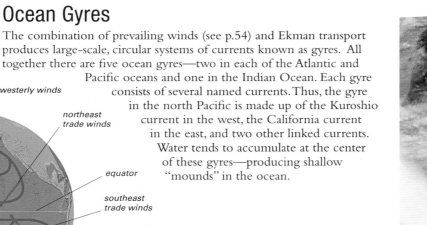

Labrador · N. Atlantic Drift · Canary

N. Equatorial · Gulf Stream · E. Greenland · Agulhas · Somali

Alaska · Oyashio

N. Pacific · Kuroshio

California

N. Equatorial

Equatorial Counter

S. Equatorial

Antarctic Circumpolar

E. Australia

W. Australia

Peru · S. Equatorial · Benguela · S. Equatorial

Brazil

MAIN CURRENTS
This map shows all
of the world's main
surface currents,
both warm and cold.

→ warm current

→ cold current

Ocean Gyres

The combination of prevailing winds (see p.54) and Ekman transport
produces large-scale, circular systems of currents known as gyres. All
together there are five ocean gyres—two in each of the Atlantic and
Pacific oceans and one in the Indian Ocean. Each gyre
consists of several named currents. Thus, the gyre
in the north Pacific is made up of the Kuroshio
current in the west, the California current
in the east, and two other linked currents.
Water tends to accumulate at the center
of these gyres—producing shallow
"mounds" in the ocean.

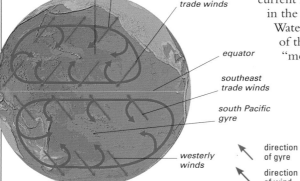

north Pacific gyre

westerly winds

northeast trade winds

equator

southeast trade winds

south Pacific gyre

westerly winds

↖ direction of gyre

↖ direction of wind

GYRE CREATION
In the north Pacific, the combination of
westerly and trade winds, always pushing
water to the right (by Ekman transport)
produces a clockwise gyre. In the south
Pacific, where winds push water to the left,
a counterclockwise gyre is created.

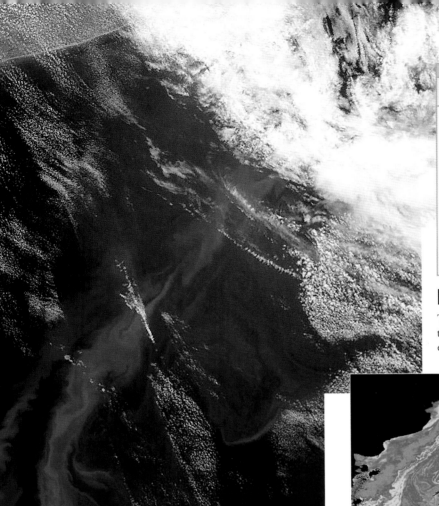

BENJAMIN FRANKLIN

The American statesman and inventor Benjamin Franklin (1706–1790) made one of the earliest studies of an ocean current, publishing a map of the Gulf Stream's course. He became interested in it after the British postal authorities asked him why American postal ships crossed the Atlantic faster than English ships. The answer was that American ships were utilizing an eastward extension of the Gulf Stream.

Boundary Currents

The currents at the edges of gyres are called boundary currents. Those on the western side of gyres are strong, narrow, and warm—they move heat energy away from the equator. Examples of these currents are the Gulf Stream and the Brazil Current in the southwestern Atlantic. Eastern boundary currents are weaker, broader cold currents that move water back toward the tropics. Examples are the Benguela Current off southwest Africa and the California Current. At the gyre boundaries close to the equator are warm, west-flowing equatorial currents. Other currents feed into or out of the main gyres. These include, for example, the warm North Atlantic Drift, an offshoot of the Gulf Stream, and cold currents that bring water down from the Arctic, such as the Oyashio and East Greenland currents.

WARM CURRENT
Satellite devices can detect phytoplankton levels in the water, which can be related to temperature. Here, yellow and red indicate high levels of plankton and the warm Brazil Current.

COLD CURRENT
In this satellite view, sea ice is visible flowing past the Kamchatka Peninsula in the cold Oyashio Current. Eddies within the current have produced spiral patterns in the sea ice.

Meeting of Currents

In a few areas, warm and cold currents meet and interact. Examples include the meeting of the warm Gulf Stream with the cold Labrador Current off the eastern seaboard of the US and Canada, and the meeting of the cold Oyashio Current with the warm Kuroshio Current to the north of Japan. At these confluences, the denser water in the cold current dives beneath the water in the warm current, usually producing some turbulence. This can trigger an upward flow of nutrient-rich waters from the sea floor, encouraging the growth of plankton, and producing good feeding grounds for fish, sea birds, and mammals.

SEA SMOKE
Dolphins cavort amid steep waves. The "sea smoke" is created when water vapor is added to cold air drifting across the boundary between cold and warm currents.

OPPOSING CURRENTS
The warm Brazil Current on the left, and the colder Falklands Current on the right, each carry differently colored populations of plankton.

Underwater Circulation

THE WATERS THAT MAKE UP EARTH'S OCEANS circulate deep below the surface. Subsurface currents are complex. Some are vertical, moving water upward and downward to and from the surface, processes called upwelling and downwelling. Surface and subsurface currents are all linked in a global pattern of deep-water circulation.

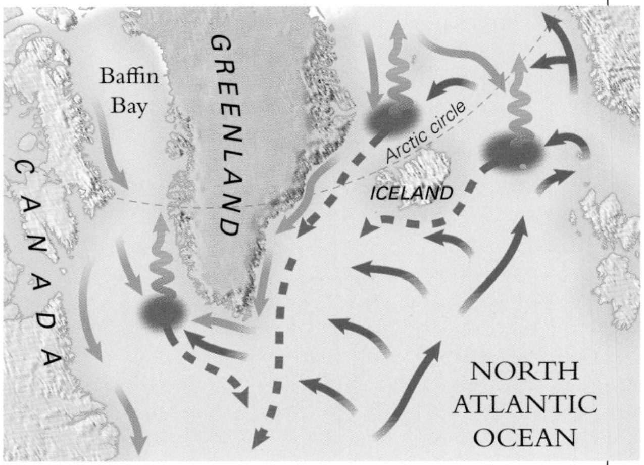
Baffin Bay
GREENLAND
CANADA
Arctic circle
ICELAND
NORTH ATLANTIC OCEAN

NORTH ATLANTIC DOWNWELLING ZONES
At the important downwelling sites shown here, warm surface water meets colder Arctic water, loses heat, become denser, and sinks.

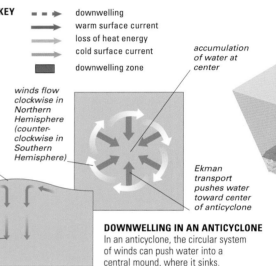

KEY
- downwelling
- warm surface current
- loss of heat energy
- cold surface current
- downwelling zone

Downwelling

The most important causes of downwelling are thermohaline processes ("thermo" means heat, and "haline" means salt), which alter either the temperature or salinity of seawater. For example, where warm, salty water is carried by a surface current into the Arctic Ocean, it rapidly cools when it meets colder, less salty, polar water. As it cools, its density increases, and it sinks down. Downwelling also occurs on some coasts. For example, winds blowing toward the equator on the western side of oceans push seawater toward land by Ekman transport (see p.58). As it reaches the coast, it is forced down. Finally, downwelling also occurs beneath the mounds of water that accumulate in the middle of anticyclones (see p.55) and ocean gyres (see p.58).

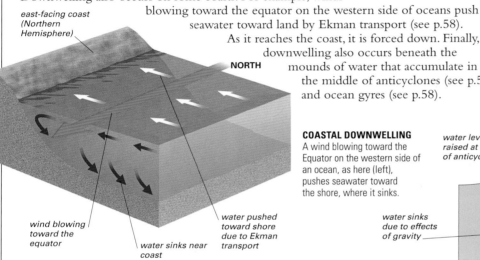

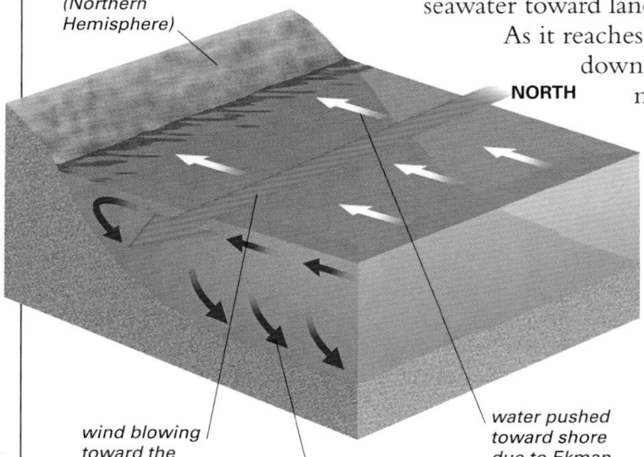
east-facing coast (Northern Hemisphere)
NORTH
wind blowing toward the equator
water sinks near coast
water pushed toward shore due to Ekman transport

COASTAL DOWNWELLING
A wind blowing toward the Equator on the western side of an ocean, as here (left), pushes seawater toward the shore, where it sinks.

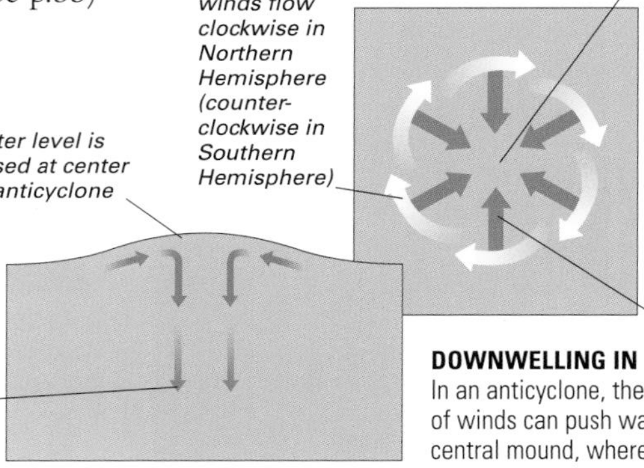
accumulation of water at center
winds flow clockwise in Northern Hemisphere (counter-clockwise in Southern Hemisphere)
water level is raised at center of anticyclone
Ekman transport pushes water toward center of anticyclone
water sinks due to effects of gravity

DOWNWELLING IN AN ANTICYCLONE
In an anticyclone, the circular system of winds can push water into a central mound, where it sinks.

Upwelling

Upwelling can occur in various situations, some of which are simply the reverse of the conditions that cause downwelling. For instance, winds blowing toward the equator on the eastern sides of oceans push seawater away from land by Ekman transport, so deeper water must upwell near the coast to replace it. Water rises toward the surface in the center of cyclones (the opposite of anticyclones, see p.55), and will also rise where surface waters tend to be pushed apart at boundaries between ocean gyres—for example, in some equatorial parts of the Pacific and Atlantic. Some seawater upwells to replace sinking, denser, water. An example occurs around Antarctica, where upwelling replaces super-dense, cold, salty water forming and sinking under developing sea ice.

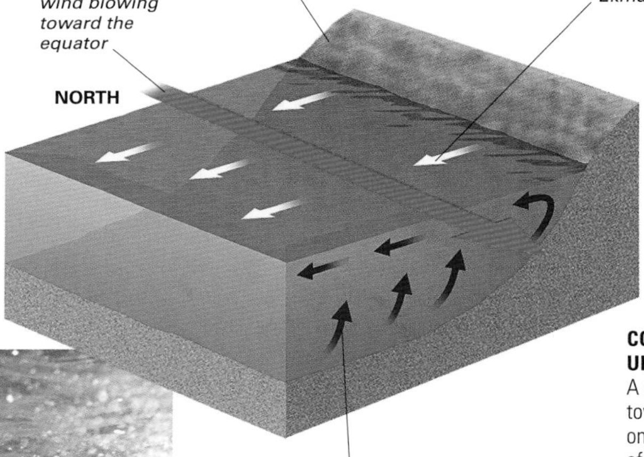
west-facing coast (Northern Hemisphere)
wind blowing toward the equator
NORTH
water moves away from shore as a result of Ekman transport
water moves upward to replace the water moving offshore at the surface

COASTAL UPWELLING
A wind blowing toward the equator on the eastern side of an ocean, as here, pushes seawater away from the shore, causing upwelling near the coast.

PLANKTON-HARVESTER
Where upwelling occurs, it brings large amounts of nutrients up from the sea floor. These encourage the growth of plankton, attracting plankton-grazers such as this manta ray as well as smaller fish, whales, and other marine life.

Deep-water Circulation

Seawater circulates slowly through the deeper parts of the oceans, driven by water sinking in major downwelling zones, such as in the north Atlantic. Analysis of seawater samples from various parts of the deep oceans indicates that there is a large-scale circulation involving all the oceans, called the global conveyor. A specific mass of seawater takes about 1,000 years to complete a lap of this circuit. Concerns have been raised that the melting of Arctic sea ice and the Greenland Ice Sheet and increased river runoff, all caused by global warming, might disrupt the global conveyor by flooding the Arctic water with fresh water, which would interfere with downwelling in the North Atlantic. This could have an effect on the climate of Europe. However, although some weakening of the conveyor in the North Atlantic has been measured, experts do not believe there will be any dramatic consequences during this century.

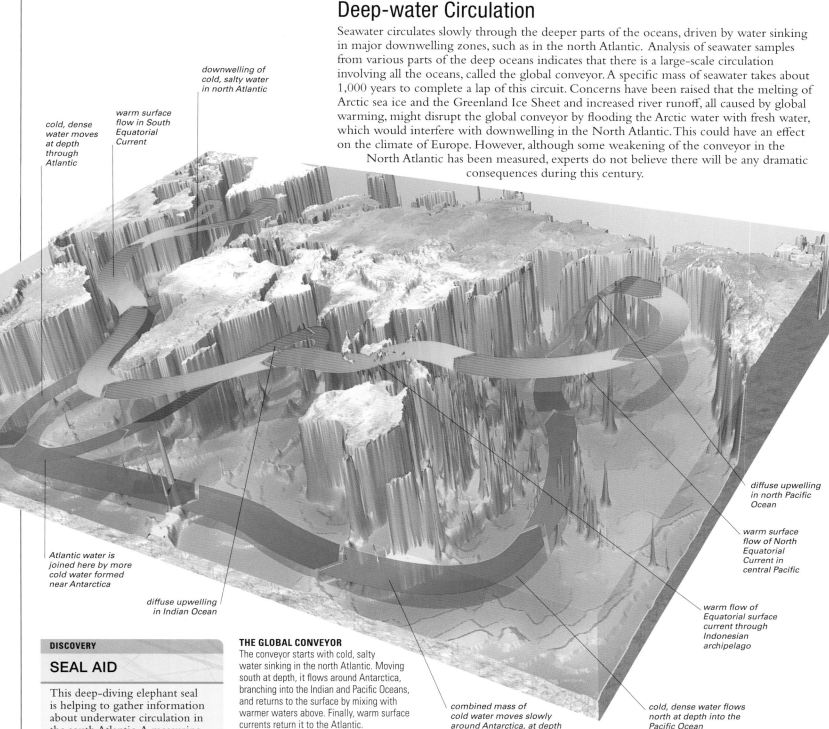

downwelling of cold, salty water in north Atlantic

warm surface flow in South Equatorial Current

cold, dense water moves at depth through Atlantic

Atlantic water is joined here by more cold water formed near Antarctica

diffuse upwelling in Indian Ocean

diffuse upwelling in north Pacific Ocean

warm surface flow of North Equatorial Current in central Pacific

warm flow of Equatorial surface current through Indonesian archipelago

cold, dense water flows north at depth into the Pacific Ocean

combined mass of cold water moves slowly around Antarctica, at depth

THE GLOBAL CONVEYOR
The conveyor starts with cold, salty water sinking in the north Atlantic. Moving south at depth, it flows around Antarctica, branching into the Indian and Pacific Oceans, and returns to the surface by mixing with warmer waters above. Finally, warm surface currents return it to the Atlantic.

DISCOVERY

SEAL AID

This deep-diving elephant seal is helping to gather information about underwater circulation in the south Atlantic. A measuring device—attached to its head with glue that sloughs off when the animal molts—collects data about temperature and salinity at varying depths. The information gained may also help to conserve elephant seal populations.

Circulation Cells

One type of circulation that affects only the upper 70 ft (20 m) of the ocean, but is more complex than either a simple horizontal or vertical flow of water, is known as Langmuir circulation. This is wind-driven and consists of rows of long, cylinder-shaped cells of water, aligned in the direction in which the wind is blowing and each rotating in the opposite direction from its neighbor—alternate cells rotate clockwise and counterclockwise. Each cell is about 30–160 ft (10–50 m) wide and can be hundreds of yards long. On the sea surface, the areas between adjacent cells where seawater converges are visible as long white streaks of foam, or congregations of seaweed, called windrows. The whole pattern of circulation was first explained in 1938 by an American chemist named Irving Langmuir, after he crossed the Atlantic in an ocean liner. It was subsequently named in his honor.

LANGMUIR WINDROWS
These long streaks of foam on the sea surface are the windrows of Langmuir circulation cells. The distance between windrow lines increases with the wind speed.

PLASTIC BAG DANGER
Loggerhead turtles often mistake plastic bags for jellyfish, one of their main foods. If a turtle ingests a bag, it may end up causing a serious obstruction in the loggerhead's digestive tract, which may eventually cause its death.

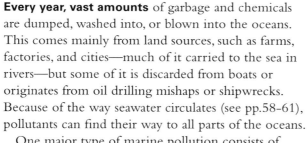

Every year, vast amounts of garbage and chemicals are dumped, washed into, or blown into the oceans. This comes mainly from land sources, such as farms, factories, and cities—much of it carried to the sea in rivers—but some of it is discarded from boats or originates from oil drilling mishaps or shipwrecks. Because of the way seawater circulates (see pp.58-61), pollutants can find their way to all parts of the oceans.

One major type of marine pollution consists of plastic objects. This floating plastic, and other debris, tends to accumulate in a few specific surface areas called garbage patches (see map, below). The plastic objects eventually fragment into vast numbers of tiny particles called microplastics. Much of each garbage patch consists of a "soup" of these particles.

Another form of marine pollution, with chemicals, can take many forms. One of these, nutrient pollution, occurs when agricultural fertilizers enter rivers and the sea. The resulting raised level of nutrients in coastal areas promotes the growth of algal blooms, which can be toxic to organisms such as corals. Blooms also remove oxygen from the water, which can cause fish to suffocate. Other pollution consists of an array of poisonous chemicals. Although diluted in the ocean, these can become concentrated if they enter and move up food chains (see p.212), eventually reaching levels that can be harmful to predators such as seals and large fish, and to humans who eat fish.

Solutions proposed for tackling marine pollution include prevention and cleanup; however, changing peoples' attitude toward plastic will also be vital.

Ocean Garbage Patch Locations

The ocean garbage patches have formed in the middle of ocean gyres—large-scale circular systems of currents (see p.58). Because ocean surface water tends to be pushed toward the centers of these gyres, floating debris also accumulates there. The largest patch, known as the Great Pacific Garbage Patch or Pacific Trash Vortex, is in the North Pacific and has distinct western and eastern parts. It contains an estimated 100 million tons (90 million metric tons) of debris. Other smaller garbage patches exist in the South Pacific, North Atlantic, South Atlantic, and Indian Oceans.

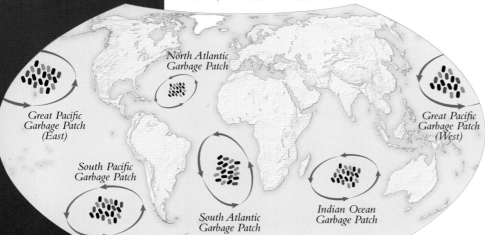

Great Pacific Garbage Patch (East)

North Atlantic Garbage Patch

South Pacific Garbage Patch

South Atlantic Garbage Patch

Indian Ocean Garbage Patch

Great Pacific Garbage Patch (West)

POLLUTION HAZARDS

MARINE PLASTICS

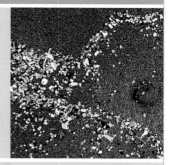

MICROPLASTICS Microplastics are plastic particles smaller than ⅕ in (5 mm) across. Today, hundreds of trillions of these particles are present in surface areas of the oceans, and some have been carried to the deepest ocean depths. These particles can cause physical harm to small marine creatures in many ways, such as by filling their stomachs and impairing their ability to feed.

ENTANGLEMENT

NET ENTRAPMENT Entanglement in items such as discarded fishing nets results in the deaths of hundreds of thousands of marine mammals, sea turtles, and birds (such as this guillemot) each year. Most occur as a result of the animal being trapped and then drowning or starving to death.

OIL SPILLS

GULF SPILL Catastrophic marine oil spills occur from time to time as a result of mishaps involving oil tankers or from drilling operations going wrong, as shown in the Gulf of Mexico in 2010. Oil can damage oceanic wildlife by fouling—for example, by coating a bird's wings so it cannot fly—or by releasing toxic chemicals into the water.

BEACH POLLUTION

HEALTH HAZARD One obvious indicator of the current level of marine pollution is the hundreds of tons of trash that wash up on beaches all over the world each day. Much of this is plastic, but it can also include items such as shoes, diapers, rubber tires, and clumps of oil, as well as raw sewage. Apart from looking extremely unsightly, polluted beaches pose a serious health risk to people and animals who visit them.

CLEANING UP

FILTER CLEANING Ridding the oceans of plastic debris and other pollutants would seem like a daunting task. But an environmental organization called The Ocean Cleanup has made a start after developing systems for filtering debris from ocean garbage patches and for stopping trash from entering the ocean. Its System 002 is shown here being tested.

The Global Water Cycle

THE WORLD'S OCEANS DO NOT FORM a self-contained system but continually exchange water with the atmosphere and landmasses through evaporation, cloud formation, precipitation, wind transport, and river flow. This complex of interconnected processes, which is ultimately driven by heat from the Sun, is called the global water cycle or hydrologic cycle. The cycle is made up of many smaller cycles, such as the formation and melting of sea ice.

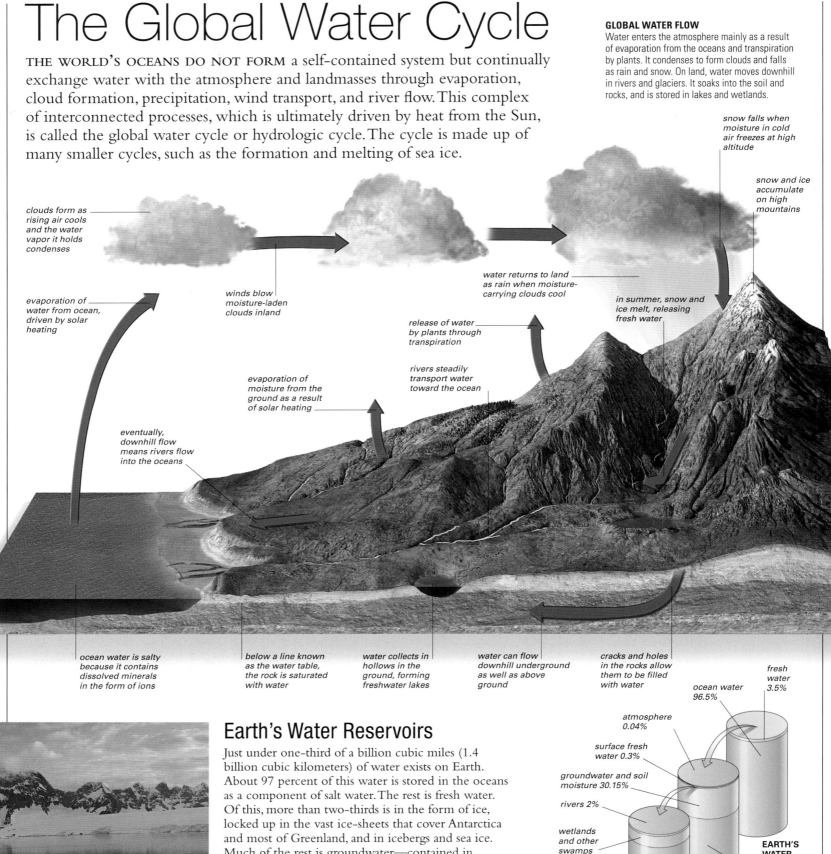

clouds form as rising air cools and the water vapor it holds condenses

snow falls when moisture in cold air freezes at high altitude

snow and ice accumulate on high mountains

evaporation of water from ocean, driven by solar heating

winds blow moisture-laden clouds inland

water returns to land as rain when moisture-carrying clouds cool

in summer, snow and ice melt, releasing fresh water

release of water by plants through transpiration

evaporation of moisture from the ground as a result of solar heating

rivers steadily transport water toward the ocean

eventually, downhill flow means rivers flow into the oceans

ocean water is salty because it contains dissolved minerals in the form of ions

below a line known as the water table, the rock is saturated with water

water collects in hollows in the ground, forming freshwater lakes

water can flow downhill underground as well as above ground

cracks and holes in the rocks allow them to be filled with water

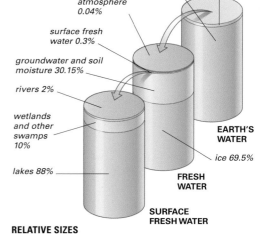

PLAYERS IN THE CYCLE
The sea, ice, mountains, and clouds all play a part in the global water cycle. This coastal scene is near Port Lockroy in Antarctica.

Earth's Water Reservoirs

Just under one-third of a billion cubic miles (1.4 billion cubic kilometers) of water exists on Earth. About 97 percent of this water is stored in the oceans as a component of salt water. The rest is fresh water. Of this, more than two-thirds is in the form of ice, locked up in the vast ice-sheets that cover Antarctica and most of Greenland, and in icebergs and sea ice. Much of the rest is groundwater—contained in underground rocks—while a tiny amount (less than 1 part in 2,000) is water vapor in the atmosphere. Fresh liquid water on the Earth's land surface, in lakes, wetlands, and rivers, makes up just 0.3 percent of all the world's fresh water, or 0.02 percent of the total water. The Earth's different water reservoirs have not always had the same relative sizes that they have today. For instance, during the ice ages, a higher proportion was locked up in ice, with less in the oceans.

ocean water 96.5%

fresh water 3.5%

atmosphere 0.04%

surface fresh water 0.3%

groundwater and soil moisture 30.15%

rivers 2%

wetlands and other swamps 10%

lakes 88%

ice 69.5%

EARTH'S WATER

FRESH WATER

SURFACE FRESH WATER

RELATIVE SIZES
Earth's ocean water (the bulk of the rear cylinder, above) hugely exceeds its reservoirs of fresh water, and the relative proportion of fresh water found on the land surface is tiny.

Ocean Evaporation and Precipitation

A total of 108,000 cubic miles (450,000 cubic kilometers) of water evaporates from the oceans per year. Of this, about 96,800 cubic miles (403,500 cubic kilometers) falls back into the sea as precipitation (rain, snow, sleet, and hail). The remainder is carried onto land as clouds and moisture. Evaporation and precipitation are not evenly spread over the surface of the oceans. Evaporation rates are greatest in the tropics and lowest near the poles. High rates of precipitation occur near the equator and in bands between the latitudes of 45° and 70° in both hemispheres. Drier regions are found on the eastern sides of the oceans between the latitudes of approximately 15° and 40°.

PEOPLE

SENECA THE YOUNGER

In his book *Natural Questions*, the Roman statesman, dramatist, and philosopher Seneca the Younger (4 BCE–65 CE) pondered why ocean levels remain stable despite the continuous input of water from rivers and rain. He argued there must be mechanisms by which water is returned from the sea to the air and land and proposed an early version of the hydrologic cycle to explain this.

Freshwater Inflow

The net loss of 11,200 cubic miles (46,500 cubic kilometers) of water from the oceans each year by evaporation and transport onto land is balanced by an equal amount returned from land in runoff. Just 20 rivers, including the Amazon and some large Siberian rivers, account for over 40 percent of all input into the oceans. Inflows from the different river systems change over time as they are affected by human activity and climate change. For instance, global warming appears to have increased the flow from Siberian rivers into the Arctic Ocean, as water frozen in the tundra melts. These inflows lower the salinity of Arctic waters and may influence global patterns of ocean circulation (see p.62).

EQUATORIAL RAINSTORM
In some areas near the equator, as here in the tropical Pacific, annual rainfall is over 120 in (300 cm), compared to under 4 in (10 cm) in the driest ocean areas.

SIBERIAN RIVER AMUR FLOODING
Climate change is thought to have contributed to severe flooding of the Amur River in Northeast Asia in recent years (below). A color-enhanced satellite image (left) shows the engorged river in black; green areas are plant-covered land.

The Sea-ice Cycle

In addition to the overall global water cycle, there is a local seasonal cycle in the amount of water locked up as sea ice. In the polar oceans, the extent of sea ice increases in winter and decreases in summer. This has important climatic consequences, because sea-ice formation releases, and its melting absorbs, latent heat to and from the atmosphere; and because the presence or absence of sea ice modifies heat exchange between the oceans and atmosphere. In winter, sea ice insulates the relatively warm polar oceans from the much colder air above, thus reducing heat loss. However, especially when covered with snow, sea ice also has a high reflectivity (albedo) and reduces the absorption of solar radiation at the surface. Overall, the sea-ice cycle is thought to help stabilize air and sea temperatures in polar oceans. Also, because it affects surface salinity, sea-ice formation helps drive large-scale circulation of water through the world's oceans (see p.61).

SEA ICE FORMING
As sea ice forms, it releases heat to the atmosphere and increases the saltiness of the surrounding water (by rejecting salt). These processes affect climate and the circulation of seawater.

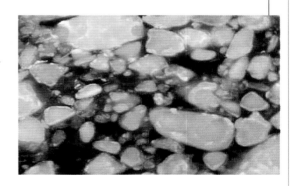

INTRODUCTION

Oceans and Climate

THE OCEANS HAVE A PROFOUND INFLUENCE on the world's climate, most strikingly in the way they absorb solar energy and redistribute it around the world in warm surface currents. Cold currents also produce local climatic effects, while alterations in currents are associated with climatic fluctuations such as El Niño (see p.68). The future behavior of the oceans is crucial to the future course of climate change, as they are an important store for carbon dioxide, the principal greenhouse gas.

SOLAR HEATING
The surface layers of the oceans absorb about half the solar energy that reaches Earth. Currents move this from the equator toward the poles at a rate of several billion megawatts.

Warm Currents

Five or six major surface currents (see p.58) carry heat away from the tropics and subtropics toward the poles, giving some temperate regions a warmer climate than they would otherwise enjoy. A prime example is the effect of the warm Gulf Stream and its extension, the North Atlantic Drift, on Europe. The North Atlantic Drift carries heat originally absorbed in the Caribbean Sea and Gulf of Mexico across the Atlantic, where it is released into the atmosphere close to the shores of France, the British Isles, Norway, Iceland, and other parts of northwestern Europe. As the prevailing westerly winds blow this warmed air over land, these countries benefit from a milder climate than equivalent regions at similar, or even lower latitudes, on the western side of the Atlantic. For example, winter temperatures are typically higher in Reykjavik, the capital of Iceland, than in New York. Similarly, in the northwest Pacific, the Kuroshio Current warms the southern part of Japan, while in the extreme southwest Pacific, the East Australian Current gives Tasmania a relatively mild climate.

BALMY BEACHFRONT
Penzance, in southwest England, has a mild climate that supports subtropical vegetation—the effect of the North Atlantic Drift is to raise temperatures here by about 9°F (5°C).

Cold Currents

In some instances, the climatic effect of cold currents is simply to produce a cooler climate than would otherwise be the case. For instance, the west coast of the US is cooled in summer by the cold California Current. Cold currents also affect patterns of rainfall and fog formation. In general, the various cold currents flowing toward the equator on the eastern sides of oceans—combined with upwellings of cold water from the depths in these regions—cool the air, reduce evaporative losses of water from the ocean, and cause downdrafts of drier air from higher in the atmosphere. Although clouds and fog often develop over the ocean in these areas (as what little moisture there is condenses over the cold water), these quickly disperse once the air moves over land. Thus, cold currents contribute to the development of deserts on land bordering the eastern sides of oceans, such as the Namib desert in southwestern Africa.

COAST OF NORTHERN CHILE
The cold Peru Current flows along the coast of northern Chile. It encourages the development of clouds and fog over the sea (visible above left in the satellite image) but also contributes to the extreme aridity of the coastal strip (left).

STRAYING NORTH

Most penguins live in Antarctica but, somewhat surprisingly, the world's most northerly-living penguins inhabit the Galápagos Islands, on the equator. The islands have a cool climate—sea-surface temperatures in most years average 9°F (5°C) less than typical temperatures in the tropics, due to the cold Peru Current that flows up the west coast of South America.

Carbon in the Oceans

The oceans contain Earth's largest store of carbon dioxide (CO_2)—the main greenhouse gas implicated in global warming. Huge amounts of carbon are held in the oceans, some in the form of CO_2 and related substances that readily convert to CO_2, and some in living organisms. The oceanic CO_2 is in balance with the atmospheric content of the same gas. For many years, the oceans have been alkaline, and acted as an important store for the excess CO_2 released by human activity. Biological and chemical processes turn some of this CO_2 into the calcium carbonate shells and skeletons of organisms, other organic matter, and carbonate sediments. However, the increasing CO_2 concentration is beginning to acidify the oceans, threatening shell and skeleton formation in marine organisms, as acid tends to dissolve carbonates. Further, some scientists fear that the rate at which the oceans can continue to absorb CO_2 will soon slow down, further aggravating global warming.

FORAMINIFERAN SHELL

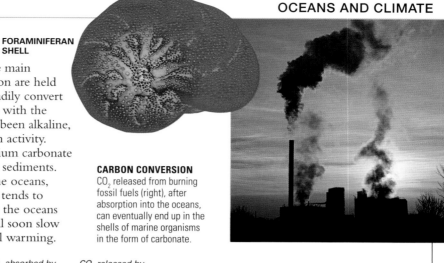

CARBON CONVERSION
CO_2 released from burning fossil fuels (right), after absorption into the oceans, can eventually end up in the shells of marine organisms in the form of carbonate.

METHANE HYDRATE DEPOSIT
This substance is found as a solid on some areas of sea floor. There are concerns that ocean warming could release this into the atmosphere as methane gas, which traps more heat than CO_2.

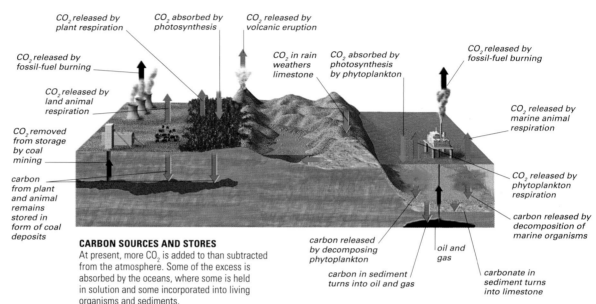

CO_2 released by plant respiration

CO_2 absorbed by photosynthesis

CO_2 released by volcanic eruption

CO_2 released by fossil-fuel burning

CO_2 released by land animal respiration

CO_2 in rain weathers limestone

CO_2 absorbed by photosynthesis by phytoplankton

CO_2 released by fossil-fuel burning

CO_2 removed from storage by coal mining

CO_2 released by marine animal respiration

carbon from plant and animal remains stored in form of coal deposits

CO_2 released by phytoplankton respiration

carbon released by decomposition of marine organisms

carbon released by decomposing phytoplankton

oil and gas

carbon in sediment turns into oil and gas

carbonate in sediment turns into limestone

CARBON SOURCES AND STORES
At present, more CO_2 is added to than subtracted from the atmosphere. Some of the excess is absorbed by the oceans, where some is held in solution and some incorporated into living organisms and sediments.

GOLDEN GATE FOG
The climate of San Francisco is influenced by exceptionally cold water, produced by upwelling, off the California coast. Fog is produced as westerly winds blow moist air over this cold water.

El Niño and La Niña

EL NIÑO AND LA NIÑA ARE LARGE climatic disturbances caused by abnormalities in the pattern of sea surface temperature, ocean currents, and pressure systems. They are in the tropical Pacific Ocean. These disturbances have important repercussions for weather throughout the Pacific and beyond. Most scientists regard El Niño and La Niña as extreme phases of a complex global weather phenomenon called the El Niño–Southern Oscillation (ENSO).

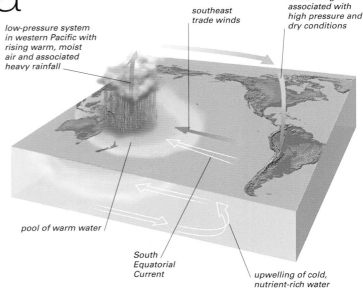

low-pressure system in western Pacific with rising warm, moist air and associated heavy rainfall

southeast trade winds

descending air associated with high pressure and dry conditions

pool of warm water

South Equatorial Current

upwelling of cold, nutrient-rich water

NORMAL PATTERN
A low-pressure system in the western Pacific draws southeast trade winds across from a high-pressure system over South America. These winds drive the South Equatorial Current, which maintains a pool of warm surface water in the western Pacific.

El Niño Events

The Spanish term *el niño* means "the little boy" or "Christ child." It originally denoted a warm current that was occasionally noticed around Christmas off Peru. Later it was restricted to unusually strong rises in temperature in the waters of the eastern Pacific, with a reduction in the upwelling of nutrient-rich waters that normally occur there. It is now used to mean a much wider shift in ocean and atmospheric conditions that affects the whole globe. El Niño events typically last from 12 to 18 months and occur cyclically, although somewhat unpredictably. On average, they occur about 30 times per century, with intervals that are sometimes as short as three years and sometimes as long as 10 years. Their underlying cause is not understood.

TEMPERATURE PATTERNS
These satellite-generated images of the Pacific compare surface temperature patterns. Red and white indicate warm water; green, blue, and purple denote cooler water.

JANUARY 1997
(NORMAL)

JANUARY 2015
(EL NIÑO)

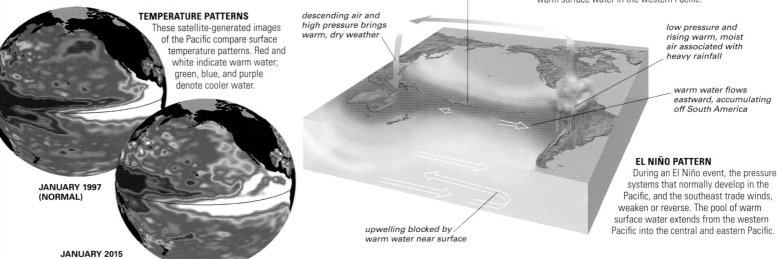

southeast trade winds reverse or weaken

descending air and high pressure brings warm, dry weather

low pressure and rising warm, moist air associated with heavy rainfall

warm water flows eastward, accumulating off South America

upwelling blocked by warm water near surface

EL NIÑO PATTERN
During an El Niño event, the pressure systems that normally develop in the Pacific, and the southeast trade winds, weaken or reverse. The pool of warm surface water extends from the western Pacific into the central and eastern Pacific.

Effects of El Niño

An El Niño event causes wetter-than-normal conditions, and floods, in countries on the western side of South America, particularly Ecuador, Peru, and Bolivia. These conditions may also extend to the southeastern United States. In other parts of the world, it causes drier conditions. Drought and forest fires become more common in the western Pacific, particularly in Indonesia and parts of Australia, but also in East Africa and northern Brazil. The warmer waters in the eastern Pacific cause a reduction in the Peru Current and reduced upwelling near the coast of South America. This reduces the level of nutrients in the seawater, which has a negative impact on fish stocks. Other effects include a quieter Atlantic hurricane season and an increase in the extent of sea ice around Antarctica. Japan, western Canada, and the western US typically experience more storms and warmer weather than normal.

GIANT WAVES
During an El Niño event, storms become more frequent and violent in the central Pacific. These storms can produce gigantic waves, up to 33 ft (10 m) high, in Hawaii, as here on the island of Oahu.

width of rings directly related to amount of growth

1746–1747 El Niño ring

EVIDENCE OF AN HISTORICAL EL NIÑO
Increased tree growth can be linked to high rainfall that occurred during historic El Niño events. One of the rings in this sample has been linked to an El Niño in 1746–1747.

CORAL BLEACHING
This small circular coral reef has suffered severe bleaching (whitening). El Niño events are often associated with bleaching caused by unusually high sea surface temperatures.

MONITORING

The tropical Pacific is regularly monitored for temperature changes. The main monitoring methods are the use of satellites, which measure sea temperatures indirectly from slight variations in the shape of the ocean surface, and an array of instrumented weather buoys.

INSTRUMENTED BUOY
Buoys such as this one, strung in an array across the equatorial Pacific, are used to make regular measurements of water temperature at varying depths.

La Niña Events

La niña is Spanish for "the little girl." A La Niña event is the reverse of an El Niño event. It is characterized by unusually cold ocean temperatures in the eastern and central equatorial Pacific, and by stronger winds and warmer seas to the north of Australia. La Niña conditions frequently, but not always, follow closely on an El Niño. Like El Niño, La Niña causes increased rainfall in some world regions and drought in others. India, Southeast Asia, and eastern Australia are lashed by rains, but southwestern US generally experiences higher temperatures and low rainfall. Meanwhile, northwestern states of the US experience colder, snowier winters. La Niña is also associated with an increase in Atlantic hurricane activity. Overall, the effects of a La Niña event often tend to be strongest during Northern Hemisphere winters.

DROUGHT CONDITIONS
This water reservoir in Texas completely dried up during a severe La Niña-related drought that affected the southwestern US in 2011.

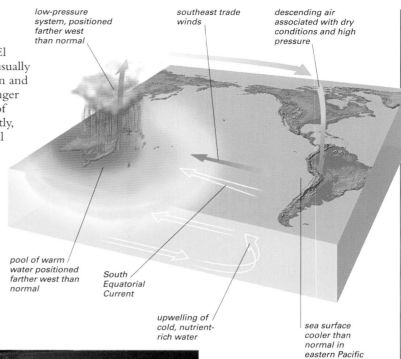

low-pressure system, positioned farther west than normal

southeast trade winds

descending air associated with dry conditions and high pressure

pool of warm water positioned farther west than normal

South Equatorial Current

upwelling of cold, nutrient-rich water

sea surface cooler than normal in eastern Pacific

LA NIÑA PATTERN
During a La Niña event, the area of low pressure in the western Pacific is farther west than normal, and the pool of warm surface water is also pushed west. Unusually cold surface temperatures develop in the eastern Pacific as the cold Peru Current strengthens off South America.

PEATLAND FIRE
A villager tries to put out a peatland fire in Kalimantan, Indonesia, in 2015. Due to abnormally dry conditions caused by a strong El Niño, such fires were frequent throughout Indonesia in 2015 and 2016.

Hurricanes and Typhoons

HURRICANES AND TYPHOONS ARE TERMS USED IN different parts of the world for very similar weather phenomena. They are characterized by violent winds moving in a circular pattern over the ocean, dense bands of clouds, and rainfall. In the Atlantic they are known as hurricanes; those in the west Pacific are called typhoons. Similar phenomena elsewhere are called severe storms or cyclones. They start as a low-pressure system (depression) over warm oceans in the tropics, between latitudes 5° and 20°, and occur mainly in late summer.

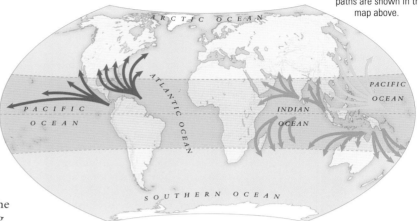

DISTRIBUTION
Severe tropical cyclones start as depressions over warm oceans in the tropics. They move across the ocean surface for several days, causing huge damage on reaching land. Their paths are shown in the map above.

Development

All tropical cyclones develop from the effects of the Sun warming the surface of a broad area of ocean and the air above it. This heating causes masses of warm, moist air to rise, creating a region of low pressure at the surface, and dense clouds above it. The low pressure sucks in more air, which spirals to the center, creating a circular wind system. As it grows stronger, becoming a tropical storm, it is pushed westward by the prevailing trade winds. In the Atlantic, a storm attains hurricane status once its winds exceed 74 mph (119 kph). Eventually, most of these violent storms move away from the equator— that is, to the north in the Northern Hemisphere. When one reaches land, it begins to lose energy, as it is no longer fed by heat from the ocean.

KEY
→ hurricanes
→ severe cyclones
→ typhoons

THREE DEVELOPMENT STAGES: HURRICANE IDA, AUGUST 2021

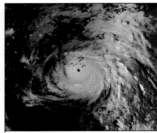
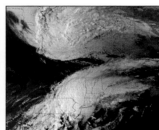
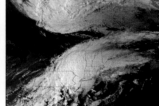

❶ On August 27, Hurricane Ida was a spiraling mass of clouds, with a distinct central nucleus over Cuba.

❷ By August 29, as it reached the coast of Louisiana, Ida had compacted and developed a central eye.

❸ By September 1, the hurricane had greatly weakened and was dissipating over the northeastern USA.

Structure

A fully developed typhoon or hurricane is usually 185–370 miles (300–600 km) in diameter and 6–9 miles (10–15 km) high. At its center is a calm region of low atmospheric pressure, called the eye. Within the rest of the cyclone, winds spiral in an counterclockwise direction in the Northern Hemisphere and clockwise in the Southern Hemisphere (the difference is due to the Coriolis effect, see p. 54). Within an area surrounding the eye, called the eyewall, the air spins upward, forming dense clouds. The eye stays calm because the winds that spiral in toward it never reach the center. Radiating out from the eye and eyewall are well-defined bands of clouds, called rainbands.

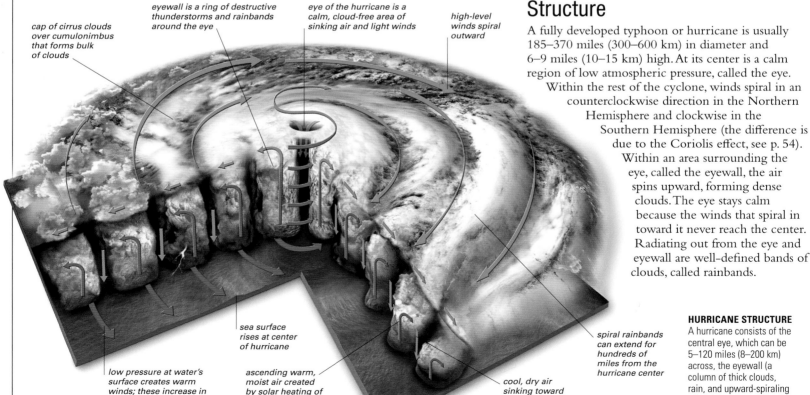

cap of cirrus clouds over cumulonimbus that forms bulk of clouds

eyewall is a ring of destructive thunderstorms and rainbands around the eye

eye of the hurricane is a calm, cloud-free area of sinking air and light winds

high-level winds spiral outward

sea surface rises at center of hurricane

low pressure at water's surface creates warm winds; these increase in speed toward the eye

ascending warm, moist air created by solar heating of the ocean surface

cool, dry air sinking toward the ocean surface

spiral rainbands can extend for hundreds of miles from the hurricane center

HURRICANE STRUCTURE
A hurricane consists of the central eye, which can be 5–120 miles (8–200 km) across, the eyewall (a column of thick clouds, rain, and upward-spiraling winds), and the rainbands.

STORM SURGE
Hurricane Frances hits Juno Beach, Florida, in September 2004. Classed as a Category 2 hurricane when it hit land, Frances caused a storm surge 6 ft (2 m) high, which ripped across highways and flooded homes and business premises.

HURRICANE CATEGORIES

A classification system called the Saffir–Simpson Scale divides hurricanes into five categories. It is used to estimate the damage and flooding to be expected along a coast impacted by the hurricane. Wind speed is the determining factor in the scale.

CATEGORY	WIND SPEED	HEIGHT OF SURGE
Tropical Storm	39–73 mph (63–118 kph)	less than 3 ft (1 m)
Category 1 hurricane	74–95 mph (119–153 kph)	3–5 ft (1–1.5 m)
Category 2 hurricane	96–110 mph (154–177 kph)	6–8 ft (2–2.4 m)
Category 3 hurricane	111–129 mph (178–208 kph)	9–12 ft (2.7–3.7 m)
Category 4 hurricane	130–156 mph (209–251 kph)	13–18 ft (4–5.5 m)
Category 5 hurricane	over 156 mph (251 kph)	over 19 ft (5.8 m)

DISCOVERY

STORM CHASERS

The United States' National Oceanic and Atmospheric Administration (NOAA) monitors Atlantic hurricanes using specially equipped aircraft. They fly into hurricanes to drop instrument packages, which radio back data.

LOCKHEED WP-3D ORION
This turboprop aircraft, equipped with a sophisticated array of instruments, is one of those used in hurricane study.

Coastal Effects

As it moves across the ocean, the low-pressure eye of a tropical cyclone sucks seawater up into a mound, which can be up to 11 ft (3.5 m) above sea level for a Category 2 hurricane or 25 ft (7.5 m) for a Category 5. When the cyclone hits land, the water in this mound surges over the coast in what is known as a storm surge. The surge may flood homes, wash boats inland, destroy roads and bridges, and seriously erode a section of coastline up to 95 miles (150 km) wide. These effects compound the devastation caused by high winds, which can topple unstable buildings, uproot trees, damage coastal mangroves, and bring down power lines. Human deaths are not uncommon, so coastal areas threatened by a severe cyclone are normally evacuated in advance. Offshore, the water movements associated with a storm surge can devastate coral reefs. In the Caribbean, branching corals that live near the surface, such as elkhorn corals, are particularly vulnerable. Healthy reefs can recover from such damage, although it can take 10–50 years depending on the extent of injury.

CORAL DAMAGE
This colony of elkhorn coral was smashed by Hurricane Gilbert on Mexico's Caribbean coast in 1988.

WATERSPOUT
Waterspouts are tornadoes (narrow, whirling masses of air) over the sea. They are quite commonly spawned around the edges of tropical cyclones.

REDUCED TO TATTERS
Survivors stand among ruined houses in Tacloban, on Leyte island in the Philippines, after its destruction by Typhoon Haiyan and the accompanying storm surge.

Typhoon Haiyan, known in the Philippines as Typhoon Yolanda, was an exceedingly powerful tropical cyclone that devastated the Philippines and some other parts of Southeast Asia between November 4 and November 11, 2013. It is the deadliest typhoon ever recorded in the Philippines, killing more than 6,000 people in that country alone. Haiyan is also the most powerful storm ever to hit land, and the fourth most intense tropical cyclone ever recorded in terms of highest sustained wind speeds.

Most of the catastrophic destruction occurred in a central group of islands within the Philippines called the Visayas. Although wind speeds were extreme, the major cause of damage and loss of life was a storm surge, particularly hitting the eastern coasts of the islands of Samar and Leyte. The city of Tacloban, with a population of over 220,000, was almost completely destroyed. On November 10 and 11, Haiyan passed over Vietnam and southern China, by which time it had weakened to a tropical storm and then to a depression. Nevertheless, there was extensive flooding in some areas, thousands of homes were destroyed, and around 50 people were killed.

Some climate scientists believe that global warming will increase the frequency of high-intensity tropical cyclones like Typhoon Haiyan. If this is the case, governments in affected countries will need to make provisions for dealing with similar-scale catastrophes more often in the future.

Track of the Typhoon

The map below shows Haiyan's path between November 5, when it reached the equivalent of a category 4 hurricane, and November 11, when it was downgraded to a tropical depression as it moved into China. Its most dangerous phase was reached on November 8, when it crossed the Philippines as a super-typhoon, equivalent to a Category 5 hurricane.

These satellite images show Haiyan's appearance as it centered over the Philippines (top) on Friday November 8, and as it moved into southern China (bottom) on Monday, November 11.

CHINA
Nov 11
Tropical depression
Nov 10–11
Tropical storm
Nov 10   Nov 9
Category 1 Category 3

PACIFIC OCEAN

LAOS

PHILIPPINES

Nov 9–10
Category 2

Nov 6–8
Category 5

VIETNAM

Nov 8–9
Category 4

PALAU
Nov 5
Category 4

PATH OF DESTRUCTION

EYE OF THE STORM
Haiyan originated around November 2 from an area of low pressure in the western Pacific. Over the next 3–4 days, it grew into a super-typhoon and developed a distinct "eye" at its center.

STORM SURGE *By November 7, Haiyan was producing sustained winds of up to 168 mph (270 kph). The next day, it made landfall in the Philippines, where a storm surge pounded coastal areas.*

EVACUATION *Warnings had been issued across a wide region to evacuate or seek safe refuge. Here, a Filipino child is being taken to board a military plane as part of an evacuation program.*

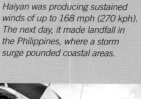

DESTROYED COMMUNITY *The destruction caused to parts of the central Philippines that Haiyan passed across was almost total. This aerial view shows the town of Guiuan in eastern Samar province.*

INFRASTRUCTURE DAMAGE *The high level of destruction in urban areas, which included downed power lines and radio masts, blocked roads and interfered with relief efforts.*

FLATTENED PALMS *The typhoon's effect on rural areas was equally terrible. This aerial view shows a destroyed coconut plantation and village on Leyte island.*

FOOD DROPS *The disaster left an estimated 1.9 million Filipinos in need of food, water, and shelter. Here, a Philippine Air Force crew drops sacks containing food supplies.*

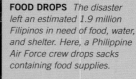

MEDICAL AID *Thousands needed medical assistance and urgent measures to combat spread of infectious diseases. Here, a child receives a measles vaccine.*

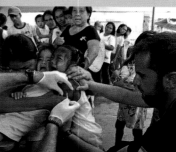

WAVES AND TIDES are two important physical phenomena that affect every area of the oceans but tend to be most noticeable, and have their main effects, on or near coasts. Ocean waves are mostly wind-generated and vary from tiny coastal ripples; to the regular, rolling swell of the open ocean; to monster breakers on world-famous surfing beaches. All waves transmit energy—when the waves reach land, this energy may be dissipated destructively, eroding coastlines, or constructively, building up features such as beaches. Tides are caused mainly by interactions between the Moon and Earth. As well as regular rises and falls in sea level, they can cause strong currents around coasts and, in some places, even more dramatic phenomena such as whirlpools and eddies.

WAVES AND TIDES

LAPPING WAVES
Waves lapping on the seashore meet a rocky stream at low tide near Kipahulu on the southeast coast of the Hawaiian island of Maui, in the Pacific.

Ocean Waves

WAVES ARE DISTURBANCES in the ocean that transmit energy from one place to another. The most familiar types of waves—the ones that cause boats to bob up and down on the open sea and dissipate as breakers on beaches—are generated by wind on the ocean surface. Other wave types include tsunamis, which are often caused by underwater earthquakes (see p.49), and internal waves, which travel underwater between water masses. Tides (see p.78) are also a type of wave.

Wave Properties

A group of waves consists of several crests separated by troughs. The height of the waves is called the amplitude, the distance between successive wave crests is known as the wavelength, and the time between successive wave crests is the period. Waves are classified into types based on their periods. They range from ripples, which have periods of less than 0.5 seconds, up to tsunamis and tides, whose periods are measured in minutes and hours. (Their wavelengths range from hundreds to thousands of miles.)

In between these extremes are chop and swell—the most familiar types of surface waves. Ocean waves behave like light rays: they are reflected or refracted by obstacles they encounter, such as islands. When different wave groups meet, they interfere—adding to, or canceling, each other.

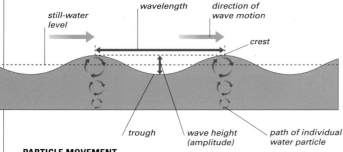

still-water level · wavelength · direction of wave motion · crest · trough · wave height (amplitude) · path of individual water particle

PARTICLE MOVEMENT
As waves pass over the surface, the particles of water do not move forward with the waves. Instead, they gyrate in little circles or loops. Underwater, the particles move in ever-smaller loops. At a depth below about half the distance between crests, they are quite still.

ROGUE WAVES
Interference between two or more large waves occasionally causes a giant or "rogue" wave. This one, recorded in the Atlantic Ocean in 1986, had an estimated height of 56 ft (17 m). It broke over the ship pictured, bending its foremast back by 20°.

Wave Generation

Wind energy is imparted to the sea surface through friction and pressure, causing waves. As the wind gains strength, the surface develops gradually from flat and smooth through growing levels of roughness. First, ripples form, then larger waves, called chop. The waves continue to build, their maximum size depending on three factors: wind speed; wind duration; and the area over which the wind is blowing, called the fetch. When waves are as large as they can get under the current conditions of wind speed and size of fetch, the sea surface is said to be "fully developed." The overall state of a sea surface can be summarized by the significant wave height—defined as the average height of the highest one-third of the waves. For example, in a fully developed sea produced by winds of about 25 mph (40 kph), the significant wave height is typically about 8 ft (2.5 m).

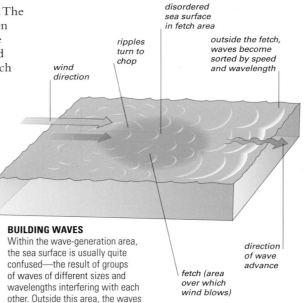

wind direction · ripples turn to chop · disordered sea surface in fetch area · outside the fetch, waves become sorted by speed and wavelength · direction of wave advance · fetch (area over which wind blows)

BUILDING WAVES
Within the wave-generation area, the sea surface is usually quite confused—the result of groups of waves of different sizes and wavelengths interfering with each other. Outside this area, the waves become sorted by speed to produce a more regular pattern, called a swell.

CAPILLARY WAVES (RIPPLES)
These tiny waves are just a few millimeters high and have a wavelength of under 1½ in (4 cm).

CHOPPY SEA
In a choppy sea, the waves are 4–20 in (10–50 cm) high and have a wavelength of 10–40 ft (3–12 m).

FULLY DEVELOPED ROUGH SEA
Wind speeds over 40 mph (60 kph) can generate very rough seas with waves more than 10 ft (3 m) high.

Wave Propagation

In the fetch, many different groups of waves of varying wavelength are generated and interfere. As they disperse away from the fetch, the waves become more regularly sized and spaced. This is because the speed of a wave in open water is closely related to its wavelength. The different groups of waves move at different speeds and so are naturally sorted by wavelength: the largest, fastest-moving waves at the fore, the smaller, slower-moving ones behind. This produces a regular wave pattern, or swell. Occasionally, groups of waves from separate storms interfere to produce unusually large "rogue" waves. As they propagate across the open ocean, wind-generated waves maintain a constant speed, which is unaffected by depth until they reach shallow water. Only with waves of extremely long wavelength—tsunamis—is the speed of propagation affected by water depth.

SWELL
A swell is a series of large, evenly spaced waves, often observed hundreds of miles away from the storm that spawned them. Wavelengths range from tens to hundreds of feet.

PLUNGING BREAKER
"Barrel" or "tube-forming" breakers like this occur when the waves reaching shore have large amounts of energy. The seabed must be firm and quite steep.

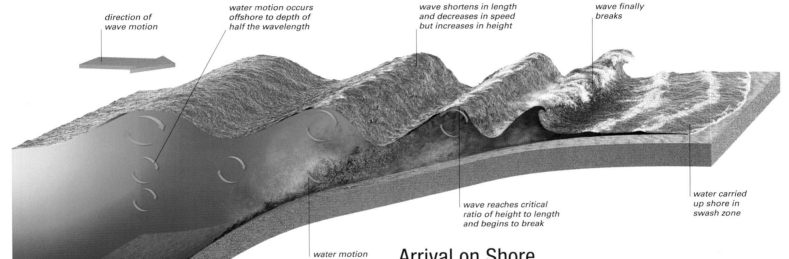

direction of wave motion

water motion occurs offshore to depth of half the wavelength

wave shortens in length and decreases in speed but increases in height

wave finally breaks

water motion caused by the wave begins to interact with the seabed and slow down

wave reaches critical ratio of height to length and begins to break

water carried up shore in swash zone

SHOALING AND BREAKING
Shoaling occurs as waves enter shallow water. Their wavelength and speed decrease, but their height increases. When the front of each wave becomes too steep, its crest curls and breaks.

INTRODUCTION

Arrival on Shore

As waves approach a shore, the motion they generate at depth begins to interact with the sea floor. This slows the waves down and causes the crests in a series of waves to bunch up—an effect called shoaling. The period of the waves does not change, but they gain height as the energy each contains is compressed into a shorter horizontal distance and eventually break.

There are two main types of breakers. Spilling breakers occur on flatter shores: their crests break and cascade down the front as they draw near the shore, dissipating energy gradually. In a plunging breaker, which occurs on steeper shores, the crest curls and falls over the front of the advancing wave, and the whole wave then collapses at once. Waves can also refract as they reach a coastline. This concentrates wave energy onto headlands (see p.93) and shapes some types of beach (see p.106).

HUMAN IMPACT

RIDING THE WAVES

When a swell reaches a suitably shaped beach, it can produce excellent surfing conditions. Small spilling breakers are ideal for novice surfers, while experts seek out large plunging breakers that form a "tube" they can ride along. For tube-riding, the break of the wave must progress smoothly either to the right or left. Here, a surfer rides a right-breaking wave in Hawaii—it is breaking from left to right behind the surfer.

WAVE REFRACTION
When waves enter a bay enclosed by headlands, they are refracted (bent) as different parts of the wave-front encounter shallow water and slow down.

Tides

TIDES ARE REGULAR RISES AND FALLS in sea level, accompanied by horizontal flows of water, that are caused by gravitational interactions between the Moon, the Sun, and Earth. They occur all over the world's oceans but are most noticeable near coasts. The daily pattern of high and low tides is caused by the Moon's influence on Earth. Variations in the range between high and low tides over a monthly cycle are caused by the combined influence of the Sun and the Moon.

High and Low Tides

Although the Moon is usually thought of as orbiting Earth, in fact both bodies orbit around a common center of mass—a point located inside Earth. As Earth and the Moon move around this point, two forces are created at Earth's surface: a gravitational pull toward the Moon, and an inertial or centrifugal force directed away from the Moon. These forces combine to produce two bulges in Earth's oceans: one toward the Moon, and the other away from it. As Earth spins on its axis, these bulges sweep over the planet's surface, producing high and low tides. The cycle repeats every 24 hours 50 minutes (one lunar day) rather than every 24 hours (one solar day), because during each cycle, the Moon moves round a little in its orbit.

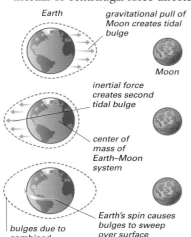

Earth

gravitational pull of Moon creates tidal bulge

Moon

inertial force creates second tidal bulge

center of mass of Earth–Moon system

Earth's spin causes bulges to sweep over surface

bulges due to combined forces

DAILY TIDES
The two ocean bulges caused by the gravitational interaction between Earth and the Moon are shown (much exaggerated) here.

Tidal Patterns

If no continents existed and the Moon orbited in the Earth's equatorial plane, the sweeping of the tidal bulges over the oceans would produce two equal daily rises and falls in sea level (a semidiurnal tide) everywhere on Earth. In practice, landmasses interfere with the movement of the tidal bulges, and the Moon's orbit tilts to the equatorial plane.

Consequently, many parts of the world experience tides that differ from the semidiurnal pattern. A few have just one high and one low tide a day (called diurnal tides), and many experience high and low tides of unequal size (known as mixed semidiurnal tides). In addition, the tidal range, or difference in sea level between high and low water, varies considerably across the globe.

KEY

diurnal

mixed

semidiurnal

..... small tidal range

.... medium tidal range

——— large tidal range

GLOBAL PATTERNS
This map shows the general pattern of tides (diurnal, semidiurnal, or mixed) and size of tidal range (average difference between high and low water) around the world.

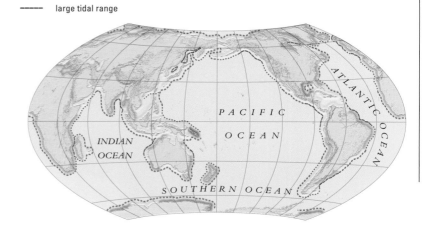

Time (hrs)
0   6   12   18   24

10ft/3m
6ft/2m
3ft/1m
0ft/0m

Tide heights

SEMIDIURNAL TIDE

0   6   12   18   24

10ft/3m
6ft/2m
3ft/1m
0ft/0m

DIURNAL TIDE

0   6   12   18   24

10ft/3m
6ft/2m
3ft/1m
0ft/0m

MIXED SEMIDIURNAL TIDE

ATLANTIC OCEAN

PACIFIC OCEAN

INDIAN OCEAN

SOUTHERN OCEAN

INTERTIDAL LIFE

Compared to permanently submerged plants and animals, organisms living in the intertidal zone have to cope with many extra stresses. They need to adapt, for instance, to the problem of becoming dried out (desiccated) when the tide is out. They may also have to endure extreme cold on frosty winter nights and even predation by land animals. Mussels, for example, often have to wait for hours between high tides to feed. At low tide, their shells close tightly to prevent desiccation and to protect against predators.

Monthly Cycle

In addition to the daily cycle of high and low tides, there is a second, monthly, cycle. In this case, the Sun and the Moon combine to drive the cycle. As with the Moon, the interaction between Earth and the Sun causes bulges in Earth's oceans, though these are smaller than those caused by the Moon. Twice a month, at the times of new and full Moon, the Sun, Moon, and Earth are aligned, and the two sets of tidal bulges reinforce each other. The result is spring tides—high tides that are exceptionally high, and low tides that are exceptionally low. By contrast, at the times of first and last quarter Moon, the effects of the Sun and Moon partly cancel out, bringing tides with a smaller range, called neaps.

SPRING TIDES

new Moon

high tide

low tide

low tide

high tide

full Moon

NEAP TIDES

first quarter Moon

low tide

high tide

high tide

low tide

last quarter Moon

ALTERNATING SPRINGS AND NEAPS
Twice a month (top), the alignment of the Sun, Moon, and Earth creates spring tides. At other times (left), when the Sun and Moon lie at right angles, it creates neap tides. The alternation between springs and neaps can be seen in the 28-day tidal graph shown below.

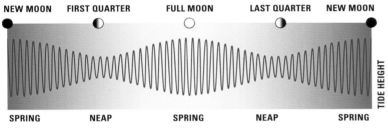

NEW MOON	FIRST QUARTER	FULL MOON	LAST QUARTER	NEW MOON

TIDE HEIGHT

SPRING | NEAP | SPRING | NEAP | SPRING

Tidal Currents

The vertical variation in sea level that occurs locally with tides can happen only through horizontal flows of water, called tidal currents. Over each daily tidal cycle, the currents typically (but not invariably) run fastest about halfway between high and low tide at that location—at intermediate times, they slow ("slack water") and then reverse direction.

The shape of a coast can have a crucial influence on current strength. Bottlenecks to water flow, such as narrow channels and promontories, are often associated with very powerful currents, called tidal races, that develop twice or four times a day. Where the flowing water meets underwater obstructions, phenomena such as whirlpools or vortices (spiraling, funnel-shaped disturbances), eddies (larger, flatter, circular currents), and standing waves may develop. Other tide-related phenomena include tide rips—turbulence caused by converging currents—and overfalls, defined as a tidal current flowing opposite to the wind direction.

TIME OF LOW WATER, WELLINGTON

Wellington

COOK STRAIT CURRENTS
These maps show the pattern of strong tidal currents in the Cook Strait, between the North and South islands of New Zealand, which occur twice a day, just over six hours apart. Water must funnel through a narrow channel in the Strait.

Wellington

TIME OF HIGH WATER, WELLINGTON

LOW TIDE AT BAMBURGH BEACH
Due to tides, large swathes of coast around the world are alternately covered and uncovered by the sea. This intertidal sandflat in Northumberland, England, has a tidal range averaging about 13 ft (4 m).

MINI-VORTEX
This mini-whirlpool, about 20 ft (6 m) wide and 16 in (50 cm) deep, would be called a "piglet" by experienced Old Sow watchers. Sometimes, several of these small vortices occur, rather than a single large whirlpool.

HUMAN IMPACT

SURVIVING THE OLD SOW

The Old Sow has been blamed for about a dozen fatalities from drowning over the past 200 years. The last recorded death was in 1943. It is unwise to go too close to the whirlpool in a small rowing or sailing boat. In recent times, a few people in powerboats have had anxious experiences when their engines have stalled. Experts advise that if caught in a whirlpool, the priority is to keep the boat on an even keel and avoid getting swamped. Most objects floating in a stable position will eventually spin clear.

ATLANTIC OCEAN NORTHWEST

The Old Sow Whirlpool

FEATURES
Tidal race, small whirlpools, occasional large whirlpool

TIMING
Four times daily

LOCATION Off the southern tip of Deer Island, New Brunswick, Canada

Situated at the southern end of Passamaquoddy Bay on the US–Canada border, the Old Sow is one of the largest whirlpools in the world, and by far the largest in the Americas. Passamaquoddy Bay is at the lower end of the Bay of Fundy, which is famous for its strong tides. The Old Sow, when it appears, is located at a spot where various tidal streams flowing through the channels

COASTAL SETTING
The Old Sow develops between Deer Island (top) and Moose Island in Maine, USA (foreground).

between different islands converge during the ebb tide or diverge during the flood tide. As they flow, these currents encounter underwater obstructions, such as ledges and small seamounts, so as they reach their maximum speed of up to 17 mph (28 kph), the whole sea surface in this area becomes rough and disordered. Typical disturbances include standing waves, troughs (long depressions in the surface), and "boils" (smooth circular areas where water spouts up from deep below). Occasionally and unpredictably, the Old Sow itself appears, forming a vortex that can be 100 ft (30 m) wide and 10 ft (3 m) deep. More often, one or several smaller vortices, known locally as piglets, appear. As with all tidal disturbances, these phenomena are more powerful during a spring tide, which occurs a day or two after a full or new moon.

ATLANTIC OCEAN NORTHEAST

Lofoten Maelstrom

FEATURES
Tidal race and large, weak eddy

TIMING
Four times daily

LOCATION Between Lofoten Point and Mosken in the Lofoten Islands, off northwest Norway

Also known as the Moskenstraumen, the Lofoten Maelstrom is a complex pattern of sea-surface disturbances caused by tidal flows of water over a broad, submerged ledge of rock between two of the Lofoten Islands. These flows result from large sea-level differences that develop four times a day between the Norwegian Sea and the Vestfjord on the eastern side of the Lofoten Islands. The word "maelstrom" originates with the tidal phenomena in this area, and is derived from the Nordic word *male*, meaning "to grind." In Norse mythology, the Maelstrom was the result of a large salt-grinding millstone on the floor of the Norwegian Sea, which sucked water into its central hole as it turned. First described by the Greek explorer Pytheas in the 3rd century BCE, the Lofoten Maelstrom is marked on many historical charts as an enormous and fearsome whirlpool. In 1997, a detailed study of tidal currents in the vicinity of the island of Mosken found that the reality is somewhat different. Although some strong tidal currents were measured, no obvious large whirlpool, with a vortex, was detected. Instead, the researchers found a weak eddy, about 4 miles (6 km) in diameter, to the north of Mosken. This eddy develops twice a day during the flood tide, when it moves in a clockwise direction, and twice on the ebb tide, when it moves slightly farther north and goes counterclockwise.

PEOPLE

JULES VERNE

The French novelist Jules Verne (1828-1905) included a somewhat fanciful description of the Lofoten Maelstrom in *Twenty Thousand Leagues Under the Sea*. At the end of the novel, Captain Nemo and his submarine, *Nautilus*, are sucked down into the whirlpool, "whose power of attraction extended to a distance of twelve miles," suffering an unknown fate.

TIDAL DISTURBANCE
For centuries, the Lofoten Maelstrom had a reputation as one of the world's most powerful tidal phenomena.

Saltstraumen

FEATURES
Tidal race and small whirlpools

TIMING
Four times daily

LOCATION Between Saltenfjord and Skjerstadfjord, northwest coast of Norway

The Saltstraumen tidal race occurs on the northwest coast of Norway and is generally acknowledged to be the strongest and most extreme tidal current in the world. It forms at a bottleneck between the Saltenfjord, an inlet from the Norwegian Sea, and the neighboring Skjerstadfjord: its driving force is a difference in sea level of up to 10 ft (3 m) that develops four times a day between the two bodies of water. The channel at the center of the bottleneck—Saltstraumen itself—is a 2-mile (3-km) long strait between two headlands, with a width of just 500 ft (150 m) and a depth that varies

from 65 to 330 ft (20 to 100 m). Twice a day, some 105 billion gallons (400 billion liters) of water roar through this strait on the flood tide, reaching maximum speeds of up to 25 mph (40 kph), as tidal forces act to fill the 30-mile (50-km) long Skjerstadfjord. Twice a day, the waters flow out again through the same channel. The flows of water, and associated whirlpools, are equally strong during the ebb as the flood tide. Despite Saltstraumen's ferocity, the channel is regularly used by shipping. For short periods every day, the tidal flows slow almost to a halt, allowing large vessels to pass safely into and out of Skjerstadfjord. Smaller vessels do remain at risk from residual underwater currents during these periods of "slack water," but many experienced pilots still venture out. Saltstraumen offers both interesting opportunities for divers and excellent angling (see panel, below). Incoming tides carry large amounts of plankton through the channel, and fish of various sizes follow.

DANGEROUS WATERS
When the tidal race flows, the spinoff vortices, which can be 33 ft (10 m) across, are capable of pulling objects down to the rocky bottom of the channel.

DISCOVERY

LIFE BENEATH THE WHIRLPOOLS

It is possible to dive into and explore the Saltstraumen, although this can safely be attempted only when the tidal streams are at a minimum. Divers have discovered rich and colorful marine life at the bottom of the channel, dominated by long strands of kelp and a variety of invertebrates, as well as fish such as lumpsuckers, coley, and wolf-fish.

TEEMING WITH LIFE
Invertebrate life at the bottom of the channel includes colorful sponges and anemones.

Corryvreckan Whirlpool

FEATURES
Tidal race, standing waves, and whirlpools

TIMING
Twice daily

LOCATION Between the islands of Jura and Scarba, west coast of Scotland, UK

The most famous tidal phenomenon in the British Isles can be found in the Gulf of Corryvreckan. Twice a day on the flood tide, strong Atlantic currents and unusual underwater topography conspire to produce an intense tidal race. As the tide enters the narrow bottleneck at Corryvreckan, currents of up to 14 mph (22 kph) develop. Underwater, these currents encounter a variety of irregular features on the

SPIN-OFF VORTEX
In the whirlpool area, massive upthrusts of water occur in pulses, producing vortices that spin away with the tidal flow.

seabed, including a conical obstruction known as the Pinnacle, which rises to within 95 ft (30 m) of the surface. The steep east face of this obstruction forces a plume of water to the surface, producing whirlpools and standing waves up to 13 ft (4 m) high, and the roar of the rushing water is sometimes heard 6 miles (10 km) away. Over the years, the Corryvreckan has caused numerous emergencies—the author George Orwell nearly drowned there in 1947.

DISTURBED SEA
An area of disturbance begins to develop in the channel north of Jura, seen here with the island of Scarba lying behind it.

Slough-na-more Tidal Race

FEATURES
Tidal race with eddies and standing waves

TIMING
Four times daily

LOCATION Between Rathlin Island and Ballycastle Bay, County Antrim, Northern Ireland, UK

The Slough-na-more Tidal Race results from strong tidal flows of billions of gallons of seawater between the Atlantic Ocean and the Irish Sea, via a narrow channel. During spring tides, the tidal stream can attain a speed of 8 mph (13 kph). Where it passes Rathlin Island, a complex of fast-moving currents, eddies, and standing waves is created. In contrast, the same sea area is usually calm during other phases of the tidal cycle. In 1915, the strength of the Slough-na-more Tidal Race forced the Irish steam coaster SS *Glentow* aground on the Irish coast, and the ship later broke up.

ATLANTIC OCEAN NORTHEAST
Needles Overfalls

FEATURES
Tidal race and overfalls

TIMING
Four times daily

LOCATION Needles Channel, northwestern coast of the Isle of Wight, England, UK

The Needles Channel is a 5-mile (7-km) long stretch of water between a line of chalk sea stacks on one side (the Needles) and an underwater reef on the other. This stretch of water is affected by short, breaking waves (overfalls) at the time of the maximum ebb or flood tide. If the wind is blowing in the opposite direction of the tidal stream, these overfalls are greatly exacerbated, producing an extremely rough sea.

ATLANTIC OCEAN EAST
Garofalo Whirlpool

FEATURES
Tidal race, small whirlpools, and overfalls

TIMING
Four times daily

LOCATION Strait of Messina, between the northeast coast of Sicily and Calabria in mainland Italy

The Strait of Messina separates the "toe" of Italy from the Mediterranean island of Sicily. It varies in width from 2 to 10 miles (3 to 16 km) and is the site of numerous complex currents and small whirlpools that vary over the tidal cycle and hamper navigation through the Strait. In Italy, the small whirlpools that form are called *garofali,* but in the English-speaking world, the whole system of tidal disturbances is known as the Garofalo Whirlpool.

PACIFIC OCEAN NORTHEAST
Yellow Bluff
Tide Rip

FEATURES
Tide rip, standing waves, and eddies

TIMING
Twice daily

LOCATION San Francisco Bay, California, US

A tide rip is a stretch of rough, turbulent water caused by a tidal current converging with, or flowing across, another current. Thus it differs from a tidal race, which occurs where a tidal stream of water accelerates through a narrow opening in a coast. An example of a tide rip occurs at a place called Yellow Bluff in San Francisco Bay, not far from the bay's entrance, the famous Golden Gate.

Four times a day, strong movements of water occur through the Golden Gate—twice flowing into the bay on the flood tide and twice flowing out on the ebb tide. These currents can reach a speed of up to 5 mph (8 kph) during spring tides. Inside the bay, the pattern of currents becomes more complex, as they either split (during the flood tide) or converge (during the ebb tide) from different parts of the bay. The currents are also modified by the varying depth of the water around the shoreline, by the shoreline's shape, and by subsurface obstructions. At Yellow Bluff, disturbances to the sea surface are most noticeable during the ebb tide, when the tidal streams are converging, and are characterized by such phenomena as extremely rough, fast-moving water, standing waves, and eddies. The spot is popular with extreme kayakers, who challenge themselves against the strong currents and surf on the standing waves.

PACIFIC OCEAN NORTHEAST
Skookumchuck
Narrows Tidal Race

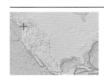

FEATURES Tidal race, small whirlpools, and standing wave on flood tide

TIMING Four times daily; flood tide twice daily

LOCATION Skookumchuck Narrows, British Columbia, Canada

One of the world's most famous tidal races occurs at the Skookumchuck Narrows on British Columbia's Sunshine Coast, not far from Vancouver (*Skookum* is a native American word for "strong" and *chuck* means "water"). Four times a day, there is a strong tidal rush of water through this 1,000-ft (300-m) wide channel, which connects two inlets into the coast—the Sechelt and Jervis inlets. A 10-ft (3-m) difference in sea level between low and high tide causes some 167 billion gallons (760 billion liters) of seawater to rush through the gap, creating turbulence and some small whirlpools. On the flood tide, when water is flowing into the Sechelt Inlet (but not the ebb tide, when it flows out), the tidal stream across an outcrop of bedrock in the channel creates a large standing wave—a mound of breaking water that remains stationary at a particular spot on the surface. At its peak, the flow rate is about 4.75 million gallons (18 million liters) per second, and current velocities can reach 20 mph (32 kph).

HUMAN IMPACT

SURF-KAYAKING

Skookumchuck Narrows is a popular destination for enthusiasts of extreme surf-kayaking. The standing wave that arises there is up to 8 ft (2.5 m) high and 23 ft (7 m) wide and is regarded as one of the world's great white-water kayaking locations. When surf-kayaking, the object is to stay in the wave as long as possible, which requires strength and skill.

POWERFUL RAPIDS
Here, water is flowing right to left, from the Sechelt Inlet into Jervis Inlet. Six hours later, it flows back in the opposite direction.

PACIFIC OCEAN NORTHWEST

Naruto Whirlpools

	FEATURES Tidal race and whirlpools
	TIMING Four times daily

LOCATION Naruto Strait, between the islands of Shikoku and Awaji, Japan

The Naruto is a spectacular system of whirlpools that develops four times a day in a channel separating the island of Shikoku (one of Japan's main islands) from Awaji Island, a much smaller island lying off Shikoku's northeastern coast. The channel, called the Naruto Strait, is one of several that join the Pacific Ocean to the Inland Sea, which is a large body of water lying between Shikoku and Japan's largest island, Honshu. Four times a day, billions of gallons of water move into and out of the Inland Sea through this channel, generated by tidal variations in sea level between the Inland Sea and the Pacific Ocean of up to 5 ft (1.5 m).

The tidal flows can reach speeds of up to 12 mph (20 kph) during spring tides (that is, twice a month, around the time of a full or new moon). They create vortices up to 65 ft (20 m) in diameter where they encounter a submarine ridge. These vortices are not stationary but tend to move with the current, persisting for 30 seconds or more before disappearing. The whirlpools can be viewed from Awaji Island, from sightseeing boats that regularly negotiate the rapids, or from a 1-mile (1.6-km) long bridge that spans the Naruto Strait.

HUMAN IMPACT

ARTISTIC INSPIRATION

The Naruto Whirlpool has existed since ancient times. It is mentioned many times in Japanese poetry and is possibly the only tidal phenomenon to feature in a well-known piece of art, namely *Whirlpool and Waves at Naruto, Awa Province*, by the 19th-century Japanese artist Utagawa Hiroshige (a fragment is shown below).

TROUBLED WATERS
A walkway built into the side of the Onaruto Bridge, which spans the Naruto Straits, provides an excellent view of the whirlpools below.

INTRODUCTION

OCEAN ENVIRONMENTS

COASTS INCLUDE SOME of the most beautiful, but also some of the most rapidly changing, places on Earth. There are a great many forms that they can take—from high cliffs composed of anything from lava to limestone, to beaches, spits, and barrier islands, river deltas, estuaries, and tidal flats. Each coast has its own unique history of formation, brought about by processes such as land rise and fall, sea-level change, glacial and volcanic action, and marine erosion and deposition. On and around these coasts, a variety of habitat types—ranging from sandy coastal dunes to salt marshes, coastal lagoons, and, in the tropics, mangrove swamps—are shaped by the interaction of tidal flows, breaking waves, discharge of river sediment, and a variety of biological and human-induced processes.

COASTS AND THE SEASHORE

SANDSTONE COAST
This dramatic Australian coastline consists of eroded sandstone strata beautifully shaped and sculpted by an azure sea.

Coasts and Sea-level Change

A COAST IS A ZONE WHERE THE LAND MEETS THE SEA—it extends from the shoreline inland to the first significant terrain change. Coastlines constantly alter in response to sea-level change, land-based processes, wave action, and tides. Coasts can be classified into many types, some of which are contrasting—for example, two opposite forms linked to sea-level change are drowned and emergent coasts. The sea-level change itself may either have been global in nature (caused by a change in the volume of ocean water, for example) or only local (stemming from regional uplift or sinking of land).

OCEAN ENVIRONMENTS

Global Sea-level Change

The most important cause of a global change in sea level is an increase or decrease in the extent of the world's ice sheets and glaciers. This is related to Earth's climate. If it cools, more water becomes locked up as ice, so there is less in the oceans. If it heats up (global warming), the ice melts and increases the volume of ocean water. Another cause of global sea-level change, which is also affected by climate, is a rise or fall in ocean temperature. Warming lowers the density of water, so if the upper layers of the oceans heat up, they expand and increase the total volume of the oceans. Any changes in the size of the ocean basins, the ocean's containers, also impact globally on sea levels. For example, a change in activity at mid-ocean ridges can have such an effect and may be important in driving long-term sea-level change.

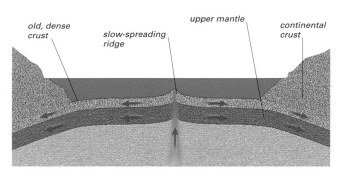

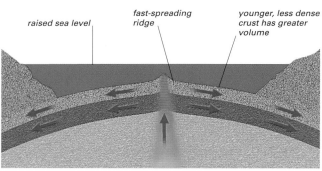

GLACIAL CYCLES
During an ice age (top) the volume of ocean water is low as water is locked up in ice sheets. When the ice melts (bottom), the oceans expand, raising sea levels globally.

OCEAN-BASIN CHANGE
A slow, global rise in sea level can occur when new crust is produced at a fast-spreading mid-ocean ridge. The relatively hot, buoyant new crust swells, pushing the ocean water upward.

Local Sea-level Change

Local sea-level change occurs when a particular area of land rises or falls relative to the general sea level. One of the main causes is tectonic uplifting of land, which occurs in regions where oceanic crust is being forced beneath continental crust (a process often associated with earthquakes). Another cause is glacial rebound, which is a gradual rise of a specific area of land after an ice sheet that once weighed it down has melted.

During the last ice age, heavy ice sheets covered much of North America and Scandinavia. Since the ice melted, these regions have risen, and they continue to do so today at rates of up to a few inches a year. In contrast, other coastal areas are slowly sinking. Often, this occurs where a heavy load of coastal sediments is pushing the underlying bedrock down. A slow subsidence is occurring, for example, on the eastern coast of the US. Many volcanic islands also start to subside soon after they form. This is due to the fact that the material from which they are created cools, compacts, and then contracts, while the sea floor under them warps downward.

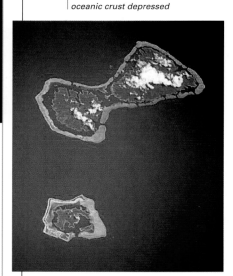

SINKING ISLANDS
These two volcanic Pacific islands, Rai'atea (top) and Bora-Bora, are subsiding. Locally, the current global rise in sea level is therefore slightly exacerbated.

Drowned Coasts

A drowned (or submergent) coast can be the result of either global or local sea-level rise. There are two types—rias and fjords. In a ria coast, the sea-level rise has drowned a region of coastal river valleys, forming a series of wide estuaries, often separated by long peninsulas. In a fjord coast, the sea-level rise has drowned one or more deep, glacier-carved valleys. Both types are characteristically irregular and indented. Due to a significant global rise in sea level over the past 18,000 years, drowned coasts are common worldwide. Ria coasts are particularly prevalent in northwestern Europe, the eastern US, and Australasia. Large numbers of fjords are present in coastal Norway, Chile, Canada, and New Zealand.

RIA COAST
The coastline around Hobart, in Tasmania, Australia, was formed by a rise in sea-level flooding a series of river valleys. Here, the Hobart Bridge spans one such drowned valley.

UPLIFTED TERRACE
This coastal region of New Zealand has experienced a localized sea-level fall in the recent geological past, as the land was significantly raised by an earthquake. What was beach is now flat clifftop.

Emergent Coasts

Emergent coasts occur where land has uplifted faster than the sea has risen since the last ice age. The causes are either activity at the edge of a tectonic plate or glacial rebound. On emergent coasts, areas that were formerly sea floor may become exposed above the shoreline, while former beaches often end up well behind the shoreline, or even on clifftops. Sometimes, staircaselike structures called marine terraces are created by a combination of uplift and waves gradually cutting flat platforms at the bases of cliffs (wave-cut platforms). Emergent coasts are typically rocky, but sometimes they have a smooth shoreline. Examples of these coasts occur on the US Pacific Coast and in Scotland, Scandinavia, New Zealand, and Papua New Guinea.

Past Change

Scientists study past sea-level changes by examining rocks and fossils near shorelines. They also analyze ocean sediments to calculate past ocean temperatures and climatic properties. Over the past 500 million years, global sea levels have fluctuated by more than 1,000 ft (300 m). About 120,000 years ago, sea level was a few meters higher than it is today, but some 20,000 years ago, it was 400 ft (120 m) below today's level. Most of the rise since then occurred prior to 6,000 years ago. From some 3,000 years ago to the late 19th century, sea level rose at about $1/254$–$1/127$ in (0.1–0.2 mm) per year. In the late 20th century, this increased to an average of about $1/12$ in (2 mm) per year.

FOSSIL MAMMOTH TOOTH

RAISED BEACH
In this bay in the Hebridean Islands, Scotland (left), the green areas behind the beach are former beaches that have been raised by glacial rebound since the end of the last ice age.

UPLIFTED CLIFFS
These marine cliffs in Crete, Greece (right), have been uplifted by tectonic activity, eroded, and finally tilted from the horizontal, also by tectonic activity.

New York
Washington D.C.

NORTH AMERICA

ATLANTIC OCEAN

Miami

THEN AND NOW
The red dotted line on this map shows where the east coast of North America was 15,000 years ago. At that time, mammoths roamed on what is now continental shelf—it is not uncommon for a mammoth tooth (above) to turn up in fishing trawls from these areas.

FLOODED PIAZZA IN VENICE
Since 1960, high tides have often flooded Venice's St. Mark's Square. The city is threatened long term by sea-level rise, though recently completed tide barriers will protect the city in the shorter term.

The measurement of global sea-level change is complex and, until satellite-based techniques were introduced in 1992, was somewhat imprecise. During the 1980s, a consensus emerged that sea level had been rising at $1/32–1/8$ in (1–3 mm) per year since 1900. Today, satellite data indicates a current average rise of about $1/6$ in (4 mm) per year, so the rise seems to have accelerated markedly. Since 1900, there has also been a rise in the average temperature at Earth's surface (global warming) of about 2.2°F (1.2°C). There are two plausible mechanisms by which the sea-level rise might be linked to temperature. First, through melting of glaciers and ice sheets, which increases the amount of water in the oceans; and second, through the expansion of seawater as it warms. Because there are no other convincing explanations of what might be causing the sea-level rise, the overwhelming scientific opinion is that global warming is the cause.

Based on different models of the future course of global warming (which most scientists now believe is linked to human activity), it is possible to make various predictions of how sea level will change in the future. For example, in 2021, the Intergovernmental Panel on Climate Change (IPCC) predicted that, by the year 2050, there will be a further sea-level rise of 6-12 in (15-30 cm). If global greenhouse gas emissions remain as they currently are, by 2100, there is likely to be a rise of about 1–3 ft (60cm–1.1 m). This rise would displace tens of millions of people living in low-lying coastal areas, and have a devastating effect on some small island nations.

Sea-level Rise in Southeastern US

The maps below indicate the areas of the southeastern US that would be threatened by sea-level rises of 3¼ ft (1 m), 6½ ft (2 m), and 20 ft (6 m). A 3¼-ft (1-m) rise is a little above the upper end of estimates for what can be expected this century. With this rise, parts of Florida and southern Louisiana would be inundated up to 18 miles (30 km) from the present coastline. A rise of 20 ft (6 m)—which would be exceeded if the Greenland Ice Sheet were to melt completely—would submerge a large part of Florida, while Louisiana would be flooded as much as 50 miles (80 km) in from the present coastline. This would appear to be highly unlikely in this century but could happen within a few hundred years if global warming continues.

3¼-ft (1-m) rise

flooded area

6½-ft (2-m) rise

20-ft (6-m) rise

EFFECTS OF GLOBAL WARMING

GLACIER RETREAT

SHRINKING ICE *Global warming is having a marked impact on the world's glaciers. The majority have shrunk since 1975, as their ice has melted faster than new ice has formed. These photographs, taken from the same viewpoint, show the extent of Briksdal Glacier in Olden, Norway, in 2002 (left) and 2011 (right).*

SUBMERGING ISLANDS

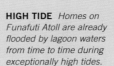

FUNAFUTI ATOLL *This atoll is part of Tuvalu, a group of small, low-lying Pacific islands whose future existence is threatened by sea-level rise.*

HIGH TIDE *Homes on Funafuti Atoll are already flooded by lagoon waters from time to time during exceptionally high tides.*

POPULATIONS AT RISK

CITY UNDER WATER *Dhaka, the capital of Bangladesh, and about three-quarters of the country's land area, is less than 27 ft (8 m) above sea level. Much of the country would be flooded by melting of the Greenland Ice Sheet. A rise of 2 ft (65 cm) would cause loss of 40 percent of productive land in southern Bangladesh. About 20 million people in coastal areas are affected by salinity in drinking water now.*

ANIMALS IN DANGER

STARVED TO DEATH *Polar bears are one of the animal species most severely threatened by global warming. The bears use Arctic sea ice as their summer hunting ground, and as the extent of sea ice diminishes, so do their opportunities for hunting and feeding.*

Coastal Landscapes

A GREAT VARIETY OF LANDSCAPES ARE FOUND along the coastlines of the world's oceans. Coasts are shaped by processes such as sea-level change and wave erosion, as well as by land-based processes such as weathering, erosion and deposition by rivers, glacier advance and retreat, the flow of lava from volcanoes, and tectonic faulting. Some coastal features are made by living organisms, including the reefs built by corals and the harbors, coastal defenses, and artificial islands built by humans.

FRINGING REEF
This reef-fringed coast, around the south Pacific island of Bora Bora, is a secondary coast, as it has been modified by the activities of living organisms, notably corals.

Classification of Coasts

Coasts can be classified as either primary or secondary. Primary coasts have formed as a result of land-based processes, such as the deposition of sediment from rivers (forming deltas), land erosion, volcanic action, or rifting and faulting in Earth's crust. Coasts formed as a result of recent sea-level change, which include drowned coasts and emergent coasts (see pp.88–89), are also usually considered primary, as are coastlines consisting mainly of wind-deposited sand, glacial till, or the seaward ends of glaciers. Coasts are considered secondary if they have been heavily shaped by marine erosional or depositional processes, or by the activities of organisms, such as corals, mangroves, or, indeed, people. A few coasts—for example, emergent coasts that have undergone significant marine erosion—display both primary and secondary features and so fit into an intermediate category.

ARTIFICIAL COAST
Singapore Harbor, in Southeast Asia, is an example of a coast that has been heavily shaped by human activity. Before human intervention, it was a mangrove-lined estuary.

VOLCANIC COAST
This land-eroded volcanic cone is in the Galápagos Islands. The entire coastline around these islands was formed by volcanic activity and so is a primary coast.

SEA ARCH
This spectacular arch in southern England is known as Durdle Door. A remnant of a once much larger headland, it is a classic feature of a marine-eroded coast.

Wave-erosion Coasts

Of all the different types of coastal landscape, perhaps the most familiar are wave-eroded cliffed coasts, a type of secondary coast. Wave erosion on these coasts occurs through two main mechanisms. First, waves hurl beach material against the cliffs, which abrades the rock. Second, each wave compresses air within cracks in the rocks, and on reexpansion the air shatters the rock. Where waves encounter headlands, refraction (bending) of the wave fronts tends to focus their erosive energy onto the headlands. At these headlands, distinctive features tend to develop in a classic sequence. First, deep notches and then sea caves form at the bases of cliffs on each side of the headland. Wave action gradually deepens and widens these caves until they cut through the headland to form an arch. Next, the roof of the arch collapses to leave an isolated rock pillar called a stack, and finally the stack is eroded down to a stump.

part of wave front opposite beach continues forward

beach

energy concentrated on headland as wave front refracts

erosion eventually divides headland into stacks

part of wave front opposite headland slows as it encounters shallower water

lobe of sediment

CONCENTRATION OF WAVE ENERGY
When a wave front reaches a shore consisting of bays and headlands, it refracts in such a way that wave energy tends to be concentrated onto the headlands.

wave front (extended crest of wave)

UNDERCUT CLIFF
Wave action has eroded a notch, and an adjoining platform, at the base of this cliff in the Caribbean.

SEA CAVE
This deep indentation and sea cave have been eroded into cliffs in the Algarve, Portugal.

SEA STACKS
The Old Harry Rocks are chalk sea stacks off Handfast Point, a headland near Swanage in southern England.

Marine-deposition Coasts

Marine depositional coasts are formed from sediment brought to a coast by rivers, eroded from headlands, or moved from offshore by waves. An important mechanism in their formation is longshore drift. When waves strike a shore obliquely, the movement of surf (swash) propels water and sediment up the shore at an angle, but backwash drags them back down at a right angle to the shore. Over time, water and sediment are moved along the shore. Where the water arrives at a lower-energy environment, the sediment settles and builds up to form various depositional features, including spits, baymouth bars, and barrier islands (long, thin islands parallel to the coast).

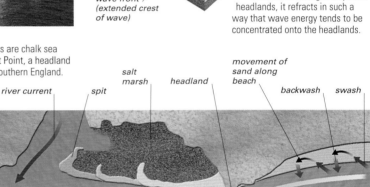

river current

spit

salt marsh

headland

movement of sand along beach

backwash

swash

second most common wind and wave direction

direction of longshore drift

prevailing wind and wave direction

BAYMOUTH BAR
Where a spit extends most or all of the way across the mouth of a bay or estuary, the result is called a baymouth bar. Here, a bar across the mouth of an estuary in Scotland, and an older spit, have created a sheltered coastal area of sandflats and salt marshes.

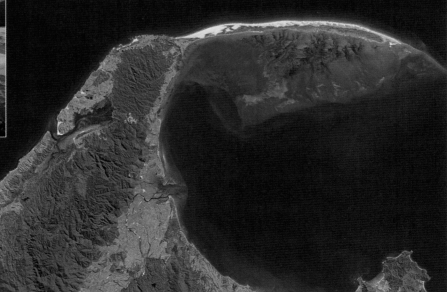

SPIT FORMATION
On this coastline, sand and water is carried past the headland by longshore drift, but the sand settles at the mouth of an estuary where the waves are opposed by the sluggish outflow from a river. There it forms a slowly growing spit—a sandy peninsula with one end attached to the land.

FAREWELL SPIT
This satellite view shows an impressive spit at Cape Farewell, at the northern tip of New Zealand's South Island. It formed as a result of a coastal current moving and depositing sand to the east of the cape.

OCEAN ENVIRONMENTS

ATLANTIC OCEAN NORTHWEST

Greenland Ice Coast

TYPE	Primary coast
FORMATION	Extension of ice-sheet to sea level in outlet glaciers
EXTENT	About 600 miles (1,000 km)

LOCATION Parts of western and eastern coasts of Greenland

An ice coast forms where a glacier extends to the sea, so that a wall of ice is in direct contact with the water. This is a common feature around the highly indented margins of Greenland, mainly at the landward end of long fjords. Together, these ice walls form an interrupted ice coast, and they are the source of enormous numbers of icebergs, many of which escape the fjords and eventually reach the Atlantic. The ice coast extends along only a fraction of the total Greenland coastline, which is an astonishing 27,500 miles (44,000 km) long.

ICE COAST NEAR CAPE YORK

ATLANTIC OCEAN NORTHWEST

Acadia Coastline

TYPE	Primary coast
FORMATION	Glaciation, then drowning by sea-level rise
EXTENT	41 miles (66 km)

LOCATION Southeast of Bangor, Maine, northeastern US

The coastline of Acadia in Maine is one of the most spectacular in the northeastern US. It forms part of Acadia National Park, most of which is located on a single large island, Mount Desert Island, and some smaller nearby islands. Sea-level rise since the last ice age has separated these islands from each other and from the mainland. The mountains that make up the basis of this coastline began to form 500 million years ago from seafloor sediments. Magma (molten rock) rising up from Earth's interior intruded into and consumed these sedimentary rocks, producing a mass of granite that was gradually eroded to form a ridge. About 2–3 million years ago, a huge ice sheet started to blanket the area, depressing the land and sculpting out a series of mountains

MOUNT DESERT ISLAND
The south-facing coast of Mount Desert Island consists of a series of fractured granitic steps that were produced by the action of glaciers some 100,000 years ago.

separated by U-shaped valleys. Since the ice sheet receded, the land has gradually rebounded upward, but global sea-level rise has caused the Atlantic to overtake the rebound at a rate of 2 in (5 cm) per century. Today, waves and tidal currents are major agents of change at Acadia, gradually eroding the cliffs and depositing rock particles mixed with shell fragments at coves around the coastline.

STAIRWAY TO THE SEA
The tops of the columns form stepping stones that first lead up from the foot of the cliff to a mound and then progress downward until they dip below the sea.

ATLANTIC OCEAN NORTHWEST

Hatteras Island

TYPE Secondary coast

FORMATION Deposition of sediment by waves and currents

EXTENT 70 miles (112 km)

LOCATION Off the coast of North Carolina, eastern US

Hatteras Island is a classic barrier island of sandy composition. It runs parallel to the mainland and is long and narrow, with an average width of 1,500 ft (450 m), and has been shaped by complex processes of deposition effected by ocean currents and waves. It is part of a series of barrier islands called the Outer Banks and has two distinct sections, which join at a promontory called Cape Hatteras. The dangerously turbulent waters in this area have resulted in hundreds of shipwrecks over the centuries.

HATTERAS SHORELINE
Hatteras Island is a typical barrier island, being low-lying with wave-straightened shorelines.

HUMAN IMPACT

CAPE HATTERAS LIGHTHOUSE

Erosion and deposition often cause shorelines to migrate. In 1999, the Cape Hatteras Lighthouse was moved because the sea had begun to lap at its base, threatening its destruction.

NEW POSITION
The lighthouse is now located about 1,500 ft (450 m) back from the shoreline.

ATLANTIC OCEAN WEST

Les Pitons

TYPE Primary coast

FORMATION Volcanic lava-dome formation followed by volcano collapse and erosion

EXTENT 3 miles (5 km)

LOCATION Southwestern coast of St. Lucia, Lesser Antilles, eastern Caribbean

The southwestern coastline on the Caribbean island of St. Lucia is rocky, highly indented, and steeply shelving. A landmark here is Les Pitons ("The Peaks"), two steep-sided mountain spires, each more than 2,430 ft (740 m)

high. These are the eroded remnants of two lava domes (large masses of lava) that formed some 250,000 years ago on the flank of a huge volcano. The volcano later collapsed, leaving behind the peaks and other volcanic features in the area. The volcanic rocks on this coast are densely vegetated, except on the very steepest parts of Les Pitons themselves. Beneath the sea are some scattered coral reefs within a series of protected marine reserves. This region was declared a World Heritage Site in 2004.

TWIN PEAKS
In this view, Petit Piton is the nearer peak, while Gros Piton, which is slightly higher and much broader, is visible in the background.

ATLANTIC OCEAN NORTHEAST

Giant's Causeway

TYPE Primary coast

FORMATION Cooling of basaltic lava flow from an ancient volcanic eruption

EXTENT ³/₅ mile (1 km)

LOCATION Northernmost point of County Antrim, Northern Ireland, UK

The Giant's Causeway is a tightly packed cluster of some 40,000 columns of basalt (a black volcanic rock). It is located at the foot of a sea cliff that rises 300 ft (90 m) on the northern coast of Northern Ireland. Although legend says the formation was created by a giant named Finn McCool, it in fact resulted from a volcanic eruption some 60 million years ago, one of a series that brought about the opening up of the North Atlantic. The eruption spewed up vast amounts of liquid basalt lava, which cooled to form the columns. They are up to 42 ft (13 m) tall and are mainly hexagonal, although some have four, five, seven, or eight sides.

OCEAN ENVIRONMENTS

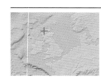

ATLANTIC OCEAN NORTHEAST

Gruinard Bay

TYPE	Primary coast
FORMATION	Ice-sheet retreat and postglacial rebound
EXTENT	8 miles (13 km)

LOCATION West of Ullapool, northwestern Scotland, UK

Around Gruinard Bay in Scotland there is evidence of a phenomenon known as postglacial rebound, in which a landmass, once pushed down by the huge weight of ice sheets during the last glacial period, rises again. In some areas, such as Scotland and Scandinavia, this upward rebound has outstripped the sea-level rise caused by the ice sheets melting. At Gruinard Bay, rebound is indicated by its raised beaches—flat, grassy areas behind the present-day beaches. Over the last 11,000 years, this part of Scotland has been moving upward relative to sea level, up to 4 in (10 cm) per century.

RAISED BEACH
The green area beyond the present-day beach, well above the line of high tide, is the remnant of an ancient beach.

ATLANTIC OCEAN NORTHEAST

Devon Ria Coast

TYPE	Primary coast
FORMATION	Former river valleys drowned by sea-level rise
EXTENT	About 60 miles (100 km)

LOCATION Between Plymouth and Torbay, southwestern coast of England, UK

Much of the south coast of the English county of Devon consists of the drowned valleys of the Dart, Avon, Yealm, and Erme Rivers and the Salcombe–Kingsbridge Estuary. The inlets, also known as rias, are separated by rugged cliffs and headlands. This beautiful coastal area was formed by the partial flooding of valleys, through which small rivers once flowed, as a result of global sea-level rise since the last ice age. The rise in sea level has been accentuated by the fact that the southern parts of the British Isles have been tipping downward since the last glacial maximum at a rate of up to 3 in (7 cm) per century.

SALCOMBE–KINGSBRIDGE ESTUARY
The highly scenic Salcombe Estuary is the largest of the five rias on the south Devon coast. Its protected waters provide ideal conditions for sailing.

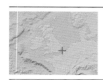

ATLANTIC OCEAN NORTHEAST

White Cliffs of Dover

TYPE	Secondary coast
FORMATION	Marine erosion of a large mass of ancient chalk
EXTENT	8 miles (13 km)

LOCATION Southeastern coast of England, to east and west of Dover, UK

One of England's most famous natural landmarks, the White Cliffs of Dover run along the northwestern side of the Strait of Dover, the narrowest part of the English Channel. They are complemented on the French side of the Strait by similar cliffs at Cap Blanc Nez. The chalk from which the cliffs are composed was formed between 100 million and 70 million years ago, when a large part of what is now northwestern Europe was underwater. The shells of tiny planktonic organisms that inhabited those seas gradually accumulated on the sea floor and became compressed into a layer of chalk that was several hundred yards thick. Subsequently, as the sea level fell during successive ice ages, this mass of chalk lay above the sea, and it later formed a land bridge between present-day England and France. However, about 8,500 years ago, the buildup of a large lake in an area now occupied by the southern North Sea caused a breach in the land-bridge. It eroded rapidly, causing flooding of the area that now forms the English Channel.

Today, the cliffs at Dover continue to be eroded at an average rate of an inch or two per year. Occasionally a large chunk detaches from the cliff edge and falls to the ground. Many marine fossils have been discovered in the cliffs, ranging from sharks' teeth to sponges and corals.

HIGH CHALK CLIFFS
Up to 360 ft (110 m) high, these cliffs owe their remarkable appearance to the almost pure chalk of which they are composed.

ATLANTIC OCEAN NORTHEAST

Cape Creus

TYPE	Primary coast
FORMATION	Land-eroded rocky coastline of schists and other metamorphic and igneous rocks
EXTENT	6 miles (10 km)

LOCATION East of Figueres, northeastern Catalonia, Spain

AN EASTERLY POINT OF THE CAPE

Cape Creus marks the point where the mountains of the Pyrenees meet the Mediterranean Sea. It has one of the most rugged coastlines in the entire Mediterranean region, with cliffs made of extremely rough-textured rocks, interspersed by small coves. Designated as a natural park in 1998, Cape Creus also boasts a varied underwater marine life, and is rich in invertebrate animals such as sponges, anemones, fan worms, and red corals. As such, it is a popular diving location. The landscape is said to have inspired the Spanish surrealist artist Salvador Dalí (1904–1989), and it features in many of his paintings, including *The Persistence of Memory*.

Western Algarve

TYPE Secondary coast	
FORMATION Erosional action of waves on ancient rock strata	
EXTENT 84 miles (135 km)	

LOCATION Southern and southwestern coast of Portugal

The western Algarve coast extends from the city of Faro in southern Portugal to Cape St. Vincent, at the southwestern tip of the Iberian Peninsula, and then for a further 30 miles (50 km) to the north. This coastline, which is bathed by the warm Gulf Stream, is notable for its picturesque, honey-colored limestone cliffs, small bays and coves, sheltered beaches of fine sand, and emerald-green water. Many stretches of this coast show typical features of marine erosion at work, including caves at the feet of cliffs, grottoes, blowholes, arches forming through headlands, and sea stacks (isolated pillars of rock set off from headlands). Although limestone is a primary component of the landscape, other rocks, including sandstones and shales, form parts of the cliffs along scattered stretches of the coast. The strikingly beautiful scenery has helped make this coast a popular vacation destination.

BALANCED STACK
At Marinha Beach near Carvoeiro, wind and waves have produced distinctive rock formations, such as this eroded sea stack balanced on the shoreline.

OCEAN ENVIRONMENTS

Amalfi Coast

TYPE Secondary coast	
FORMATION Marine erosion of folded and inclined limestone rock strata	
EXTENT 43 miles (69 km)	

LOCATION Southern side of the Sorrento Peninsula, south of Naples, southern Italy

Stretching along the southern edge of the Sorrento Peninsula, south of Naples, the Amalfi Coast is famous for its steep cliffs punctuated by caves and grottoes, and for its picturesque coastal towns, some of which are built into the cliffs. The inclined layers of limestone rock that form the cliffs lie at the foot of the Lattari Mountains and were formed between 100 and 70 million years ago.

CLIFFS AT SANT'ELIA, AMALFI COAST

Nile Delta

TYPE Primary coast	
FORMATION Deposition of sediment at mouth of Nile River	
EXTENT 150 miles (240 km)	

LOCATION North of Cairo, northern Egypt

The Nile Delta is one of the world's largest river deltas. As with all deltas, its shoreline is classed as a primary coast because it formed as a result of sediment deposition from a river, a land-based process. The flow of water that formed it and nourishes it has been reduced significantly by the Aswan Dam in Upper Egypt and by local water usage. The sand belt at the delta's seaward side, which prevents flooding, is currently eroding, and anticipated future rises in sea level pose a severe threat to its agriculture, freshwater lagoons, wildlife, and reserves of fresh water. Farming has already had to be abandoned in some areas.

PROTECTIVE BELT
In this satellite view, the sand belt at the front of the Nile Delta is clearly visible. The protrusions through the sand belt mark the mouths of two Nile tributaries.

ATLANTIC OCEAN SOUTHEAST

Skeleton Coast

TYPE	Secondary coast
FORMATION	Wind-formed desert dunes
EXTENT	310 miles (500 km)

LOCATION Extending northwest from the city of Swakopmund on the coast of Namibia

The Skeleton Coast is an arid coastal wilderness in southwestern Africa, where the northern part of the Namib Desert meets the South Atlantic. Some stretches of this coast are dominated by sand dunes that extend to the sea; others consist of low gravel plains. An important influence on this largely straight coastline is the Benguela Current, a surface current that flows in a northerly direction offshore, bringing cool waters from the direction of Antarctica. Prevailing southwesterly winds blow onto the coast from the Atlantic, but as they cross the cold offshore water, any moisture in the air condenses. This leads to an almost permanent fog bank and allows strange desert plants such as *Welwitschia mirabilis*, a species that survives for hundreds of years, to thrive. The coast is home to a large seal colony at Cape Fria in the north and includes many salt pans.

SHIPWRECKS

The Skeleton Coast is aptly named. Its frequent fogs, onshore winds, and pounding surf have made it a graveyard for both ships and sailors. Behind the coast is a steep mountain escarpment, so before the days of rescue parties, the escape route for shipwrecked mariners was a long march along the coast through an arid desert.

WOODEN SKELETON
This wreck of a wooden vessel is one of many ships that have foundered on this treacherous coast.

HIGH DUNES AND POUNDING SURF
The coast's high dunes present ever-changing contours as they are blown by strong southwesterly winds. Below the dunes, waves pound the beaches.

INDIAN OCEAN NORTHWEST

Red Sea Coast

TYPE	Primary coast
FORMATION	Faulting and sinking of land
EXTENT	1,200 miles (1,900 km)

LOCATION Coasts of Egypt, Sudan, Eritrea, and Saudi Arabia, from gulfs of Suez and Aqaba to Djibouti

The Red Sea was created as a result of a rifting process that has been gradually separating Africa from the Arabian Peninsula for the past 25 million years. Rifting is the splitting of a region of Earth's crust into two parts, which then move apart, creating a new tectonic plate boundary. This process begins when an upward flow of heat from Earth's interior stretches the continental crust, causing it to thin, and eventually it may fracture, or fault. Sections of crust may sink, and if either end of the rift connects to the sea, flooding will occur, creating new coasts. On both sides of the Red Sea, there is evidence of the downward movement of blocks of crust, in the form of steep escarpments (lines of mountains). The Red Sea shoreline itself shelves steeply in many parts. On the land side, the coast is sparsely vegetated because of the region's hot, dry climate, but underwater there are many rich and spectacular coral reefs.

SEA MEETS DESERT
The steep Sarawat mountain escarpment, which runs the length of the eastern coast of the Red Sea, can be seen in the distance in this view from part of the Sinai Peninsula.

INDIAN OCEAN NORTHWEST

Tigris-Euphrates Delta

TYPE Primary coast

FORMATION Sediment deposition from Tigris, Euphrates, and Karun

EXTENT 95 miles (150 km)

LOCATION Parts of southeastern Iraq, northeastern Kuwait, and southwestern Iran

The Tigris-Euphrates delta is a broad area of marshes and alluvial plain at the northern head of the Arabian Gulf, formed from sediment deposited by three major rivers. An important wildlife haven, the delta suffered great ecological damage between the 1970s and 2003 from various drainage and damming schemes carried out for military and political purposes. Fisheries and several animal species became threatened. However, a substantial recovery has occurred since 2003.

SATELLITE VIEW
The delta's seaward edge has advanced by about 150 miles (250 km) in the past 3,000 years.

INDIAN OCEAN SOUTHEAST

The Twelve Apostles

TYPE Secondary coast

FORMATION Wave erosion of cliffs producing large sea stacks

EXTENT 2 miles (3 km)

LOCATION Near Port Campbell, southwest of Melbourne, Victoria, southeastern Australia

One of Australia's best-known geological landmarks is a group of large sea stacks formed through the erosion of 20-million-year-old limestone cliffs. Known as the Twelve Apostles, even though there were originally only nine of them, the stacks are up to 230 ft (70 m) tall. In 2005, one of the stacks collapsed, leaving just eight. Collapses such as this are quite common and are an integral part of the erosion process.

ONGOING EROSION
The effects of wave erosion can clearly be seen at the bases of the remaining Apostles.

INDIAN OCEAN NORTHEAST

Krabi Coast

TYPE Primary coast

FORMATION Chemical erosion of limestone followed by drowning

EXTENT 100 miles (160 km)

LOCATION Andaman Sea coast of southwestern Thailand

The area around Krabi on the western coast of southern Thailand is notable for its fantastic-looking formations of partially dissolved limestone, known as karst. This limestone was originally formed about 260 million years ago. At that time, a shallow sea covered what is now south Asia and slowly built up deposits of shells and coral that sediments washed in from the land subsequently buried. These formed layers of limestone, which were later thrust upward and tipped over at an angle when India began to collide with mainland Asia some 50 million years ago. Around Krabi and Phang Nga Bay to its north, chemical erosion of these limestone strata by rainwater, followed by sea-level rise, has created thousands of craggy karst hills and islands. These include a number of isolated cone- and cylinder-shaped karst towers that rise out of the sea to heights of up to 700 ft (210 m) and groups of towers that sit on broad masses of limestone. Many of these karst formations are elongated in a northeast-southwest direction, reflecting the axis (or strike line) around which the original layers of limestone were tipped.

KOH TAPU ISLAND
Some of the karst formations along this coast have been weathered into unusual shapes, as in these examples at Koh Tapu Island in Phang Nga Bay to the north of Krabi.

PACIFIC OCEAN WEST

Hong Kong Harbor

TYPE Secondary coast

FORMATION Artificial coast built around various natural harbors and nearby islands

EXTENT 25 miles (40 km)

LOCATION Southeast of Guangzhou, on the South China Sea coast of southeastern China

A number of natural harbors surround Hong Kong Island, which is the best-known part of the Hong Kong region of China. Together, these harbors make up one of the world's busiest ports, both in terms of number of ship movements and amount of cargo handled. The largest, naturally deepest, and most sheltered of the harbors is Victoria Harbour, which has an area of over 16 square miles (42 square km) and is situated between Hong Kong Island and Kowloon Peninsula. Other smaller harbors include Aberdeen Harbor, which separates Hong Kong Island from a satellite island, Ap Lei Chau. The margins of all these harbor areas have been artificially modified by the construction of concrete piers, seawalls, jetties, and other structures. This coastline can be classified as a secondary coast because it has been modified by living organisms, in this case, humans. In the whole of the Hong Kong region, more than 60 miles (100 km) of coastline have been artificially constructed or modified.

VICTORIA HARBOUR
This nighttime view shows Victoria Harbour with Hong Kong Island on the left and Kowloon on the right. The harbor is visited by more than 200,000 ships per year.

OCEAN ENVIRONMENTS

THE SKELETON COAST
On Africa's desolate Skeleton Coast, the huge, steep-faced dunes of the Namib Desert meet the cold waters of the southern Atlantic. On this part of the coast, the remains of an ancient shipwreck can be seen close to the shoreline. Above it, fog hangs in the air from condensation of moisture blown in by southwesterly winds.

PACIFIC OCEAN WEST

Ha Long Bay

TYPE	Primary coast
FORMATION	Chemical dissolution and drowning of limestone formations
EXTENT	75 miles (120 km)

LOCATION On the Gulf of Tonkin, east of Hanoi, Northeastern Vietnam

Ha Long Bay is a distinctive region on the coast of Vietnam, within the Gulf of Tonkin. It consists of a body of water filled with nearly 2,000 islands composed of karst (limestone partially dissolved by rainwater). This landscape, which covers an area of just over 585 square miles (1,500 square km), was created by sea-level rise and flooding of a region with a high concentration of karst towers. Several of the islands are hollow and contain huge caves, and a few have been given distinctive names, such as Ga Choi ("Fighting Cocks") Island, Man's Head Island, and the Incense Burner, as a result of their unusual shapes. Most are uninhabited. The Bay's shallow waters are biologically highly productive and sustain hundreds of species of fish, mollusks, crustaceans, and other invertebrates, including corals. Designated a World Heritage Site in 1998, Ha Long Bay is currently under threat from destruction of mangroves and from pollution caused by urban development and mining activities nearby. A further problem has been a high level of plastic jettisoned from tourist boats into the bay.

TOWERING LIMESTONE
Several large karst islands, each topped with thick tropical vegetation, tower over a central area of Ha Long Bay. These islands rise up to 660 ft (200 m) above sea level.

PACIFIC OCEAN WEST

Huon Peninsula

TYPE	Primary coast
FORMATION	Uplift of fossil coral reefs as a result of tectonic plate movement
EXTENT	50 miles (80 km)

LOCATION Eastern Papua New Guinea, north of Port Moresby

For hundreds of thousands of years, the Huon Peninsula has been forced upward at a rate of about 10 in (25 cm) per century by movements of Earth's crust at a tectonic plate boundary. This activity has pushed coastal coral reefs above the shoreline to form a series of terraced reefs on land. The oldest of these are hundreds of yards back from the coast. By studying them, scientists have learned much about changes in sea level and climate over the past 250,000 years.

AERIAL VIEW OF THE PENINSULA

PACIFIC OCEAN NORTHEAST

Puget Sound

TYPE	Primary coast
FORMATION	Glacier-carved coastal channels and bays
EXTENT	90 miles (150 km)

LOCATION North and south of Seattle, Washington State, northwestern US

Puget Sound, with its numerous channels and branches, was created primarily by glaciers. About 20,000 years ago, a glacier from present-day Canada advanced over the area, covering it in thick ice. Over the next 7,000 years, glaciers advanced and retreated several times. When they finally withdrew, they left behind many deeply gouged channels and thick layers of mud, sand, and gravel deposited by meltwater. Waves and weather have since reworked the deposits, molding landforms and shoreline, and forming beaches, bluffs, spits, and other sedimentary features.

SOUND SETTLEMENTS
Much of the shoreline around Puget Sound has now been settled. The town of Tacoma is seen here, with Mount Rainier in the distance.

HUMAN IMPACT

ENVIRONMENTAL STRESSES

The human population living around the shores of Puget Sound has increased several-fold since the 1950s, creating a number of environmental stresses. These include runoff of pollutants into the waters of the Sound in stormwater, a decrease in the area of habitats such as salt marshes and seagrass beds, and overfishing. These stresses have in turn led to a severe decline in the numbers of many marine species that were once abundant in the region, such as the harbour porpoise, pinto abalone (a mollusk), copper rockfish, and various species of scoter (variety of sea duck). A number of restoration plans are now in operation in an attempt to reverse the degradation that has taken place.

OCEAN ENVIRONMENTS

PACIFIC OCEAN NORTHEAST

Big Sur

TYPE Intermediate coast

FORMATION Tectonic uplift combined with rapid wave erosion

EXTENT 90 miles (145 km)

LOCATION Southeast of San Francisco, coast of California, US

The Big Sur coastline of central California, where the rugged Santa Lucia Mountains descend steeply into the Pacific Ocean, is one of the most spectacular in the US. Like much of the west coast of North America, Big Sur is an emergent shoreline, in that the coast has risen up faster than sea level since the end of the last ice age. This uplift has resulted from interactions at the nearby boundary between the Pacific and North American tectonic plates—this region is crisscrossed by a complex system of faults in Earth's crust and is subjected to frequent earthquakes. At Big Sur a combination of tectonic uplift and relentless wave erosion has produced steep cliffs and partially formed marine terraces (platforms cut at the base of cliffs by waves and then lifted up). The coast is susceptible to landslides as a result of wave action, the weakening of the cliffs by faulting and fracturing, the destruction of vegetation by summer fires, and heavy winter rainfall.

RAISED PLATFORM
In this view of part of Big Sur, a grassed-over marine terrace (the green area) is visible above the present-day cliff, with a raised ancient cliff behind it.

PACIFIC OCEAN CENTRAL

Hawaiian Lava Coast

TYPE Primary coast

FORMATION Lava flow into the sea from an active volcano

EXTENT 14 miles (20 km)

LOCATION Southeastern coast of the Big Island of Hawaii, US

One of the fastest ways for a coast to change shape is as a result of lava flow to the sea. On southeastern Big Island, new coast has been added intermittently since 1969 as a result of lava flows from satellite craters of the active volcano Kilauea. From 1983 to 2018, lava from the Pu'u O'o crater flowed 9 miles (15 km) to the sea, where it cooled and hardened. The landscape created by these flows is a primitive scene of black beaches and dark cliffs made of rough, fractured lava. Since 2018, Kilauea's eruption has shifted to an area a few miles to the east.

STEAM PLUMES
As red-hot lava enters the sea, it solidifies amid huge plumes of steam. Newly forming shoreline sometimes collapses to reveal ripped-open lava tubes.

PACIFIC OCEAN SOUTHEAST

Chilean Fjordlands

TYPE Primary coast

FORMATION Deep glacier-carved valleys flooded by sea-level rise

EXTENT 950 miles (1,500 km)

LOCATION Pacific coast of southern Chile from Puerto Montt to Punta Arenas

The Chilean fjordlands are a labyrinth of fjords, islands, inlets, straits, and twisting peninsulas, lying to the west of the snow-capped peaks of the southern Andes. The fjordlands extend for most of the length of southern Chile, as far south as Tierra del Fuego, and their total area is some 21,500 square miles (55,000 square km).

Some 10,000 years ago, this region was covered in glaciers, but these have largely retreated into large ice-filled areas within the mountains on the Chile–Argentina border called the Northern and Southern Patagonian Ice Fields. The glaciers left behind a network of long, deeply gouged valleys, which were filled by glacier meltwater and then flooded by the sea to form today's fjords. Rainfall here is heavy, and clear skies are rare because the moisture-laden Pacific air cools and forms clouds as it rises to cross the Andes. On the edges of the fjords, waterfalls cascade down steep granite walls, while hundreds of species of birds nest and feed around the often mist-shrouded coast and islands. Mammals that live along this coast include sea lions, elephant seals, and marine otters.

ICE-CHOKED FJORD
The calving ends of outlet glaciers, which choke the waters with icebergs, are found at the landward end of some fjords.

BATTERED BY THE SEA
The Eastern Scheldt storm-surge barrier in the Netherlands is one of the world's largest sea defenses. Its 62 sliding steel gates are held between concrete piers.

Coastal defense refers to various types of engineering techniques aimed at protecting coasts from the sea. The threats posed by the sea fall into two main categories. First is the danger of flooding of low-lying coastal areas during severe storms. Second is the continuous gradual erosion of some coasts. There are a number of different approaches to coastal defense. To prevent flooding of low-lying regions, one solution is to build a large-scale system of dams and tidal barriers. Another is to encourage the development of natural barriers, such as salt marshes, around coasts, and to conserve existing areas of this type. A third possibility is managed retreat. Instead of trying to hold back the sea, some areas of coast are allowed to flood. The idea is that, in time, the flooded land will turn into a marsh, providing natural protection.

To slow coastal erosion, various "hard" engineering techniques are commonly employed, such as the building of sea walls, breakwaters, or groynes. These methods can be effective for a while (they usually have to be rebuilt after a few decades), but are expensive and can increase erosion on neighboring areas of coast by interfering with longshore movement of sediment. "Soft" engineering techniques are more environmentally friendly. They include the temporary solution of beach nourishment (see panel, right), which has to be repeated every few years, and encouraging the development of coastal dunes.

Crumbling Coasts

Many coasts around the world are eroding. In England, stretches of the east coast are eroding by up to 6 ft (1.8 m) a year—the highest rate in Europe—while parts of the Jurassic coast in the south have frequent landslides. The one pictured below, near Seatown in Dorset, occurred in April 2021, with the collapse of a 4,460-ton (4,050-metric-ton) chunk of cliff. Some eroding cliffs have doomed houses situated on top. In the US, for example, it is estimated that tens of thousands of coastal homes are at risk of falling into the sea by 2060. Coastal defenses can slow coastal erosion temporarily, but in the long run, maintenance becomes prohibitively expensive, and the sea wins out.

TYPES OF DEFENSES

HARD ENGINEERING

SEA WALL *A sea wall is designed to reflect wave energy. Modern walls have a curved top that prevents water from spraying over the wall in storms. A wall protects the land behind it for some years but usually increases erosion of the beach in front of it.*

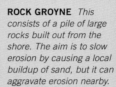

ROCK GROYNE *This consists of a pile of large rocks built out from the shore. The aim is to slow erosion by causing a local buildup of sand, but it can aggravate erosion nearby.*

SOFT ENGINEERING

DUNE STABILIZATION *Coastal dunes provide valuable protection against erosion if they can be stabilized and prevented from shifting. This is usually achieved by planting with grasses.*

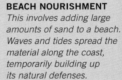

BEACH NOURISHMENT *This involves adding large amounts of sand to a beach. Waves and tides spread the material along the coast, temporarily building up its natural defenses.*

MODERN SOLUTION

GEOTEXTILE TUBE *Also called a geotube, a geotextile tube is a long, cylindrical container made of a type of durable but permeable textile and filled with sand and water. The tube can be laid along the top of a beach or partially embedded at the base of a dune, where it reduces coastal erosion. This tube formed part of a marsh restoration project at Barren Island, Maryland.*

LARGE-SCALE PROTECTION

DAMS AND STORM-SURGE BARRIERS *The Netherlands has invested in an extensive series of engineering works to protect a large region of the country from future marine flooding. Known as the Deltaworks, it includes many dams and movable storm-surge barriers. The works were initiated in 1953 after a serious storm and floods killed a total of 1,835 people.*

OCEAN ENVIRONMENTS

Beaches and Dunes

BEACHES ARE DEPOSITS OF SEDIMENTARY MATERIAL, ranging in size from fine sand to rocks, that commonly occur on coasts above the low-tide line. Sources of beach material include sediment brought to a coast by rivers, or eroded from cliffs or the sea floor, or biological material such as shells. This material is continually moved on and off shore and around coasts, by waves and tides. Wind can also influence beach development and is instrumental in forming coastal dunes.

DISSIPATIVE BEACH AND DUNE
Dissipative beaches are usually made up of fine sand, and they slope at an angle of less than 5°.

Beach Anatomy

A typical beach has several zones. The foreshore is the area between the average high- and low-tide lines. On the seaward side of the foreshore is the nearshore, while behind it is the backshore; the latter is submerged only during the very highest tides and usually includes a flat-topped accumulation of beach material called a berm. The sloping area seaward of the berm, making up most of the foreshore, is the beach face. At the top end of the beach face there are sometimes a series of crescent-shaped troughs, called beach cusps. The swash zone is the part of the beach face that is alternately covered and uncovered with water as each wave arrives. Seaward of the swash zone, extending out to where the waves break, is the surf zone. The shape of a beach often alters as wave energy changes over the year.

SWASH
The surge of water and sediment up a beach when a wave arrives is called swash. If waves reach a beach at an angle, the combined effect of swash and backwash moves material along the beach.

Types of Beaches

The level of wave energy, the direction the waves arrive from, and the geological makeup of a coast all affect the type of beach that will form. Dissipative beaches are gently sloping and absorb wave energy over a broad area, while reflective beaches are steeper and shorter, and consist of coarser sediment. If a cliffed coast contains a mixture of both easily eroded and erosion-resistant rock, headlands tend to form, with crescent-shaped beaches within the bays (embayed beaches) or smaller "pocket" beaches. Both of these tend to be "swash-aligned"—the waves arrive parallel to shore and do not transport sediment along the beach. Many long, straight beaches are "drift-aligned"—the waves arrive at an angle and sediment is moved along the beach by longshore drift.

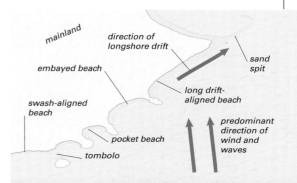

RANGE OF BEACHES
This imaginary coast (right) shows several beach types, ranging from a tombolo (a sand deposit between the mainland and an island) to a drift-aligned beach.

mainland
direction of longshore drift
embayed beach
sand spit
swash-aligned beach
long drift-aligned beach
pocket beach
tombolo
predominant direction of wind and waves

NEARSHORE   FORESHORE   BACKSHORE

average low-tide line | surf zone | swash zone | average high-tide line | beach face | beach cusp | berm | berm crest | foredune

BEACH PARTS AND ZONES
This photograph (above) shows the main zones on a beach and the locations of the berm, beach face, and beach cusps. It was taken when the sea was approaching low tide.

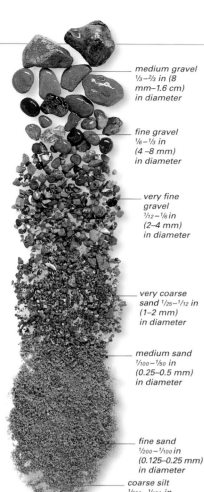

medium gravel
$\frac{1}{3}$–$\frac{2}{3}$ in (8 mm–1.6 cm) in diameter

fine gravel
$\frac{1}{6}$–$\frac{1}{3}$ in (4–8 mm) in diameter

very fine gravel
$\frac{1}{12}$–$\frac{1}{6}$ in (2–4 mm) in diameter

very coarse sand $\frac{1}{25}$–$\frac{1}{12}$ in (1–2 mm) in diameter

medium sand
$\frac{1}{100}$–$\frac{1}{50}$ in (0.25–0.5 mm) in diameter

fine sand
$\frac{1}{200}$–$\frac{1}{100}$ in (0.125–0.25 mm) in diameter

coarse silt
$\frac{1}{850}$–$\frac{1}{400}$ in (0.03–0.06 mm) in diameter

GRAIN SIZES
The silts, sands, and gravels that make up most beaches tend to become sorted by the action of waves, with material of different sizes deposited on different parts of the beach.

Beach Composition

The composition of a beach at any particular location depends on the material available and on the energy of the arriving waves. Most beaches are composed of sand, gravel, or pebbles produced from rock erosion. Sand consists of grains of quartz and other minerals, such as feldspar and olivine, typically derived from igneous rocks such as granite and basalt. Other common beach-forming materials, seen particularly in the tropics, include the fragmented shells and skeletons of marine organisms. In general, higher wave energies are associated with coarser beach material, such as gravel or pebbles, rather than fine sand. Occasionally, large boulders are found on beaches—usually they have rolled down to the shore from local cliffs, but some boulders have ended up on beaches as a result of glacial transport or even backwash from tsunamis.

PEBBLES AND SHELLS
This high-energy beach (left) contains many large pebbles. Mollusk shells in the beach below reflect favorable offshore feeding conditions for the live mollusks.

MARRAM GRASS
This grass is a common colonizer of embryo dunes. It develops deep roots that allow it to tap into deep groundwater stores. The roots bind the sand together, while the grass traps more blown sand, assisting in foredune development.

Coastal Dunes

Coastal dunes are formed by wind blowing sand off the dry parts of a beach. Dunes develop in the area behind the backshore, which together with the upper beach face supplies the sand. For dunes to develop, this sand has to be continually replaced on the beach by wave action. The actual movement of sand to form dunes occurs through a jumping and bouncing motion along the ground called saltation. Some coastal areas have more than one set of vegetated dunes that run parallel to the shoreline. The dunes closest to shore are called foredunes; behind them is a primary dune ridge, secondary dune ridge, and so on. These anchored, vegetated dunes are important for the protection they provide against coastal erosion. On some coasts, nonvegetated, mobile dunes occur; these move in response to the prevailing winds. They can often be anchored by planting with grasses.

REFLECTIVE BEACH
This beach in the Seychelles is an example of a reflective beach because of its quite steeply shelving face. It has a distinct berm and berm crest.

OCEAN ENVIRONMENTS

ATLANTIC OCEAN WEST

Pink Sands Beach

TYPE Dissipative beach, protected by reefs	
COMPOSITION Sand mixed with broken shells and skeletons	
LENGTH 2.5 miles (4 km)	

LOCATION Harbor Island, off Eleuthera, northeast of Nassau, Northern Bahamas

Pink Sands in the Bahamas is a gently sloping beach that faces east onto the Atlantic Ocean. It is protected from ocean currents by an outlying reef. The pale pink color of the sand comes from small, single-celled organisms called foraminiferans, in particular, the species *Homotrema rubrum*, also known as the sea strawberry. The shells of these organisms are bright red or pink due to the presence of an iron salt. In parts of the Bahamas they are abundant, living on the underside of reefs. When they die, they fall to the seafloor, where they are broken by wave action and mixed with other debris, such as the white shells of snails and sea urchins, as well as mineral grains. This mixture is then finely pulverized and washed up on the shore as pink-colored sand by wave action.

GENTLE SLOPE
Pink Sands is an example of a dissipative beach, on which waves break some distance from the shore, then slowly roll in, dissipating their energy across a broad surf zone.

ATLANTIC OCEAN SOUTHWEST

Copacabana Beach

TYPE Embayed, dissipative beach	
COMPOSITION White sand	
LENGTH 2 miles (3.5 km)	

LOCATION Rio de Janeiro, southeastern Brazil

One of the most famous beaches in the world, Copacabana Beach is a wide, gently curving stretch of sand between two headlands. Behind the beach lies the city of Rio de Janeiro, with green, luxuriant hills in the hinterland. The beach is crowded much of the year and is known for its beach sports and New Year's Eve firework displays. The sea area off the beach is not always recommended for swimming, due to strong currents.

COPACABANA LOOKING NORTHEAST

ATLANTIC OCEAN NORTHEAST

St. Ninian's Tombolo

TYPE Tombolo	
COMPOSITION Yellow and white sand	
LENGTH 1/3 mile (500 m)	

LOCATION West coast of southern Mainland, the main island of the Shetland Isles, off Scotland, UK

St. Ninian's Isle in the Shetland Isles provides a classic example of a tombolo, or ayre—a short spit of sedimentary material that connects an island to a nearby land mass or mainland. A tombolo is formed by waves curving around the back of an island so that they deposit sediment on a neighboring land mass, at the point directly opposite the island. Over time, these sediments gradually build up into a tombolo, which typically projects at right angles to the coast and has a beach on each side. St. Ninian's Tombolo has been in existence for at least 1,000 years, and its permanence may be due to a cobble base underlying the sand. This tombolo tends to become lower and narrower during storms as a result of destructive wave action, while during calmer weather the waves build it up again with sand carried from offshore or the nearshore. The sediment that forms a tombolo may come from the mainland, the island, the sea floor, or a combination. Scientists have deduced that a tombolo will usually form when the island's distance from the shore is less than two-thirds of its length parallel to the shore (the distance of St. Ninian's Isle from the shore is only about one third of its length). In other cases, a feature called a salient may form—a sand spit that extends toward the island but does not quite reach it.

NARROW CONNECTION
A slim, sandy tombolo extends from the Shetland island of Mainland in the foreground, to St. Ninian's Isle.

North Jutland Dunes

TYPE	Coastal dunes
COMPOSITION	Yellow sand, marram grass
LENGTH	155 miles (250 km)

LOCATION North and northwest coast of Jutland, Denmark

Much of the northern coastline of Denmark's Jutland Peninsula consists of sand dunes, which cover several thousand square miles of coast. These dunes are "active" in that they have a natural tendency to migrate along the coast, carried by wind (sand drift) and wave erosion. In some areas, attempts have been made to restrict this dune drift, to prevent sand from inundating summer houses. Some early attempts were fruitless. For example, sand fences were built into the dunes during World War II, but the dunes have since moved behind them, leaving the fences on the beach. More recently, many dune areas have been stabilized more successfully by planting with grasses and conifer trees.

SHIFTING SANDS
Many sand dunes on the peninsula have marram grass growing in them, which helps constrain their movement.

Porthcurno Beach

TYPE	Pocket beach
COMPOSITION	Yellow-white sand, composed mainly of shell fragments
LENGTH	500 ft (150 m)

LOCATION Southwest of Penzance, Cornwall, southwestern England, UK

Porthcurno is a typical pocket beach located near Land's End at England's southwesternmost tip. Like all pocket beaches, it nestles between two headlands that protect the sandy cove from erosion by winter storms and strong currents. Pocket beaches are common where cliffs made of different types of rock are subject to strong wave action. Rock that is especially hard and resistant to erosion forms headlands, while intervening areas of softer rock are worn down to form pocket beaches. Unlike other beaches, pocket beaches exchange little or no sand or other sediment with the adjacent shoreline, because the headlands prevent longshore drift. The sea at Porthcurno is a distinctive turquoise, possibly due to the reflective qualities of the sand, which is made mainly of shell fragments.

GRANITE HEADLANDS
The headlands on either side of the beach are formed from 300-million-year-old granite.

Chesil Beach

TYPE	Storm beach on tombolo
COMPOSITION	Gravel of flint and chert
LENGTH	18 miles (29 km)

LOCATION West of Weymouth, Dorset, southern England

Chesil Beach forms the seaward side of the Chesil Bank, a remarkably long, narrow bank of sedimentary material that connects the coast of Dorset in southern England to the Isle of Portland. Behind the bank is a tidal lagoon called the Fleet. Running parallel to the coast, Chesil Bank looks like a barrier island. However, because it connects the mainland to an island, it is classified as a tombolo. How Chesil Bank and its beach originally formed is debated—the most widely accepted theory is that it originally formed offshore and was then gradually moved to its current location by waves and tides. The beach is classified as a storm beach, as it is affected by strong waves because it faces southwest toward the Atlantic and the prevailing winds. Like most storm beaches, it is steep, with a gradient of up to 45 degrees, and is made of gravel.

CHESIL BANK
The bank is about 560 ft (170 m) wide and 50 ft (15 m) high along its entire length. The beach (left) is on its seaward side.

DISCOVERY

GRADED PEBBLES

Chesil Beach's pebbles change in size progressively from potato-sized at one end to pea-sized at the other. This reveals the differences in wave energy along its length—at one end, strong waves wash smaller pebbles offshore; at the other, weaker waves wash them onshore.

ATLANTIC OCEAN NORTHEAST

Cap Ferret

TYPE Coastal dunes on a spit	
COMPOSITION Sand, grasses, forest	
LENGTH 7¹/₂ miles (12 km)	

LOCATION Coast of Aquitaine, southwest of Bordeaux, southwestern France

Cap Ferret lies at the southern end of a long sand spit in western France. It separates the Arcachon Lagoon from the Atlantic Ocean and forms part of the spectacular Aquitaine coast, which at 143 miles (230 km) is the longest sandy coast in Europe. This region is characterized by a series of straight, sandy beaches backed by longitudinal sand dunes, which are the highest dune formations in Europe. They include the highest individual European sand dune, the Dune du Pilat, which rises to about 350 ft (107 m) above sea level.

Behind the main dune area is a forest, originally planted in the 18th century to try to prevent the dunes from shifting. Unfortunately, this coast is undergoing serious erosion, of more than 33 ft (10 m) a year in some places, mainly because excessive urban development has degraded the vegetation cover.

SAND MOUNTAINS
Along the coast to the north and south of Cap Ferret, mini-mountains of pale, rippling sand are backed by an extensive vegetation cover.

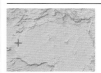

ATLANTIC OCEAN EAST

Banc d'Arguin

TYPE Coastal dunes and tidal flats	
COMPOSITION Yellow sand	
LENGTH 100 miles (160 km)	

LOCATION Between Nouakchott and Nouadhibou on the northwest coast of Mauritania, West Africa

The Banc d'Arguin National Park is a vast region of dunes, islands, and shallow tidal flats covering more than 4,600 square miles (12,000 square km) of the Mauritanian coast. The dunes, which consist mainly of windblown sand from the Sahara, are concentrated in the southern region of the Park. Banc d'Arguin contains a variety of plant life and is a major breeding or wintering site for many migratory birds, including flamingos, pelicans, and terns. It was declared a UNESCO World Heritage Site in 1987.

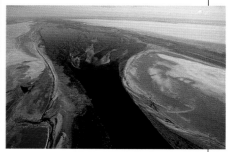

SAND BANKS AT BANC D'ARGUIN

INDIAN OCEAN SOUTHWEST

Jeffreys Bay

TYPE Series of gently sloping, dissipative beaches	
COMPOSITION Sand	
LENGTH 9 miles (15 km)	

LOCATION West of Port Elizabeth, eastern Cape Province, South Africa

Jeffreys Bay is famous both as a highly popular surfing spot and for the large numbers of beautiful seashells that wash up on its shores. It consists of a series of wide beaches strung out along a southeast-facing stretch of the South African coastline.

As a surfing destination, Jeffreys Bay is regularly ranked among the top five beaches in the world by those seeking the "perfect wave." The most acclaimed surfing spot or wave "break" is known as Supertubes. Here, the combination of shoreline shape, bottom topography, and direction of wave propagation regularly generates waves that form huge, glassy-looking hollow tubes as they break. Other nearby wave breaks in Jeffreys Bay have been given such colorful names as Boneyards, Magna Tubes, and Kitchen Windows. Some of these waves can carry a skilled surfer several hundred yards along the beach on a single ride. The same waves that attract surfers are also responsible for the vast numbers and wide variety of seashells that are washed up onto the beach with each tide.

Conchologists have identified the shells of over 400 species of marine animals, including various gastropods, chitons, and bivalves, making the bay the most biologically diverse natural coastline in South Africa. Dolphins, whales, and seals are also seen.

HEADING FOR SUPERTUBES
The waves at Supertubes may be 10 ft (3 m) high and invariably break right-to-left as viewed from the shore.

HUMAN IMPACT

HIDDEN DANGERS

Every surfing spot, including Jeffreys Bay, has dangers that would-be surfers should know about. The most important are rip currents. The enormous volume of seawater washed up on shore by the waves tends to pool at specific points on the beach and is then funneled back out to sea in swift currents. These move rapidly away from the beach, straight out through the surf zone, and can sweep unsuspecting swimmers out to sea. They can be escaped by swimming parallel to the shore. At Jeffreys Bay, there have also been rare reports of surfers being bitten by sharks, most often by the sand tiger or ragged-tooth shark.

INDIAN OCEAN NORTH

Anjuna Beach

TYPE Series of embayed beaches	
COMPOSITION Yellow sand	
LENGTH 1 mile (1.5 km)	

LOCATION On the Arabian Sea coast, northwest of Panaji, southwestern India

Anjuna Beach is one of the most scenic and popular of the renowned string of beaches that lie on the coast of the Indian State of Goa. The beach has an undulating shape and is broken up into several sections by rocky outcrops that jut into the sea. By reducing rip currents and cross-currents, these outcrops help to make Anjuna one of the safest swimming beaches on the Goa coast. During the monsoon season, from June to September, much of the beach sand is stripped away and carried offshore by heavy wave action, but after the monsoons, calmer seas restore the sand deposits.

PICTURESQUE SETTING
With its calm seas and sand crescents backed by swaying palms and low, rocky hills, Anjuna has been a favored vacation destination since the 1960s.

INDIAN OCEAN NORTH

Cox's Bazar Beach

TYPE Dissipative coastal plain beach	
COMPOSITION Yellow sand	
LENGTH 75 miles (120 km)	

LOCATION South of Chittagong, southeastern Bangladesh

Cox's Bazar Beach lies on a northeastern stretch of the Bay of Bengal and is the second-longest unbroken natural beach in the world—Ninety Mile Beach (see p.112) in Australia is the longest. It fronts a range of dunes and, at its southern end, a spit of land. The dunes, spit, and beach have been built up over hundreds of years through a combination of wave action and deposition of sediment from the Bay of Bengal. This is a gently sloping beach that offers safe swimming and surfing and is also popular among collectors of conch shells.

SATELLITE VIEW OF BEACH (BOTTOM RIGHT)

INDIAN OCEAN SOUTHEAST

Shell Beach

TYPE Embayed beach	
COMPOSITION Shells of a species of cockle	
LENGTH 70 miles (110 km)	

LOCATION Northwest of Perth, Western Australia

Shell Beach, in Western Australia's Shark Bay, has a unique composition, consisting almost entirely of the white shells of *Fragum erugatum*, a species of cockle (a bivalve). The beach lies in a partially enclosed area of Shark Bay known as L'Haridon Bight. This cockle thrives here because its predators cannot cope with the high salinity of the seawater. On the foreshore of Shell Beach, the layer of shells reaches a depth of 26–30 ft (8–9 m). The shells also form the sea floor, stretching for hundreds of yards from the shoreline. On the backshore, away from the water line, many of the shells have become cemented together, in some areas leading to the formation of large, solid conglomerations. These were formerly mined to make decorative wall blocks.

SHELL BANK
Individual shells in the beach are about ½ in (1 cm) wide. Accumulations of these shells over about 4,000 years has led to the formation of a long bank along the seashore.

OCEAN ENVIRONMENTS

PACIFIC OCEAN SOUTHWEST

Ninety Mile Beach

TYPE	Dissipative coastal plain beach
COMPOSITION	Yellow sand
LENGTH	90 miles (145 km)

LOCATION Southeast of Melbourne, Victoria, southeastern Australia

Australia's Ninety Mile Beach, on the coast of Victoria, has a solid claim to be the world's longest uninterrupted natural beach. The beach runs in a southwest to northeasterly direction and fronts a series of dunes. Waves generally break too close to the beach for good surfing, and strong rip currents make the conditions hazardous for swimmers. In its northeastern part, several large lakes and shallow lagoons, known as the Gippsland Lakes, lie behind the dunes. Beneath the sea, vast plains of sand stretch in every direction and are home to a large variety of small invertebrate life, including crustaceans, worms, and burrowing mollusks.

AERIAL VIEW
Facing out onto the Bass Strait, Ninety Mile Beach is subject to strong waves during the winter months.

PACIFIC OCEAN SOUTHWEST

Moeraki Beach

TYPE	Embayed beach
COMPOSITION	Dark sand and large boulders
LENGTH	2 miles (3 km)

LOCATION Northeast of Dunedin, southeastern New Zealand

The beach north of Moeraki on New Zealand's South Island is strewn with groups of large, near-spherical boulders. Scientists believe they are mineral concretions that formed over a few million years within 60-million-year-old mudstones—thick layers of sedimentary rock making up the sea floor. These mudstones were later uplifted and now form a cliff at the back of the beach. There, gradual erosion exposes and releases the boulders, which eventually roll down onto the beach.

SUPERSIZED BOULDERS
The boulders are up to 7 ft (2.2 m) in diameter, and some weigh several tons. Some are half-buried in the sand.

PACIFIC OCEAN CENTRAL

Punalu'u Beach

TYPE	Pocket beach
COMPOSITION	Black sand
LENGTH	1/3 mile (500 m)

LOCATION Northeast of Naalehu, on the south coast of Big Island, Hawaii

Punalu'u Beach on Hawaii's Big Island is a steeply shelving pocket beach. It is best known for its dramatic-looking black sand, which is composed of grains of the volcanic rock basalt. The sand has been produced by wave action on local cliffs of black basaltic lava. Punalu'u, in common with about half the land area of Hawaii, lies on the flank of Mauna Loa, the world's most massive volcano. Lava produced by the volcano dominates the local landscape—although no lava has reached Punalu'u from Mauna Loa or the nearby active volcano Kilauea for several hundred years.

The beach is a popular location for swimming and snorkeling, but underwater springs that eject cold water into the sea close to the beach can cause discomfort. Punalu'u Beach is also visited by green turtles, which come to eat seaweed off rocks at the edge of the beach and bask on the warm, heat-absorbing sand.

BLACK AND BLUE
The sand on Punalu'u is almost perfectly black, contrasting with the deep blue Pacific waters. Removal of the sand is prohibited.

PACIFIC OCEAN NORTHEAST

Columbia Bay

TYPE	Series of embayed beaches
COMPOSITION	Gravel and rocks
LENGTH	31 miles (50 km)

LOCATION Southwest of Valdez, southern Alaska, US

Many beaches in southern Alaska, and other beaches at high latitudes in the Northern Hemisphere, consist of gravel, small rocks, and boulders. These materials come from coarse glacial till—mixtures of clay, silt, sand, gravel, and rocks that were carried to a location by ancient glaciers, remaining there when the glaciers melted. The till has usually been reworked by wave action, with the lighter material (clay, silt, and sand) washed away and the heavier gravel and rocks sorted by size and deposited in different areas along the shoreline. Such is the case in Columbia Bay, a region within Alaska's Prince William Sound. Many of the beaches in this area have old tidal lines visible above the present ones, the result of a huge earthquake in 1964 that raised the land by 8 ft (2.4 m).

BEACH AND BAY
The backshore area visible here, which has been colonized by plants, was foreshore prior to the 1964 earthquake.

PACIFIC OCEAN NORTHEAST

Oregon National Dunes

TYPE	Coastal dunes
COMPOSITION	Yellow sand, grasses, conifers
LENGTH	40 miles (64 km)

LOCATION Southwest of Portland, Oregon, northwestern US

Oregon National Dunes is the largest area of coastal sand dunes in North America, extending along the coast of Oregon between the Siuslaw and Coos Rivers. These dunes have been created through the combined effects of coastal erosion and wind transport of sand over millions of years and extend up to 2½ miles (4 km) inland, rising to 500 ft (150 m) above sea level. A continuum of dry and wet conditions extends through the dune area. Close to the beach are low foredunes of sand and driftwood stabilized by marram grass. Behind these are hummocks where sand collects around vegetation. Water accumulates around the hummocks seasonally, giving them the appearance of floating islands. Behind the hummocks are further distinct regions, ranging from densely vegetated areas that become marshlike in winter to completely barren, wind-sculpted high dunes. The dunes are a popular location for various recreational activities, including riding all-terrain vehicles (ATVs) and dune buggies.

HUMAN IMPACT

DUNE DESTABILIZATION

The use of dune buggies and ATVs, especially when raced in large numbers, may destroy the grass on the dunes, making them susceptible to wind scour. This may in turn lead to self-propagating breaches in the dune ridges. To protect the dunes, ATV usage is restricted.

SEA OF DUNES
The wind molds the sand of the dunes into wave shapes, with crests at right angles to the wind direction.

PACIFIC OCEAN NORTHEAST

Dungeness Spit

TYPE	Sand spit
COMPOSITION	Sand
LENGTH	5½ miles (9 km)

LOCATION Northwest of Seattle, Washington State, northwestern US

Dungeness Spit, one of the world's longest natural sand spit, juts out from the Olympic Peninsula in Washington State. It is part of the Dungeness National Wildlife Refuge and is as little as 100 ft (30 m) wide in places. In addition to its great length, the spit has a complex shape, the result of seasonal changes in wind and wave direction. During part of the year, these bring sandy sediments from the northwest, and at other times from the northeast. The resulting pattern of sedimentation has created a large sheltered coastal area, providing refuge for many shorebirds and waterfowl, which nest along the beach, and for Pacific harbor seals. The tidal flats nourish a variety of shellfish, and the inner bay is an important nursery habitat for several salmon species.

GROWING SPIT
The spit grows at about 15 ft (4.5 m) a year. It provides shelter for a large inner bay and an area of tidal flats.

PACIFIC OCEAN EAST

Tamarindo Beach

TYPE	Embayed beach
COMPOSITION	Yellow sand
LENGTH	2 miles (3 km)

LOCATION Northwest of San José, northwestern Costa Rica

Tamarindo Beach is a curved, gently shelving crescent of sand situated close to a mangrove-lined estuary and backed by modern dwellings within

COASTAL SETTING
In this view, the main part of Tamarindo Beach is in the background, with the entrance to an estuary that curves around behind the beach on the right.

a scattered forest. It faces directly onto the Pacific, with its enormous fetch (wave-generation area), and so benefits from a strong year-round incoming swell, making the beach a popular surfing location. To the north and south of the main beach are two further beaches, which together form the Las Baulas National Marine Park. From October to May, they are important nesting sites for the leatherback turtle (see p.371).

(see p.371)

OCEAN ENVIRONMENTS

Estuaries and Lagoons

ESTUARIES AND COASTAL LAGOONS ARE BOTH semienclosed, coastal bodies of water. An estuary typically connects to the open sea, is quite narrow, and receives a significant input of fresh water from one or more rivers. This fresh water mixes with the salt water to a varying degree, depending on river input and tides. Many estuaries are simply the seaward, tidally affected ends of large rivers. Coastal lagoons are usually linked to the sea only by one or more narrow channels, through which water flows in and out; sometimes these channels open only at high tide.

Estuary Formation

Estuaries form in four main ways. First, sea level may rise and flood an existing river valley on a coastal plain, such as in Chesapeake Bay in the US. Second, sea level can rise to flood a glacier-carved valley, forming a fjord. Estuaries formed in this way are deeper than other types, but have shallow sills at their mouths that partially block inflowing seawater. Third, coastal wave action can also create an estuary, by building a sand spit or bar across the open end of a bay fed by a stream or river (see p.93). Fourth, estuaries result from movement at tectonic faults (lines of weakness) in Earth's crust, where downward slippage can result in a surface depression. This becomes an estuary if seawater later floods in.

CONGO RIVER ESTUARY
Formed by flooding of a river valley, this estuary is the world's second largest (after the Amazon) in terms of discharge rate.

FORMATION PROCESSES
An estuary can form when sea-level rise causes the seaward end of a river valley to flood (top) or inundates a glacier-carved valley to create a fjord (middle), or when a spit extends across a bay (bottom).

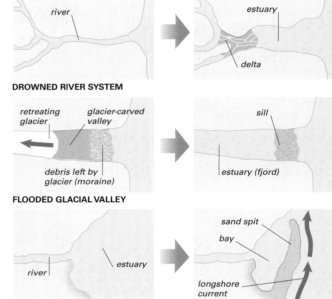

DROWNED RIVER SYSTEM
river — estuary — delta

FLOODED GLACIAL VALLEY
retreating glacier — glacier-carved valley — sill — debris left by glacier (moraine) — estuary (fjord)

SPIT ACROSS A BAY
river — estuary — sand spit — bay — longshore current

Types of Estuaries

The way in which fresh and salt water mixes in an estuary determines its classification. A strong river inflow usually means minimal mixing—the less-dense fresh water flows over the denser salt water, which forms a wedge-shaped intrusion into the bottom of the estuary. This is a salt-wedge (river-dominated) estuary. In partially mixed and fully mixed (tide-dominated) estuaries, there is considerable mixing, producing turbulence and increased salinity in the fresh water. In each case, this is balanced by a strong, tidally influenced influx of salt water from the sea: this influx brings sediments from offshore, which are deposited as mud in the estuary.

strong flow of fresh water — minimal mixing of salt and fresh water — slightly salty water flows out — fresh water — wedge of sea water — small tidal countercurrent

SALT-WEDGE ESTUARY
In a salt-wedge estuary (left), there is a strong flow of fresh river water over a wedge of salt water, with little mixing between the two layers.

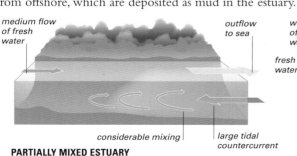

medium flow of fresh water — outflow to sea — considerable mixing — large tidal countercurrent

PARTIALLY MIXED ESTUARY
In this type of estuary, there is considerable mixing between fresh and salt water. The saltiness of the water increases with depth in all parts of the lower estuary.

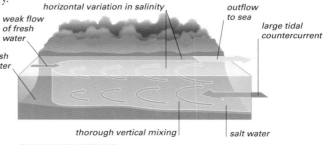

horizontal variation in salinity — weak flow of fresh water — outflow to sea — large tidal countercurrent — fresh water — thorough vertical mixing — salt water

FULLY MIXED ESTUARY
In a fully mixed (or tide-dominated) estuary, the fresh and salt water are well-mixed vertically, but there is some horizontal variation in saltiness.

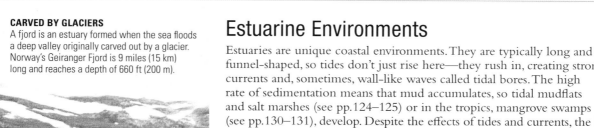

CARVED BY GLACIERS
A fjord is an estuary formed when the sea floods a deep valley originally carved out by a glacier. Norway's Geiranger Fjord is 9 miles (15 km) long and reaches a depth of 660 ft (200 m).

Estuarine Environments

Estuaries are unique coastal environments. They are typically long and funnel-shaped, so tides don't just rise here—they rush in, creating strong currents and, sometimes, wall-like waves called tidal bores. The high rate of sedimentation means that mud accumulates, so tidal mudflats and salt marshes (see pp.124–125) or in the tropics, mangrove swamps (see pp.130–131), develop. Despite the effects of tides and currents, the high turbidity that reduces plant photosynthesis, and fluctuations in salinity and temperature, most estuaries are biologically highly productive. This is partly due to the high concentration of nutrients in river water, and because estuaries are well oxygenated. Although only a limited range of organisms, such as mussels, cope with living in estuaries, populations are often huge.

COMMON EUROPEAN OYSTER

RICH FOOD SOURCE
Estuaries attract waders and other shorebirds because of the high concentrations of small animals (such as worms and shrimp) that live in the mud deposits. These lapwings and an egret are congregating to feed in the Thames estuary, UK.

ESTUARY DWELLER
Various species of starfish tolerate the estuarine environment, where they feed on mussels, crustaceans, and worms. This common starfish is in an estuary in Brittany, France.

Coastal Lagoons

Coastal lagoons occur worldwide, and are different from the lagoons found at the centers of coral atolls (see p.152). Calmer and usually shallower than estuaries, most lagoons are connected to the sea by tidal channels. Although fresh water does not usually flow into coastal lagoons, some do receive a significant river inflow. So, as well as saltwater lagoons, there are also some partly, or predominantly, freshwater lagoons.

In hot climates, some lagoons are hypersaline (saltier than ocean water), due to high evaporative losses. Although some coastal lagoons are severely polluted, the cleaner ones are often well stocked with fish, crustaceans, and other marine life, and frequently attract large numbers of shorebirds. Some provide feeding or breeding areas for sea turtles and whales.

LAGOON AND CHANNELS
Matagorda Bay is a lagoon on the coast of Texas, separated from the Gulf of Mexico by a long, narrow peninsula. Two channels, located near the southwest corner of the lagoon, connect it to the gulf.

OCEAN ENVIRONMENTS

ATLANTIC OCEAN NORTHWEST

St. Lawrence Estuary

TYPE	Salt-wedge (river-dominated) estuary
AREA	Approximately 10,000 square miles (25,000 square km)

LOCATION Quebec, eastern Canada

The St. Lawrence Estuary is one of the world's largest estuaries. Some 500 miles (800 km) long, it discharges about 2.2 million gallons (10 million liters) of water into the Gulf of St. Lawrence each second. The estuary is rich in marine life. In its wide middle and lower reaches, the icy Labrador Current flows 1,000 ft (300 m) below the surface in the opposite direction of the main estuarine flow. In one section, near the mouth of a fjord that branches off the estuary, the current's nutrient-rich waters rise abruptly and mix with warmer waters above. This upwelling of nutrients encourages plankton growth, providing the base of a food chain that involves many species of fish and birds, and a small population of beluga whales.

WINTER SCENE
In winter, much of the estuary becomes iced over. A stretch of the estuary is seen here at dusk.

ATLANTIC OCEAN NORTHWEST

Chesapeake Bay

TYPE	Partially mixed estuary
AREA	4,480 square miles (11,600 square km)

LOCATION Surrounded by Maryland and parts of eastern Virginia, US

Chesapeake Bay is the largest estuary in the US. Its main course, fed by the Susquehanna River, is over 185 miles (300 km) in length. It has numerous subestuaries, and more than 150 rivers and streams drain into it. This body of water was created by sea-level rise drowning the valley of the Susquehanna and its tributaries over the last 15,000 years. Once famous for its seafood, such as oysters, clams, and crabs, the bay is now far less productive, though it still yields more fish and shellfish than any other estuary in the US. Industrial and farm waste running into the bay causes frequent algal blooms, which block sunlight from parts of its bed. The resulting loss of vegetation has lowered oxygen levels in some areas, severely affecting animal life. The depletion of oysters, which naturally filter water, has had a particularly harmful effect on the bay's water quality.

MAIN CHANNEL FLOWING THROUGH DELTA

BAY BRIDGE
A major bridge in the upper bay connects Maryland's rural eastern shore to its urban western shore.

DISCOVERY

IMPACT CRATER

In the 1990s, drilling of the seabed at Chesapeake Bay led to the discovery of a meteorite impact crater 53 miles (85 km) wide under its southern region. The 35-million-year-old crater helped shape today's estuary.

SHOCKED QUARTZ
Evidence for the crater included the discovery of grains of shocked quartz, which forms when intense pressure alters its crystalline structure.

ATLANTIC OCEAN WEST

Mississippi Estuary

TYPE	Salt-wedge (river-dominated) estuary
AREA	25 square miles (60 square km)

LOCATION Southeastern Mississippi Delta, southeastern Louisiana, US

The Mississippi Estuary is about 30 miles (50 km) long and lies at the seaward end of the Mississippi River, where the river flows through its own delta. The estuary consists of a main channel and several subchannels. Together, these discharge an average of some 3.5 million gallons (13 million liters) of water per second into the Gulf of Mexico. The main channel is a classic example of a salt-wedge estuary—its surface waters contain little salt, but they flow over a wedge of salt water, which extends deep down for several miles up the estuary.

OCEAN ENVIRONMENTS

Laguna Madre

TYPE
Hypersaline coastal lagoon

AREA
14,400 square miles (3,660 square km)

LOCATION Southern Texas, US, and northeastern Mexico, along the coast of the Gulf of Mexico

The Laguna Madre is a shallow lagoon in two distinct parts extending about 285 miles (456 km) along the coast of the Gulf of Mexico. Its northern part, in Texas, is separated from the Gulf by a long, thin barrier island, Padre Island. The southern part, in Mexico, is similarly cut off by a string of barrier islands. The entire lagoon connects with the Gulf only via a few narrow channels, and it is less than 3 ft (1 m) deep in most parts. It is saltier than seawater because it receives no input of river water and lies in a hot, dry region, leading to high rates of evaporation. Seagrass meadows and several species of crustaceans and fish thrive in the lagoon, which also supports many wintering shorebirds and waterfowl. Threats to the lagoon's health include dredging, overfishing, and algal blooms.

FLY-FISHING FOR REDFISH
The sale of licenses for fly-fishing—for trout and redfish—in the Laguna provides funds for protecting its water quality and wildlife.

Lagoa dos Patos

TYPE
Tidal coastal lagoon

AREA
3,900 square miles (10,000 square km)

LOCATION South of the city of Porto Alegre, southern Brazil

Lagoa dos Patos ("Lagoon of Ducks") is the world's largest coastal lagoon. Its name is said to have been given to it by Jesuit settlers in the 16th century, who bred waterfowl on its shores. It is a shallow, tidal body of water, 180 miles (290 km) long and up to 40 miles (64k m) wide. A sand bar separates it from the Atlantic, with which it connects at its southern end via a short, narrow channel that disgorges a large plume of sediment into the ocean. Marine animals use this channel to access the lagoon; sea turtles are found in the lagoon in spring and summer.

At its northern end, the lagoon receives an inflow of fresh water from the Guaíba Estuary, formed from the confluence of the Rio Jacui and three smaller rivers. Along its inner side are a number of distinctive wavelike "cusps" that have been caused by the accumulation and erosion of sediments driven by tidal action and winds. The salinity of the lagoon varies. It consists mainly of fresh water at times of high rainfall, but there is considerable saltwater intrusion at its southern end at times of drought. Lagoa dos Patos is one of Brazil's most vital fishing grounds. However, runoff from rice fields and pastureland, industrial effluents, and increasing population have led to concerns for the lagoon's ecosystem.

TWO LAGOONS
In this aerial view, Lagoa dos Patos is the pale central area. Below it, the darker Lagoa Mirim extends to the Brazil–Uruguay border.

ATLANTIC OCEAN SOUTHWEST

Amazon Estuary

TYPE Salt-wedge (river-dominated) estuary

AREA Approximately 7,800 square miles (20,000 square km)

LOCATION Northern Brazil

The Amazon Estuary is a stretch of the Amazon River that extends more than 190 miles (300 km) inland from the river's mouth to an area southwest of the city of Macapá. Varying in width from 15 to 190 miles (25 to 300 km), the estuary is partly filled by numerous low-lying, forested islands.

The Amazon Estuary has by far the largest water output of any estuary in the world, discharging an average of 45 million gallons (200 million liters) per second into the Atlantic. The sheer magnitude of this discharge means that, almost uniquely among estuaries, there is very little saltwater intrusion into it. Instead, nearly all of the mixing between the river's discharge and seawater occurs outside the estuary, on an area of continental shelf. Despite the relative lack of seawater intrusion, the whole estuary is significantly affected by twice-daily tides, which cause inundation (by river water) of most of the islands in the estuary.

MARAJÓ ISLAND
The Amazon Estuary is so enormous that the biggest of the forested islands lying within it, Marajó Island, has its own river system.

HUMAN IMPACT

POROROCA SURF

Tidal bores, locally called *pororocas*, occur on large spring tides in several of northern Brazil's river estuaries. Some of these bores attain heights of 10 ft (3 m) and can be surfed for several miles. This sport is rather hazardous, however, because the waters through which the *pororocas* surge are home to dangerous snakes, fish, and crocodiles.

ATLANTIC OCEAN SOUTHWEST

River Plate

TYPE Salt-wedge (river-dominated) estuary

AREA 13,500 square miles (35,000 square km)

LOCATION On the Argentina–Uruguay border, east of Buenos Aires and southwest of Montevideo

The Plate River, or Rio de la Plata, is not a river but a large, funnel-shaped estuary formed by the confluence of the rivers Uruguay and Paraná. These rivers and their tributaries drain about one-fifth of the land area of South America. At 180 miles (290 km) long and 136 miles (220 km) wide at its mouth, the Plate discharges about 6.5 million gallons (25 million liters) of water per second into the Atlantic Ocean. As well as transporting this vast amount of water, the estuary receives about 2 billion cubic feet (57 million cubic meters) of silt each year from its input rivers. This mud accumulates in great shoals, so that the water depth in most of the estuary is less than 10 ft (3 m). Constant dredging is therefore needed to maintain deep-water channels to the ports of Buenos Aires, which lies near the head of the estuary, and Montevideo, which is close to its mouth. Surface salinity varies uniformly through the estuary, from close to zero in its upper parts to a value just below average ocean salinity near its mouth. Deep down, a wedge of salt water penetrates deep into the estuary. Biologically, the Plate is highly productive, yielding large annual masses of plankton, which support large numbers of fish and dense beds of clams. It is also a habitat for the La Plata Dolphin, which has an IUCN Vulnerable conservation status.

SATELLITE VIEW
The main flow of river water over the sediments on the estuary bed is visible here, as well as the Paraná River at top left and the Uruguay River at top center.

ATLANTIC OCEAN NORTHEAST
Curonian Lagoon

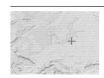

TYPE
Freshwater coastal lagoon

AREA
610 square miles
(1,580 square km)

LOCATION On the Baltic Sea coasts of Lithuania and the Kaliningrad Oblast (part of Russia)

The Curonian Lagoon is a nontidal lagoon on the southeastern edge of the Baltic Sea, with an average depth of just 12 ft (3.8 m). The Neman River flows into the lagoon's northern (Lithuanian) section, which discharges into the Baltic via a narrow channel, the Klaipeda Strait. While most of the lagoon consists of fresh water, seawater sometimes enters its northern part via the Klaipeda Strait following storms. In the past, the lagoon has suffered heavy pollution from sewage and industrial effluents, but attempts are now being made to address this problem.

The lagoon is separated from the Baltic by the narrow, curved Curonian Spit, which is 60 miles (98 km) long. The spit is notable for its mature pinewoods and drifting barchans (sand dunes), some reaching a height of 200 ft (60 m), which extend for 20 miles (31 km) along the spit. The sandy beaches on the spit, together with vistas over the lagoon, woods, and drifting dunes, make it a tourist attraction, and in 2000 the entire spit was designated a UNESCO World Heritage site.

DUNES AND LAGOON
A white stork in a quiet corner of the lagoon, near Morskoe village. Migrating birds use the lagoon and nearby Neman Delta for vital rest breaks.

ATLANTIC OCEAN NORTHEAST
Humber Estuary

TYPE Fully mixed (tide-dominated) estuary

AREA Approximately 80 square miles (200 square km)

LOCATION West and southeast of Kingston-upon-Hull, eastern England, UK

This large estuary on Great Britain's eastern coastline is formed from the confluence of the Ouse and Trent Rivers. On average, it discharges about

READ'S ISLAND
This low-lying island, in the upper part of the estuary, is a breeding ground for birds such as pied avocets and is managed as a nature reserve. The view here is looking downstream.

66,000 gallons (250,000 liters) of water per second into the North Sea, the largest input from any British river into this sea. After the end of the last ice age, when sea levels were much lower, the Humber was a river that flowed up to 30 miles (50 km) past the present coastline before reaching the sea.

About 3.6 million cubic feet (100,000 cubic meters) of sediment are deposited in the estuary every year, mainly from offshore by tidal action. Shifting shoals formed by this sediment can obstruct shipping. The estuary's intertidal areas are productive ecosystems that support a wide range of mollusks, worms, crustaceans, and other invertebrates. These are vital sources of food for birds, especially waders. The estuary also supports a colony of gray seals, and many lampreys pass through it every year.

ATLANTIC OCEAN NORTHEAST
Hardanger Fjord

TYPE Highly stratified estuary; fjord

AREA Approximately 290 square miles (700 square km)

LOCATION Southeast of Bergen, southwestern Norway

Like all fjords, the Hardanger Fjord in Norway is much deeper than a typical coastal-plain estuary, with a maximum depth of some 2,600 ft (800 m). Near its mouth is a sill just 500 ft (150 m) deep. At 114 miles (183 km) long, it is the third-longest fjord in the world.

UPPER FIORD
Hardanger Fjord is around 2.5 miles (4 km) wide on average, and is surrounded by dramatic snow-dusted mountains.

Hardanger Fjord was formed about 10,000 years ago, when a large glacier that had carved out and occupied a deep U-shaped valley in the area began to melt and retreat. As it did so, seawater flooded into the valley to create the fjord. Today, the fjord continues to receive a large input of fresh water from glacier melt. Throughout much of its length, the fjord is stratified into a lower layer of salt water, which moves into the fjord during flood tide, and an upper layer of fresher water that flows outward to the sea on the ebb tide.

ATLANTIC OCEAN NORTHEAST
Eastern Scheldt Estuary

TYPE
Former estuary, now a sea-arm

AREA
140 square miles (365 square km)

LOCATION Southwest of Rotterdam, southwestern Netherlands

The Eastern Scheldt Estuary is a tidal body of water 25 miles (40 km) long, with a salinity similar to that of seawater. Since the late 1980s, it has been cut off from its input of fresh water from the Scheldt River by dams, leading to its reclassification as a sea-arm rather than an estuary.

It has also been defended against seawater flooding by a storm-surge barrier (see p.104). This was originally to have been a fixed dam to prevent any ingress of seawater at all, but there were fears that, with a dam of this type, the estuary would gradually lose its salinity, producing an adverse effect on its fauna and flora—in particular, there were concerns that it would end the large-scale mussel and oyster farming in the area and degrade the tidal flats and salt marshes that form an important habitat for birds. The government of the Netherlands therefore commissioned a movable barrier, the construction of which was completed in 1986.

STORM BARRIER GATES
The gates are usually raised, allowing tidal water in and out of the Eastern Scheldt Estuary. They are lowered about twice a year, during stormy weather.

Gironde Estuary

TYPE	Fully mixed (tide-dominated) estuary
AREA	Approximately 245 square miles (635 square km)

LOCATION North of Bordeaux, western France

The Gironde Estuary, formed by the confluence of the Garonne and Dordogne Rivers, is the largest estuary in Europe at almost 47 miles (75 km) long and up to 7.5 miles (12 km) wide. The estuary's average discharge rate into the Atlantic is 265,000 gallons (1 million liters) per second. It has a large tidal range, of up to 16 ft (5 m) during periods of spring tide, and the strong tidal currents in the estuary, as well as numerous sand banks, tend to hamper navigation. One of the Gironde's most impressive features is its tidal bore—a large, wall-like wave at the leading edge of the incoming tide—known locally as the Mascaret. Occurring with each flood tide at the time of spring tides (that is, twice daily for a few days every two weeks), the bore surges from the Gironde upstream into its narrower tributaries. On the Garonne, the Mascaret sometimes forms a barreling wave, which can reach a height of 5 ft (1.5 m) and tends to break and reform.

The Gironde is an important artery of the Bordeaux wine region and a rich source of eels and a wide variety of shellfish, which feature on local restaurant menus. Wild sturgeon (the source of caviar) were once also plentiful in the estuary, and although their numbers have declined due to overfishing, they are still farmed in small numbers.

THE MASCARET
When it reaches the Dordogne River, the Mascaret, or Gironde tidal bore, turns into a series of waves, which may travel up to 20 miles (30 km) upstream.

Venetian Lagoon

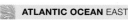

TYPE	Saltwater coastal lagoon
AREA	210 square miles (550 square km)

LOCATION On the Adriatic coast of northeastern Italy

The Venetian Lagoon is a very shallow, crescent-shaped coastal lagoon off the northern part of the Adriatic Sea. It is the largest Italian wetland and a major Mediterranean coastal ecosystem.

In addition to Venice, which sits on a small island at the center, the lagoon contains many other islands, most of which were marshy but have now been drained. Its average depth is just 28 inches (70 cm), so most boats cross the lagoon only via dredged navigation channels, and four-fifths of its area consists of salt marshes and mudflats. It takes in both riverine fresh water and seawater, and its tides have a range of up to 3 ft (1 m). In the past, Venice and other islands in the lagoon have suffered many serious floods during periods of spring tides, but a much-delayed project to prevent these, called MOSE, was successfully tested in 2020. Marine life in the lagoon includes many species of fish (from anchovies to eels, mullet, and sea bass) and invertebrates. Seabirds, waterfowl, and waders proliferate on the many uninhabited islands. In 2021, cruise ships were banned from entering the lagoon due to damage caused to Venice's foundations and to the lagoon's ecosystem by the large waves generated by the ships.

WATERY GEM
In the center of this photograph, taken from the International Space Station, is the fish-shaped main island of Venice. Below it is one of the lagoon's three protective barrier islands.

JAMES ISLAND

Gambia Estuary

TYPE	Salt-wedge (river-dominated) estuary
AREA	Approximately 400 square miles (1,000 square km)

LOCATION East of Banjul, The Gambia, West Africa

The Gambia Estuary is the western half of the Gambia River, which runs 700 miles (1,120 km) through West Africa. The estuary is tidal throughout and discharges about 528,000 gallons (2 million liters) per second into the Atlantic during the rainy season, but only 528 gallons (2,000 liters) in the dry season. It contains abundant stocks of fish and shellfish, including catfish, barracuda, and shrimp. Kunta Kinteh Island, or James Island, some 20 miles (30 km) from the estuary's mouth, was formerly a slave-collection point and is now a UNESCO World Heritage Site.

ATLANTIC OCEAN EAST
Ebrié Lagoon

TYPE
Coastal lagoon of variable salinity

AREA
200 square miles (520 square km)

LOCATION West of Abidjan, Ivory Coast, West Africa

The Ebrié Lagoon is one of three long, narrow lagoons that line the shores of the West African state of Ivory Coast. With a length of 81 miles (130 km) and an average width of 2½ miles (4 km), it is the largest lagoon in West Africa. Its average depth is 16 ft (5 m). Near its

TIAGBA VILLAGE
In the village of Tiagba, on the outskirts of a small island in the Ebrié Lagoon, the buildings are raised up on wooden piles.

eastern end, it connects to the Atlantic via a narrow artificial channel, the Vridi Canal, opened in 1951. Abidjan, the largest city in Ivory Coast, stands on several converging peninsulas and islands in an eastern part of the lagoon; other communities situated on or in the lagoon include the town of Dabou and the village of Tiagba (see below). The Komoé River provides the main input of fresh water. In winter the lagoon becomes salty, but it turns to fresh water during the summer rainy season. Levels of pollution in the lagoon have been high for some years due to runoff of agricultural fertilizers and the discharge of industrial effluents and sewage from nearby urban areas.

INDIAN OCEAN NORTH
Kerala Backwaters

TYPE Chain of coastal saltwater lagoons

AREA Approximately 1,780 square miles (4,600 square km)

LOCATION Southeast of Cochin, Kerala State, southwestern India

The backwaters of Kerala in southern India are a labyrinth of lagoons and small lakes, linked by 560 miles (900 km) of canals. The lagoons are

VEMBANAD LAKE
Vembanad, the largest Kerala coastal lagoon, is listed as a Wetland of International Importance under the Ramsar Convention.

shielded from the sea by low barrier islands and spits that formed across the mouths of the many rivers flowing down from the surrounding hills. During the summer monsoon rains, the lagoons overflow and discharge sediments into the sea, but toward the end of the rains, the seawater rushes in, altering salinity levels. The aquatic life in the backwaters, which includes crabs, frogs, otters, and turtles, is well adapted to this seasonal variation.

LAKES AND LAGOON
In this satellite view, the Coorong Lagoon is the narrow blue strip behind the yellow sand dunes. Above are the lakes Alexandrina (left) and Albert (right).

PELICANS IN DECLINE

The Coorong is home to a large breeding colony of Australian pelicans, which inhabit a string of islands in the center of the lagoon. Since the 1980s, however, their numbers have fallen significantly due to reduced flows of fresh water into the Coorong from the Murray River. The resultant higher salt levels in the lagoon have reduced the growth of an aquatic weed that is a major part of the food chain.

AUSTRALIAN PELICANS
This pelican, one of eight species worldwide, is widespread in Australia, where it lives on freshwater, brackish, and saltwater wetlands.

INDIAN OCEAN SOUTHEAST
Coorong Lagoon

TYPE
Saltwater coastal lagoon

AREA
80 square miles (200 square km)

LOCATION Southeast of Adelaide on the southeastern coast of South Australia

The Coorong Lagoon is a wetland that lies close to the coast of South Australia. It is famous as a haven for birds, ranging from swans and pelicans

to ducks, cranes, ibis, terns, geese, and waders such as sandpipers and stilts. The lagoon is separated from the Great Australian Bight (considered part of the Indian Ocean) by the Younghusband Peninsula, a narrow spit of land covered by sand dunes and scrubby vegetation. The lagoon is about 81 miles (130 km) long, with a width that varies from 3 miles (5 km) to just 330 ft (100 m). At its northwestern end, the lagoon meets the outflow from Australia's largest river, the Murray, after the river has passed through Lake Alexandrina. In this region, called the Murray Mouth, both river and lagoon meet the sea,

and the Coorong can receive both fresh and salty water. The lagoon was once freely connected to the lake, from which it received a much larger supply of fresh water. In 1940, however, barrages were built between the lagoon and the lake to prevent seawater from reaching the lake and the lower reaches of the Murray River. The salinity of the lagoon's waters increases naturally with distance from the sea due to evaporative losses. However, reduced water flows from the Murray, due to a combination of barrage construction and extraction of water for irrigation projects, has caused a gradual further increase in salinity throughout the

lagoon. There is ample evidence that this has adversely affected the lagoon's ecosystem. In particular, several species of plants have become less abundant or disappeared, many fish species have declined, and migratory bird numbers have fallen. Further, the reduced flow from the Murray may result in the eventual closure of the channel joining the lagoon to the ocean, which would prevent migration of fish and other animals between the two.

PACIFIC OCEAN WEST
Pearl River Estuary

TYPE
Salt-wedge
(river-dominated) estuary

AREA
450 square miles
(1,200 square km)

LOCATION Northwest of Hong Kong, Guangdong, southeastern China

The bell-shaped Pearl River Estuary receives and carries most of the outflow from the Pearl River, the common name for a complex system of rivers in the southern Chinese province of Guangdong. The estuary is nearly 37 miles (60 km) long, and its width increases from 12 miles (20 km) at its head to about 30 miles (50 km) at its mouth. To the north and west of the estuary is a delta, formed from the confluence of the Xi Jiang and other rivers of the Pearl River system. Together, these rivers discharge an average of 2.2 million gallons (10 million liters) of water per second into the South China Sea. Mostly less than 30 ft (9 m) deep, but containing some deeper dredged channels, the Pearl River Estuary has a tidal range of 3–6ft (1–2 m). It drains water from one of the most densely urbanized regions in the world and is severely polluted as a result of billions of tons of sewage and industrial effluent entering it each year. One result of this has been the increasing occurrence of algal blooms that threaten local fishing. Pollution is also a threat to a dwindling population of Indo–Pacific humpback white dolphins (less than 1,000) that live in the estuary. Only since 2008 have efforts begun to reduce pollution through the building of more water treatment plants.

GUANGZHOU
Formerly known as Canton, this large and busy port city lies on a northerly extension of the Pearl River Estuary.

INDIAN OCEAN SOUTHEAST
Northern Spencer Gulf Estuary

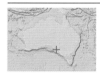

TYPE
Inverse estuary

AREA
Approximately
2,000 square miles
(5,000 square km)

LOCATION Northwest of Adelaide, South Australia

The estuary in the north of Australia's Spencer Gulf is classified as an inverse estuary, owing to its unusual pattern of salt distribution and water circulation.

DEEP GULF
Spencer Gulf is the larger wedge-shaped coastal indent visible in this satellite image. The desert around its head helps produce the estuary's unusual circulation pattern.

In a reverse of the usual pattern, this estuary's waters become saltier toward its head, away from its mouth. This is because its head is surrounded by hot desert and loses more water to evaporation than enters it from rivers. The head's high salinity means that it draws in from the mouth ocean water of lower salinity than the water drawn in by a typical estuary. The estuary is surrounded by extensive tidal flats, seagrass banks, and mangroves.

PACIFIC OCEAN WEST
Yangtze Estuary

TYPE
Partially mixed
estuary

AREA
1,000 square miles
(2,500 square km)

LOCATION Northwest of Shanghai, eastern China

The Yangtze Estuary is the lower, tide-affected part of the Yangtze (or Changjiang)—the longest river in Asia and the third longest in the world. The estuary occupies 430 miles (700 km) of the river's 3,900-mile (6,300-km) length. Near its mouth, it splits into three smaller rivers and numerous streams that run through a delta. Here, silt deposition continually creates new land, which is used for agriculture.

The estuary carries an average of 6.6 million gallons (30 million liters) of water per second into the East China Sea; its average depth is 23 ft (7 m), and the average tidal range at its mouth is 9 ft (2.7 m). It supports large numbers of fish and birds, although fish stocks have declined over the past 20 years due to overfishing and pollution. A species of river dolphin that used to live in the estuary, the Baiji or Yangtze River dolphin, is now thought to be extinct.

In winter, salt water intrudes a significant distance upstream, making the water unfit for drinking and irrigation. Recently, this intrusion has occurred more frequently due to reduced river flow—a reduction exacerbated by the Three Gorges Dam project farther upstream. Reduced flows have worsened the acute water shortage in Shanghai on the estuary's southern shore, as well as affected the dispersion and dilution of pollutants around the estuary. Silt deposition in the delta is also likely to fall, reducing the rate of new land creation.

SHANGHAI YANGTZE RIVER BRIDGE
This bridge across the Yangtze Estuary, situated very close to its mouth, is 6 miles (10 km) long. It opened in 2009.

PACIFIC OCEAN SOUTHWEST

Doubtful Sound

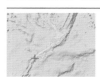

TYPE
Highly stratified estuary; fjord

AREA
30 square miles
(70 square km)

LOCATION West of Dunedin, southwestern South Island, New Zealand

Doubtful Sound is one of 14 major fjords that were formed 15,000 years ago in a scenic part of New Zealand's South Island. Some 25 miles (40 km) long and opening onto the Tasman Sea, it is surrounded by steep hills from which hundreds of small waterfalls descend during the rainy season. Its name originated in 1770, during the first voyage to New Zealand by the English explorer Captain James Cook (1728–1779). He called the fjord Doubtful Harbour because he was skeptical of being able to sail out again if he entered it. Doubtful Sound is the second-longest and the deepest of the New Zealand fjords, with a maximum depth of 1,380 ft (421 m). It receives fresh water from a hydroelectric power station at its head and from a huge 236 in (600 cm) of rainfall annually. Like all fjords, it contains fresh water in its top few yards and a much denser, colder, saltier layer below. There is little mixing between the two. Doubtful Sound is home to bottlenose dolphins; New Zealand fur seals; and many species of fish, starfish, sponges, and sea anemones.

SOUND VIEW

This view of the head of Doubtful Sound, looking toward the open ocean, is from the hills of the south-central region of New Zealand's South Island.

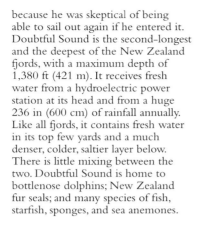

PACIFIC OCEAN NORTHEAST

San Francisco Bay

TYPE
Partially mixed tectonic estuary

AREA
460 square miles
(1,200 square km)

LOCATION Central California, western USA

San Francisco Bay, the largest estuary on North America's west coast, consists of four smaller, interconnected bays. One of these, Suisun Bay, receives fresh water drained from about 40 percent of California's land area. This water flows into San Pablo Bay and then Central Bay, where it mixes with salt water that has entered at depth from the Pacific Ocean via the Golden Gate channel. From Central Bay, there is little flow of fresh water to the largest body of water, South San Francisco Bay, but there is some surface outflow of brackish water to the Pacific. San Francisco Bay is a tectonic estuary—one caused by movement at tectonic faults (lines of weakness) in the Earth's crust, of which there are several in the area, notably the San Andreas Fault.

During the past 150 years, human activity has resulted in the loss of 90 percent of the bay's surrounding marshy wetland, a greatly reduced flow of fresh water (which has been diverted for agricultural purposes), and contamination by sewage and effluent. Nevertheless, the bay remains an important ecological habitat. Its waters are home to large numbers of economically valuable marine species, such as dungeness crab and California halibut, and millions of geese and ducks annually use the bay as a refuge.

OAKLAND BAY BRIDGE

Thick fog surrounds the lower half of the San Francisco–Oakland Bay Bridge, one of five bridges that cross the bay.

PACIFIC OCEAN EAST

Laguna San Ignacio

TYPE
Hypersaline coastal lagoon

AREA
140 square miles
(360 square km)

LOCATION On the Pacific coast of the Baja California Peninsula, Mexico, southeast of Tijuana

The Laguna San Ignacio is a coastal lagoon in northwestern Mexico best known as a sanctuary and breeding ground for Pacific gray whales. Latin America's largest wildlife sanctuary, it is also an important feeding habitat for four endangered species of sea turtle. The lagoon, which is 25 miles (40 km) long and on average 6 miles (9 km) wide, receives only occasional inflows of fresh water, and its evaporative losses are high. Its salinity is therefore significantly higher at its head than at its mouth, where it connects to the sea. Apart from whale watching, the main human activities in the area are small-scale fisheries and oyster cultivation. In 1993, the lagoon was designated a World Heritage Site.

LAGOON BEACH

Waves break on the shore at San Ignacio Lagoon, which is surrounded by a landscape of sparse desert scrub.

HUMAN IMPACT

WHALE WATCHING

The Laguna San Ignacio is a popular whale-watching site. Between January and March, large numbers of gray whales can be found there. The whales, which often approach boats, use the upper part of the lagoon for giving birth, while the lower lagoon is where males and females look for mates. Females swim with their calves in the middle part of the lagoon.

Salt Marshes and Tidal Flats

A SALT MARSH IS A VEGETATED AREA OF COAST that is partly flooded by the sea at high tide and completely flooded by the highest spring tides. Many areas of salt marsh are bordered by tidal flats. These are broad areas of mud or sand, mainly without vegetation, that are uncovered at low tide and covered as the tide rises. Salt marshes and tidal flats are depositories for large amounts of organic material, derived from decaying plants and animals. This provides the base for an extensive food chain.

Formation and Features

Tidal flats occur on low-energy sheltered coasts, such as estuaries and enclosed bays, where sediment held in the water settles out and builds up. The most extensive flats occur where there is a high tidal range. Tidal flats may consist either of sand (sandflats) or mud (mudflats), or a mixture of these. Mudflats contain a higher concentration of the decaying remains of dead organisms than sandflats and are also the first stage in the development of salt marshes. These develop on the landward side of mudflats. As various salt-tolerant plants grow, their roots trap sediment and stabilize the mud. As the vegetated flat builds up, different types of plants become established. The result is a salt marsh, consisting of blocks of flat, low-growing vegetated areas of mud, broken up by sinuous channels.

DISTRIBUTION
Salt marshes and tidal flats occur only north of the latitude of 32°N and south of 38°S. In latitudes nearer the equator, they are replaced by mangrove swamps.

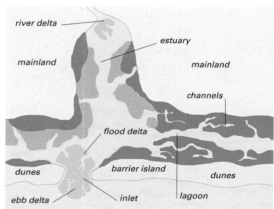

river delta

estuary

mainland

mainland

channels

flood delta

dunes

barrier island

dunes

ebb delta

inlet

lagoon

COASTAL SETTING
Salt marshes commonly develop in coastal lagoons or in estuarine areas that are sheltered from the sea by spits or barrier islands. The channels transport salt water, plankton, nutrients, sediment, and plant detritus into and out of the marsh.

KEY

- salt marsh
- tidal flats

BAY OF FUNDY
In this small subestuary of Canada's Bay of Fundy, an area of salt marsh is visible in the background. In the foreground is a broad intertidal area of mud and gravel.

Zones and Evolution

Salt marshes have two main zones. The parts flooded by every high tide are called low marsh, while the areas that are only occasionally flooded are termed high marsh. Each zone is colonized by distinct species of salt-tolerant plants. Each species, of which there are many, has developed special mechanisms to deal with the high levels of salt they are exposed to: some possess salt-excreting glands, for instance, while others have storage systems for collecting the salt until they can dilute it with water. Salt marshes and adjoining mudflats usually evolve over time. As sediment builds up, the mud surface in the marsh, the adjoining flats, and the bay or estuary as a whole tends to rise. As it does so, areas of low marsh become high marsh and areas of mudflat are colonized by plants, turning into low marsh.

SEA LAVENDER
Sea lavenders are common high-marsh colonizers. They bloom in summer, producing purple or lavender flowers.

SALT-MARSH CORDGRASS
Also called smooth cordgrass, this species is the dominant low-marsh plant throughout the Atlantic coast of North America. Strands of this grass grow to 7 ft (2 m) high.

SALT-MARSH ZONES
The low marsh is the part flooded once or twice a day at high tide, while the high marsh is the area above the mean high-tide level—it is flooded only occasionally, by the highest spring tides. Each zone has distinctive vegetation.

pool

highest spring tide

mean high tide

upper high marsh

lower high marsh

mean sea level

upland

high marsh

low marsh

mudflat

ALGAE-COVERED MUDFLATS
Some mudflats, such as these in Alaska, become heavily encrusted with green algae. The algae is often itself colonized by large numbers of tiny marine snails.

CONSERVING SALT MARSHES

Salt marshes are threatened worldwide through being built on, converted to farmland, or even used as waste dumps. Over half of the original salt marshes in the US, for example, have been destroyed. This is regrettable, as salt marshes are valuable wildlife habitats and centers of biodiversity.

MARSH HOUSING DEVELOPMENT
This coastal development in Myrtle Beach, South Carolina, has been built on top of a drained salt marsh. However, the adjoining area of marsh has been carefully preserved.

Animal Life

Measured by the amount of organic matter (the base material for food chains) that they produce, salt marshes are extremely productive habitats. Most of this material comes from decaying plant material. When plants die, they are partially decomposed by bacteria and fungi, and the resulting detritus is consumed by animals such as worms, mussels, snails, crabs, shrimp, and amphipods living in the marsh, and zooplankton living in the salt water. These in turn provide food for larger animals. Salt marshes provide nursery areas for many species of fish, and feeding and nesting sites for birds such as egrets, herons, harriers, and terns. Tidal flats are home to many types of crustaceans, worms, and mollusks, which either feed on the surface or burrow beneath it. These in turn provide food for enormous numbers of wading birds.

NATTERJACK TOAD
This toad, found in parts of western and northern Europe, inhabits upper salt marsh habitats (just below the high marsh), where it uses shallow ponds to breed.

GREAT EGRET
A common inhabitant of salt marshes in the USA and parts of east Asia, the great egret, and closely related eastern great egret, feeds on small fish, invertebrates, and small mice.

LUGWORM CASTS
Lugworms live in burrows some 8–16 in (20–40 cm) deep in tidal flats. They feed by taking in sand or mud, digesting any organic matter, and excreting the rest as a cast.

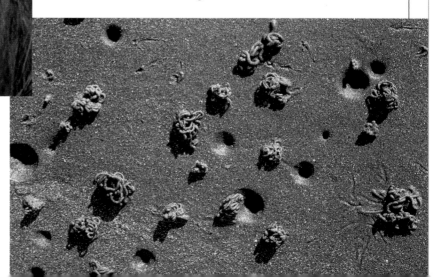

MARSH AT LOW TIDE
Patches of salt marsh surround the basin, together with tidal flats that can extend for up to 3 miles (5 km) from the shore at low tide.

ATLANTIC OCEAN NORTHWEST

Minas Basin

TYPE Tidal sandflats and mudflats, and salt marshes

AREA 445 square miles (1,150 square km)

LOCATION Eastern part of Bay of Fundy, Nova Scotia, Canada

The Minas Basin is a semienclosed inlet of the Bay of Fundy. It consists of a triangular area of tidal mudflats and sandflats, surrounded by patches of salt marsh, most of which have been diked and drained for agriculture. Twice a day, the sea fills and empties the basin, rising and falling by an average of 48 ft (14.5 m) during spring tides, which is the largest tidal range in the world. No other coastal marine area has such a large proportion of its floor exposed at low tide. Sediments in the basin range from coarse sand to fine silt and clay. The tidal flats formed by these sediments contain high densities of a marine amphipod, the Bay of Fundy mud-shrimp, which provides food for huge numbers of migrating shorebirds, including sandpipers and plovers. The numbers peak from July to October, and for some species exceed 1 percent of the world population.

SEMIPALMATED SANDPIPER
Hundreds of thousands of semipalmated sandpipers stop off in the Bay of Fundy each year on their way from Arctic regions to South America.

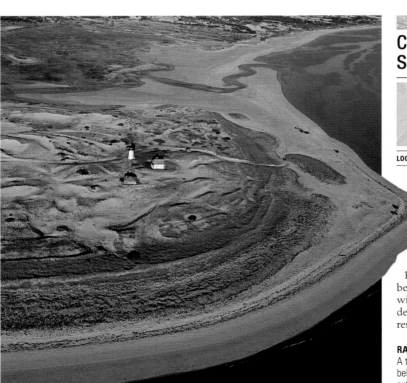

ATLANTIC OCEAN NORTHWEST

Cape Cod Salt Marshes

TYPE Salt marshes

AREA 20 square miles (Less than 50 square km)

LOCATION Cape Cod, eastern Massachusetts, US

Salt marshes are the dominant type of coastal wetland around Cape Cod, although more than a third of the region's marshes have been lost or severely degraded within the past 100 years. These salt marshes occur behind barrier beaches or spits and within estuarine systems, and have developed over the past 3,000 years in response to sea-level rise. They mainly consist of high marsh, where the dominant plant species is saltmeadow cordgrass, with some scattered areas of low intertidal marsh, dominated by smooth cordgrass. The low marsh areas are flooded twice daily and the high marsh twice a month, during the highest spring tides. The largest individual marsh is the Great Salt Marsh to the west of the town of Barnstable. With deep channels running through it, this is a popular area to explore by kayak.

The marshes around Cape Cod serve as a breeding and foraging habitat for a diversity of animals such as the northern harrier, saltmarsh sparrow, least tern, and two reptiles with a vulnerable conservation status, the diamond-backed terrapin and eastern box turtle. Restoring degraded salt marshes on Cape Cod is regarded as a top priority for many regional and national conservation organizations. Restoration will allow these wetlands to regain their function as a barrier protecting the coastline from storm surges and as a natural sponge that filters pollutants and excess nutrients from the water runoff in the region.

RACE POINT
A typical area of salt marsh can be seen here behind dunes at Race Point, at the northern extreme of Cape Cod.

SALT-MARSH MOSAIC
The edges of the Wadden Sea are a mosaic of marsh patches broken up by shallow tidal channels.

ATLANTIC OCEAN NORTHWEST

South Carolina Low Country

TYPE
Salt marshes and tidal mudflats

AREA
630 square miles (1,600 square km)

LOCATION South Carolina coast, southwest and northeast of Charleston, US

The Low Country contains one of the most extensive systems of salt marsh and tidal flats in the United States. Its size results from the broad, gently sloping, sandy coast of the US eastern seaboard, coupled with a moderately high tidal range of 5–7 ft (1.5–2 m).

CORDGRASS MEADOWS
A tidal channel weaves its way through stands of smooth cordgrass, the dominant plant species in the lower marsh areas.

Each day, two high tides inundate a vast area of the coastal zone, maintaining a system of channels, creeks, and rivers. The influence of both fresh and salt water here results in some diverse ecological communities. Smooth cordgrass is the dominant grass in the lower marshes, where the ground stays wet and muddy as a result of the tides. From late spring to fall, darker dead-looking sections of a grass called needle rush can also be seen. These two grasses are replaced toward higher ground by sea oxeye and the similar but taller marsh elder.

In the lower marshes and the bordering tidal flats, mud snails, crabs, shrimp, worms, and tiny inhabitants burrow into the mud, while attached and clinging to the stalks of the grasses are ribbed mussels and marsh periwinkles. Among the fish living in the silty tidal wash are croaker, menhaden, and mullet. Birds living here include marsh wrens and clapper rails.

ATLANTIC OCEAN NORTHEAST

Wadden Sea

TYPE Tidal mudflats and sandflats, salt marshes, and islands

AREA 4,000 square miles (10,000 square km)

LOCATION North Sea coast from Esbjerg, Denmark, along northern Germany, to Den Helder, Netherlands

The Wadden Sea is an extensive body of shallow water and associated tidal flats, salt marshes, and low-lying islands in northwestern Europe. Straddling the shores of Denmark, Germany, and the Netherlands, the Wadden Sea has been formed by storm surges and sea-level rise inundating an area of coast, combined with the deposition of fine silt by rivers. It is an important nursery for North Sea fish species such as European plaice and common sole, and its extensive mudflats are home to a number of mollusks and worms. The salt marshes provide a habitat for more than 1,500 species of insects and are important feeding and breeding grounds for many species of birds. Unfortunately, these marshes are threatened by intensive farming, industrial development, and climate change. Since 2014, the whole of the Wadden Sea has been a UNESCO World Heritage Site.

ATLANTIC OCEAN NORTHEAST

Morecambe Bay

TYPE Tidal mudflats and sandflats, and salt marshes

AREA 120 square miles (310 square km)

LOCATION Northwest England, UK

Formed from the confluence of five estuaries, those of the Kent, Keer, Leven, Lune, and Wyre Rivers, Morecambe Bay is the largest continuous area of tidal flats in the UK. Broad, shallow, and funnel-shaped, the bay has a large tidal range, of up to 35 ft (10.5 m). During periods of spring tides, the sea can ebb as far as 7 miles (12 km) back from the high-water mark. The flood tide comes up the bay faster than a person can run, and parts of the bay are also affected by quicksand, posing dangers for anyone who does not know the area well.

The bay's extensive mudflats support a rich and diverse range of invertebrate animals, including cockles and mussels, snails, shrimp, and lugworms, as well as one of the largest populations of shorebirds in the UK. The bay regularly hosts 170,000 wintering waders, with several species present in internationally significant numbers, including oystercatchers, curlews, dunlins, and knots. The tidal flats are surrounded by extensive salt marshes, which make up about 5 percent of the total salt marsh in the UK and support a number of rare plants. Much of this marsh area is grazed by sheep and cattle.

The bay is an important location for commercial fishing; the fish species most commonly caught include bass, cod, whitebait, and plaice. However, Morecambe Bay has not escaped the problems of pollution common to many coastal areas of northwestern Europe. Oil, chemicals, and plastic are among the more common pollutants of this ecosystem.

MORECAMBE MUDFLATS
The ebbing tide reveals half of the bay's total area as undulating expanses of mud and sand, meandering channels, and tidal pools.

HUMAN IMPACT

COCKLING

Morecambe Bay has many rich cockle beds. The cocklers use planks of wood called jumbos to soften the sand, which helps draw the cockles to the surface. Because of the fast-moving tides, cockling has to be carried out with an eye on safety. In February 2004, a total of 23 Chinese migrant workers drowned after being cut off by the tides.

OCEAN ENVIRONMENTS

127

OCEAN ENVIRONMENTS

The Wash

TYPE
Salt marshes, tidal sandflats, and mudflats

AREA
240 square miles (620 square km)

LOCATION Northeast of Peterborough, England, UK

The Wash is a large, roughly square-shaped estuary on the eastern coast of England, surrounded by extensive areas of tidal sandflats, some mudflats, and salt marshes. It is fed by four main rivers: the Great Ouse, Nene, Welland, and Witham. The sandflats of the Wash range from extensive fine sands to drying banks of coarse sand and are home to large communities of bivalve mollusks, crustaceans, and polychaete worms. The extensive salt marshes

TERRINGTON MARSHES
Located close to the mouth of the Nene River, these marshes form part of the Wash National Nature Reserve.

comprise the largest single area of this habitat in Britain and are growing in extent. The main plant species making up these salt marshes, which are traditionally used as grazing lands by farmers, are cordgrass and glasswort in roughly equal amounts.

The Wash is one of the most important sites in the UK for wild birds, its sheltered tidal flats providing a vast feeding ground for migrating birds, such as geese, ducks, and waders.

These come to spend the winter in the Wash in huge numbers, with an average total of about 300,000 birds, from as far away as Greenland and Siberia. In addition, the Wash is an important breeding area for common terns and a feeding area for marsh harriers. It is a Site of Special Scientific Interest (SSSI).

In 2000, parts of the artificial coastal defenses on the western side of the Wash were deliberately breached

to increase the area of salt marsh in the region. This has taken pressure off other nearby sea defenses, because the newly establishing area of salt marsh soaks up wave energy, acting as a natural sea defense. This is a relatively novel approach to coastal management that employs "soft engineering" techniques to defend against the erosive power of the sea. It also has the added environmental advantage of providing additional habitat for wildlife.

EDGE OF THE MARSHES
The dominant plant species in the non-exploited areas of salt marsh, such as at La Turballe in the northern part of the marshes, include sea-blite, cordgrass, and glasswort.

Guérande Salt Marshes

TYPE Salt marshes, artificial salt pans, and tidal mudflats

AREA 7 square miles (50 square km)

LOCATION Northwest of St. Nazaire, Atlantic coast of France

The region of salt marshes close to the medieval town of Guérande is most famous for its salt production but is also a noted ecological site, important for its role as a feeding and resting site

for large numbers of birds. The salt marshes came to exist in their present state through a combination of geology, climatic factors, and human intervention. Around the coast near Guérande, a system of spits and coastal dunes developed thousands of years ago, cutting off an area of shallow water, which was nevertheless subject to tides—seawater could flow in through two inlets in the dune belts. Over the centuries, marshes and tidal flats developed in this basin. During the past 1,000 years or so, these have been artificially converted into a mosaic of salt pans, separated by clay walls, although some areas remain unexploited. During the flood tide, seawater is allowed to flow through

channels into the pans, and during the warm summer months, when the rate of evaporation is high, sea salt is skimmed from the surface of the pans by an army of salt-farmers (*paludiers*).

The areas of marsh surrounding the salt pans are made up of various salt-tolerant plants. More than 70 different species of birds nest and breed in the area, and many species spend the winter here in large numbers. For many years, the salt-farmers and the French ornithological society, the LPO, have jointly organized exhibitions and guided tours in the Guérande Salt Marshes, which are themed on the economics of salt production, the ecology of the marshes, and their need for protection.

SALT HARVEST

The Guérande region has had salt pans for over 1,000 years. Today, more than 200 salt-farmers work in the area, one of the few places in France where salt continues to be produced in a traditional manual way. The average annual harvest is about 10,000 tons of natural, mineral-rich sea salt, which is sold unrefined, with nothing added and nothing removed. The salt has a light gray color because of its content of fine clay from the salt pans.

PACIFIC OCEAN NORTHWEST

Saemangeum Wetlands

TYPE
Mudflats, sandflats, and salt marshes

AREA
155 square miles
(400 square km)

LOCATION South of Seoul, on the west coast of South Korea

Situated at the confluence of the Mangyeung and Dongjin river estuaries, on South Korea's Yellow Sea coast, the Saemangeum Wetlands is a shorebird staging site of great importance. Its tidal flats and shallows support many bird species, some of which are considered to be globally threatened. However, the South

SPOON-BILLED SANDPIPER
This extremely rare species is one of the shorebirds most threatened by the reclamation project.

Korean government built a 21-mile (33-km) sea wall across the mouth of the estuaries to convert the wetlands into a large freshwater reservoir and some dry land for development, agriculture, and tourism. Since the completion of the seawall in 2010,

SEA WALL AND RESERVOIR
In this view, the wall extends top to bottom at left of center. There are calls for sluice gates to be built into it—some seawater could then enter, with potential to restore the wetlands.

the wetlands' ecosystem has been devastated, with the deaths of vast numbers of shellfish, a dramatic decline in migratory birds, and the destruction of the local fishing industry.

PACIFIC OCEAN NORTHWEST

Yatsu-Higata Tidal Flat

TYPE
Tidal mudflat

AREA
1/6 square mile
(0.4 square km)

LOCATION Narashino City, at the northern part of Tokyo Bay, Japan

Yatsu-Higata is a tiny rectangular mudflat at the northern end of Tokyo Bay, and is unusual because it is almost completely surrounded by a dense urban area. Once open shoreline, Yatsu-Higata now sits 3/5 mile (1 km) inland. Twice daily, it experiences a tidal inflow and outflow of water from Tokyo Bay via two concrete channels. When the tide comes in, the mudflat fills with about 3 ft (1 m) of water. When it flows out, a variety of resident and migrant shorebirds congregate to feed on the lugworms, crabs, and other marine animals that live within the fine silt that remains. Yatsu-Higata is an important stopover point for migrating birds flying from Siberia to Australia and Southeast Asia.

PACIFIC OCEAN NORTHEAST

Alaskan Mudflats

TYPE
Tidal mudflats

AREA
4,000 square miles
(10,000 square km)

LOCATION Various coastal inlets of southern and western Alaska, US

Many areas on the coast of southern and western Alaska are fringed by mudflats that appear at low tide. They are formed of a finely ground silt that in some areas is several hundred yards deep. This silt has originated from the action of Alaska's numerous glaciers, which have been grinding away at the surrounding mountains for thousands of years. As these glaciers melt, the silt is carried to the coast in meltwater and deposited as sediment

when it reaches the sea. Because tidal ranges around Alaska are generally high, the total area of mudflats exposed at low tide is huge. These mudflats are an important stopover for migrating shorebirds. Various species of burrowing worms and bivalve mollusks are an important source of food for these waders and for the waterfowl that feed on the mudflats through the winter. Harbor seals also use the mudflats as rest areas.

The mudflats are dangerous for human visitors, because in some areas they behave like quicksand. Even mud that at first seems firm enough to support a person may in reality be treacherous. A number of people have become stuck and some have even drowned.

DRYING MUDFLATS
These mudflats are at the edge of a large delta on the southwest coast of Alaska, formed by the Yukon and Kuskokwim Rivers.

DIGGING FOR CLAMS

Brown bears are occasional visitors to some Alaskan mudflats, where they dig for Pacific razor clams buried in the mud. They probably find the clams by looking for the small holes they leave on the surface as they burrow down. Extracting them is tricky, since when disturbed, they burrow down further.

ALASKAN BROWN BEAR
This large adult bear is digging on the coast of Katmai National Park, at the eastern end of the Alaskan Peninsula.

Mangrove Swamps

A MANGROVE SWAMP IS A COLLECTION of salt-tolerant evergreen trees, thriving in an intertidal environment in the tropics or subtropics. Mangrove swamps line about 7 percent of the world's coastlines, where they filter pollutants from river runoff and help prevent the silting up of adjacent marine habitats. They also protect coastlines against erosion and provide a home for fish, invertebrates, and many other animals.

AERIAL ROOTS
Many mangrove species have aerial roots. These prop the tree up and take in oxygen, which is usually not available in the mud that most mangroves grow in.

Formation

Mangrove swamps develop in coastal areas protected from direct wave action. These areas often fringe estuaries and coastal lagoons (see p.114). Most mangroves develop in fine muds or sandy sediments that form in these environments. As the lower parts of the mangrove roots develop in the sediment, aerial roots form a tangled network above it. This traps silt and other material carried there by rivers and tides. Land is built up, and then colonized by other types of vegetation.

DISTRIBUTION
Mangrove swamps occur only between latitudes 32°N and 38°S. Salt marshes and tidal flats (see p.124) replace them elsewhere.

Plants

Some 54 species of trees and shrubs are classified as "true" mangroves, occurring only in mangrove habitats. Each has evolved special adaptations to the conditions they grow in, such as salty water. For example, some mangroves can excrete salt in their leaves. On most mangrove shorelines, there are two or three zones, each dominated by different mangrove species. In the Americas, just four main species are found. The area closest to the sea is dominated by red mangroves. Landward of this are black mangroves—the roots of this and some other species develop pencil-like breathing tubes, called pneumatophores. White and button mangroves grow farther landward. The mangrove swamps on the coasts of the rest of the tropics contain greater species richness.

PNEUMATOPHORES
These vertical tubes grow up out of the sand or mud as extensions of horizontal roots. When exposed to the air, they take in oxygen.

RED MANGROVE
This mangrove species can grow in deep water by means of its numerous prop roots, which often have a reddish tint. It also has a particularly high salt tolerance.

Animal Life

Mangrove swamps are rich centers of biodiversity. Mangrove trees produce enormous amounts of leaf litter, as well as twigs and bits of bark, which drop into the water. Some of this immediately becomes food for animals such as crabs, but most is broken down by bacteria and fungi, which turn it into food for fish and shrimp. These in turn produce waste, which, along with the even smaller mangrove litter, is consumed by mollusks, amphipods, marine worms, small crustaceans, and brittlestars. Some of these become food for larger fish, and the various fish species provide food for larger animals.

Across the world, mangrove swamps are home to an enormous number and diversity of birds and several endangered species of crocodiles. Other types of animals found in great numbers and diversity in mangrove swamps include frogs, snakes, insects, and mammals ranging from swamp rats to tigers.

BANDED ARCHERFISH
This little fish inhabits mangrove swamps in the Indian and Pacific Oceans. It is known as an archerfish because it feeds mainly on flying insects, which it knocks out of the air and into the water by spitting at them.

MANGROVE BRITTLESTAR
This scavenger is one of the few echinoderms found in mangrove swamps. It is highly mobile, using its long arms to pull itself along.

JABIRU STORK
This large stork inhabits mangrove swamps and other wetlands throughout the tropical Americas, feeding on a range of prey, including snakes.

SHELTER FROM PREDATORS
Cardinalfish, sheltering here in a mangrove swamp in Papua New Guinea, are one of the many types of small tropical fish that use mangrove roots for protection from predators.

HUMAN IMPACT

SHRIMP FARMING

More than a third of the world's mangrove swamps have been destroyed since 1980 and have been built on or turned into commercial enterprises such as shrimp farms (including the one shown here, in Vietnam). Unfortunately, intensive shrimp farming often has devastating environmental effects. Typically, the effluent from shrimp ponds pollutes nearby coastal waters, destroying more mangroves as well as coral reefs along the coastline.

OCEAN ENVIRONMENTS

MANGROVE-LINED CHANNEL
Here, parallel stands of red mangrove line a shallow offshoot channel of Florida Bay in the southern part of the Everglades National Park.

ANHINGA
This diving bird hunts fish, frogs, and baby alligators in the Everglades mangroves.

ATLANTIC OCEAN WEST

Everglades

PRINCIPAL SPECIES
Red, black, and white Mangroves

AREA Mangroves only: 600 square miles (1,500 square km)

LOCATION Southwestern Florida, US

Mangroves occupy a large, roughly triangular area at the southwestern tip of southern Florida, where a maze of islands along the coast is intersected by mangrove-lined channels. Here, where the salt water of the Gulf of Mexico and Florida Bay meets fresh water that has traveled from Lake Okeechobee in central Florida, is the largest area of mangrove swamps in North America.

The dominant species along the edges of the sea and the numerous channels is the red mangrove—water within the channels is normally stained brown from tannin contained in the leaves of this species. In addition to their role in stabilizing shorelines with their large prop roots, red mangroves are crucial to the Everglades ecosystem, acting as a nursery for many species of fish, as well as shrimp, mussels, sponges, crabs, and other invertebrates. The other principal mangrove species in the Everglades are the black mangrove and white mangrove. Both of these grow closer to the shore than red mangroves, so they are in contact with seawater only at high tide. The Everglades swamps provide a feeding

---

ATLANTIC OCEAN WEST

Alvarado Mangrove Coast

PRINCIPAL SPECIES
Red, white, and black mangroves

AREA 200 square miles (500 square km)

LOCATION To the southwest of Veracruz, Mexico, on southwestern Gulf of Mexico

The Alvarado Mangroves Ecoregion in southern Mexico is an extensive area of mangrove swamps mixed in with other habitats such as reed beds and palm forests. The mangroves grow on flat coastal land interspersed with brackish lagoons fed by several small rivers. The swamps are brimming with life, from rays gliding in the calm waters to snails climbing the mangrove roots, whose tangled network protects many fish and invertebrates from predators. Bird life in and around the swamps includes the keel-billed toucan, reddish egret, wood stork, and several species of herons and kingfishers, while the mammalian inhabitants include spider monkeys and West Indian manatees. Some large areas of mangroves in the region have been destroyed, and those that remain are under pressure from logging, agricultural expansion, oil extraction, and frequent oil spills.

ATLANTIC OCEAN WEST

Sian Ka'an Biosphere Reserve

PRINCIPAL SPECIES
Red, black, white, and button Mangroves

AREA 1,080 square miles (2,800 square km)

LOCATION Eastern coast of Yucatán Peninsula, eastern Mexico, 90 miles (150 km) south of Cancún

Stretching for 75 miles (120 km) along Mexico's Caribbean coast, the Sian Ka'an Biosphere Reserve contains a mixture of mangrove swamps, lagoons, and freshwater marshes; it was declared a World Heritage Site by UNESCO in 1987. The mangroves are protected from the energy of the Caribbean Sea by a barrier reef growing along the coast. However, the reserve's terrestrial part is between 20 and 75 percent flooded, depending on season. Sian Ka'an's mangrove systems are some of the most biologically productive in the world and their health is critical for the survival of many species in the western Caribbean region. Hidden between the massive mangrove roots live oysters, sponges, sea squirts, sea anemones, hydroids, and crustaceans. Bird species found here include roseate spoonbills, pelicans, greater flamingos, jabiru storks, and 15 species of heron. The swamps are also home to West Indian manatees and two crocodilian species: the American crocodile and Morelet's crocodile. The explosion of tourism in the nearby resort of Cancún poses several threats to the area. Unregulated development has increased pollution and altered the distribution and use of water in Sian Ka'an, compromising the health of the mangroves.

BOAT TOUR
Because a large part of Sian Ka'an is flooded for much of the year, there are few roads into the area, so much of it can be reached and explored only by boat.

and nesting site for several mammals, including swamp rats, and numerous bird species, such as herons, egrets, gallinules, anhingas, and brown pelicans. While much of the region has an abundant alligator population, the swamps are also home to a small but increasing population of American crocodiles. Also occasionally spotted in the channels between the mangroves are West Indian manatees (sea cows). The

Florida mangroves are sometimes damaged by the hurricanes that hit the region, hurricanes Katrina and Wilma in 2005 being examples. Hurricanes damage mangroves in two ways: strong winds may defoliate them, and storm surges harm them by depositing large quantities of silt on their roots. Fortunately, mangrove forests are resilient ecosystems, and they usually regenerate fully from hurricane damage within a few years.

CICHLID INVASION

Since 1983, the Mayan cichlid, an exotic fish species from Central America, has been spreading rapidly through the mangrove swamps and other wetland areas of the Everglades. No one yet knows what effect it may have on the region's ecosystem. There are worries that it may displace native fish species; alternatively, it could be occupying a new "niche" that no other fish species has filled.

ATLANTIC OCEAN WEST

Zapata Swamp

PRINCIPAL SPECIES
Red, black, white, and button mangroves

AREA
1,680 square miles (4,350 square km)

LOCATION Western Cuba, 100 miles (160 km) southeast of Havana

The Zapata Swamp is a mosaic of mangrove swamps and freshwater and saltwater marshes that form the largest and best-preserved wetland in the Caribbean. The swamp was designated a Biosphere Reserve in 1999 and forms a vital preserve for Cuban wildlife, a spawning area for commercially valuable fish, and a crucial wintering territory for millions of migratory birds from North America. More than

GREATER FLAMINGOS
Large numbers of these colorful birds live in the swamp, feeding off algae, shrimp, mollusks, and insect larvae that inhabit the mud at the bottom of the shallow waters.

900 plant species have been recognized in the swamp, and all but three of the 25 bird species endemic to Cuba breed there. All together, about 170 bird species have been identified in the swamp, including the common black-hawk, the greater flamingo, and the world's smallest bird, the bee hummingbird. It also contains most of the remaining few thousand Cuban crocodiles, which are critically endangered. Mammalian residents include the Cuban hutia, a gopher-like rodent. The manjuari, or Cuban gar, is an unusual fish found only in the swamp. Adjacent to the swamp is the Bay of Pigs, where millions of land crabs breed each spring.

ATLANTIC OCEAN WEST

Belize Coast Mangroves

PRINCIPAL SPECIES
Red, black, white, and button mangroves

AREA
600 square miles (1,500 square km)

LOCATION Eastern Belize, on the western margins of the Caribbean Sea

The mangrove swamps of Belize are a nursery for many fish species associated with the huge Belize Barrier Reef. By filtering runoff from rivers and trapping sediment, the mangroves also protect the clarity of the coastal waters, helping the coral reef to survive. Numerous cays—small islands composed largely of coral or sand—along the coast are covered with mangroves and form a habitat for birds. In all, more than 250 bird species share the swamps with West Indian manatees and a variety of reptiles, including boa constrictors, American crocodiles, and iguanas.

MANGROVE ROOTS
A tangled maze of mangrove roots extends beneath the water's surface all along this coast, providing refuge for a variety of juvenile fish.

INDIAN OCEAN WEST

Madagascar Mangroves

PRINCIPAL SPECIES Gray, yellow, long-fruited stilt, and large-leafed orange mangroves

AREA 770 square miles (2,000 square km)

LOCATION Scattered areas around the coast of Madagascar, off the eastern coast of Africa

Mangroves occur in a wide range of environmental conditions on Madagascar, fostered by a high tidal range, extensive low-lying coastal areas, and a constant supply of fresh river water, which brings a high silt load. They occupy about 600 miles (1,000 km) of the island's coastline and are often associated with coral reefs, which protect them from ocean swell. The mangroves, in turn, capture river sediment that otherwise would threaten both reefs and seagrass beds. Up to nine different mangrove species

have been recorded in Madagascar, although only six are widespread. Several of Madagascar's endemic birds, including the Madagascar heron, Madagascar teal, and Madagascar fish-eagle, use the mangroves and associated wetland habitats. Dugongs (relatives of manatees) glide through the waters, feeding on sea grasses, while huge quantities of invertebrates and fish swim freely among the fingerlike roots of the mangroves. These provide an abundance of food for animals such as the Nile crocodile,

sharks, and aquatic and wading birds, such as herons, spoonbills, and egrets. Many of the fish and bird species here are found nowhere else in the world. Unfortunately, the mangroves are threatened by urban development, overfishing, and the development of land for rice and shrimp farming.

MANGROVE MAZE
This area of coastal mangroves, bisected by numerous channels, is located on the northeastern coast of Madagascar, at the mouth of the Ambodibonara River.

INDIAN OCEAN NORTH

Pichavaram Mangrove Wetland

PRINCIPAL SPECIES Gray, milky, stilted, small-fruited orange, and yellow mangroves

AREA 4 square miles (11 square km)

LOCATION Tamil Nadu, southeastern India, 90 miles (150 km) south of Chennai

The Pichavaram Mangrove Wetland lies on a delta between the Vellar and Kollidam estuaries in southeastern India. It consists of a number of small and large mangrove-covered islets intersected by numerous channels and creeks. Fishing villages, croplands, and aquaculture ponds surround the area. This small, carefully preserved wetland is thought to have saved many lives during the 2004 Indian Ocean tsunami. When the tsunami struck, six villages that were physically protected by the mangroves incurred no damage, while other, unprotected villages were totally devastated. The wetland may have reduced the tsunami's impact partly by slowing the onward rush of the sea through frictional effects and partly by absorbing water into its numerous canals and creeks.

INDIAN OCEAN NORTH

Sundarbans Mangrove Forest

PRINCIPAL SPECIES Sundri, milky mangrove, yellow mangrove, Indian mangrove, keora

AREA 3,200 square miles (8,000 square km)

LOCATION Southwestern Bangladesh and northeastern India, between Calcutta and Chittagong

This forest, a World Heritage Site since 1997, is the largest continuous mangrove ecosystem in the world. It is part of a

huge delta formed by sediments from the Ganges, Brahmaputra, and Meghna Rivers. The region contains thousands of mangrove-covered islands intersected by an intricate network of waterways. The Bengal tiger swims here from island to island, hunting prey such as spotted deer and wild boar. Other inhabitants include fishing cats, rhesus macaque monkeys, water monitor lizards, hermit crabs, the Ganges River dolphin, and various sharks and rays. Habitat destruction threatens this region: more than half of the original mangroves have been cut down.

GAVIAL

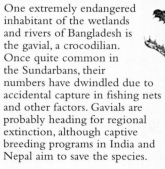

One extremely endangered inhabitant of the wetlands and rivers of Bangladesh is the gavial, a crocodilian. Once quite common in the Sundarbans, their numbers have dwindled due to accidental capture in fishing nets and other factors. Gavials are probably heading for regional extinction, although captive breeding programs in India and Nepal aim to save the species.

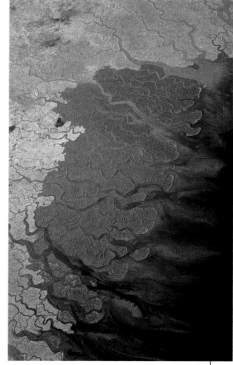

SATELLITE VIEW
In this satellite view of part of the Ganges–Brahmaputra–Meghna delta, the Sundarbans Mangroves form the area that appears dark red. On the right is the Bay of Bengal.

PACIFIC OCEAN WEST

Kinabatangan Mangroves

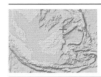

PRINCIPAL SPECIES Stilt mangrove, long-fruited stilt mangrove, gray mangrove, nipa palm

AREA 300 square miles (800 square km)

LOCATION Southeast of Sandakan, eastern Sabah, Malaysia

Mangrove swamps occupy a coastal region of the Kinabatangan River delta, within eastern Sabah in the northern part of the island of Borneo. The mangrove swamps in this area form a complex mosaic with other types of lowland forest (including palm forest) and open reed marsh. They are home to dozens of species of saltwater fish, invertebrates such as shrimp and crabs, otters, and some 200 species of birds including various species of fish eagle, egret, kingfisher, and heron.

Irrawaddy dolphins are also occasionally spotted in the region, while other spectacular inhabitants include Borneo's indigenous proboscis monkey and the saltwater crocodile (the world's largest crocodile species), which was almost hunted to extinction but whose numbers are now recovering. Over the past 30 years, there has been extensive clearance of mangroves in the Kinabatangan delta for purposes of timber and charcoal production. The mangroves have either been replaced by oil palms or the cleared land has been developed for shrimp farming. Inevitably, the wildlife has suffered, but the government of Sabah is now engaged in a large-scale mangrove replanting operation.

MANGROVE MONKEY
A female proboscis monkey, able to both swim and walk upright, is seen here with an infant, leaping across a waterway in the Kinabatangan mangroves. Her long tail helps to stabilize her movement through the air.

AERIAL ROOTS AT LOW TIDE
The mangroves are anchored in the soft mud by a dense network of roots that also provide a habitat for many animals.

PACIFIC OCEAN EAST

Darien Mangroves

PRINCIPAL SPECIES Red, black, button, white, mora, and tea mangroves

AREA 360 square miles (900 square km)

LOCATION Southeast of Panama City on the Pacific coast of eastern Panama

The Darien mangrove swamps lie around estuaries in eastern Panama in the Darien National Park, adjacent to the Gulf of Panama. Here, the roots of mangroves create a haven for mollusks, crustaceans, and many fish species. Shrimp are particularly abundant—the larvae hatch offshore, migrate to the mangrove "nursery" for a few months, and then return to sea as adults. Some of the mangrove swamps in this region have been converted to shrimp ponds or farmland. Other threats include urbanization and pollution.

PACIFIC OCEAN WEST

New Guinea Mangroves

PRINCIPAL SPECIES Gray, long-fruited stilt, tall-stilted, and cannonball mangroves

AREA 10,000 square miles (26,000 square km)

LOCATION Scattered areas around the island of New Guinea in the western Pacific

Mangrove swamps occur in extensive stretches on New Guinea's coastline. The longest and deepest stretches are found on the south side of the island, around the mouths of large rivers such as the Digul, Fly, and Kikori Rivers. Mangrove communities here are the most diverse in the world—more than 30 different species of mangroves have been found in a single swamp—and they form a vital habitat for a variety of animals living on the water's edge. Underwater, over 200 different fish species, ranging from cardinal fish and mangrove jacks to seahorses and anchovies, have been recorded in either their adult or juvenile stages. Mudskippers (species of fish that can leave the water and climb trees), snails, and crabs climb the mangrove roots, while saltwater crocodiles patrol the channels between the mangrove stands. Although there are many species of fish and mangrove in these swamps, terrestrial animal diversity is relatively low. Two endemic species of bats and a species of monitor lizard are found here.

Ten bird species are endemic, including the New Guinea flightless rail, two species of lory, the Papuan swiftlet, red-breasted paradise-kingfisher, and red-billed brush-turkey. Although largely intact, the mangrove regions in the western part of New Guinea have recently come under threat of pollution from the rapidly expanding oil and gas industries.

SEAHORSE
This small seahorse is adopting the yellow color of fallen mangrove leaves.

BLACK MANGROVE
This mangrove forest is at the outlet of the Rio Mogue on Panama's Pacific coast.

NEW GUINEA MANGROVES
This young saltwater crocodile is feeding among mangrove roots. Fully grown, this species is the largest of all crocodiles, growing up to 23 ft (7 m) long. Despite its name, it prefers fresh water, and adults compete fiercely for control of prime channels in swamps, often forcing juveniles into marginal rivers or out to the open sea.

THE SHALLOW SEAS that cover the continental shelves around Earth's landmasses nurture an extraordinary diversity of life. Energy from the Sun and nutrients from the land and sea ensure good conditions for plant growth, on which all marine life depends. The Moon is also a key player. Its gravitational pull drives the tides, which uncover the seashores each day, creating tidal currents that distribute plant nutrients and bring food to waiting animals. Each area of seabed provides a specific habitat for marine life that has adapted to the local conditions. The shallow seas comprise those parts of the oceans with which we are most familiar; however, we are only just beginning to understand the complexity of life there and its importance to the overall health of the planet.

SHALLOW SEAS

CORAL REEFS
From polar seas to the tropics, the reflective surface of the sea hides a realm populated with unfamiliar life forms. Here in the tropics, animals look like plants, and plants hide inside the tissues of corals.

Continental Shelves

CONTINENTAL SHELVES ARE ESSENTIALLY the flooded edges of continents, inundated by sea-level rise after the last ice age. The shelf seabed is now approximately 600 ft (200 m) below the surface, and its width varies, occasionally extending to hundreds of miles. The shelf seabed and water quality are influenced by land processes. Rivers bring fresh water and nutrients, making shelf waters very productive ecologically, while river-borne material settles on the seabed as sediment. The continental shelf has a huge diversity of marine life and habitats, but it is also the area of the sea that suffers most from pollutants.

DISCOVERY
FIORDS

Fiords are deep, sheltered sea inlets originally gouged out by glaciers and then flooded by the sea. They often extend many miles inland and are made up of deep basins, separated from the open sea by shallow sills.

This basin-and-sill structure has a huge influence on marine life. In this sheltered environment, still, dark salt water lies beneath peaty fresh water. This mimics the marine conditions off the continental shelf, and animals normally confined to much deeper water, such as cold-water corals, inhabit water shallow enough for divers to explore.

Fertile Fringes

The coastal fringes have the greatest diversity of life in the oceans. Light penetration is highly variable, from turbid basins to clear tropical waters. In many places, enough light reaches the shallow seabed for good growth of photosynthetic organisms. Seaweeds, seagrasses, and phytoplankton thrive here, fed by solar energy, nutrients from land, and sediments stirred up by winds and currents. The coastal fringes are much more productive than the open oceans. Combined with diverse habitats, this results in complex marine communities, making rich feeding and nursery grounds for animals from deeper water. In higher latitudes, seasonal variations in the Sun's strength stimulate an annual cycle of plankton and seaweed growth. In the tropics, where seasons are less pronounced, seagrasses and seaweeds grow year-round.

SHALLOW SEAWEED
Seaweeds grow best on shallow, sunlit rocks, thrive in strong water movement, and provide food and shelter for many small animals.

Productive Plains

Much of the continental shelf is covered with deep sediments. Sand, gravel, and pebbles are deposited in shallow water, while fine mud is carried into deeper water offshore. An important part of shelf sediments is biogenic (made from the remains of living organisms). It consists of carbonates (chemical compounds containing carbon) derived from, for example, coral skeletons, and microscopic plankton.

At first sight, sediment plains appear barren. However, many different animals live hidden beneath the surface, either permanently or emerging from burrows and tubes to feed and reproduce. Shifting sand and gravel is a difficult place to live, but more stable sediments occur on deeper seabeds. Varying particle size makes it suitable for constructing burrows and tubes, and it can contain huge numbers of animals, providing a rich food source. These animal communities are all sustained by plankton falling from the continental-shelf surface waters, and by the products of decomposition of seagrasses and seaweeds.

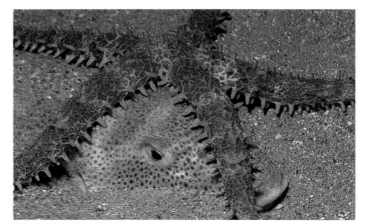

SEDIMENT PREDATORS
Fish and starfish are top predators on sediments, eating the many different animals on the surface or buried beneath. Fish catch a wide range of creatures, while starfish capture slower-moving prey.

Shelf Fisheries

The waters and seabed of the continental shelf support most of the world's major fisheries. In coastal waters, there is planktonic food for larvae and cover for juveniles, and this is where 90 percent of the world's total seawater catch reproduces. Demersal fish (living on or just above the seabed) such as cod and haddock feed on seabed life. Pelagic (open water) shoaling fish such as sardines and herring feed on zooplankton. They are also important food for larger fish such as mackerel and sharks, as well as for cetaceans and seabirds. Commercially important invertebrates such as shrimp are caught in shelf waters. Worldwide, coastal communities are sustained by small-scale, inshore fisheries, which catch a wide range of marine life.

JUVENILE SHELTER
These baby cod are feeding in horse mussel beds, before moving offshore as adults.

QUEEN SCALLOP
Scallops feed by filtering seawater, and can be collected by diving, or farmed, with no damage to the marine environment.

SHELF DEPOSITS
The Mississippi River flows into the sea through a network of channels. As its silt-laden waters reach the sea, sediments fall out along the continental shelf.

Geology of the Continental Shelf

Shelf deposits can be extremely thick. For example, those off eastern North America are up to 9 miles (15 km) deep, and have been accumulating and compacting for millions of years. A cross-section here reveals ancient sediments other than those deposited by rivers and glaciers, including carbonates, evaporites, and volcanic materials. Carbonates are largely produced by marine life in shallow tropical seas. Evaporites are salts resulting from seawater evaporation in shallow basins or on arid coastlines. Evaporite deposits create domes in overlying sedimentary rocks, trapping oil and gas.

DREDGED TREASURE
Metals such as gold, tin, rare earth elements, and aggregates for the building industry are extracted by dredging continental shelves.

HUMAN IMPACT

COASTAL POLLUTION

For many years, coastal seas have been used as a convenient dump for human waste. Even the most remote seashores are now littered with plastic. More insidious is invisible pollution: nutrients and pathogens from sewage; heavy metals, organohalogens, and other toxins from industrial and agricultural effluents; radioactive waste from power stations; and hydrocarbons from effluents, oil spills, and other sources.

Rocky Seabeds

FROM THE WARM TROPICS TO COLD POLAR SEAS, many distinctive communities of marine life develop on the rocky floors of shallow seas. Underwater rocks provide points of attachment for both seaweeds and marine animals and are often covered with life. Seaweeds thrive in the sunlit shallows and provide a sheltered environment for animal communities. Firmly attached animals extend arms and tentacles to catch planktonic food from water currents, or pump water through their bodies to filter out nutrients. Mobile animals graze seaweeds or prey on fixed animals or each other. The life on a rocky reef depends on many environmental factors.

ROCKY SLOPE
These underwater rocks in British Columbia, Canada, are covered with marine life. A sunstar and leather star search for prey among pink soft corals and sponges, while urchins graze below.

The Seaweed Zone

Seaweeds rely on sunlight for growth, and thrive only on the shallowest rocks. The depth in which they can grow depends on water clarity, from a few yards in turbid seas, to more than 330 ft (100 m) in the clearest waters. In colder waters, huge forests of kelp and other large brown seaweeds dominate the shallows, with smaller seaweeds in deeper water. Large seaweeds are often scarce on rocks in the tropics—instead, the Sun's energy is harnessed by tiny unicellular algae inside coral tissues. Seaweeds harbor a plethora of associated animals. Some live permanently in the seaweed zone, while others use it as a breeding ground or nursery before moving into deeper water.

FOOD SOURCE
Energy from sunlight captured by seaweeds is used by grazing animals. Here, green seaweeds cover rocks in Orkney, Scotland.

BALLAN WRASSE
In summer, adult ballan wrasses lay eggs in nests built of seaweed, secured in rock crevices. Young wrasses are often patterned, providing camouflage.

ROCK GRAZERS

Sea urchins are highly successful marine invertebrates, well defended by sharp spines. They graze the seabed, eating virtually everything except hard-shelled animals and coralline seaweed crusts. They have a profound effect on sea bed communities. If urchins are abundant, they can seriously reduce the diversity of life on the sea bed, leaving urchin "barrens." Conversely, where urchins are sparse, they can increase diversity, by clearing spaces for new life to settle.

Animal-dominated Deeps

In deeper water, light levels are too low for most seaweeds, although encrusting red seaweeds need little light and grow farther down. Much of the plantlike growth in deeper water actually consists of fixed animals, which are most abundant in places with strong tidal currents. For mobile animals living here, the seabed is a minefield of toxic substances, released by fixed animals to deter predators. Below 160 ft (50 m), water movement from waves is much less, and fragile animals such as sponges and sea fans can grow to a large size. Here, and in places more sheltered from water movement, a smothering layer of fine silt continually settles on the rock surfaces, restricting the animal life to forms that can hold themselves above the rock or can remove the silt. On the most heavily silted rocks, animals may grow only on vertical or overhanging surfaces.

ROCKY-BOTTOM PREDATOR
Stonefish have a textured skin and irregular shape, making them difficult to spot. A huge mouth engulfs prey, while the dorsal spines contain venom that can be fatal.

protruding eye used when hiding in sediment

dorsal spines with venom glands

camouflage skin color and texture

large mouth

tail fin

Vertical Rock

Underwater cliffs are often more heavily colonized with invertebrates than gently sloping rocks. In shallow water exposed to strong waves, various mobile seabed animals, particularly grazing sea urchins and predatory starfish, find it harder to cling to vertical and overhanging surfaces, and are knocked off by waves in rough weather. Vertical walls receive less sunlight, and are harder places for seaweed spores to settle, so there is less competition from seaweeds here than on horizontal rock. At sheltered sites, upward-facing rock is often covered with silt and has few animals, but vertical and overhanging rock, by contrast, is silt-free and may have abundant life. Ledges and crevices in underwater cliffs provide safe refuges for fish and crustaceans.

JEWEL ANEMONES
Multicolored jewel anemones carpet vertical, wave-exposed rocks, with tentacles outstretched to catch food from the currents.

Crevices and Caves

Irregularities in underwater rock features can provide additional habitats for marine life. Crevices and small caves provide shelter for nocturnal fish that hide during the day and are active at night. Elongated fish are well shaped to live in crevices, while fish that are active by day need holes to hide in at night and when predators approach. Deep, dead-ended caves contain a range of habitats, from sunlit, wave-exposed entrances to dark, still inner waters and sheltered sediments. Shrimp and squat lobsters occupy cave ledges, while animals that actively pump water to feed live in the quiet water inside the cave and coat the walls. Flashlight fish hiding in caves during the day signal to each other with light produced by bacteria in organs beneath their eyes. Small crevices are important because they form a refuge for small animals from sea urchins.

TAKING REFUGE
The flattened body of this spiny squat lobster enables it to retreat far into narrow crevices if threatened, and the spines help to wedge it in small spaces.

Storms and Scour

Shallow rocky reefs take the full force of waves during storms, but rock-living animals and seaweeds on open, exposed coasts are firmly attached and are generally well-adapted to cope with pounding waves. Larger seaweeds and animals will be torn from shallower rocks, making space for new life to settle, while many seaweeds and colonial animals can regrow from holdfasts or basal parts. However, few animals or plants survive on rolling boulders or on bedrock scoured by nearby sand and pebbles. Where rock meets sand, there is often a band of bare, sandblasted rock. Just above, tough-shelled animals such as keelworms survive, together with patches of hard encrusting calcareous red seaweeds. Above this, fast-growing colonial animals such as sponges and barnacles can colonize in the intervals between storms.

KEELWORMS
Keelworms have a hard, calcareous shell that protects their bodies from sand scour.

CORALLINE ALGAE
Like a coating of hard pink paint, encrusting coralline algae can withstand considerable scouring from nearby sand and pebbles.

Sandy Seabeds

MOST OF THE CONTINENTAL SHELF is covered with thick sediments, accumulated from millennia of land and coast erosion. The calcareous remains of marine life are continually added to the mix. Unlike deep-sea sediments (see pp.180–181), shelf sediments are stirred up by waves during storms, resuspending nutrients and profoundly affecting marine life and productivity. Sediments are largely the domain of animals, as seagrasses and seaweeds grow only in limited, shallow areas. Buried beneath the surface of a sandy seabed, there may be vast numbers of animals hiding from, or waiting for, prey.

Gravel and Sand

The coarsest sediments from coastal and land erosion are usually deposited inshore by rivers and glaciers as they enter the sea. Frequently shifted by waves and tides, clean, coarse sand and gravel make a difficult habitat; typical inhabitants include tough-shelled mollusks, sea cucumbers, burrowing urchins, and crabs. A wider range of organisms live in the more stable sand and gravel, where purple-pink beds of maerl can be found. This unattached, calcareous seaweed is made up of coral-like nodules. The open structure of live maerl twiglets is ideal for sheltering tiny animals, newly settled from the plankton, while the dead maerl gravel underneath supports burrowing animals. Beds of seagrass and green seaweeds thrive in shallow sand, harboring a wide range of life. Embedded shells and stones provide anchors for various seaweed species. Many fish have adapted to life on sandy seabeds, the most familiar being flatfish. Shallow-water anglerfish wave their fishing lures to tempt prey within striking distance of their huge mouths, while garden eels live permanently in sand burrows, partly emerging to eat plankton. Sand eels and cleaver wrasse dive into the sand to avoid predators.

coarse bristles (chaetae) on sides

feltlike dorsal chaetae

SANDY HABITAT
A marine segmented worm, the sea mouse lives in muddy sand.

GRAVEL DWELLER
This flame shell lives in a nest of gravel, pebbles, and shells. It pumps seawater through the nest, extracting food with its sticky, acidic tentacles.

EXPLOITING SANDY SEABEDS
Stingrays are among the many animals that hide in the sand—this southern stingray does so both to escape predators and to ambush prey.

Soft Mud

In sheltered waters in enclosed bays, estuaries, and fiords and in the deeper parts of the continental shelf, the finest particles of sediment settle as soft mud. Easily stirred up, the fine particles smother newly settled larvae and clog gills. There is little oxygen just below the mud surface, so buried animals must find ways to obtain oxygen from seawater. Despite these challenges, mud can be very productive. Bacteria and diatoms are often abundant on the mud surface, providing food for hoovering animals such as echiuran worms. Stable burrows are more easily built in mud than in sand or gravel. Animals such as sea pens and burrowing anemones anchor themselves in the mud, raising sticky polyps and tentacles to catch the raining plankton or to ensnare a passing fish or crustacean.

ANCHORED IN MUD
This sea pen's branches are covered with small polyps that feed on the plankton.

Mixed Sediments

Most sediments on the continental shelf are a mix of coarse and fine materials. An important part of these are calcareous fragments, derived from hard-shelled animals. Mixed sediments offer a wider range of building materials for tubes and burrows than sand or mud and are easier to traverse, so a far greater variety of animals live here. Seaweeds and hydroids cover the bed, attached to shells and pebbles. Visible life includes tube worms, brittlestars, and burrowing anemones; most of these withdraw into the sediment if threatened. Below the surface, hidden animals, including bivalves and crustaceans, provide a rich source of food for animals that can find and excavate it, such as starfish, crabs, and rays.

LIFE ON THE SEDIMENT
Its mouth fringed by tentacles, this half-buried sea cucumber (left) and a hermit crab inhabit these mixed sediments.

SEABED STABILIZERS

This seabed owes its luxuriant growth, including hydroids, soft corals, and brittlestars, to the many flame shells and horse mussels hidden under the surface. These mollusks bind the shifting sediments with strong threads, creating a stable, complex surface that many other animals can colonize. Flame shell nests join together to form extensive reefs, with holes for water exchange, so many other organisms can live inside and beneath the nests.

SEDENTARY HABIT
This Norway lobster lives in a U-shaped burrow with two exits and is mainly nocturnal.

DISCOVERY

SEA WRECKS

The complex shape and hard surfaces of shipwrecks such as the one shown here (the *Eagle,* off Florida) attract sedentary invertebrates and fish. A new wreck may take some time to become colonized, depending on the material from which it is made. Small hydroids, barnacles, and keelworms often settle first, paving the way for other animals and seaweeds to grow on their hard shells. Filter-feeders thrive in enhanced currents on the superstructure, while the spaces inside offer hiding places for fish and octopuses.

Beneath the Surface

Wave-disturbed sand and gravel creates a mobile, well-oxygenated environment. Animals that live here, such as crustaceans and echinoderms, move through the shifting sand without building permanent homes. Animals that disturb sediments in this way, or by ingesting and defecating it, are called bioturbators and are important recyclers of nutrients. Less-disturbed sediments are inhabited by sediment stabilizers. These sedentary animals, many living in permanent burrows or tubes, can cope with oxygen depletion and being covered over. Some strengthen their burrows by lining them with substances such as mucus and draw in seawater to supply food and oxygen. Others filter seawater or hoover sediment by extending their siphons to the surface. Microscopic creatures (the meiofauna) live in between the sand grains.

Seagrass Beds and Kelp Forests

SEAGRASS BEDS AND KELP FORESTS are very different habitats, but both are highly productive and contribute significantly to the total primary production of inshore waters. Seagrasses are the only fully marine flowering plants. They thrive in shallow, sunlit water on sheltered, sandy seabeds, primarily in warm water. Kelps are large brown seaweeds that grow as dense forests on rocks of the lower shore and subtidal zone, preferring cold water. Both of these ecosystems have a complex structure and provide shelter for a wide range of associated animals and seaweeds, some of them found nowhere else.

COLD-WATER KELP
Kelps are large brown seaweeds that live mainly in shallow subtidal zones.

ENDANGERED GRAZERS

Seagrasses are the primary food of green turtles, and the only food of manatees and dugongs. Globally, these animals are now endangered or vulnerable, threatened by the destruction of their feeding grounds. The coastal areas in which seagrass beds are found are often vulnerable to pollution. Runoff of nutrients and sediments from land affects water clarity, and is probably the biggest threat worldwide.

Seagrass Beds

Seagrasses are the only flowering plants (angiosperms) that live entirely in the sea, and they grow best in shallow, sandy lagoons or enclosed bays, where the water clarity is good. They are also tolerant of variable salinity. Unlike seaweeds, seagrasses have roots, which they use to absorb nutrients from within the sediments, thus recycling nutrients that would otherwise be locked up below the surface. Their intertwined rhizomes and roots help to stabilize the sand, protecting against erosion and encouraging the buildup of sediments. The productivity and complex physical structure of seagrasses attract a considerable diversity of associated species, some of which are only found in seagrass beds. A variety of seaweeds and sedentary animals, including species of hydroids, bryozoans, and ascidians, grow on the leaves. Seagrasses are also a critically important food for animals such as manatees, dugongs, green turtles, and many aquatic birds.

SEAGRASS MEADOWS
Seagrass meadows help to protect shallow sandy seabeds against erosion.

NATURAL CAMOUFLAGE
This greater pipefish's elongated shape and drab color make it hard to spot among seagrass leaves.

Kelp Forests

The term "kelp" was originally used to refer to the residue resulting from burning brown seaweeds, which was used in soap-making. It is now used more generally to refer to the many kinds of large brown seaweeds of the order Laminariales. Kelp forests grow best in colder waters, on shallow rocks with good water movement. The top edge of some kelp beds is visible at the lowest tides. Kelps grow densely on rock slopes down to around 30–70 ft (10–20 m) deep, depending on water clarity. In deeper water, there is less light for photosynthesis and kelps grow more sparsely; in most coastal waters they cannot survive below 80 ft (25 m). In exceptionally clear water, kelps can grow at 160 ft (50 m). Many kelp species have gas-filled floats, which hold the fronds up to the light and away from grazers. Within the kelp forest, waves are subdued and many organisms live in its shelter. Although kelp habitats support rich marine communities, only about 10 percent of kelp is eaten directly by animals; the rest enters the food chain as detritus or dissolved organic matter.

DISTRIBUTION MAP
Seagrass beds flourish in the tropics, while kelp forests thrive in cold, nutrient-rich waters, extending into the polar regions.

■ kelp forests
■ seagrass beds

COASTAL DEFENSES
A band of giant kelp can help to protect coasts from severe storms by absorbing wave energy.

Kelp Communities

Many kelps are treelike in shape, with a branched holdfast for attachment and a long stem (stipe), sometimes with floats, supporting a palmlike frond. This makes a kelp forest a multilayered environment in which different organisms live at different levels. Small spaces in the holdfast can harbor hundreds of small animals from predators. Some kelps have rough stipes covered with red seaweeds, although sea urchins and limpets may graze these in calm weather and in deeper water. Actively growing kelp fronds exude slime, which deters most animals from settling, but as growth slows later in the season, the fronds may become covered with a few species, particularly bryozoans, hydroids, and tube worms. These animals reduce the light reaching the fronds, and some kelps shed their fronds to get rid of unwanted settlers before growing new ones. The sea floor beneath the kelps may be covered with marine growth, or relatively barren if heavily grazed by sea urchins.

KELP ANEMONE
This large anemone is unusually mobile, and crawls or drifts up onto seaweed fronds to catch floating prey.

BLUE-RAYED LIMPET
At the end of the growing season, these limpets move down into the holdfast to avoid being discarded with the old frond.

Nurseries and Refuges

Seagrass beds and kelp forests are important refuges for young fish that need to hide from predators until they reach maturity. Many fish, such as the lumpsucker and swell shark, do not live among seagrasses or kelps as adults, but come into these habitats to spawn, giving their young a greater chance of survival. Small fish need small prey, and they find an abundance of food in the form of tiny worms, crustaceans, and mollusks among the seagrasses and in the sediment beneath, or in the undergrowth of kelp forests. These young fish are often unlike their parents, usually camouflaged in shades of green and brown to avoid detection. Some herbivorous fish from surrounding reefs come into seagrass beds only at night. Seagrass beds are important nurseries for some commercial invertebrates, including shrimp and cuttlefish.

LUMPSUCKER
This baby lumpsucker is very vulnerable. However, it is well camouflaged on kelp fronds, to which it attaches itself with a sucker.

DENSE KELP FOREST
Giant kelp is the world's biggest seaweed. Its stipes can be more than 100 ft (30 m) long, and it can grow as fast as 20 in (50 cm) per day.

OCEAN ENVIRONMENTS

OCEAN ENVIRONMENTS

Laguna de Términos

COASTAL TYPE Shallow lagoon	
WATER TYPE Tropical	
PRIMARY VEGETATION Seagrasses, seaweeds, and mangroves	

LOCATION In the southwest of the Yucatán Peninsula, Campeche State, Mexico

SATELLITE VIEW, WITH LAGOON AT TOP

Two channels connect this sediment-laden lagoon to the Gulf of Mexico, while three rivers feed in fresh water, producing a pronounced change in salinity. The seagrasses *Thalassia testudinum*, *Syringodium filiforme*, and *Halodule wrightii* cover 29 percent of the lagoon. With 448 recorded animal species, Términos is the most species-rich of Mexico's four large lagoons.

Sound of Barra

COASTAL TYPE Island chain with sounds	
WATER TYPE Cool	
PRIMARY VEGETATION Seagrasses, maerl, and kelp	

LOCATION Between South Uist, Eriskay, and Barra, Outer Hebrides, Scotland, UK

Strong tidal currents flow through the Sound of Barra, and its clear, shallow waters and sandy sea floor provide an ideal habitat for the eelgrass *Zostera marina*. The eelgrass beds, together with beds of maerl (see p.245), are home to many species of small animals. Such rich, current-swept communities in this part of Scotland are threatened by the building of rock causeways across the sounds, which cut off the nutrient-bearing currents that are essential for healthy growth. Eelgrass also grows in nearby brackish lagoons, together with the tasselweed *Ruppia maritima*, which is regarded by some scientists as a type of seagrass. Forests of the kelp *Laminaria hyperborea* grow on rocks at the edges of the sound, and these are home to abundant sea squirts and sponges.

EELGRASS STANDS
Healthy stands of eelgrass now thrive in the current-swept sound. Almost 90 percent of western Europe's eelgrass was lost to a wasting disease in the 1930s.

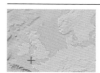

Falmouth Bay

COASTAL TYPE Rocky with inlets	
WATER TYPE Cool	
PRIMARY VEGETATION *Laminaria hyperborea* kelp, eelgrass	

LOCATION Southwest Cornwall, England, UK

The coastline of Falmouth Bay includes two drowned river valleys (rias), the Fal and Helford, which are now long, sheltered sea inlets. Because of their rich marine life, these inlets, together with part of Falmouth Bay, have been designated as a European marine Special Area of Conservation. Beds of the eelgrass *Zostera marina* and maerl (see p.245) in the inlets are home to a wide variety of animals, including the rare Couch's goby.

On the wave-exposed rocky coasts outside the inlets, the kelp *Laminaria hyperborea* grows in dense forests that support many associated seaweeds and animals. This kelp has a stiff stipe, which raises the frond off the seabed and means that the forest has developed well vertically. On the rock beneath the kelp, there is competition for space among anemones, sponges, and smaller seaweeds, while other animals hide in kelp holdfasts. The kelp stipes have a rough surface and provide effective attachment points for red seaweeds, bryozoans, soft corals, and other types of encrusting animals. On the fronds, tiny blue-rayed limpets graze, and colorful sea slugs eat small hydroids and lacy bryozoans.

In deeper water, the kelp *Laminaria ochroleuca* grows, close to its northern limit in Europe. This kelp is similar to *Laminaria hyperborea*, but has a smooth stipe on which little can grow. Two other kelps are found in the area, the sugar kelp (*Saccharina latissima*), which has a crinkled frond, and furbelows (*Saccorhiza polyschides*), which has a large, hollow holdfast and grows up to 13 ft (4 m) long in just one season.

CORNISH KELP FOREST
This forest of *Laminaria hyperborea* kelp has many different plants and animals living on the rocks beneath it, and on the kelp itself.

SOUTH ATLANTIC KELP FOREST
On the west coast of South Africa, sea bamboo is the largest of the local kelps. It can grow as tall as 50 ft (15 m).

Saldanha Bay

COAST TYPE Rocky and sandy bay with lagoon	
WATER TYPE Cool currents	
PRIMARY VEGETATION Kelp and eelgrass	

LOCATION Western Cape, South Africa

The cold Benguela Current flowing northward along the west coast of South Africa brings nutrient-rich water that is ideal for kelp growth, and sea bamboo (*Ecklonia maxima*) is abundant in Saldanha Bay. The split-fan kelp (*Laminaria pallida*) becomes dominant in deeper water. South Africa is famous for its diversity of limpets, and the kelp limpet *Cymbula compressa* is found only on sea bamboo. Its shell fits neatly around the stipe, where it grazes. The highly endangered limpet *Siphonaria compressa* occurs only in the bay's Langebaan Lagoon, grazing on the endemic eelgrass *Zostera capensis*.

INDIAN OCEAN WEST

Gazi Bay

COASTAL TYPE Shallow bay and fringing reef

WATER TYPE Tropical

PRIMARY VEGETATION Seagrasses and mangroves

LOCATION 30 miles (50 km) south of Mombasa, Kenya

Gazi Bay's shallow, subtidal mud and sand flats are sheltered by fringing coral reefs. Twelve species of seagrass grow on the mudflats, and these seagrass beds cover about half of the bay's 6 square miles (15 square km). Mangrove-lined creeks flow into the bay, and this unusual proximity of mangrove, seagrass, and coral reef systems has led to scientific studies on how they interact. The seagrass beds proved to be important in trapping particles washed into the bay from the creeks. Most were trapped within 1¼ miles (2 km) of the mangroves. The seagrass beds provide food directly for shrimp larvae, zooplankton, shrimp, and oysters, and they are the main feeding grounds of all the fish in the bay, making them very important to the health of the local fisheries.

MANGROVES AND SEAGRASS
Unusually for mangrove-lined creeks, the water here is clear and stands of seagrasses are able to flourish in the waterways leading into Gazi Bay.

INDIAN OCEAN EAST

Lombok

COASTAL TYPE Semisheltered bays on rocky coast

WATER TYPE Tropical

PRIMARY VEGETATION Seagrasses

LOCATION Lesser Sunda Islands, Indonesia

At least 11,600 square miles (30,000 square km) of seabed around Indonesia is covered by seagrasses. In the warm, shallow lagoons and bays, 12 species of seagrass flourish. Gerupuk Bay in the south of the island of Lombok contains 11 of the 12 Indonesian seagrass species, with *Enhalus acoroides* and *Thalassodendron ciliatum* forming dense stands. Analyses of the gut contents of fish that live among seagrass in Lombok's waters revealed that crustaceans were the dominant food source. However, a species of *Tozeuma* shrimp found there avoids the attention of predators by having an elongated body colored green with small white spots, a perfect camouflage against seagrass leaves. At low tide, local people use sharp iron stakes to dig for intertidal organisms. This damages the seagrass leaves and roots, thereby threatening the survival of the beds.

HUMAN IMPACT

THREAT FROM TOURISM

The islands of Southeast Asia contain the greatest diversity of seagrasses in the world, but human activity threatens them in many places. Tourism is a means of bringing a much-needed boost to many local economies and this necessitates the building of hotels and other tourist facilities in previously unspoiled areas.

HOTEL DEVELOPMENT
Future tourist development in the region may threaten the Lombok seagrass beds as a result of pollution and loss of habitat through the building of beach facilities, such as marinas.

BAY OF PLENTY
The seagrass beds in Lombok's bays are a source of seaweeds, sea urchins, sea cucumbers, mollusks, octopus, and milkfish for the area's inhabitants.

OCEAN ENVIRONMENTS

PACIFIC OCEAN WEST

Sea of Japan/East Sea

COASTAL TYPE
Mainly rocky

WATER TYPE
Warm to cold

PRIMARY VEGETATION
Kelp and seagrasses

LOCATION Off the west coast of the island of Hokkaido, northern Japan

The Sea of Japan/East Sea is influenced by the warm Tsushima Current from the south and the cold Liman Current from the north, so its marine flora is a rich mix of temperate and cold-water species because of the wide range of water temperatures in different parts of the coast. The mixing of these currents also provides plentiful nutrients for plant growth. Seagrass diversity is moderate, but eelgrasses are particularly well represented with seven species, several of them endemic to the area. Kelps are also diverse, with species of *Undaria*, *Laminaria*, and *Agarum* thriving in the colder waters in the north. Kelp is highly nutritious, and Hokkaido is the traditional center of kelp harvesting.

INVASIVE KELP

Since 1981, Asian kelp (*Undaria pinnatifida*) has spread from its indigenous sites in Japan, China, and Korea to four continents.

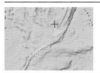
FEMALE RED PIGFISH IN KELP FOREST

PACIFIC OCEAN SOUTHWEST

Poor Knights Islands

COASTAL TYPE Offshore islands

WATER TYPE Temperate

PRIMARY VEGETATION Kelp and other brown seaweeds

LOCATION Off the east coast of Northland, North Island, New Zealand

In 1981, a marine reserve was set up around the Poor Knights Islands, extending 2,600 ft (800 m) out from the shore. The area is popular with divers for its caves and kelp forests. In the most exposed places, the kelp *Lessonia variegata* is predominant, while at more sheltered sites *Ecklonia radiata* is more abundant, together with the large brown seaweed *Carpophyllum flexuosum*. Large numbers of sea urchins dominate in some places.

INDIAN OCEAN EAST

Shark Bay

COASTAL TYPE Shallow, semienclosed bay

WATER TYPE Tropical; high salinity

PRIMARY VEGETATION Seagrasses

LOCATION Inlet of the Indian Ocean, north of Perth, Western Australia

Shark Bay is a UNESCO World Heritage Site containing one of the largest, most diverse seagrass beds in the world. Its 12 species of seagrass, which include *Amphibolis antarctica* and *Posidonia australis*, dominate the subtidal zone to depths of about 40 ft (12 m). The vast seagrass beds provide

SEAGRASS BANKS

Shark Bay has one of the world's largest seagrass beds, covering about 1,500 square miles (4,000 square km).

food for one of the world's largest populations of dugongs (see p.419), which are preyed on by sharks. The adjacent Hamelin Pool is too salty for seagrasses, but it is well known for the growth of stromatolites (see p.232).

POSIDONIA AUSTRALIS

PACIFIC OCEAN NORTHEAST

Izembek Lagoon

COASTAL TYPE Rocky coast and lagoon

WATER TYPE Cold; low salinity

PRIMARY VEGETATION Eelgrass and kelp

LOCATION On the northern side of the Alaskan Peninsula, Alaska, US

Izembek Lagoon covers 150 square miles (388 square km) of the Izembek State Game Refuge and is the site of one of the world's largest eelgrass beds. The eelgrass *Zostera* grows in dense beds here, both subtidally and on intertidal flats, where it is grazed by wading birds at low tide. Over half a million geese, ducks, and shorebirds stop over at the lagoon during migration to refuel on the eelgrass. On the rocky, open coasts outside the lagoon, kelp forests thrive in the cold water. The most common forest-forming kelp here is bull kelp (*Nereocystis luetkeana*), which can grow to 130 ft (40 m) in length. Bull kelp is an annual, which means that it reaches maturity within a single year. It grows quickly, at a rate of up to 5 in (13 cm) per day. The huge fronds, which have many long, strap-shaped blades, are supported by gas-filled bladders (pneumatocysts) that are up to 6 in (15 cm) in diameter.

FEEDING GROUNDS

After raising their young farther north, thousands of Brant geese graze on eelgrass in Izembek Lagoon in the fall before flying south to Baja California, Mexico.

PACIFIC OCEAN EAST

Monterey Bay Kelp Forest

COASTAL TYPE
Rocky and sandy

WATER TYPE
Cool to warm

PRIMARY VEGETATION
Kelp

LOCATION South of San Francisco, California, US

The California coast is famous for its beds of giant kelp (*Macrocystis pyrifera*), the largest seaweed on the planet (see p.238). It forms dense forests just offshore, and in Monterey Bay it outcompetes bull kelp for sunlight in many places, but the latter dominates in more exposed areas. Inshore of these giant species, other smaller kelps thrive. The kelp forests provide a unique habitat. Critically endangered giant sea bass (*Stereolepis gigas*) can be found in the kelp forest. This apex predator captures fish, crustaceans, and squid by suddenly opening its mouth and sucking its prey into the vacuum created. Sea otters (see p.402), which live among the kelp forests and eat sea urchins, are thought to be important in controlling the urchins, which graze on the kelp. Seagrasses of the genus *Phyllospadix* are also found in Monterey Bay. Unusually for seagrasses, they can attach to rock, and grow in the surf zone or in intertidal pools on rocky coasts. Each year, over 140,000 tons of giant kelp are harvested in California for the extraction of alginates, which are used in the textile, food, and medical industries.

SUNLIT FOREST
In exceptional circumstances, giant kelp can be 265 ft (80 m) long. The forests are at their thickest in late summer and decline during the dark winter months.

OCEAN ENVIRONMENTS

Coral Reefs

CORAL REEFS ARE SOLID STRUCTURES built from the remains of small marine organisms, principally a group of colony-forming animals called stony (or hard) corals. Reefs cover about 108,000 square miles (280,000 square km) of the world's shallow marine areas, growing gradually as the organisms that form their living surfaces multiply, spread, and die, adding their limestone skeletons to the reef. Coral reefs are among the most complex and beautiful of Earth's ecosystems, and are home to a fantastic variety of animals and other organisms; but they are also among the most heavily utilized and economically valuable. Today, the world's reefs are under pressure from numerous threats to their health.

Types of Reefs

Coral reefs fall into three main types: fringing reefs, barrier reefs, and atolls. The most common are fringing reefs. These occur adjacent to land, with little or no separation from the shore, and develop through upward growth of reef-forming corals on an area of continental shelf. Barrier reefs are broader and separated from land by a stretch of water, called a lagoon, that can be many miles wide and dozens of yards deep.

Atolls are large, ring-shaped reefs, enclosing a central lagoon; most atolls are found well away from large landmasses, such as in the South Pacific. Parts of the reef structure in both atolls and barrier reefs often protrude above sea level as low-lying coral islands—these develop as wave action deposits coral fragments broken off from the reef itself.

Two other types of reefs are patch reefs—small structures found within the lagoons of other reef types—and bank reefs, comprising various reef structures that have no obvious link to a coastline.

CORAL DIVERSITY
In this seascape off a Fijian island, groups of shoaling sea goldies hover over diverse species of coral, sponges, and other reef organisms.

FRINGING REEF
A fringing reef directly borders the shore of an island or large landmass, with no deep lagoon.

BARRIER REEF
A barrier reef is separated from the coast by a lagoon. In this aerial view, the light blue area is the reef and the distant dark blue area is the lagoon.

ATOLL
An atoll is a ring of coral reefs or coral islands enclosing a central lagoon. It may be elliptical or irregular in shape.

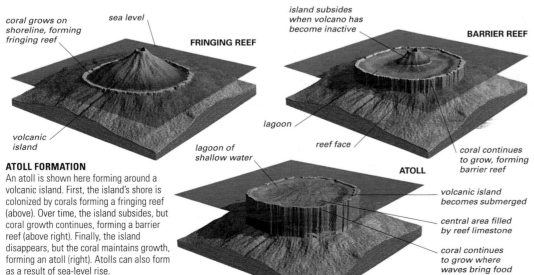

FRINGING REEF

coral grows on shoreline, forming fringing reef

sea level

volcanic island

island subsides when volcano has become inactive

BARRIER REEF

lagoon

reef face

coral continues to grow, forming barrier reef

ATOLL FORMATION
An atoll is shown here forming around a volcanic island. First, the island's shore is colonized by corals forming a fringing reef (above). Over time, the island subsides, but coral growth continues, forming a barrier reef (above right). Finally, the island disappears, but the coral maintains growth, forming an atoll (right). Atolls can also form as a result of sea-level rise.

lagoon of shallow water

ATOLL

volcanic island becomes submerged

central area filled by reef limestone

coral continues to grow where waves bring food

Reef Formation

The individual animals that make up corals are called polyps. The polyps of the main group of reef-building corals, stony corals, secrete limestone, building on the substrate underneath. The polyps also form colonies that create community skeletons in a variety of shapes. An important contributor to the life of these corals is the presence within the polyps of tiny organisms called zooxanthellae, which provide much of the polyps' nutritional needs. Other organisms that add their skeletal remains to the reef include mollusks and echinoderms.

Grazing and boring organisms also contribute, by breaking coral skeletons into sand, which fills gaps in the developing reef. Algae and other encrusting organisms help bind the sand and coral fragments together. Most reefs do not grow continuously but experience spurts of growth interspersed with quieter periods, which are sometimes associated with recovery from storm damage.

STONY CORAL
This group of branching hard corals is growing at a depth of about 16 ft (5 m) off the coast of eastern Indonesia. Individual stony corals can grow up to a few inches per year.

OPEN POLYPS
At the center of each polyp is an opening, the mouth, which leads to an internal gut. The tissue around the gut secretes limestone, which builds the reef.

Distribution of Reefs

Stony corals can grow only in clear, sunlit, shallow water where the temperature is at least 64°F (18°C), and preferably 77–84°F (25–29°C). They grow best where the average salinity of the water is 36 ppt (parts per thousand) and there is little wave action or sedimentation from river runoff. These conditions occur only in some tropical and subtropical areas. The highest concentration of coral reefs is found in the Indo-Pacific region, which stretches from the Red Sea to the central Pacific. A smaller concentration of reefs occurs around the Caribbean Sea. In addition to warm-water reefs, awareness is growing about other corals that do not depend on sunlight, and form deep, cold-water reefs—some of them outside the tropics (see p.178).

WARM-WATER REEF AREAS
The conditions needed for the growth of warm-water coral reefs are found mainly within tropical areas of the Indian, Pacific, and Atlantic Oceans. The reefs are chiefly in the western parts of these oceans, where the waters are warmer than in the eastern areas.

COLD-WATER CORAL
This species, *Lophelia pertusa*, is one of a few of the reef-forming corals that grow in cold water, at depths to 1,650 ft (500 m).

see p.178

HUMAN IMPACT

CORAL BLEACHING

Bleaching, or color loss, in corals occurs when the tiny organisms called zooxanthellae, which give corals their colors, are ejected from polyps. This can sometimes lead to the coral's death. Various stresses can cause bleaching, including ocean temperature rise (the main cause), chemicals in sunscreens, and ocean acidification (see p.67). In the past 20 years, many severe bleaching events have affected reefs over wide areas.

OCEAN ENVIRONMENTS

Parts of a Reef

Distinct zones exist on coral reefs, each with characteristic levels of light intensity, wave action, and other parameters. Each zone's characteristics determine the organisms that live there. The reef slope, or forereef, is the part that faces the sea. The upper parts of the reef slope are dominated by branching coral colonies and intermediate depths by massive forms. These are the areas of the reef with the greatest diversity of species.

At the top of the reef slope is the reef crest. This takes the brunt of the wave action and is subject to high light levels. Shoreward of the reef crest is the reef flat, a shallow, relatively flat expanse of limestone, sand, and coral fragments that may become exposed at high tide. The number of corals decreases toward the shore. Barrier reefs and atolls have a final zone, the lagoon area.

Species Diversity

In addition to reef-building corals, the warm, sunny waters of a reef are populated by a huge variety of other animals as well as seaweeds. The richest and healthiest reefs are home to thousands of species of fish and other marine vertebrates, such as turtles, while all the major groups of invertebrate animals are also represented.

These include sponges, worms, anemones, and non–reef-building corals (such as sea fans), crustaceans, mollusks (which include snails, clams, and octopuses), and echinoderms (sea urchins and relatives). Every nook and cranny of a reef is used by some animal as a hiding place and shelter. All the organisms in the reef are part of a complex web of relationships. Many organisms are also involved in mutualistic partnerships with other organisms, in which both species benefit.

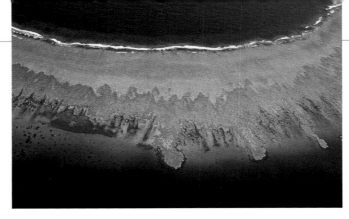

REEF CREST
In front of the reef crest (the uppermost, seaward part of a reef), spurs of coral sometimes grow out into the sea, separated by grooves.

sea urchin

crinoid

elkhorn coral

staghorn coral

maze coral

tube sponge

sea fan

star coral

lettuce coral

platelike star coral

finger coral

sea whip

QUEEN ANGEL FISH
One of hundreds of fish species found on the Caribbean reefs, this juvenile angelfish feeds on small crustaceans and algae.

TUBE SPONGES
Different species of sponges are found in many parts of the reef, including caves and cavities, as well as on the open reef slope.

REEF ZONES
The structure of a typical fringing reef, including forereef, reef crest, and reef flat, and some of the sea life that inhabits it, are shown here. The forereef has three zones, which are dominated by different coral forms: branching coral, massive coral, and platy coral. Individual corals are not shown to scale.

PLATY CORAL ZONE
Corals in this deep, dark part of the forereef expand horizontally to capture maximum sunlight, forming platelike colonies.

beach

small brain coral

seagrass

golf ball coral

sea anemone

SAND AND ALGAL ZONE
This area is dominated by sand and seagrass, which may harbor small marine life.

REEF FLAT
The animals living here must be able to endure high temperatures and salinity.

REEF CREST
The corals inhabiting this zone are invariably robust, as they must withstand energetic wave action.

BRANCHING CORAL ZONE
This zone is just below the reef crest and is dominated by corals with branching forms, such as staghorn coral.

MASSIVE CORAL ZONE
This central part of the forereef is usually dominated by massive corals—that is, colonies with rounded shapes.

SUBMARINE STUDY
Here researchers record the frequency of algal species on a reef in the Hawaiian Islands, using a camera, a frame for delineating areas of reef, and underwater writing implements.

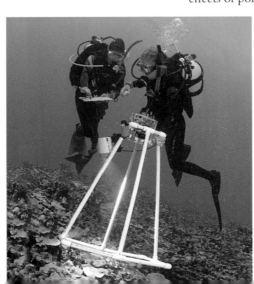

SEA URCHIN
Sea urchins graze on algae and are important in preventing algal overgrowth on coral reefs.

The Importance of Reefs

Coral reefs are of inestimable value for many reasons. First, they provide a protective barrier around islands and coasts: without the reefs, these would erode away into the ocean. Second, reefs are highly productive, creating more living biomass than any other marine ecosystem and providing an important food source for many coastal peoples. Third, they support more species per square unit area than any other marine environment. In addition to known coral-reef species, scientists estimate that there may be several million undiscovered species of organisms living in and around coral reefs. This biodiversity may be vital in finding new medicines for the 21st century—many reef organisms contain biochemically potent substances that are being studied as possible cures for arthritis, cancer, and other diseases. Finally, because of their outstanding beauty, reefs contribute to local economies through tourism, particularly attracting snorkelers and scuba-diving enthusiasts (see p.474).

REEF FISHING
Small-scale fishing using hand nets, often transported to a suitable site by canoe, is common throughout the Indian and Pacific Oceans, as shown here off Pantar Island in eastern Indonesia.

GOLDEN CRINOID
Crinoids, or feather stars, are related to starfish. They usually live in a hole or other shelter on the reef, extending their elegant arms to catch food.

crinoid arm

CORAL POISONING

One of the most destructive fishing practices, liable to kill corals over wide areas of reef, involves the use of poison to help catch tropical fish for the aquarium trade. This is practiced in parts of Southeast Asia such as the Philippines. The young boy photographed below, swimming at a depth of about 70 ft (20 m), carries a catch bag, net, and a squirt bottle containing a solution of sodium cyanide. The cyanide is used to immobilize selected reef fish, making them easier to catch, but kills all the living corals that it comes in contact with, taking a terrible toll on the health of the reef.

Vulnerable Reefs

Many types of stress can damage reefs and are doing so on a massive scale; the extent of the world's reefs has declined by about 50 percent since 1950. Some of the harm has been caused directly by human activity, including the effects of pollution (see pp.62-63), destructive fishing practices, and mining of reefs for materials. Natural disturbances include tropical storms and mass die-offs of animals that help to maintain reef health. The burning of fossil fuels indirectly damages reefs by increasing atmospheric levels of carbon dioxide (CO_2), causing a warming of the atmosphere and raising ocean temperatures. Atmospheric CO_2 is also absorbed by the oceans, leading to their acidification (see p.67). Reefs can recover from periodic natural traumas, but if they are subjected to multiple sustained stresses, they perish. It has recently been estimated that two-thirds of the world's warm-water reefs are at risk of disappearing in the near future.

OCEAN ENVIRONMENTS

Bermuda Platform

TYPE	Atoll with fringing and patch reefs
AREA	150 square miles (370 square km)
CONDITION	Healthy with only minor bleaching

LOCATION Northwest Atlantic, extending west and north of the islands of Bermuda

The Bermuda Platform is the elliptical, flattened summit of a huge volcanic submarine mountain (seamount) in the northwest Atlantic. Its surface lies 13–60 ft (4–18 m) below sea level and is covered in a thick layer of limestone, formed over millions of years from the

BOILER REEFS
These small reefs, close to the surface, are called "boilers" after their frothy appearance when waves break on them.

remains of corals and other organisms growing on the platform. Along the platform's southern and eastern edges, limestone sand has gradually built up to form the Bermuda islands. Coral reefs are present around the other edges of the platform, forming an atoll, while patch reefs grow on its central surface. The diversity of reef flora and fauna here is less than that associated with the reefs in the Caribbean Sea to the south.

Nevertheless, 21 different species of stony coral, 17 species of soft (non-reef-building) coral, including many spectacular purple sea fans, and about 120 different species of fish have been recorded here.

Lighthouse Reef

TYPE	Atoll with patch reefs
AREA	120 square miles (300 square km)
CONDITION	Partially impaired

LOCATION Western Caribbean, 60 miles (80 km) east of central Belize

Lighthouse Reef is an atoll lying 35 miles (55 km) east of the huge Belize barrier reef, off the coast of central Belize. It is roughly oval-shaped, about 23 miles (38 km) long, and 5 miles (8 km) wide on average.

Florida Reef Tract

TYPE	Barrier reef, patch reefs
AREA	400 square miles (1,000 square km)
CONDITION	Degraded

LOCATION From east of Soldier Key, Biscayne Bay, to south of the Marquesas Keys, Florida, USA

This system of coral reefs is 160 miles (260 km) long and curves to the east and south of the Florida Keys. Some geologists classify it as a barrier reef, others as a barrierlike collection of bank reefs. It is the largest area of coral reefs in the US and has a high biodiversity, being home to more than 40 species of stony coral, 500 species of fish, and hundreds of mollusk species. The reefs' health has declined

over the past 40 years due to human impact. Live coral cover has decreased, inhabitants that were once common (such as the queen conch) have largely disappeared, and mats of algae have expanded. Since 2014, there has been an ongoing outbreak of an affliction

termed stony coral tissue loss disease. Causes of this degradation include overfishing, fertilizer runoff from south Florida, increase in ocean temperature and sea level, and sewage pollution from boats. Other contributing factors include hurricane damage and declines in algae-grazing sea urchins. In an attempt to restore the health of the reefs, a number of coral nurseries have been established in the region.

CARYSFORT REEF
Carysfort Reef, part of the Florida Reef Tract, lies close to Key Largo and is the site of many ancient shipwrecks.

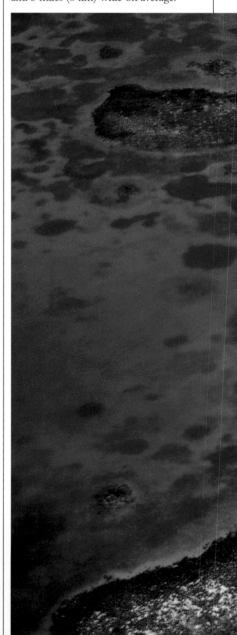

Bahama Banks

TYPE	Fringing reefs, patch reefs, barrier reef
AREA	1,200 square miles (3,150 square km)
CONDITION	Fair to Impaired

LOCATION Bahamas, southeast of Florida, US, and northeast of Cuba

The Bahamas is an archipelago of some 700 islands scattered over two limestone platforms, the Little Bahama and Great Bahama Banks, in the West Indies. The platforms have been accumulating for at least 70 million years—the Great Bahama Bank is over 15,000 ft (4,500 m) thick—yet their surfaces remain 33–80 ft (10–25 m) below sea level. Many of the islands have fringing coral reefs; there are also many patch reefs on the Banks and a

barrier reef near the island of Andros. The reefs are home to a range of corals and coral reef-dwelling animals that is typical for the western tropical Atlantic. A 2020 report categorized the condition of most of the Bahamas reefs as fair and the rest as impaired. The leading threats to the reefs were listed as hurricanes, coral bleaching events, coastal development, unregulated fishing practices, and stony coral tissue loss disease.

HARD AND SOFT CORALS
This diverse group of corals, including a large purple sea fan, was photographed off the island of New Providence.

Like all atolls, it is bounded by a ringlike outer structure of coral formations, many of which break the surface. These form a natural barrier against the sea and surround a lagoon, which sits on top of a mass of limestone. The lagoon is relatively deep but contains numerous patch reefs along with six small, sandy, low-lying islands, or cays (one containing a dive center). At its center is Lighthouse Reef's most remarkable feature—a large, almost circular sinkhole in the limestone, known as the Great Blue Hole. Approximately 407 ft (124 m) deep, this feature formed some 18,000 years ago during the last ice age, when much of Lighthouse Reef was above sea level. At that time, freshwater erosion

produced a complex of air-filled caves and tunnels in the limestone. At some point, the ceiling of one of the caves collapsed, producing what is now the entrance to the Blue Hole. Later, as sea level rose, the cave complex flooded.

Elsewhere, the atoll boasts some areas of healthy reef, but others have been significantly damaged by unregulated fishing and a severe coral bleaching event that occurred in 2016. As well as patch reefs within the atoll, around its margins are many coral-encrusted walls (dropoffs) that descend to depths of several hundred yards. Lighthouse Reef exhibits a biological diversity typical of the region; it is home to some 200 fish species and 60 species of stony corals.

HUMAN IMPACT

DIVING THE GREAT BLUE HOLE

The Great Blue Hole is one of the world's most exciting dive sites. It is not recommended for the fainthearted (as sharks are commonly encountered) or for novice divers (because perfect buoyancy control is needed). At 125 ft (38 m) depth, an array of impressive ancient stalactites can be seen hanging from the slanting walls of the hole. The entrance to a system of caves and tunnels lies a few yards farther down.

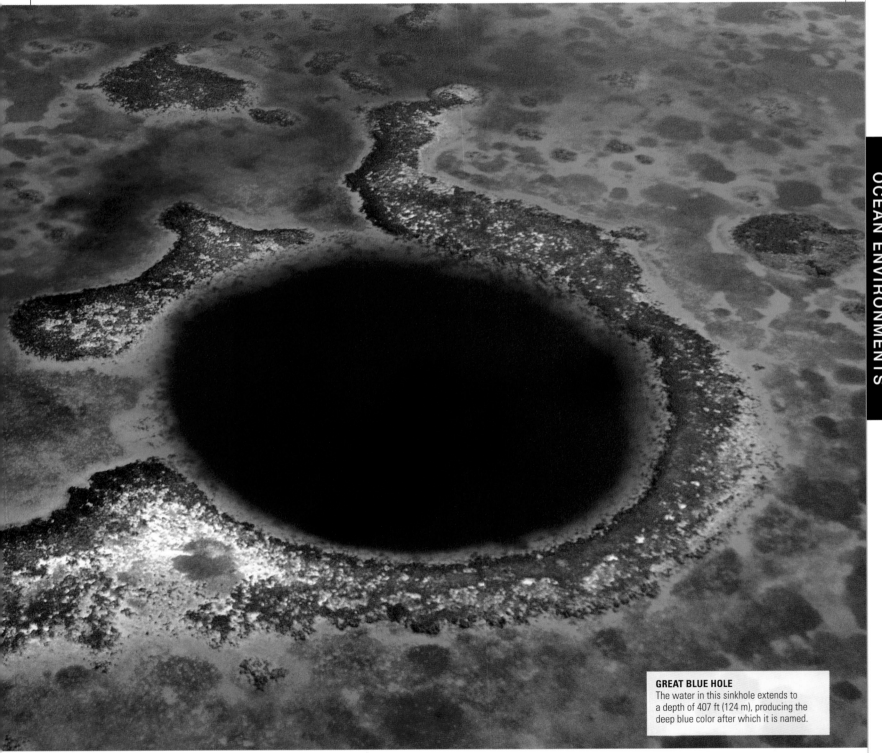

GREAT BLUE HOLE
The water in this sinkhole extends to a depth of 407 ft (124 m), producing the deep blue color after which it is named.

OCEAN ENVIRONMENTS

OCEAN ENVIRONMENTS

 INDIAN OCEAN NORTHWEST

Red Sea Reefs

TYPE	Fringing, patch, and barrier reefs; atolls
AREA	6,300 square miles (16,500 square km)
CONDITION	Localized areas of severe damage

LOCATION Red Sea coasts of Egypt, Israel, Jordan, Saudi Arabia, Sudan, Eritrea, and Yemen

The Red Sea contains arguably the richest, most biologically diverse, and most spectacular coral reefs outside Southeast Asia. The coral reefs in the northern and southern areas of the sea differ considerably. In much of the northern section, the coasts shelve extremely steeply and there are few offshore islands. The coral reefs here are mainly narrow fringing reefs, with reef flats typically only a few yards wide, and slopes that plunge steeply toward the sea floor. In the south, off Eritrea and southwestern Saudi Arabia, is a much wider area of shallow continental shelf. Many of the reefs in this area surround offshore islands, and there are fewer steep dropoffs. The southern Red Sea also receives a continuous inflow of water from the Gulf of Aden to its south that is high in nutrients and plankton, making the waters more cloudy, which restricts reef development. Live coral cover throughout the Red Sea reefs is generally high, at about 60–70 percent, as is the diversity of corals, fish (including the famous Red Sea lionfish), and other reef organisms.

Although many reef areas are healthy, others have been damaged by intensive diving tourism and deposition of untreated sewage. A large-scale tourism project (including a new city named Neom) being planned in Saudi Arabia will put further pressure on part of the reef system. Against that, in 2020, it was discovered that some coral species in the northernmost part of the Red Sea possess a natural tolerance to rising sea temperatures.

GULF OF AQABA REEF
Groups of little red fish of the genus *Anthias* fluttering around hard coral heads, or colonies, are a familiar sight on Red Sea reefs.

 INDIAN OCEAN NORTHWEST

Aldabra Atoll

TYPE	Atoll
AREA	60 square miles (155 square km)
CONDITION	Excellent, although it has suffered some coral bleaching

LOCATION Western extremity of the Republic of Seychelles archipelago, northwest of Madagascar

At 20 miles (34 km) long and 9 miles (14.5 km) wide, Aldabra is the largest raised coral atoll in the world. The term "raised" refers to the fact that the limestone structures forming its rim, which originated from coral reefs, have grown into four islands that protrude as much as 27 ft (8 m) above sea level. Situated on top of an ancient volcanic pinnacle, the islands enclose a shallow lagoon, which partially empties and then fills again twice a day with the tides. Because of its remote location, and its status as a Special Nature Reserve and (since 1982) UNESCO World Heritage Site, Aldabra has escaped the worst of the stresses that human activities have placed on most of the world's coral reefs. Although, in common with many Indian Ocean locations, the atoll was affected by coral bleaching events in 1997–1998 and 2015–2016, its external reefs are in a near-pristine state. They are rich in marine life, featuring large schools of reef fish, green and hawksbill turtles, forests of yellow, pink, and purple sea-fans, groupers, hammerhead sharks, and barracuda. The atoll's inner lagoon contains numerous healthy patch reefs, is fringed by mangrove swamps, and is inhabited by turtles, parrotfish, and eagle rays.

On land, Aldabra is famous for its giant tortoises; rare exotic birds such as a subspecies of white-throated rail; and giant coconut crabs, which can grow up to 3 ft (1 m) in length.

MUSHROOM ROCK
Strong tidal flows of ocean water into and out of Aldabra's lagoon have sculpted some raised clumps of old reef into mushroom-shaped islets known as champignons.

 INDIAN OCEAN WEST

Bazaruto Archipelago

TYPE	Fringing reefs, patch reefs
AREA	60 square miles (150 square km)
CONDITION	Generally good; some damage

LOCATION Southeastern coast of Mozambique, northeast of Maputo

The Bazaruto Archipelago is a chain of sparsely populated islands on the coast of Mozambique, formed where sand was deposited over hundreds of thousands of years by the Limpopo River. A Marine National Park, established in 2001, covers most of the archipelago, protecting its impressive fringing reefs and kaleidoscopic range of marine life. More than 2,000 fish species, 100 species of stony corals, and 27 dazzling soft-coral species, including unusual "green tree" corals, are found on Bazaruto's reefs, as well as eagle rays, manta rays, and five species of turtles. The archipelago is also a refuge for one of the remaining populations of dugongs (see p.419) in the western Indian Ocean.

REEF SAFARI
A peaceful way of visiting the shallow, crystal-clear waters around the Bazaruto reefs is on a dhow, as part of a reef safari.

INDIAN OCEAN CENTRAL
Diego Garcia Atoll

TYPE Atoll

AREA 17 square miles (44 square km)

CONDITION Recovering from a severe coral bleaching event in 2016

LOCATION Chagos Archipelago, central Indian Ocean, southwest of Sri Lanka

This atoll, best known as a US military base, is also home to one of the world's largest populations of breeding sea birds. The reefs around the atoll's edges and within its central lagoon are home to 220 species of stony coral. In 2010, the Chagos Archipelago, of which Diego Garcia forms a part, was declared a "no-take" marine reserve, making it the largest marine protected area in the world.

WESTERN SIDE OF DIEGO GARCIA

INDIAN OCEAN CENTRAL
Maldives

TYPE Atolls, fringing reefs

AREA 3,500 square miles (9,000 square km)

CONDITION Recovering from coral bleaching

LOCATION Off southern India, southwest of Sri Lanka, in the Indian Ocean

The Maldives are a group of 26 atolls, many of them very large, in the Indian Ocean. The majority are composed of numerous separate reefs and coralline islets (some 1,200 in all), arranged in ringlike structures. Within most of the atoll lagoons, which are 60–180 ft (18–55 m) in depth, there are usually many patch reefs and numerous structures called faros, which are rare outside the Maldives. These look like mini-atolls and consist of roughly elliptical reefs with a central lagoon.

Most of the Maldivian atolls are themselves arranged in a large, elliptical ring, some 500 miles (800 km) long and 60 miles (100 km) wide. The reefs that fringe all the Maldivian atolls, islets, and faros contain more than 200 species of colorful stony coral, more than 1,000 different fish species, and are abundant in other marine life. Groupers, snappers, and sharks, for example, are frequently encountered.

The coral reefs of the Maldives suffered severe coral bleaching events in 1998 and 2016. After the 1998 event, more than 90 percent of stony corals died, and it took 16 years for a full recovery. Some recovery has now been reported from the 2016 event.

ATOLLS WITHIN ATOLLS
The numerous ringlike structures in this aerial view are faros—mini-atolls within a larger Maldivian atoll.

HUMAN IMPACT
ATOLL CITY

Male, the Maldives' capital city, covers the entire surface area of a coral island that forms part of an atoll rim. Its reef has been mined to provide building materials for artificially extending the island.

The partly dismantled reef leaves the island poorly protected from storms, so a sea wall has been built around much of its perimeter, preventing major damage during the 2004 Indian Ocean tsunami.

INDIAN OCEAN NORTHEAST
Andaman Sea Reefs

TYPE Fringing reefs

AREA 2,000 square miles (5,000 square km)

CONDITION Some areas poor due to coral bleaching, diver damage

LOCATION Andaman Sea coasts: Thailand, Myanmar, Andaman and Nicobar Islands, Malaysia, Sumatra

Most Andaman Sea reefs are fringing reefs around islands off the coasts of Thailand and Myanmar or, in the northwest, off the eastern coasts of the Andaman and Nicobar islands—the site of the largest continuous area of reefs in south Asia. About 200 coral species and more than 500 fish species have been recorded here. Repeated coral bleaching events in 1993, 1998, 2010, and 2016 due to abnormally high sea surface temperatures have caused significant damage to the region's reefs, and they are only slowly recovering. The 2004 Indian Ocean tsunami caused relatively little damage. Other threats to these reefs include collection of marine life for aquariums, destructive fishing techniques, siltation caused by badly managed deforestation on some of the islands, and anchor damage from dive boats.

SOFT CORAL COLONIES
These soft corals and glass fish, which are almost transparent, were photographed off southwest Thailand.

Shiraho Reef

TYPE	Fringing reef
AREA	4 square miles (10 square km)
CONDITION	Recovering from severe coral bleaching event in 2016

LOCATION Southeast coast of Ishigaki Island, at the southwestern extremity of Japanese archipelago

Shiraho Reef, off Ishigaki Island, part of the Japanese archipelago, came to notice in the 1980s as an outstanding example of biodiversity, with some 120 species of coral and 300 fish species concentrated in a few square miles. The reef also contains the world's largest colony of rare Blue Ridge Coral (*Heliopora coerulea*). For decades, environmentalists battled to save the reef from a proposal to build an airport on top of it. This battle was eventually won in the 1990s, but since then, the health of the reef has been put under pressure from typhoon damage and severe coral bleaching events in 1998, 2007, and 2016.

BLUE RIDGE CORAL
Despite its name, the color of this coral varies from violet through blue, turquoise, and green to yellow-brown. Its branching vertical plates can form massive colonies.

Tubbataha Reefs

TYPE	Atolls
AREA	130 square miles (330 square km)
CONDITION	Suffered some mild to moderate coral bleaching in 2020

LOCATION Central Sulu Sea, between the Philippines and northern Borneo

The Tubbataha Reefs lie mainly around two atolls in the center of the Sulu Sea and are famous for the many large pelagic (open ocean) marine animals attracted to them—such as sharks, Manta Rays, turtles, and barracuda. The steeply shelving reefs here are also rich in smaller life, including many species of crustaceans, colorful nudibranchs (sea slugs), and more than 350 species of stony and soft coral.

In 1988, the Philippines government declared the area a National Marine Park, and since 1993, it has also been a UNESCO World Heritage Site. For many decades, scuba divers have rated the reefs at Tubbataha as among the top 10 dive sites in the world, and they have generally remained in excellent condition. Their health has been attributed partly to the enforcement of measures such as a prohibition on fishing and a ban on boats anchoring on the reefs. (Craft must use mooring buoys.) A setback occurred in 2013, however, when a US Navy minesweeper ran aground on an area of reef, damaging over 21,500 square ft (2,000 square m). A further setback occurred in 2020, when mild to moderate coral bleaching was observed in some reef areas.

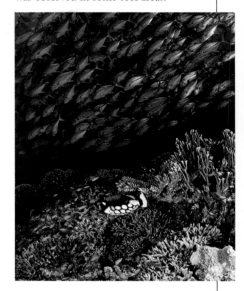

CORAL DROP-OFF
In this photograph of a steeply shelving reef slope, several species of soft coral are visible, together with a shoal of Bigeye Trevally.

Nusa Tenggara

TYPE	Fringing reefs, barrier reefs
AREA	2,000 square miles (5,000 square km)
CONDITION	Damaged by fishing practices

LOCATION Southern Indonesia, from Lombok in the west to Timor in the east

Nusa Tenggara is a chain of around 500 coral-fringed islands in southern Indonesia. The northern islands are volcanic in origin, while the southern islands consist mainly of uplifted coral limestone. Many of the reefs have been only rarely explored. However, what surveys have been carried out indicate an extremely high diversity of marine life in this region. For example, a single large reef can contain more than 1,200 species of fish (more than in all the seas in Europe combined), and 500 different species of reef-building coral. Common animals here include Eagle Rays, Manta Rays, Humphead Parrotfish, and various species of octopuses and nudibranchs (sea slugs). Major threats to the reefs in Nusa Tenggara include pollution from land-based sources, removal of fish for the aquarium trade, and reef destruction by blast fishing. Coral bleaching affected some reefs in 2010.

REEF FLAT OFF PANTAR ISLAND
This shallow reef area, featuring numerous species of stony coral and a starfish, is in east Nusa Tenggara.

PACIFIC OCEAN SOUTHWEST

Great Barrier Reef

TYPE Barrier reef

AREA 14,300 square miles (37,000 square km)

CONDITION Significantly degraded by factors such as a series of coral bleaching events

LOCATION Parallel to Queensland coast, northeastern Australia

Australia's Great Barrier Reef, which stretches 1,250 miles (2,010 km), is the world's largest coral reef system. Often described as the largest structure ever made by living organisms, it in fact consists of some 3,000 individual reefs and small coral islands. Its outer edge ranges from 18 to 155 miles (30 to 250 km) from the mainland, and its biological diversity is high. The reef contains about 350 species of stony coral and many of soft coral. Its 1,500 species of fish range from 45 species of butterflyfish, to several shark species, including silvertip, hammerhead, and whale sharks. In 1975, the Great Barrier Reef Marine Park was established, and in 1981, it was

REEF CHANNEL
In this view of a central area of the reef, a deep, meandering channel separates two reef platforms. The region's high tidal range drives strong currents through such channels.

declared a World Heritage Site. Sadly, the reef has lost more than half of its corals since 1995 due to stresses such as pollution, predation by the crown-of-thorns starfish, and several coral bleaching events. A steep decline in the reef's health came after widespread bleaching in 2016–2017, and more damage occurred in 2020.

THE WORLD'S SMALLEST MARINE FISH?

One of the tiniest residents of the Great Barrier Reef, at just less than ⅓ in (7–8 mm) long from snout to tail, is the Stout Infantfish. When discovered in 2004, the Infantfish was declared to be the world's smallest vertebrate. That title has since been claimed for a type of frog found in Papua New Guinea, but the infantfish still vies with a species of Indonesian cyprinid for the title of world's smallest fish.

PACIFIC OCEAN SOUTHWEST

Marshall Islands

TYPE Atolls

AREA 2,400 square miles (6,200 square km)

CONDITION Recovering from severe coral bleaching event in 2016

LOCATION Micronesia, southwest of Hawaii, western Pacific

The Marshall Islands consist of 29 coral atolls and five small islands in the western Pacific. The atolls lie on top of ancient volcanic peaks that are thought to have erupted from the ocean floor 50-60 million years ago. They include Kwajalein, the largest atoll in the Pacific at 1,000 square miles (2,500 square km), and Bikini and Enewetak atolls, which were used by the US for testing nuclear weapons between 1946 and 1962. Direct human pressures on these two remote, evacuated atolls have been minimal during the past 60 years, and marine life has recovered, but the islands did not escape the effects of a worldwide coral bleaching event in 2016.

MAJURO ATOLL
As with many Pacific atolls, the rim of Majuro Atoll consists partly of shallow submerged reef and partly of small, low-lying islands.

PACIFIC OCEAN SOUTHWEST

Society Islands

TYPE Fringing reefs, barrier reefs, atolls

AREA 600 square miles (1,500 square km)

CONDITION Some coral bleaching in 2019

LOCATION French Polynesia, northeast of New Zealand, south-central Pacific

The Society Islands comprise a chain of volcanic and coral islands in the South Pacific, including islands with barrier reefs (such as Rai'atea), islands with both fringing and barrier reefs (such as Tahiti), and atolls or near-atolls (such as Maupihaa and Maupiti). The reefs' biological diversity is moderate compared with the reefs of Southeast Asia, although more than 160 coral species, 800 species of reef fish, 1,000 species of mollusks, and 30 species of echinoderm have been recorded. The reefs' health is generally good, but some reefs around the busy holiday destination islands of Tahiti, Moorea, and Bora Bora have been affected by construction, sediment runoff, and coral bleaching.

MOOREA
A wide fringing reef almost completely surrounds the shoreline of mountainous Moorea, part of which is visible in this view.

PACIFIC OCEAN CENTRAL

Hawaiian Archipelago

TYPE Fringing reefs, atolls, submerged reefs

AREA 450 square miles (1,180 square km)

CONDITION Widespread coral bleaching in 2019

LOCATION North-central Pacific

The Hawaiian Archipelago consists of the exposed peaks of a huge undersea mountain range. These mountains have formed over tens of millions of years as the Pacific Plate moves northwest over a hotspot in the Earth's mantle. Coral reefs fringe some coastal areas of the younger, substantial islands at the southeastern end of the chain, such as Oahu and Molokai. To the northwest, located on the submerged summits of older, sunken islands, are several near-atolls (such as the French Frigate Shoals) and atolls (such as Midway Atoll). These reefs are highly isolated from all other coral reefs in the world, and although their overall biological diversity is relatively low, many new species have evolved on them. The more remote reefs are healthy, but in 2013, a serious coral disease was reported affecting reefs on Oahu and Kauai.

FRENCH FRIGATE SHOALS
Reef fish, including Longfin Bannerfish, Milletseed Butterflyfish, and Bluestripe Snappers, swim around a table coral.

OCEAN ENVIRONMENTS

THE GREAT BARRIER REEF
The warm, clear waters of the reef support an astonishing variety of life. Here, fairy basslets can be seen shoaling over vividly colored soft corals. Bright coloration can serve several purposes for reef fish, including helping members of a species recognize each other and acting as a warning to predators.

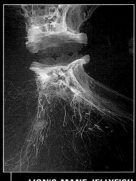

LION'S-MANE JELLYFISH
This daunting giant of the plankton can grow up to 6 ft (2 m) across, with 200-ft (60-m) tentacles.

The Pelagic Zone

THE PELAGIC ZONE IS THE WATER COLUMN ABOVE the continental shelf (although the term is also used to refer to the water column of the open ocean). It is a vast environment, and temperature and salinity variations within it result in distinct water masses. These are separated by "fronts" and characterized locally by different plankton. Coastal and shelf waters are more productive than the open ocean. When calm, the water stratifies, cutting off the surface plankton from essential nutrients in the layers below. Storms cause the layers to mix, stimulating phytoplankton blooms. High latitudes have seasonal plankton cycles; in warmer waters, seasonal upwelling of nutrient-rich deeper water triggers phytoplankton growth.

Microscopic Productivity

Much of the primary productivity in the world's oceans and seas occurs over the continental shelves. Tiny phytoplankton floating in the surface waters harness the Sun's energy through photosynthesis to produce living cells. Some of the tiniest algae (picoplankton) are thought to supply a considerable amount of primary production. As well as sunlight, nutrients and trace metals are needed for phytoplankton growth. These are often in short supply in the open ocean, but shelf waters benefit from a continual input from rivers, mixing by waves and, on some coasts, the upswelling of nutrient-rich water.

PRIMARY PRODUCTION
This satellite map shows variations in primary production, indicated by the concentration of the pigment chlorophyll a in the oceans and the amount of vegetation on land.

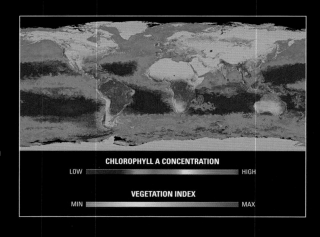

CHLOROPHYLL A CONCENTRATION
LOW   HIGH

VEGETATION INDEX
MIN   MAX

The Plankton Cycle

In temperate and polar seas, optimal phytoplankton growth occurs in both spring and summer. There are long daylight hours and maximum nutrient levels after winter storms have mixed the water column and resuspended dissolved nutrients from the seabed. The well-known spring blooms can rapidly turn clear seawater into pea soup, or a variety of other colors, depending on the organism. Typically, there is a succession of phytoplankton species with short blooms. Responding to abundant food and increasing temperatures, tiny zooplankton begin grazing the phytoplankton and reproducing. Bottom-living coastal animals release clouds of larvae to feed in the nutritious broth, before taking up life on the seabed. Spawning fish also contribute a mass of eggs and larvae. Eventually, the phytoplankton is grazed down, nutrients are exhausted, and productivity drops off, in an annual cycle that will be renewed again next spring.

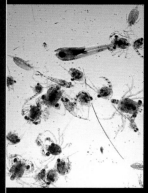

DRIFTING ZOOPLANKTON
Continental shelf zooplankton contains many larvae of seabed animals that then drift away to new areas.

Riding the Currents

From tiny algae to giant jellyfish, the animals and plants of the plankton either float passively or swim weakly. This is mainly to keep them up in the sunlit surface waters, where most production occurs; these drifters must go wherever the currents take them. On most continental shelves, there is a residual drift in a particular direction, although wind-driven surface currents, where most of the plankton live, can move in any direction for short periods. Some animals go on long migrations to spawn, relying on residual currents to bring their larvae back to areas suitable for their growth into adults; for example, conger eel larvae take around two years to drift back from their spawning grounds far off the continental shelf. The larvae of the majority of coastal animals, including those of barnacles, mussels, hydroids, and echinoderms, spend much shorter periods in the plankton—just long enough to disperse to new areas of coast. However, the plankton is a dangerous place, full of hungry mouths and tentacles, and though millions of eggs and larvae are released, the vast majority of planktonic feeders will die; only a lucky few find a suitable place to settle and grow.

COLONIZED ROPE
This rope was colonized over the course of a year by sea squirts, feather stars, fan worms, and anemones, their planktonic larvae having been transported by ocean currents.

RED TIDE

A rapid increase in a population of marine algae is called a bloom. This bloom on the Scottish coast, known as a red tide, was caused by the dinoflagellate *Noctiluca scintillans*. Sometimes blooms poison marine life. Often, the sheer numbers of organisms clog fish gills, suffocating them. Dense blooms occur naturally, but man-made pollution from nutrient runoff into the sea may also feed these blooms, making them more frequent and extensive.

MIGRATORY SHOALS
Pelagic fish such as these mackerel move around the ocean in response to temperature changes. They are among the pelagic zone's larger predators.

Active Swimmers

The animals of the plankton, especially small crustaceans such as copepods and krill, are eaten by fish, mainly small, shoaling species such as herring, sand eels, sardines, and anchovies. Most of these fish live permanently in midwater, using the seabed only to spawn or to avoid predators. They are strong swimmers (nekton), using speed to catch prey and evade predators. They can travel long distances against residual currents to feed and also to reach their spawning grounds. Small, shoaling fish are, in turn, food for larger predators, such as squid, tuna, cetaceans, and sharks. Whale sharks, basking sharks, and baleen whales are among the largest of the marine animals, yet they feed directly on plankton, consuming vast quantities.

NEKTONIC INVERTEBRATE
Squid are the only invertebrates that swim strongly enough to be classed as nekton. They catch a variety of prey including fish and planktonic crustaceans.

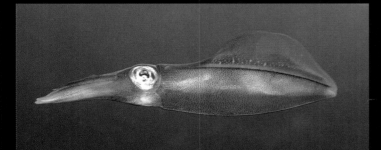

Pelagic Fisheries

Continental shelf waters support massive quantities of pelagic fish, ultimately sustained by abundant plankton. The most important fisheries are for herring, sardines, anchovies, pilchards, mackerel, capelin and jackfish. Squid are also fished commercially. Many fish stocks are under severe pressure as boats and nets get bigger and the technology to pinpoint shoals becomes ever more sophisticated. Pelagic fish and squid are caught in drift nets, which hang down from the surface. In the north Pacific, many tens of thousands of miles of drift net are used by major fisheries; unfortunately, these nets also trap dolphins, sea lions, turtles, and diving birds. Drifting longlines are used for tuna and swordfish; these also catch juvenile fish, sharks, turtles, and seabirds. Midwater trawls capture vast quantities of shoaling fish such as herring, mackerel, and sardines. Small-scale fisheries for a wide variety of other pelagic species are important in sustaining local coastal communities worldwide.

FOOD CHAIN THREAT
Sand eels are food for sea-birds (such as this Arctic tern), seals, cetaceans, and larger fish. Despite their importance at the base of many food chains, vast quantities are taken by fisheries for feeding to livestock and farmed fish, and are burned as fuel oil.

THE STEEL-BLUE WAVES of the open ocean conceal an extraordinary landscape, where the continents plunge down to a vast, undulating, muddy plain. Here, the ocean water column supports layer upon layer of life, from the surface zone, powered by sunlight, to the crushing pressure of the darkest depths. In places, the abyssal plain is broken by underwater volcanoes and mountain ranges high enough to rival the Himalayas. Springs of super-hot water emerge from these mountainsides, supporting living communities unlike any others on the planet. Elsewhere, Earth's vast tectonic plates collide, ripping trenches in the ocean floor. Incredibly, few people have explored these mysterious depths, but the integrity of the deep sea is threatened by plans for human exploitation.

THE OPEN OCEAN AND OCEAN FLOOR

BENEATH THE WAVES
In the deepest ocean, an underwater Mount Everest could be hidden beneath these waves—and still leave plenty of space to put the Burj Khalifa (the world's tallest building) on top. As a result, we have better maps of the Moon than of the deep seabed.

Zones of the Open Ocean

CONDITIONS IN THE OCEAN vary greatly with depth. Light and temperature changes occur quickly, while pressure increases incrementally. Although many of these changes are continuous, the ocean can be divided into a series of distinct depth zones, each of which produces very different conditions for living things.

The Surface Layer

The top three feet of the ocean is the richest in nutrients. This upper layer is sometimes called the neuston, although this term is also used for the animals that live there, such as jellyfish. Amino acids, fatty acids, and proteins excreted by plants and animals float up into this surface layer, as do oils from the decomposing bodies of dead animals. These produce a rich supply of nutrients for phytoplankton.

The top three feet of seawater is also the interface where gas exchange takes place between the ocean and the atmosphere. This is vitally important to all life on Earth, as half of the oxygen animals need for survival comes from the ocean. Not surprisingly, phytoplankton gathers in this surface zone in daylight, as do the animals that feed on them. This zone is also highly susceptible to chemical pollution and floating litter, which can be deadly for marine life.

NOCTURNAL AND DIURNAL DISTRIBUTION
Only a small proportion of marine life inhabits the deep zone; the majority live above 3,300 ft (1,000 m). The sunlit zone is dangerous for animals—many stay in the twilight zone by day and only go upward at night. The sunlit zone is much emptier by day.

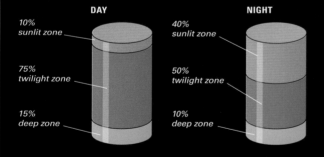

DAY
10% sunlit zone
75% twilight zone
15% deep zone

NIGHT
40% sunlit zone
50% twilight zone
10% deep zone

HUMAN IMPACT

FREE DIVING

When divers breathe compressed air underwater, excess nitrogen dissolves in their blood, and they risk the bends if they surface too fast. Free divers avoid this by holding their breath underwater. Pressure squeezes their lungs, but the surrounding blood vessels swell to protect them, and blood nitrogen levels stay safe. Trained free divers can hold their breath long enough to reach 660 ft (200 m), using aids to help them descend and ascend.

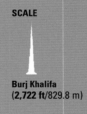

OCEAN ZONES

SUNLIT ZONE 0–660 ft
Seawater rapidly absorbs sunlight, so only one percent of light reaches 660 ft (200 m) below the surface. Phytoplankton use the light to photosynthesize, forming the base of food chains. This zone drives all ocean life.

TWILIGHT ZONE 660–3,300 ft
Too dark for photosynthesis, but with just enough light to hunt by, many animals move from this zone into the sunlit zone at night.

DARK ZONE 3,300 ft–13,100 ft
Almost no light penetrates below 3,300 ft (1,000 m). From here to the greatest depths, it is dark, so no plants can grow, and virtually the only source of food is the "snow" of waste from above. Temperatures down here are a universally chilly 35–39°F (2–4°C), and the pressures so extreme that only highly adapted animals can survive.

The dark zone is defined as continuing down to the abyssal plain, below 13,100 ft (4,000 m). Technically, all the water below 3,300 ft (1,000 m) is a dark zone, where the only light comes from bioluminescent animals (see p.224). However, for convenience, the waters below the dark zone can be further subdivided.

ABYSSAL ZONE 13,100–19,700 ft
Beyond the continental slope, the seabed flattens out. In many areas, it forms vast plains at depths below 13,100 ft (4,000 m). Some areas drop deeper to a sea floor that undulates down to depths of 19,700 ft (6,000 m). Around 30 percent of the total seabed area lies between these depths. Animals living here move up and down through a narrow column above the seabed, called the abyssal zone.

HADAL ZONE 19,700–36,100 ft
The sea floor plunges below 19,700 ft (6,000 m) in only a few deep ocean trenches. This hadal zone makes up less than 2 percent of the total seafloor area. As of 2021, only 22 human beings have reached the foot of the deepest ocean canyon, in the Mariana Trench, in four manned submersibles (see p.173).

As yet, little is known about life at these depths, although anemones and jellyfish have been observed at a depth of 27,000 ft (8,221 m) and a fish has been dredged from a depth of 27,500 ft (8,370 m). Amphipods, as well as amoebae and various other microbes, have been found living at the bottom of the Mariana Trench, the deepest point in the oceans.

SCALE

Burj Khalifa
(2,722 ft/829.8 m)

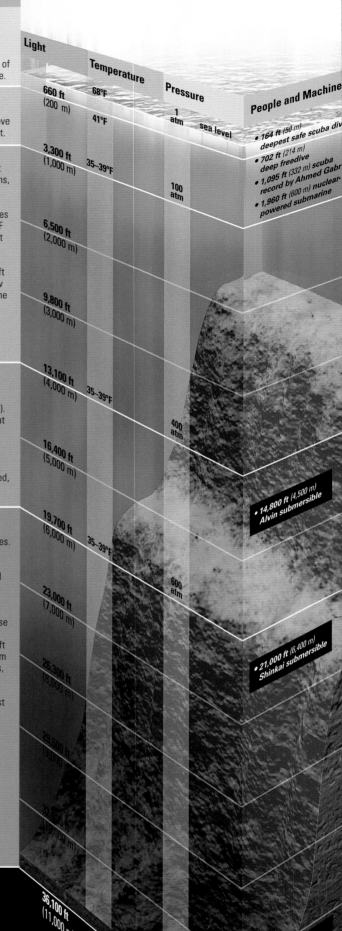

OCEAN ZONES

Light   Temperature   Pressure   People and Machines

660 ft (200 m)   68°F   1 atm   sea level
        41°F
                • 164 ft (50 m) deepest safe scuba dive
3,300 ft (1,000 m)   35–39°F
                • 702 ft (214 m) deep freedive
                100 atm   • 1,095 ft (332 m) scuba record by Ahmed Gabr
6,500 ft (2,000 m)
                • 1,960 ft (600 m) nuclear-powered submarine
9,800 ft (3,000 m)

13,100 ft (4,000 m)   35–39°F
                400 atm
16,400 ft (5,000 m)
                • 14,800 ft (4,500 m) Alvin submersible
19,700 ft (6,000 m)   35–39°F

                600 atm
23,000 ft (7,000 m)
                • 21,000 ft (6,400 m) Shinkai submersible
26,300 ft (8,000 m)

29,500 ft (9,000 m)

32,800 ft (10,000 m)

36,100 ft (11,000 m)

deep trenches plunging to almost 36,100 ft (11,000 m) cover just 2 percent of the seabed

• 35,853 ft (10,928 m) DSV Limiting Factor

The column below shows the average depth (yellow band) and greatest depth (red band) of the oceans and some of the world's seas.

Marine Life

859 ft (201 m) Dolphin

1,148 ft (350 m) King penguin

2,230 ft (680 m) Great white shark

4,199 ft (1,280 m) Leatherback turtle

6,677 ft (2,035 m) Sperm Whale

7,835 ft (2,388 m) Southern Elephant Seal

North Sea
average depth **308 ft** (94 m)

Baltic Sea
greatest depth **1,473 ft** (449 m)

North Sea
greatest depth **2,296 ft** (700 m)

Arctic Ocean
average depth **3,953 ft** (1,205 m)

Mediterranean Sea
average depth **4,921 ft** (1,500 m)

Caribbean Sea
average depth **8,685 ft** (2,647 m)

Almost one third of the total seabed area is made up of abyssal plains at around 14,800 ft (4,500 m).

Southern Ocean
average depth **10,728 ft** (3,270 m)

Atlantic Ocean
average depth **11,962 ft** (3,646 m)

Indian Ocean
average depth **12,274 ft** (3,741 m)

Pacific Ocean
average depth **13,386 ft** (4,080 m)

Mediterranean Sea
greatest depth **16,762 ft** (5,109 m)
(Calypso Deep)

Arctic Ocean
greatest depth **18,599 ft** (5,669 m)
(Molloy Deep)

Indian Ocean
greatest depth **23,917 ft** (7,290 m)
(Java Trench)

Southern Ocean
greatest depth **24,140 ft** (7,358 m)
(South Sandwich Trench)

Caribbean Sea
greatest depth **25,213 ft** (7,685 m)
(Caymen Trench)

Beyond the abyssal plains, undulating, rocky seabed stretches down to around 19,700 ft (6,000 m). Only the ocean trenches reach deeper.

Atlantic Ocean
greatest depth **27,585 ft** (8,408 m)
(Puerto Rico Trench)

Pacific Ocean
greatest depth
35,853 ft (10,928 m)
(Challenger Deep)

ZONES OF LIFE
The different zones of life in the deep ocean are shown here, together with the maximum depths reached by humans and a selection of marine animals. Most life is concentrated above 3,300 ft (1,000 m), where there is some light.

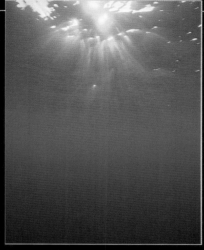

CRYSTAL WATERS
Crystal-clear tropical waters look idyllic, but the clarity indicates that there are few nutrients and therefore few phytoplankton in the water. As a result, feeding for animals is quite poor.

The Sunlit Zone

The sunlit zone is the range in which there is enough sunlight for photosynthesis. The ocean absorbs different wavelengths of sunlight to differing extents (see p.36). Nearly all red light is absorbed within 30 ft (10 m), so red animals look black below this depth. Green light penetrates much deeper in clear water, down to around 330 ft (100 m), and blue light to twice that. Due to the presence of chlorophyll, phytoplankton preferentially absorb the red and blue portions of the light spectrum (for photosynthesis) and reflect green light. They can photosynthesize down to about 660 ft (200 m) in clear water.

In cloudy water, the sunlit zone is shallower, because light is absorbed more quickly. The accumulations of phyto- and zooplankton in fertile waters absorb sunlight, reducing the depth of the sunlight zone. Phytoplankton must stay in the sunlit layer during daylight to photosynthesize. Zooplankton follows them there to feed, along with animals that feed on zooplankton. This zone is dangerous because light makes animals conspicuous to their hunters.

FEEDING IN THE SUNLIT ZONE
The phytoplankton of the sunlit zone is the food of zooplankton. Larger animals, such as these shrimp, in turn feed on zooplankton. Phytoplankton is at the bottom of most ocean food chains.

Living in the Sunlit Zone

Phytoplankton must remain in the sunlit zone if it is to catch enough sunlight for photosynthesis. This zone is the warmest and richest in the nutrients needed for growth. It would be counterproductive to expend huge amounts of energy to stay in this zone, so phytoplankton has developed a wide range of mechanisms to help it hang there effortlessly. Buoyancy bubbles, droplets of oil, or stores of light fats keep some species afloat. Others are covered in spines, which increase their surface area and help buoy them up. Some phytoplankton forms colonial chains, which produce more drag in water and slow the rate at which the phytoplankton sink. One group, called dinoflagellates, have threadlike flagellae that let them swim weakly. In this highly productive zone, phytoplankton produces half of the oxygen in the atmosphere. In temperate regions, phyto-plankton proliferates in summer, sometimes forming dense blooms.

DIATOM
Diatoms are a very prolific type of phytoplankton. Some grow colonially, attached to rocks in chains or mats. Each year, six billion tons of phytoplankton grow in the oceans worldwide.

ZOOPLANKTON
This sample of zooplankton, collected in a net, includes an echinoderm (bottom left), a radiolarian, and a crab larva (center), with a fish egg (bottom right).

Plankton and Nekton

In spring, as phytoplankton blooms begin to develop, zooplankton start multiplying. They follow the phytoplankton into the sunlit zone to feed. Most are herbivores that feed on phytoplankton; some are carnivores that hunt other zooplankton. Many are classed as meroplankton—the young of animals like crabs, lobsters, barnacles, and some fish—which have a planktonic larval stage and use the currents to spread. By taking advantage of the summer phytoplankton feast, they avoid competing for food with adults of their own kind. While plankton drift with the currents, many free-swimming animals (collectively called nekton) gather to feed on them: fish, squid, marine mammals, and turtles. These, in turn, are food for predatory fish and seabirds. Some larger animals, such as basking sharks, also feed on zooplankton and nekton.

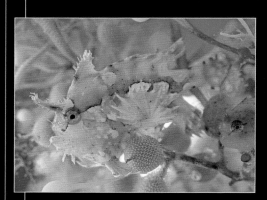

SARGASSUMFISH
Here, two *Sargassumfish* are hiding in *Sargussum* seaweed, floating on the surface of the Sargasso Sea.

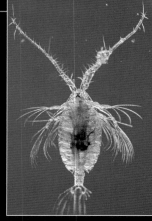

COPEPOD
Copepods are herbivores. They make up 70 percent of the total zooplankton population, with thousands in a cubic yard.

The Twilight Zone

In the twilight zone, there is just enough light for animals to see—and be seen. As a result, predators and prey are in constant battle. Many species are almost totally translucent, to avoid casting even a faint shadow. Others are reflective, to disguise themselves against the light from above, or have wafer-thin bodies that reduce their silhouette. To cope with dim light, many animals in this zone have large eyes.

The main source of food here is detritus. Many animals therefore migrate upward into the sunlit zone, where food is plentiful, at night, returning to the twilight zone as the Sun rises. Millions of tons of animals, equivalent to around 30 percent of the total marine biomass, make this daily trek—by far the largest migration of life on Earth. The length of the journey is a matter of scale. Small planktonic animals measuring less than $1/25$ in (1 mm) in length may only migrate through 70 ft (20 m), but some larger shrimp travel 2,000 ft (600 m) each way, every day.

GIANT FILTER-FEEDER
More than 36 ft (11 m) long, basking sharks like this one scoop up shoals of plankton, then filter them from the water with the white gill rakers inside their jaws.

The Dark and Abyssal Zones

The waters below the twilight zone are all dark, cold, subject to high pressure, and impoverished in food. For animals adapted to these deep zones, pressure is not a problem: their liquid-filled bodies are almost incompressible, compared to the gas-filled bodies of surface-living birds and mammals, which are much more easily compressible and subject to the effects of pressure. Most fish use gases in their swim bladders to maintain buoyancy, and these are susceptible to pressure change. Many deep-water fish therefore have no functional swim bladders.

For most deep-water species, lack of food is the biggest problem: only about 5 percent of the energy that plants produce at the surface filters down to these depths. Animals of the deep are typically slow-moving, slow-growing, and long-lived. They conserve energy by waiting for food to come to them. Many therefore have massive mouths and powerful teeth. Others use tricks to catch prey: anglerfish dangle lures, and some species even harbor luminescent bacteria or use chemical processes to make these lures and other structures glow.

SQUID OF THE OPEN OCEAN
Squid live in several zones of the ocean, from the sunlit zone, where this big-fin reef squid is found, to the deep zone. Deep-sea squid are difficult to photograph and are often photographed only as dead specimens.

SPOOKFISH
The brownsnout spookfish is found at depths of up to 3,300 ft (1,000 m), on the boundary of the dark zone. Its bones are so thin that it is almost transparent, and its large eyes look upward to spot predators attacking from above. It feeds mainly on copepods, and gives birth to live young that float in the plankton.

mucus on body attract bacteria, which protect it from heat

red tentacles around head gather food and provide sensory information

HEAT-TOLERANT WORM
This polychaete worm (a type of segmented worm) was discovered by the *Alvin* submersible in 1979—and named *Alvinella pompejana* in its honor. It is the most heat-tolerant animal on Earth, living near water emerging from hydrothermal vents at 570°F (300°C).

THE *CHALLENGER*

Many oceanographic discoveries were made by HMS *Challenger,* a converted British warship that made a 68,900-mile (110,900-km) voyage around the oceans from 1872 to 1876, collecting depth soundings as it went.

In March 1875, near Guam, it dropped a sounding line to a depth of 26,850 ft (8,184 m) and collected clay to prove this was the seabed. By good luck, the ship was over the Mariana Trench, close to the deepest spot in the ocean, now appropriately called the Challenger Deep.

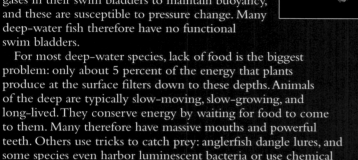

DEEP SQUEEZE
A polystyrene cup attached to the outside of a submersible resurfaces at a fraction of its original size, illustrating the effects of pressure in the deep ocean.

ALVIN SUBMERSIBLE
Alvin is designed to withstand the extreme pressure of the deep zone and has enabled scientists to make many important discoveries during over 4,300 dives.

The Hadal Zone

Few deep-water species have been observed in their natural environment of the hadal zone and even fewer photographed. Many species are known only from samples dredged up in nets, and most photographs are of dead specimens (including the fangtooth on the right). Sometimes deep-sea animals can be studied in aquariums, but many species cannot survive temperature and pressure changes when brought to the surface.

Although many of the animals here hunt each other, the food chain must begin with a supply of food from above. Whereas animals on the seabed can patrol large areas to find food particles accumulated there over weeks and months, animals in midwater must grab food particles in the short time when they float downward past them, which is much trickier. Only a small proportion of the detritus from above is harvested in midwater, so food is always scarce.

Scientists observing this zone often see the same species repeatedly. The environment of this zone is remarkably uniform worldwide, and there are few physical or ecological barriers to block the movement of species. Many deep-water species therefore are widely distributed, and several are found in every ocean. As a result, species diversity is low: only around a few dozen of the known 35,000 fish species live at this depth.

FANGTOOTH FISH
The fangtooth has been recorded at depths of 16,380 ft (4,992 m). Like many deep-water fish, it has a large head and massive teeth. Sensory organs along its body detect prey movement in the dark.

GLOBAL EXPLORER
The *Global Explorer* is an ROV controlled by cable from a mother ship, involved in surveys of the Arctic Ocean and Celebes Sea near the Philippines between 2002 and 2007.

Submersibles are underwater vehicles, smaller than submarines, used mainly for exploration, scientific study of the oceans, and recreation. First developed in the 1960s, they have opened up the deep ocean to exploration. Modern submersibles include manned vehicles and smaller, unmanned vessels called Remotely Operated Vehicles (ROVs).

A famous manned submersible is *Alvin*, operated by the Woods Hole Oceanographic Institute (US). In 1977, *Alvin*'s crew discovered the first hydrothermal vents (see pp.188–189) and, in 1986, they explored the wreckage of the *Titanic*. The vessel was completely redesigned in 2011–2013 and is still in operation after more than 4,300 dives. Along with *Shinkai 6500* (Japan), *Jiaolong* (China), and others, it belongs to a class known as Deep Submergence Vehicles or DSVs. These are mostly used for scientific research.

Without humans on board, ROVs can be smaller and more adaptable. They can drill cores in the seabed and make sonar surveys, as well as record images and collect specimens. The first models relied on buoyancy tanks and ballast for depth control, but modern ones use technologies designed for flight. Most ROVs are directed by a tether from surface vessels, but some are now semiautonomous, guided by Artificial Intelligence (AI) systems.

Into the Deep

Most DSVs and ROVs have maximum depths varying from 3,300 ft (1,000 m) to 23,000 ft (7,000 m), but a few can go to the deepest spot in the oceans, the Challenger Deep of the Mariana Trench. As of 2021, only four manned vehicles have done this: Jacques Piccard's bathyscape *Trieste* in 1960 (see p.183), *Deepsea Challenger* in 2012, and *DSV Limiting Factor* in 2019, and *Fendouzhe* in 2020. Two ROVs have also reached these depths: *Kaiko* (Japan) in 1995 and *Nereus* (US) in 2009.

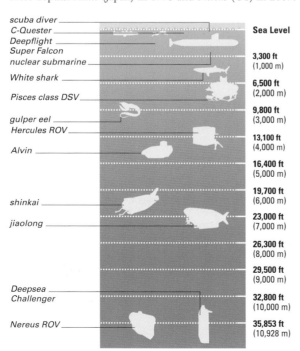

scuba diver	**Sea Level**
C-Quester	
Deepflight Super Falcon	
nuclear submarine	**3,300 ft** (1,000 m)
White shark	**6,500 ft** (2,000 m)
Pisces class DSV	
gulper eel	**9,800 ft** (3,000 m)
Hercules ROV	
	13,100 ft (4,000 m)
Alvin	
	16,400 ft (5,000 m)
shinkai	**19,700 ft** (6,000 m)
jiaolong	**23,000 ft** (7,000 m)
	26,300 ft (8,000 m)
	29,500 ft (9,000 m)
Deepsea Challenger	**32,800 ft** (10,000 m)
Nereus ROV	**35,853 ft** (10,928 m)

TYPES OF SUBMERSIBLES

SHALLOW EXPLORATION

DEEPFLIGHT SUPER FALCON
Designed by American engineer Graham Hawkes, Super Falcon is an underwater vessel intended mainly for recreational exploration from yachts or resorts. It "flies" through the water and can take a pilot and one or two passengers down to 330 ft (100 m).

C-QUESTER
Developed by Netherlands-based company U-boat Worx, C-Quester submersibles allow two or three people to explore down to 160 ft (50 m).

MULTIPERSON DEEPWATER RESEARCH

PISCES IV *An example of a Deep Submergence Vehicle (DSV) used in scientific research, Pisces IV is owned and operated by the Hawaii Undersea Research Laboratory. It carries three people and can operate down to 6,600 ft (2,000 m).*

SHINKAI 6500
Launched in 1989 by the Japan Marine Science and Technology Center, Shinkai 6500 is one of the deeper-diving DSVs. In June 2013, it transmitted the world's first live broadcast from 16,500 ft (5,000 m).

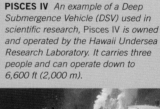

REMOTELY OPERATED VEHICLES (ROVS)

HERCULES ROV *A fairly typical ROV, Hercules can descend to a depth of 13,500 ft (4,000 m) and take high-definition images. It is equipped with six thrusters that allow it to "fly" in any direction, like a helicopter. Slightly positively buoyant, it will gently float up to the surface if its thrusters stop turning.*

RACE TO INNER SPACE

DSV LIMITING FACTOR
In April 2019, this Deep Submergence Vessel (DSV), with its pilot Victor Vescovo, reached the foot of the Challenger Deep in the Mariana Trench (see p.467) at a new record depth of 35,853 ft (10,928 m). This was 66 ft (20 m) deeper than the previous dive there in 2012.

OCEAN ENVIRONMENTS

Seamounts and Guyots

SEAMOUNTS ARE TOTALLY submerged, undersea mountains that rise at least 3,300 ft (1,000 m) from the sea floor; smaller ones are called sea knolls. Guyots are seamounts that once rose above sea level—as a result, they have a flat top caused by erosion. Often isolated in deep ocean, seamounts and guyots provide a habitat for marine life adapted to shallower water. The obstruction of a seamount forces nutrient-rich, deep-sea currents to rise closer to the surface, forming eddies above the seamount. These trap nutrients and support plankton, which in turn attract shoals of fish.

PEOPLE

HENRY GUYOT

Arnold Henry Guyot (1807–1884) was the first professor of geology at Princeton University. He set up a system of weather observatories that led to the formation of the US Weather Bureau. Guyots were named in his honor by a later Princeton geology professor, Harry Hass. Hass discovered guyots using echo-sounding equipment during World War II.

Geological Origins

Seamounts start as undersea volcanoes, where a rift in the seabed allows volcanic eruptions. Many arise at rifts on the crest of mid-ocean ridges, formed by the movement of tectonic plates (see p.185). Because these rifts are generally linear, seamounts tend to be elliptical or elongated in shape. They are made of volcanic basalt rock, but a thin layer of marine sediment accumulates over time. Seamounts often occur in chains or elongated groups, either because there are several weak spots along a rift, or because a series of seamounts originated sequentially at a single, stationary volcanic hotspot. Sometimes volcanic eruptions break above the ocean surface to form island chains, and these may continue out to sea as a line of guyots, or tablemounts. Newly formed volcanic rock is easily eroded, so over time, the above-water peak of the volcanic island is eroded down to a flat top. Then, as the ocean plates carry it away from the zone of volcanic activity, the flat-topped guyot sinks beneath the surface.

SEAMOUNT FORMATION
A seamount forms from an underwater volcanic eruption. Erosion here is slower than on land, so it remains conical.

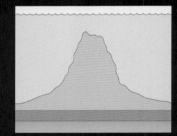

EVOLUTION OF A GUYOT

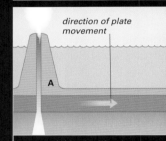

direction of plate movement

A

1 A guyot (A) begins life when a volcano erupts above a "hotspot," creating a small volcanic island.

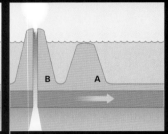

B A

2 Over millennia, erosion reduces the island to a flat top at sea level, while it (A) moves away from the hotspot. A new island (B) forms.

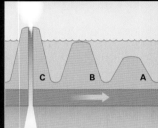

C B A

3 As the island moves farther, it sinks and forms a guyot. New islands (B and C) erupt from the hotspot.

Upwellings

The open ocean is mainly barren, because cold, nutrient-rich currents are confined to deep water, far beneath the reach of plankton. Seamounts—which stand up to 13,000 ft (4,000 m) above the seabed—form a major obstruction to these currents, diverting them and pushing them upward. This brings an upwelling of nutrients into the sunlit zone, and allows phytoplankton to flourish. As these nutrient-rich currents rush over the top of the seamount, they split in two and sweep around it. This makes the water above the seamount rotate, encircling a cylindrical column of still water that extends high above the height of the seamount. This "virtual" cylinder is called a Taylor Column. Above a seamount, it forms an area of back-eddies and still water in which nutrients accumulate and plankton get trapped. This creates a zone of incredible richness and productivity above the seamount—an "oasis" in the nutrient desert of the open ocean.

World Distribution

There may be more than 45,000 seamounts and guyots in the oceans, but many still have to be mapped or studied, and the total number is unknown. Seamounts may occur either singly or in clusters or chains, reflecting zones of past volcanic activity. The Pacific, with its Ring of Fire, is the most volcanically active ocean, containing at least 14,700 seamounts and guyots. Pacific chains typically form in a northwesterly direction, matching the direction of plate movement, with 10 to 100 seamounts in each chain, sometimes connected by an undersea ridge. In the Atlantic and Indian Oceans, by contrast, seamounts mostly occur singly.

DISTRIBUTION MAP OF SEAMOUNTS
Some seamounts and guyots arise over volcanic hotspots, often in chains. Others form singly along mid-ocean ridges. Total numbers are unknown.

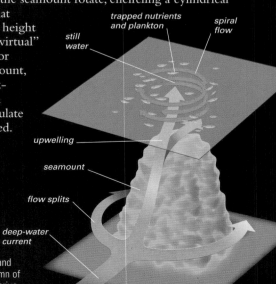

trapped nutrients and plankton

spiral flow

still water

upwelling

seamount

flow splits

deep-water current

WATER COLUMNS
The currents spiraling around and over a seamount create a column of still water above it. Plankton thrive on the nutrients trapped there.

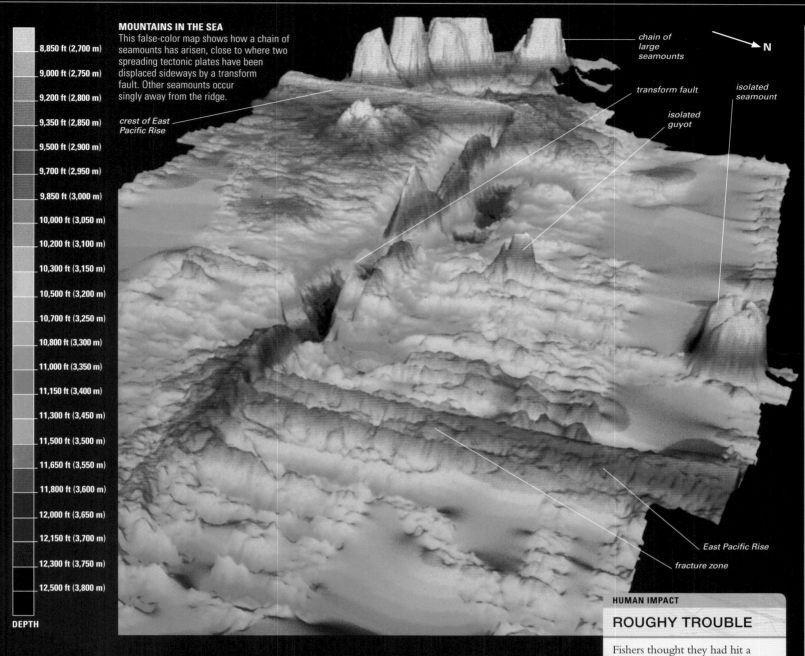

DEPTH

8,850 ft (2,700 m)	
9,000 ft (2,750 m)	
9,200 ft (2,800 m)	
9,350 ft (2,850 m)	
9,500 ft (2,900 m)	
9,700 ft (2,950 m)	
9,850 ft (3,000 m)	
10,000 ft (3,050 m)	
10,200 ft (3,100 m)	
10,300 ft (3,150 m)	
10,500 ft (3,200 m)	
10,700 ft (3,250 m)	
10,800 ft (3,300 m)	
11,000 ft (3,350 m)	
11,150 ft (3,400 m)	
11,300 ft (3,450 m)	
11,500 ft (3,500 m)	
11,650 ft (3,550 m)	
11,800 ft (3,600 m)	
12,000 ft (3,650 m)	
12,150 ft (3,700 m)	
12,300 ft (3,750 m)	
12,500 ft (3,800 m)	

MOUNTAINS IN THE SEA
This false-color map shows how a chain of seamounts has arisen, close to where two spreading tectonic plates have been displaced sideways by a transform fault. Other seamounts occur singly away from the ridge.

crest of East Pacific Rise

chain of large seamounts

N

transform fault

isolated seamount

isolated guyot

East Pacific Rise

fracture zone

Life on a Seamount

Some seamounts were first detected when fishers discovered large shoals of fish in the area. The nutrient-rich waters trapped above seamounts support dense concentrations of phytoplankton as well as the zooplankton that feed on them. Free-swimming animals are attracted by this feast, including fish at densities found nowhere else in the open ocean. Predators such as sharks and seals also gather to feed. Seamount rock is colonized by suspension feeders—animals that catch plankton and detritus as it floats past. Only about one in a thousand seamounts has been explored underwater. However, in studies of 25 seamounts in the Tasman and Coral seas, 850 species (some previously thought extinct) were recorded. Seamounts are important biodiversity hotspots, with up to one-third of species found there restricted to a single seamount or group of seamounts.

PRIMNOID CORAL THREAT
Scientists fear some primnoid coral species may be wiped out by bottom-trawl fishing before they have even been named.

SEAMOUNT FEEDER
Found in the tropics, this octocoral is a colony of soft corals. The feeding polyps, lined up along the branches, catch food from currents sweeping over the seamount.

FEEDING FROM THE CURRENTS
This squat lobster, or pinch bug, is a scavenger living on rock faces. Currents welling over the Bowie Seamount in the northeast Pacific supply rich pickings.

HUMAN IMPACT

ROUGHY TROUBLE

Fishers thought they had hit a bonanza when they discovered huge shoals of fish living over seamounts. By the late 1980s, nearly 99,000 tons (90,000 metric tons) of Orange Roughy (see p.353) were caught globally each year. New Zealand alone caught 60,000 tons (54,000 metric tons). The roughy is slow-growing and does not produce eggs until it is 20–30 years old, so such heavy fishing was unsustainable. New Zealand now carefully manages stocks, restricting roughy catch to 6,600–7,700 tons (6,000–7,000 metric tons) a year.

OCEAN ENVIRONMENTS

The Continental Slope and Rise

THE CONTINENTAL SLOPE AND RISE are areas of sloping sea floor that lead from the continental shelf to the abyssal plain. Beyond a point on the shelf called the shelf break, the seabed begins to drop more steeply. This is the continental slope, which leads into the open ocean. It sweeps down to 9,800–14,800 ft (3,000–4,500 m), where the seabed flattens out. In places, the slope is broken by submarine canyons. Sediments wash down these canyons, and accumulate at the base of the slope in a gentler gradient, forming the continental rise.

Continental Slope

The rock of the continental slope is blanketed by sediments washed from the land that have accumulated over millions of years. Crustaceans, echinoderms, and many other animals live in, or on, these sediments. The slope is dissected by deep canyons. These have been cut by an abrasive mix of sediment and water, called turbidity currents, which flow down the gorges at 50–60 mph (80–100 kph). Some submarine canyons are massive: the Grand Bahama Canyon in the Caribbean has cliffs rising 14,060 ft (4,285 m) from the canyon floor. Many canyons are seaward extensions of great rivers. At the canyon end, the sediment is deposited as a spreading outwash fan, extending far out onto the abyssal plain.

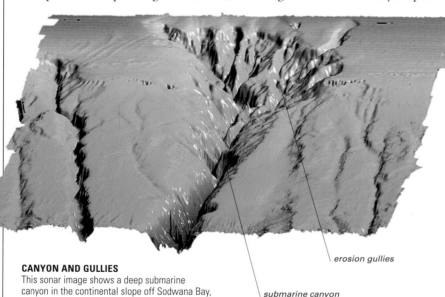

CANYON AND GULLIES
This sonar image shows a deep submarine canyon in the continental slope off Sodwana Bay, in KwaZulu Natal, South Africa.

erosion gullies

submarine canyon

CONTINENTAL MARGIN
A typical continental margin is shown here, including the transition from a shoreline to the abyssal plain via the continental shelf, slope, and rise. The continental slope is about 87 miles (140 km) wide, and the continental rise is about 60 miles (100 km) wide. The vertical scale has been exaggerated: the continental slope actually has a gentle gradient, of about 1 in 50 (2 percent); and the rise is even gentler, at about 1 in 100 (1 percent).

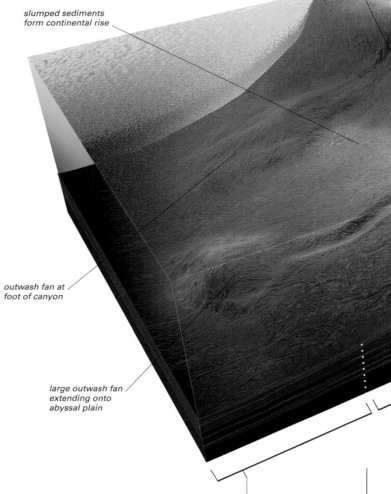

submarine canyon

shelf break— around 660 ft (200 m) below surface

slumped sediments form continental rise

outwash fan at foot of canyon

large outwash fan extending onto abyssal plain

ABYSSAL PLAIN
This flat plain is formed by a deep accumulation of sediments. It typically lies at a depth of 15,000 ft (4,500 m).

CATCHING SABLEFISH
Sablefish are caught with longlines, ⅔ mile (1.2 km) long, that reach down toward the continental slope.

Life on the Continental Slope

Like the shelf, the continental slope is enriched by nutrients washed off the land. This helps support both midwater (pelagic) and bottom-dwelling (demersal) fish. Fish stocks over most continental-shelf regions have declined dramatically in recent decades, as a result of overexploitation and poor management, driving more fishers to seek deeper-water species over the continental slope. Unfortunately for fisheries, although deep-water species are long-lived, they breed slowly, and stocks take a long time to recover. So many fisheries are now in serious decline.

SABLEFISH
Sablefish breed slowly, and it takes 14 years to replace each fish caught. Fish farms (right) may be a better option.

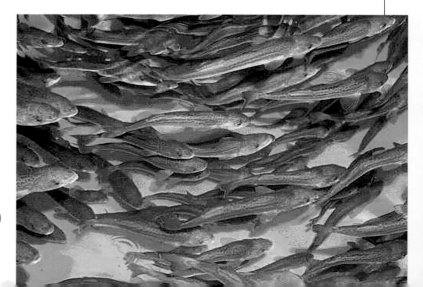

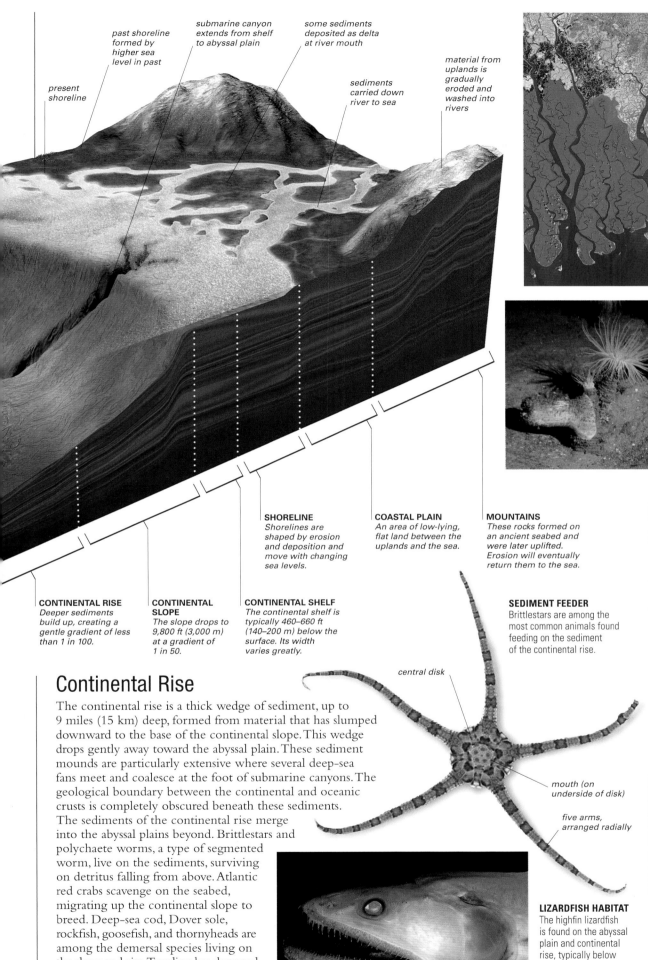

present
shoreline

past shoreline
formed by
higher sea
level in past

submarine canyon
extends from shelf
to abyssal plain

some sediments
deposited as delta
at river mouth

sediments
carried down
river to sea

material from
uplands is
gradually
eroded and
washed into
rivers

SHORELINE
Shorelines are
shaped by erosion
and deposition and
move with changing
sea levels.

COASTAL PLAIN
An area of low-lying,
flat land between the
uplands and the sea.

MOUNTAINS
These rocks formed on
an ancient seabed and
were later uplifted.
Erosion will eventually
return them to the sea.

CONTINENTAL RISE
Deeper sediments
build up, creating a
gentle gradient of less
than 1 in 100.

**CONTINENTAL
SLOPE**
The slope drops to
9,800 ft (3,000 m)
at a gradient of
1 in 50.

CONTINENTAL SHELF
The continental shelf is
typically 460–660 ft
(140–200 m) below the
surface. Its width
varies greatly.

SEDIMENT FEEDER
Brittlestars are among the
most common animals found
feeding on the sediment
of the continental rise.

central disk

mouth (on
underside of disk)

five arms,
arranged radially

Continental Rise

The continental rise is a thick wedge of sediment, up to
9 miles (15 km) deep, formed from material that has slumped
downward to the base of the continental slope. This wedge
drops gently away toward the abyssal plain. These sediment
mounds are particularly extensive where several deep-sea
fans meet and coalesce at the foot of submarine canyons. The
geological boundary between the continental and oceanic
crusts is completely obscured beneath these sediments.
The sediments of the continental rise merge
into the abyssal plains beyond. Brittlestars and
polychaete worms, a type of segmented
worm, live on the sediments, surviving
on detritus falling from above. Atlantic
red crabs scavenge on the seabed,
migrating up the continental slope to
breed. Deep-sea cod, Dover sole,
rockfish, goosefish, and thornyheads are
among the demersal species living on
the slope and rise. Trawling has damaged
many of these habitats, but the deeper
canyons remain havens of biodiversity.

GANGES DELTA
The Ganges River
carries 2 billion tons
of sediment a year.
Some is deposited in this
massive delta. More is
carried out to sea where
it forms a deep-sea fan
over the Bay of Bengal.

TUBE ANEMONES
These sea anemones bury
their bodies in sediment, at
depths of 13,100 ft (4,000 m),
feeding with their tentacles.

DISCOVERY

STAKING A CLAIM

Although climate change calls
into question the need for new
oil fields, oil companies have
begun test-drilling in waters as
deep as 11,800 ft (3,600 m) on the
continental rise. Deep-sea mining
for minerals is also being actively
promoted, and deep-sea fisheries
are being extended. Under current
maritime law, the rights of a
coastal state over certain resources,
such as oil, extend out to the
continental margin—essentially to
the boundary between continental
rise and abyssal plain—or to 200
nautical miles from the coast,
whichever is the greater (but never
exceeding 350 nautical miles).

LIZARDFISH HABITAT
The highfin lizardfish
is found on the abyssal
plain and continental
rise, typically below
about 6,600 ft (2,000 m),
in water colder than
39°F (4°C).

OCEAN ENVIRONMENTS

COLD-WATER COMMUNITY
A squat lobster shelters among the polyps of the cold-water stony coral *Lophelia pertusa*, or tuft coral, in a Norwegian fjord.

Deep-sea corals were first discovered in 1869, but it took the advent of sonar and deep-sea submersibles to reveal the size and abundance of the reefs that they build. Although less well studied than their tropical counterparts, these cold-water reefs are just as rich in life. The stony corals that form deep-water reefs flourish in water temperatures of 39–55°F (4–13°C). Unlike tropical corals, they can live in total darkness because they do not rely on zooxanthellae (p.153) living inside them to produce nourishment by photosynthesis in sunlight. Instead, they survive by filtering food from the water. Some scientists have suggested that there may be a link between the existence of deep-water reefs and the seepage of certain substances, such as methane, from the seafloor. Methane may provide energy for bacteria at the bottom of a food chain, which are then filtered from the water by the coral polyps.

In the northeast Atlantic, cold-water reefs are mainly formed by the coral *Lophelia pertusa* (see p.270), at depths of around 3,300 ft (1,000 m). One of the largest of these reef complexes is the Røst Reef at the continental margin of northern Norway, southwest of the Lofoten Islands. A few hundred individual reefs there cover an overall area of around 37 square miles (100 square kilometers). Other *Lophelia* reefs occur at similar depths on many seamounts in the Atlantic, and also in shallow cold water in Norway's fjords. Several other coral species form cold-water reefs elsewhere in the world. For example, in the Pacific, the main reef species on seamounts and oceanic banks around Tasmania and New Zealand are *Goniocorella dumosa* and *Solenosmilia variabilis*. Over 1,300 species of animals have been recorded on deep-sea reefs, and they are important nursery grounds for many commercial fish species.

Location of Deep-sea Reefs

The map below shows the global distribution of cold-water reefs. Some of these reefs are small, while others cover up to 770 square miles (2,000 square km), although the map dots exaggerate their extent. The many reefs detected in the north Atlantic probably reflect the intensity of surveying there, particularly in the search for oil. More detailed surveys of other oceans are likely to reveal the existence of further deep-sea reefs.

LIFE IN COLD WATER

DEEP-SEA CORALS

LOPHELIA REEF This reef lies deep in the Atlantic off the west coast of Ireland, where it can be studied only by means of a submersible. Lophelia can also be viewed in water as shallow as 128 ft (39 m) in some Norwegian fjords.

GONIOCORELLA CORAL This deep-sea coral thicket is made mostly of Goniocorella dumosa, a species that is restricted to the Southern Hemisphere. It forms reefs at depths to 5,000 ft (1,500 m).

ASSOCIATED MARINE LIFE

SQUAT LOBSTER This tiny squat lobster is sitting on Madrepora oculata coral polyps, 1,290 ft (390 m) down in the Bay of Biscay, north of Spain.

CHIROSTYLUS CRABS Many animals live among the coral. These long-limbed crabs are crawling over a black coral in the northeast Atlantic.

THREATS FROM DEEP-SEA TRAWLING

DAMAGED REEF Fishing gear has snagged on this reef west of Ireland, tearing off chunks of living reef that could be up to 8,500 years old. Fishing has been banned near the reef, called the Darwin Mounds, since 2005.

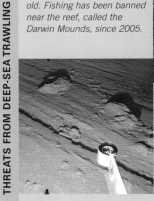

TRAWL MARK Even before scientific surveys discovered them, many deep-water reefs had been severely damaged by trawls dragged across the seabed to catch bottom-living fish. The scarred seabed shown here is at a depth of 2,900 ft (885 m).

PACIFIC OCEAN

ATLANTIC OCEAN

INDIAN OCEAN

PACIFIC OCEAN

SOUTHERN OCEAN

Ocean Floor Sediments

OVER VAST AREAS OF THE SEABED, THE UNDERLYING landforms are hidden beneath deep layers of sediments. Made up of silts, muds, or sands that have built up over 200 million years, they now form a blanket that is several miles thick in places. The sediments have various origins. One group, terrigenous sediments, come from land, mainly from fragments of eroded rock that are carried down rivers into the sea, then down the continental slope to form the continental rise and abyssal plain beyond. Other sediments are biogenic, formed from the hard remains of dead animals and plants. A few, called authigenic sediments, are made up of chemicals precipitated from seawater. There are even cosmogenic sediments, which come from outer space as particles in space dust and meteors. All accumulate to form extensive, flat plains. Various animals feed here and burrow into the sediments for shelter.

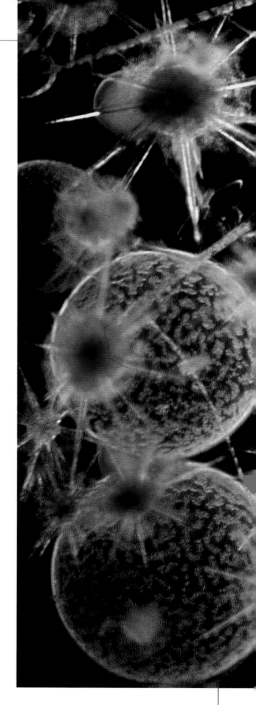

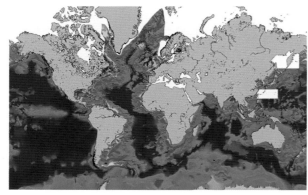

SEDIMENT THICKNESS

| 0 | 1,650 ft (500 m) | 3,300 ft (1,000 m) | 3 miles (5 km) | 6 miles (10 km) | 12 miles (20 km) |

MAPPING SEDIMENTS
Ocean sediment depths can be measured and mapped using echo-sounding. Some areas (white on the map above) are still unsurveyed. Sediment is thickest near land. Glaciers also carry many sediments into the oceans.

Deep-sea Sediments

The average thickness of sediments on the ocean floor is 1,500 ft (450 m), but in the Atlantic Ocean and around Antarctica, sediments can be up to 3,300 ft (1,000 m) deep. Closer to the continents—along the continental rise—sediments washed from the land accumulate more rapidly, and can be up to 9 miles (15 km) deep. In the open ocean, further from the source of terrigenous sediments, the buildup rate is very slow: from a fraction of an inch to a few inches in a thousand years. That is slower than the rate at which dust builds up on furniture in an average house. The accumulated sediments tell scientists a great deal about the last 200 million years of Earth's history. Their form and arrangement provide a vivid snapshot of sea-floor spreading, the evolving varieties of ocean life, alterations in Earth's magnetic field, and changes in ocean currents and climate.

WHITE CLIFFS OF DOVER
These chalk cliffs originated on the sea bed from a biogenic ooze, formed from algal scales (coccoliths) that built up to form layers hundreds of yards thick. They are now raised above sea level.

PTEROPOD OOZE

Pteropods are small winged snails that float in midwater. When they die, their internal shells of aragonite (calcium carbonate) sink to the seabed, contributing to biogenic oozes. The presence of pteropod remains in samples collected from deep in the ooze reveal changes over millennia in water temperatures and sea levels.

Sediments Derived from the Land

Most terrigenous sediments come from the weathering of rock on land and are swept into the oceans, mainly by rivers but also by glaciers, ice sheets, and wind. Coastal erosion adds to these sediments. Often, they are washed down through submarine canyons to the deeper ocean. Sometimes, the route from land to sea is more indirect: volcanic eruptions eject material into the upper atmosphere before it falls as "rain" into the ocean.

In the deepest ocean floors, below about 13,000 ft (4,000 m), the main sediment is red clay, composed mostly of fine-grained silts that have washed off the continents and accumulated incredibly slowly—about $1/32$ in or 1 mm per thousand years. These clays may include up to 30 percent of fine, biogenic particles and have four main mineral components—chlorite, illite, kaolinite, and montmorillonite. Clay types depend on origin and climate. Unfortunately, human debris is also increasing in the deep sea. In 2021, microplastics were found in 28 out of 30 deep-sea sediment cores from the Antarctic and Southern Oceans.

DUST STORM RESULTS IN SILT
Winds from arid regions, such as North Africa (shown in this satellite image) carry dust far out to sea, where it sinks to form silts.

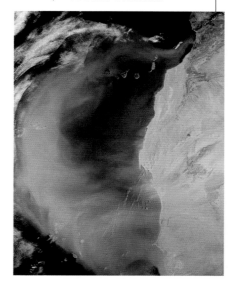

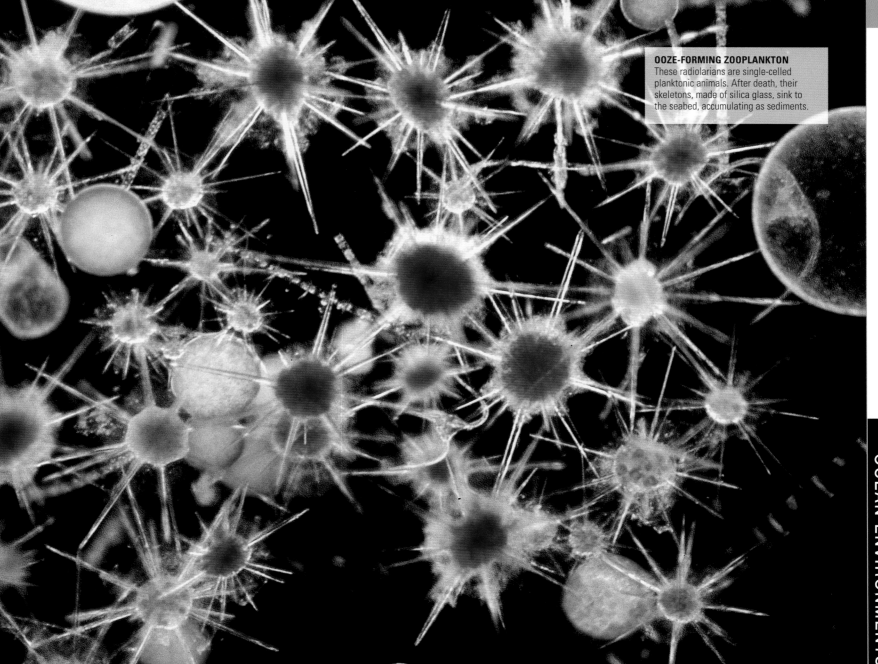

OOZE-FORMING ZOOPLANKTON
These radiolarians are single-celled planktonic animals. After death, their skeletons, made of silica glass, sink to the seabed, accumulating as sediments.

Biogenic Oozes

Biogenic sediments are formed mainly from the shells and skeletons of microscopic organisms that sink to the seabed after death. The decaying remains of larger organisms, such as mollusks, corals, calcareous algae, and starfish, add to this accumulation. Oozes are calcareous if derived from the calcium carbonate shells of foraminifera, pteropods, and coccolithophores (microscopic algae), or siliceous if derived from the silica shells of single-celled radiolarians or diatoms. Because silica dissolves rapidly in seawater, siliceous oozes only build up beneath zones of high primary production. As calcareous shells and skeletons sink, they reach a depth (around 15,000 ft/4,500 m) where the water becomes more acidic; this, combined with pressure, means calcareous remains are dissolved rapidly in seawater at depth. Calcareous oozes therefore occur only above this "calcium carbonate compensation depth," beneath which the seabed consists mainly of terrigenous red clays.

COCCOLITHOPHORE
When this coccolithophore dies, its platelets will add to the calcareous ooze.

FORAMINIFERA
The tiny shells of dead foraminiferans add to the biogenic oozes.

Feeding on the Ooze

The "snow" of calcareous and siliceous remains from the upper levels accumulate on the ocean floor, providing the main source of food for animals living in or on the sediments. Bacteria live in the ooze, where they break down organic remains. In turn, they—along with other organic matter—are consumed by multitudes of tiny foraminiferans. Nematodes, roundworms, isopods, and small bivalve mollusks live and feed in the mud. Brittlestars feed on the ooze by sweeping food off its surface with their arms. Sea pens, crinoids, and glass sponges, which are anchored to the seabed, filter organic particles from the water column.

SEA CUCUMBER FEEDING
Sea cucumbers wander widely over the seabed, sucking up the sediment and then extracting its organic content.

tube feet enable animal to traverse sediment while foraging

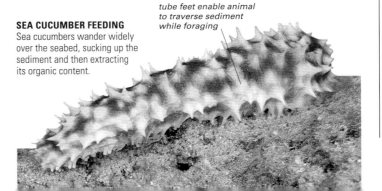

Abyssal Plains, Trenches, and Mid-ocean Ridges

OVER VAST AREAS, THE SEABED IS COVERED BY a flat expanse of accumulated sediments. The sparse life here relies on food falling from above. In places, the abyssal plains are disrupted by more dramatic features, created by tectonic shifts. Where tectonic plates diverge, magma wells up through the gap to create mid–ocean ridges, at which new seabed is constantly being formed. At the other extreme, where plates collide, one plate is dragged downward, opening up a trench.

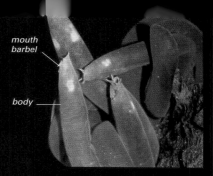

SEABED SCAVENGERS
Hagfish feed on animal corpses that fall to the abyssal plain. Blind and jawless, these primitive fish are attracted by smell. They bore into corpses, using their horny teeth, and secrete clouds of mucus to deter other scavengers.

mouth
barbel

body

Abyssal Plains

Over large areas of the ocean floor, sediments have built up a blanket several miles thick, obscuring the underlying topography. This produces vast, gently undulating abyssal plains, typically at depths around 14,800 ft (4,500 m). However, there are depth differences between some plains, with barriers between them. This leads to submarine waterfalls, where water spills over the barrier and down into the plain below, at rates of up to 5 mph (8 kph). Occasional abyssal storms also occur, stimulated by instabilities at the ocean surface resulting from conditions. Most animals in this zone are scavengers with a body temperature close to that of the surrounding water. They move and grow slowly, reproduce infrequently, and live longer than their relatives at the surface. Even these depths are not free from human threats. Licenses have been granted for pilot schemes to mine manganese nodules found, most notably, in a vast stretch of the Pacific abyssal plain between Hawaii and Mexico. The nodules are rich in rare minerals and metals, but mining them threatens the pristine environment of the seafloor.

MANGANESE NODULE
In places, the abyssal plain is littered with potato-sized nodules of manganese, often contaminated with other valuable metals such as nickel, cooper, and cobalt.

ABYSSAL FLOOR
A recent study off the east coast of North America revealed 798 species buried in a small sediment sample from the seabed.

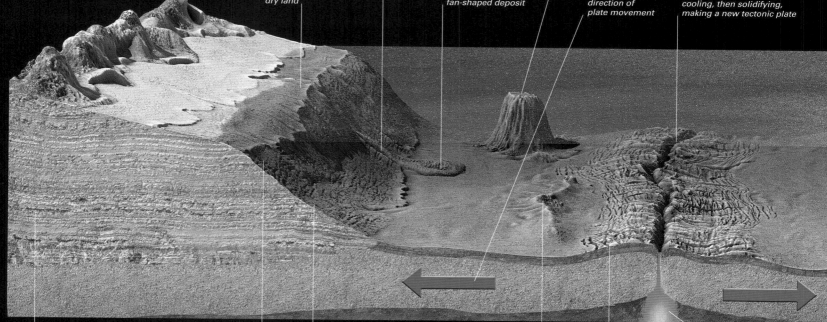

the continental shelf is the flooded edge of a continent, which was once dry land

ocean currents carve a deep gorge, called a submarine canyon, in the continental slope

silt carried down a canyon spreads out at the bottom as a fan-shaped deposit

when a volcanic island sinks, it eventually becomes a flat-topped seamount, or "guyot"

direction of plate movement

at the mid-ocean ridge, two plates pull apart and magma rises up between them, cooling, then solidifying, making a new tectonic plate

Earth's outer layer of rock in continental areas is called continental crust

the steep continental slope goes down to about 10,000 ft (3,000 m)

the gently sloping continental rise is a region that extends down from the continental slope

an underwater plateau is a large, flat-topped mound caused by a few million years of underwater volcanic eruptions

each tectonic plate is made of crust and the top layer of the mantle

melted rock is called magma when it occurs beneath Earth's surface, and lava when it is found above Earth's surface

Ocean Trenches

Ocean trenches are created by a process called subduction. Where oceanic and continental tectonic plates collide, the denser but thinner oceanic plate is forced down beneath the thicker but less dense continental plate, and plunges to its destruction in the mantle deep below. Where two oceanic plates collide, the older plate is subducted beneath the younger. The buckling where the plates collide causes a deep depression at the point of impact—an ocean trench. Trenches are the deepest points in the ocean, and only one person has visited the deepest trench in each ocean: the American private equity investor Victor Vescovo. In 2018–2019, on board *DSV Limiting Factor* (see p.173), he reached the foot of the Puerto Rico Trench (Atlantic Ocean); South Sandwich Trench (Southern Ocean); Java Trench (Indian Ocean); Molloy Deep (Arctic Ocean); and, deepest of all, the Mariana Trench (Pacific Ocean).

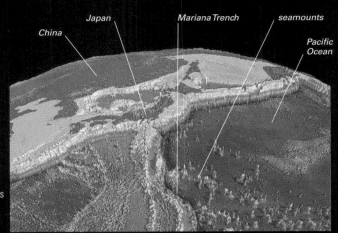

China · Japan · Mariana Trench · seamounts · Pacific Ocean

MARIANA TRENCH
The Mariana Trench is roughly 1,600 miles (2,500 km) long and 40 miles (70 km) wide. It lies in the western Pacific, around 1,000 miles (1,600 km) to the south and east of Japan.

THE DEATH OF A WHALE

Occasionally, a dead whale sinks to the abyssal plain and provides a feast. Scientists have counted 12,000 animals of 43 species feeding on the bones of a single whale. It may take them 11 years to strip the flesh from a blue whale. Later, bacteria invade and decompose the remaining bones. This process leaches out sulfides that sustain a complex community of seafloor life.

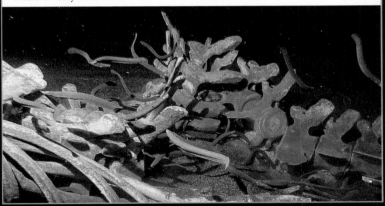

Life in the Ocean Trenches

Animals have been found at great depths in the ocean trenches. The depth record for a fish belongs to a cuskeel, *Abyssobrotula galathea*. This was dredged in 1970 from a depth of 27,453 ft (8,370 m) in the Puerto Rico Trench. In 1998, the unmanned Japanese submersible *Kaiko* collected some large amphipods (shrimplike crustaceans) called *Hirondellea gigas* from the bottom of the Mariana Trench. These were later found to harbor wood–dissolving enzymes in their gut, suggesting they can digest woodfall (tree debris swept into the ocean that eventually sinks to the bottom). *Kaiko* also collected sediment samples that contained 432 different species of foraminiferans and a range of bacteria.

Since 2010, some giant unicellular organisms more than 4 in (10 cm) across, belonging to a class called xenophyophores (a form of foraminiferan), have been observed in the Mariana Trench and elsewhere. Cameras on board the *Deepsea Challenger* that descended to the bottom of the Mariana Trench in 2012 detected sea cucumbers and a jellyfish, as well as xenophyophores and amphipods.

GELATINOUS BLINDFISH
A small number of these curious fish have been collected from the seabed in the Atlantic, Pacific, and Indian Oceans, at depths of at least 10,000 ft (3,000 m). Like many deep-water fish, they are almost transparent, with tiny eyes.

the abyssal plain is a flat expanse of mud that covers a vast area of seafloor

volcanic islands form an arc parallel to the ocean trench

each volcanic island is the above-water part of a huge undersea volcano

THE OCEAN FLOOR
The seafloor lies about 2.3 miles (3.7 km) below the sea surface. It is made of a layer of dark-colored rock, called oceanic crust, which is covered in muddy sediment. Tectonic plates are generally made of this oceanic crust and continental crust, along with part of Earth's deeper mantle layer. Features such as volcanic islands and seamounts are caused by erupting magma.

oceanic crust is thinner than continental crust, and made of dark-colored rock

DISCOVERY

THE *TRIESTE* EXPEDITION

In 1960, two oceanographers, Don Walsh and Jacques Picard, dived to 35,797 ft (10,911 m) in the Challenger Deep section of the Mariana Trench in the bathyscaphe *Trieste*. It took five hours to descend to that depth, and after just 20 minutes hanging there, the crew began their return to the surface. Later submersibles reached 56 ft (17 m) deeper.

The Ring of Fire

All around the margins of the Pacific Ocean, tectonic plates are colliding. This produces a belt of intense volcanic and earthquake activity encircling the Pacific, known as the Ring of Fire. It extends for 18, 600 miles (30,000 km) in a series of arcs, from New Zealand, through Japan, and down the west coast of the Americas to Patagonia. About three-quarters of the Pacific lies over a single plate, the Pacific Plate, which is colliding around its edges with the North American, Australian, and various minor plates. In the eastern Pacific, the smaller Cocos and Nazca plates are colliding with the Caribbean and South American plates. As the edges of the Pacific, Cocos, and Nazca plates subduct (move down) beneath the younger, less dense edges of neighboring plates, massive slabs of rock shatter explosively along faults, producing earthquakes.

A series of deep ocean trenches, arranged in arcs around the Ring of Fire, mark the boundaries where the subducting plates move beneath neighboring plates. Parallel to these trenches— typically at a distance of 100 miles (160 km) and always on the side of the overriding plate—are arcs of often highly active volcanoes, taking either the form of volcanic islands or (on the eastern side of the Pacific) forming lines of volcanoes on land, such as the volcanoes of Central America.

SAFE HAVEN

Mid-ocean-ridge islands offer protected breeding places for many sea birds, with rich feeding provided by upwelling currents offshore. The sooty tern is found in all tropical seas. It nests on oceanic islands. Cats and rats halved the Sooty Tern population on one of these islands, Ascension, but numbers have increased since cats were eradicated in 2004.

HOTSPOTS OF VOLCANIC ACTIVITY
Red on this map shows areas of volcanic activity around the Pacific Ocean, highlighting the Ring of Fire. These volcanoes form on continental plates as oceanic plates are thrust below.

THE RIDGE ON LAND
For most of its vast length, the Mid-Atlantic Ridge is hidden deep beneath the ocean. However, at Iceland, where both the Eurasian and North American plates are separating, it rises above the surface.

WHAKAARI
Whakaari (White Island), an active volcano off the east coast of North Island, New Zealand, is part of the Ring of Fire. In December 2019, it erupted explosively, killing 22 visiting tourists.

Mid–ocean Ridges

New seabed is produced wherever tectonic plates diverge. As plates move apart, they create a rift. Magma wells up through this rift from the Earth's mantle, forming volcanoes and creating an underwater mountain chain, called a mid-ocean ridge. The lava cools as it meets the seawater, and solidifies in vertical basalt dikes or fields of pillow lava (see p.42).

Mid-ocean ridges are assembly lines along which new ocean floor is being produced. The ridges and lava fields remain visible for some time before sediments accumulate over them. Sometimes the volcanoes extend above sea level, producing islands such as Iceland. Some mid-ocean ridges spread slowly, allowing deep rift valleys to form down their centers—others are much faster-spreading but lack rift valleys. Sometimes the ridges are disrupted sideways by transform faults.

As the new seabed spreads outward, tensions are created, making it crack. Water seeps into these cracks and reemerges from hydrothermal vents (see p.188). The oceanic ridge system is the third largest feature on the Earth's surface, after the oceans and continents.

PILLOW LAVA
Under the high pressure of the deep ocean, lava oozes slowly from the mid-ocean crests. When it meets cold seawater, it cools rapidly to form globular masses, called pillow lavas due to their shape. About 1.4 square miles (3.5 square km) of new sea floor is formed each year along mid-ocean ridges.

ASCENSION ISLAND
Ascension Island arises where the Mid-Atlantic Ridge protrudes above sea level in the south Atlantic. It covers 35 square miles (90 square km) and ascends to 2,817 ft (859 m) on Green Mountain. Sooty terns and sea turtles breed around its shores.

OCEAN WANDERERS
Macquarie Island, on the Macquarie Ridge, south of New Zealand, provides a nesting site for the black-browed albatross. Outside of the breeding season, it wanders the Southern Ocean.

Ridges of the World

The longest mid-ocean ridge occurs where the Eurasian and African plates are diverging from the North and South American plates. The Mid-Atlantic Ridge runs along this boundary for 10,000 miles (16,000 km), from the Arctic Ocean to beyond the southern tip of Africa, rising 6,000–13,000 ft (2,000–4,000 m) above the sea floor.

A chain of volcanoes runs down its length, most famously in Iceland. An eruption close to Iceland in 1963 created a new volcanic island, Surtsey. Ascension Island lies very close to the ridge, and the Azores straddle it, while St. Helena and Tristan da Cunha arise from isolated volcanoes, displaced from it. A valley, 15 miles (25 km) wide, extends along the ridge crest. In the Pacific the main ridge system is the East Pacific Rise. This is Earth's fastest-spreading system, separating at 5–6 in (13–16 cm) per year. A series of mid-ocean ridges encircle Antarctica, along the divergent boundaries between the Antarctic Plate and its neighbors, and the Carlsberg Ridge runs down the center of the Indian Ocean.

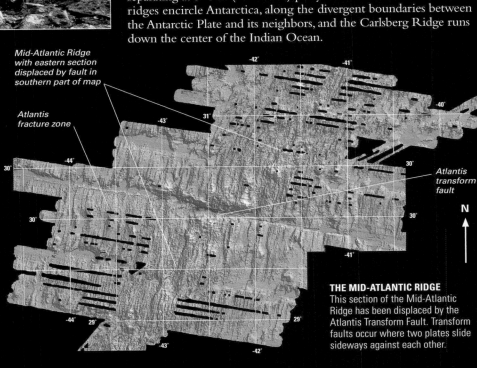

Mid-Atlantic Ridge with eastern section displaced by fault in southern part of map

Atlantis fracture zone

Atlantis transform fault

N

THE MID-ATLANTIC RIDGE
This section of the Mid-Atlantic Ridge has been displaced by the Atlantis Transform Fault. Transform faults occur where two plates slide sideways against each other.

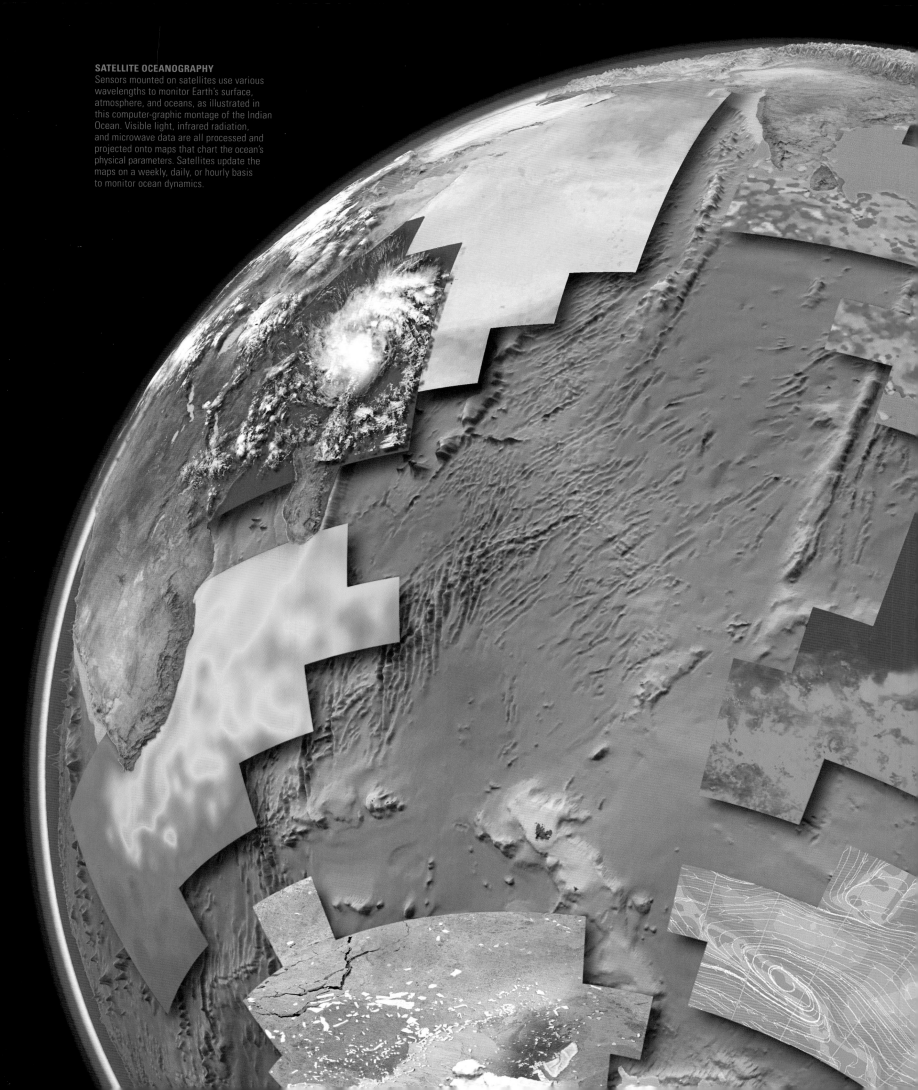

SATELLITE OCEANOGRAPHY
Sensors mounted on satellites use various
wavelengths to monitor Earth's surface,
atmosphere, and oceans, as illustrated in
this computer-graphic montage of the Indian
Ocean. Visible light, infrared radiation,
and microwave data are all processed and
projected onto maps that chart the ocean's
physical parameters. Satellites update the
maps on a weekly, daily, or hourly basis
to monitor ocean dynamics.

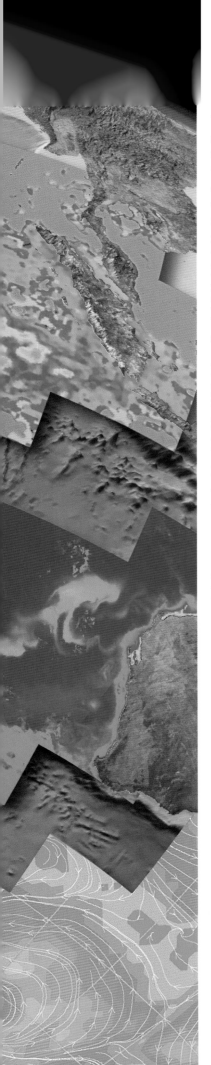

The world's oceans are too vast to be adequately studied using ships alone. Even if all of the depth soundings that were taken during the 20th century were to be plotted, the resultant map would provide only sparse information on the sea floor and would even be blank in large areas. The advent of satellite remote sensing in the 1960s brought a revolution in oceanography. For the first time, it was possible to take a picture showing an entire ocean basin. Hurricane tracking and warning was one of the first benefits to accrue from early weather satellites. Eventually, a large range of sensors were developed to probe the physical attributes of the ocean surface and the atmosphere above.

Ocean color, temperature and salinity, sea-level height, and surface roughness are among the parameters that can be monitored in some detail. Satellite-derived information is a vital component of practical applications such as weather forecasting, commercial fishing, oil prospecting, and ship routing. In some cases, 40 years of continuous observations have been built up, helping scientists to track seasonal and long-term changes in the ocean environment and understand its effects on the global climate.

Measuring Ocean Depth from Space

Satellites cannot directly measure the depth of the seafloor, but it can be derived from the height of the sea surface. The sea is not flat. Water piles up above gravity anomalies caused by ocean-floor features such as seamounts, producing variations in the surface that are much larger than those produced by tides, winds, and currents. By comparing the height of the sea surface against a reference height, the depth of the sea floor can be estimated.

SENTINEL-6 SATELLITE
Sentinel-6 carries a radar altimeter, which is similar to the instruments on aircraft that measure their height above the surface of Earth.

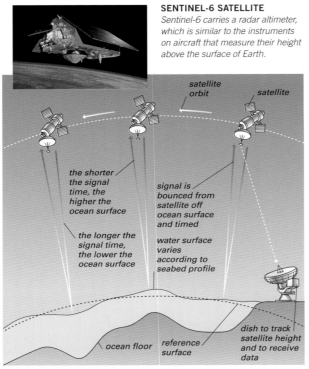

satellite orbit

satellite

the shorter the signal time, the higher the ocean surface

the longer the signal time, the lower the ocean surface

signal is bounced from satellite off ocean surface and timed

water surface varies according to seabed profile

ocean floor

reference surface

dish to track satellite height and to receive data

FEATURES STUDIED FROM SPACE

OCEAN CHEMISTRY

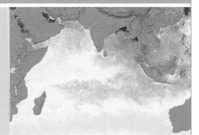

SALINITY *Microwave emissions from the ocean surface tell us about its salinity. Salty seas, such as the Red Sea and Persian Gulf, are shown in red on this map, and relatively fresh water is shown in blue.*

WEATHER

RAINFALL *The Tropical Rainfall Measuring Mission uses a microwave radiometer to see through clouds and detect the presence of liquid water in the atmosphere. Rainfall measures are used in computer models of the climate and ocean.*

PLANT LIFE

CHLOROPHYLL *Ocean color cameras use wavelengths of visible light to measure the concentration of chlorophyll, which is present in phytoplankton. This information is used for water-quality assessment, finding fish, and in various aspects of marine biology.*

WIND SPEED

MICROWAVE SCATTEROMETER *Surface wind speed and direction are measured by satellites that bounce radio beams off the surface of the ocean. Wind-induced ocean waves modify the return signal, and the data can be used for meteorology and climate research.*

ICE COVER

SYNTHETIC APERTURE RADAR *Imaging radar systems, such as the one carried by Radarsat, penetrate clouds and can operate through the dark of the extended polar night to monitor ice shelves, sea ice, and icebergs all year round.*

TEMPERATURE

SURFACE TEMPERATURE *Infrared radiometers can measure the temperature of the sea surface precisely. Shifts in ocean currents, cold-water upwelling, and ocean fronts can be monitored for ocean and climate research.*

THERMAL GLIDER *A new generation of instrument platforms is being developed to sample the vast subsurface volume of the world's oceans. Autonomous Underwater Vehicles, or "sea gliders," can undertake long cruises, surfacing every day to return their data via satellite communication links.*

Vents and Seeps

HYDROTHERMAL VENTS ARE SIMILAR to hot springs on land. Located near ocean ridges and rifts, at an average depth of 7,000 ft (2,100 m), they spew out mineral-rich, superheated seawater. Some have tall chimneys, formed from dissolved minerals that precipitate when the hot vent water meets cold, deep-ocean water. The mix of heat and chemicals supports animal communities around the vents—the first life known to exist entirely without the energy of sunlight. Elsewhere, slower, cooler emissions of chemicals called hydrocarbons occur from sites known as cold seeps.

DISCOVERY
DISCOVERING WHITE SMOKERS

The first hydrothermal vents that scientists observed from *Alvin*, a submersible, in 1977, were black smokers. Scientists then explored other sites near mid-ocean ridges and found more vent systems. Some looked different: their fluids were white, cooler, and emerged more slowly from shorter chimneys. These were called white smokers (see right).

Hydrothermal Vents

Hydrothermal vents always form close to mid-ocean ridges and rifts (see p.185), where new ocean crust is forming and spreading, and where magma from the Earth's mantle lies relatively close to the surface. Seawater seeps into rock cracks opened up by the spreading sea floor. It penetrates several miles into the newly formed crust, close to the hot magma below. This heats the water to 660–750°F (350–400°C). The high pressure at these depths stops it from boiling, and it becomes superheated, dissolving minerals from the rocks that it is passing through, including sulfur which forms hydrogen sulfide. The hot water rises back up through cracks and erupts out of the vents as a hot, shimmering haze, complete with its load of minerals.

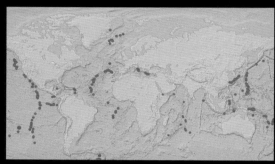

DISTRIBUTION OF VENTS AND RIDGES
Since their discovery in 1977, hydrothermal vents have been found in the Pacific and Indian Oceans, in the mid-Atlantic, and even in the Arctic, always near mid-ocean ridges and rifts.

Black and White Smokers

As superheated water erupts from a hydrothermal vent, it meets the colder water of the ocean depths. This causes hydrogen sulfide in the vent water to react with the metals dissolved in it, including iron, copper, and zinc, which then come out of solution in the form of sulfide particles. Sometimes these form pools on the seabed. However, if the water is particularly hot, it spouts up a little before being chilled by the surrounding seawater, and the metal sulfides form a cloud of black, smokelike particles. Some of these minerals form a crust around the "smoke" plumes, building up into chimneys that can reach more than 100 feet in height. Such vents are called black smokers. Subsequently, a different form of vent was discovered. In these, the black sulfides come out of solution as solids well beneath the sea floor, but other minerals remain in the vent water. Silica and a white mineral called anhydrite form the "smoke" from these chimneys, which, because of their color, are called white smokers.

THE FORMATION OF A SMOKER
Water, heated by magma deep beneath the seabed, dissolves minerals from the rocks. When it erupts through vents, the water is chilled by the surrounding sea. This makes minerals precipitate as smoky clouds, which can be white or black; other minerals are deposited to form chimneys.

SMOKING CHIMNEYS
The minerals from black smokers, like this one, can increase the height of a chimney by an incredible 12 in (30 cm) a day. However, the chimneys are fragile, and they collapse when they get too high.

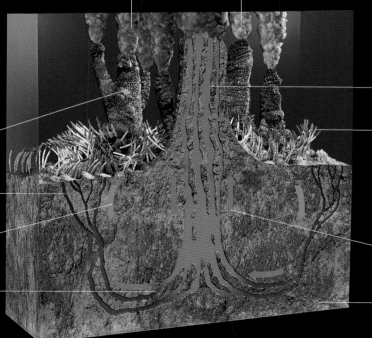

black cloud of metal sulfide particles

white cloud of silica and anhydrite particles

mineral chimney, which can grow at up to 12 in (30 cm) per day

shaft or conduit

unique ecosystems, including tube worms, develop in clusters around some smokers

seepage of seawater down through cracks in the oceanic crust

cold water descends through cracks

heated water rises through vents

superheated water reaching temperatures above 750°F (400°C)

hot rock, or magma, heats water that has seeped into crust

Life Without Sunlight

The first biologists to explore hydrothermal vents were amazed at the life they saw. Masses of limpets, shrimps, sea anemones, and tube worms cluster close to the vents, beside unusually large clams and mussels. White crabs and a few fish, such as the eelpout, scrabble among them. Not every vent system is the same: in the Atlantic, there are no tube worms, clams, or mussels, but lots of white shrimps. Some animals that live in darkness depend on sunlit waters for their food supply but vent animals are remarkable in that they do not need sunlight for energy. White mats of bacteria around vents are the key. They oxidize sulfides from the vent water to make energy, and are the vent animals' food source. Some animals have the bacteria living inside their bodies.

VENT FISH
This fish, called an eelpout, feeds on mussels, shrimps, and crabs living around vents.

GHOSTLY CRAB
The hydrothermal vent crab is one of many vent creatures. Scientists continue to document further new species living around vents.

DIFFERENT ANIMAL COMMUNITIES
Animal communities vary between vent systems. Vents on the Mid-Atlantic Ridge are inhabited by swarms of rift shrimps (shown here), feeding on sulfide-fixing bacteria, but there are no giant clams.

Cold Seeps

The discovery of hydrothermal vents proved that not all deep-sea life depends on sunlight for energy. Soon, other seabed communities were found that could survive in the dark. In the Gulf of Mexico, diverse animal colonies live in shallow waters near where oil companies drill for petroleum. Here, seeps of methane and other hydrocarbons (compounds containing carbon and hydrogen) ooze up from rocks beneath the sea. Mats of bacteria feed on these cold seeps, providing energy for a food chain that includes soft corals, tube worms, crabs, and fish. Other animal communities in deep-sea trenches off the coasts of Japan or between California and Washington, USA, rely on methane, which is released by tectonic activity. Cold-seep communities may be more common than first thought at depths below 1,800 ft (550 m), although there is often no obvious seepage. Such communities may instead rely on chemical-rich sediments exposed by undersea landslides or currents.

OCEAN SMOKER
This black smoker, seen from *Alvin*, is similar to the one that scientists first observed in 1977, spewing out dark fluids from deep in the ocean crust.

WORM WITHOUT A MOUTH

The vent tube worm (below) can be 6 ft (2 m) long and as thick as a human arm. It has no apparent way of feeding. However, its body sac contains an organ called a trophosome, filled with grapelike clusters of bacteria. The worm's crimson plumes collect sulfides from vent water, and the bacteria use these to produce organic material, which the worm absorbs as food.

LIFE ON A SEEP
Mussels containing methane-fixing bacteria live alongside tube worms, soft corals, crabs, and an eelpout at this cold seep, 9,800 ft (3,000 m) down on the seabed near Florida.

THE TWO POLAR OCEANS are the Arctic Ocean in the Northern Hemisphere and the Southern Ocean, which surrounds the continent of Antarctica, in the Southern Hemisphere. They differ from other oceans in several respects, not least in the sheer quantity of ice that floats on them. This includes sea ice, which is frozen seawater, and icebergs and ice shelves, which are frozen fresh water. The polar oceans contain fewer temperature layers than other oceans, being uniformly cold, and they have different circulation patterns, which are partly wind-driven but also influenced by such factors as river inflow (in the Arctic Ocean) and sea ice formation. The edges of the sea ice are biologically productive zones where plankton blooms occur in summer, attracting many fish, birds, and mammals.

POLAR OCEANS

PENGUINS UNDER THE ICE
These emperor penguins are swimming in a break in the sea ice off the coast of Antarctica. They can dive to 1,750 ft (530 m), staying down for up to 20 minutes.

Ice Shelves

AN ICE SHELF IS A HUGE FLOATING ice platform, formed where a glacier, or group of glaciers, extends from a continental ice sheet over the sea. The landward side of an ice shelf is fixed to the shore, where there is a continuous inflow of ice from glaciers or ice streams that flow down from the ice sheet. At its front edge, there is usually an ice cliff, from which massive chunks of ice break off (calve) periodically, forming icebergs. Ice shelves are almost entirely an Antarctic phenomenon, with only a few small ones in the Arctic.

SIR JAMES CLARK ROSS

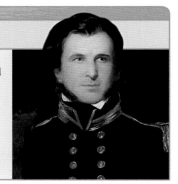

The British naval officer Sir James Clark Ross (1800–1862) spent his early adulthood exploring the Arctic. In 1839, he set off to find the south magnetic pole, and on January 11, 1840 reached Antarctica, near the western side of what is now called the Ross Sea. Later, Ross and his crew discovered an ice cliff 165 ft (50 m) high. This was later named the Ross Ice Shelf.

Antarctic Ice Shelves

Ice shelves surround about 44 percent of the continent of Antarctica and cover an area of some 600,000 square miles (1.5 million square km). The largest is the Ross Ice Shelf, also called the Great Ice Barrier, discovered by Sir James Clark Ross (see panel, above). It is as large as mainland France, with an area of about 188,000 square miles (487,000 square km) and is fed by seven different ice streams. The second largest, the Filchner–Ronne Ice Shelf, covers about 166,000 square miles (430,000 square km). About 15 or so other ice shelves are dotted around the edge of the continent. Since 1995, a few of the smaller ice shelves around the Antarctic Peninsula, including parts of the Larsen Ice Shelf, have disintegrated, most probably as a result of ocean warming (see p.487).

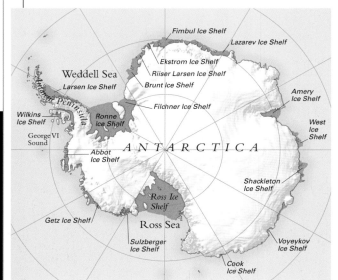

ICE-SHELF LOCATIONS
The two largest ice shelves—the Ross and Filchner–Ronne ice shelves—sit on either side of west Antarctica.

Structure and Behavior

Every ice shelf is anchored to the sea floor (ending at a point called the grounding line) and has a front part that floats. The front part is usually 330–3,300 ft (100–1,000 m) thick, though only about one-ninth protrudes above water. The back of an ice shelf is fixed while the front part moves up and down with the tides, creating stresses that can lead to the formation of cracks. Overall, there is a gradual movement of ice from the rear to the front of an ice shelf, from where large tabular icebergs occasionally calve. There is sometimes also a slow upward migration of ice, due to seawater freezing to the bottom of a shelf and the ice on the upper surface melting and evaporating in summer. Even deposits from the sea floor under an ice shelf are sometimes brought to the surface by this mechanism.

CALVING SHELF
The front part of an ice shelf will sometimes break up and the pieces drift off as tabular icebergs. Each piece visible here has a surface area of several square miles.

GAINS AND LOSSES
An ice shelf gains ice from glaciers flowing into its landward end, from new snowfall, and from seawater freezing to its undersurface. It loses ice by iceberg calving, by some summer melting of its upper surface and through evaporation, and by melting of part of on its undersurface.

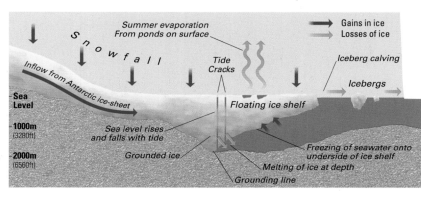

Surface and Interior

The upper surfaces of Antarctic ice shelves are inhospitable places. For most of the year, cold air streams called katabatic winds blow down from the Antarctic Ice Sheet and over the ice shelves. The surface of the ice is not flat, but is shaped by the winds into a series of ridges and troughs, called sastrugi. These are typically covered in a snow blanket. In some areas, the surface is littered with rocks from the input glacier or glaciers, or even with material that has been carried upward from the sea floor by vertical movement. In summer, small ponds form on some ice shelves and provide a home for various types of microscopic organisms. Internally, an ice shelf usually contains some tide-induced cracks and crevasses.

CAVE INSIDE AN ICE SHELF
In summer, the internal cracks and crevasses in an ice shelf may enlarge to form caves as some of the ice melts.

Beneath the Ice Shelves

Underneath the Antarctic ice shelves are extensive bodies of water that are some of the least explored regions on Earth. Seawater is thought to circulate constantly here, caused partly by new ice formation underneath and around the ice shelves. As new ice forms, it "rejects" salt, making the surrounding seawater denser. This causes the seawater to sink, and helps drive the circulation. Recent attempts have been made to explore these areas, using robotic submarines to take measurements. Little is known about the organisms that live here, although in 2005 a community of clams and bacterial mats was found on the sea floor under the Larsen B Ice Shelf after it broke up (see p.487).

LIFE UNDER THE ICE
Organisms such as starfish and worms live in shallow water around the edge of Antarctica, and possibly under the ice shelves.

SEAWATER CIRCULATION
A continuous circulation of seawater is thought to occur under large ice shelves, driven by sea ice formation on its undersurface and partial melting at depth.

Diagram labels: annually reforming fast ice; marine ice is found beneath sea level; ice shelf; melting zone; new ice; ice platelets rise as density decreases; low-salinity water; high-salinity water; grounding line; ice pump driven by salt rejection

ICE CLIFF
This massive ice cliff was photographed at the seaward edge of the Riiser-Larsen Ice Shelf. In front of it, emperor penguins line up to enter the water at Atka Bay, on the Weddell Sea.

OCEAN ENVIRONMENTS

Icebergs

ICEBERGS ARE HUGE, FLOATING chunks of ice that have broken off, or been calved, from the edges of large glaciers and ice shelves. These chunks range from car-sized objects to vast slabs of ice that are bigger than some countries. It is estimated that each year 40,000 to 50,000 substantial icebergs are calved from the glaciers of Greenland. A smaller number of gigantic icebergs break off the ice shelves around Antarctica. Surface currents carry icebergs away from their points of origin into the open ocean, where they drift and slowly melt. They can last for years and are a considerable danger to shipping.

ICEBERG PROPORTIONS
Because pure ice is 90 percent as dense as seawater, an iceberg made entirely of ice will have only 10 percent of its mass visible above water.

Iceberg Properties

Icebergs consist principally of frozen fresh water, with no salt content. This is because they originate not from seawater but from glaciers or ice shelves (floating glaciers), and glaciers themselves come from compacted snow. Typically, an iceberg has a temperature of about 5 to -4°F (-15 to -20°C) at its core and 32°F (0°C) at its surface. In addition to ice, some icebergs contain rock debris. This is material that has fallen onto the parent glacier from surrounding mountains, or frozen to the glacier's edges, and eventually becomes incorporated into the ice. An iceberg's rock load affects its buoyancy. An iceberg with a high rock content may float up to 93 percent submerged.

Sizes and Colors

Icebergs include pieces of ice that are hundreds of square miles in area, down to ones the size of houses (bergy bits) or cars (growlers). Tabular icebergs may rise to a height of up to 200 ft (60 m) above the sea surface and extend underwater to a depth of up to 1,000 ft (300 m). Most icebergs appear white because of the light-reflecting properties of air bubbles trapped in the ice. Those made of dense, bubble-free ice absorb all but the shortest (blue) light wavelengths and so have a vivid blue tint. Occasionally, icebergs roll over and expose a previously submerged section to view, which appears aqua green because of algae growing in the ice.

RANGE OF SHAPES
Icebergs come in a range of shapes including tabular (flat-topped), domed, pinnacled or pyramidal, wedge-shaped, and various irregular shapes, as shown here.

PINNACLED

IRREGULAR

TABULAR

DOMED

North Atlantic Icebergs

Most icebergs seen in the north Atlantic begin as snow falling on Greenland. This snow eventually becomes ice, which over thousands of years is transported from the Greenland ice sheet down to the sea as glaciers. Icebergs calved from the glaciers on the west coast of Greenland (and many from the east coast) move into Baffin Bay. The Labrador Current carries these icebergs southeast, past Newfoundland, into the north Atlantic. There, most of the icebergs rapidly melt, but a few reach as far south as 40°N—around the same latitude as New York and Lisbon.

ARCTIC OCEAN

ORIGINS AND DISTRIBUTION
Most north Atlantic icebergs are calved by glaciers in west Greenland, such as the Jakobshavn and Hayes glaciers.

ICEBERG DETECTION

Because of their threat to shipping, north Atlantic icebergs are monitored by the US Coast Guard. Information on iceberg sightings, obtained by aircraft and ships, is fed into a computer along with ocean-current and wind data. The future movements of the icebergs are then predicted so that ships can be warned. The southernmost iceberg ever spotted in the Atlantic was only 170 miles (275 km) from Bermuda at 30°N and 62°W.

ICELANDIC ICEBERGS
These massive icebergs were calved from Breidamerkurjökull, a glacier in Iceland, and form a surreal tourist attraction, drifting in the glacial lagoon.

Southern Ocean Icebergs

All Southern Ocean icebergs have broken off one of the ice shelves that surround Antarctica (see p.192). Most start off as extremely large, tabular icebergs—satellite monitoring of their drift tracks has provided useful information about Southern Ocean currents. After calving, these icebergs drift westward around Antarctica in a coastal current (the East Wind Drift). A few are carried in an eastward direction by the Antarctic Circumpolar Current. In extreme cases, they drift further, reaching as far north as 42°S in the Atlantic Ocean. The largest iceberg ever recorded, seen near Antarctica in 1956, was 208 miles (335 km) long and 60 miles (97 km) wide. However, as this was before satellite imagery, the accuracy of these measurements has been questioned.

DISTRIBUTION
The approximate limit of iceberg drift from Antarctica is shown by the red dotted line. Most Southern Ocean icebergs remain close to the Antarctic Circle at 67°S.

Ice Rafting

Icebergs that contain rock debris gradually release this material as they melt, and the debris sinks to the sea floor. Thus rock fragments can be transported from Greenland, for example, to the bottom of the north Atlantic. The process is called ice rafting. By examining sediment samples taken from the ocean floor, scientists can often identify rock fragments that have been transported in this way. Such studies can provide clues about past patterns of iceberg calving and iceberg distribution. For example, they have shown that there were short cold periods during the last ice age, called Heinrich Events, when vast armadas of icebergs were calved and crossed the Atlantic eastward from the coast of Labrador.

DIRTY ICEBERG
The fact that this iceberg contains considerable amounts of rock and dust is plain from its "dirty" appearance. This rock will end up on the sea floor as ice-rafted material.

OCEAN ENVIRONMENTS

WRECK OF THE TITANIC
The *Titanic*'s bow section, of which the upper deck and railings are seen here, is mostly intact, although deeply embedded in the seafloor.

The sinking of the ocean liner *Titanic* on April 15, 1912, in the north Atlantic ranks as one of the worst peacetime maritime disasters in history. It is also arguably the most famous sinking of all time, partly because the ship had been considered unsinkable. A total of more than 1,500 people died in the disaster, while just over 700 survived.

The exact sequence of blunders by which the *Titanic* came to collide with an iceberg has never been fully explained. It is known that during the 12 hours preceding the disaster, messages were sent from other ships that large icebergs lay in the *Titanic*'s path. However, these messages may not have reached the ship's bridge. When the collision occurred, the iceberg did not hit the *Titanic* head-on, but brushed the starboard side. However, this was enough to buckle the hull and dislodge rivets below the waterline, creating leaks into five of the ship's hull compartments. Although lifeboats were deployed, there were not enough to hold everyone. Furthermore, some were launched before they were full. As a result, about 1,500 people were still on the ship when it sank. Most are thought to have died of hypothermia in the ice-cold waters.

In 1985, the wreck of the *Titanic* was located by an American–French team, by means of an underwater vehicle with a video camera and lights attached. A notable discovery was that the ship had split in two before sinking—the bow and stern were found lying 2,000 ft (600 m) apart, facing in opposite directions.

First and Last Voyage

The *Titanic* left England on April 10, 1912, bound for New York. After crossing the English Channel, the ship took on additional passengers in France, and also stopped in Ireland the next day, before continuing on its journey. Three days later, on April 14, the ship's captain altered course slightly to the south, possibly in response to iceberg warnings received over the radio. However, at 11:40 p.m., lookouts spotted a large iceberg directly in front of the ship. Despite a frantic avoiding maneuver, the *Titanic* hit the iceberg, and by 2:20 a.m., the ship had sunk.

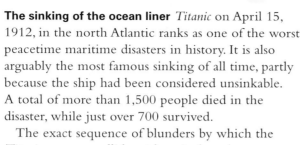

WAS THIS THE ICEBERG?
This photograph, taken six days later in the vicinity of the disaster, shows an iceberg that closely accorded with descriptions provided by survivors.

CANADA

departs Southampton 10th April

Queenstown

Cherbourg

sinks 15th April

New York

ATLANTIC OCEAN

THE HISTORY OF THE *TITANIC*

SETTING SAIL

LEAVING SOUTHAMPTON *At the time of its launch, the Titanic was the world's largest passenger liner and also the most opulent. When the ship left Southampton docks on April 10, 1912, it was carrying about 900 crew and 1,300 passengers, including some of the world's richest and most prominent people.*

MEDIA SENSATION *The sinking of the Titanic caused shock around the world. This Chicago newspaper dates from April 16, 1912.*

DISASTER

THE UNSINKABLE SINKS
In the early morning of April 15, about 2.5 hours after colliding with an iceberg, the Titanic's stern rose out of the water as the ship sank.

BOB BALLARD *Along with French scientist Jean-Louis Michel, American oceanographer Bob Ballard led the team that discovered the wreck of the Titanic on September 1, 1985, at a depth of 12,500 ft (3,800 m).*

DISCOVERY OF THE WRECK

LIFEBOAT WINDLASS *This piece of deck machinery was barely recognizable under a covering of rusticles (nodules containing a mixture of iron compounds and microbes that feed on wrought iron).*

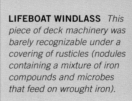
BANKNOTES
Banknotes in surprisingly good condition have been retrieved, including this $5 bill found in the purser's bag.

CHINA DISHES *Rows of dishes were found lying on the seafloor. Such diverse items as books, watches, and wireless messages have also been retrieved, along with a bronze cherub and hundreds of other objects.*

ARTIFACTS

Sea Ice

SEA ICE IS SEAWATER THAT HAS FROZEN at the ocean surface and floats on the liquid seawater underneath. It includes pack ice—ice that is not attached to the shoreline and drifts with wind and currents—and fast ice, which is frozen to a coast. Sea ice formation and melting influences the large-scale circulation of water in the oceans. It has important stabilizing effects on the world's climate, since it helps control the movement of heat energy between the polar oceans and atmosphere. Sea ice strongly reflects solar radiation, so in summer it reduces heating of the polar oceans. In winter, it acts as an insulator, reducing heat loss. Today, scientists are concerned about shrinking sea ice in the Arctic because of its possible effects on climate and wildlife.

Formation

Seawater starts to freeze when it reaches a temperature of 28.8°F (−1.8°C), slightly cooler than the freezing point of fresh water. Sea ice formation starts with the appearance of tiny needlelike ice crystals (frazil ice) in the water. Salt in seawater cannot be incorporated into ice, and the crystals expel salt. The developing sea ice gradually turns into a thick slush and then, under typical wave conditions, into a mosaic of ice platelets called pancake ice. Subsequently, it consolidates into a thick, solid sheet, through processes such as "rafting" (in which the ice fractures and one piece overrides another) and "ridging" (where lines of broken ice are forced up by pressure). Where ridging occurs, each ridge has a corresponding structure, a keel, that forms on the underside of the ice. Newly formed, compacted sheet ice is called first-year ice and may be up to 12 in (30 cm) thick. It continues to thicken through the winter. Any ice that remains through to the next winter is called multiyear ice.

TESTING THE ICE
Pancake ice, consisting of ice platelets, can be up to 4 in (10 cm) thick. Waves and wind have caused these platelets to collide, hence their curled-up edges.

OCEAN ENVIRONMENTS

HOW ICE FORMS
The stages of sea ice formation vary according to whether the sea surface is calm or affected by waves. A typical sequence in an area of moderate wave action is shown below.

GREASE ICE
Fine ice spicules, called frazils, appear in the water. These coagulate into a viscous soup of ice crystals, called grease ice.

PANCAKE ICE
Wave action causes the grease ice to break into slushy balls of ice, called shuga. These clump into platter shapes called pancakes.

FIRST-YEAR ICE
The ice pancakes congeal, consolidate, and thicken through processes such as rafting and ridging to form a continuous sheet of ice.

MULTIYEAR ICE
Further thickening, for a year or more, produces multiyear ice. This has a rough surface and may be several yards thick.

Extent and Thickness

The extent of sea ice in the polar oceans varies over an annual cycle. About 85 percent of the winter ice that forms in the Southern Ocean melts in summer, and on average this ice only reaches a thickness of a yard or two. In the Arctic, some of the ice lasts for several seasons, and this multiyear ice attains a greater thickness—on average 7–10 ft (2–3 m). In winter, ice covers most of the Arctic Ocean. In summer, it shrinks in area by more than two-thirds. Since around 2005, the summer retreat has been more pronounced, raising fears that summer ice coverage may disappear altogether by 2050 or earlier.

ARCTIC SEA ICE COVERAGE
Coverage varies from a winter high of 6 million square miles (15 million square km) to a summer low of about 1.4–1.8 million square miles (3.7–4.7 million square km).

☐ year-round ice

▨ winter sea ice

USS NAUTILUS

In 1958, a US submarine, the USS *Nautilus*, crossed the Arctic Ocean underneath its cover of sea ice, passing the North Pole on August 3. The crossing proved that there is no sizable land mass in the middle of the Arctic Ocean. The submarine traversed the Arctic from the Beaufort Sea to the Greenland Sea in four days at a depth of about 500 ft (150 m).

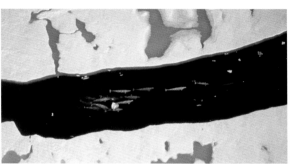

ICE LEAD
An ice lead forms when an area of sea ice shears. Stresses from winds and water currents are thought to be the cause. Here, a group of beluga whales swims along a lead.

Gaps in the Ice

Even in parts of the polar oceans that are more or less permanently ice-covered, gaps and breaks sometimes appear or persist in the ice. These openings vary greatly in size and extent and have different names. Fractures are extremely narrow ruptures that are usually not navigable by boats of any size. An ice lead is a long, straight, narrow passageway that opens up spontaneously in sea ice, making it navigable by surface vessels and some marine mammals. Polynyas are persistent regions of open water, up to a few hundred square miles in area and often roughly circular in shape. They sometimes develop where there is upwelling of warmer water in a localized area, or near coasts where the wind blows new sea ice away from the shore as it forms.

Life Around the Ice

Life thrives around sea ice. One reason for this is that as ice forms, salt is expelled into the seawater, causing it to become denser and sink. This forces nutrient-laden water to the surface. In summer, the combination of nutrients and sunlight encourages the growth of phytoplankton, which provide a rich food source. These organisms form the base of a food chain for fish, mammals, and birds. In the Arctic, sea ice provides a resting and birthing place for seals and walruses and a hunting and breeding ground for polar bears and Arctic foxes. In the Antarctic, it supports seals and penguins. Breaks in the ice are vital to this wildlife. Seals, penguins, and whales rely on them for access to the air, while polar bears hunt near them. Decreases in Arctic sea ice would drastically shrink some habitats, pushing them toward extinction.

ANTARCTIC KRILL
These crustaceans form an important part of the food chain in the Southern Ocean, where they congregate in dense masses.

WEDDELL SEAL
The Weddell seal, found only in the Antarctic, is one of nine seal species that inhabit polar oceans. Weddell seals never stray far from sea ice.

ICEBREAKERS

Icebreakers are ships designed for moving through ice-covered environments. An icebreaker has a reinforced hull and a bow shape that causes the ship to ride over sea ice and crush it as it moves forward. The shape of the vessel clears ice debris to the sides and under the hull, allowing steady progress. The most powerful modern icebreaker can advance through sea ice that is 9 ft (2.8 m) thick.

Polar Ocean Circulation

THE ARCTIC AND SOUTHERN OCEANS each have their own unique patterns of water flow, which link in with the rest of the global ocean circulation. These flows are driven partly by wind and partly by various factors that influence the temperature and salinity of the surface waters in these oceans—including seasonal variations in air temperature and sea ice coverage, and large inflows of fresh water from rivers. Although driven by similar influences, the significantly different water-flow patterns of these two oceans are largely due to the fact that the Arctic Ocean is encircled by land, whereas the Southern Ocean surrounds a frozen continent.

MOUTH OF THE LENA RIVER
The Lena flows across Siberia and discharges 100 cubic miles (420 cubic km) of water into the Arctic Ocean every year.

Arctic Surface Circulation

The upper 170 ft (50 m) of the Arctic Ocean is affected by currents that keep it in constant motion. There are two main components to this circulation (see pp.424–425). In a large area north of Alaska, there is a slow, circular motion of water called the Beaufort Gyre. This clockwise movement is wind-generated and completes one rotation every four years. The second component, the Transpolar Drift, is driven by water discharged into the Arctic Ocean from Siberian rivers. It carries water and ice from a region north of eastern Siberia toward northeastern Greenland.

CIRCULATION AND FEEDING
The Southern Ocean meets warmer water at the Antarctic Convergence, creating a biologically rich feeding area for whales, including these humpbacks.

OCEAN ENVIRONMENTS

Arctic Deep-water Circulation

In the deeper waters of the Arctic, there is a slow circulation of cold, dense water. This circulation is restricted by the structure of the Arctic Ocean, which consists of a central deep basin (the Arctic Basin), bisected by several underwater ridges and surrounded on most sides by shallow continental shelf. Only on the Atlantic side is there a connection between the deep waters of the Arctic and deep ocean waters to the south. On the opposite side, the connection with the Pacific is via the shallow and narrow Bering Strait. What little circulation of water occurs in the Arctic Basin largely involves influxes of mostly relatively warm, salty Atlantic water at various depths to the north of Russia via the Barents Sea and outflows of cooler, denser (and thus sinking) water to the west—particularly, to the east of Greenland.

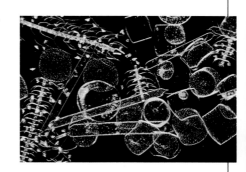

MISTY SEAS
The seas around the Antarctic Convergence are prone to mists. Here, a cruise liner approaches a channel just south of the Convergence.

The Antarctic Convergence

The Antarctic Convergence is a region of the Southern Ocean encircling Antarctica, located roughly at latitude 55°S (but deviating from this in places), where cold, northward-flowing waters from Antarctica sink beneath the relatively warmer waters to the north.

At the Convergence, there is a sudden change in surface ocean temperature of 5–9°F (3–5°C) as well as alterations in the chemical composition of seawater. As a result, the Convergence forms a barrier to the movement of animal species, and the groups of marine animals found on either side of it are quite different. This is a turbulent area. The meeting of different water masses brings dissolved nutrients from the seabed to the ocean surface. This acts as a fertilizer, encouraging the growth of plankton during the Southern Hemisphere summer.

ARCTIC BASIN CIRCULATION
Atlantic water enters to the north of Russia and Pacific water via the Bering Strait. As it cools, some of the Atlantic water becomes denser and dips far below the sea ice cover, where it flows around slowly. The main outflow is to the east of Greenland.

→ atlantic water
→ pacific water

PHYTOPLANKTON
In summer, massive blooms of phytoplankton occur around the Antarctic Convergence, forming the base of a productive food chain.

Southern Ocean Circulation

In the Southern Ocean, surface waters move under the influence of two wind-driven currents. Off the coast of Antarctica, the Antarctic Coastal Current carries water from east to west around Antarctica. Several hundred miles north, the Antarctic Circumpolar Current (ACC) moves water in the opposite direction, from west to east, and pushes the Antarctic waters northward. The ACC is a major ocean current that connects the Pacific, Atlantic, and Indian Oceans and isolates Antarctica from the warmer ocean currents to the north. In the Southern Ocean, an important movement of water also occurs deep down. In an area near Antarctica, masses of dense, salty water form as salt is rejected from seawater as it freezes. This cold water sinks and moves north into the southern Atlantic.

ALBATROSSES
These black-browed and gray-headed albatrosses inhabit the biologically productive Southern Ocean.

PEOPLE

FRIDJTOF NANSEN

The Norwegian explorer and scientist Fridjtof Nansen (1861–1930) is most famous for his Arctic voyage of 1893–1895 on a specially built wooden ship called the *Fram*. Nansen deliberately allowed the *Fram* to drift across the Arctic Ocean locked in ice, and in doing so proved the existence of the surface current now called the Transpolar Current. In 1895, setting off with one companion from the *Fram*, Nansen walked and skied to within 400 miles (640 km) of the North Pole, closer than anyone else up to that time.

OCEAN LIFE

BY FAR THE LARGEST ECOSYSTEM on Earth, the ocean contains a great range of habitats and environments as disparate as mangrove swamps and deep-sea vents. Living organisms have found a place to take hold in every ocean environment, even the deepest trenches, more than 6 miles (10 km) beneath the surface. Ocean life teems with greatest abundance and variety in the sunlit surface waters. Here, microscopic plants and plantlike organisms, the phytoplankton, fuel productive communities of organisms right up to top predators. Life most likely began in the ocean, and saltwater was where many groundbreaking steps in evolution took place. Tracing the history of this evolution puts into context the astonishing variety of today's marine life.

INTRODUCTION
TO OCEAN LIFE

KELP FOREST COMMUNITY
Ocean life exists within many different characteristic communities, according to physical and biological conditions. Here in cool, shallow water in the UK, a canopy of kelp towers above an undergrowth of smaller seaweeds, while sea urchins graze among them and on the kelp stipes (stems).

Classification

BY CLASSIFYING ORGANISMS AND FITTING them into a universally accepted framework, scientists have created a massive reference system that accommodates all forms of life. Over 2 million organisms have been described, of which about 20 percent live in the oceans. The marine proportion is likely to increase because many new species continue to be discovered annually, particularly in the deep ocean.

LINNAEAN HIERARCHY

Linnaeus used a hierarchy of ranked categories of increasing exclusiveness. Today's expanded system includes many ranks, from domain down to species. Below is an example of a series of ranked categories, illustrating those that classify the common dolphin.

DOMAIN Eucarya
Includes all Eukaryotes—organisms that have complex cells with distinct nuclei. Only bacteria and archaea fall outside this domain.

KINGDOM Animalia
Includes all animals—multicellular eukaryotes that need to eat food for energy. All animals are mobile for at least part of their lives.

PHYLUM Chordata
Includes all chordates—animals possessing a notochord. In most cases, the notochord is replaced before birth by the backbone.

CLASS Mammalia
Includes all mammals—air-breathing chordates that feed their young on milk. The jaw is made up of a single bone.

ORDER Cetartiodactyla
Includes all cetaceans (whales and dolphins)—marine mammals that have a tail with boneless, horizontal flukes for propulsion.

FAMILY Delphinidae
Includes all dolphins (a subgroup of toothed cetaceans) with beaks and 50–100 vertebrae. The skull lacks a crest.

GENUS *Delphinus*
Includes a few colorful, oceanic dolphins with 40–50 teeth on each side of the jaw. These dolphins form large social groups.

SPECIES *Delphinus delphis*
Specifies a single type of dolphin with a V-shaped black cape under the dorsal fin and criss-cross hour-glass patterning on its sides.

Principles of Classification

Classification helps us make sense of the natural world by grouping organisms on the basis of features that they share. It gives scientists a clear and accurate understanding of the diversity of life, and because everyone uses the same system, the knowledge is accessible on a worldwide basis. The hierarchical system devised by the Swedish scientist Carolus Linnaeus (see panel, left) in the 18th century still forms the basis of today's classification. Each species is identified with a unique two-part scientific name (made up of the genus and species names), then categorized in a series of ever-larger groupings. However, as our knowledge increases, it is often necessary to revise the groups. Sometimes, this leads to subdivision of categories, for example phylum Arthropoda has been split into four subphyla. Many new species are discovered each year but describing and publishing them is laborious and time-consuming.

WHAT IS A SPECIES?

A species is the basic unit of classification. One commonly accepted definition of a species is a group of living organisms that have so many features in common that they can interbreed and exchange genes in natural conditions. This definition cannot be applied to fossil species. It also does not work for Bacteria and Archaea and there can be no one universal definition. Other factors such as geographical isolation and DNA (see below) are also important.

The Evidence

In the past, scientists could identify and classify organisms only by studying anatomy, by looking at form, function, and embryological development (animals only), and by examining the fossil record. Recently, scientists have also been able to investigate organisms by looking at their

DETAILED ANATOMY
By making a detailed anatomical examination of material in museum collections, scientists can distinguish between similar organisms and classify them according to shared characters.

proteins and their DNA. DNA is a complex molecule whose sequential structure is unique to each organism. The relatedness of organisms can be determined by comparing these DNA molecules for shared features. This molecular evidence has led to many revisions of classification.

Cladistics

By the 1950s, although most people used the same system of classification, the criteria they used for placing organisms in categories were often neither measurable nor repeatable. The idea emerged to analyze many characters using an automatic, computerlike process, not only to classify organisms, but also to trace their evolution. This process became known as cladistics, and it is a widely used technique today. A cladistic analysis examines a wide selection of characters shared by a study group of organisms. It finds the most likely pattern of evolutionary changes that link the organisms, involving the least number of steps (evolutionary branching points). It then arranges the organisms in a tree diagram (cladogram) that reflects their relationships. A cladogram is made up of nested groups called clades. A clade encompasses all the descendants of the group's common ancestor.

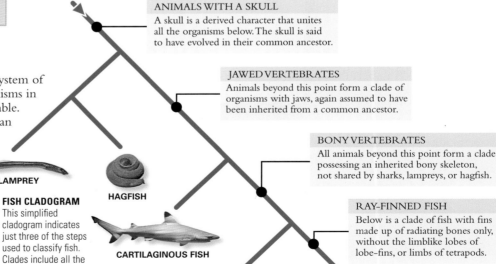

ANIMALS WITH A SKULL
A skull is a derived character that unites all the organisms below. The skull is said to have evolved in their common ancestor.

JAWED VERTEBRATES
Animals beyond this point form a clade of organisms with jaws, again assumed to have been inherited from a common ancestor.

BONY VERTEBRATES
All animals beyond this point form a clade possessing an inherited bony skeleton, not shared by sharks, lampreys, or hagfish.

RAY-FINNED FISH
Below is a clade of fish with fins made up of radiating bones only, without the limblike lobes of lobe-fins, or limbs of tetrapods.

LAMPREY

FISH CLADOGRAM
This simplified cladogram indicates just three of the steps used to classify fish. Clades include all the descendants of a common ancestor, so some new groups, such as "lobe-finned fish and tetrapods" result, since all tetrapods (land vertebrates) share a common ancestor.

HAGFISH

CARTILAGINOUS FISH

LOBE-FINNED FISH AND TETRAPODS

RAY-FINNED FISH

Marine Life

THE CLASSIFICATION FRAMEWORK USED in this book is shown on the following three pages. In this framework, all living things are divided into three domains. Within domains, only the marine groups are shown, although the numbers of classes and species cited include all living organisms within the group whether they are marine or not. Some groupings, such as fish, are shown in dotted lines because although they are useful categories, they are not true taxonomic groups. Others, such as bottom-living phyla and planktonic phyla, are ecological groupings and do not reflect taxonomy or evolutionary history.

BACTERIA
DOMAIN Bacteria **SPECIES** Many millions

ARCHAEA
DOMAIN Archaea **SPECIES** Probably millions

EUKARYOTES
DOMAIN Eucarya **KINGDOMS** At least 8 **SPECIES** 2 million

EUKARYOTES

THIS DOMAIN INCLUDES ALL ORGANISMS that have cells with a nucleus and other complex structures not seen in prokaryotes (bacteria, archaea). The eukaryotes comprise protists, chromists, plants, fungi, and animals.

Chromists

THE CLASSIFICATION OF SINGLE-CELLED ORGANISMS is complex, difficult, and constantly in flux. Many important marine plankton groups are collectively referred to as chromists. Others are instead plants or protozoans (Kingdom Protozoa).

Dinoflagellates
PHYLUM Myzozoa **INFRAPHYLUM** Dinoflagellata **CLASSES** 2 **SPECIES** 2,931

Ciliates
PHYLUM Ciliophora **CLASSES** About 11 **SPECIES** 3,039

Radiolarians
PHYLUM Radiozoa **CLASSES** 5 **SPECIES** 467

Foraminiferans
PHYLUM Foraminifera **CLASSES** 4 **SPECIES** 8,195

Coccolithophorids
PHYLUM Haptophyta **CLASSES** 3 **SPECIES** 391

Ochrophyta
PHYLUM Ochrophyta **CLASSES** 15 **SPECIES** 9,839

> #### DIATOMS
> **CLASS** Bacillariophyceae **ORDERS** 35 **SPECIES** 5,494
>
> #### GOLDEN YELLOW ALGAE
> **CLASS** Chrysophyceae **ORDERS** 8 **SPECIES** 738
>
> #### BROWN SEAWEEDS
> **CLASS** Phaeophyceae **ORDERS** 20 **SPECIES** 2,364

+ SEVERAL MORE PHYLA

Plants

KINGDOM Plantae **DIVISIONS** 7 **SPECIES** 315,000

PLANTS COMPRISE SEVEN DIVISIONS, only three of which have truly marine species and are included in this book. Mosses (Bryophyta) are additionally included because a few of them live in the intertidal zone.

RED SEAWEEDS
DIVISION Rhodophyta **CLASSES** 7 **SPECIES** 7,895

GREEN SEAWEEDS AND ALGAE
DIVISION Chlorophyta **CLASSES** About 10 **SPECIES** 7,371

> #### GREEN MICROALGAE
> **CLASS** Prasinophyceae **ORDERS** 1 **SPECIES** 36

GREEN SEAWEEDS
CLASS Ulvophyceae **ORDERS** 8 **SPECIES** About 2,180

+ EIGHT MORE CLASSES OF MAINLY GREEN MICROALGAE

MOSSES
DIVISION Bryophyta **CLASSES** 8 **SPECIES** 20,000

VASCULAR PLANTS
DIVISION Trachaeophyta **CLASSES** 7 **SPECIES** 350,000

> #### FLOWERING PLANTS
> **CLADE** Angiospermae **ORDERS** 64 **SPECIES** 325,000

Fungi
KINGDOM Fungi **PHYLA** 5 **SPECIES** 148,000

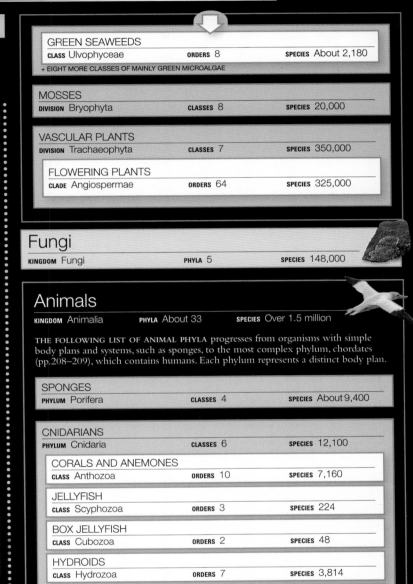

Animals
KINGDOM Animalia **PHYLA** About 33 **SPECIES** Over 1.5 million

THE FOLLOWING LIST OF ANIMAL PHYLA progresses from organisms with simple body plans and systems, such as sponges, to the most complex phylum, chordates (pp.208–209), which contains humans. Each phylum represents a distinct body plan.

SPONGES
PHYLUM Porifera **CLASSES** 4 **SPECIES** About 9,400

CNIDARIANS
PHYLUM Cnidaria **CLASSES** 6 **SPECIES** 12,100

> #### CORALS AND ANEMONES
> **CLASS** Anthozoa **ORDERS** 10 **SPECIES** 7,160
>
> #### JELLYFISH
> **CLASS** Scyphozoa **ORDERS** 3 **SPECIES** 224
>
> #### BOX JELLYFISH
> **CLASS** Cubozoa **ORDERS** 2 **SPECIES** 48
>
> #### HYDROIDS
> **CLASS** Hydrozoa **ORDERS** 7 **SPECIES** 3,814
>
> #### STALKED JELLYFISH
> **CLASS** Staurozoa **ORDERS** 1 **SPECIES** 49

PLANKTONIC PHYLA

THE FOLLOWING THREE PHYLA FLOAT with the ocean currents in the plankton and are grouped here on this basis. The Ctenophora and Chaetognatha contain so few species that they are known as minor phyla.

COMB JELLIES
PHYLUM Ctenophora **CLASSES** 2 **SPECIES** 205

ARROW WORMS
PHYLUM Chaetognatha **CLASSES** 1 **SPECIES** 132

ROTIFERANS
PHYLUM Rotifera **CLASSES** 2 **SPECIES** About 2,000

FLATWORMS

PLATYHELMINTHES		
PHYLUM Platyhelminthes	SUBPHYLA 2	SPECIES 21,274

XENACOELMORPHA		
PHYLUM Xenacoelomorpha	SUBPHYLA 2	SPECIES 456

RIBBON WORMS		
PHYLUM Nemertea	CLASSES 4	SPECIES 1,350

SEGMENTED WORMS		
PHYLUM Annelida	CLASSES 2	SPECIES 16,000

BOTTOM-LIVING PHYLA

MEMBERS OF THE FOLLOWING PHYLA all live in or on the ocean floor. The list is not comprehensive—the following phyla are among those not included: Entoprocta, Acanthocephala, and Placozoa.

SPOON WORMS		
PHYLUM Echiura	CLASSES 1	SPECIES 173

LAMP SHELLS		
PHYLUM Brachiopoda	CLASSES 3	SPECIES 414

HORSESHOE WORMS		
PHYLUM Phoronida	CLASSES 1	SPECIES 13

PEANUT WORMS		
PHYLUM Sipuncula	CLASSES 2	SPECIES 158

CYCLIOPHORANS		
PHYLUM Cycliophora	CLASSES 1	SPECIES 2

GASTROTRICHS		
PHYLUM Gastrotricha	CLASSES 1	SPECIES 866

ROUND WORMS		
PHYLUM Nematoda	CLASSES 3	SPECIES 20,000

PRIAPULA WORMS AND MUD DRAGONS		
PHYLA Priapulida and Kinorhyncha	CLASSES 2	SPECIES 330

WATER BEARS		
PHYLUM Tardigrada	CLASSES 3	SPECIES 1,000

PTEROBRANCH WORMS AND ACORN WORMS		
PHYLUM Hemichordata	CLASSES 3	SPECIES 130

MOLLUSKS

PHYLUM Mollusca	CLASSES 8	SPECIES 79,000

CAUDOFOVEATES		
CLASS Caudofoveata	ORDERS 1	SPECIES 131

SOLENOGASTRES		
CLASS Solonogasters	ORDERS 4	SPECIES 273

MONOPLACOPHORANS		
CLASS Monoplacophora	ORDERS 1	SPECIES 30

TUSK SHELLS		
CLASS Scaphopoda	ORDERS 1	SPECIES 576

BIVALVES		
CLASS Bivalvia	ORDERS 17	SPECIES 9,209

GASTROPODS		
CLASS Gastropoda	ORDERS 16	SPECIES 67,000

CEPHALOPODS		
CLASS Cephalopoda	ORDERS 9	SPECIES 822

CHITONS		
CLASS Polyplacophora	ORDERS 3	SPECIES 1,026

ARTHROPODS

PHYLUM Arthropoda	SUBPHYLA 4	SPECIES About 1.25 million

CRUSTACEANS		
SUBPHYLUM Crustacea	CLASSES 6	SPECIES 67,411

WATER FLEAS AND RELATIVES		
CLASS Branchiopoda	ORDERS 9	SPECIES 1,534

BARNACLES AND COPEPODS		
CLASS Thecostraca	ORDERS 17	SPECIES 16,211

MUSSEL SHRIMPS		
CLASS Ostracoda	ORDERS 5	SPECIES 6,429

MALACOSTRACANS		
CLASS Malacostraca	ORDERS 18	SPECIES 43,237

MANTIS SHRIMPS		
ORDER Stomatopoda	FAMILIES 17	SPECIES 491

ISOPODS		
ORDER Isopoda	FAMILIES 94	SPECIES 10,541

AMPHIPODS		
ORDER Amphipoda	FAMILIES 119	SPECIES 10,333

KRILL		
ORDER Euphausiacea	FAMILIES 2	SPECIES 86

LOBSTERS, CRABS, AND SHRIMPS		
ORDER Decapoda	FAMILIES 105	SPECIES 17,032

+ 10 MORE MINOR ORDERS

CHELICERATES		
SUBPHYLUM Chelicerata	CLASSES 13	SPECIES c. 71,500

SPIDERS, SCORPIONS, TICKS, AND MITES		
CLASS Arachnida	ORDERS 12	SPECIES 70,000

HORSESHOE CRABS		
CLASS Merostomata	ORDERS 1	SPECIES 4

SEA SPIDERS		
CLASS Pycnogonida	ORDERS 1	SPECIES 1,368

HEXAPODS		
SUBPHYLUM Hexapoda	CLASSES 4	SPECIES About 1.11 million

INSECTS		
CLASS Insecta	ORDERS 29	SPECIES 1.1 million

+ 1 OTHER SUBPHYLUM, MILLIPEDES AND CENTIPEDES (MYRIAPODA)

BRYOZOANS

PHYLUM Bryozoa	CLASSES 3	SPECIES 6,530

ECHINODERMS

PHYLUM Echinodermata	CLASSES 5	SPECIES 7,483

SEA LILIES AND FEATHER STARS		
CLASS Crinoidea	ORDERS 4	SPECIES 672

STARFISH		
CLASS Asteroidea	ORDERS 9	SPECIES 1,921

BRITTLESTARS		
CLASS Ophiuroidea	ORDERS 6	SPECIES 2,111

SEA URCHINS		
CLASS Echinoidea	ORDERS 16	SPECIES 1,015

SEA CUCUMBERS		
KINGDOM Holothuroidea	ORDERS 7	SPECIES 1,764

CHORDATES

PHYLUM Chordata	**SUBPHYLA** 3	**SPECIES** About 66,000

THE VERTEBRATES DOMINATE PHYLUM CHORDATA. The remaining two, much smaller, subphyla are united with vertebrates by the presence of the rodlike notochord, which becomes the backbone before birth in vertebrates.

TUNICATES (SEA SQUIRTS AND RELATIVES)

SUBPHYLUM Tunicata	**CLASSES** 3	**SPECIES** 3,091

LANCELETS

SUBPHYLUM Cephalochordata	**CLASSES** 1	**SPECIES** 30

VERTEBRATES

SUBPHYLUM Vertebrata	**CLASSES** 12	**SPECIES** About 63,000

HAGFISH HAVE AT TIMES been excluded from the vertebrates because they have only vestiges of a vertebral column. However, recent molecular studies confirm they are related to the other jawless fish, the lampreys. Reptiles, birds, and mammals (as well as amphibians, of which there are no marine species) are informally grouped together as tetrapods (Tetrapoda) within the larger group of jawed vertebrates (Gnathostomata), which also include fish.

FISH

"FISH" IS AN INFORMAL TERM for four classes of animals. Similarly, "jawless fish," "cartilaginous fish," and "bony fish" are informal groupings.

JAWLESS FISH (AGNATHANS)

HAGFISH

CLASS Myxini	**ORDERS** 1	**SPECIES** 92

LAMPREYS

CLASS Cephalaspidomorphi	**ORDERS** 1	**SPECIES** 50

SHARKS, RAYS, AND CHIMAERAS

SHARKS, SKATES, AND RAYS

CLASS Elasmobranchii	**ORDERS** 13	**SPECIES** 1,303

SHARKS		
ORDERS 9	**FAMILIES** 35	**SPECIES** 563

SKATES AND RAYS		
ORDERS 4	**FAMILIES** 19	**SPECIES** 740

CHIMAERAS

CLASS Holocephali	**ORDERS** 1	**SPECIES** 57

BONY FISH

LOBE-FINNED FISH

SUBCLASS Sarcopterygii	**ORDERS** 2	**SPECIES** 8

RAY-FINNED FISH

SUBCLASS Actinopterygii	**ORDERS** 45	**SPECIES** 31,282

STURGEONS AND PADDLEFISHES			SALMONS		
ORDER Acipenseriformes	**SPECIES** 27		**ORDER** Salmoniformes	**SPECIES** 242	
TARPONS AND TENPOUNDERS			LIGHTFISH AND DRAGONFISH		
ORDER Elopiformes	**SPECIES** 9		**ORDER** Stomiiformes	**SPECIES** 449	
BONEFISH			GRINNERS		
ORDER Albuliformes	**SPECIES** 13		**ORDER** Aulopiformes	**SPECIES** 300	
EELS			LANTERNFISH AND RELATIVES		
ORDER Anguilliformes	**SPECIES** 1,030		**ORDER** Myctophiformes	**SPECIES** 278	
SWALLOWERS AND GULPERS			VELIFERS, TUBE-EYES, RIBBONFISH		
ORDER Saccopharyngiformes	**SPECIES** 28		**ORDER** Lampriformes	**SPECIES** 27	
HERRINGS AND RELATIVES			COD FISH AND RELATIVES		
ORDER Clupeiformes	**SPECIES** 435		**ORDER** Gadiformes	**SPECIES** 652	
MILKFISH			TOADFISH AND MIDSHIPMEN		
ORDER Gonorhynchiformes	**SPECIES** 6		**ORDER** Batrachoidiformes	**SPECIES** 5	
CATFISH AND KNIFEFISH			CUSK EELS		
ORDER Siluriformes	**SPECIES** 4,163		**ORDER** Ophidiiformes	**SPECIES** 563	
SMELTS AND RELATIVES			ANGLERFISH		
ORDER Osmeriformes	**SPECIES** 46		**ORDER** Lophiiformes	**SPECIES** 99	

CLINGFISH			GOBIES AND RELATIVES		
ORDER Gobiesociformes	**SPECIES** 186		**ORDER** Gobiiformes	**SPECIES** 29	
NEEDLEFISH			PIPEFISH AND SEAHORSES		
ORDER Beloniformes	**SPECIES** 283		**ORDER** Syngnathiformes	**SPECIES** 4	
SILVERSIDES			PERCHLIKE FISH		
ORDER Atheriniformes	**SPECIES** 393		**ORDER** Perciformes	**SPECIES** Many thousands	
SQUIRRELFISH AND RELATIVES			FLATFISH		
ORDER Holocentriformes	**SPECIES** 92		**ORDER** Pleuronectiformes	**SPECIES** 825	
DORIES AND RELATIVES			PUFFERS AND FILEFISH		
ORDER Zeiformes	**SPECIES** 35		**ORDER** Tetraodontiformes	**SPECIES** 47	

+ 16 MORE ORDERS

REPTILES

CLASS Reptilia	**ORDERS** 4	**SPECIES** 11,570

TURTLES AND TORTOISES		
ORDER Testudines	**FAMILIES** 12	**SPECIES** 360

SNAKES AND LIZARDS		
ORDER Squamata	**FAMILIES** 44	**SPECIES** 10,980

CROCODILES		
ORDER Crocodylia	**FAMILIES** 3	**SPECIES** 26

+ 1 NONMARINE ORDER: THE TUATARAS (SPHENODONTIDA)

BIRDS

CLASS Aves	**ORDERS** 40	**SPECIES** 10,650

IN THIS CLASSIFICATION, the birds have been divided into 40 orders. Some scientists consider birds to be grouped within the reptiles.

WATERFOWL (DUCKS, GEESE, AND SWANS)		
ORDER Anseriformes	**FAMILIES** 2	**SPECIES** 170

PENGUINS		
ORDER Sphenisciformes	**FAMILIES** 1	**SPECIES** 18

DIVERS (LOONS)		
ORDER Gaviiformes	**FAMILIES** 1	**SPECIES** 5

ALBATROSSES AND PETRELS		
ORDER Procellariiformes	**FAMILIES** 4	**SPECIES** 143

GREBES		
ORDER Podicipediformes	**FAMILIES** 1	**SPECIES** 20

TROPICBIRDS		
ORDER Phaethontiformes	**FAMILIES** 1	**SPECIES** 3

PELICANS AND RELATIVES		
ORDER Pelicaniformes	**FAMILIES** 5	**SPECIES** 111

CORMORANTS AND RELATIVES		
ORDER Suliformes	**FAMILIES** 4	**SPECIES** 60

BIRDS OF PREY		
ORDER Accipitriformes	**FAMILIES** 5	**SPECIES** 265

WADERS, GULLS, AND AUKS		
ORDER Charadriiformes	**FAMILIES** 18	**SPECIES** 381

KINGFISHERS AND RELATIVES		
ORDER Coraciiformes	**FAMILIES** 9	**SPECIES** 180

+ 29 NONMARINE ORDERS

MAMMALS

CLASS Mammalia	**ORDERS** 27	**SPECIES** 6,431

THREE PARTLY OR WHOLLY MARINE mammal orders are listed here. The pinnipeds (seals, sea lions, and walruses), until recently classified as order Pinnipeda, do not form a natural group, and have been placed within order Carnivora (cats, dogs, bears, otters, and relatives). The 27 mammal orders includes new orders formerly classified as marsupials.

CARNIVORES		
ORDER Carnivora	**FAMILIES** 16	**SPECIES** 302

WHALES AND DOLPHINS		
ORDER Cetacea	**FAMILIES** 13	**SPECIES** 93

SEA COWS		
ORDER Sirenia	**FAMILIES** 2	**SPECIES** 4

+ 23 MORE NONMARINE ORDERS

RED SEA REEF
The Red Sea is one of the world's top 18 coral hotspots. Its colorful reefs are home to an abundance of marine life, including the venomous red lionfish.

Many people have heard of biodiversity hotspots, particularly in the context of documentaries about ocean life. These sites are very popular with filmmakers for the variety of life they exhibit. However, the term is a slight misnomer. Strictly speaking, such sites are "species diversity hotspots," places where the largest number of species are concentrated in a small area. Identifying such hotspots helps conservationists to decide where protected areas should be set up. However, places where species diversity is low, such as the ocean trenches, are also important because of the remarkable animals that live there.

Many countries have produced inventories of biodiversity in their national waters and begun to set up networks of marine protected areas. International assessments have been made of coral reefs, seagrass beds, and mangrove swamps. Hotspots have been identified for seabirds, sea turtles, sharks, and open-ocean fish. Progress on the deep ocean has been slower, but data has accumulated on biodiversity around seamounts, shelf breaks, and hydrothermal vents. In 2020, the Pew Charitable Trusts funded a project in which scientists from 13 universities worldwide, led by the University of California Santa Barbara, analyzed 22 billion "data points" and produced the first map of marine biodiversity in the open ocean. As of 2021, over 100 countries had signed up to an international target to protect at least 30 percent of the world's ocean by 2030.

Riches of the Deep

An international team of marine scientists (see above) mapped the biodiversity of the open ocean waters for which individual states have responsibility. They assessed species richness in these waters, especially for species threatened by human activities. They also considered the variety of seabed habitats, including seamounts and hydrothermal vents, and they identified areas of high primary production that best support ocean food chains.

They mapped areas with fewer features of interest, through to those with the most features of interest and so having the most urgent need of protection. From this, the team selected 10 representative areas with especially rich biodiversity, highlighted on the map below.

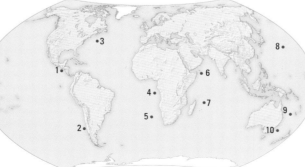

1. Costa Rica Dome
2. Salas y Gómez and Nazca Ridges
3. Sargasso Sea and Corner Rise Seamounts
4. Gulf of Guinea
5. Walvis Ridge
6. Arabian Sea
7. Mascarene Plateau
8. Emperor Seamount Chain
9. Lord Howe Rise
10. South Tasman Sea

TYPES OF HOTSPOTS

HIDDEN HOTSPOT

LOCH CARRON *The northwest Highlands of Scotland may be scenic, but very little biodiversity is found in the harsh, rocky landscape that surrounds Loch Carron.*

BENEATH THE SURFACE *Underwater, however, Loch Carron is as full of life as any tropical coral reef. Animals include soft corals, dahlia anemones, and brittlestars.*

CARIBBEAN TREASURE CHEST

SABA BANK *A 2006 survey of this atoll in the Caribbean Netherlands found a high species biodiversity. The area was declared a national park in 2012.*

NEW TO SCIENCE *The Saba Bank study discovered a seven-spined goby living on the seabed. It is a new species, and probably a new genus.*

SEAMOUNT COMMUNITIES

FISH HAVEN *Soldierfish and snappers gather at a seamount in the Indian Ocean. The upwelling of nutrient-rich currents around seamounts makes them "oases" of the ocean.*

GUADALUPE SEAMOUNT *Up to a third of species of seaweed, plants, and animals on isolated seamounts may be unique (or endemic) to that seamount, having evolved there over millions of years.*

BURIED RICHES

SAMPLING SEDIMENTS *Mud cores collected from the seabed can contain thousands of species of microorganisms. Deep-sea sediment was once thought to be like a desert but this hidden biodiversity proves otherwise.*

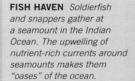

Cycles of Life and Energy

ALL LIFE DEPENDS ON ORGANISMS that harness energy from chemicals or the Sun to produce food. These organisms, whether phytoplankton, seaweeds, chromists, or bacteria, are called primary producers and form the first link of a food chain. This first link is just one point in a cycle that processes chemical energy and nutrients through the entire community of life in an ecosystem, into the physical environment, and back again.

Energy Flow

As each organism in an ecosystem is eaten in turn by the next organism in the food chain, food energy flows from prey to consumer. The primary producers—the organisms such as diatoms and bacteria at the beginning of the food chain—are eaten by organisms called primary consumers, which are eaten by secondary consumers, and so on to top predators—animals not preyed upon by anything else. In land-based ecosystems, the total mass of organisms at each succeeding food-chain level decreases, leaving very few top predators. However, in marine ecosystems with phytoplankton as producers, the mass is greatest at the primary consumer level. This is possible because phytoplankton grow so rapidly that they provide great turnover despite having little mass.

FOOD-ENERGY PYRAMID
At each level of a food chain, energy is lost as heat, so less is available to the next consumer. The diminishing energy at each level can be represented by a pyramid (below) and accounts for the scarcity of top predators.

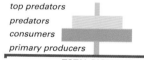

top predators
predators
consumers
primary producers

TOTAL ENERGY

BIOMASS PYRAMID
The biomass pyramid (below) for a system with plankton producers is partly inverted, because the producers have low total mass. Despite this, the rapid reproduction of the plankton keeps the food chain supplied.

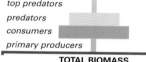

top predators
predators
consumers
primary producers

TOTAL BIOMASS

Recycling

All living things need a supply of chemical nutrients, such as nitrates, phosphates, and silicates, to grow and reproduce. They are taken up by primary producers then passed along the food chain. Although some nutrients are available from seawater, most are derived ultimately from the sea floor. When an organism dies, any parts that are not eaten by other animals gradually sink to the sea floor, where they are broken down by bacteria and other decomposers. Fecal matter also ends up on the seabed and is processed by detritus feeders or decomposers. Eventually, the nutrients are released into the environment in their mineral, nonliving forms. They may then remain at depth, or they may be returned to surface waters by circulating water currents within an ocean basin (see upwelling, opposite).

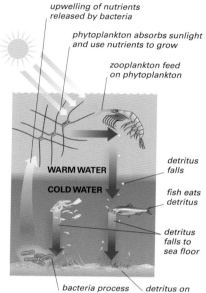

upwelling of nutrients released by bacteria

phytoplankton absorbs sunlight and use nutrients to grow

zooplankton feed on phytoplankton

WARM WATER

COLD WATER

detritus falls

fish eats detritus

detritus falls to sea floor

bacteria process detritus

detritus on sea floor

NUTRIENT CYCLE
Small particles of organic matter, or detritus, are found in the water column. They may be eaten by scavengers or broken down still further by bacteria present in the water. However, many of them rain down on the ocean floor where they decompose, releasing nutrients. The nutrient cycle is completed by upwelling water currents that then carry the nutrients back to the surface where they can be utilized by the phytoplankton.

OVERHARVESTING

These fishermen are harvesting Pacific cod. Cod populations have drastically declined and many important stocks have collapsed because too many are being caught for human consumption before they can reproduce successfully. The imposition of quotas by governments has not solved the problem, although numbers are starting to recover in some areas.

COD FISHING
More than 1.5 million tons of cod (Atlantic and Pacific) were caught in 2010 using various methods, including trawls and longlines.

FOOD-ENERGY PYRAMID / **FOOD WEB**

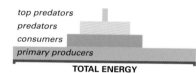

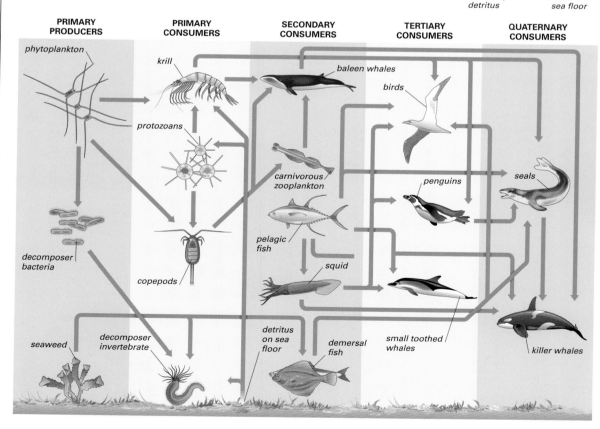

PRIMARY PRODUCERS | PRIMARY CONSUMERS | SECONDARY CONSUMERS | TERTIARY CONSUMERS | QUATERNARY CONSUMERS

phytoplankton

krill

baleen whales

birds

protozoans

carnivorous zooplankton

penguins

seals

pelagic fish

decomposer bacteria

copepods

squid

small toothed whales

killer whales

seaweed

decomposer invertebrate

detritus on sea floor

demersal fish

FOOD WEB
Many food chains have been combined to form this complex food web, extending from primary producers to quaternary consumers (top predators) for a Southern Ocean ecosystem. Each arrow shows the flow of food energy from prey to predator, grazer, or decomposer. It shows how organisms depend on one another for food. Some animals feed on organisms from several different levels of the food chain, adding to its complexity. Food webs are delicately balanced and easily upset by human interference.

Productivity

Throughout the world's oceans, the abundance of marine life varies dramatically. The ocean is more productive in some places and at some times than others. The amount of sunlight is a major influence on productivity and changes with latitude and time of year. The supply of nutrient-rich water from the sea floor and light for photosynthesis is affected by changing water movements and day length, affecting plankton levels. Temperature also affects productivity as it influences the rate of photosynthesis.

CLEAR, TROPICAL OCEAN
Tropical waters are not mixed seasonally, so few nutrients are returned to the surface, and little plankton growth is possible. Here, a solitary turtle cruises in crystal-clear surface waters near Hawaii.

RICH, MURKY TEMPERATE SEA
In coastal and temperate areas, water turbulence circulates nutrient-rich water that supports a variety of algae, such as this kelp forest in the Pacific.

Upwelling

The open-ocean surface water can become impoverished, as nutrients are constantly absorbed by phytoplankton and fall with detritus to the sea floor. Nutrient-rich water can be restored to the surface on a large scale by vertical ocean currents in a process called upwelling (see p.60).

Near land, coastal upwelling is caused by surface currents, such as the Humboldt Current off South America (see p.58). In the equatorial waters of the Pacific and Atlantic, midocean upwelling occurs when water masses are driven north and south by the trade winds, and cooler, nutrient-rich water rises to take their place. Polar upwelling can happen where winter storms cause intense water movement. When upwelling occurs and there is sufficient sunlight, phytoplankton multiply rapidly to support a vast number of organisms, creating the most productive ocean waters in the world.

NUTRIENT-RICH WATERS
Where there is upwelling, large numbers of small fish gather to feed on the plankton. They, in turn, attract larger predators like these copper sharks feeding on sardines off the coast of South Africa.

Swimming and Drifting

MOST OF THE OCEAN'S LIVING SPACE IS NOT ON THE SEABED but in the water column and out in the open ocean—areas known as the pelagic zone. Salt water provides support, as well as the nutrients that allow many plants and animals to live in the water column without ever going near the seabed. Some animals live at the interface between ocean and air, or alternate between both environments, because it is more energy-efficient. The water surface, water column, and seabed are all interconnected, and many animals move between these habitats.

Plankton

The sunlit, surface layers of the ocean are home to many tiny plants and animals (plankton) that drift with the water currents. Phytoplankton consist of bacteria, microalgae (see p.248), and plantlike chromists (see p.234) that can photosynthesize and make their own food. Along with fixed seaweeds and seagrasses, phytoplankton form the basis of ocean food webs. Zooplankton consists of animals, most of which are very small and feed on the phytoplankton. However, jellyfish can grow to a huge size. Many deep-sea forms have strange shapes and soft bodies that are very delicate. Some zooplankton, such as arrow worms, comb jellies, and copepods, live permanently in the plankton, hunting and grazing (holoplankton), while others are simply the larval and dispersal stages of animals, including crabs, worms, and cnidarians (meroplankton) that will spend part or all of their adult lives on the seabed.

Many planktonic organisms have elegant spines, long legs, or feathery appendages that help them float. Tropical zooplankton generally have more of these than their temperate or polar equivalents because warm water tends to be less dense and viscous, and so provides less support.

PLANKTONIC LARVA
The eggs of the common shore crab hatch into floating, spiny zoea larva.

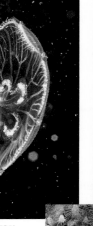

TEMPORARY PLANKTON
Most temporary zooplankton are the larvae of animals that, as adults, live on the seabed. The common jellyfish, however, has a planktonic adult stage (shown above), and a fixed, asexual, juvenile stage (right).

Nekton

Fish and most other free-living marine animals can all swim, even if only for short distances, over the seabed. However, many animals spend their whole lives swimming in the open ocean and are collectively called nekton. This group includes the majority of fish and all whales, dolphins, and other marine mammals, turtles, sea snakes, and cephalopods. There are also some representatives from other groups such as swimming crabs and shrimp. Most nektonic animals are streamlined, and there is a remarkable similarity in shape between some dolphins and open-ocean nektonic fish such as tuna.

TYPICAL NEKTON FEATURE
Dusky dolphins are typical of nektonic animals, most of which are vertebrates (animals with backbones).

The Ocean–air Interface

Some animals live at the interface between air and water, either floating at the surface or alternating between the two environments. Oceanic birds such as albatrosses, petrels, gannets, and tropic birds spend their whole lives out at sea. They eat, sleep, preen, and even mate on the ocean surface. Large rafts of such seabirds are particularly vulnerable to oil spillages. Other diving seabirds, such as terns, alternate between hunting at sea and resting on land. Just as these birds plunge down into the water to catch fish, so some sharks lunge out of the water to catch birds and turtles. Flying fish launch into the air to escape their predators. Some planktonic animals live permanently at the water surface with part of their body projecting into the air. The by-the-wind sailor is a small, colonial cnidarian that is supported by a sail-like float and transported by wind blowing against its vertical sail. Drifting with it on a raft of mucous bubbles is the violet sea snail, which also feeds on it. There are even surface-living insects, such as the genus *Halobates*, that drift the oceans.

DRIFTING AT THE INTERFACE
The large gas-filled float of the Portuguese man-of-war supports the whole colony at the water surface.

FLYING AND DIVING
The brown pelican is one of several species that dive or dip down from the air into the water to catch fish. It uses its capacious beak as a scoop.

FLOATING COMMUNITY
Ocean sunfish often drift at the ocean's surface. They will investigate floating rafts of seaweed and logs for potential food such as small fish and crustaceans.

Drifting Homes

Many pelagic fish species are attracted to floating objects that provide shelter from predators, currents, and even sunlight. Floating logs and seaweed also provide a meeting point. Fishers have exploited this tendency by using fish-attracting devices (FADs, see panel, right) to concentrate fish in one area to increase their catch. These FADs vary from simple rafts with hanging coconut palm leaves to complex technological devices.

Mini-ecosystems often develop on and around large drifting logs. Seaweeds and goose barnacles settle, providing shelter and food for crabs, worms, and fish. Shipworms bore into the wood, and their tunnels provide further refuge. Occasionally reptiles, insects, and plant seeds survive and drift on logs, and may eventually be washed ashore to colonize new places, including new volcanic islands.

SARGASSO HAVEN
Floating *Sargassum* seaweed provides a safe haven for the sargassumfish. More than 50 animal species have been recorded in this habitat.

LURING FISH

Fish-attracting devices (FADs) have increased catches in many areas by making fish stocks easier to exploit. However, they make no contribution to biological productivity because they simply gather fish together, and so may contribute to overexploitation. Artificial reefs also attract fish, but provide safe breeding sites, too.

FISH-ATTRACTING DEVICE
Even simple FADs, such as this floating buoy in Hawaii, will attract fish. Juvenile jacks and endemic Hawaiian damselfish can be seen sheltering under this one.

Bottom-living

ANIMALS LIVING ON THE OCEAN FLOOR or within its sand and mud, either moving over it or firmly attached, are called benthic animals. On land, plants provide a structural habitat within which animals live. In the ocean, this is rarely the case, except in shallow, sunlit areas dominated by kelp, seaweeds, or seagrasses. Instead, wherever areas of hard seabed provide a stable foundation, a growth of benthic animals develops, fixed to the seabed and often resembling plants. A seabed of shifting sediments is no place for fixed animals. Here, a community of burrowers develops instead.

Fixed Animals

Many benthic animals such as sponges, sea squirts, corals, and hydroids spend their entire adult lives fixed to the seabed, unable to move around. On land, animals must move around in search of food, whether they are grazers, predators, or scavengers. In the ocean, water currents carry an abundant supply of food in the form of plankton and floating dead organic matter. Fixed animals can take advantage of this by simply catching, trapping, or filtering their food directly from the water, without having to move from place to place. When it is time to reproduce, they simply shed eggs and sperm into the water, where the eggs are fertilized and grow into planktonic larvae. Sometimes, they retain their larvae or eggs, and release them only when the young are well developed. Water currents distribute the offspring to new areas, where they can settle and grow.

REEF-FORMING TUBE WORM
In some Scottish sea lochs, the chalky cases of tube worms form substantial reefs.

Mobile Animals

Dense growths of seaweeds or fixed animals provide shelter and food for many mobile animals. Grazers, such as sea urchins, crawl through the undergrowth, eating both seaweeds and fixed animals. Meanwhile, crabs, lobsters, and starfish scramble and swim around, hunting and scavenging for food. Sea slugs are specialist predators, each species feeding on one, or a few, types of bryozoans, hydroids, or sponges. Sea slugs therefore live in close association with their prey and rarely stray far. Kelp holdfasts provide a safe haven for small, mobile animals such as worms.

BENEATH THE SEAWEED
Below the seaweed-dominated zone around northern European coasts, on seabeds too deep and dark for photosynthesis, dead man's fingers, sponges, and tube worms typically grow attached to subtidal rocks.

SEABED IN THE SUN
Seaweeds anchor in the tidal zone of rocky shores and on rocky reefs, such as this one in the Canary Islands. On sunlit, temperate seabeds, it is seaweeds that provide the community structure.

FIXED TO THE BOTTOM
Christmas-tree worms live attached to the bottom in hard tubes that they cement into coral reefs. They feed by filtering plankton from the water, using their beautiful double spiral of tentacles.

FISH IN DISGUISE
Scorpionfish live on the seabed among the seaweeds and fixed animals. Their intricate skin-flaps blend in with this habitat.

Burrowing and Boring

Much of the seafloor is covered in soft sediments, such as sand and mud. Living on the surface of the sediment is both difficult and dangerous, and most animals burrow below or build tubes in which to live and hide. Bivalve mollusks and segmented worms cope especially well in this habitat, and many different species can be found in sediments all over the world. Safe under the sediment surface, a bivalve draws in oxygen-rich water and plankton through one of its two long siphons, expelling waste through the other. It never has to come out to feed or breathe. Piddocks and shipworms bore into rocks and wood, then use their siphons in a similar way. Sediment is not a completely safe home—predatory moon snails dig through sand and bore into bivalve shells, eating the contents. Ragworms are also active predators, hunting through the sediment for other worms and crustaceans. Some worms build flexible tubes from sand grains, their own secretions, or both. The tubes stick out of the sand, and they feed by extending feathery or sticky tentacles from the tube to catch plankton. If danger threatens, they can withdraw rapidly. A similar strategy is adopted by tube anemones and sea pens.

REPLACING SIPHONS
The siphon tops of buried bivalve mollusks are sometimes nipped off by flatfish but can regrow.

BORING INTO ROCK
The boring sponge uses chemicals to dissolve tunnels in calcareous shells and rocks, creating a living space for itself.

Symbiosis

Bottom-living is a challenge for marine organisms. A safe crevice on a coral reef, for instance, is valuable, but fiercely fought over. The solution to finding a home is often to enter an intimate relationship with a different organism—a situation called symbiosis. When only one partner benefits, the relationship is called commensal, and often involves one animal providing a home for the other. Small pea crabs live inside mussels, gaining shelter and food, while the mussel merely tolerates their presence. Symbiosis in which both partners benefit is called mutualism. Many tropical gobies live in such relationships with blind or nearly blind shrimp. The shrimp digs and maintains a sandy burrow they both live in, while its sharp-eyed partner goby acts as a lookout. Some anemones stick to the shells of hermit crabs, gaining from the crab's mobility and access to its food scraps. The crab is protected, in return, by the anemone's stinging tentacles. The third type of symbiosis is parasitism, in which one partner, the host, is harmed. The crustacean *Sacculina* spreads funguslike strands through its host crab's body to extract nutrients, weakening or killing it.

MUTUAL RELATIONSHIP
The banded coral shrimp earns its place in the moray eel's well-defended crevice by cleaning the teeth of its host.

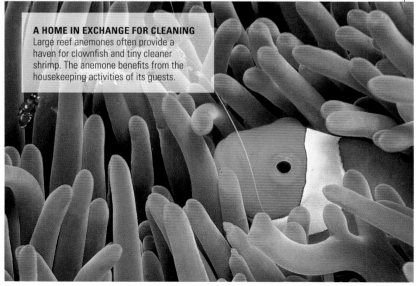

A HOME IN EXCHANGE FOR CLEANING
Large reef anemones often provide a haven for clownfish and tiny cleaner shrimp. The anemone benefits from the housekeeping activities of its guests.

Zones of Ocean Life

NO PART OF THE OCEAN IS DEVOID of organisms, from polar seas to the tropics and from coasts and the seashore to the deepest depths. The seabed and the water column above it both support a huge variety of life. However, marine organisms are distributed unevenly both horizontally and vertically. As on land, climate (mainly temperature) and food play a large part in determining distributions and biodiversity. In the harsh environment at the poles, there is less coastal life than in the warm tropics, but beneath the surface, Antarctic seas support rich marine communities. Although there is life at every depth, most creatures can only survive within particular depth zones at particular pressures, temperatures, and light regimens.

Geographical Zones

Seawater temperatures are much more stable than those on land because water loses and gains heat more slowly than does air. However, the distribution of marine coastal and continental shelf communities still follows a global pattern, with distinct polar, temperate, and tropical ecosystems. Coastal salt marsh in temperate parts is replaced in the tropics by mangroves. Kelp forests only grow in cool waters but extend into the tropics in places where cold water upwells from the deep, such as off the coast of Oman on the Arabian peninsula. Planktonic species and bottom-living species with planktonic larvae might be expected to occur anywhere that ocean currents take them. However, a boundary between water masses with different physical characteristics may present as effective a barrier in an ocean as mountains do on land. Below a certain depth, there are fewer such barriers, and conditions are stable and similar worldwide, so deep-sea animals often have very wide distributions.

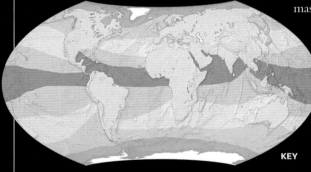

CLIMATIC ZONES
The shape and tilt of our planet results in differences in the amount of solar radiation reaching land and ocean at different latitudes. This produces large-scale climatic zones that ring Earth.

KEY

■	equatorial	■	temperate
■	tropical	■	subpolar
■	subtropical	■	polar

Endemic Species

Some marine organisms, especially pelagic species, have a wide global distribution, since there are few barriers to their dispersal. Others live in restricted geographical ranges and are said to be endemic to a particular sea, island, or country. The most remote patches of habitat, such as small oceanic islands, tend to have the most endemic species. This is because animals in their dispersive stages, such as eggs and larvae, may survive only for short periods and so never reach distant shores. The Red Sea holds many endemic fish species. It is connected to the Indian Ocean only by a narrow channel and so is effectively isolated. Endemic fish are often those that cannot or do not swim far. Anemonefish, for example, lay

GALÁPAGOS PENGUIN
This penguin species lives only

MALDIVES ANEMONEFISH
This endemic fish is not a strong swimmer. It does not have planktonic larvae and lives only in the Maldives and Sri Lanka in the Indian Ocean. Its host anemone has a wider distribution, because its larvae disperse on ocean currents.

Depth Zones

As depth increases, so does pressure, while light, temperature, and food supply decrease. These changes impose limits on the types of marine organisms that can survive and prosper at different depths. The areas on and over continental shelves around the world are rich in life as they are well-supplied with nutrients from river discharge and stirred-up sediments. Schooling fish, such as herring, feed on plankton sustained by the nutrients. Most commercial fisheries are over continental shelves. Below the continental shelf, no phytoplankton or seaweeds grow. Pelagic animals either eat each other or make daily feeding migrations into the upper layers. Rocky areas support a diverse fauna including coldwater coral reefs, sponge reefs, and hydrothermal vent communities. Fine sediments cover the immense, flat abyssal plains at the foot of the continental slope. While microorganisms abound, large animals are relatively scarce.

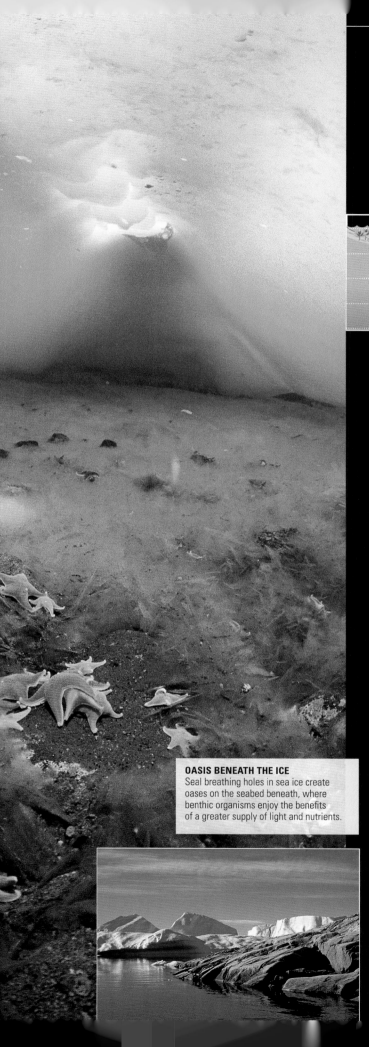

OASIS BENEATH THE ICE
Seal breathing holes in sea ice create oases on the seabed beneath, where benthic organisms enjoy the benefits of a greater supply of light and nutrients.

BARREN POLAR SHORE
This polar shore in Greenland supports little life due to the grinding action of winter ice, though below the reach of the ice, rich communities may develop.

SEABED IN THE SUNLIT ZONE

- seaweeds
- sponge
- starfish

0 ft	
160 ft (50 m)	
330 ft (100 m)	
500 ft (150 m)	
650 ft (200 m)	

- phytoplankton
- zooplankton
- whale shark
- Portuguese man-of-war

660 ft (200 m)

SUNLIT ZONE
On seabed, high biodiversity—seaweeds, corals, sessile animals; in water, rich plankton, abundant fish, cetaceans

- mackerel
- tuna
- jellyfish
- salp

3,300 ft (1,000 m)

- shark
- squid

TWILIGHT ZONE
On seabed, crinoids, sponges, sea fans, sea pens, sea cucumbers, Greenland shark; in water, zooplankton, squid, shrimp, predators—sperm whale, silvery fish with large eyes, such as hatchet fish and lanternfish

6,500 ft (2,000 m)

- hatchetfish
- comb jelly
- crinoid
- sponge
- deep-sea anglerfish

9,800 ft (3,000 m)

- hagfish
- black swallower

DARK ZONE
On seabed, similar to twilight zone; in water, mostly small, dark-colored fish with large mouths and stomachs, gulper eels, rattails, anglerfish, red shrimp, deep-sea jellies

ABYSSAL ZONE
On seabed, few large animals, rattails, hagfish, and sea cucumbers, very diverse protists, nematode worms, bacteria; in water, some deep-sea fish

13,100 ft (4,000 m)

VERTICAL LIFE ZONES
Environmental conditions change gradually as depth increases, but zones can be recognized based on both physical and biological parameters. The types of marine life in each zone are shown here.

HADAL ZONE
Little-known region, but some large organisms found in deepest depths; deepest fish caught at 26,200 ft (8,000 m)

16,400 ft (5,000 m)

19,700 ft (6,000 m)

- cusk eel

Ocean Deserts

Some areas of ocean are similar to deserts on land and support few species. Clear, blue surface water over the deep oceans often supports only small amounts of plankton, because it is very poor in the nutrients and minerals needed by phytoplankton to grow. This is especially the case in areas where there are few storms to stir the water and bring nutrients up from deep water. The nutrient iron can be a limiting factor, and experiments in which areas were seeded with iron have shown greatly increased phytoplankton production. The ocean floor in abyssal depths can support only a few large animals and was once considered to be a virtual desert. However, recent work on deep-sea sediments has shown the opposite. If all the bacteria and tiny animals living between the sediment particles are counted, then this habitat is as diverse as a tropical rainforest.

Ocean Migrations

FEEDING AND BREEDING ARE THE MAIN REASONS that animals migrate. They move from one place to another, often at the same time of day or year, and usually follow the same, well-defined routes. Migratory species include many of the larger marine animals, such as whales and turtles, but smaller creatures, such as squid and plankton, also make spectacular journeys in order to survive and reproduce. Animal migration in the oceans is more complicated than on land because animals can move both horizontally and vertically through the water column.

Types of Migration

The driving force behind any animal migration is survival. Individuals must eat to live, and some will travel long distances to find food. Such journeys often coincide with peak production times of plankton and other food sources in particular places, such as sites of seasonal upwellings. A shorter, more regular feeding migration is made daily by plankton and active swimmers such as squid (see p.221).

A species' survival depends on reproductive success. Gathering together and breeding in a few places at the same time optimizes conditions for offspring survival. Breeding grounds where food is abundant and conditions favorable are used repeatedly, with individuals often returning to their birthplace to breed. Some shore animals also migrate up and down the beach, following the outgoing tide to feed and avoiding immersion by returning before the tide turns.

MARINE MIGRATORY CYCLE
Some marine organisms migrate to a specific spawning site to release their eggs. The eggs hatch into larvae that join the plankton and drift in the currents to another nursery area, where they feed and mature before joining the adult population.

TRACKING

Until recently, any journey made by a marine animal, such as the leatherback turtle shown below, was poorly understood because tracking devices used on land were inappropriate for use in water. This changed when satellite-tracking devices became available. Attaching one to a turtle does not impede or harm it in any way but it can still pose problems. Yet turtles are threatened in the wild, so knowing where a female goes after laying her eggs is vital to conservation work.

SATELLITE TRACKING
The device on this turtle's shell transmits data to an orbiting satellite whenever the turtle surfaces to breathe. The information is then relayed to the scientists who are tracking the turtle's movements.

ARCTIC TERN MIGRATION
This small bird flies from the Antarctic to the Arctic to breed and then returns south, a round trip of nearly 22,000 miles (35,500 km). Terns spend 90 days at the nesting grounds each year. The rest of the time is spent mostly on the wing.

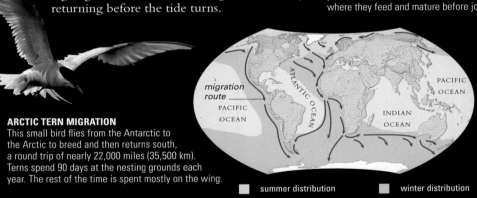

summer distribution   winter distribution

LOBSTER MIGRATION
Caribbean spiny lobsters migrate in single file across the sea floor in winter, seeking warmer water, and return to shallow water in summer.

Migrating between Salt and Fresh Water

Although some marine species can cope with a great range of salinity and temperature, only a few move between fresh and salt water at particular stages of their lives. Some, such as salmon, start and finish their lives in fresh water and spend the rest of their time in the ocean. Such fish are described as anadromous. Eels, on the other hand, start and finish their lives in the sea, but spend 10 to 14 years in fresh water while maturing. Fish such as these are termed catadromous.

At maturity, both of these fish return to their birthplace to breed, after which they die. Changing from fresh to salt water or vice versa would be fatal to most fish, but various physiological adaptations, including the way their kidneys function, allow both anadromous and catadromous fish to make the transition without experiencing any ill effects.

SALMON RETURNS TO FRESHWATER SPAWNING GROUNDS

1 At three to five years of age, a coho salmon is ready to return to the river where it was born to spawn. Some mature at only two years, returning to their home river as "jacks."

2 In migrating upstream to its spawning grounds, the salmon may swim 2,175 miles (3,500 km) against the water current, negotiating several waterfalls and rapids.

3 As soon as the female has deposited her eggs in the gravel on the riverbed, the male swims over them and releases his sperm, optimizing the chance of fertilization.

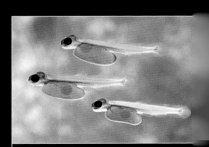

4 Newly hatched salmon live among the gravel until they absorb their yolk sacs and become fry. They then begin their journey downstream to the sea.

HORIZONTAL AND VERTICAL TRAVEL
Some animals migrate horizontally and vertically. Here, longfin squid migrate to spawn in May. They also move up and down the water column each day to feed.

Navigation

While satellite tracking provides information on migration routes and confirms that many individuals travel the same path, how animals navigate over vast distances is still poorly understood.

Salmon return to their spawning grounds by smelling the unique chemical composition of the water, but they first have to get close enough to pick up this scent. Some aquatic species may use water currents to guide them, while others use the Earth's magnetic field to navigate.

Animal migration is amazingly accurate, in terms of both direction and timing. Oceanic birds, turtles, and mammals can also navigate using the sun, the stars, and familiar landmarks.

BELUGA WHALES MIGRATING
Belugas, like these in Lancaster Sound, Canada, live in the Arctic and subarctic, but some migrate to warmer waters in summer.

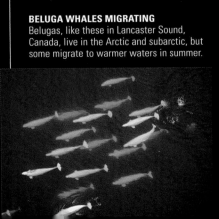

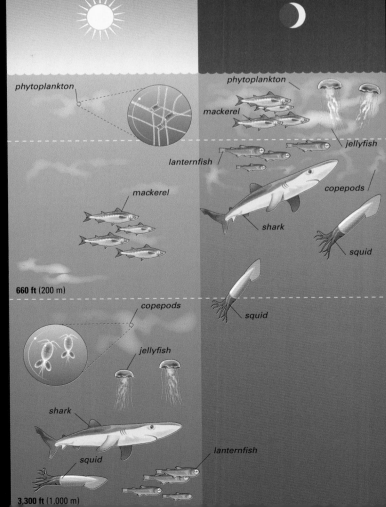

DAY

NIGHT

phytoplankton

phytoplankton

mackerel

jellyfish

lanternfish

mackerel

copepods

shark

squid

660 ft (200 m)

copepods

squid

jellyfish

shark

squid

lanternfish

3,300 ft (1,000 m)

Vertical Migrations

In temperate and tropical regions, zooplankton migrates to the ocean surface at night and then moves down again during the hours of daylight. In a single day, this vertical movement may range from 1,310 to 3,300 ft (400 to 1,000 m), depending on the size and type of animal involved. In polar regions, where darkness lasts for several months, zooplankton migrates up and down on a seasonal basis, being at the surface during summer and at depth in winter.

It is thought that zooplankton rises to feed on the phytoplankton that lives in the surface waters, but then retreats to depth for safety, or possibly because it expends less energy in cooler water. Maturing planktonic larvae of animals such as crabs will eventually migrate to the sea floor and become benthic.

EARTH'S GREATEST MASS MIGRATION
During the day, when many animals remain in the depths, out of sight of predators, the phytoplankton utilizes the Sun's energy to produce food. At night, the biomass of the surface waters (the sunlit zone) increases by as much as 30 percent, as zooplankton comes up to feed on phytoplankton and is, in turn, eaten by various fish and other animals. This regular movement of animals up and down the water column is the greatest mass migration on Earth.

Living Down Deep

THE DEEP-SEA ENVIRONMENT APPEARS INHOSPITABLE—cold, dark, and with little food. However, it is remarkably stable: temperatures remain between 35 and 39°F (2 and 4°C) year-round, salinity is constant, and the perpetual darkness is overcome by novel communication methods (see pp.224–225). Although deep-sea pressures are immense, most marine animals are unaffected, since they have no air spaces, while animals living below about 5,000 ft (1,500 m) show subtle adaptations. Species diversity of large animals decreases with depth, but there is a huge diversity of small organisms living within deep-sea sediments.

Pressure Problems

Deep-ocean animals experience huge pressures, but problems arise only in gas-filled organs such as the lungs of diving mammals and the swim bladders of fish. Sperm whales, Weddell seals, and elephant seals all make deep dives, and so have flexible ribcages and exhale before diving. While underwater, they use oxygen stored in blood and muscles. Deep-sea fish can cover a large vertical range because pressure changes at depth are proportionately less, per foot, than near the surface, so the pressure or size of their swim bladders does not change radically. In oceanic trenches, the pressure is so great that it affects the operation of biological molecules, such as proteins. Pressure-loving bacteria in this habitat have specialized proteins—they cannot grow or reproduce when brought to

DEEP-SEA ADAPTATIONS
Anglerfish have a lightweight skeleton and muscles for neutral buoyancy. This specimen's muscles have been "cleared" to show the bone, which is stained red.

SPERM WHALE
Sperm whales can dive to at least 3,300 ft (1,000 m), where the pressure is 100 times

Finding Food

The major problem of deep-sea living is finding enough food. With the exception of communities based around hydrothermal vents and cold seeps (see pp.188–189), animals living in the deep ocean and on the deep-ocean floor are ultimately reliant on food production in the sunlit layer, thousands of feet above. In the depths, it is too dark for photosynthetic plankton to live and to provide food. Sometimes, large mammal or fish carcasses reach the sea bed, but most food arrives as tiny food fragments, slowly sinking from above. Much is eaten before it reaches the sea floor, but much is also added in the form of skins, shed from midwater crustaceans and salps. Bacteria grow on such material, helping it to clump together and so fall more rapidly.

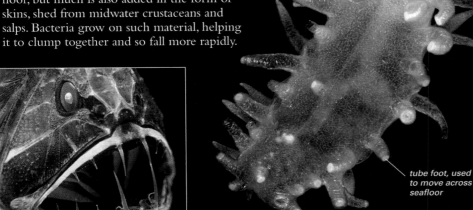

mouth surrounded by modified tube feet

tube foot, used to move across seafloor

MIDWATER FEEDER
The fangtooth lives at midwater depths of about 1,600–6,500 ft (500–2,000 m). Food is scarce, so its large mouth and sharp teeth help it to catch all available prey.

SEABED CONSUMER
Sea cucumbers vacuum up organic remains from the sea floor. At high latitudes, more food rains down in spring, following surface phytoplankton blooms; these rains may trigger sea cucumbers to reproduce.

A WINDOW ON DEEP-SEA LIFE
Deep Rover is a two-person submersible capable of diving to 3,300 ft (1,000 m), launched from a semisubmersible platform. The occupants can see all the way around through the acrylic hull.

Scavenging Giants

Many deep-sea animals are smaller than their relatives in shallow water. This is an evolutionary response to the difficulties of finding food in the deep ocean. However, some scavengers survive by growing much larger than their shallow-water counterparts. For example, amphipod and isopod crustaceans that measure only about ½ in (1 cm) long are common in shallow water, where they scavenge on rotting seaweed and other debris. Carrion in the deep sea is sparse, but it comes in big, tough lumps such as whale carcasses. Some deep-sea amphipods grow to a length of 4–6 in (10–15 cm), more than ten times larger than shallow-water species, and so are able to tackle such a bonanza. In the low temperature of the deep ocean, these animals move and grow slowly and reproduce infrequently, but live much longer than their shallow-water counterparts. Sea urchins, hydroids, seapens, and other animals also have giant deep-sea forms. Similar giants are found in cold Antarctic waters.

DEEP-SEA GIANT
The widespread deep-sea scavenger amphipod *Eurythenes* grows to over 3 in (8 cm).

Staying Aloft

Huge areas of the deep-sea floor are covered in soft sediments many yards thick, called oozes (see p.181). Seabed animals need ways of staying above these sediments so that they can feed and breathe effectively. Many sedentary filter-feeding animals, such as sea lilies, sea pens, and some sponges, have long stalks, enabling them to keep their feeding structures above the sediment. Some sea cucumbers have developed stiltlike tube feet that help them walk over the sediment surface, instead of having to plow through it. The tripodfish props itself up on its elongated pectoral and anal fin rays. One species of sea cucumber, *Paelopatides grisea*, has an unusually flattened shape that allows it to lift itself off the sea bed with slow undulations of its body.

SEA LILIES ANCHORED IN THE OOZE
To catch food, sea lilies reach up into the current on stalks up to 2ft (60cm) high. The stalk extends deep into the sediment to provide an anchor.

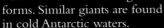

OCEAN LIFE

Bioluminescence

BIOLUMINESCENCE IS A COLD LIGHT produced by living organisms. On land, only a few nocturnal animals, such as fireflies, produce light, but in the ocean, thousands of species do so. Deep-water fish and squid use bioluminescence extensively, but there are many other light producers, such as species of bacteria, dinoflagellates, sea pens, jellyfish, mollusks, crustaceans, and echinoderms. Evidence suggests that marine organisms use bioluminescence for defense (as camouflage or distraction), for finding and luring prey, and for recognizing and signaling to potential mates.

USING LIGHT TO COMMUNICATE
Many bioluminescent marine organisms use their light in communication. This bristlemouth fish can signal to its own kind with its specific photophore pattern.

Light Production

Bioluminescence is produced by a chemical reaction in special cells known as photocytes, usually contained within light organs called photophores. A light-producing compound called luciferin is oxidized with the help of an enzyme called luciferase, releasing energy in the form of a cold light. Most bioluminescent light is blue-green, but some animals can produce green, yellow, or, more rarely, red light.

A range of light-producing structures is found in different animals. The hydroid *Obelia* has single photocytes scattered in its tissues, while certain fish and squid have complex photophores with lenses and filters. Some animals, including flashlight and eyelight fish, some anglerfish, ponyfish, and some squid, adopt a different strategy. They culture symbiotic, bioluminescent bacteria in special organs. The bacteria produce their light, while their host provides them with nutrients and a safe place in which to live.

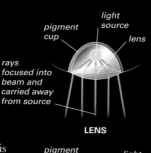

light source
pigment cup
lens
rays focused into beam and carried away from source
LENS

pigment cup
light source
light pipe
LIGHT PIPE

pigment cup
light source
deep-red pigment filter
filter allows only deep-red light to pass
COLOR FILTER

PHOTOPHORE TYPES
Photophores often feature a pigment cup and a lens that directs the light into a parallel beam. With a light pipe, light can be channeled from the photophore, which might be buried in the animal's body. Color filters in front of the light source fine-tune the color of the emitted light.

HUNTING WITH A SPOTLIGHT
The dragonfish produces a beam of red light, from a photophore beneath its eye, to spotlight its prey. Red light is invisible to most deep-sea animals.

body covered with tiny, flashing photophores

Light Disguise

Animals using bioluminescence to attract prey, or to signal to each other, risk alerting their own predators to their presence. However, lights can also be used for camouflage. Hatchetfish live at depths where some surface light is still dimly visible. To keep their silhouette from being seen from below, they manipulate the light they emit from photophores along their belly, to mimic the intensity and direction of the light coming from above. Bioluminescence is also used to confuse potential predators. Flashlight fish turn their cheek lights on and off. Some squid, shrimp, and worms eject luminous secretions or break off luminous body parts that act as decoys, while they escape.

MANIPULATING LIGHT
The silvery, vertical flanks of hatchetfish reflect downwelling light, and their photophores shine downward, camouflaging their silhouette from below.

organs producing downward-directed beams of light

LUMINOUS SMOKESCREEN
A firefly squid presents a predator with a myriad of confusing pinprick lights emitted from its body. It can also secrete a cloud of luminous particles into the water to act as a smokescreen, allowing it to escape.

squid's ink is bioluminescent

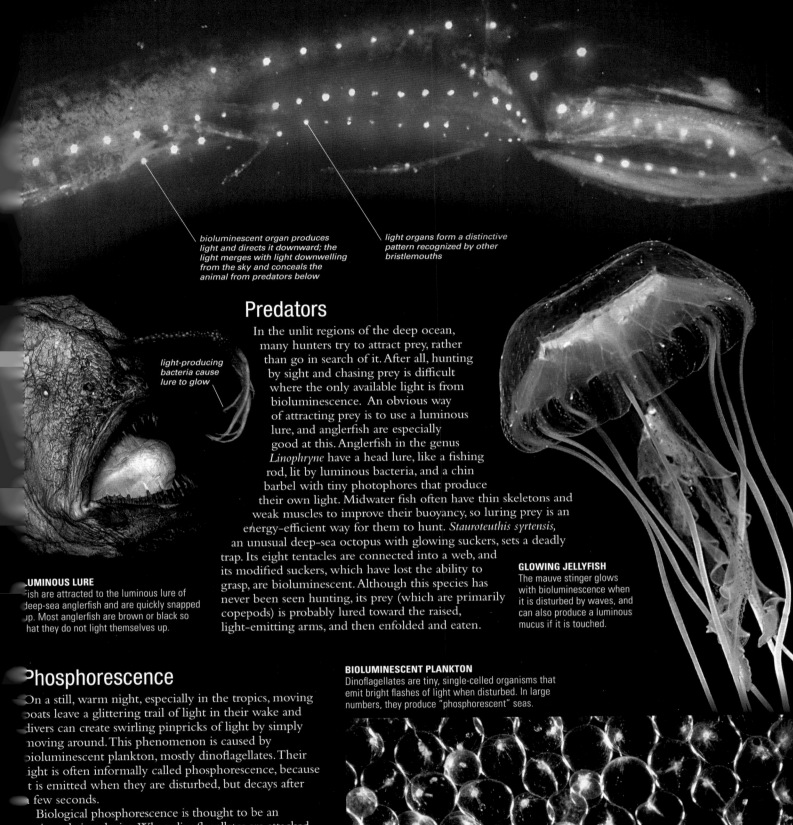

bioluminescent organ produces
light and directs it downward; the
light merges with light downwelling
from the sky and conceals the
animal from predators below

light organs form a distinctive
pattern recognized by other
bristlemouths

Predators

In the unlit regions of the deep ocean, many hunters try to attract prey, rather than go in search of it. After all, hunting by sight and chasing prey is difficult where the only available light is from bioluminescence. An obvious way of attracting prey is to use a luminous lure, and anglerfish are especially good at this. Anglerfish in the genus *Linophryne* have a head lure, like a fishing rod, lit by luminous bacteria, and a chin barbel with tiny photophores that produce their own light. Midwater fish often have thin skeletons and weak muscles to improve their buoyancy, so luring prey is an energy-efficient way for them to hunt. *Stauroteuthis syrtensis,* an unusual deep-sea octopus with glowing suckers, sets a deadly trap. Its eight tentacles are connected into a web, and its modified suckers, which have lost the ability to grasp, are bioluminescent. Although this species has never been seen hunting, its prey (which are primarily copepods) is probably lured toward the raised, light-emitting arms, and then enfolded and eaten.

light-producing
bacteria cause
lure to glow

LUMINOUS LURE
Fish are attracted to the luminous lure of deep-sea anglerfish and are quickly snapped up. Most anglerfish are brown or black so that they do not light themselves up.

GLOWING JELLYFISH
The mauve stinger glows with bioluminescence when it is disturbed by waves, and can also produce a luminous mucus if it is touched.

Phosphorescence

On a still, warm night, especially in the tropics, moving boats leave a glittering trail of light in their wake and divers can create swirling pinpricks of light by simply moving around. This phenomenon is caused by bioluminescent plankton, mostly dinoflagellates. Their light is often informally called phosphorescence, because it is emitted when they are disturbed, but decays after a few seconds.

Biological phosphorescence is thought to be an anti-predation device. When dinoflagellates are attacked by planktonic copepods, they flash. This alerts nearby shrimp and fish to the copepods' presence, and the copepods themselves may then become prey. Some dinoflagellates, such as *Gonyaulax polyedra,* only produce light at night, so they do not waste energy on light production when it cannot be seen.

Deep-sea jellyfish may use a similar anti-predator strategy. The jellyfish light up only when disturbed by vibrations, which indicate an approaching predator. Often, a series of erratic flashes travels over the entire body surface. Such lights may serve to distract the predator

BIOLUMINESCENT PLANKTON
Dinoflagellates are tiny, single-celled organisms that emit bright flashes of light when disturbed. In large numbers, they produce "phosphorescent" seas.

The History of Ocean Life

LIFE HAS BEEN PRESENT IN THE OCEANS for over 3,500 million years. The great diversity of today's marine life represents only a minute proportion of all species that have ever lived. Evidence of early life is hard to find, but it is seen in a few ancient sedimentary rocks. The fossil record has many gaps, but it is the only record of what past life looked like. Fortunately, many marine organisms have shells, carapaces, or other hard body parts, such as bones and teeth. They are more likely to be preserved than entirely soft-bodied creatures, although in exceptional circumstances these have also been fossilized. Using fossils, and information from the sediments in which they are preserved, scientists can reconstruct the history of marine life.

PEOPLE

A. I. OPARIN

In 1924, Russian biochemist Aleksandr Oparin (1894–1980) theorized that life originated in the oceans. He suggested that simple substances in ancient seas harnessed sunlight to generate organic compounds found in cells. These compounds eventually evolved into a living cell.

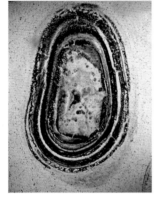

EARLY MICROFOSSILS
This micrograph of a section of chert (a form of silica) from the Gunflint Formation, Canada, includes 2,000 million-year-old microfossil remains. These microfossils contain the oldest and best-preserved fossil cells known.

3,800–2,200 Million Years Ago
The Origin of Life

When Earth formed, it was totally unsuitable for life. The atmosphere changed, however, and the oceans formed and cooled (see pp.42–43), so that by 3,800 million years ago, conditions allowed biochemical reactions to take place. It is thought that simple, water-soluble organic compounds called amino acids accumulated in the water, eventually forming chains and creating proteins. These combined with other organic compounds, including self-replicating DNA, to form the first living cells.

Earth's atmosphere was further developed by mats of algae and cyanobacteria called stromatolites, whose fossil record stretches from over 3,500 million years ago to the present day. Stromatolites could perform photosynthesis, and their growth eventually flooded the atmosphere with oxygen. Cyanobacteria are single-celled organisms with DNA but no nucleus or complex cell organelles. It was not until at least 2,200 million years ago that cells with nuclei and complex organelles (eukaryote cells) appeared.

620–542 MYA Precambrian Life

Ancient life, though soft-bodied, fossilizes under certain conditions, offering rare glimpses of early multicellular life. About 620 million years ago, a community of soft-bodied animals known as the Ediacaran fauna left their body impressions and trackways in a shallow seabed. The seabed now forms the sandstone of the Ediacaran Hills in Australia, where the fauna was discovered in the 1940s.

The ancient sea was inhabited by strange, multicellular animals. Some resembled worms and jellyfish, but others were thin, flat, and unfamiliar, making it difficult to know if they are related to existing animals or a separate, extinct, evolutionary line. These animals are the only link between the single-celled organisms that preceded them and the rapid diversification of life that followed. Additional Ediacaran fauna are also found in Namibia, Sweden, Eastern Europe, Canada, and the UK.

EDIACARAN FOSSILS
These are typical examples of Ediacaran fossils preserved as impressions in rock. Mawsonites (above) is believed to be a complex animal burrow, although scientists do not agree. Spriggina (left) may be an arthropod, or a new life form.

541–520 MYA
Cambrian Explosion

Over the course of 20 million years, around the start of the Cambrian Period, many life-forms made a sudden appearance. Indeed, most of today's major animal groups (phyla) abruptly appear in the fossil record. The Cambrian Explosion of evolution may have been caused by the creation of new ecological niches as the coastline increased, due to the breakup of the Rodinia supercontinent. Further niches arose as a rise in sea level produced large expanses of warm, shallow water. The Cambrian seas were dominated by arthropods, chiefly trilobites, but there were also foraminiferans, sponges, corals, bivalves, and brachiopods. All readily fossilize, as they each have some sort of mineralized "skeleton."

BRACHIOPODS
Brachiopods may resemble bivalve mollusks, but they are unrelated life-forms and were among the first animals to appear in the Cambrian Period. Over 6,000 genera have been described. Only about 400 species survive today.

FIRST REEFS
The Cambrian reefs were built by extinct sponges called archaeocyathids. They resembled tube sponges (above), having a similar shape and a calcareous skeleton.

ARTHROPOD TRAILBLAZERS
Trilobites evolved a multitude of different body forms and remained a ubiquitous arthropod group for the next 100 million years. They became extinct during the Permian Period.

LIVING MARINE STROMATOLITES
Built by Earth's oldest type of organism, stromatolites are now found in only a few places such as here, in the hyper-saline water of Hamelin Pool, Australia.

419–359 MYA The Age of Fish

The earliest vertebrate fossils known are primitive jawless fish that lived some 520 million years ago. Jawed fish appeared in the Silurian Period, following the development of massive coral and sponge reefs that provided them with a multitude of habitats in which to diversify. The now-extinct acanthodians, with their prominent spines on the leading edges of their fins, were among the earliest of these. Having hinged jaws allowed fish to feed more efficiently, and paired fins gave them the speed and manoeuvrability to hunt. The following Devonian Period (419–359 million years ago) saw an evolutionary radiation that could be called the "Age of Fish." Armored fish called placoderms dominated Devonian seas, some reaching maximum lengths of around 30 ft (9 m). Ray-finned fish, sharks, and lobe-finned fish also appeared at this time and have survived to the present day, although marine lobe-finned fish are known only from the Coelacanth. Lobe-finned fish are important in the fossil record because one group gave rise to early tetrapods (limbed vertebrates).

EARLY JAWLESS FISH
Jawless fish first evolved in the ocean, later spreading into brackish and freshwater habitats. The bony head shield and dorsally situated eyes of this *Cephalaspis* suggest it is a bottom-dweller.

EVOLUTIONARY INSIGHT
This lobe-finned fish, *Tiktaalik roseae*, has gills and scales like a fish, but has tetrapodlike limbs and joints. This "missing link" helps to reveal how animals moved from the oceans onto land.

OCEAN LIFE

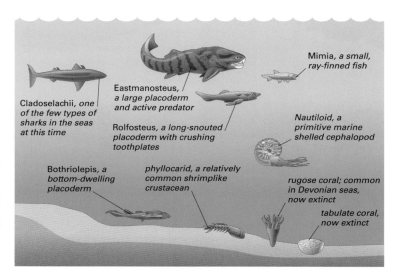

DEVONIAN COMMUNITY
The Devonian reef fauna (right) from Gogo, Australia, is typical of the time. It is dominated by a wide variety of armored placoderms, but ray- and lobe-finned fish, and a shark, have also been found there.

Cladoselachii, *one of the few types of sharks in the seas at this time*

Eastmanosteus, *a large placoderm and active predator*

Rolfosteus, *a long-snouted placoderm with crushing toothplates*

Mimia, *a small, ray-finned fish*

Nautiloid, *a primitive marine shelled cephalopod*

Bothriolepis, *a bottom-dwelling placoderm*

phyllocarid, *a relatively common shrimplike crustacean*

rugose coral; *common in Devonian seas, now extinct*

tabulate coral, *now extinct*

OCEAN LIFE

252–65 MYA
Giant Marine Reptiles

During Triassic, Jurassic, and Cretaceous times, evolution of reptiles, similar to that of the dinosaurs on land, occurred in the oceans. Between 252 and 227 million years ago, three groups appeared—turtlelike placodonts, lizardlike nothosaurs, and dolphinlike ichthyosaurs. Of these, only ichthyosaurs survived until the Cretaceous.

In the Triassic, ichthyosaurs greater than 66 ft (20 m), possibly up to 98 ft (30 m) ruled the oceans. The Jurassic oceans teemed with life. Modern fish groups were well represented, as were ammonites, echinoderms, squid, and modern corals.

The gap left by the extinction of the placodonts and nothosaurs was filled by long-necked plesiosaurs. Those with a short body and tail and a small head lived in shallow water, while larger forms, called pliosaurs, probably lived in deep water. It is also likely that some of the flying reptiles, called pterosaurs, lived on coastal cliffs and survived by eating fish caught at the water's surface. During the Cretaceous Period, reptiles remained the largest marine carnivores, (plesiosaurs now coexisting with mosasaurs, distant relatives of monitor lizards,) but none survived the mass extinction that occurred 66 million years ago.

FOSSILIZED ICHTHYOSAUR
The dolphinlike features of this ichthyosaur are evident from its fossilized remains. The powerful tail was half-moon-shaped, but here only the down-turned backbone is preserved.

dorsal vertebrae with attachment points for long ribs

rib

long neck comprising 30 vertebrae

PLESIOSAUR
Cryptoclidus eurymerus is a mid-Jurassic plesiosaur. It has a small head, long neck, and short tail, which is typical of shallow-water forms. Its sharp teeth indicate that it ate small fish or shrimplike crustaceans.

pointed, interlocking teeth trap prey

bones of shoulder girdle are flattened into thin plates

large bones of pelvic girdle

paddlelike hind flipper

TIMELINE OF EARTH HISTORY

Million years ago	4,100 MYA	4,000 MYA	3,500 MYA	3,000 MYA	2,500 MYA	2,200 MYA
CRYPTOZOIC EON 4,500–542 MYA						

first organic molecules

first stromatolites

first microfossils

first eukaryotes and multicellular algae

50–14 MYA Return to Water

Following the mass extinction that saw the demise of marine reptiles, some mammals that had evolved on land began returning to the water. Around 50 million years ago, the oceans started to resemble modern oceans in terms of their geographical positions and fauna. The ancestors of whales, however, were unlike their modern counterparts. The earliest whales, such as *Pakicetus*, were close relatives of the hoofed mammals (ungulates). *Ambulocetus*, which means "walking whale," is another early form. It had few adaptations for living in water and probably still spent much time on land.

The productivity of the oceans increased, whales diversified, and other marine mammals appeared. Whales similar to today's toothed whales appeared first, and a few million years later, baleen whales evolved. By 24 million years ago, baleen whales had reached today's giant sizes, suggesting that plankton was present in vast numbers for them to feed on. By 15 million years ago, pinnipeds and sirenians (dugongs and manatees) were commonly found in the oceans. It is thought that pinnipeds arose from a family of carnivores not unlike otters. Their present-day forms are seals, sea lions, and walruses.

ANCIENT WHALE SKELETON
This skeleton has been exposed in a desert in Saccao, Peru. Whales evolved over the last 50 million years, so this area must have been an ancient sea at some point in this period.

SKULL WITH A BLOWHOLE
The nostrils of *Prosqualodon davidis* are positioned on top of the head, forming a blowhole. This feature proves that this fossil skull is from a primitive whale.

Today: Life in Modern Oceans

We know much more about life in today's oceans because, as well as having entire organisms to study, we can also observe life cycles, locomotion, and behavior. Each of the five oceans supports a wide variety of life. Some species are very specialized and are restricted to a small area, while others are migratory or generalists and have a wider distribution. Sometimes, closely related species live in the same habitat in different oceans, separated by land or other physical barriers (see right).

By studying living organisms and the characteristics of the water they live in, scientists can also better understand ancient ocean environments and organisms. The deep ocean is still poorly known, but it contains an ecosystem that could be crucial to our understanding of life—black and white smokers (see p.188). Isolated from sunlight and from the surrounding water by a steep thermal gradient, it is possible that this is the type of environment in which life first evolved around 3.5 billion years ago.

GRAY REEF SHARK
Like its close relative the Caribbean reef shark (below), this shark lives in warm, shallow waters, near coral atolls and in adjacent lagoons. It is found in the Indian and Pacific Oceans, but it is cut off from the Atlantic.

CARIBBEAN REEF SHARK
Like the gray reef shark (above), this species lives in shallow water near coral reefs. Its range is isolated from the Indo-Pacific by the deep, cold ocean around South Africa, so it is restricted to warm parts of the Atlantic, from the Caribbean to Uruguay.

OCEAN LIFE

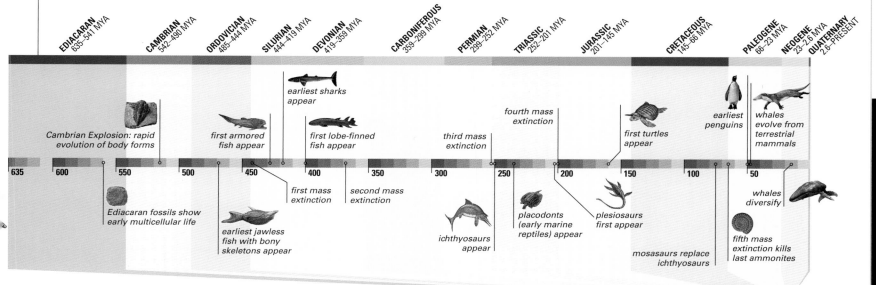

EDIACARAN 635–541 MYA
CAMBRIAN 542–490 MYA
ORDOVICIAN 485–444 MYA
SILURIAN 444–419 MYA
DEVONIAN 419–359 MYA
CARBONIFEROUS 359–299 MYA
PERMIAN 299–252 MYA
TRIASSIC 252–201 MYA
JURASSIC 201–145 MYA
CRETACEOUS 145–66 MYA
PALEOGENE 66–23 MYA
NEOGENE 23–2.6 MYA
QUATERNARY 2.6–PRESENT

earliest sharks appear

fourth mass extinction

earliest penguins

whales evolve from terrestrial mammals

Cambrian Explosion: rapid evolution of body forms

first armored fish appear

first lobe-finned fish appear

third mass extinction

first turtles appear

635 600 550 500 450 400 350 300 250 200 150 100 50

Ediacaran fossils show early multicellular life

first mass extinction

second mass extinction

whales diversify

earliest jawless fish with bony skeletons appear

ichthyosaurs appear

placodonts (early marine reptiles) appear

plesiosaurs first appear

fifth mass extinction kills last ammonites

mosasaurs replace ichthyosaurs

2,000 MYA   1,500 MYA   1,000 MYA   700 MYA   635 MYA   PRESENT DAY

PHANEROZOIC EON 541 MYA–PRESENT

first fossil evidence of microscopic mineralized skeletons

beginning of the Ediacaran period, which soon features the first multicellular life

Mass Extinctions

The history of life is punctuated by five mass major extinctions—catastrophic events in which many life forms died out. The first occurred 444 million years ago, when prominent marine invertebrates disappeared from the fossil record. About 368 million years ago, global cooling and an oxygen shortage in shallow seas caused about 21 percent of marine families to disappear, including corals, brachiopods, bivalves, fishes, and ancient sponges. At the end of the Permian Period, 252 million years ago, the cooling and shrinking of oceans killed over half of all marine life. Another mass-extinction event at the end of the Triassic Period, 201 million years ago, caused major losses of cephalopods, especially the ammonites. The fifth extinction, 66 million years ago, caused the demise of the dinosaurs; in the oceans, it caused the giant marine reptiles to disappear. The next mass extinction is likely to be a result of human activity.

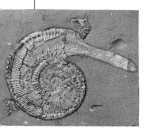

AMMONITE FOSSIL
Many ammonite species, such as this *Kosmoceras*, thrived in the Jurassic and Cretaceous.

VOLCANIC ARMAGEDDON
Volcanic activity in the western Ghats of India is now thought to have been a factor in the most recent mass extinction. The eruptions would have caused destruction and climate change on a global scale.

LIFE ON EARTH was once thought to fall into five great kingdoms—the animals; plants; fungi; and the two microscopic kingdoms, the protists and the bacteria. Scientists are now looking at life ever more closely, and each discovery expands our perspective on life's vast variety. Many experts now consider that the familiar life-forms, plants and animals, represent just 2 of 30 or more kingdoms. The oceans are the ancestral home of life and are still home to all major groups of animals. Although plants are far more diverse on land, their place is taken in the oceans by a range of seaweeds and microorganisms. The following section showcases the entire range of ocean life. It is organized into kingdoms and further divided into the smaller units used by scientists to order and understand nature.

KINGDOMS OF OCEAN LIFE

SUCCESS IN WATER
This violet-spotted reef lobster is a flamboyant example of one of the marine success stories of the animal kingdom—the varied and abundant crustaceans. The crustaceans as a group include crabs, crayfish, shrimp, and some of the most common members of the zooplankton.

Bacteria and Archaea

DOMAINS	Bacteria
	Archaea
KINGDOMS	13
SPECIES	Many millions

THE SMALLEST ORGANISMS ON EARTH are the bacteria and their relatives, the archaea. Bacteria occupy virtually all oceanic habitats, whereas many archaea are confined to extreme environments, such as deep-sea vents. Bacteria and archaea play vital roles in the recycling of matter. Many are decomposers of dead organisms on the ocean floor. Others are remarkable in being able to obtain their energy from minerals in the complete absence of light.

Anatomy

Bacteria and archaea are single-celled organisms that are far smaller than any other, even protists. Most have a cell wall, which, in bacteria, is made from a substance called peptidoglycan. None has a nucleus or any of the other cell structures of more complex organisms (eukaryotes). Some bacteria and archaea can move by rotating threads called flagella; others have no means of propulsion.

Scientists separated the Archaea and Bacteria groups on the basis of chemical differences in their cell make-up. All living cells contain tiny granules (ribosomes), which help to make proteins, but those in archaea are differently shaped to those in bacteria. The oily substances that make up their cell membranes are also different. Additionally, archaea have special molecules associated with their DNA that protect them in the harsh environments in which they live.

Scientists now think that their chemical differences are sufficiently important to rank Archaea as a distinct evolutionary branch of life. Initially considered to be primitive, the archaea are now thought to be closer to the ancestors of eukaryotes than are the bacteria.

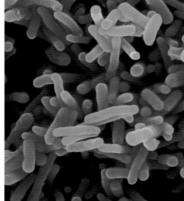

THRIVING IN THE RIGHT CONDITIONS
The bacterium *Nitrosomonas* forms colonies wherever there is enough ammonia and oxygen in the water.

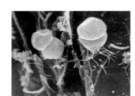

HEAT-LOVING ARCHAEA
Most archaea can adapt to extreme conditions. This heat-loving example, GRI, was ejected from the sea floor in an undersea eruption.

Habitats

Bacteria are found throughout the ocean environment, because nearly all habitats provide them with the materials necessary to obtain energy. Most bacteria get energy by breaking down organic matter. Much of this matter accumulates on the ocean floor and provides excellent conditions for the decomposer bacteria. Bacteria are also found in large numbers in the water column, feeding on suspended matter.

A few kinds of bacteria, such as cyanobacteria, can photosynthesize and so live nearer to the surface, in brightly lit waters. Some form colonies and build huge structures, called stromatolites, near the shore.

Many archaea and bacteria can live in extreme conditions, such as high temperatures, water with high acidity, or low oxygen levels. For example, archaea and bacteria live around deep-sea vents, getting their energy from chemical reactions of methane and sulphide compounds ejected by the vents. Others survive in the very high concentrations of salt on some sea shores.

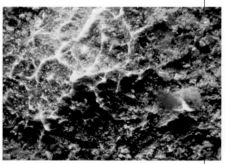

LIVING ON THE SEA BED
Bacterial mats form on the sea bed where oxygen supply is low. This mat of *Beggiatoa* sp. is at the mouth of the Mississippi, USA.

HYPER-SALINE CONDITIONS
The hyper-saline water of Hamelin Pool, west Australia, is one of only three places where stromatolites are found. The rocks are formed by the cyanobacteria cementing sediment particles together.

DOMAIN BACTERIA

Oscillatoria willei

SIZE Filament length 0.13 mm

DISTRIBUTION Tropical waters

Once known as blue-green algae, cyanobacteria are bacteria that are able to use photosynthesis to make foods in a similar way to plants. *Oscillatoria willei* and other related cyanobacteria occur in rows of similarly sized cells that form filaments called trichomes. Many trichomes are enveloped in a firm casing, but in *Oscillatoria*, the casing is thin or may be absent altogether, which allows the filaments to glide quickly forward or backward or even to rotate. Some species of *Oscillatoria* can fix nitrogen but, unlike *Trichodesmium* (below), they may not have cells specialized for the purpose. Fragments of filaments, called hormogonia, which consist of dozens of cells, sometimes break off and glide away to establish new colonies. These

bacteria may cause skin irritations in humans who come into contact with them in tropical waters.

DOMAIN BACTERIA

Trichodesmium erythraeum

SIZE 1–10 mm per colony

DISTRIBUTION Tropical and subtropical seas worldwide

Individual filamentous colonies of the cyanobacteria *Trichodesmium erythraeum* are just visible to the naked eye, and these bacteria have traditionally been known as sea sawdust by mariners. Under warm conditions, the bacteria is able to multiply extremely rapidly to create massive blooms that may have such an extent that they are visible from space. This is a prolific nitrogen-fixing bacterium that harnesses about half of the nitrogen passing through oceanic food chains. The bacteria form long colonial

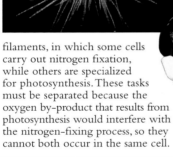

filaments, in which some cells carry out nitrogen fixation, while others are specialized for photosynthesis. These tasks must be separated because the oxygen by-product that results from photosynthesis would interfere with the nitrogen-fixing process, so they cannot both occur in the same cell.

DOMAIN BACTERIA

Vibrio fischeri

SIZE 0.003 mm cell length

DISTRIBUTION Worldwide

Many marine organisms, particularly those in the deep sea, make use of bioluminescence, the biochemical emission of light. Many of these creatures depend on bacteria, such as the rod-shaped *Vibrio fischeri*, to generate the light, and in these cases the bacteria live within the body of their host in a mutually beneficial relationship. The bacteria produce light using a chemical reaction that takes place inside their cells. *Vibrio fischeri* also occurs as a free-living organism, moving through water by means of a flagellum and feeding on dead organic matter. The distinctive, comma-shaped cells seen in *Vibrio fischeri*, below, are characteristic of the genus. Other *Vibrio* species (which are not luminescent) are responsible for the potentially fatal disease cholera.

DOMAIN BACTERIA

Calothrix crustacea

SIZE Filament length 0.15 mm

DISTRIBUTION Worldwide

Forming single filaments or small bundles, bacteria of the genus *Calothrix* are widespread in oceans everywhere. Unlike those of *Oscillatoria* and *Trichodesmium* (left), the filaments of *Calothrix crustacea* have a broad base and a pointed tip that ends in a transparent hair. The filament has a firm or jellylike coating, which is often made up of concentric layers that may be colorless or yellow-brown.

EYE LIGHTS

Eyelightfish (such as the one shown below) have light-emitting organs called photophores under each eye. The light is produced by colonies of *Vibrio fischeri* living in the photophores. The light organs can be covered and uncovered and may be used as an aid to recognition and communication between fish of this species. The ability to emit light may also play a part in prey capture and the avoidance of predators.

Unusually, the filament grows in much the same way as a plant root, its growth being confined to a special region just behind the tip, called a meristem. Sometimes, the filament sheds the tapering tip above the growth region, enabling *Calothrix* to reproduce asexually by casting off fragments called hormogonia from the meristem. These fragments are able to form new filaments far away from the parent. These kinds of cyanobacteria often form slimy coatings on coastal rocks and seaweeds. At least one species of *Calothrix* is known to make up the photosynthetic part of some rocky shore lichens, such as *Lichina pygmaea* (p.255).

DOMAIN ARCHAEA

Halobacterium salinarium

SIZE 0.001–0.006 mm

DISTRIBUTION Dead Sea and other hypersaline areas of the world

Archaea that have adapted to live in waters with exceptionally high salt concentrations are called halophiles. One example of this type of organism is *Halobacterium salinarium*, which is rod-shaped, produces pink pigments called carotenoids and forms extensive areas of pink scum on salt flats. The cell membranes of halophiles contain substances that make them more stable than other types of cell membrane, preventing them from falling apart in the high salt concentrations in which they live. Their cell walls are also modified for the same function.

These archaea obtain nourishment from organic matter in the water.

In addition, their pigments absorb some light energy, which the bacteria then use for fueling processes within the cells.

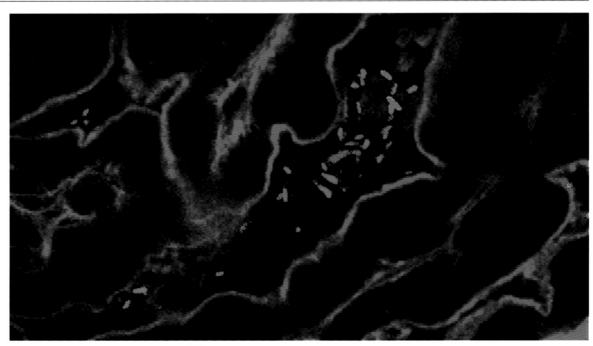

Chromists

DOMAIN	Eucarya
PHYLA	At least 11
SPECIES	31,200

MOST CHROMISTS are microscopic in size, including many of the important photosynthetic plankton groups such as diatoms. However, brown seaweeds, which can grow to a huge size, are also chromists and unlike green seaweeds are not considered to be true plants. Like true plants, chromists make their own food by photosynthesis, but they use different pigments to capture the sun's energy.

Anatomy

Brown seaweeds and the various kinds of microscopic chromists all have very different shapes and structures. Their shared characteristics are at the cellular level. Chromists use an additional type of chlorophyll (a pigment) for photosynthesis and, unlike true plants, the food they manufacture is not stored as starch. They also have other colored pigments, which give brown seaweeds their color. Like green and red seaweeds, brown seaweeds do not need roots to absorb water and nutrients. Some have stiff stalks (stipes) and gas–filled bladders (pneumatophores) that hold up their leaflike fronds to the light. Some microscopic chromists, such as diatoms, have a rigid outer skeleton and float passively. Others, including dinoflagellates, can propel themselves along using whiplike threads called flagella.

BROWN SEAWEED
In place of roots, brown seaweeds have a holdfast, which acts as an anchor. In this one, called *Laminaria hyperborea*, the holdfast is a small disk.

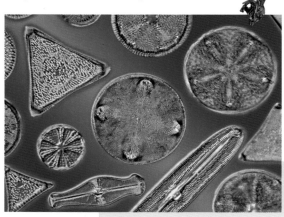

DIFFERENT SHAPES
There are thousands of species of diatoms. Each has a differently shaped silica skeleton.

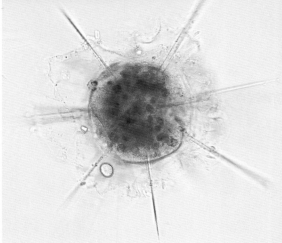

DRIFTING PLANKTON
Radiolarians use their delicate arms to trap food particles and aid buoyancy as they float in the plankton.

GIANT CHROMISTS
The largest chromists are brown seaweeds called kelps. Firmly attached to rocks, they are seen flourishing in this clear, shallow water along the coast of South America.

CHROMIST OR PROTIST?

Scientists used to place all single-celled eukaryotes, including microalgae, protozoans, and chromists, in one taxonomic group: the Protista. Today the term protist is used informally to indicate any of these diverse unicellular organisms. But there is no final agreement as to whether all the organisms included as chromists in this book really belong together in this kingdom.

Habitats

Chromists live in every ocean in the world. Some need sunlight to photosynthesize, so the microscopic forms, most of which are phytoplankton, are only found in the well-lit surface layers. Others, such as foraminiferans, can live on the seabed, along with brown seaweeds, which often dominate rocky seashores in cooler climates. Great underwater forests of brown kelps grow in colder waters, but they and most other brown seaweeds do not usually grow below about 66 ft (20 m). Some brown seaweeds grow unattached in sheltered lagoons and sea lochs, and a few grow in salt marshes anchored in mud. The brown sargassum or gulfweed, *Sargassum natans*, is unusual in that it floats at the surface of the open ocean, forming the basis for a unique ecosystem (see p.238).

PLANKTON BLOOM
The milky bloom in this satellite image off Cornwall, UK, is of the coccolithophore *Emiliania huxleyi*, which has multiplied rapidly in favorable conditions.

Life Strategies

Most single-celled chromists drift in the ocean surface layers as phytoplankton. They reproduce by splitting into two, and some can do so very rapidly (see above). They can also reproduce sexually, forming the equivalent of egg and sperm cells. Their number increases dramatically with the warmth and longer days of spring and summer. Many seashore brown seaweeds produce slippery mucus, both to protect from drying out and to deter grazers. While some brown seaweeds are annuals, living for only a year, large species are usually perennial.

NEW KELP GROWS FROM OLD
A new, yellow frond is growing from the top of this kelp stipe. The old frond, which will drop off, is covered in white animals called bryozoans, which block vital photosynthesis.

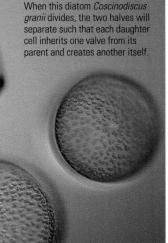

upper valve of pillbox-shaped diatom

DIATOM DIVISION
When this diatom *Coscinodiscus granii* divides, the two halves will separate such that each daughter cell inherits one valve from its parent and creates another itself.

HUMAN IMPACT

FARMING THE OCEANS

Seaweeds are harvested wild, but more are now grown in ocean nurseries and farms, especially in Asia. They are used for food, and seaweed extracts are used in a wide range of products—for example, in gels as a stabilizer, in cosmetics and pharmaceuticals, in beer-making, and as a fertilizer.

SEAWEED HARVEST IN ZANZIBAR
Many seaweeds grow readily on floating rafts, as seen here. They flourish in strong light, away from grazing invertebrates, and provide vital income for coastal communities.

OCEAN LIFE

Noctiluca scintillans

DIAMETER Up to ¹/₁₆ in (2 mm)

HABITAT Surface waters

DISTRIBUTION Worldwide

Also known as sea sparkle, *Noctiluca scintillans* is a large, bioluminescent dinoflagellate that floats near the surface of the ocean, buoyed up by its oily cell contents. It is one of the naked dinoflagellates, which do not have a protective outer theca (shell). Like all dinoflagellates, it has two flagella, but one is tiny. This species feeds on other plankton, and its second large flagellum helps sweep food particles toward it, which it then engulfs. Other dinoflagellates also feed in this way, but there are many that are photosynthetic.

BIOLUMINESCENCE

Floating just below the surface of the water at night, dinoflagellates, and in particular *Noctiluca scintillans*, are the most common cause of bioluminescence in the open ocean. Millions of *Noctiluca scintillans* cells twinkle in the waves, hence the common name sea sparkle. The blue-green light is emitted from small organelles within the cells and is generated by a chemical reaction. Unlike many bioluminescent fish, it does not depend on light-emitting bacteria.

Gymnodinium pulchelum

DIAMETER 0.025 mm

HABITAT Surface waters

DISTRIBUTION Temperate and tropical waters above continental shelves, and Mediterranean

Some red-tide organisms such as *Gymnodinium pulchellum* produce toxins that affect the nervous system and the clotting properties of the blood, causing high mortality among fish, as well as invertebrates. The cause of red tides is not well understood, but some scientists think they may be influenced by coastal pollution providing nutrients that might otherwise be in short supply and so normally limit the population size. Rapid reproduction by simple cell division results in huge numbers of *Gymnodinium pulchellum* being present in the water, turning it a characteristic brown-red color, as shown here in the seas around Hong Kong. Unlike many other types of dinoflagellate, this species lacks a test and also produces food by photosynthesis.

Neoceratium tripos

LENGTH 0.2–0.35 mm

HABITAT Surface waters

DISTRIBUTION Worldwide

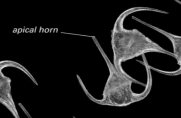

apical horn

lateral horns
aid flotation

The unique three-pronged shape of the dinoflagellate *Ceratium tripos* makes it easy to identify among the phytoplankton, where it is one of the dominant organisms. Although this species is usually solitary, several individuals may be seen together, attached to each other by the single apical horn. This occurs when a cell divides and the daughter cells remain linked in short chains. *Ceratium tripos* is sometimes parasitized by other protists.

Strombidium sulcatum

DIAMETER 0.045 mm

HABITAT Surface waters

DISTRIBUTION Atlantic, Pacific, and Indian Oceans

Organisms such as *Strombidium sulcatum* are classified as ciliates because the cell membrane has many hairlike projections, called cilia, that are used in locomotion. In *Strombidium sulcatum*, the cilia are restricted to a collar at one end of its spherical body, which has no shell. Ciliates are the most complex of all protists and have two nuclei in their single cell, a macronucleus and a micronucleus. Most of the time, *Strombidium sulcatum* reproduces asexually by splitting both nuclei and the cell into two. Periodically, it must undergo a type of sexual reproduction called conjugation. Two individuals join, exchange part of their micronucleus, and then separate. The new micronucleus divides in two, one forming a new macronucleus as the original one disintegrates. The ciliate then divides in two. Ciliates and dinoflagellates share some cell characteristics, and both belong to a group known as the alveolates.

PHYLUM RADIOZOA
Cladococcus viminalis

DIAMETER 0.08 mm

HABITAT Surface waters

DISTRIBUTION Mediterranean

Radiolarians produce extremely complex silica tests of spines and pores that are laid down in a well-defined geometric pattern. The spines aid buoyancy and the pores provide outlets for cell material, called pseudopodia, which engulf any food that becomes trapped on the spines and carry it to the center of the cell to be digested. *Cladococcus viminalis* is a polycystine radiolarian, which are the most commonly fossilized radiolarians and are frequently found in chalk and limestone rocks.

PHYLUM HAPTOPHYTA
Emiliania huxleyi

DIAMETER 0.006 mm

HABITAT Surface waters

DISTRIBUTION Atlantic, Pacific, and Indian Oceans

Emiliana huxleyi is a protist belonging to a group of haptophytes commonly known as coccolithophores. The name comes from a covering of intricately sculptured calcite plates called coccoliths with patterns that are unique to each species. Like some other protists, *E. huxleyi* can multiply very quickly in favorable conditions and form blooms that can cover areas of up to 38,600 square miles (100,000 square km). These blooms are visible from space because the coccoliths act like tiny mirrors and reflect sunlight so the water they are in appears a milky white. By reflecting light and heat and by "locking up" carbon in their calcite coccoliths, they help reduce ocean warming. They have been found worldwide in chalk deposits dating from 65 million years ago. The famous white cliffs of Dover in the UK are mainly formed from coccolith plates.

PHYLUM FORAMINIFERA
Hastigerina pelagica

LENGTH ¼ in (6 mm)

HABITAT Warm waters at depth of 660 ft (200 m)

DISTRIBUTION Subtropical and tropical waters of North Atlantic and western Indian Ocean

Foraminiferans are unicellular organisms found only in marine habitats. *Hastigerina pelagica* is one of the larger forms. It is often pinkish red and has a calcareous test with several globular-shaped chambers from which radiate calcite spines covered with cytoplasmic strands (pseudopodia) for collecting food. *Hasterigina pelagica* is unique in surrounding its test with a gelatinous capsule of tiny frothy bubbles, which is thought to aid buoyancy. Dinoflagellates sometimes live on the surface of the capsule and up to 79 have been counted on a single individual, although 6–10 is more usual. The relationship between the two organisms is not clearly understood because *Hasterigina pelagica* is carnivorous, yet the dinoflagellates are unharmed.

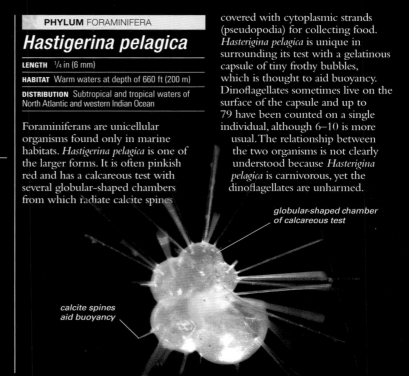

globular-shaped chamber of calcareous test

calcite spines aid buoyancy

PHYLUM OCHROPHYTA
Ethmodiscus rex

DIAMETER 1/16–1/8 in (2–3 mm)

HABITAT Warm, nutrient-poor water

DISTRIBUTION Open ocean worldwide

Ethmodiscus rex is the largest of all diatoms and can be seen with the naked eye. It is a single cell with a rigid cell wall, called a test, which is impregnated with silica and covered in regular rows of pits. The test is made up of two disk-shaped halves, called valves, which fit tightly together. Because each diatom has a unique test, *Ethmodiscus rex* can be easily identified in the fossil record. It is found in rocks that date from the Pliocene, and the fossils can be up to 5 million years old. The cells need to remain near the water surface in order to utilize the Sun's energy for food, which they do by transforming the products of photosynthesis into oily substances that increase their buoyancy. *Ethmodiscus rex* can reproduce sexually but, if conditions are favorable, it multiplies rapidly, simply by dividing into two (see p.235). Over a 10-day period, one individual that divides three times a day can theoretically have more than 1.5 billion descendants.

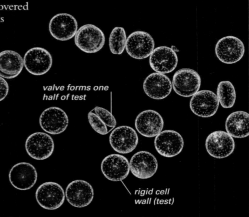

valve forms one half of test

rigid cell wall (test)

PHYLUM OCHROPHYTA
Chaetoceros danicus

LENGTH 0.005–0.02 mm

HABITAT Surface waters

DISTRIBUTION Worldwide

First described in 1844, *Chaetoceros* is one of the largest and most diverse genera of marine diatoms, containing nearly 200 species. *Chaetoceros danicus* is a colonial form, and groups of seven cells are not uncommon (as shown here). It is easily recognized because it has highly distinctive long, stiff hairs, called setae, which project perpendicularly from the margins of its test. and have prominent secondary spines along their length. Chloroplasts, which contain pigments used in photosynthesis, are numerous and found inside both the cell and the setae. The setae are easily broken, and if large quantities lodge in the gills of a fish, they may kill it. The secondary spines anchor the setae to the sensitive gill tissue, causing irritation, and the fish reacts by producing mucus. Eventually, it dies from suffocation.

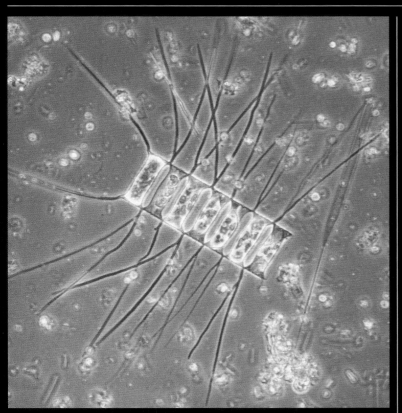

PHYLUM OCHROPHYTA
Dictyocha fibula

LENGTH 0.045 mm

HABITAT Surface waters

DISTRIBUTION Atlantic, Mediterranean, Baltic Sea, and eastern Pacific off coast of Chile

The golden yellow pigments visible in this image of *Dictyocha fibula* are typical of two groups of golden algae known as Chrysophyceae and Dictyochophyceae (this species). The word Dictyocha means "net" and refers to the large windows in the silica test. Fewer than 20 species of *Dictyocha* are alive today. They are all that remains of a group of organisms that flourished more than 5 million years ago. Their fossils are abundant in some Miocene deposits.

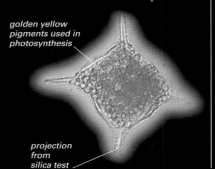

golden yellow pigments used in photosynthesis

projection from silica test

OCEAN LIFE

PHYLUM OCHROPHYTA

Limey Petticoat

Padina gymnospora

HEIGHT	Up to 4 in (10 cm)
HABITAT	Rock pools and shallow subtidal rocks
WATER TEMPERATURE	68–86°F (20–30°C)

DISTRIBUTION Coasts in tropical and subtropical areas worldwide

Padina is the only genus of brown seaweeds to have calcified fronds, hence this species' common name of limey petticoat. The reflective chalk shows as bright white concentric bands on the upper surface of the fan-shaped fronds. The fronds are only 4–9 cells thick and curled inward. Older fronds may become split into wedge-shaped sections. This species is widespread in tropical seas, often growing in masses on shallow subtidal rocks, and on old coral and shells.

PHYLUM OCHROPHYTA

Giant Kelp

Macrocystis pyrifera

LENGTH	150 ft (45 m)
HABITAT	Rocky seabeds, occasionally sand
WATER TEMPERATURE	41–68°F (5–20°C)

DISTRIBUTION Temperate waters of southern hemisphere and northeastern Pacific

Giant kelp (pictured on pp.240–241) is the largest seaweed on Earth. It can grow at the rate of 24 in (60 cm) per day in ideal conditions, and reaches lengths of over 100 ft (30 m) in a year.

PHYLUM OCHROPHYTA

Oyster Thief

Colpomenia peregrina

DIAMETER	Up to 4 in (10 cm)
HABITAT	Intertidal and subtidal rocks and shells
WATER TEMPERATURE	49–83°F (6–28°C)

DISTRIBUTION Coasts of western North America, Japan, and Australasia; introduced in Atlantic

The oyster thief gets its unusual name from its habit of growing on shells, including commercially grown oysters. The frond is initially spherical and solid, but as it grows, it becomes irregularly lobed and hollow and fills with gas. Sometimes, this can make it sufficiently buoyant to lift the oyster,

Giant kelp normally grows at a depth of 30–100 ft (10–30 m), but it can grow much deeper in very clear water. The huge branched holdfast, which is about 24 in (60 cm) high and wide after three years, is firmly attached to the seabed. From it, a number of stalks (or stipes) stretch toward the surface, bearing many straplike fronds, each buoyed by a gas-filled bladder. The stem and fronds continue to grow on reaching the surface, floating as a dense canopy. giant kelp has a two-phase life cycle. Fronds (sporophylls) at the base of the kelp produce spores that develop into tiny creeping filaments. The filaments produce eggs and sperm, which combine to produce embryonic kelp plants.

which is not fixed to the seabed, and they may both be carried away by the tide. This seaweed has a thin wall with only a few layers of cells. The outer layer is made of small, angular cells which contain the photosynthetic pigments that give the oyster thief its brown color.

PHYLUM OCHROPHYTA

Landlady's Wig

Desmarestia aculeata

LENGTH	Up to 6 ft (1.8 m)
HABITAT	Subtidal rocks, and kelp forests
WATER TEMPERATURE	32–64°F (0–18°C)

DISTRIBUTION Near coasts in temperate, cold, and polar regions

This large seaweed has narrow brown fronds with many side-branches. Its bushy appearance is the reason for its common name of landlady's wig. The smallest branches are short and spinelike, hence the species name *aculeata*, which means "prickled." In summer, the whole plant is covered with delicate branched hairs. This species is particularly abundant on boulders and in kelp forests disturbed by waves.

PHYLUM OCHROPHYTA

Sea Palm

Postelsia palmaeformis

LENGTH	Up to 24 in (60 cm)
HABITAT	Wave-exposed shores
WATER TEMPERATURE	46–64°F (8–18°C)

DISTRIBUTION West Coast of North America

Sea palms are kelps, which are large brown seaweeds that belong to the order Laminariales, as does giant kelp (above). Unusually for a kelp, sea palm grows on the midshore, where it forms dense stands on wave-exposed coasts. It has a branched holdfast, and a stout, hollow stalk, which stands erect when the tide is low. The top of the stalk is divided into many short, cylindrical branches, each of which bears a single frond up to 10 in (25 cm) long, with toothed margins and deep grooves running down both faces. Spores are released into the grooves and drip off the frond tips onto the holdfasts and nearby rocks at low tide, so that the developing seaweeds grow as dense clumps. Some sea palms attach to mussels and are later ripped off during storms, making more rock available for other sea palms to grow.

OCEAN LIFE

CROFTER'S WIG

In very sheltered bays and sea lochs, detached pieces of "normal" knotted wrack will continue to grow, lying loose on the seabed. In situations where the fronds are alternately covered by salt and fresh water, they divide repeatedly to form a dense ball that has no bladders or reproductive structures. This unattached form, which is known as crofter's wig, appears very different to the attached form, even though it is genetically identical.

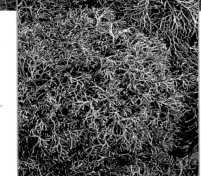

PHYLUM OCHROPHYTA

Knotted Wrack
Ascophyllum nodosum

LENGTH Up to 10 ft (3 m)

HABITAT Sheltered seashores

WATER TEMPERATURE 32–64°F (0–18°C)

DISTRIBUTION Coasts of northwestern Europe, eastern North America, and north Atlantic islands

Knotted wrack belongs to a group of tough brown seaweeds that often dominate rocky seashores in cooler climates. It is firmly attached to the rocks by a disk-shaped holdfast, from which arise several narrow fronds that often grow to 3 ft (1 m) in length, and exceptionally to 9 ft (3 m) in very sheltered situations. Single oval bladders grow at intervals down the frond. The fronds produces about one bladder a year, so the seaweed's age can be roughly estimated by counting a series of bladders. The bladders hold the fronds up in the water so that they gain maximum light, which is an advantage in the often turbid waters where knotted wrack grows. This also makes it harder for grazing snails to reach the fronds when the tide is in.

The dark brown fronds may be bleached almost to yellow in summer. Reproductive structures that look like swollen raisins are borne on short side-branches, and orange eggs can sometimes be seen oozing from them.

PHYLUM OCHROPHYTA

Neptune's Necklace
Hormosira banksii

LENGTH Up to 12 in (30 cm)

HABITAT Lower shore and subtidal rocks

WATER TEMPERATURE 50–68°F (10–20°C)

DISTRIBUTION Coasts of southern and eastern Australia and New Zealand

Neptune's necklace is one of the many brown seaweeds endemic (unique) to New Zealand and the cooler waters around Australia. Its distinctive fronds, which look like a string of brown beads, are made up of chains of ovoid, hollow segments joined by thin constrictions in the stalk. Small reproductive structures are scattered over each "bead."

Dense mats composed almost entirely of this one species can be found on seashore rocks. The fronds are attached to the rock by a thin, disk-shaped holdfast. Neptune's necklace also lives unattached among mangrove roots. The shape of its segments varies according to habitat. They are spherical and about ¾ in (2 cm) wide in fronds growing on sheltered rocks, mussel beds on tidal flats, or in mangrove swamps. Fronds growing on subtidal rocks on moderately exposed coasts have smaller segments that are just ¼ in (6 mm) long.

PHYLUM OCHROPHYTA

Japweed
Sargassum muticum

LENGTH 6–33 ft (2–10 m)

HABITAT Intertidal and subtidal rocks and stones

WATER TEMPERATURE 41–79°F (5–26°C)

DISTRIBUTION Coasts of Japan, introduced in western Europe and western North America

Japweed can reproduce all year round and forms dense stands in quiet waters. Native to Japan (hence its common name), it was accidentally introduced to western North America and Europe, and is steadily expanding its range in these areas. It outcompetes other seaweeds and in these regions is regarded as an invasive species. This long, bushy seaweed has numerous side-branches, which have many leaflike fronds up to 4 in (10 cm) long. The fronds bear small, gas-filled bladders, either singly or in clusters.

GIANT KELP
This enormous seaweed can grow at a rate of 20 in (50 cm) per day in favorable conditions, such as the relatively cold water off California (shown here). Air bladders help keep the kelp's blades afloat as they grow upward toward the surface, where there is an enhanced supply of light and nutrients.

OCEAN LIFE

Plant Life

DOMAIN	Eucarya
KINGDOM	Plantae
SPECIES	315,000

PLANTS FORM A GREAT kingdom of life-forms, all of which use the pigment chlorophyll to fix carbon dioxide from the atmosphere into organic molecules, using energy from sunlight. Most organisms in the plant kingdom are vascular plants, which evolved on land and remain land-based. Of these, several unrelated families of flowering plants (see p.250) have since returned to the sea or taken up residence on the coast. The plant kingdom, as defined in this book, also encompasses more primitive nonvascular organisms, including mosses, the microscopic green algae (microalgae), and the green and red seaweeds (macroalgae). Brown seaweeds appear very similar but may not be closely related to true plants and are classified in this book in the phylum Chromista (see p.234).

Marine Plant Diversity

Plants are united by their use of chlorophyll for photosynthesis. Vascular plants encompass several major land-based groups, including ferns and conifers. Because they evolved on land, they are adapted to life in air and to fresh water. They have tissues bearing vessels that transport water and food.

The largest group of vascular plants are the flowering plants, some of which live in marine habitats. They mostly inhabit the coastal fringes, but the seagrasses are fully marine. Green seaweeds and microscopic green algae (microalgae) lack stems and roots, have neither woody tissues nor transport vessels, and are mostly aquatic.

TEMPERATE MARINE PLANTS
Temperate seas are rich in phytoplankton, including microscopic green algae. Green and red seaweeds algae commonly grow on rocks, and green seaweeds also on mud flats. Seagrasses have true roots and live in sediment in the shallow subtidal and intertidal zones, and in brackish lagoons. Above high water, cliffs, sand dunes, and salt marshes are home to flowering plants, mosses and lichens.

TROPICAL MARINE PLANTS
Seagrasses are abundant in tropical lagoons, and green seaweeds include calcified species. Mangroves line estuaries and creeks, and other flowering plants, including shrubs and trees, colonize the back of sandy beaches. Although not shown here, seaweeds may grow seasonally on rocky shores.

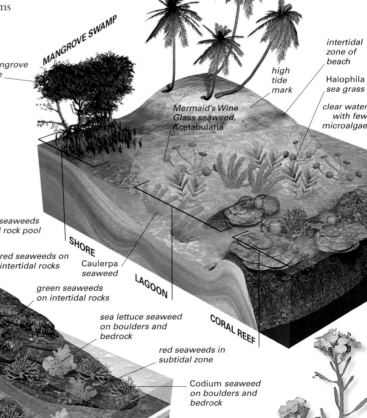

- sea campion
- SEA CLIFF
- lichens
- green seaweeds in tidal rock pool
- DUNES
- marram grass
- thrift
- SALT MARSH
- red seaweeds on intertidal rocks
- SHORE
- Caulerpa seaweed
- green seaweeds on intertidal rocks
- LAGOON
- sea lettuce seaweed on boulders and bedrock
- red seaweeds in subtidal zone
- brackish lagoon
- Ulva seaweed in channel
- sea rocket above high tide
- intertidal zone
- seagrass in sediment in subtidal zone
- surface water thick with green microalgae
- SHELTERED SHORE
- SEMISHELTERED SHORE
- EXPOSED SHORE
- Codium seaweed on boulders and bedrock
- CORAL REEF
- MANGROVE SWAMP
- mangrove tree
- BEACH
- Coconut Palm
- high tide mark
- Mermaid's Wine Glass seaweed, Acetabularia
- intertidal zone of beach
- Halophila sea grass
- clear water with few microalgae

SEA ROCKET
A European member of the brassica family, sea rocket can grow on fine sand, just above high tide, where it traps the sand, forming small foredunes. Its waxy leaves repel sea spray, while its stubby fruit pods are dispersed by the tide.

ABOVE AND BELOW THE TIDE LINE
Marine flowering plants range from coastal trees growing above the high tide mark, such as the coconut palm, to seagrasses, the only wholly marine flowering plants, below it.

PLANTS OF SHIFTING SHORES
Sea mayweed, with daisylike white flowers, and oyster plants, with their dark blue-green leaves, are salt-tolerant flowering plants that grow on semisheltered shores of shingle.

Above High Water

Above the reach of the highest tides, the environment is essentially terrestrial, but its proximity to the sea makes life hard for all but a few specialized flowering plants and lichens. In places, the coast is covered with dunes of sand blown from the seashore. Dunes are often alkaline, being rich in calcium carbonate from marine shells and maerl (see p.245). With few nutrients, dunes only support hardy colonizers tolerant of infertile and salty soil. Acidic dunes made from sand with little shell support many lichens and mosses such as golden dune moss. Marram, a grass with a fast-growing root and rhizome network stabilizes dunes and takes the first steps toward soil formation. Plants that can fix nitrogen in root nodules, such as the casuarina tree, have an advantage here.

On rocky cliffs plants are safe from grazing animals but must withstand salty spray, drying winds and scanty soil. Plants such as sea campion have very long roots that reach deep into rock crevices for water draining from the land. Plants with succulent leaves and waxy cuticles can store the water they find.

Between the Tides

Plants between the tides have to live both in and out of water in variable conditions sometimes hot, dry, and salty, and at other times drenched with cold, fresh rainwater. Intertidal zones support green and red seaweeds, seagrasses, and mangroves. Many green seaweeds are tolerant of brackish water and grow down the upper shore in strips following freshwater seeps and streams. Lower down, delicate red seaweeds thrive in pools or in the damp beneath tough brown seaweeds. Many are ephemeral, growing quickly in fair conditions and dispersing many spores. They may quickly cover tropical coasts where monsoons bring humid conditions, but then dry up and blow away when the sun returns. Seagrasses grow on lower intertidal flats; here the sediments retain moisture until the tide returns. In the tropics, mangroves colonize intertidal sediments, but only their roots are regularly submerged, while the rest of the plant remains in the air. In colder climates, salt-marsh vegetation develops on mudflats.

SUBMERGED SALT MARSH
The salt-tolerant thrift, or sea pink, often grows in salt marshes, which develop on sheltered coasts in temperate regions. Growing at higher levels, the sea pink is submerged only in the highest tides.

Aquatic Plants

Only seaweeds, seagrasses, and microalgae live permanently submerged in seawater. All seaweeds can absorb nutrients and gases over their whole body surface, so they do not need the transport systems of land plants, and their holdfast simply attaches them to the seabed. Seagrasses have a land-plant anatomy, so they need extra structures such as air spaces (lacunae) to aid gas exchange. Plants living in seawater can only thrive in the top few yards, because enough light for photosynthesis cannot penetrate beyond this. Many marine plants are tough to deter grazing animals or produce toxic or distasteful chemicals.

SEAWEED BED
Green seaweeds in the English Channel include *Codium* (in the foreground) and sea lettuce (lower right). They are growing here with an unrelated brown seaweed called serrated wrack.

Red Seaweeds

DOMAIN	Eucarya
KINGDOM	Plantae
DIVISION	Rhodophyta
CLASSES	2 or more
SPECIES	6,394

RED SEAWEEDS are attractive marine plants and are found in shallow seas around the world. Their red color comes from the extra pigments they have in addition to green chlorophyll. While red seaweeds appear to be plants, they differ in some cellular details and metabolism. Some scientists still do not consider them to be true plants, but the consensus is that they are.

Anatomy

The shape and form of red seaweeds is highly diverse but in general they are relatively small and delicate. Like green and brown seaweeds most have a holdfast, stipes (stems) and fronds that absorb water, nutrients, and sunlight. Coralline seaweeds have a heavily calcified frond, which is too hard for most grazers to eat. They look more like pink crusts or small corals than plants. Maerl forms unattached nodules that resemble twiggy coral lying on the seabed. Some red seaweeds have two phases, growing as a tiny long-lived crust and as an annual bushy frond. These look so different they were first described as separate species.

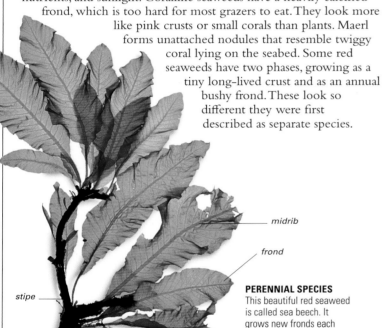

midrib

frond

stipe

holdfast

PERENNIAL SPECIES
This beautiful red seaweed is called sea beech. It grows new fronds each year from a perennial stipe, and reproduces from spores in winter.

Habitats and Distribution

On the shore, red seaweeds mostly live at the lowest levels where they are less likely to dry out. In deeper water, their additional pigments allow them to flourish in the dim blue light that remains and they generally extend deeper than brown and green seaweeds. Red (and brown) seaweeds are less abundant in tropical waters, an exception being the red coralline encrusting seaweeds, which play a very important role in cementing coral reefs.

DULSE
Red seaweeds rely on moving water to bring them nutrients and oxygen. The fingerlike fronds of this dulse increase its surface area while preventing it from tearing in rough conditions.

BARBED COLONIZER
This red seaweed has barbed branches enabling detached fragments to hook onto other organisms and even ships' hulls. It disperses in this way and has been transported outside its native range.

DIVISION RHODOPHYTA

Small Jelly Weed

Gelidium foliaceum

LENGTH	2 in (5 cm)
HABITAT	Intertidal rocks
WATER TEMPERATURE	50–68°F (10–20°C)

DISTRIBUTION Coasts of southern Africa and southern Japan

There are many species of *Gelidium* worldwide, and they are difficult to identify because the plants can look very different depending on their habitat and whether they have been grazed by seashore animals such as limpets. Ongoing work on molecular sequencing is gradually resolving some of these problems, and *Gelidium foliaceum* is one species that has been reclassified. It has a flattened, much lobed and curled frond, which grows in dense clumps on rocky seashores. The fronds are tough and cartilaginous, and the seaweed is attached to the rock at frequent intervals by small hairlike structures, or rhizoids, from a creeping stem, or stolon. This creeping habit is probably the main method of spreading, but some species of *Gelidium* also reproduce sexually. Some *Gelidium* species are a source of the gelatinous substance agar, which is used in cooking and microbiology.

DIVISION RHODOPHYTA

Black Laver
Porphyra dioica

LENGTH	Up to 20 in (50 cm)
HABITAT	Intertidal rocks
WATER TEMPERATURE	43–64°F (6–18°C)

DISTRIBUTION Coasts of northeastern and western Europe and Mediterranean around Italy

This species of red seaweed has only recently been separated from the very similar *P. purpurea* on the basis of how they reproduce. *P. dioica* is dioecious (male and female reproductive cells are on separate fronds), while *P. purpurea* is monoecious (male and female reproductive cells are on the same frond). *P. dioica* grows on intertidal, sandy rocks and is most abundant in the spring and early summer. The membranous frond is only one cell thick and is olive-green to purple-brown or blackish. This species appears to have a limited distribution in western Europe, but the genus is widespread throughout the world. All species of *Porphyra* are edible and are often harvested for food worldwide, especially in Japan where they are cultivated and known as *nori*. In the UK, wild laver is collected and made into the Welsh delicacy laverbread.

DIVISION RHODOPHYTA

Grape Pip Weed Moss
Mastocarpus stellatus

LENGTH	7 in (17 cm)
HABITAT	Lower shore and subtidal rocks
WATER TEMPERATURE	32–77°F (0–25°C)

DISTRIBUTION Coasts of northeastern North America, northwestern Europe, and Mediterranean

This tough red seaweed is common on exposed shores, often forming a dense turf on the lower shore. Its frond is attached to rock by a disk-shaped holdfast, from which arises a narrow stipe (stalk) that gradually expands into a divided blade, which is slightly rolled to form a channel with a thickened edge. Reproductive structures produced in small nodules on the blade's surface develop into a very different seaweed in the form of a thick black crust (it was originally named *Petrocelis cruenta* because it was thought to be an entirely different species). Spores from this crust grow back into the erect form, in a typical two-phase life history. *Mastocarpus stellatus* and the similar *Chondrus crispus* are both also called Irish moss or carrageen moss and are collected on an industrial scale on both sides of the north Atlantic to produce the gelling agent carrageenan.

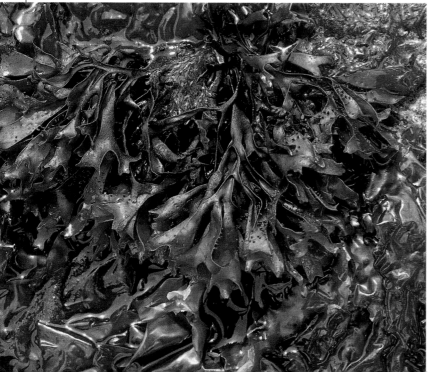

DIVISION RHODOPHYTA

Maerl
Phymatolithon calcareum

DIAMETER	Up to 2¾ in (7 cm)
HABITAT	Subtidal seabed sediments
WATER TEMPERATURE	32–77°F (0–25°C)

DISTRIBUTION Coasts of Atlantic islands, northern and western Europe, and Mediterranean

The term "maerl" describes various species of unattached coralline seaweeds that live on seabeds. *Phymatolithon calcareum* forms brittle, purple-pink, branched structures that look more like small corals than seaweed. It grows as spherical nodules at sheltered sites, or as twigs or flattened medallions at more exposed sites. In places with some water movement from waves and tides, but not enough to break the maerl nodules, extensive beds can develop. Maerl is as much a habitat as a species, and both the living maerl and the maerl-derived gravel beneath it harbor many small animals. Maerl grows slowly and the beds are vulnerable to damage from bottom trawlers.

DIVISION RHODOPHYTA

Coral Weed
Corallina officinalis

LENGTH	Up to 4¾ in (12 cm)
HABITAT	Rock pools and shallow subtidal rocks
WATER TEMPERATURE	32–77°F (0–25°C)

DISTRIBUTION Coasts worldwide except for far north and Antarctica

Coral weed belongs to a group of red seaweeds known as coralline seaweeds, which have chalky deposits in the cell walls that give them a hard structure. Coral weed fronds have rigid sections that are separated by flexible joints. The branches usually lie in one plane, forming a flat, featherlike frond, but the shape is very variable. On the open shore, the fronds are often stunted, forming a short mat a few inches high in channels and rock pools and on wave-exposed rocks. These mats often harbor small animals, and other small red seaweeds attach to the hard fronds. Subtidally, the fronds grow much longer. The color of coral weed varies from dark pink when it lives in the shade to light pink in sunny locations. When the seaweed dies, its hard skeleton breaks up and becomes part of the sand.

DIVISION RHODOPHYTA

Cotton's Seaweed
Kappaphycus alvarezii

LENGTH	20 in (50 cm)
HABITAT	Intertidal and shallow subtidal rocks
WATER TEMPERATURE	50–86°F (10–30°C)

DISTRIBUTION Coasts of Africa, southern and eastern Asia, and Pacific islands

Formerly called *Eucheuma cottonii*, this is a much-branched, cylindrical red seaweed that is farmed extensively in the Philippines for extraction of carrageenan, a gelling agent similar to agar. In the wild, it grows attached to rocks or lies loose in sheltered places. Like some other red seaweeds, its fronds are often shades of green and brown rather than red.

DIVISION RHODOPHYTA

Iridescent Drachiella
Drachiella spectabilis

LENGTH	Up to 2½ in (6 cm)
HABITAT	Subtidal rocks at 6–100 ft (2–30 m)
WATER TEMPERATURE	46–64°F (8–18°C)

DISTRIBUTION Off western coasts of Scotland, UK, Ireland, France, and Spain

This colorful seaweed is rarely seen, except by divers, because it normally grows in relatively deep water and is rarely washed ashore. It also grows in shallower water within kelp forests. It has a thin, fan-shaped frond, split into wedges, that spreads out over the rock and reattaches with small rootlike structures called rhizoids. Young plants have a purple-blue iridescence, which is lost as the seaweed ages. Sexual reproduction is unknown in this species and spores are produced asexually.

Green Seaweeds

DOMAIN	Eucarya
KINGDOM	Plantae
DIVISION	Chlorophyta
CLASSES	About 8
SPECIES	5,426

GREEN ALGAE LARGE ENOUGH to be seen with the naked eye are known as green seaweeds. They are classified with the microscopic green algae, or microalgae (see p.248). True plants of the sea, they have pigments and other features in common with higher plants. They can be abundant in tropical lagoons, and proliferate seasonally on many temperate seashores. *Ulva* (sea lettuce) is grown for food.

frond

large, fibrous holdfast

SEAWEED BODY PARTS
Green seaweeds have a simple structure, with an erect frond and a disk-shaped or fibrous holdfast. This tropical *Udotea* species has calcified fronds with many branched siphons.

Habitats

Green seaweeds often attach to rocks on rocky coasts, particularly in temperate and cold waters, and are ephemeral colonizers in seasonally disturbed tidal and shallow subtidal habitats. *Ulva* species, such as sea lettuce, dominate in high-level rock pools, or where fresh water seeps over the shore, since they can withstand changes in saltiness and temperature. The more delicate *Cladophora* and *Bryopsis* species live in rock pools or among red and brown seaweeds in the shallow subtidal zone. Green seaweeds also thrive in shallow, tropical lagoons, where species of *Caulerpa*, *Udotea*, and *Halimeda* are often abundant. *Caulerpa* species have runners (stolons), which creep through sand or cling to rock, while the bases of *Udotea* and *Halimeda* are a bulbous mass of fibers that anchor in sand. *Halimeda* (cactus seaweed) is heavily encrusted with calcium carbonate, which breaks up when the plant dies, contributing to the lagoon sand.

FLEXIBLE SEAWEED
Able to handle fluctuations in salinity and temperature, *Ulva* species thrive in this freshwater stream as it flows across the seashore.

Anatomy

The body structure of green seaweeds lacks stems and roots. Green seaweed shapes range from threadlike (filamentous) to tubes, flat sheets, and more complex forms. Their texture varies from slimy to rigid and smooth to bristly. Their bright green color is due to the fact that their chlorophyll is not masked by additional pigments, unlike red and brown seaweeds. Many of the features of green seaweeds and green algae, including their types of chlorophyll, are shared by terrestrial "higher" plants, including mosses, liverworts, and vascular plants.

FRAGILE FRONDS
This delicate *Bryopsis plumosa* has coenocytic fronds, meaning its fronds do not have the crosswalls common in other green seaweeds.

CODIUM FOREST
This mini-forest of *Codium fragile* is growing on shallow rocks in a sheltered bay in Scotland. The fronds are buoyant, holding the plants up to the light.

Flaccid Green Seaweed

Ulothrix flacca

SIZE	Up to 4 in (10 cm)
HABITAT	Intertidal on various shore types
HABITAT	32–68°F (0–20°C)

DISTRIBUTION Northern Atlantic, Mediterranean, waters off South Africa, Pacific

This seaweed is made up of many unbranched green filaments, which themselves consist of strings of cells. The filaments form soft, woolly masses or flat green layers that stick to intertidal rocks. Each filament is attached to its rock by a single cell called a basal cell, which may be given additional anchorage by outgrowths called rhizoids. This seaweed reproduces by releasing up to a hundred gametes, each with two flagellae, from some of the cells. In another phase of its life cycle it is a single globular cell.

Sea Lettuce

Ulva lactuca

SIZE	Up to 40 in (100 cm)
HABITAT	Intertidal and shallow subtidal
WATER TEMPERATURE	32–86°F (0–30°C)

DISTRIBUTION Coastal waters worldwide

Sea lettuce is common worldwide on seashores and in shallow subtidal areas, growing in a wide range of conditions and habitats. Its frond is a bright green, flat sheet, which is often split or divided, and has a wavy edge. The plant is very variable in shape and size, ranging from short, tufted plants on exposed shores to

sheets over a yard long in sheltered, shallow bays, especially where extra nutrients are available in polluted harbors. Sea lettuce reproduces by releasing gametes from some cells, and it can also spread vegetatively by regeneration of small fragments. Large fronds lying on the seabed may be full of holes made by grazing animals. It is a also popular food for humans in many parts of the world.

CLASS ULVOPHYCEAE
Sailor's Eyeball
Valonia ventricosa

SIZE	Up to 1½ in (4 cm)
HABITAT	Rock and coral to 100 ft (30 m)
WATER TEMPERATURE	50–86°F (10–30°C)

DISTRIBUTION Western Atlantic, Caribbean, Indian and Pacific Oceans

This odd seaweed looks like a dark green marble, and consists of a single large cell attached to the substratum (which is often coral rubble) by a cluster of filaments called rhizoids. Younger plants have a bluish sheen, but older ones become overgrown with encrusting coralline red seaweeds. Sailor's eyeball has an unusual way of reproducing vegetatively: daughter cells are formed within the parent, which then degenerates, releasing the young plants in the process.

CLASS ULVOPHYCEAE
Cactus Seaweed
Halimeda opuntia

SIZE	Up to 10 in (25 cm)
HABITAT	Rock and sand
WATER TEMPERATURE	68–86°F (20–30°C)

DISTRIBUTION Red Sea, Indian Ocean, and western Pacific

The heavily calcified skeletons of species of *Halimeda* contribute much of the calcareous sediment in the tropics. The plant consists of strings of flattened, kidney-shaped, calcified segments, linked by uncalcified, flexible joints. By day, its chloroplasts are in the outer parts of the frond; at night they withdraw deep into the plant's skeleton. This, along with sharp crystals of aragonite, and the presence of toxic substances in the frond, protects them from nocturnal grazing.

CLASS ULVOPHYCEAE
Velvet Horn
Codium tomentosum

SIZE	Up to 8 in (20 cm)
HABITAT	Intertidal pools, shallow subtidal rocks
WATER TEMPERATURE	46–86°F (8–30°C)

DISTRIBUTION Coastal waters worldwide

The spongy fronds of velvet horn are made up of interwoven tubes, arranged rather like a tightly packed bottlebrush, with each tube ending in a swollen bulb. Many of these bulbs packed together make up the outside of the frond, which is usually repeatedly branched in two. Many short, fine hairs cover the seaweed, giving it a fuzzy appearance when in water. The plants are attached to rocks by a spongy holdfast.

Although this seaweed is present year-round, its maximum development is in winter, and it also reproduces during the winter months. Velvet horn, like all *Codium* species, is often grazed by sacoglossans, small sea slugs that suck out the seaweed's contents, but can keep the photosynthetic chloroplasts alive and use them to make sugars inside their own tissues. The chloroplasts color the sea slugs green, which helps to disguise them from predators. There are about 160 species of *Codium*.

CLASS ULVOPHYCEAE
Giant Cladophora
Lychaete mirabilis

LENGTH	Up to 40 in (100 cm)
HABITAT	Subtidal rocks and kelp
WATER TEMPERATURE	50–59°F (10–15°C)

DISTRIBUTION Southern Atlantic off southwest Africa

CLASS ULVOPHYCEAE
Sea Grapes
Caulerpa racemosa

HEIGHT	Up to 12 in (30 cm)
HABITAT	Shallow sand and rock
WATER TEMPERATURE	59–86°F (15–30°C)

DISTRIBUTION Warm waters worldwide

This seaweed has creeping stolons (stems) that anchor it to rocks or in sand, and from which arise upright shoots covered with round sacs, or vesicles, hence the common name sea grapes. Each plant is a single huge cell. Old plants may become densely branched and entangled, growing to 6 ft (2 m) across. There are many varieties of sea grapes, and around 100 species of *Caulerpa* worldwide.

A giant among *Cladophora* species, *C. mirabilis* grows to 40 in (1 m) long. It is bluish green and filamentous, with many straggly side-branches. It is made up of strings of cells, but individual cells in the main axis may be ½ in (12 mm) long. The plant attaches using a disk made of interwoven extensions of its basal cell, and often has red algae growing on it. It has a very limited distribution in South Africa, but other related species are common worldwide.

KILLER SEAWEED

A strain of *Caulerpa taxifolia* that is widely used in marine aquariums is an invasive species. It is toxic to grazers, grows rapidly, and forms a dense, smothering carpet on the seabed. In 1984 it was discovered in the Mediterranean off Monaco, and has since spread rapidly along the coast, altering native marine communities.

CLASS ULVOPHYCEAE
Mermaid's Wineglass
Acetabularia acetabulum

SIZE	1¼ in (3 cm)
HABITAT	Shallow subtidal rocks
WATER TEMPERATURE	50–77°F (10–25°C)

DISTRIBUTION Eastern Atlantic off North Africa, Mediterranean, Red Sea, Indian Ocean

This curious little green seaweed grows in clusters on rocks or shells covered with sand in sheltered parts of rocky coasts within its range. Although it grows to 1½ in (3 cm), it consists of just one cell. Its calcified frond appears white because it is encrusted with calcium carbonate, and it terminates in a small cup. The cup is made up of fused rays that produce reproductive cysts. The cysts are released after the remainder of the plant has decayed, and they then require a period of dormancy in the dark before they begin to germinate.

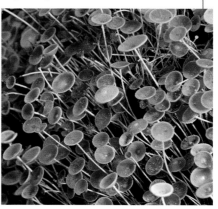

Green Microalgae

DOMAIN	Eucarya
KINGDOM	Plantae
DIVISION	Chlorophyta
CLASSES	9
SPECIES	At least 5,400

THESE MICROSCOPIC, MOSTLY single-celled plants live in the surface layers of the ocean and fresh waters in immense numbers, and they form an important part of the phytoplankton (see p.212). Like seaweeds and microscopic chromists (see p.234), they produce their own food through photosynthesis. Along with the microscopic chromists, these tiny, drifting organisms are sometimes referred to as "grasses of the sea." The vast majority of green microalgae live in fresh water, with only about 500 known marine species. Green seaweeds that are visible to the naked eye are sometimes called macroalgae (see p.246), as are the red and the brown seaweeds.

Habitats

With a few exceptions, marine microalgae swim and float, in their millions, in the sunlit layers of the ocean so that photosynthesis can occur. They are more numerous in nutrient-rich waters, such as those with coastal runoff. In temperate coastal waters, green microalgae and photosynthetic chromists multiply rapidly each spring in response to rising nutrient and light levels, creating abundant food for zooplankton. Such population explosions, or blooms, can reduce the water clarity for weeks. Some green microalgae live inside the bodies of animals and inside chromist plankton—in the appendages (rhizopoda) of radiolarians (see p.237) and inside food vacuoles within the green form of the dinoflagellate *Noctiluca* (see p.236).

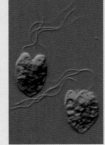

HALOSPHAERA
These microalgae (shown greatly enlarged) are green with chlorophyll and bear hairlike swimming appendages called flagellae.

GREEN TIDE
Green microalgae grow quickly and are first to respond in spring when nutrients become available. While grazer levels are low, the algae are free to multiply until their density turns the ocean green.

Anatomy

Marine microscopic green algae belong to several different classes of algae. Each consists of a single living cell that is generally too small to be visible to the naked human eye. Even the larger species, such as members of the genera *Halosphaera* (Pyramimonadophyceae) and *Pterosperma* (Prasinophyceae), measure just 0.1–0.8 mm across, so appear as no more than a speck. Some green microalgae can swim, using hairlike structures, called flagella, to move through the water. Others lack flagella and cannot propel themselves. Several groups of these plants have a two-stage life history, including both swimming (motile) and nonswimming forms. All green algae possess chloroplasts—structures that contain the green pigment chlorophyll that plants use in photosynthesis.

GREEN BEACHES

A few green microalgae and flatworms form symbiotic partnerships, in which both species gain. The beach-living worms ingest algae, giving them a green color. At low tide, they move up through the sand to pools on the surface, where the algae photosynthesize. In return, the worms absorb food from the algae. Vast numbers of the worms tinge beaches green.

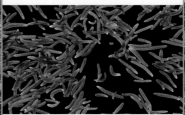

ANIMAL–ALGA PARTNERSHIP
When young, these marine flatworms ingest green algae, which may multiply until there are 25,000 algal cells living in each worm. The adult worms obtain all their nutrition from the algae.

Halosphaera viridis

SIZE
1 mm
Swimming (motile) phase
20–30 micrometers

DISTRIBUTION
Northeastern Atlantic, eastern Pacific

Halosphaera viridis is a tiny, nearly cylindrical cell with four swimming flagella at one end. Each motile cell reproduces by splitting into two, allowing numbers to build up rapidly. Some of these develop into globular cysts (shown here), which increase greatly in size, and the contents then divide into small disks. Before being released from the cyst, each disk develops into another flagellated, motile cell. There can be thousands of cysts per square foot in the open ocean, and such large phytoplankton are an important food source for larger zooplankton.

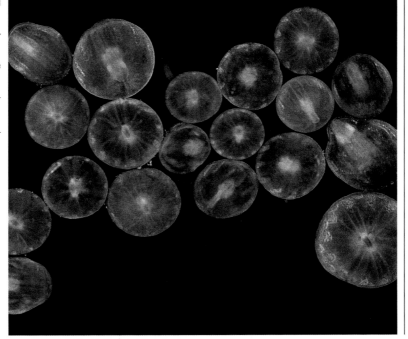

Tetraselmis convolutae

SIZE
10 micrometers

DISTRIBUTION Northeastern Atlantic, off the western coasts of Britain and France

Although it can survive free-living, the tiny cells of *Tetraselmis convolutae* often live inside a worm host (see box, above) in a symbiotic relationship. The worm provides them with shelter and a constant environment inside its body. The worm's light-seeking behavior gives the algae ideal conditions for photosynthesis, which in turn provides both algae and worm with nutrients and energy.

Mosses

DOMAIN	Eucarya
KINGDOM	Plantae
DIVISION	Bryophyta
SPECIES	13,365

MOSSES ARE LOW-GROWING, nonvascular plants that thrive in damp habitats on land, where they may carpet the ground or rocks. They dislike salty environments and only a few species manage to live in the intertidal zone of coasts, mainly in cooler climates. A much wider variety of mosses can be found slightly farther inland, away from the direct effects of sea spray but within range of moisture-laden sea mists.

Habitats

Mosses generally prefer moist, shady places and are most numerous in the cooler and damper climates of temperate regions. This is because they lack the thick cuticle that enables other types of plant to retain moisture. Without the protection of this skinlike surface, mosses soon shrivel up in dry conditions. However, some mosses have an amazing capacity to recover quickly when wetted after a long period of drought. A few species grow in salt marshes or among the lichens at the top of rocky shores; on sheltered coasts, where there is little salt spray, they may live only just above the high tide level. Sand-dune mosses grow rapidly to keep pace with accumulating sand, and blown fragments of moss can colonize new areas of dunes. Many more moss species grow on sea cliffs and in damp gullies away from the intertidal zone.

SYNTRICHIA RURALIFORMIS
This moss grows in coastal sand dunes. Its leaves curl up when dry (left of picture) but unfurl a few minutes after wetting (on right).

Anatomy

Most mosses have a recognizable structure of stems and leaves, which, as in other plants, gather sunlight and perform photosynthesis. However, unlike flowering plants (see p.250), they do not have woody tissues for support, and they also lack the conducting tissues that transport water and nutrients. Mosses have a very thin outer layer of cells, or cuticle, that can absorb (and lose) water, nutrients, and gases over their entire surface. Their "roots" are simple strands called rhizoids, which anchor the plant to its growing surface. Mosses reproduce sexually by means of wind-blown spores, or asexually by spreading across the ground.

SPORE PRODUCTION
Mosses have low-growing leaves, but sprout taller structures with bulbous tips called capsules, from which spores are released.

CLASS BRYOPSIDA

Sand-hill Screw-moss
Syntrichia ruralis

SIZE	½–1½ in (1–4 cm)
FORM	Yellow-green to orange-brown cushions and carpets
HABITAT	Mobile dunes

DISTRIBUTION Eastern Pacific, northwestern Atlantic, Mediterranean

This is one of the first mosses to colonize mobile dunes. It often forms extensive colonies that cover many square yards of sand, giving the sand a golden tinge. Its leaves are covered by hundreds of small papillae that enable swift absorption of water. The leaves gradually taper into long white hair points. This moss is able to establish new plants from fragments dispersed by the wind.

CLASS BRYOPSIDA

Salt Marsh Moss
Hennediella heimii

SIZE	⅛ in (3 mm)
FORM	Single green plants
HABITAT	Salt marshes, other coastal areas

DISTRIBUTION Patchy distribution on temperate and cool waters worldwide

This tiny moss is a halophyte, meaning it can grow in salt-laden soil. It is mainly found on the coast, but not right by the sea. One of the few mosses that may be regularly found in salt marshes, it grows on patches of bare ground between the other vegetation in the upper parts of the salt marsh. It also grows in various other coastal habitats, including the banks of creeks and behind sea walls. Although small, the plants may be abundant and may appear conspicuous from a distance, due to their prolific number of stout, dark, rusty-brown capsules, which are borne on short stalks less than ½ in (1 cm) tall. These have a little cap with a long point, which lifts to allow spores to escape, but remains attached to the capsule by a central stalk. The Salt marsh Moss has a wide distribution in colder climates.

CLASS BRYOPSIDA

Seaside Moss
Schistidium maritimum

SIZE	1 in (2 cm)
FORM	Dark blackish green, compact cushions
HABITAT	Hard, acidic rocks, salt marshes

DISTRIBUTION Western and eastern coasts of North America, coasts of western Europe

This moss grows as small, dark green cushions on hard, acidic rocks, with seashore lichens, just above high-tide mark. It also occurs in salt marshes. It is often soaked by salt spray and occasionally covered by the highest tides. Seaside moss appears to be a true halophyte, functioning normally even after immersion in sea water for a few days, and growing only in saline conditions; in Britain, it is found no farther than 1,300 ft (400 m) from the sea. Its leaves curl when dry. In winter, it produces small brown capsules on short stalks.

CLASS BRYOPSIDA

Southern Beach Moss
Orthotrichum crassifolium

SIZE	1–5 in (2–13 cm)
FORM	Black cushions and mats
HABITAT	Rocks

DISTRIBUTION Southern tip of South America, islands in the Southern Ocean

This moss is the southern version of seaside moss (see above), growing on coastal rocks in the usually lichen-dominated splash zone, where it is often inundated by the sea in stormy weather. It grows in southern Chile and on subantarctic islands, where it can become dominant. On Heard Island, for example, a salt spray community of plants found on exposed coastal lava rock, at elevations of less than 16 ft (5 m), is dominated by southern beach moss, which has also colonized derelict buildings.

Flowering Plants

DOMAIN	Eucarya
KINGDOM	Plantae
DIVISION	Trachaeophyta
CLASS	Angiospermae
SPECIES	260,000

PLANTS CONQUERED LAND, and then land-based flowering plants grew to be among the most abundant and diverse life-forms on Earth. However, relatively few have adapted to the poor soil, salt spray, and drying wind of coastal dunes and cliffs. These few include some fascinating plants found nowhere else. Few flowering plants have returned to the sea: salt-marsh plants and mangroves get wet at high tide, but only the seagrasses live fully submerged.

Anatomy

Flowering plants, technically called magnoliophytes or angiosperms, uniquely possess fruit and flowers, unlike mosses, seaweeds, and other algae. They are adapted to life in air, absorbing fresh water through their roots. If they take salt water into their vascular system, water from their own cell sap is attracted to the more concentrated salts and sucked out by osmosis. This is fatal to cells, but mangroves cope by excreting the salt, while succulents partition it within their cells. Seagrasses have fully adapted by matching their cells' salt concentration to that of seawater. Most seagrasses have a similar form, with thin, grasslike blades that allow easy exchange of nutrients and gases. Mangroves grow in mud that lacks oxygen by growing aerial roots to assist gas exchange in their underground roots. Many angiosperm seeds are killed by seawater, but those of the coconut can stay viable at sea for long periods inside a waterproof case. Seagrasses are water-pollinated, and to increase the chances of a pollen grain catching onto a female stigma, pollen is released as a sticky string.

waterproof seed case

germinating plant

GERMINATING SEED
This coconut is a fruit—a defining characteristic of flowering plants. The coconut has the marine adaptations of buoyancy and a waterproof case.

Seawater Plants

Seagrasses are monocotyledons (the group of flowering plants with thin, straplike leaves), but are not true grasses, and they do not share a single evolutionary origin. There are 60 species in 4 or 5 families, although the Ruppiaceae, living mainly in brackish water, is not always accepted as a seagrass family. Salt-marsh plants are mainly small and herbaceous, with early colonizers including the salt-excreting cord grass and small succulents, such as common glasswort. Further salt-tolerant flowering plants grow farther up the shore in established salt marshes, forming a dense, grassy turf. Salt marshes (see p.124) form in cooler climates, and are replaced in tropical seas by mangroves—trees with characteristic aerial roots. At least 65 species from 16 diverse families are classed as mangrove plants. They do not have a single origin, so the mangrove habit evolved separately, several times.

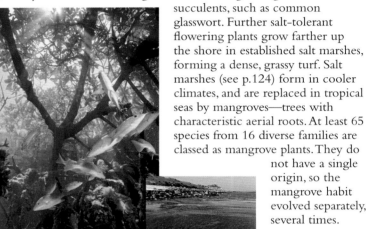

MANGROVES SUBMERGED
When the tide is in, mangroves form a mini-jungle of arching roots where small fish hide.

ZOSTERA
Seagrasses, such as this *Zostera* off the Isles of Scilly, can be found from the cold waters of Alaska to tropical seas.

Coastal plants

A greater variety of flowering plants can grow above the high-water mark. Salt-tolerant grasses are important constituents of the upper parts of salt marshes, and at the seaward edge of sand-dune systems, grasses are often the first to stabilize the shifting sand. In sand and sheltered gravel at the top of the shore, a few deep-rooted plants grow. A much wider variety of flowers and a few mosses colonize sand dunes and slacks just inland from the coast. Here they are subjected to salt spray but never inundated by tides. Nitrogen fixers thrive in these poor, sandy soils. In warmer climates, annuals bloom like desert flowers after seasonal rains. On cliff-tops, plants may be fertilized by sea-bird guano, stimulating lush growth.

COASTAL FLOWERS
The beautiful flower heads of sea pink, or thrift, transform rocky seashores and salt marshes in late spring. The sea pink's compact cushions resist wind and cold.

DUNE-SLACK FLORA
Here, in dune slacks behind a beach in the Canary Islands, annual plants bloom for a short period after rain. They flower and produce seeds quickly before drying up in the summer sun.

ORDER NAJADALES
Neptune Grass
Posidonia oceanica

TYPE	Perennial
HEIGHT	12 in (30 cm)
HABITAT	Rocks and sand

DISTRIBUTION Mediterranean

Neptune grass (also known as Mediterranean tapeweed) forms meadows from shallow water to a depth of about 150 ft (45 m) in the clearest waters. It grows on both rock and sand, has a tough, fibrous base, and persistent rhizomes (stems) that grow both horizontally and vertically. These build up into a structure known as "matte," which can be several meters high and thousands of years old. Around the island of Ischia, Italy, more than 800 species have been associated with Neptune grass beds.

ORDER HYDROCHARITALES
Paddle Weed
Halophila ovalis

TYPE	Perennial
HEIGHT	2½ in (6 cm)
HABITAT	Sand

DISTRIBUTION Coasts of Florida, USA, East Africa, Southeast Asia, Australia, and Pacific islands

Members of the genus *Halophila* look quite unlike other seagrasses, having small, oval leaves that are borne on a thin leaf stalk. As its scientific name indicates, paddle weed is particularly tolerant of high salinities (*halophila* means "salt-loving"). Pollination takes place underwater, and the tiny, oval pollen grains are released in chains, which assemble into rafts like floating feathers. This is thought to increase the chances of pollination of a female flower. Despite its small size, paddle weed is an important food for the dugong (see p.419). An adult can eat more than 88 lb (40 kg) of it a day.

ORDER POALES
Marram Grass
Ammophila arenaria

TYPE	Perennial
HEIGHT	1½–4ft (0.5–1.2 m)
HABITAT	Coastal sand dunes

DISTRIBUTION Western Europe and Mediterranean (natural occurrence); introduced elsewhere

Marram grass is a tall, spiky grass that plays a key role in binding coastal sand and building sand dunes. Its underground stems (rhizomes) spread through loose sand, and upright shoots develop regularly along their length. When the tangle of stems and leaves impede onshore breezes, sand carried in the wind is deposited. Progressively, the sand builds up, the stems grow up through the sand, and a sand dune is formed. In dry weather, the leaves curl into a tube. The underside of the leaf then forms the outer surface and its waxy coating helps to reduce water loss from the plant. Marram grass is widely planted to stabilize eroded dunes, and has been introduced for this purpose to North America (where it is known as European beach grass), Chile, South Africa, Australia, and New Zealand.

ORDER CARYOPHYLLALES
Common Glasswort
Salicornia europaea

TYPE	Annual
HEIGHT	4–12 in (10–30 cm)
HABITAT	Coastal mudflats and salt marshes

DISTRIBUTION Coasts of western and eastern North America, western Europe, and Mediterranean

Glasswort, also known as marsh samphire, is an early colonizer of the lower levels of salt marshes and mudflats, where plants are inundated twice a day by the tide. It is a small, cactuslike plant with bright green stems that later turn red. The tiny flowers and scalelike leaves are sunk into depressions in the fleshy stem. Glasswort is protected externally from salt water and moisture loss by a thick, waxy skin. It is able to prevent the salt absorbed through its roots from doing any damage by locking it away in vacuoles (small cavities) within its cells. The plant stores water inside its succulent stems, hence its cactuslike shape. For centuries, glasswort was gathered and burnt to produce an ash rich in soda (impure sodium carbonate). The ash was then baked and fused with sand to make crude glass—hence its common name. Glasswort can also be eaten boiled or pickled in vinegar.

ORDER CARYOPHYLLALES

Common Sea Lavender

Limonium vulgare

TYPE	Perennial
HEIGHT	8–20 in (20–50 cm)
HABITAT	Muddy salt marshes

DISTRIBUTION Coasts of western Europe, Mediterranean, Black Sea, and Red Sea

This showy plant, which flowers in late summer, often forms dense colonies in salt marshes, particularly along the sides of muddy creeks. Several closely related species of sea lavender are highly localized in their distribution. For example, two species are confined to two rocky peninsulas in Wales, while others are only found in parts of Sicily or Corsica. Varieties of sea lavender, often called statice, are grown commercially as "everlasting" flowers. The colored, papery "flowers" are actually what remains after the true flowers have fallen.

ORDER PAPAVERALES

Yellow Horned Poppy

Glaucium flavum

TYPE	Biennial or perennial
HEIGHT	20–36 in (50–90 cm)
HABITAT	Shingle, sometimes sand

DISTRIBUTION Coasts of western Europe, Mediterranean, and Black Sea

Also known as yellowhorn poppy, this plant has leaves covered in a waxy coating to protect it from salt spray and reduce water loss. Its taproot penetrates deep into shingle in search of water beneath. It blooms through most of the summer, producing flowers that are up to 3½ in (9 cm) across.

ORDER SOLANALES

Beach Morning-glory

Ipomoea imperati

TYPE	Perennial
LENGTH	Up to 16 ft (5 m)
HABITAT	Coastal beaches and grasslands

DISTRIBUTION Widespread on many coasts and islands with tropical or warm temperate climates

This pioneer plant helps to stabilize coastal sands, creating a habitat into which other species move. It can endure low nutrient levels, high soil temperatures, abrasion and burial by blown sand, and occasional frosts, but not hurricanes, according to studies in Texas. It also grows occasionally on disturbed ground inland. Beach morning-glory is recorded on six continents and many isolated islands.

ORDER MALPIGHIALES

Pacific Stilt-mangrove

Rhizophora stylosa

HABIT	Woody perennial
HEIGHT	Commonly 16–26 ft (5–8 m), but can be up to 130 ft (40 m)
HABITAT	Intertidal mudflats

DISTRIBUTION Coasts of northern Australia, Southeast Asia, and South Pacific islands

The aerial roots of Pacific stilt-mangroves arch down from the main trunk, with secondary roots coming off the primary ones before they reach the ground to form a tangle of roots growing in all directions. When the tide is in, these form a sheltered refuge for many small fish. This species of mangrove can tolerate a wide range of soils, but thrives best in the fine, muddy sediments of river estuaries. Its roots absorb water selectively, so much of the damaging salt is not taken up, but it still has to excrete some salt through the leaves.

PROFILE BRASSICALES

Scurvy-grass

Cochlearia officinalis

HABIT	Biennial or perennial
HEIGHT	4–16 in (10–40 cm)
HABITAT	Coastal rocks and salt marshes

DISTRIBUTION Coasts of northern Europe and Asia and northern North America

The thick, fleshy leaves of this coastal plant help it to store water in an environment where fresh water soon drains away (scurvy-grass plants found on mountains have thinner leaves and may belong to a different species). Scurvy-grass leaves are rich in vitamin C. They were once eaten, or pulped and drunk, to prevent scurvy—a disease caused by vitamin C deficiency to which sailors were prone ("grass" is Old English for any green plant).

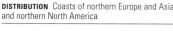

ORDER CARYOPHYLLALES

Grand Devil's-claw

Pisonia grandis

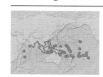

TYPE	Woody perennial
HEIGHT	46–98 ft (14–30 m)
HABITAT	Coastal and island forests

DISTRIBUTION Coasts and islands in Indian Ocean, Southeast Asia, and South Pacific

The grand devil's-claw is typically found on small tropical islands and its distribution is associated with sea bird colonies. It can grow as tall as 98 ft (30 m), the trunk can be up to 6½ ft (2 m) in diameter, and it is often the dominant tree in coastal forests that are undisturbed by humans. The trees provide nesting and roosting sites for many species of sea bird, whose guano is an important fertilizer on isolated islands. The branches break easily, and can root in the ground.

BIRD-KILLING TREE

The seeds of grand devil's-claw are produced in clusters of 50–200 and exude a resin that makes them extremely sticky. They attach to the feathers of sea birds and may subsequently be flown to remote islands. This is an effective means of dispersal, but the seeds are so sticky that small birds often become completely entangled and die.

ORDER CASUARINALES

Casuarina

Casuarina equisetifolia

TYPE	Woody perennial
HEIGHT	66–98 ft (20–30 m)
HABITAT	Coastal and island forests

DISTRIBUTION Southeast Asia, eastern Australia, and islands in southeast Pacific

Casuarina has many common names, including Beach she-oak, beefwood, ironwood, and Australian pine. It is typically found at sea level, but also grows inland to 2,600 ft (800 m). Casuarina is fast-growing, reaching a height of 65 ft (20 m) in 12 years. It is drought-tolerant, and can grow in poor soils because it can fix nitrogen in nodules on its roots. Its wood is very hard and is used as a building material and as firewood. The bark is widely used in traditional medicines.

ORDER ARECALES

Coconut Palm

Cocos nucifera

HABIT	Woody perennial
HEIGHT	66–72 ft (20–22 m)
HABITAT	Coastal rocky, sandy, and, coralline soils

DISTRIBUTION Tropical and subtropical coasts worldwide

The coconut palm was once the mainstay of life on Pacific islands. It provided food, drink, fuel, medicine, timber, mats, domestic utensils, and thatching for roofs. It remains an important subsistence crop on many Pacific islands today. Its original habitat was sandy coasts around the Indo-Malayan region, but it is now found over a much wider area, assisted by its natural dispersal mechanism, and deliberate planting by humans. The fibrous husk of the coconut fruit is a flotation aid that enables the seeds to be carried vast distances by ocean waves and currents. The coconut palm cannot develop viable fruits outside of the tropics and subtropics.

COCONUT FRUIT
The fruit of the coconut palm weighs 2¼–4½ lb (1–2 kg). It contains one seed, which is rich in food reserves and is part solid (flesh) and part liquid (coconut milk).

fibrous husk

edible flesh

COCONUT TREE
The coconut palm can live as long as 100 years, a mature tree producing 50–80 coconuts a year. The trunk is ringed with annual scars left by fallen leafbases.

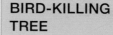

Fungi

DOMAIN	Eucarya
KINGDOM	Fungi
PHYLA	5
SPECIES	About 148,000

FUNGI FORM A GREAT KINGDOM OF SINGLE-CELLED and filamentous life forms, including yeasts and molds. Some organize their filaments into complex fruiting structures, such as mushrooms. Few marine fungi have been described, but it appears that they can live in almost every ocean habitat. Fungi are, however, abundant on shorelines, but only in close association with certain algae, cyanobacteria, or a combination of both. The organisms grow in partnership in a kind of symbiotic, compound organism called a lichen. Lichens proliferate in the hostile, wave-splashed zone of bare rock just above high tide.

LICHEN ENCRUSTATION
Fungi thrive on the coast if they grow in association with algae, in an intimate symbiosis called lichen. Here, encrusting and foliose lichens cover sandstone cliffs in the Shetland Isles, Scotland.

Anatomy

A lichen's body (thallus) is composed mainly of fungal filaments called hyphae. The cells of the fungus's photosynthesizing partner are restricted to a thin layer below the surface, where they are protected and cannot dry out and the fungus can access the nutrients they produce. Fungal hyphae also form rootlike structures, called rhizines, that anchor the lichen to the substrate. Lichens grow in one of four ways: bushy (fruticose); leaflike (foliose); tightly clustered (squamulose); or lying flat (crustose).

Marine funguslike organisms, such as slime nets (labyrinthulids) and thraustochytrids, are microscopic, usually transparent, and encased in a network of slimy threads. The cells move up and down within the threads and react positively toward food. They are now classed as protists, however, rather than fungi.

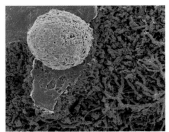

LICHEN COMPOSITION
This false-color micrograph of a lichen (below) shows the smooth surface of the thallus, to the left, and fungal hyphae, to the right.

Habitats

Most lichens require alternating dry and wet periods, but marine lichens can withstand continuous drought or dampness. On most rocky shores, yellow and gray lichens dominate surfaces splashed by waves at high tide (the splash zone). They endure both the drying Sun and wind, and the salt spray of the sea. Below, in the tidal zone, the brightly colored lichens give way to black encrusting lichen, such as *Verrucaria maura*, which covers the bedrock and any large, stable boulders. *Verrucaria serpuloides* lives yet further down the shore and is the only lichen to survive permanent immersion in sea water.

Marine fungi exists on living and nonliving substrates in marine and estuarine environments. Some are protected from the dehydrating effects of salt water because they live as parasites within seagrasses or green algae. Some live on driftwood and are decomposers; others live on creatures such as sponges.

BELOW THE SPLASH ZONE
Some lichens, such as this crustose black *Verrucaria*, live below the splash zone, and may be surrounded by seaweeds.

PHYLUM ASCOMYCOTA

Sea Ivory

Ramalina siliquosa

LENGTH (BRANCHES)	1–4 in (2–10 cm)
HABITAT	Hard siliceous rocks above the splash zone

DISTRIBUTION Northeast and southwest Atlantic, coasts of Japan and New Zealand

Nutrient-poor siliceous rocks are the favorite habitat of gray lichens, such as sea ivory. This lichen is usually gray-green in color, with a brittle, bushlike (fruticose) structure and disk-shaped fruiting bodies, called apothecia, at its branch tips. Sea ivory cannot withstand being trampled or extensively grazed, and so it grows best on vertical rock faces, to which it sticks by a single basal attachment.

PHYLUM ASCOMYCOTA

Black Shields

Tephromela atra

WIDTH	Up to 4 in (10 cm)
HABITAT	In and above the splash zone

DISTRIBUTION Polar coasts, coast of California, US, Gulf of Mexico, Mediterranean, Indian Ocean

Crustose lichens such as black shields, which form a crust over the rock, attach themselves so firmly using fungal filaments that they cannot be easily removed from it. Over time, these anchoring filaments break down the rock as they alternately shrink when dry and swell when moist. Black shields is a thick, gray lichen with a rough, often cracked, surface from which project a number of characteristic black fruiting bodies.

PHYLUM ASCOMYCOTA

Yellow Splash Lichen

Xanthoria parietina

WIDTH	Up to 4 in (10 cm)
HABITAT	Splash zone; favors surfaces high in nitrogenous compounds

DISTRIBUTION Temperate Atlantic, Gulf of Mexico, Indian and Pacific oceans

On most rocky shores, different species of lichen have a marked vertical territory related to their tolerance of salt exposure. The yellow splash lichen is found in the splash zone and forms a bright orange band across the shore, with gray lichens above it and black lichens below. It has a leaflike (foliose) form, with slow-growing, leafy lobes held more or less parallel to the rock on which it lives. Usually bright orange in color, it tends to become greener if in shade. Lichens are widely used to monitor air pollution because they simply disappear when conditions deteriorate. The yellow splash lichen is particularly sensitive to sulfur dioxide, a by-product of industrial processes and of burning fossil fuels.

PHYLUM ASCOMYCOTA

Black Tar Lichen

Verrucaria maura

THICKNESS	1/32 in (1 mm)
HABITAT	Intertidal

DISTRIBUTION Temperate and polar coasts, Indian Ocean, Japan

This smooth, black, crustose lichen covers large areas of bedrock or stable boulders in a thin layer, making them appear as though they have been covered with dull black paint. Many types of lichen accumulate heavy metals, and the black tar lichen is no exception, having been found to have levels of iron that are about 2.5 million times more concentrated than the surrounding seawater. That may be an adaptation to deter grazers, such as gastropods, from eating it.

PHYLUM ASCOMYCOTA

Black Tufted Lichen

Lichina pygmaea

WIDTH (LOBES)	To 1/2 in (1.5 cm)
HABITAT	Lower littoral fringe to middle shore, regularly covered by the tide

DISTRIBUTION Northeast Atlantic from Norway to northwest Africa

Typically found on exposed sunny rock faces, this lichen looks rather like a seaweed, being fruticose (bushlike) in form with branching, brownish black, flattened lobes. Its fruiting bodies form in small swellings at its branch tips. It is often seen growing in association with barnacles but does not tolerate algal (seaweed) growth. Its compact growth and rigid branches provide a refuge for several mollusks, particularly *Lasaea rubra*, a small, pink-shelled gastropod. All *Lichina* species are limited to coastal habitats.

PHYLUM ASCOMYCOTA

Gray Lichen

Pyrenocollema halodytes

SIZE	Not recorded
HABITAT	Upper shore on rocks and on shells of some sedentary invertebrates

DISTRIBUTION Temperate northeast and southwest Atlantic

Seen on hard, calcareous rocks, where it forms small, black-brown patches, gray lichen is unusual in being an association of three organisms— a fungus, a cyanobacteria, and an alga. The fungus anchors the lichen to the rock; the cyanobacteria and the alga contain chlorophyll and make food by photosynthesis. The cyanobacteria can also utilize nitrogen, a process that uses a lot of energy, and this comes from the sugar made during photosynthesis.

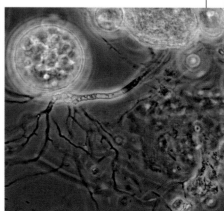

Animal Life

DOMAIN	Eucarya
KINGDOM	Animalia
PHYLA	About 33
SPECIES	Over 1.5 million

ANIMAL LIFE FIRST APPEARED IN THE OCEAN over one billion years ago. It has since diversified into a vast array of different organisms. The range of scale among marine animals is immense: the smallest invertebrates are over half a million times smaller than the largest whales. Despite this huge disparity, animals all share two key features. First, they are heterotrophs, meaning they obtain energy from food. Second, they are multicellular, which distinguishes them from single-celled life forms.

INVERTEBRATE
This yellow tube sponge, from the sea off Belize, is a typical sessile invertebrate. Instead of moving to find food, it filters out particles of food by pumping water through its pores.

Marine Animal Diversity

Animals are classified into 30 or more major groups (phyla), all of which include at least some marine animals. Twenty-nine of these phyla are composed of animals without backbones (invertebrates), each phylum representing a completely different body plan. Only one phylum, the chordates, contains animals with backbones (vertebrates). In salt water, vertebrates include fish, reptiles, birds, and ammals—animals that are often described as the dominant forms of ocean life. However, in terms of abundance and diversity, invertebrates have a stronger claim to this title.

Invertebrates exist in all ocean habitats and outnumber marine vertebrates by a million to one. They include an array of fixed (sessile) animals, such as corals and sponges. They also form most of the zooplankton, a drifting community of tiny organisms and animal eggs and larvae.

CHANGING SHAPE
Most invertebrates change shape as they develop. Feather stars start as drifting larvae, which eventually attach themselves to corals or rock before changing into swimming adults.

VERTEBRATE
Active predators, such as this barracuda, need sharp senses and rapid reactions to catch prey. Unlike invertebrates, they have fast-acting nerves and well-developed brains.

Support and Buoyancy

On land, most animals have hard skeletons to counteract gravity's pull. Life is different in the sea, because water is denser than air. It buoys up soft-bodied animals, such as jellyfish, enabling them to grow large. They use internal pressure to keep their shape, the same principle that works in balloons. Animals with hard body parts, such as fish and mollusks, are often denser than water, and would naturally sink. To combat this, many have a buoyancy device. Bony fishes have an adjustable gas-filled swim bladder, while squid have an internal float made of chalky material, containing many gas-filled spaces. Some surface dwellers, such as the violet sea snail, have gas-filled floats that prevent them from sinking.

BUBBLE RAFT
The violet sea snail stays afloat by producing bubbles of mucus. The mucus slowly hardens, forming a permanent raft.

Groups and Individuals

Among marine animals, there is a social spectrum from species that live on their own to those that form permanent groups. Whale sharks are solitary apart from when they mate or form eating aggregations. They can do this because their huge size means they have few natural predators. Smaller fish often form schools, which reduce each fish's chances of being singled out for attack. Many invertebrates, from corals to tunicates, live in permanent groups, known as colonies. In most coral colonies, the individual animals, or polyps, are anatomically identical and function as independent units, even though they are joined. Other animal colonies, such as the Portuguese man-of-war, are made of individuals with distinct forms. Each form carries out a different task, like parts of a single animal.

COLONY ON THE MOVE
A diver films a pyrosome colony in the sea off Florida. It consists of thousands of tiny soft-bodied animals called tunicates, joined together to form a tube.

LONE GIANT
The normally solitary whale shark (below) forms feeding groups of up to 100 individuals in coastal areas where food is abundant.

SAFETY IN NUMBERS
Crowded together in a ball, gregarious striped catfish (right) make a confusing target for predators.

Reproduction

Animals reproduce in two ways. In asexual reproduction, which occurs in many marine animals from flatworms to sea anemones, a single parent divides in two, or grows (buds off) parts that become independent. In sexual reproduction, the eggs of one parent are fertilized by the sperm of another. Sessile animals, such as corals and clams, usually breed by shedding their eggs and sperm into the water, leaving them to meet by chance. In some fish, all mammals, and birds, fertilization is internal, which means that the two parents have to mate. Marine animals vary greatly in reproductive potential. Most whales have a single calf each time they breed, but an ocean sunfish can produce over 300 million eggs in a single spawning event.

SINGLE PARENT
This sea anemone is budding off young that will eventually take up life on their own. Asexual reproduction is quick and simple, but it does not produce genetic variation, making it more difficult for a species to adapt to change.

COURTSHIP
Two waved albatrosses display to each other in the Galápagos Islands. Complex courtship rituals like this ensure that each parent finds a partner of the right species and the right sex, and they cement the bond once breeding begins.

SEA SYMPHONY
A wrasse feeds among coral in the Red Sea. Coral reefs contain the greatest diversity of animal life in the oceans, and are one of the few habitats that are actually created by animals.

OCEAN LIFE

Sponges

DOMAIN	Eucarya
KINGDOM	Animalia
PHYLUM	Porifera
CLASSES	4
SPECIES	About 9,400

THIS ABUNDANT AND diverse group of often colorful invertebrates lives permanently attached to the sea floor. Naturalists once thought they were plants, but they are now known to be very simple and ancient animals. Sponges live by drawing water into their bodies through tiny holes called pores, filtering it for food and oxygen and pushing it out again. Many species are found on coral reefs or rocks, and a few live in fresh water.

Anatomy

The body plan of a sponge is based on a system of water canals lined with special cells known as collar cells. Collar cells are unique to sponges. They draw water into the sponge through pores, by each beating a long, whip-like flagellum. A ring of tiny tentacles around the base of the flagellum traps food particles, and the water and waste material then flows out of the sponge through larger openings. Rigidity is provided by a skeleton made up of tiny splinters (spicules) of silicon dioxide or calcium carbonate throughout the body.

osculum
central cavity
collar cell
flagellum
spicule
pore

BODY SECTION
A sponge has specialized cells, but no organs. Water enters the sponge through pore cells and exits via openings called osculae.

Habitats

Most sponges need a hard surface for attachment, but some can live in soft sediment; a few species are able to bore into rocks and shells. Sponges are common on rocky reefs, shipwrecks, and coral reefs in a wide range of temperatures and depths. The largest populations occur where there are strong tidal currents, which bring extra food. Animals such as crabs and worms sometimes live inside sponges, but little manages to settle and grow on their surface. This is because sponges produce chemicals to discourage predators.

CHANGING SHAPE
Many sponges grow different shapes in different habitats. This sponge develops fingers in strong currents (above), but has an encrusting form (right) when it grows in wave-exposed sites.

CARNIVOROUS SPONGES
A few exceptional deep-sea sponges ensnare live prey. Pictured is the harp sponge *Chondrocladia lyra*, which has small hooks on its vertical branches.

Reef-forming Sponge

Heterochone calyx

HEIGHT	Up to 5 ft (1.5 m)
DEPTH	300–800 ft (100–250 m)
HABITAT	Deep hard seabed

DISTRIBUTION Deep cold waters of north Pacific

The reef-forming sponge not only looks like a delicate glass vase, but its skeleton spicules are made from the same material as glass, silica. Each spicule has six rays, hence the scientific name of its class, Hexactinellida. Many glass sponges grow very large—off Canada's British Columbian coast, the reef-forming sponge forms huge mounds nearly 65 ft (20 m) high spread over several miles. Other members of their class also contribute to these reefs, which may have started forming nearly 9,000 years ago. Like coral reefs, sponge reefs provide a home for many other animals.

CLASS DEMOSPONGIAE

Barrel Sponge

Xestospongia testudinaria

HEIGHT	Up to 6 ft (2 m)
DEPTH	6–165 ft (2–50 m)
HABITAT	Coral reefs

DISTRIBUTION Tropical waters of western Pacific

These gigantic sponges grow large enough to fit a person inside. Their hard surface is deeply ridged, but their rim is thin and delicate. The barrel sponge belongs to the *Demospongiae*, the largest class of sponges, containing about 95 percent of sponge species. The skeleton of sponges in this class is made from both scattered spicules of silica and organic collagen called spongin. An almost identical barrel sponge, *Xestospongia muta*, occurs in the Caribbean.

CLASS HOMOSCLEROMORPHA

Flesh Sponge
Oscarella lobularis

HEIGHT About 0.3 in (1 cm)

HABITAT Sublittoral rock

DISTRIBUTION Mediterranean and south to Senegal

The blue color of this species is unusual among sponges but this species can also be green, violet, or brown. It grows as irregular lobules that look and feel smooth and soft because it has no spicules. It doesn't have many other skeletal fibers either and collapses when out of water. This genus of sponges, along with six others, have recently been separated out from the demosponges and placed in their own class. A similar yellow sponge that occurs around the UK is called by the same name but the two forms are thought to be different species.

CLASS DEMOSPONGIAE

Breadcrumb Sponge
Halichondria panicea

WIDTH To more than 12 in (30 cm)

DEPTH Shore to sublittoral zone

HABITAT Hard surfaces

DISTRIBUTION Temperate coastal waters of northeastern Atlantic and Mediterranean

The appearance of this soft encrusting sponge varies from thin sheets to thick crusts and large lumps. On wave-exposed shores, it usually grows under ledges as a thin, green crust, its osculae opening at the tops of small mounds. Its green color is produced by photosynthetic pigments in symbiotic algae in the sponge's tissues. In deeper, shaded waters, the sponge is usually a creamy yellow. In waters with strong currents, this sponge may cover large rocky areas and kelp stems.

CLASS DEMOSPONGIAE

Tube Sponge
Haliclona fascigera

HEIGHT Up to 3 ft (1 m)

DEPTH Below 33 ft (10 m)

HABITAT Coral reefs

DISTRIBUTION Tropical reef waters of western Pacific; likely to be more widespread than shown

The elegant, tubular branches of this beautiful sponge are easily torn, and so it occurs only on deeper reef slopes, where wave action is minimal.

It sometimes grows as a single tube, but it is more often seen as bunches of tubes joined at the base. The tips of the tubes are translucent and slightly rolled in. The color of this sponge is usually pinkish violet, although some specimens are pinkish blue. When this sponge releases sperm, it resembles smoking chimneys.

The taxonomic status of this species and its relationship to other species in the same family has not been fully determined, and it is listed under various names in different sources. Such uncertainties are not unusual in the study of sponges and mean that the exact distribution of this and many other species is yet to be established.

CLASS DEMOSPONGIAE

Mediterranean Bath Sponge
Spongia officinalis

WIDTH Up to 14 in (35 cm)

DEPTH 3–165 ft (1–50 m)

HABITAT Rocks

DISTRIBUTION Mediterranean, especially the eastern part

The Mediterranean bath sponge, as its name suggests, is collected and processed for use as a bath sponge. It grows as rounded cushions and mounds, and is usually dull gray to black outside but yellowish white inside. It can be used as a sponge because it has no sharp skeletal spicules, just a network of tough fibers made from an elastic material called spongin. Huge numbers were once harvested, but today they are rare.

CLASS CALCAREA

Lemon Sponge
Leucetta chagosensis

WIDTH Up to 8 in (20 cm)

DEPTH Shallow

HABITAT Steep coral reef and rock slopes

DISTRIBUTION Tropical reef waters of western Pacific

The lemon sponge is a beautiful, bright yellow color and is easy to spot underwater. It grows in the form of sacs, which may have an irregular, lobed shape. Each sac has a large opening—the osculum—through which used water flows out of the sponge. Through the osculum, entrances to the water-intake channels that run throughout the sponge can be seen. The lemon sponge belongs to a small class of sponges in which the mineral skeleton is composed entirely of calcium carbonate spicules, most of which have three or four rays. The densely packed spicules give the sponge a solid texture. Like all sponges, this sponge is hermaphroditic. It incubates its eggs inside and releases them as live larvae through the osculum. Each larva is a hollow ball of cells with flagella for swimming.

CLASS DEMOSPONGIAE

Coralline Sponge
Vaceletia crypta

SIZE Not recorded

DEPTH At least 65 ft (20 m)

HABITAT Dark reef caves

DISTRIBUTION Not fully known, but includes tropical waters of western Pacific

Vaceletia crypta, discovered in the 1970s, is a living member of the coralline sponges group, most of which are known only from fossils. Coralline sponges have a massive skeleton made of calcium carbonate, as well as silica spicules and organic fibers. They were the dominant reef-building organisms before the stony corals of modern reefs evolved. They were once put in a separate class (the Sclerospongiae) but are now part of the Demospongiae.

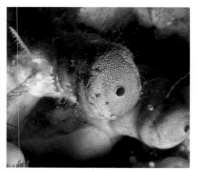

Cnidarians

DOMAIN	Eucarya
KINGDOM	Animalia
PHYLUM	Cnidaria
CLASSES	5
SPECIES	10,886

THIS ANCIENT GROUP OF AQUATIC ANIMALS emerged in Precambrian times, about 600 million years ago. Most are marine, and the group includes reef-building corals, jellyfish, anemones, and hydroids. Cnidarians have a radially symmetrical body shaped like a simple sac, with stinging tentacles around a single opening that serves as both mouth and anus. There are two body forms: the polyp form, typified by sea anemones, which is fixed to a solid surface and has an upward-facing mouth and tentacles; and the medusa, best known in adult jellyfish, which can swim and has a downward-facing mouth and tentacles.

Anatomy

Corals and anemones exist only as polyps, whereas other cnidarians can be either polyps or medusae at different stages of their life cycle. The body wall of both polyps and medusae consists of two types of tissue. On the outside is the epidermis, which acts like a skin to protect the animal. The inner tissue layer, lining the body cavity, is the gastrodermis, which carries out digestion and produces reproductive cells. Separating and connecting these two layers is a jellylike substance called the mesoglea. The tentacles have stinging cells called cnidocytes, which are unique to this phylum and give it its name. A simple nervous system responds to touch, chemicals, and temperature.

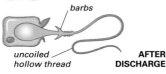

TENTACLE ARRANGEMENT
The number of tentacles on coral polyps varies from one group to another. The polyps of all soft corals (above) have eight tentacles, hence their alternative name of octocorals. Hexacorals (right) have tentacles arranged in multiples of six.

POLYP

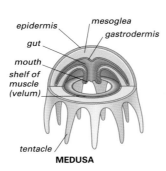

tentacle
cnidocyte
epidermis
mesoglea
gastrodermis
budding juvenile
gut
basal disk

POLYP AND MEDUSA
Polyps are essentially a tube, closed at one end, that attaches to a hard surface by a basal disk. They live singly or in colonies. Medusae are bell-shaped and usually have a thicker mesoglea; some also have a shelf of muscle for locomotion.

epidermis
gut
mouth
shelf of muscle (velum)
mesoglea
gastrodermis
tentacle

MEDUSA

cnidocyte
coiled thread
cnidocyst

BEFORE DISCHARGE

barbs

uncoiled hollow thread

AFTER DISCHARGE

STINGING CELLS
Each cnidocyte contains a bulblike structure, called a cnidocyst, which houses a coiled, barbed thread. When triggered by touch or chemicals, the thread explodes outward and pierces the prey's skin. The animal's tentacles are then used to haul the victim in.

SCLERITES
Small slivers of calcium carbonate called sclerites are scattered through the tissues of soft corals and sea fans. Here, they are visible as white shards under the skin of this soft coral.

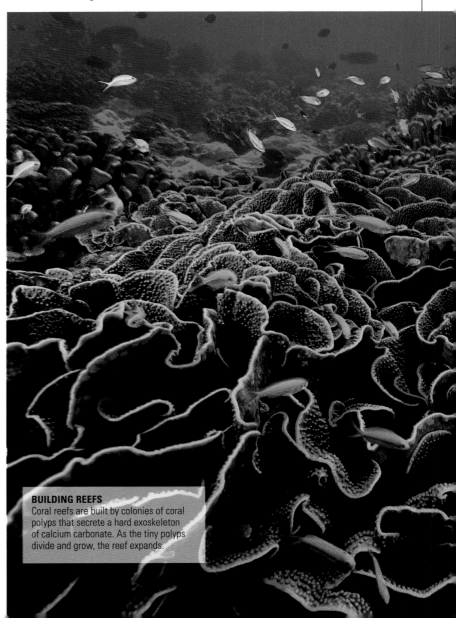

BUILDING REEFS
Coral reefs are built by colonies of coral polyps that secrete a hard exoskeleton of calcium carbonate. As the tiny polyps divide and grow, the reef expands.

Locomotion

The most mobile cnidarians are free-living jellyfish and medusae, which mainly drift in water currents but also swim actively using a form of jet propulsion. Most colonial cnidarians, such as corals and sea fans, cannot move from place to place. However, they can expand and contract their polyps to feed or avoid danger, and some sea pens can withdraw the whole colony below the surface of the sediment in which they live. Unattached mushroom corals may move slowly, or even right themselves if overturned. Anemones can creep slowly over the seabed on their muscular basal disk, and a few species swim if attacked.

JELLYFISH SWIMMING
A jellyfish swims by using muscles to contract its bell, forcing water out and pushing it along. The muscles then relax and the bell opens again.

bell relaxed and flattened, ready to propel forward

bell begins to contract and force water out

bell fully contracted, with little water remaining inside

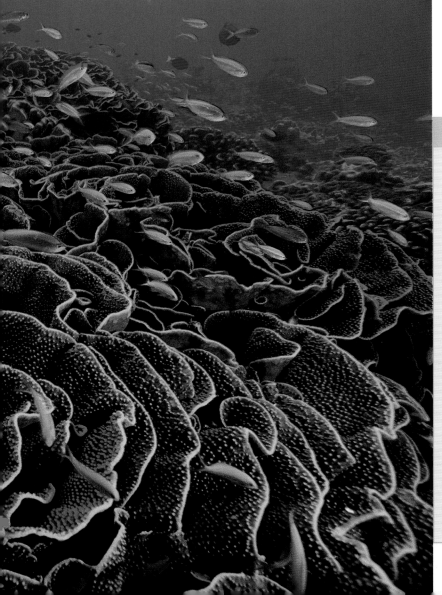

Reproduction

Members of the class Anthozoa, such as corals and anemones, reproduce by asexual budding. A genetically identical copy of the adult grows on the polyp's body wall. This budding juvenile drops off or stays attached to form a colony. Anthozoans also reproduce sexually, producing eggs and sperm within the polyps. Fertilized eggs develop into hairy, oval larvae (planulae), which either swim free or are brooded internally and then released. Hydrozoans have a two-stage life cycle. Their polyps release tiny free-swimming medusae into the water which, when mature, shed eggs and sperm. The resulting fertilized eggs develop into planulae that settle in a new area to grow into polyps. In contrast, the medusa form of true jellyfish (Scyphozoa) is usually much larger than the fixed polyp form and the polyps bud asexually.

BUDDING JELLYFISH POLYPS
Jellyfish polyps are minuscule, and their sole function is to reproduce asexually by budding off baby jellyfish.

Zooxanthellae

The massive skeletons secreted by reef-building corals require energy for their construction. Corals cannot catch enough plankton in clear tropical waters to provide this energy. Instead, they rely on tiny, symbiotic single-celled algae, called zooxanthellae, living in their cells. These algae manufacture organic matter by photosynthesis, and make more food than they need, so the excess is used by the coral. The algae benefit from a safe place to live and obtain "fertilizer" from the coral by using its nitrogenous waste products. If stressed by disease or high temperatures, corals expel their zooxanthellae, in a process called coral bleaching, and may die of starvation.

ZOOXANTHELLAE
In this image of coral polyps, the green patches are zooxanthellae living in the corals' tissues. The zooxanthellae give color to the otherwise colorless polyps.

CNIDARIAN CLASSIFICATION

Cnidarians are divided into five classes and a large number of orders and families. This phylum used to be called the Coelenterata, a name still used by some authorities. Many species remain undescribed.

BOX JELLYFISH
Class Cubozoa

48 species
These jellyfish have a cube-shaped bell with four flattened sides and a domed top. There are four tentacles or clusters of tentacles, one at each corner. Most are virulent stingers.

HYDROIDS
Class Hydrozoa

3,814 species
These colonial cnidarians mostly resemble plant growths attached to the seabed. A few have hard skeletons and resemble corals, and some colonies float at the surface like jellyfish. Most species have a free-living medusa stage.

STALKED JELLYFISH
Class Staurozoa

49 species
Less than an inch high, stalked jellyfish have a bell-shaped body with eight clusters of short, knobbed tentacles around the rim of the bell. They attach to seaweeds by a stalk that extends from the "top" of the bell.

ANTHOZOANS
Class Anthozoa

7,160 species
These colonial or solitary polyps are diverse in shape and have no medusa phase. Octocorals (soft corals, sea fans, and sea pens) have polyps with eight feathery tentacles; hexacorals (including hard corals and anemones) have polyps with multiples of six simple tentacles; ceriantipatharians have polyps with unbranched tentacles.

JELLYFISH
Class Scyphozoa

224 species
These mostly free-swimming medusae are shaped like a bell or saucer with a fringe of stinging tentacles. The edges of the mouth, located on the underside, are drawn out to form trailing mouth tentacles or oral arms.

MYXOZOANS
Class Myxozoa

790 species
Microscopic internal parasites.

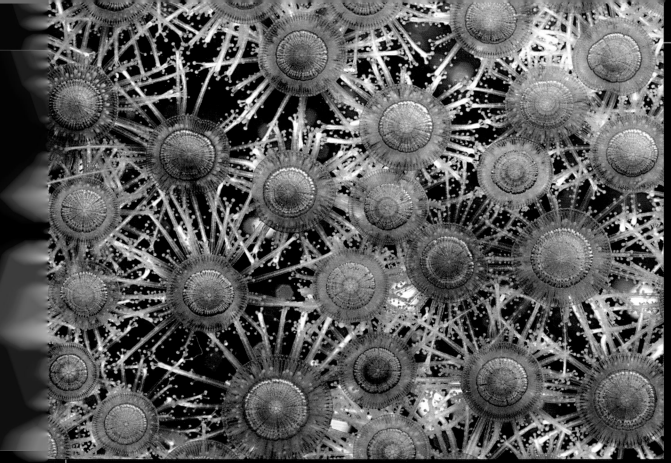

Blue Buttons
Porpita porpita

DIAMETER	³/₄ in (2 cm)
DEPTH	Surface
HABITAT	Surface waters

DISTRIBUTION Worldwide in warm waters

At first sight, blue buttons could be mistaken for a small jellyfish or even a piece of blue plastic. In fact, it is a hydrozoan colony that is modified for a free-floating existence. Swarms of these unusual creatures can be seen drifting on the water's surface or can sometimes be found washed up on the shore. The animal is kept afloat by a buoyant circular disk. Around the edge hang protective stinging polyps modified as knobbed tentacles. In the center underneath hangs a large feeding polyp that acts as the mouth for the whole colony. In between this and the tentacles are circlets of reproductive polyps. Unlike the Portuguese man-of-war (see p.214) to which it is related, blue buttons do not have a powerful sting.

CLASS HYDROZOA

Stinging Hydroid
Pachyrhynchia cuppressina

HEIGHT	Up to 16 in (40 cm)
DEPTH	10–100 ft (3–30 m)
HABITAT	Coral reefs

DISTRIBUTION Tropical reefs in Indian Ocean and southwestern Pacific

While most hydroids are harmless to touch, the stinging hydroid has a powerful sting. The colonies look like clumps of feathers or ferns dotted around among the corals on a reef. Individual polyps are arranged along one side of the smallest branches and extend their stinging tentacles to catch small planktonic animals. The sting is not usually dangerous to humans, but it results in an itchy rash that can irritate for up to a week.

CLASS SCYPHOZOA

Stalked Jellyfish
Haliclystus auricula

HEIGHT	Up to 2 in (5 cm)
DEPTH	0–50 ft (0–15 m)
HABITAT	On seaweed or seagrass

DISTRIBUTION Coastal waters of north Atlantic and north Pacific

Most jellyfish drift and swim freely in the water, but stalked jellyfish spend their lives attached by a stalk. Its body is shaped like a tiny funnel, urn, or goblet made up of eight equally spaced arms joined together by a membrane. Each arm ends in a cluster of tentacles on the funnel rim, and in some species there is an extra anchor-shaped tentacle between these. They cannot swim, but those with anchors can move by bending over their stalk and turning "head-over-heels," using the anchor tentacles to attach temporarily to the seabed, flip over, and reattach the adhesive disk.

Stalked jellyfish can be found attached to seaweed or seagrass in the intertidal zone and shallow water, where they feed by catching prey, such as small shrimp and fish fry, with their tentacles and passing it to the mouth inside the funnel. Undigested remains are expelled from the mouth

CLASS SCYPHOZOA

Deep-sea Jellyfish
Periphylla periphylla

HEIGHT	8–14 in (20–35 cm)
DEPTH	3,000–23,000 ft (900–7,000 m)
HABITAT	Open water

DISTRIBUTION Deep water worldwide, except Arctic Ocean

This jellyfish belongs to a group called coronate jellyfish, which are shaped like a ballet tutu. The upper part of the bell is a tall, stiff cone and the lower part a wider, soft, crown-shaped base with a scalloped edge. The 12 thin tentacles are often held in an upright position. The insides of the deep-sea jellyfish are a deep red color, and this may hide the bioluminescent light given out by its ingested prey. The jellyfish itself can squirt out a bioluminescent secretion that may help to confuse any predators. Unlike many jellyfish, the deep-sea jellyfish does not develop from a fixed bottom-living stage.

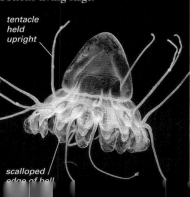

tentacle held upright

scalloped edge of bell

CLASS SCYPHOZOA

Moon Jellyfish
Aurelia aurita

DIAMETER	Up to 12 in (30 cm)
DEPTH	Near surface
HABITAT	Open water

DISTRIBUTION Worldwide; polar distribution unknown

The moon jellyfish is possibly the most widespread of all jellyfish and can be found in almost every part of the ocean except polar regions, where other *Aurelia* species take its place. It exists mainly in coastal waters and is sometimes cast ashore in large numbers because it is not a strong swimmer and lives near the surface. The body is shaped like a saucer with fine, short tentacles, which it uses to catch plankton. It can also trap plankton in sticky mucus on its bell and slide this down into its mouth on the underside. The gonads show through the translucent bell as four opaque horseshoe shapes.

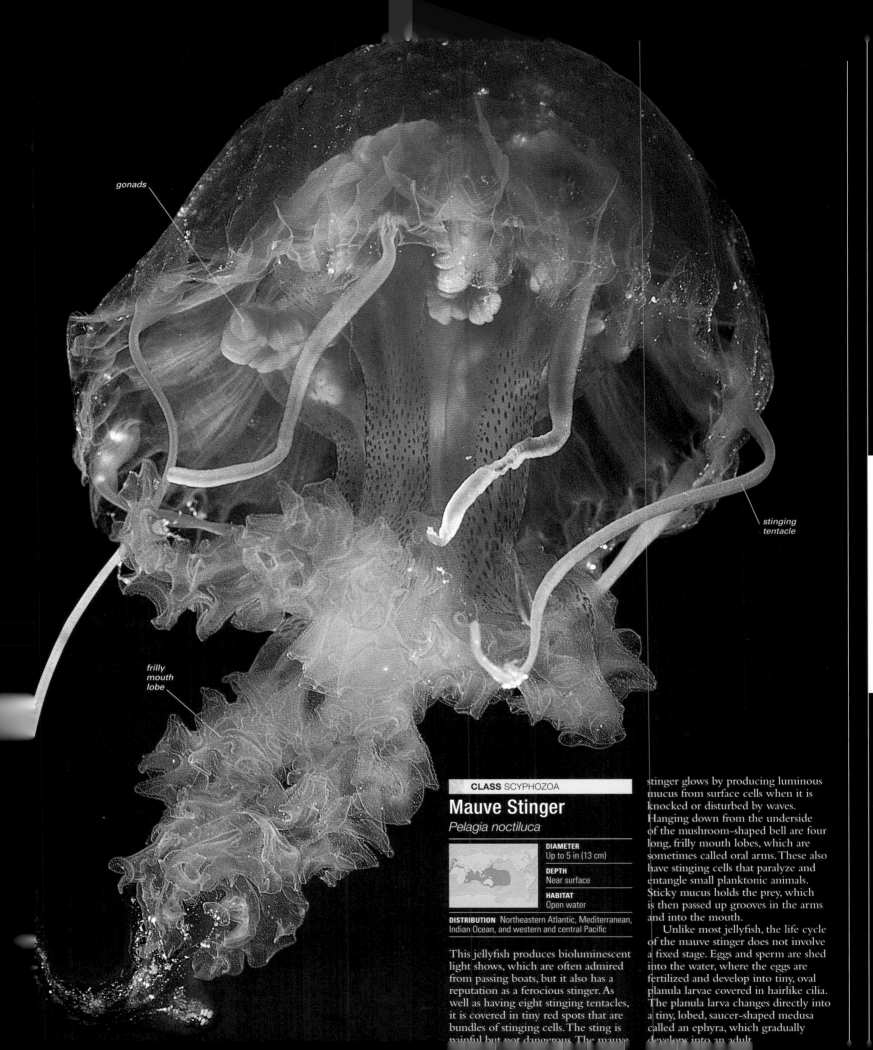

gonads

stinging
tentacle

frilly
mouth
lobe

CLASS SCYPHOZOA

Mauve Stinger

Pelagia noctiluca

DIAMETER
Up to 5 in (13 cm)

DEPTH
Near surface

HABITAT
Open water

DISTRIBUTION Northeastern Atlantic, Mediterranean, Indian Ocean, and western and central Pacific

This jellyfish produces bioluminescent light shows, which are often admired from passing boats, but it also has a reputation as a ferocious stinger. As well as having eight stinging tentacles, it is covered in tiny red spots that are bundles of stinging cells. The sting is painful but not dangerous. The mauve

stinger glows by producing luminous mucus from surface cells when it is knocked or disturbed by waves. Hanging down from the underside of the mushroom-shaped bell are four long, frilly mouth lobes, which are sometimes called oral arms. These also have stinging cells that paralyze and entangle small planktonic animals. Sticky mucus holds the prey, which is then passed up grooves in the arms and into the mouth.

Unlike most jellyfish, the life cycle of the mauve stinger does not involve a fixed stage. Eggs and sperm are shed into the water, where the eggs are fertilized and develop into tiny, oval planula larvae covered in hairlike cilia. The planula larva changes directly into a tiny, lobed, saucer-shaped medusa called an ephyra, which gradually develops into an adult.

CLASS SCYPHOZOA

Upside-down Jellyfish

Cassiopea xamachana

DIAMETER	Up to 12 in (30 cm)
DEPTH	0–33 ft (0–10 m)
HABITAT	Coastal mangroves

DISTRIBUTION Tropical waters of Gulf of Mexico and Caribbean

Divers who find this jellyfish upside-down on the seabed often think they have found a dying specimen. However, the upside-down jellyfish lives like this, floating with its bell pointing downward and its eight large,

branching mouth arms held upward. The mouth arms have elaborate fringes consisting of tiny bladders filled with minute single-celled algae called zooxanthellae. The algae need light to photosynthesize, and the jellyfish behaves as it does in order to ensure its passengers can thrive. Excess food manufactured by the algae is used by the jellyfish, but it can also catch planktonic animals with stinging cells on the mouth arms. Its bell pulsates to create water currents that bring food and oxygen. When it wants to move, the upside-down jellyfish turns the right way up with the bell uppermost. A very similar jellyfish, *Cassiopeia andromeda*, is found in the tropical Indian and Pacific Oceans and may actually be the same species.

CLASS CUBOZOA

Australian Box Jellyfish

Chironex fleckeri

DIAMETER	Up to 10 in (25 cm)
DEPTH	Near surface
HABITAT	Open water

DISTRIBUTION Tropical waters of southwest Pacific and eastern Indian Ocean

A sting from this jellyfish can kill a person in only a few minutes, and this small animal is considered one of the most venomous in the ocean. At each corner of its box-shaped, transparent body is a bunch of 15 tentacles. When it is hunting prey such as shrimp and small fish in shallow water, the tentacles extend up to 10 ft (3 m), and swimmers can be stung without ever seeing the jellyfish. In the middle of each flattened side is a collection of sense organs including some remarkably complex eyes. The exact range of this jellyfish in the Indo-Pacific region north of Australia is not known, but other smaller, less dangerous box jellyfish also occur in the Indian and Pacific Oceans. Some sea turtles can eat the box jellyfish without being affected by its sting.

HUMAN IMPACT

LETHAL VENOM

The sting of a box jellyfish causes excruciating pain and skin damage and can leave permanent scars. In severe cases, death may occur from heart failure or drowning following loss of consciousness. A box jellyfish antivenin is available in Australia. In northern parts of the country, some beaches are closed to the public for periods between November and April when the jellyfish are most abundant.

CLASS ANTHOZOA

Organ Pipe Coral

Tubipora musica

DIAMETER	Up to 20 in (50 cm)
DEPTH	15–65 ft (5–20 m)
HABITAT	Tropical reefs

DISTRIBUTION Tropical reefs of Indian Ocean and western Pacific

Although the organ pipe coral has a hard skeleton, it is not a true stony coral. Instead, it belongs to a group of cnidarians called octocorals, which includes soft corals and sea fans. Its beautiful red skeleton is made up of parallel tubes joined by horizontal links, and bits of this animal's skeleton are often found washed up on tropical shores. A single polyp extends from the end of each tube, and when the polyps expand their eight branched tentacles to feed, the skeleton cannot be seen.

CLASS ANTHOZOA

Mushroom Leather Coral

Sarcophyton species

DIAMETER	Up to 5 ft (1.5 m)
DEPTH	0–165 ft (0–50 m)
HABITAT	Rocks and reefs

DISTRIBUTION Tropical waters of Red Sea, Indian Ocean, and western and central Pacific

This distinctive soft coral has a conspicuous bare stalk topped by a wide, fleshy cap covered in polyps. When the colony is touched or is resting, the polyps are withdrawn into the fleshy body, and it looks and feels like leather. Within this genus there are many similar species.

CLASS ANTHOZOA

Dead Man's Fingers

Alcyonium digitatum

HEIGHT	Up to 8 in (20 cm)
DEPTH	0–165 ft (0–50 m)
HABITAT	Rocks and wrecks

DISTRIBUTION Temperate and cold waters of northeastern Atlantic

This soft coral's strange name comes from its appearance when thrown ashore by storms. It is shaped like a thick lump with stubby fingers, which can, with a little imagination, resemble a corpse's hand. When alive, it grows attached to rocks in shallow water and often covers large areas, especially where strong currents bring plenty of planktonic

food. With the polyps extended, the colonies have a soft, furry look. Most dead man's fingers colonies are white but some, like those shown below, are orange with white polyps. Over the fall and winter, the colony retracts its polyps and becomes dormant. In the spring, the outer skin is shed, along with any algae and other organisms that have settled on it.

CLASS ANTHOZOA
Carnation Coral
Dendronephthya species

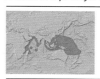

HEIGHT	Up to 12 in (30 cm)
DEPTH	33–165 ft (10–50 m)
HABITAT	Coral reefs

DISTRIBUTION Tropical reefs of Red Sea, Indian Ocean, and western Pacific

Carnation corals are among the most colorful of all reef animals. They grow as branched and bushy colonies and often cover steep reef walls with pink, red, orange, yellow, and white patches. They prefer to live where there are fast currents. When the current is running, they expand to full size and the polyps, which are on the branch ends, extend out to feed. With little or no current, they often hang down as flaccid lumps. In some species, such as the one shown here, small slivers of colored calcium carbonate show through the body tissue. These are called sclerites and help to give the soft branches some strength. Species and even genera are difficult to identify visually, and many species have not yet been described.

CLASS ANTHOZOA
Pulse Coral
Xenia species

HEIGHT	Up to 2 in (5 cm)
DEPTH	15–165 ft (5–50 m)
HABITAT	Coral reefs

DISTRIBUTION Tropical reefs of the Red Sea, Indian Ocean, and western Pacific

The most notable feature of this soft coral is the way the feathery tentacles of the polyps rapidly and continually open and close. A reef covered in pulse coral is alive with movement. The colonies have a stout trunk with a dome-shaped top covered with long polyps. Unlike the mushroom leather coral (see opposite), pulse coral polyps cannot retract and disappear. The pulsating movements of the polyps may help to oxygenate the colony as well as bring food within range of their tentacles.

CLASS ANTHOZOA
Common Sea Fan
Gorgonia ventalina

HEIGHT	Up to 6 ft (2 m)
DEPTH	15–65 ft (5–20 m)
HABITAT	Coral reefs

DISTRIBUTION Caribbean Sea

Sea fans grow attached to the seabed and look like exotic plants. Unlike soft corals, they have a supporting skeleton that provides a framework and allows them to grow quite large. It is made mainly of a flexible, horny material called gorgonin and consists of a rod that extends down the inside of all except the smallest branches. In the common sea fan, the branches are mostly in one plane and form a mesh that is aligned at right angles to the prevailing current. This increases the amount of planktonic food brought within reach of the polyps, which are arranged all around the branches. Fishing nets dragged over the reef can damage common sea fans and, as they grow quite slowly, they take a long time to recolonize. They are also collected, dried, and sold as souvenirs.

CLASS ANTHOZOA
White Sea Whip
Junceella fragilis

HEIGHT	Up to 6 ft (2 m)
DEPTH	15–165 ft (5–50 m)
HABITAT	Coral reefs

DISTRIBUTION Southwestern Pacific

Sea whips have a very similar structure to sea fans but grow up as a single tall stem. They have a very strong central supporting rod containing a lot of calcareous material as well as a flexible, horny material called gorgonin. The small polyps have eight tentacles and are placed all around the stem. White sea whips are often found in groups because they can reproduce asexually. As the whip enlarges, the fragile tip breaks off and drops onto the seabed, where it attaches and grows.

CLASS ANTHOZOA

Mediterranean Red Coral

Corallium rubrum

HEIGHT	Up to 20 in (50 cm)
DEPTH	165–650 ft (50–200 m)
HABITAT	Shaded rocks and caves

DISTRIBUTION Mediterranean and warm waters of eastern Atlantic

Often called precious coral, Mediterranean red coral has been collected and its skeleton made into jewelry for centuries. In spite of its name, it is not a true stony coral but instead is in the same group as sea fans (see p.265). Like them, its branches are covered in small polyps, each of which has eight branched tentacles. However, the supporting skeleton is made mainly from hard calcium carbonate colored a deep red or pink. This coral is now scarce in places that are easily accessible to collectors.

CLASS ANTHOZOA

Orange Sea Pen

Ptilosarcus gurneyi

HEIGHT	Up to 20 in (50 cm)
DEPTH	33–1,000 ft (10–300 m)
HABITAT	Sediment

DISTRIBUTION Temperate waters of northeastern Pacific

Unlike the majority of anthozoans, sea pens live in areas of sand and mud. They get their name from their resemblance to an old-fashioned quill pen. The orange sea pen consists of a central stem with branches on either side. The basal part of the stem is bulbous and anchors the colony in the sediment. Single rows of polyps extend their eight tentacles into the water from each leaflike branch, giving the front of the sea pen a downy appearance. The colony faces toward the prevailing current to maximize the flow of plankton over the feeding polyps. When no current is flowing, the colony can retract down into the sediment. Although they tend to stay in one place, colonies can relocate and reanchor themselves if necessary. Predators of sea pens include sea slugs and starfish.

CLASS ANTHOZOA

Slender Sea Pen

Virgularia mirabilis

HEIGHT	Up to 24 in (60 cm)
DEPTH	33–1,300 ft (10–400 m)
HABITAT	Sediment

DISTRIBUTION Temperate waters of northeastern Atlantic and Mediterranean

The muddy bottoms of sheltered sea lochs in Scotland and Norway are often carpeted in dense beds of slender sea pens. This species has a structure similar to the orange sea pen (see below, left) but has a much thinner central stalk and thin branches. Almost half the stalk is buried in the sediment and the colony can withdraw into the sediment if disturbed.

CLASS ANTHOZOA

Beadlet Anemone

Actinia equina

DIAMETER	Up to 2¾ in (7 cm)
DEPTH	0–65 ft (0–20 m)
HABITAT	Hard surfaces

DISTRIBUTION Coastal waters of Mediterranean, northeastern and eastern Atlantic

Most anemones cannot survive out of water, but the beadlet anemone can do so provided it stays damp. At low tide, this anemone can be found on rocky shores with its tentacles retracted, looking like a blob of red or green jelly. The top of the anemone's body is ringed with blue beads called acrorhagi. These contain numerous stinging cells, which the anemone uses to repel any close neighbors. Leaning over, it will sting any anemone within reach, and the defeated anemone will move slowly out of the victor's territory. The beadlet anemone broods its eggs and young inside the body and ejects them through its mouth.

CLASS ANTHOZOA

Giant Anemone

Condylactis gigantea

DIAMETER	Up to 12 in (30 cm)
DEPTH	10–165 ft (3–50 m)
HABITAT	Coral reefs and rocks

DISTRIBUTION Tropical waters of Caribbean Sea and western Atlantic

The long, purple-tipped tentacles of this large anemone bring a splash of color to Caribbean reefs. Its columnar body is usually tucked away between rocks or corals, leaving only the stinging tentacles exposed. Several small reef fish (mainly blennies) can live unharmed among the tentacles, where they gain protection from predators. The giant anemone can move slowly along on its basal disk if it wants to find a better position on the reef.

acrorhagi containing stinging cells

CLASS ANTHOZOA
Plumose Anemone
Metridium senile

HEIGHT	Up to 12 in (30 cm)
DEPTH	0–330 ft (0–100 m)
HABITAT	Any hard surface

DISTRIBUTION Temperate waters of north Atlantic and north Pacific

This tall anemone resembles an ornate piece of architecture. It has a long column, topped by a collarlike ring and a wavy disk with thousands of fine tentacles. The most common colors are white or orange, but it can also be brown, gray, red, or yellow. Fragments from the base of large anemones can grow into tiny new anemones. Large ones can be found on pier pilings and wrecks projecting into the current, but they deflate when the water is still.

CLASS ANTHOZOA
Cloak Anemone
Adamsia palliata

DIAMETER	2 in (5 cm)
DEPTH	0–650 ft (0–200 m)
HABITAT	Hermit crab shells

DISTRIBUTION Temperate waters of northeastern Atlantic and Mediterranean

The cloak anemone lives with its wide base wrapped around the shell of a hermit crab and its tentacles trailing beneath the crab's head. In this position, the tentacles are ideally placed to pick up food scraps. The enveloping column of the anemone is off-white with distinct pink spots. Neither partner thrives without the other, though young cloak anemones can be found on rocks and shells between the tidemarks waiting to find a host.

ARMORED VEHICLE

The hermit crab *Pagurus prideaux* is always seen with its protective anemone cloak. It does not have to find a bigger shell as it grows because the cloak anemone secretes a horny extension. The anemone on the crab on the left has thrown out pink stinging threads, called acontia, to repel another hermit crab.

CLASS ANTHOZOA
Antarctic Anemone
Urticinopsis antarctica

SIZE	Not recorded
DEPTH	15–740 ft (5–225 m)
HABITAT	Rocky seabeds

DISTRIBUTION Southern Ocean around Antarctica and South Shetland Islands

Like many other Antarctic marine animals, the Antarctic anemone grows to a large size, but rather slowly. It has long tentacles with powerful stinging cells and is capable of catching and eating starfish, sea urchins, and jellyfish much larger than itself. As there are often many anemones living close together, two or more may hold a large jellyfish. As in most anemones, stinging cells on the tentacles fire barbed threads into the prey to hold it and to paralyze or kill it.

OCEAN LIFE

CLASS ANTHOZOA
Jewel Anemone
Corynactis viridis

DIAMETER	½ in (1 cm)
DEPTH	0–260 ft (0–80 m)
HABITAT	Steep rocky areas

DISTRIBUTION Temperate waters of northeastern Atlantic and Mediterranean

Jewel anemones often cover large areas of shaded underwater cliff faces, creating a spectacular display. Individuals can be almost any color, and they reproduce by splitting in half, making two new identical anemones. This results in dense patches of different colors. Each anemone has a small saucer-shaped disk circled by stubby translucent tentacles. The tentacles have knobbed tips that are often a contrasting color to the tentacle shafts, disk, and column of the anemone. The color combination shown here is one of the most common. Jewel anemones are not true anemones but belong to a group of anthozoans called coralliomorphs. These closely resemble the polyps of hard corals but have no skeleton. Coralliomorphs are found in all oceans but are most common in the tropics.

CLASS ANTHOZOA

Table Coral
Acropora hyacinthus

DIAMETER	Up to 10 ft (3 m)
DEPTH	0–33 ft (0–10 m)
HABITAT	Coral reefs

DISTRIBUTION Tropical waters of Red Sea, Indian Ocean, and western and central Pacific

The magnificent flat plates of table coral are ideally shaped to expose as much of their surface as possible to sunlight. Like most hard corals, the cells of table coral contain zooxanthellae that need light to photosynthesize and manufacture food for themselves and their host. Table coral is supported on a short, stout stem that is attached to the seabed by a spreading base. The horizontal plates have numerous branches that mostly project upward from the surface, so each plate, or table, resembles a bed of nails. Each of these branches is lined by cup-shaped extensions of the skeleton called corallites, from which the polyps extend their tentacles in order to feed, mainly at night.

The usual color of table coral is a dull brown or green, but it is brightened up by the numerous reef fish that shelter under and around its plates. However, the shade the plates cast means that few other corals can live underneath a table coral. There are many other similar species that are also called table coral, but *Acropora hyacinthus* is one of the most abundant and widespread.

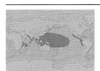

CLASS ANTHOZOA

Hump Coral
Porites lobata

DIAMETER	Up to 20 ft (6 m)
DEPTH	0–165 ft (0–50 m)
HABITAT	Coral reefs

DISTRIBUTION Tropical waters of Red Sea, Persian Gulf, and Indian and Pacific Oceans

It can be difficult to tell that hump coral is a living coral colony because it looks just like a large, lumpy rock. Closer inspection will show that the coral grows as a series of large lobes formed into a dome. The living polyps are tiny, with tentacles that are only about $\frac{1}{32}$ in (1 mm) long, and during the day, they are hidden in their shallow skeleton cups. At night, they extend their tentacles to feed and the colony takes on a softer appearance. Hump coral is an important reef-building species.

CLASS ANTHOZOA

Daisy Coral
Goniopora djiboutiensis

DIAMETER	Up to 3 ft (1 m)
DEPTH	15–100 ft (5–30 m)
HABITAT	Turbid reef waters

DISTRIBUTION Tropical waters of Indian Ocean and western Pacific

In most corals it is difficult to see the tiny polyps, but the daisy coral has polyps that are a few inches long. The head of each polyp is dome-shaped with the mouth in the middle, surrounded by a ring of about 24 tentacles. These are arranged rather like the petals of a daisy. Unlike the majority of corals, the polyps extend to feed during the day, though they will quickly withdraw if touched. Daisy coral grows as a rounded lump, but the shape is difficult to see when the polyps are extended. While most corals need clear water to survive, this species often covers large areas where the water is made turbid by disturbed sediment.

CLASS ANTHOZOA
Mushroom Coral
Fungia species

DIAMETER	Up to 6 in (15 cm)
DEPTH	0–80 ft (0–25 m)
HABITAT	Coral reefs

DISTRIBUTION Tropical waters of Red Sea, Indian Ocean, and western Pacific

Mushroom corals are unusual in that they live as single individuals rather than a colony. Juveniles start life as a small stalked disk attached to dead coral or rock. By the time they reach about 1½ in (4 cm) in diameter, they become detached. Most species feed at night and the tentacles are withdrawn during the day, leaving the skeleton clearly visible, with the mouth at the center of the disk. The skeleton resembles the gills of a mushroom. They use their tentacles to turn the right way up if overturned by waves.

CLASS ANTHOZOA
Giant Brain Coral
Colpophyllia natans

DIAMETER	Up to 16 ft (5 m)
DEPTH	3–180 ft (1–55 m)
HABITAT	Seaward side of coral reefs

DISTRIBUTION Tropical waters of Gulf of Mexico and Caribbean

This huge coral grows as giant domes or extensive thick crusts and can live for more than 100 years. The surface of the colony is a convoluted series of ridges and long valleys, as in other species of brain coral, and this is what

CLASS ANTHOZOA
Dendrophyllid Coral
Dendrophyllia species

HEIGHT	Up to 2 in (5 cm)
DEPTH	10–165 ft (3–50 m)
HABITAT	Steep rock faces

DISTRIBUTION Tropical waters in Indian Ocean and from western Pacific to Polynesia

With their large, flamboyant polyps, corals of the genus *Dendrophyllia* look more like anemones than corals. Dendrophyllids belong to a group called cup corals. They grow as a

gives it its name. The valleys and ridges are often differently colored and the ridges have a distinct groove running along the top. Typically, the valleys are green or brown and the ridges are brown. The polyp mouths are hidden in the valleys and the tentacles are only extended at night. Since 2014, stony coral tissue loss disease (SCTLD) has killed brain corals and many others in Florida and has now spread to parts of the Caribbean. Particularly large colonies are popular tourist attractions in islands such as Tobago and attract both divers and fish, including some gobies that live permanently on the coral.

low-branching colony with each tubular individual distinct, and they do not develop the massive skeleton of reef-building corals. They have no zooxanthellae and grow in shaded parts of reefs such as below overhangs and especially on steep cliff faces. During the day, the polyps are entirely withdrawn and the coral looks like a dull orange–yellow lump. As darkness falls, the polyps expand their orange tentacles to feed on plankton and make a spectacular display that often covers large areas. This genus of coral is very difficult to identify to species level and can also be confused with cup corals belonging to the genus *Tubastrea*.

CLASS ANTHOZOA
Devonshire Cup Coral
Caryophyllia smithii

DIAMETER	1¼ in (3 cm)
DEPTH	0–330 ft (0–100 m)
HABITAT	Rocks and wrecks

DISTRIBUTION Northeastern Atlantic and Mediterranean

While most corals grow as large reef-forming colonies in tropical waters, this and other cup corals grow as single individuals in temperate parts of the ocean. Devonshire cup coral grows with its cup-shaped skeleton attached to a rock or even a shipwreck. When the tentacles are expanded, these tiny corals look just like anemones, with each tapering, transparent tentacle ending in a small knob. Devonshire cup coral occurs in a variety of colors, and the white skeleton can usually be seen through the translucent, living tissue.

CLASS ANTHOZOA
Lophelia Coral
Lophelia pertusa

DIAMETER	At least 33 ft (10 m)
DEPTH	165–10,000 ft (50–3,000 m)
HABITAT	Deep-sea reefs

DISTRIBUTION Atlantic, eastern Pacific, and western Indian Ocean; distribution not fully known

Lophelia (now considered by some to be *Desmophyllum*) lives in dark, cold water, so has no zooxanthellae to help build its white, branching skeleton. It therefore grows very slowly, and large reefs are many hundreds of years old. Each polyp has many tentacles, which it uses to capture planktonic prey, such as arrow worms and small crustaceans,

from the passing current. Stinging cells render the prey immobile and it is then transferred to the mouth. Although individual colonies grow up to 33 ft (10 m) across, they can merge to form much larger reefs. Some reefs more than 8 miles (13 km) long and 98 ft (30 m) high have been found off the coast of Norway.

Ancient mounds of this coral are often called "bioherms," the main body of which is a rocklike material formed over millennia from dead coral, with living coral spread over the top. The gaps in well-developed reefs and bioherms are colonized with sponges, soft corals, and tube worms. Mobile crustaceans, mollusks, and fish add to the biodiversity. Many reefs have been badly damaged by trawlers trying to catch deep-sea fish. In the UK, some areas of reef are now protected.

OCEAN LIFE

White Zoanthid

Parazoanthus anguicomus

HEIGHT	1 in (2.5 cm)
DEPTH	65–1,300 ft (20–400 m)
HABITAT	Shaded rocks, wrecks, and shells

DISTRIBUTION Temperate waters of northeastern Atlantic

Most zoanthids are found in tropical waters, but the white zoanthid is common in the north Atlantic. Its white polyps arise from an encrusting base and it has two circles of tentacles around the mouth. One circle is usually held upward while the other lies flat. As well as covering rocks and wrecks, this species also encrusts worm tubes and *Desmophyllum* reefs (see p.179).

Whip Coral

Cirrhipathes species

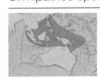

LENGTH	Up to 3 ft (1 m)
DEPTH	10–165 ft (3–50 m)
HABITAT	Coral reefs

DISTRIBUTION Tropical waters of eastern Indian Ocean and western Pacific

Whip corals, or wire corals, belong to a group of anthozoans called antipatharians to which the black corals (see right) also belong. Whip coral grows as a single unbranched colony that can be either straight

or coiled as in the species belonging to the genus *Cirripathes* shown here. (Whip corals are difficult to identify and many species remain undescribed.) The feeding polyps of whip corals and black corals can be seen easily because, unlike sea fans, they cannot retract their short, pointed tentacles. Gobies live among the tentacles, hanging onto the coral with suckerlike pelvic fins.

Feather Black Coral

Plumapathes pennacea

HEIGHT	Up to 5 ft (1.5 m)
DEPTH	15–1,100 ft (5–330 m)
HABITAT	Coral reefs

DISTRIBUTION Tropical waters of Gulf of Mexico, Caribbean Sea, and western Atlantic

Feather black coral grows as a bushy colony with branches shaped like large bird feathers. There are many different species of black corals, and they get their name from the strong black skeleton that strengthens their branches. Made of a tough, horny material, the skeleton is valuable as it can be cut and polished to make jewelry, although this particular species is not widely used for this.

Mediterranean Tube Anemone

Cerianthus membranaceus

HEIGHT	14 in (35 cm)
DEPTH	33–330 ft (10–100 m)
HABITAT	Muddy sand

DISTRIBUTION Mediterranean and northeast Atlantic

The long, pale tentacles make a spectacular display, but at the slightest disturbance, the anemone will retreat down its tube in an instant. Tube Anemones are closely related to black corals (see above). They live in tubes made of sediment-encrusted mucus that can be up to 3 ft (1 m) long even though the animals are only about a third of this length. The slippery lining of the tube allows it to withdraw rapidly. As well as the long, slender outer tentacles, the animal has an inner ring of very short tentacles surrounding the mouth. The outer tentacles may look dangerous, but the tube anemone feeds only on plankton and organic debris.

Flatworms

DOMAIN	Eucarya
KINGDOM	Animalia
PHYLUM	Platyhelminthes
	Xenacoelamorpha
CLASSES	8
SPECIES	20,430

POSSESSING VERY thin, sometimes transparent bodies, flatworms are among the simplest of animals. Marine species mostly belong to a colorful group called polyclad flatworms—leaf-shaped animals, common on coral reefs. Others are found in fresh water, and many are parasitic. In the oceans, parasitic flukes and tapeworms are common in fish, mammals, and birds. Most belong to the phylum Platyhelminthes, but acoel flatworms (see below) are now separated into the phylum Xenacoelomorpha (or Acoelomorpha).

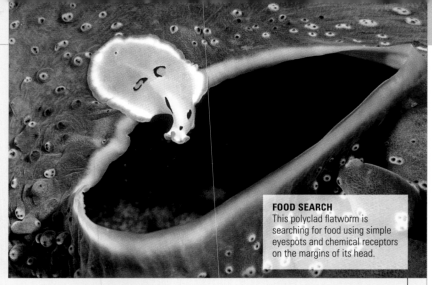

FOOD SEARCH
This polyclad flatworm is searching for food using simple eyespots and chemical receptors on the margins of its head.

Anatomy

Flatworms have a simple, solid structure with no internal cavity. They are so thin that oxygen can diffuse in from the water, and there are no blood or circulatory systems. The head end contains sense organs; advanced species have primitive eyes. The gut opens to the outside at one end, the opening serving as both mouth and anus. In polyclad flatworms, this opening is in the middle underside of the body. When feeding, they extend a muscular tube (pharynx) out of the mouth to grasp their food. Polyclad flatworms are covered in tiny hairs (or cilia) which, together with simple muscles, help them to glide over almost any surface. The anatomy of tapeworms and flukes is adapted to suit their parasitic lifestyle.

BODY SECTION
In flatworms, the space between the internal organs is filled with soft connective tissue crisscrossed by muscles.

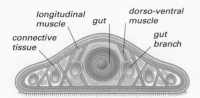

longitudinal muscle — gut — dorso-ventral muscle — connective tissue — gut branch

Reproduction

Most flatworms are hermaphrodites, so every individual has both ovaries and testes. The reproductive system is complex for such a primitive animal and includes special chambers and tubules where the ripe eggs are fertilized. When two polyclad flatworms meet, they may briefly touch heads and bodies in a short ritual before mating. After mating, the eggs are released into the water, laid in sand, or stuck to rocks. In some flatworms, the eggs develop directly into juvenile worms, but in others, they develop initially into an eight-lobed planktonic larva. Called Müller's larva, this swims for a few days and then settles onto the seabed and flattens out into a young flatworm.

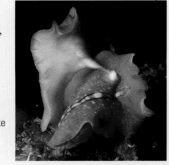

COMPLEX APPARATUS
Some flatworms undertake "penis fencing," where each tries to stab the other and inject sperm.

SUBPHYLUM ACOELOMORPHA

Acoel Flatworm

Waminoa species

LENGTH	Less than ¼ in (5 mm)
DEPTH	Not recorded
HABITAT	On bubble coral (*Pleurogyra sinuosa*)

DISTRIBUTION Tropical Indian and Pacific Oceans

These diminutive flatworms look like colored spots on the bubble coral on which they live. Their ultra-thin bodies glide over the coral surface as they graze, probably eating organic debris trapped by coral mucus. Acoel flatworms have no eyes and instead of a gut, they have a network of digestive cells. They are able to reproduce by fragmentation, each piece forming a new individual. The genus is difficult to identify to species level and the distribution is uncertain.

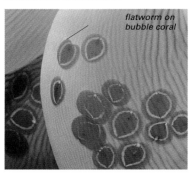

flatworm on bubble coral

SUBPHYLUM ACOELOMORPHA

Green Acoel Flatworm

Symsagittifera roscoffensis

LENGTH	Up to ½ in (1.5 cm)
DEPTH	Intertidal
HABITAT	Sheltered sandy shores

DISTRIBUTION Northeastern Atlantic; probably more widespread than shown

Although difficult to see individually, these flatworms show up as green patches when they collect together in puddles on sandy shores at low tide. Their bodies harbor tiny, single-celled algae (p. 248) that color them bright green. In warm, sunlit pools the algae can photosynthesize and pass some of the food they make to their host. These flatworms are very sensitive to vibrations and quickly disappear down into the sand if footsteps approach.

SUBPHYLUM RHABTITOPHORA

Candy Stripe Flatworm

Prostheceraeus vittatus

LENGTH	Up to 2 in (5 cm)
DEPTH	0–100 ft (0–30 m)
HABITAT	Muddy rocks

DISTRIBUTION Temperate waters of northeastern Atlantic and Mediterranean

Most brightly colored flatworms are found on tropical reefs, but the candy stripe flatworm is an exception and can be found as far north as Norway. Generally a cream color, it is marked with reddish brown, lengthwise stripes. The head end of its flattened, leaf-shaped body has a pair of distinct tentacles and groups of primitive eyes. As it crawls along, the flatworm pushes the edges of its body up into folds; it is also able to swim using sinuous movements of the body. Usually found in rocky areas, it has also been recorded on sand and mud.

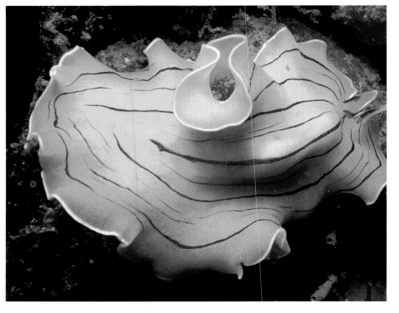

Exquisite Lined Flatworm

Pseudobiceros bedfordi

LENGTH	Up to 3 in (8 cm)
DEPTH	Not recorded
HABITAT	Coral reefs

DISTRIBUTION Tropical waters of Indian and western Pacific Oceans

Divers frequently come across this beautiful flatworm on coral reefs. Its striking pattern of pinkish transverse stripes and white dots against a black background make it easily recognizable. It is usually seen crawling over rocks in search of tunicates and crustaceans, but it is also a fairly good swimmer. Sometimes, the head end is reared up and a pair of flaplike tentacles can be seen.

Divided Flatworm

Pseudoceros dimidiatus

LENGTH	Up to 3 in (8 cm)
DEPTH	Not recorded
HABITAT	Coral reefs

DISTRIBUTION Tropical waters of Indian and western Pacific Oceans

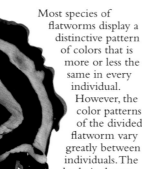

Most species of flatworms display a distinctive pattern of colors that is more or less the same in every individual. However, the color patterns of the divided flatworm vary greatly between individuals. The body is always black with an orange margin, but the width and arrangement of the yellow or white lateral stripes, zebralike bars, or narrow and wide longitudinal stripes is highly variable. These highly contrasting colors act as a warning to predators that divided flatworms are not good to eat. Like other flatworms, this species has numerous photo- and chemosensitive cells in its head region, which help the worm to find food and avoid danger.

GOOD IMITATION
The imitating flatworm has a creamy gray background color and black reticulations surrounding pale pustules.

Imitating Flatworm

Pseudoceros imitatus

LENGTH	Up to 1 in (2 cm)
DEPTH	Not recorded
HABITAT	Coral reefs

DISTRIBUTION Waters around New Guinea and northern Australia, perhaps more extensive

Unlike the majority of polyclad flatworms, which have a relatively smooth skin, the imitating flatworm has a bumpy surface covered in small pustules. This appearance is an

SOURCE OF IMITATION
Phyllidiella pustulosa is a common and widespread sea slug found on Indo-Pacific reefs in water about 15–130 ft (5–40 m) deep.

imitation of the skin of the sea slug *Phyllidiella pustulosa*, and the flatworm's color pattern is also almost identical to that of the sea slug. The sea slug secretes a noxious chemical to deter potential predators, and it may be that the imitating flatworm gains protection by looking and feeling to the touch like the distasteful sea slug.

Thysanozoon Flatworm

Thysanozoon nigropapillosum

LENGTH	Up to 3 in (8 cm)
DEPTH	3–100 ft (1–30 m)
HABITAT	Coral reef slopes

DISTRIBUTION Tropical waters of Indian and western Pacific Oceans

The highly convoluted edge of the very thin thysanozoon flatworm is prominently displayed with a white outline. The rest of the upper side

of the body is black and covered in short papillae, or protuberances, each of which ends in a yellow tip. This gives the flatworm the appearance of being peppered with yellow spots. As is the case with most tropical reef flatworms, little is known of the biology of this species, but the thysanozoon flatworm has been found in association with colonial tunicates and is thought to feed on these and other colonial animals. It has been observed to swim well, rhythmically undulating its wide body. Much of what is known about this and other tropical reef flatworms has come from observations made by recreational divers and photographers. A similar species, *Thyanozoon flavomaculatum*, is found on Red Sea coral reefs.

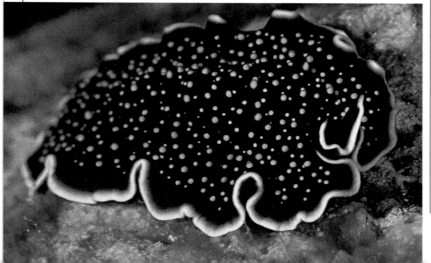

Giant Leaf Worm

Kaburakia excelsa

LENGTH	Up to 4 in (10 cm)
DEPTH	Intertidal
HABITAT	Under coastal rocks

DISTRIBUTION Temperate waters of northeastern Pacific

This large, oval flatworm crawls around rocks, stones, and undergrowth on the Pacific shores of North America. Its color is reddish-brown to tan, marked with darker spots, and when it is fully spread out, the branches of its digestive system may

Broad Fish Tapeworm

Dibothriocephalus latus

LENGTH	Up to 33 ft (10 m)
DEPTH	Dependent on host
HABITAT	Parasitic

DISTRIBUTION Probably worldwide, dependent on host species

be seen through the skin. It feeds in the same way as most polyclad flatworms, by everting its pharynx over its prey. Most intertidal flatworms in this region are only about 1 in (2 cm) long, making this species easy to identify. It is common on floating docks and in mussel beds.

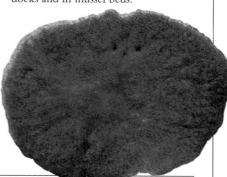

Some flatworms, including tapeworms (Class Cestoda), have become highly modified and live as parasites. This tapeworm has a complex life. It begins as a fertilized egg that is eaten by tiny freshwater crustaceans, inside which the larvae hatch. Freshwater, estuarine, and migratory marine fish (such as salmon) become infected by the larvae when they eat either the crustaceans or other infected fish. The adult tapeworm lives in fish-eating mammals and may infect humans who eat raw fish. Other tapeworm species live in the guts of marine fish.

Ribbon Worms

DOMAIN	Eucarya
KINGDOM	Animalia
PHYLUM	Nemertea
CLASSES	4
SPECIES	1,350

ALSO CALLED NEMERTEAN worms, ribbon worms can reach great lengths, although many are small and inconspicuous. *Lineus longissimus* (see below) is thought to be the longest animal on Earth and can stretch out to at least 160 ft (50 m). Such species are cylindrical and are often called bootlace worms. The majority of ribbon worms live in the ocean under rocks, among undergrowth or in sediment. A few species live inside the shells of mollusks or are parasitic.

Anatomy

Ribbon worms have a long, unsegmented body covered with mucus. Strong muscles in the body wall can shorten the worm to a fraction of its full length. Unlike flatworms, ribbon worms have blood vessels and a complete gut with mouth and anus. It is often difficult to distinguish between the front and rear end of the worm, but most species have many simple eyes at the front. The most characteristic feature of these worms is a strong, tubular structure called a proboscis that lies in a fluid–filled sheath above the gut. It can be everted either through the mouth or a separate opening, and is used to capture prey. In some species, the proboscis and mucus contain potent toxins such as tetrodotoxin.

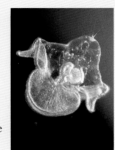

stylet · proboscis · nerve ganglion · nerve · excretory organs · proboscis sheath · blood vessel · ovary · gut

BODY SECTION
Ribbon worms have no body cavity or gills; a simple circulatory system carries oxygen around the body.

WARNING PATTERN
Some ribbon worms have bright patterns that may serve as a warning to predators that they are toxic. Drab-colored species only emerge at night to hunt.

SWIMMING LARVA
Some ribbon worms develop from a planktonic larva called a pilidium. It is able to swim by beating hairlike structures, called cilia.

Reproduction

Most marine ribbon worms have separate sexes and their numerous, simple gonads produce either eggs or sperm. These are usually shed into the sea through pores along the sides of the body. Some species cocoon themselves together in a mucous net where the eggs are duly fertilized. In some types of ribbon worms, the eggs develop directly into juvenile worms, while others initially hatch into various types of larvae. The long, fragile bodies of ribbon worms tend to break easily but they have the useful ability to regenerate any lost parts. Some species even use regeneration as a method of asexual reproduction, where the body breaks up into several pieces and each piece develops a new head and tail.

OCEAN LIFE

 CLASS PALAEONEMERTEA

Football Jersey Worm
Tubulanus annulatus

LENGTH Up to 30 in (75 cm)	
DEPTH 0–130 ft (0–40 m)	
HABITAT Gravel, stones, and sediment	

DISTRIBUTION Cold and temperate waters of north Atlantic and north Pacific

One of the most strikingly colored ribbon worms, the football jersey worm has a patterning of longitudinal white lines and regularly spaced white rings. It may be found lying in an untidy pile beneath stones on the lower shore and may also be seen scavenging when the tide is out. More usually it lives below the shore on almost any type of seabed, including mud, sand, and shell gravel. To camouflage itself, it secretes a mucous tube that becomes covered in surrounding sediment.

 CLASS PILIDIOPHORA

Bootlace Worm
Lineus longissimus

LENGTH Up to 180 ft (55 m)	
DEPTH Intertidal	
HABITAT Sediments and stones	

DISTRIBUTION Temperate waters of northeast Atlantic

The bootlace worm makes up for its rather drab brown color by its incredible length. Only a fraction of an inch in diameter, it commonly reaches 33 ft (10 m) in length, and is one of the longest animals known. On the shore, it appears as a writhing pile of loops and curves, often lying on muddy sediment beneath boulders. This rather unattractive worm is difficult to pick up, because it exudes large amounts of mucus, which helps it to slide over the seabed.

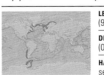 **CLASS** HOPLONEMERTEA

Ribbon Worm
Nipponnemertes pulcher

LENGTH Up to 3½ in (9 cm)	
DEPTH 0–1,900 ft (0–570 m)	
HABITAT Coarse sediments	

DISTRIBUTION Temperate and cold waters of Arctic, Atlantic, Pacific and Southern Oceans

This worm belongs to a class of nemertean worms in which the proboscis is tipped with one or more sharp stylets. *Nipponnemertes pulcher* has a short, stout body with a width of up to ¼ in (5 mm) that tapers to a pointed tail. The coloration varies from pink to orange or deep red and is paler beneath. This species has a distinctive, shield-shaped head with numerous eyes along its edges. The number of eyes increases with age. It is usually seen when dredged up by scientists from the coarse sediments in which it lives, but is sometimes found beneath stones on the lower shore. Its full distribution is unknown.

Segmented Worms

DOMAIN	Eucarya
KINGDOM	Animalia
PHYLUM	Annelida
CLASSES	2
SPECIES	15,000

SEGMENTED WORMS include terrestrial earthworms, leeches, and various other freshwater worms, grouped in the class Clitellata. Species from a diverse second class, the bristleworms or Polychaeta, are found mostly in the ocean. These include burrowing lugworms, free-living predatory ragworms, and tube-dwelling worms. All segmented worms have a long, soft body, divided into a series of almost identical, linked segments.

BRISTLEWORM
Fire worms have long, sharp bristles on each body section. These break off if the worm is attacked and can cause severe skin irritation.

Anatomy

Each body segment is called a metamere and, except for the head and tail tip, all are virtually indistinguishable from each other. In bristleworms, flattened lobes (parapods) project from the sides of each segment, and are reinforced by strong rods made of chitin. The worm uses parapods for locomotion, and projecting bundles of bristles help it to grip. Internally, the segments are separated by partitions and filled with fluid. The gut, nerve cord, and large blood vessels run all along the body.

BODY SECTIONS
Most segments contain their own organs, including excretory and reproductive organs, and branches from the main blood vessels and ventral nerve cord.

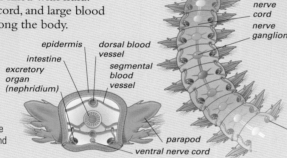

epidermis
intestine
excretory organ (nephridium)
dorsal blood vessel
segmental blood vessel
parapod
ventral nerve cord

parapod
ventral nerve cord
nerve ganglion

JAWS OF A PREDATOR
This bobbit worm seizes prey using a proboscis tipped with sharp mandibles, which it shoots out from the mouth.

excretory organ (nephridium)

Reproduction

In most polychaete worms, the sexes are separate and the eggs and sperm are shed into the water. Spawning is usually seasonal, especially at temperate latitudes. In many species, the fertilized egg develops into a larva (trochophore) that resembles a tiny spinning top. It floats and swims in the plankton, propelled by the beating of hairlike cilia around its middle. Eventually, the larva elongates and constricts into segments as it turns into an adult. Some species brood their eggs until the larvae are well developed. Many polychaete worms change shape as they become sexually mature, becoming little more than swimming bags of eggs or sperm. Known as epitokes, they swarm, burst open to release the eggs or sperm, then die.

epitoke

READY TO BURST
The egg- or sperm-laden epitoke of a palolo worm separates from the front segments, and bursts open.

CLASS POLYCHAETA

Lugworm
Arenicola marina

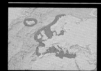

LENGTH	Up to 8 in (20 cm)
DEPTH	Shore and just below
HABITAT	Muddy sand

DISTRIBUTION Temperate shores of northeastern Atlantic, Mediterranean, and western Baltic

One of the most familiar sights on western European beaches is the neat, coiled casts of undigested sand deposited by lugworms. The worm itself is rarely seen, remaining hidden in its U-shaped tube beneath the surface of the sand. The entrance to the tube is marked by a shallow, saucer-shaped depression in the sand. The worm may be pink, red, brown, black, or green. The first six segments of its front section are thick with bristles, while the next thirteen segments have red, feathery gills. The rear third of the body is thin, with no gills or bristles.

Lugworms feed by eating sand, extracting organic matter from it, and expelling the waste. These fleshy worms are a favorite food of many wading birds and are also used by fishermen as bait. They are most abundant at midshore level in sediments containing reasonable amounts of organic matter.

CLASS POLYCHAETA

Green Paddle Worm
Eulalia viridis

LENGTH	Up to 6 in (15 cm)
DEPTH	Shore and shallows
HABITAT	Rocky areas under stones, in crevices

DISTRIBUTION Temperate coastal waters of northeastern Atlantic

Although this beautiful green worm is usually found crawling over rocks, it can also swim well. The name paddle worm comes from the large, leaf-shaped appendages called parapodia that are attached to the side of each body segment and aid in swimming. The head has two pairs of stout tentacles on each side, a single tentacle on top, and four short, forward-pointing tentacles at the front. These tentacles and two simple black eyes help the worm in its hunt for food. The green paddle worm is attracted to dead animals, especially mussels and barnacles, but will also hunt for live prey. However, unlike the king ragworm (opposite), it does not have jaws to tackle large prey. Instead, carrion and debris sticks to its proboscis and is wiped off inside the mouth.

During spring, the green paddle worm lays gelatinous green egg masses about the size of a marble on the shore and in shallow water, attaching them to seaweeds and rocks.

CLASS POLYCHAETA

Sea Mouse
Aphrodita aculeata

LENGTH	Up to 8 in (20 cm)
DEPTH	Shallow to moderate
HABITAT	Sand, muddy sand

DISTRIBUTION Temperate coastal waters of northeastern Atlantic and Mediterranean

The segmented structure of this pretty worm can be seen only if it is turned over, because its back is disguised by a thick felt of hairs that mask its segments. Running along each side of its body are numerous stiff, black bristles and a fringe of beautiful, iridescent hairs that glow green, blue, or yellow. The bristles can cause severe irritation if they puncture the skin. The sea mouse is so called because it looks like a bedraggled mouse when washed up dead on the seashore.

CLASS POLYCHAETA
King Ragworm
Alitta virens

LENGTH
Up to 20 in (50 cm)

DEPTH
Shore and shallows

HABITAT
Muddy sand

DISTRIBUTION Temperate coastal waters of northeastern and northwestern Atlantic

This large worm has strong jaws that are easily capable of delivering a painful bite to a human. The jaws are pushed out on an eversible proboscis and are used for pulling food into its mouth, as well as for defending itself. The king ragworm lives in a mucus-lined burrow in the sand, and waits for the tide to come in before coming out to feed. It swims well by bending its long body into a series of S-shaped curves. Fishermen collect it for bait.

CLASS POLYCHAETA
Magnificent Feather Duster
Sabellastarte magnifica

LENGTH
Up to 6 in (15 cm)

DEPTH
3–65 ft (1–20 m)

HABITAT
Coral reefs

DISTRIBUTION Shallow waters of the western Atlantic and Caribbean

crown of spines in three concentric rings

fingerlike gills on each body segment

CLASS POLYCHAETA
Honeycomb Worm
Sabellaria alveolata

LENGTH Up to 1½ in (4 cm)

DEPTH Shore and shallows

HABITAT Mixed rock and sand areas

DISTRIBUTION Intertidal areas of northeastern Atlantic and Mediterranean

Although honeycomb worms are tiny, the sand tubes they build may cover many yards of rock in rounded hummocks up to 20 in (50 cm) thick. The worms build their tubes close together, and the tube openings give the colony a honeycomb appearance. This worm's head is crowned by spines and it has numerous feathery feeding tentacles around the mouth, which it uses to trap plankton. The body ends in a thin, tubelike tail with no appendages.

The only part of this worm that is normally visible is a beautiful fan of feathery tentacles. The worm's segmented body is hidden inside a soft, flexible tube that it builds tucked beneath rocks or in a coral crevice or buried in sand. The tentacles are in two whorls and are usually banded brown and white. They are normally extended into the water to filter out plankton, but at the slightest vibration or disturbance, such as the exhalation of a scuba diver, the worm instantly retracts the tentacles down into the safety of the tube.

CLASS POLYCHAETA
Christmas Tree Worm
Spirobranchus giganteus

LENGTH Up to 1¼ in (3 cm)

DEPTH 0–100 ft (0–30 m) or more

HABITAT Living coral heads

DISTRIBUTION Shallow reef waters throughout the tropics

Many large coral heads in tropical waters are decorated with Christmas tree worms, which occur in a huge variety of colors. The worm lives in a calcareous tube buried in the coral and extends neat, twin spirals of feeding tentacles above the coral surface. If disturbed, the worm pulls back into its tube in a fraction of a second. For added safety, the worm can also plug its tube with a small plate called an operculum.

WORM REEFS

Honeycomb worms build their tubes by gluing together sand grains stirred up by waves. The glue is a mucus secreted by the worm, which uses a lobed lip around its mouth to fashion the tube. As new worms settle out from the plankton to build their own tubes, a reef develops and expands sideways and upward, provided there is a good supply of sand. These structures provide a home to many other species.

LIVE REEF
Live reefs will survive for many years provided new larvae settle and grow to replace wave-damaged areas and dead worms.

CLASS POLYCHAETA
Pompeii Worm
Alvinella pompejana

LENGTH Up to 4 in (10 cm)

DEPTH 6,500–10,000 ft (2,000–3,000 m)

HABITAT Hydrothermal vent chimneys

DISTRIBUTION Eastern Pacific

This extraordinary worm lives in thin tubes massed together on the sides of chimneys of deep-sea hydrothermal vents. The tubes are close to the chimneys' openings, where water from deep inside Earth pours out at temperatures of up to 660°F (350°C). The worms can survive a temperature of at least 108°F (42°C). At its head end, the Pompeii worm has a group of large gills and a mouth surrounded by tentacles. Each of the worm's body segments has appendages on the side called parapodia. The posterior parapodia have many hairlike outgrowths that carry a mass of chemosynthetic bacteria. The bacteria manufacture food that the worm absorbs, and the worm also eats some of the bacteria.

OCEAN LIFE

Mollusks

DOMAIN	Eucarya
KINGDOM	Animalia
PHYLUM	Mollusca
CLASSES	8
SPECIES	over 79,000

AMONG THE MOST SUCCESSFUL of all marine animals, mollusks display great diversity and a remarkable range of body forms, allowing them to live almost everywhere from the ocean depths to the splash zone. They include oysters, sea slugs, and octopuses. Most species have shells and are passive or slow-moving; some lack eyes. Others are intelligent, active hunters with complex nervous systems and large eyes. Filter-feeding mollusks, such as clams, are crucial to coastal ecosystems, as they provide food for other animals and improve water quality and clarity. Many mollusks are commercially important for food, pearls, and their shells.

Anatomy

Most mollusks have a head, a soft body mass, and a muscular foot. The foot is formed from the lower body surface and helps it move. Mollusks have what is called a hydrostatic skeleton—their bodies are supported by internal fluid pressure rather than a hard skeleton. All mollusks have a mantle, a body layer that covers the upper body and may or may not secrete a shell. The shell of bivalves (clams and relatives) has two halves joined by a hinge; these can be held closed by powerful muscles while the tide is out, or if danger threatens. Mollusks other than bivalves have a rasping mouthpart, or radula, which is unique to mollusks. Cephalopods (octopuses, squid, and cuttlefish) also have beaklike jaws, as well as tentacles, but most lack a shell, while most gastropods (slugs and snails) have a single shell. This is usually a spiral in snails, but can be cone-shaped in other forms, such as limpets.

REEF-DWELLING GOLIATH
The tropical giant clam is the largest bivalve and may measure more than 3 ft (1 m) across and weigh over 440 lb (220 kg).

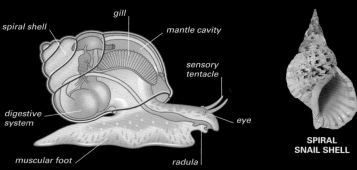

gill
spiral shell
mantle cavity
sensory tentacle
digestive system
eye
muscular foot
radula

GASTROPOD ANATOMY
The body plan (far left) of gastropods (slugs and snails) features a head, large foot, and usually a spiral shell (left). In shelled forms, all the soft body parts can be withdrawn into the shell for protection, or to conserve moisture while uncovered by the outgoing tide.

SPIRAL SNAIL SHELL

BIVALVE ANATOMY
Bivalves are housed within a shell of two halves (right) from which the siphons and muscular foot can be extended. The shell is opened and closed by the adductor muscles, labeled in the body plan (far right).

BIVALVE SHELL

hinge ligament
shell
mantle cavity
digestive system
muscular foot
siphon
gill
adductor muscle

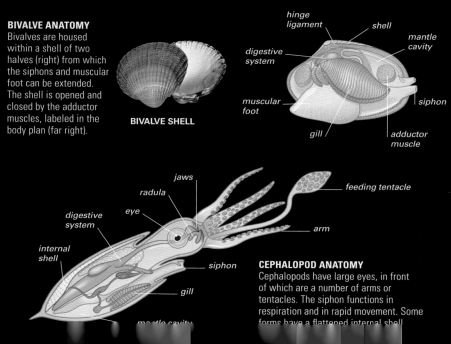

jaws
radula
feeding tentacle
digestive system
eye
arm
internal shell
siphon
gill

CEPHALOPOD ANATOMY
Cephalopods have large eyes, in front of which are a number of arms or tentacles. The siphon functions in respiration and in rapid movement. Some forms have a flattened internal shell

mantle cavity

Sense Organs

Touch, smell, taste, and vision are well developed in many mollusks. The nervous system has several paired bundles of nervous tissue (ganglia), some of which operate the foot, and interpret sensory information such as light intensity. Photoreceptors range from the simple eyes (ocelli) seen along the edges of the mantle or on bivalve siphons, to the sophisticated image-forming eyes of cephalopods. Cephalopods are also capable of rapidly changing their color.

PIGMENTED SKIN CELLS HELP CUTTLEFISH TO CHANGE COLOR

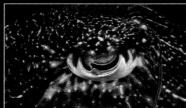

1 The giant cuttlefish's color change is due to skin cells called chromatophores. It is pale when pigment is confined to a small area of each cell.

2 When the cuttlefish passes over a darker background, it disperses the colored pigments throughout each of its chromatophores, and the animal darkens.

MOLLUSCAN BEAUTY
Displaying fabulous warning colors, this nudibranch is a shell-less example of the many thousands of marine species of gastropods (slugs and snails).

GRAFTING OYSTERS

Pearls form in oysters when a grain of sand or other irritant lodges in their shells. The oyster coats the grain with a substance called nacre, forming a pearl. Today many pearls are cultured artificially: the shell is opened just enough to introduce an irritant into the mantle cavity.

SEEDING AN OYSTER
The best-shaped artificial pearls are produced by "seeding" oysters with a tiny pearl bead and a piece of mantle tissue from another mollusk .

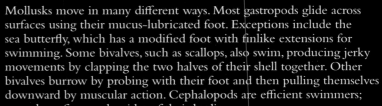

Movement

Mollusks move in many different ways. Most gastropods glide across surfaces using their mucus-lubricated foot. Exceptions include the sea butterfly, which has a modified foot with finlike extensions for swimming. Some bivalves, such as scallops, also swim, producing jerky movements by clapping the two halves of their shell together. Other bivalves burrow by probing with their foot and then pulling themselves downward by muscular action. Cephalopods are efficient swimmers; some have fins on the sides of their bodies that let them hover in the water, and they can accelerate rapidly by squirting water out through their siphons.

REDUCING DRAG
Swimming backward reduces drag from the tentacles. The siphon, used for jet propulsion, is clearly visible in this Humboldt squid.

siphon

AIDED BY MUCUS
Muscular contractions ripple through the fleshy foot of this marine snail. It secretes a lubricating mucus that helps it to move on rough surfaces.

Respiration

Most mollusks obtain oxygen from water using gills, called ctenidia, which are situated in the mantle cavity. These are delicate structures with an extensive capillary network and a large surface area for gaseous exchange. In species that are always submerged, water can continually be drawn in and over the gills. Those living in the intertidal zone are exposed to the air for short periods and must keep their gills moist. At low tide, bivalves clamp shut and some gastropods close their shell with a "door" (called an operculum) to retain moisture. Pulmonate snails have a simple lung formed from the mantle cavity instead of ctenidia and are mostly terrestrial but others live on the seashore and can absorb oxygen through their skin when immersed. The respiratory pigment in most molluscan blood is a copper compound called hemocyanin. It is not as efficient at taking up oxygen as hemoglobin and gives mollusks' blood a blue color.

external gills
(ctenidia)

COLOR CODING
Nudibranchs (sea slugs) have feathery external gills toward the rear of their bodies. The warning coloration of this species includes the bright orange gills.

Feeding

The ways in which mollusks feed are almost as varied as their anatomy. Sedentary mollusks, such as many bivalves including clams and oysters, create water currents through tubular outgrowths of their mantle (siphons). They filter food from the moving water with their mucus-covered gills. Suitably sized particles are then selected and passed to the mouth by bristly flaps called palps. Sea slugs, chitons, and many sea snails graze algae from hard surfaces using their rasplike radula. Radulae have toothlike structures called denticles, many of which are reinforced with an iron deposit for durability. Larger mollusks feed on crustaceans, worms, fish, and other mollusks, which they locate either by scent or, in the case of some cephalopods such as octopuses, by sight. Cephalopods use their suckered arms or the tips of their feeding tentacles to capture prey and their parrotlike beak to crush and dismember it. Some squid even appear to hunt in packs and swim in formation over reefs looking for prey.

SPECIES-SPECIFIC DENTICLES
The denticles on a mollusk's radula are often species-specific. This electron micrograph shows the distinctive radula of the gastropod *Sinezona rimuloides*.

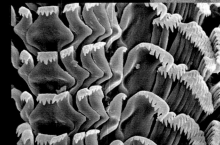

FEEDING TRAIL
Limpets continually graze the same area as the algae and bacterial film on which they feed regrow rapidly. The abrasive radula of the limpet scrapes a trail on the rock surface, as shown above.

Reproduction

In many mollusks, reproduction simply involves releasing sperm or eggs (gametes) into the water. Fertilization is external and there is no parental care. Individuals may be of separate sexes or hermaphrodites (having both male and female reproductive organs). Hermaphrodites may function as male or female at different times or, as in nudibranchs, produce both eggs and sperm, although eggs can be fertilized only by cross-fertilization. Some species, such as slipper limpets, change sex with age, while oysters can change sex several times in a breeding season. Among cephalopods, males court females, fertilization is internal, and in some species, the eggs are protected by the females until they hatch.

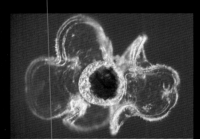

LIMPET CHAIN
Slipper limpets change from male to female as they grow. This chain of four such limpets has a female at the bottom and smaller males above her.

DEVELOPING EMBRYOS
In 4 months, Australian giant cuttlefish eggs develop into mini-replicas of the adults.

OYSTER DEMAND

Oysters have long been harvested as a food source. Their high market value and increasing demand have led to overexploitation of wild stocks. In the North Sea, the European flat oyster has vanished from much of its former range, and today most oysters are commercially farmed.

SLOW RECOVERY PERIOD
Relatively long-lived and reproducing only sporadically, the European flat oyster (right) takes a long time to recover from overexploitation.

READY AND WAITING FOR PREY
This cuttlefish hovers with its arms outstretched. When prey comes within reach, the two feeding tentacles, currently contracted and set above the two lower arms, will shoot forward to grab the prey.

Lifecycles

Most mollusks produce eggs that either float or are deposited in clusters, anchored to the substrate. Most forms have eggs that hatch into shell-less larvae, which live in the plankton. The larvae are called ciliated trochophores due to their bands of hairlike cilia, used in swimming. In gastropods, bivalves, and scaphopods, the trochophore larvae change into veliger larvae, which have larger ciliated bands, and sometimes adult features such as a mantle or a rudimentary shell or both. As they approach maturity, the larvae float down from the surface and, on reaching the seabed, change into adults. Only those that land in a suitable environment survive to reach sexual maturity. Cephalopod eggs hatch into active predators. Some resemble mini-adults; others live in the plankton and initially look and behave differently from the adults.

PLANKTONIC LARVA
The visible bands of this veliger larva of the common limpet beat with tiny hairlike cilia, which are used in locomotion and feeding.

SECURING EGG CLUSTERS
This female bigfin reef squid produces up to 400 egg capsules containing about 2,500 eggs. Here, she is securing egg capsules to a solid substrate.

each finger-shaped egg capsule holds up to seven eggs

MOLLUSK CLASSIFICATION

The phylum Mollusca is the second largest animal phylum, comprising more than 73,000 species, and their diverse form has led to the identification of eight different classes. The majority of species live in marine habitats, but freshwater and terrestrial species are also numerous.

CAUDOFOVEATES
Class Caudofoveata

131 species
These are marine, shell-less, wormlike organisms of deep-water sediments. Their horny outer layer is covered with spines.

SOLENOGASTERS
Class Solenogaster

273 species
Another marine class of shell-less, wormlike organisms, solenogasters live in or on the ocean floor. Some lack a radula.

MONOPLACOPHORANS
Class Monoplacophora

About 30 species
These deep-sea mollusks lack eyes but have a radula and a conelike shell. They are more abundant as fossils than as living species.

TUSK SHELLS
Class Scaphopoda

576 species
These animals have a tubular, tapering shell, open at both ends. The head and foot project from the wider end and dig in soft sediments.

BIVALVES
Class Bivalvia

9,733 marine species
Bivalves, or clams and their relatives, have a hinged shell of two halves, but no radula. Most are sedentary and marine. Siphons create a water current through the shell, aiding feeding and respiration. Sexes are usually separate.

GASTROPODS
Class Gastropoda

About 67,000 species
Familiar as slugs and snails, these mollusks are marine, freshwater, and terrestrial. They have a spiral shell and a large, muscular foot. The body is twisted 180° so the mantle cavity lies over the head. Many species can retract into their shell; hermaphrodite species are common.

CEPHALOPODS
Class Cephalopoda

822 species
Squid, octopuses, and cuttlefish are all cephalopods—fast-moving and intelligent, with a complex nervous system and large eyes. The shell is internal or absent in all but the nautiloids. The central mouth has a parrotlike beak and a radula. The sexes are separate.

CHITONS
Class Polyplacophora

1,026 species
Chitons have a repeating structure with a series of plates (usually 8) on their backs enclosed by an extension of the mantle. The underside is dominated by the foot.

CLASS BIVALVIA
Common Mussel
Mytilus edulis

LENGTH	4–6 in (10–15 cm)
HABITAT	Intertidal zones, coasts, estuaries

DISTRIBUTION North and southeastern Atlantic, northeastern and southwestern Pacific

Also called the blue mussel, this edible, black-shelled bivalve attaches itself in large numbers to various substrates using tough fibers called byssus threads. These fibers are extremely strong and prevent the mussels from being washed away. The fibers increase in strength in fall, possibly to cope with storms. When the mussel opens its shell, water is drawn in over the gills, or ctenidia, which absorb oxygen into the tissues and also filter food particles out of the water.

Common mussels are very efficient filter feeders—they process about 10–15 gallons (45–70 liters) of water per day and consume almost everything they trap. The sexes are separate and so grouping together in "beds" helps to ensure that their eggs are fertilized. After hatching, the planktonic larvae are dispersed by the ocean currents. After some months the larvae settle, attach, and metamorphose, but can resorb their byssus and move to a better spot.

CLASS BIVALVIA
Great Scallop
Pecten maximus

WIDTH	Up to 7½ in (17 cm)
HABITAT	Sandy seabeds, at 16–500 ft (5–150 m), commonly 33 ft (10 m)

DISTRIBUTION Northeastern Atlantic

Also known as the king scallop, the great scallop is usually found partly buried in sand. It is one of the few bivalves capable of rapid movement through water, which it achieves using a form of jet propulsion. It claps the two halves of its shell together, which pushes water out of the mantle cavity close to the hinge. It moves forward with its shell gape first, producing jerky movements as it takes successive "claps" of water. These movements are a useful strategy to escape from predators. These edible bivalves are now farmed to meet growing demand.

CLASS BIVALVIA
Black-lip Pearl Oyster
Pinctada margaritifera

LENGTH	Up to 12 in (30 cm) diameter
HABITAT	Hard substrata of inter- and subtidal zones; reefs

DISTRIBUTION Gulf of Mexico, western and eastern Indian Ocean, western Pacific

CLASS BIVALVIA
Atlantic Thorny Oyster
Spondylus americanus

LENGTH	Up to 4½ in (11 cm)
HABITAT	Rocks to a depth of 460 ft (140 m)

DISTRIBUTION Southeast coast of USA, Bahamas, Gulf of Mexico, Caribbean

The Atlantic thorny oyster's spiny shell protects it from predators. The oyster pictured here is covered with an encrusting red sponge, which provides camouflage. This species is unusual in having a ball-and-socket type hinge joining the two halves of its shell,

Black-lip pearl oysters begin life as a male before changing into a female two or three years later. Females produce millions of eggs, which are fertilized randomly and externally by the males' sperm, before hatching into free-swimming larvae. The mobile larvae pass through various larval stages for about a month before eventually settling on the seafloor, after metamorphosing into the sessile (immobile) adult form. This species is famous and much sought-after because it occasionally produces prized black pearls.

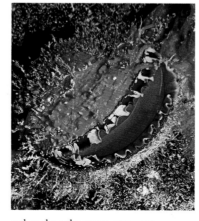

rather than the more common toothed hinge seen in many other bivalves. The Atlantic thorny oyster cements itself directly to rocks rather than using byssal threads.

CLASS BIVALVIA
Shipworm
Teredo navalis

LENGTH	Up to 23 in (58 cm)
HABITAT	Wood burrows in high-salinity seas and estuaries

DISTRIBUTION Coastal waters off North, Central, and South America, and Europe

Despite its wormlike appearance, the shipworm is a type of clam that has become elongated as an adaptation to its burrowing lifestyle. Its bivalve shell, situated at the anterior end, is very small and ridged. The shipworm uses it with a rocking motion to bore into wooden objects. Outside the shell, its body is unprotected, except for a calcareous tube it secretes to line the burrow. These worms damage wooden structures, such as piers, irreparably and in the past caused many ships to sink. The burrow entrance is only about the size of a pinhead, but the burrow itself may be over ½ in (1 cm) wide, so the extent of an infestation is often underestimated until it is too late.

Shipworms change from male to female during their lifetime, and the female form produces many eggs, from which free-swimming larvae hatch. When they mature and settle on a suitable piece of wood, the larvae quickly metamorphose into the adult form and start burrowing.

CLASS BIVALVIA

Common Piddock

Pholas dactylus

LENGTH
Up to 6 in (15 cm) across

HABITAT
Lower shore to shallow sublittoral

DISTRIBUTION South and east coasts of UK, Severn estuary in UK, west coast of France, Mediterranean

This mollusk has a pronounced "beak" covered in toothlike projections at the front end of its shell. It uses this feature for boring holes into relatively soft substrates, such as mud, chalk, peat, and shale. Like the shipworm (opposite), this piddock relies on its burrows for protection from predation, because the shell does not encase all of its body—its two fused siphons (tubes for eating, breathing, and excretion) trail out behind it. The shell is fragile, elliptical, and covered in a pattern of concentric ridges and radiating lines. If disturbed, the common piddock has an unusual defense strategy: it squirts a luminous blue secretion from its outgoing, or exhalant, siphon. Such bioluminescence is very rare in mollusks, seen in only a few species.

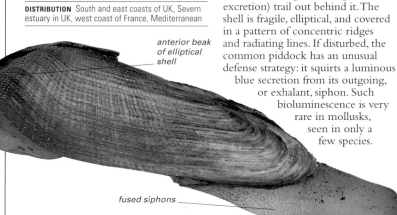

anterior beak of elliptical shell

fused siphons

CLASS BIVALVIA

Common Edible Cockle

Cerastoderma edule

LENGTH
Up to 2 in (5 cm)

HABITAT
Middle and lower shore, 2 in (5 cm) below surface of sand or mud

DISTRIBUTION Barents Sea, eastern north Atlantic from Norway to Senegal, West Africa

This edible bivalve has a robust, ribbed shell and burrows in dense populations just below the surface of sand or mud, filtering organic matter such as plankton from the water. Cockles are an important commercial species but also a vital food source for wading birds, such as oystercatchers.

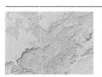

CLASS BIVALVIA

Razor Shell

Ensis directus

LENGTH
Up to 10 in (25 cm)

HABITAT
Sandy and muddy shores and shallows

DISTRIBUTION Atlantic coast of North America, introduced to North Sea

So-called because their shells resemble a straight razor, razor shells live in deep, vertical burrows on muddy and soft, sandy shores. They are native to the northeast coast of North America, and the free-swimming larval stage is thought to have been introduced to the North Sea in 1978, when a ship emptied its ballast tanks outside the German port of Hamburg. This clam has subsequently spread along the continental coast. In places, it affects local polychaete worm populations, but it is not considered a pest.

CLASS BIVALVIA

Giant Clam

Tridacna gigas

LENGTH
Up to 5 ft (1.5 m)

HABITAT
Reefs, reef flats and shallow lagoons to 65 ft (20 m)

DISTRIBUTION Tropical Indo-Pacific from south China seas to northern coasts of Australia, and Nicobar Islands in the west to Fiji in the east

The largest and heaviest of all mollusks is the giant clam. Like other bivalves, it feeds by filtering small food particles from the water using its ingoing, or inhalant, siphon, which is fringed with small tentacles. However, it differs in obtaining most of its nourishment from zooxanthellae (unicellular algae that live within its tissues)—a type of relationship also associated with coral polyps. The algae have a constant and safe environment in which to live, and the clams obtain carbon-based products (sugars) made by the algae during photosynthesis. In fact, so dependent is the giant clam on these algae that it will die without them.

The adult is sessile (immobile) and its inhalant and exhalant (outgoing) siphons are the only openings in its mantle. Although the scalloped edges of their shell halves are mirror images of one another, larger individuals may be unable to close their shells fully, so their brightly colored mantle and siphons remain constantly exposed.

Many giant clams appear iridescent due to an almost continuous covering of purple and blue spots on their mantles, while others look more green or gold, but all have a number of clear spots, or "windows," that allow sunlight to filter into the mantle cavity. Fertilization is external and the eggs hatch into free-swimming larvae before settling onto the seabed. The exhalant siphon expels water and at spawning time provides an exit point for the eggs or sperm.

SPAWNING

Reproduction in giant clams is triggered by chemical signals that synchronize the release of sperm and eggs into the water. Giant clams start life as males and later become hermaphroditic, but during any one spawning event, they release either sperm or eggs in order to avoid self-fertilization. A large clam can release as many as 50 million eggs in 20 minutes.

GIANT CLAM
The giant clam grows to its huge size with the help of photosynthetic microorganisms (zooxanthellae) living in its colorful mantle tissue, which share the sugars they produce with the clam. Collected for its meat and shell, it is now rare throughout its range, and international trade is restricted.

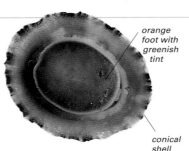

orange foot with greenish tint

conical shell

MUSCULAR FOOT
The common limpet's muscular foot, seen here from below, holds it firmly to its rock, regardless of the strength of the waves.

CLASS GASTROPODA
Common Limpet
Patella vulgata

DIAMETER
2½ in (6 cm)

HABITAT
Rocks on high shore to sublittoral zone

DISTRIBUTION Northeastern Atlantic from Arctic Circle to Portugal

Abundant on rocks from the high to the low water mark, the common limpet is superbly adapted to shore life. A conical shell protects it from predators and the elements. Limpets living at the low water mark are buffeted by the waves and so require smaller, flatter shells than those living at the high water mark, where wider, taller shells allow for better water retention during periods of exposure. Limpets travel slowly during low tide covering up to 5 ft (1.5 m) when feeding, using contractions of their single foot. They graze on algae from rocks using a radula (a rasplike structure), which has teeth reinforced with iron minerals.

RETURNING HOME
Limpets gradually grind a "scar" into their anchor spot on the rock, to aid their grip and help retain water. A mucus trail leads them back to the spot.

CLASS GASTROPODA
Zebra Nerite
Puperita pupa

LENGTH
Up to ½ in (1.3 cm)

HABITAT
Rocky tide pools

DISTRIBUTION Caribbean, Bahamas, Florida

The small, rounded, smooth, black-and-white striped shell of the zebra nerite is typical of the species, but in examples from Florida the shell is sometimes more mottled or speckled with black. These gastropods are most active during the day, when they feed on microorganisms such as diatoms and cyanobacteria, but if they become too hot or they are exposed at low tide, they cluster together, withdraw into their shells, and become inactive. This may be a mechanism for preventing excessive water loss.

Unusually for gastropods, there are separate males and females of zebra nerites and fertilization of the eggs occurs internally. The males use their penis to deposit sperm into a special storage organ inside the female. Later, she lays a series of small white eggs that hatch into planktonic larvae.

CLASS GASTROPODA
Dog Whelk
Nucella lapillus

LENGTH
Up to 2 in (5 cm)

HABITAT
Middle and lower rocky shores

DISTRIBUTION Northwestern and northeastern Atlantic

One of the most common rocky shore gastropods, the dog whelk has a thick, heavy, sharply pointed spiral shell. The shell's exact shape depends on its exposure to wave action, and its color depends on diet. Dog whelks are voracious predators, feeding mainly on barnacles and mussels. Once the prey has been located, the whelk uses its radula to bore a hole in the shell of its prey before sucking out the flesh.

CLASS GASTROPODA
Top Shell
Tectus niloticus

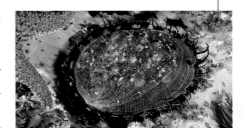

LENGTH
6 in (16 cm)

HABITAT
Intertidal and shallow subtidal areas, reef flats to 23 ft (7 m)

DISTRIBUTION Eastern Indian Ocean, western and southern Pacific

Easily distinguished from most other gastropods by the conical shape of its spiral shell, the top shell moves slowly over reef flats and coral rubble, feeding on algae. Demand for its flesh and pretty shell has led to declining numbers, especially in the Philippines, due to unregulated harvesting. It has, however, been successfully introduced elsewhere in the Indo-Pacific, such as French Polynesia and the Cook Islands, from where some original sites are being restocked.

CLASS GASTROPODA
Red Abalone
Haliotis rufescens

LENGTH
Up to 10 in (26 cm)

HABITAT
Rocks from low tide mark to 100 ft (30 m)

DISTRIBUTION East Pacific coasts from southern Oregon, US to Baja California, Mexico

The largest of the abalone species, the red abalone is so called because of the brick-red color of its thick, roughly oval shell. There is an arc of

three to five clearly visible holes in the shell, through which water flows for respiration. These are filled and replaced with new holes as the abalone increases in size. Sea otters are one of the red abalone's main predators, along with human divers.

CLASS GASTROPODA
Venus Comb
Murex pecten

LENGTH
Up to 6 in (15 cm)

HABITAT
Tropical warm waters to 650 ft (200 m)

DISTRIBUTION Eastern Indian Ocean and western Pacific

The tropical carnivorous snail known as the Venus comb has a unique and spectacular shell. There are rows of long, thin spines along its longitudinal ridges, which continue onto the narrow, rodlike, and very elongated siphon canal. The exact function of these spines is unknown, but they are thought to be either for protection or to prevent the snail from sinking into the soft substrate on which it lives. Its body is tall and columnar so that it can lift its cumbersome shell above the sediment to move in search of food.

HIDING FROM VIEW

There are times when the Venus comb buries itself just below the surface of the sea floor, displacing the sand with movements of its muscular foot. However, it leaves the opening of its tubular inhalant siphon above the sand's surface so that it can draw water into its mantle cavity to obtain oxygen and to "taste" the water for the presence of prey.

HALF BURIED
The spines of this Venus comb can be seen sticking out of the sand. The siphon is visible to the right of the picture.

CLASS GASTROPODA

Tiger Cowrie
Cypraea tigris

LENGTH	Up to 6 in (15 cm)
HABITAT	Low tide to 100 ft (30 m) on coral reefs and flats

DISTRIBUTION Indian Ocean, western Pacific

One of the largest cowrie species, the tiger cowrie has a shiny, smooth, domed shell with a long, narrow aperture, and is variously mottled in black, brown, cream, and orange. The cowrie's mantle (its body's outer, enclosing layer) can extend to cover parts of the exterior of the shell. These extensions have numerous projections, or papillae, whose exact function is unknown, but which may increase the surface area for oxygen absorption or provide camouflage of some sort. Tiger cowries are nocturnal creatures, hiding in crevices among the coral during the day and emerging at night to graze on algae. The sexes are separate and fertilization occurs internally. Females exhibit some parental care in that they protect their egg capsules by covering them with their muscular foot until they hatch into larvae, which then enter the plankton to mature.

CLASS GASTROPODA

Giant Triton
Charonia tritonis

LENGTH	Up to 20 in (50 cm)
HABITAT	Coral reefs, mostly in subtidal zones

DISTRIBUTION Indian Ocean, western and central Pacific

This gastropod is one of the very few animals that eats the crown-of-thorns starfish, itself a voracious predator and destroyer of coral reefs. The giant triton is an active hunter that will chase prey, such as starfish, mollusks, and sea stars, once it has detected them. It uses its muscular single foot to hold its victim down while it cuts through any protective covering using its serrated, tonguelike radula; it then releases paralyzing saliva into the body before eating the subdued prey.

CLASS GASTROPODA

Common Periwinkle
Littorina littorea

LENGTH	Up to 1 in (3 cm)
HABITAT	Upper shore to sublittoral rocky shores, mud flats, estuaries

DISTRIBUTION Coastal waters of northwest Europe; introduced to North America

The common periwinkle has a black to dark gray, sharply conical shell and slightly flattened tentacles, which in juveniles also have conspicuous black banding. The sexes are separate and fertilization occurs internally. Females release egg capsules, containing two or three eggs, directly into the water during the spring tides. The eggs hatch into free-swimming larvae that float in the plankton for up to six weeks. After settling and metamorphosing into the adult form, it takes a further two to three years for the adult to fully mature. It feeds mainly on algae, which it rasps from the rocks. In the 19th century, the common periwinkle was accidentally introduced into North America, where its selective grazing of fast-growing algal species has considerably affected the ecology of some rocky shores.

CLASS GASTROPODA

Flamingo Tongue
Cyphoma gibbosum

LENGTH	1–1½ in (3–4 cm)
HABITAT	Coral reefs at about 50 ft (15 m)

DISTRIBUTION Western Atlantic, from North Carolina to Brazil; Gulf of Mexico, Caribbean Sea

The off-white shell of the flamingo tongue cowrie is usually almost completely hidden by the two fleshy, leopard-spotted extensions of its body's outer casing, or mantle. When threatened, however, its distinctive coloration quickly disappears as it withdraws all its soft body parts into its shell for protection. This snail feeds almost exclusively on gorgonian corals, which dominate Caribbean reef communities. Although these corals release chemical defenses to repulse predators, the flamingo tongue cowrie is apparently able to degrade these bioactive compounds and eat the corals without coming to any harm. After mating, the female strips part of a soft coral branch and deposits the egg capsules on it. Each egg capsule contains embryos that will hatch into free-swimming planktonic larvae.

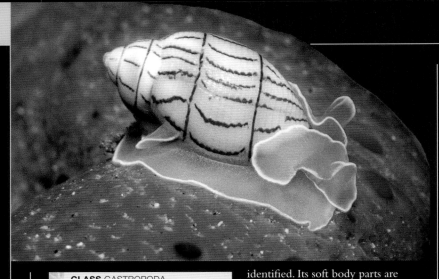

CLASS GASTROPODA

Bubble Shell

Bullina lineata

LENGTH	1 in (2.5 cm)
HABITAT	Sand, reefs to 65 ft (20 m), mainly intertidal; subtidal at range limits

DISTRIBUTION Tropical and subtropical waters of Indian Ocean and west Pacific

The pale spiral shell of the bubble shell (also known as the red-lined bubble shell) has a distinctive pattern of pinkish red lines by which it can be identified. Its soft body parts are delicate and translucent with a fluorescent blue margin and, in form, reminiscent of the Spanish dancer (opposite), which is a close relative that has lost its shell. If threatened, the bubble shell quickly withdraws into its shell and at the same time regurgitates food, possibly as a defense mechanism to distract predators. The bubble shell is itself a voracious predator, feeding on sedentary polychaete worms. This mollusk is hermaphroditic and produces characteristic spiral white egg masses.

CLASS GASTROPODA

Three-tooth Cavoline

Cavolinia tridentata

LENGTH	½ in (1 cm)
HABITAT	330–6,500 ft (100–2,000 m); carried in ocean currents

DISTRIBUTION Warm oceanic waters worldwide

This species of sea butterfly has a small, almost transparent, spherical shell with three distinctive, posterior projections. The shell also has two slits through which large extensions of the mantle pass. These brownish "wings" are ciliated and so can create weak water currents, as well as aid buoyancy. Sea butterflies are unusual among shelled mollusks in that they can live in open water. Like other members of this group, the three-tooth cavoline produces a mucus web very much larger than itself, which traps planktonic organisms, such as diatoms and the larvae of other species. It eats the web and the trapped food at intervals, then produces a new one. During their lifetime, sea butterflies change first from males into hermaphrodites and then into females.

CLASS GASTROPODA

Sea Hare

Aplysia punctata

LENGTH	Up to 8 in (20 cm)
HABITAT	Shallow water

DISTRIBUTION Northeast Atlantic and parts of the Mediterranean

The sea hare, a type of sea slug, has tentacles reminiscent of a hare's ears. It has an internal shell about 1½ in (4 cm) long that is visible only through a dorsal opening in the mantle. If disturbed, it releases purple or white ink. It is not known if this response is a defense mechanism.

CLASS GASTROPODA

Polybranchid

Cyerce nigricans

LENGTH	Up to 1½ in (4 cm)
HABITAT	Reefs

DISTRIBUTION Western Indian Ocean, western and central Pacific

This colorful sea slug is a herbivore that browses on algae. It has no need of camouflage or a protective shell, as it has two excellent alternative defense strategies. First, it can secrete distasteful mucus, by utilizing substances in the algae it feeds on and secreting them from small microscopic glands over the body. Second, its body is covered with petal-like outgrowths called cerata, spotted and striped above and spotted below, that can be shed if it is attacked by a predator, in the same way as a lizard sheds its tail. This ability to cast off body parts to distract predators is called autotomy.

The cerata are also used in respiration, their large collective surface area allowing efficient gas exchange with the surrounding water. The head carries two pairs of sensory organs— the oral tentacles near the mouth and, further back, the olfactory organs (rhinophores). These are retractile and subdivide as the polybranchid matures. They are used to assist in finding food and mates. There is some debate as to whether this sea slug is a separate species or is simply a color variation of a similar mollusk, *Cyerce nigra*,

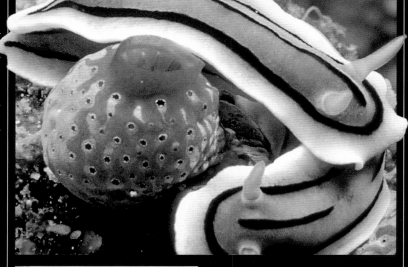

CLASS GASTROPODA

Chromodorid Sea Slug

Chromodoris lochi

LENGTH	1½ in (4 cm)
HABITAT	Reefs

DISTRIBUTION Tropical and subtropical western and central Pacific

Protected from predators by its bright warning coloration and unpleasant taste, the chromodorid sea slug forages in the open, rather than hiding away in cracks and crevices. Since it cannot swim, it glides over the tropical reefs on which it lives on its muscular foot, secreting a mucus trail much as terrestrial slugs do. The different species of the genus *Chromodoris* are distinguished by the pattern of black lines on their backs and the plain color of their gills and rhinophores (a pair of olfactory organs at the head end).

The two chromodorid sea slugs pictured here are possibly about to mate. As they are hermaphrodites, they can both produce sperm and eggs, but the way in which they mate decides which will be male. They each move their penis toward the other. The first to penetrate the other's body becomes the dominant male.

CLASS GASTROPODA

Hermissenda Sea Slug

Hermissenda crassicornis

LENGTH	Up to 3 in (8 cm)
HABITAT	Mud flats, rocky shores

DISTRIBUTION Northwest and northeast Pacific

This sea slug, usually known simply as Hermissenda, has an unusual way of deterring predators. It separates the stinging cells from any organism it eats and stores them in the orange-red tips of petal-like tentacles, or cerata, that cover its back. Any creature that touches the cerata is stung. Unlikely though it seems, Hermissenda is used extensively by scientists conducting memory experiments. The animal has an excellent sense of smell that enables it to find its way around mazes to locate food, and it can be "taught" to respond to simple stimuli.

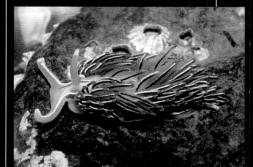

Spanish Dancer

Hexabranchus sanguineus

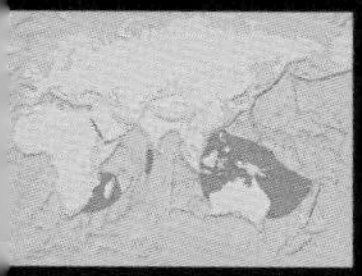

LENGTH
Up to 24 in (60 cm)

HABITAT
Shallow water on coasts
and reefs

DISTRIBUTION Parts of tropical Indian Ocean,
west Pacific

The largest of the nudibranchs is the
Spanish dancer—so called because
when it swims, the undulating
movements of its flattened body
are reminiscent of a flamenco
dancer. Adults are brightly
but variably colored,
generally in shades of red,
pink, or orange,
sometimes mixed
with white or
yellow. While
resting, crawling,
or feeding, the
lateral edges of its mantle
are folded up over its back,
displaying the less colorful
underside. If disturbed, it will
escape by swimming away,
exposing its bright colors and possibly
startling potential predators. Spanish
dancers are specialist predators that
feed only on sponges, particularly
encrusting species, from which they
modify and concentrate certain
distasteful compounds in their skin to
use as another defense against
predation. They have external gills for
respiration, which are extensively
branched and attached to the body
wall in distinct pockets and which
cannot be retracted. Like all
nudibranchs, the Spanish dancer
is hermaphroditic, but it requires
a partner in order to reproduce.

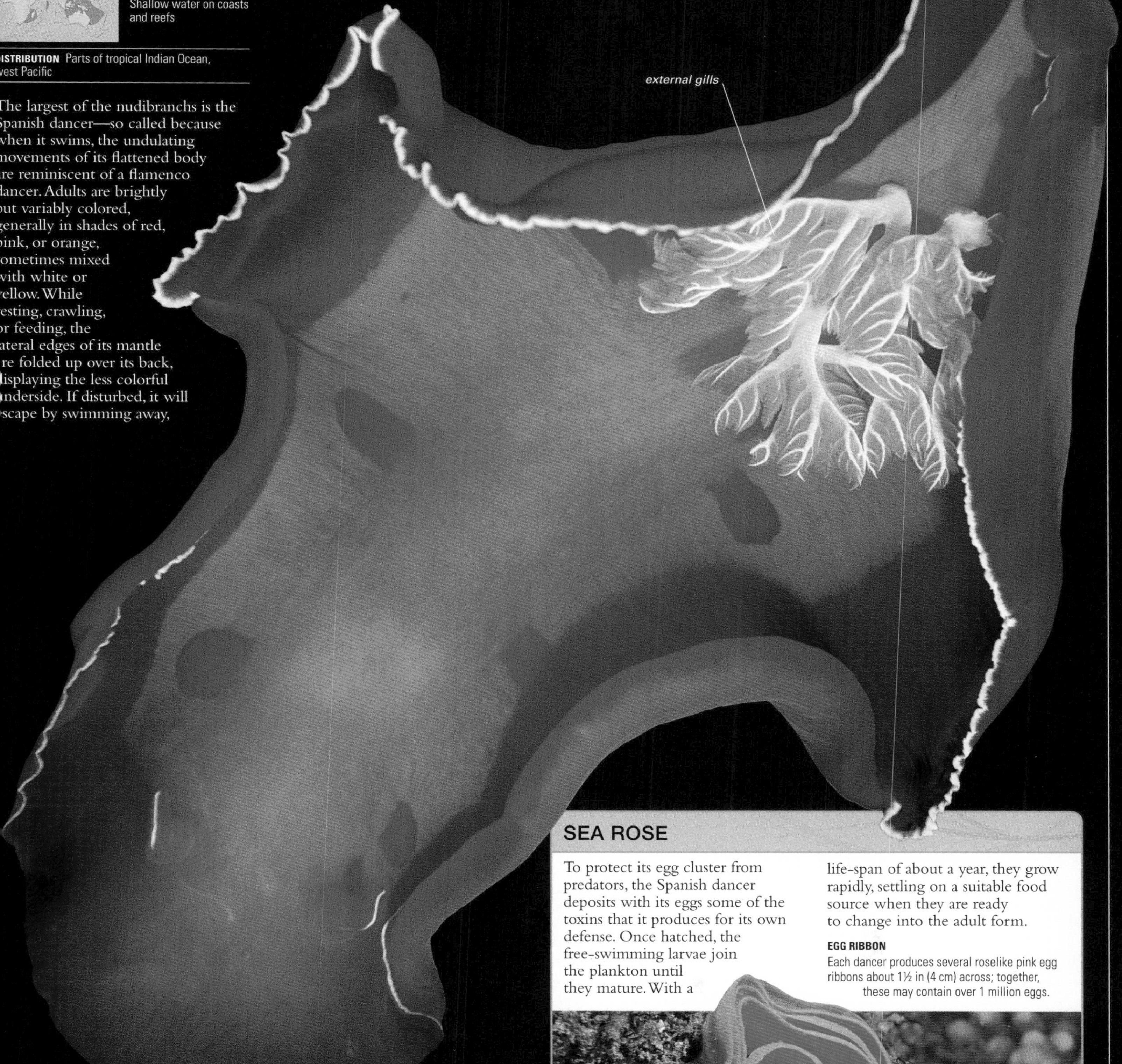

external gills

bright coloration

SEA ROSE

To protect its egg cluster from
predators, the Spanish dancer
deposits with its eggs some of the
toxins that it produces for its own
defense. Once hatched, the
free-swimming larvae join
the plankton until
they mature. With a
life-span of about a year, they grow
rapidly, settling on a suitable food
source when they are ready
to change into the adult form.

EGG RIBBON
Each dancer produces several roselike pink egg
ribbons about 1½ in (4 cm) across; together,
these may contain over 1 million eggs.

CLASS CEPHALOPODA

Nautilus

Nautilus pompilius

WIDTH
Shell up to 9 in (22 cm)

HABITAT
Tropical open waters to
1,600 ft (500 m)

DISTRIBUTION Eastern Indian Ocean, western Pacific, and Australia to New Caledonia

Nautilus and *allonautilus* are the only cephalopods with an external shell that protects them from predation as they swim in the open waters of the Indo-Pacific. Gas trapped in the shell's inner chambers provides buoyancy.

The head protrudes from the shell and has up to 90 suckerless tentacles, which are used to capture prey such as shrimp and other crustaceans; the head also features a pair of rudimentary eyes that lack a lens and work on a principle similar to a pinhole camera. The nautilus swims using jet propulsion, drawing water into its mantle cavity and expelling it forcefully through a tubular siphon, which can be directed to propel the nautilus forward, backward, or sideways. Unlike most other cephalopods, nautiluses mature late, at about ten years of age, and produce only about twelve eggs per year. They are often referred to as "living fossils," as they appear little changed from the fossil nautiloids that lived in prehistoric oceans over 200 million years ago.

CLASS CEPHALOPODA

Dumbo Octopus

Grimpoteuthis sp.

LENGTH
Up to 11 in (28 cm)

HABITAT
Deep water, to 6,500 ft
(2,000 m)

DISTRIBUTION Northwest Atlantic

Little is known about the Dumbo octopus, as only a few have been recorded. Its common name derives from a pair of unusual, earlike flaps extending from the mantle above its eyes. It has a soft body, an adaptation to its deep-water habitat, and eight arms connected to each other almost to their tips by "webbing." It is a detritus feeder, so it eats organic debris.

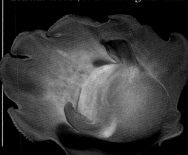

CLASS CEPHALOPODA

Southern Blue-ringed Octopus

Hapalochlaena maculosa

LENGTH
4–9 in (10–22 cm)

HABITAT
Shallow water,
rock pools

DISTRIBUTION Tropical west Pacific and Indian Ocean (all species of Hapalochlaena)

The most dangerous cephalopod is the southern blue-ringed octopus, which produces highly toxic saliva powerful enough to kill a human. To catch prey, it either releases saliva into the water and waits for the poison to take effect, or catches, bites, and injects prey directly. Its bright coloring is unusual for an octopus, and the numerous blue rings covering its body become more iridescent if it is disturbed.

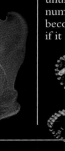

CLASS CEPHALOPODA

Giant Octopus

Enteroctopus dofleini

LENGTH
Up to 15 ft (4.5 m)

HABITAT
Bottom-dwellers,
30–2,500 ft (9–750 m)

DISTRIBUTION Temperate northwest and northeast Pacific

The North Pacific giant octopus is one of the largest invertebrates and also one of the most intelligent. It can solve problems, such as negotiating a maze by trial and error, and remember the solution for a long time. It has large, complex eyes with color vision and suckers that can distinguish between objects by touch alone. It changes color rapidly by contracting or expanding pigmented areas in cells called chromatophores, enabling it to remain camouflaged regardless of background. It also uses its color to convey mood, becoming red if annoyed and pale if stressed. Most cephalopods show little parental care, but female giant octopuses guard their eggs for up to eight months until they hatch. They do not eat during that time, and siphon water over the eggs to keep them clean and aerated.

DEFENSE MECHANISM

When threatened, a giant octopus squirts a cloud of purple ink out through its siphon into the water and at the same time moves backward rapidly using jet propulsion. Potential predators are left confused and disoriented in a cloud of ink. The octopus can repeat this process several times in quick succession.

A QUICK GETAWAY
This giant octopus is making a rapid retreat, expelling an ink jet as a defense mechanism. The jet also propels the octopus backward forcefully.

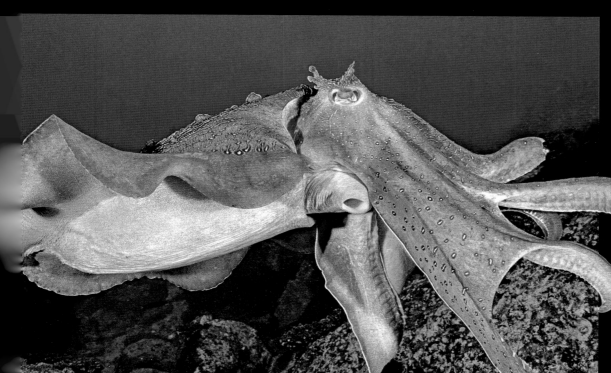

CLASS CEPHALOPODA

Australian Giant Cuttlefish

Sepia apama

LENGTH
Up to 5 ft (1.5 m)

HABITAT
Shallow water over reefs

DISTRIBUTION Coastal Australian waters

Of over 100 cuttlefish species, the Australian giant cuttlefish is the largest. Like all cuttlefish, it has a flattened body and an internal shell, known as the cuttle and familiar to many as budgerigar food. This species lives for up to three years and gathers in huge numbers to breed. Males have elaborate courtship displays, which involve hovering in the water while making rapid, kaleidoscopic changes of color, as the male shown here is doing. When a female is receptive, the male deposits a sperm package in a pouch under her mouth. This later bursts, releasing sperm and fertilizing her 100–300 golf-ball-sized eggs, which she then deposits on a hard substrate. The eggs hatch into miniature adults after several months.

CLASS CEPHALOPODA

Common Squid

Loligo vulgaris

LENGTH
Up to 17 in (42 cm)

HABITAT
60–800 ft (20–250 m)

DISTRIBUTION Eastern Atlantic, Mediterranean

A tubular body and a small, rodlike internal skeleton are characteristic features of all species of squid. They also have very large eyes relative to body size. The common squid is an inshore, commercially important species that has been harvested for centuries and is probably the best known of all cephalopods. It is a fast swimmer that actively hunts its prey, such as crustaceans and small fish. Once caught, the squid passes the prey to its mouth, where it is dismembered by powerful, beaklike jaws.

CLASS CEPHALOPODA

Glass Squid

Teuthowenia pellucida

LENGTH
Up to 8 in (20 cm)

HABITAT
5,250–7,900 ft (1,600–2,400 m)

DISTRIBUTION Circumglobal in southern temperate waters

Like many mollusks, juvenile glass squid live in the plankton, then descend to deeper, darker levels as they mature. The presence of light organs, called photophores, in the tips of their arms and in the eye may help in locating a mate. Sexually mature females are also thought to produce a chemical attractant, or pheromone.

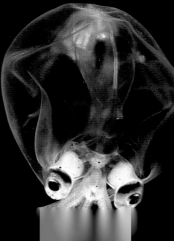

CLASS CEPHALOPODA

Vampire Squid

Vampyroteuthis infernalis

LENGTH
12 in (30 cm)

HABITAT
330–9,800 ft (100–3,000 m); usually 3,000–3,600 ft (900–1,100 m)

DISTRIBUTION Tropical and temperate oceans worldwide

This is the only squid that spends its entire life in deep, oxygen-poor water. Like many deep-living creatures, the vampire squid is bioluminescent and has light organs, or photophores, on the tips of its arms and at the base of its fins. If threatened, it flashes these lights and writhes around in the water, finally ejecting mucus that sparkles with blue luminescent light. When the lights go out, the vampire squid will have vanished. Its predators include deep-diving whales.

CLASS POLYPLACOPHORA

Lined Chiton

Tonicella lineata

LENGTH
Up to 2 in (5 cm)

HABITAT
Intertidal and subtidal zones, common on rocky surfaces

DISTRIBUTION Temperate waters of northeast and northwest Pacific

Chitons are mollusks with shells made up of eight arching and overlapping plates. The lined chiton is so called because of a series of zigzagging blue or red lines on its shell. The shell is usually pinkish in color, which provides good camouflage as this chiton grazes from rocks that are covered with encrusting pink coralline algae. The lined chiton's mantle extends around the shell on all sides, forming an unusually smooth, leathery "girdle" that helps to hold its eight shell-plates together. It has a large, muscular foot, which it uses to move over rocks and, when still, to grip on to them in much the same way as limpets do. At low tide, it remains stationary to avoid water loss. Its head is small and eyeless. The sexes are separate and it reproduces by releasing its gametes into the water.

Arthropods

DOMAIN	Eucarya
KINGDOM	Animalia
PHYLUM	Arthropoda
SUBPHYLA	4
SPECIES	About 1.25 million

ARTHROPODS HAVE ACHIEVED the greatest animal diversity on Earth, with insects by far the most numerous. Estimates of species vary, and possibly millions more remain to be discovered. Most marine arthropods are crustaceans, such as crabs, shrimp, and barnacles. Crustaceans, including both adults and larval stages, form most of the ocean's zooplankton—the community of tiny, drifting life forms that support all oceanic food chains. Like land arthropods, all marine forms have an external skeleton, segmented body, and jointed appendages. These features permit some, such as robber crabs, to live on land, as well as in water. Fully marine insects are rare but some live on seashores and coasts.

APPENDAGES
This spotted cleaner shrimp has jointed walking appendages. Two furthers pairs of jointed appendages, which are located on its head, are modified into sensory antennae.

Anatomy

Although arthropods may look very different from one another, they all have an external skeleton (exoskeleton), which is either thin and flexible or rigid and toughened by deposits of calcium carbonate. The body is segmented and has a variable number of jointed appendages—some are used for walking and swimming, while others are modified into claws and antennae or adapted for feeding. Muscles are attached across the joints to facilitate movement. Most of the body cavity is hollow. This space, called the hemocoel, contains the internal organs and a fluid—hemolymph—that is the equivalent to vertebrate blood. This is pumped around the body by the heart in an open circulatory system. Most marine forms use gills for respiration and have well-developed sense organs.

SEASHORE INSECTS
Springtails are wingless relatives of true insects. Marine species can be found in the upper regions of seashores and will spring into the air if disturbed.

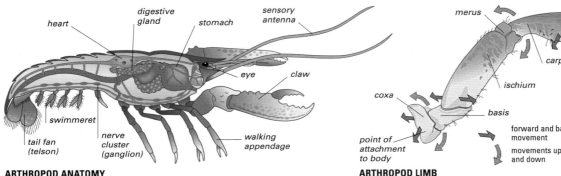

ARTHROPOD ANATOMY
heart, digestive gland, stomach, sensory antenna, eye, claw, swimmeret, tail fan (telson), nerve cluster (ganglion), walking appendage

ARTHROPOD ANATOMY
Lobsters have a protective shell carapace covering the head and thorax, large pincers, and well-developed walking appendages. The hemocoel contains the internal organs.

ARTHROPOD LIMB
merus, dactylus, carpus, propodus, ischium, coxa, basis, point of attachment to body, forward and backward movement, movements up and down

ARTHROPOD LIMB
Walking appendages, such as this crab's leg, comprise rigid sections linked with movable joints. The joints move in different planes, allowing versatile movement.

FILTER-FEEDING LIMBS
At high tide, barnacles feed by extending their long, feathery appendages from their "shell" and sweeping the water for plankton and detritus.

Feeding

Among crustaceans, feeding is extremely varied. Many crabs are scavengers that feed on dead and decaying organic matter. They are therefore vital in helping to recycle nutrients. Others are hunters and have robust claws to stun (mantis shrimp) or crush (lobsters) their prey before tearing it apart and consuming it. Many small planktonic crustaceans, such as copepods, are filter feeders that make effective use of various appendages, including long antennae, to create water currents that waft food particles toward their mouths. Barnacles, which are attached to rocks and unable to move, feed in a similar way, using their limbs to collect plankton. A few crustaceans are parasitic (some isopods, copepods, and the barnacle *Sacculina*) and obtain all their nourishment from their host. The shoreline is an ideal place for insects, such as kelp flies, that feed by releasing enzymes onto rotting seaweed and then taking in the resultant digested material. Farther inland, among the dunes, there is more vegetation, so spiders and pollen- and nectar-feeding insects start to appear.

SCAVENGING IN THE SAND
As the tide retreats, this sand bubbler crab emerges to feed on the organic material contained in the sand.

SPIDER FEATURES
Although it is called a crab and has a hinged carapace, this horseshoe crab is a close relative of spiders, ticks, and mites. Like them, it has piercing mouthparts.

Growth

Crustaceans can only develop and grow by molting and replacing their exoskeleton with a larger one. The molting process, called ecdysis, is controlled by hormones and occurs repeatedly during adult life. The exoskeleton is produced from the layer of cells situated immediately below it. Before a molt starts, the exoskeleton detaches from this cell layer and the space in between fills with molting fluid. Enzymes within this fluid weaken the exoskeleton so that it eventually splits at the weakest point, often somewhere along the back. The new exoskeleton is soft and wrinkled, so it needs to expand and then harden. Marine arthropods absorb water rapidly after molting to expand their new protective covering. Molting individuals are vulnerable until their exoskeleton hardens, so they often remain hidden for hours or days.

HUMAN IMPACT

KRILL DECLINE

Antarctic krill are only 2½ in (6 cm) long but are among the most abundant crustacean arthropods. Numbers in the Southern Ocean have fallen over the past few decades, partly due to rising water temperatures and melting of ice. Krill feed on algae that grow beneath and within the sea ice, and shelter under the ice to avoid predation. Overharvesting of krill for human and animal feed also poses a significant threat with potential to disrupt the Antarctic food web (see p.295).

THE MOLTING SEQUENCE

This sequence shows a harlequin shrimp molting. The exoskeleton has split just behind the neck joint, allowing the shrimp to pull out its head. The rest of its body quickly follows as the split enlarges. It takes only a few minutes for the shrimp to free itself completely, after which it rests for a few seconds.

The new exoskeleton is soft, since it must be flexible to buckle up to fit inside the older, smaller skeleton. It stretches to accommodate the increased size of the shrimp. Complete hardening of the new exoskeleton will take about two days.

1 The old exoskeleton splits along the back behind the harlequin shrimp's head. It eases out backward.

2 The shrimp emerges further and struggles to free itself from the old exoskeleton.

3 Molting is complete, and the old exoskeleton lies beside the shrimp, as the animal rests.

HARD SKELETON AND JOINTED LIMBS
These tiny porcelain crabs, less than 1 in (2.5 cm) wide, are filter feeders. They have typical arthropod features, such as a hard exoskeleton covering a segmented body, and jointed limbs.

Lifestyles

All marine arthropods are free-living for at least part of their lives. Some, such as crabs, have planktonic larvae that sink to the sea floor and become bottom-living, or benthic, as they mature. Crabs and their relatives tend to live alone unless seeking a partner to breed with, and may defend territory. Others, such as krill and copepods, live in vast swarms, traveling hundreds of yards up and down the water column each day to feed (see p.221). Adult barnacles remain anchored to one spot (that is, they are sessile), often settling in large aggregations on rocky shores where living conditions are most favorable. Deep-sea arthropod species are not well known, but many have cryptic red or black coloration to make themselves invisible, while others such as krill, have light organs and exhibit bioluminescence. A few arthropods live in close association with other species. Sometimes both partners benefit from a relationship (mutualism), sometimes only one has an advantage (commensalism), and sometimes one gains at some cost to the other (parasitism).

COMMENSALISM
This crab is camouflaged by a sea squirt in a commensal relationship. The crab benefits but the sea squirt neither gains nor loses.

SEA SPIDERS

As a group, sea spiders are typical of many problematic organisms whose classification is changing as information becomes available. In line with current thinking, they form a class (pycnogonids) within the subphylum that also contains spiders and horseshoe crabs (chelicerates). However, continuing research indicates they may form a completely separate group of arthropods.

DECEPTIVE APPEARANCE
Sea spiders are so called because of their resemblance to land spiders, but their exact relationship to spiders is still not clear.

PARASITISM
Some arthropods live closely with other species. This fish is being parasitized by an isopod, which is related to woodlice. There are two isopods, one under each eye, feeding on tissue fluid to the detriment of the fish.

Reproduction and Life Cycles

In most crustaceans, the sexes are separate, fertilization is often internal, and the eggs must be laid in water. Some females store sperm and then let it flow over their eggs as they release them. Others protect their eggs by carrying them around, and keep them healthy by continually wafting water over them. On hatching, the larvae join the zooplankton, and pass through various stages before maturing into adults. Barnacles are both male and female (hermaphroditic) but only function as one sex at a time. The male has a long, extendable penis and mates with all neighboring females within reach. In horseshoe crabs, fertilization occurs externally. Males and females pair up, the males fertilize the eggs as the females lay them in the sand, and then both sexes abandon them.

egg mass on underside of female crab

CRAB MOTHER AND LARVA
A velvet crab (above) carries eggs beneath her body until they hatch. The hatchlings enter a planktonic larval stage called a zoea (left). This molts four to seven times before it becomes a megalops larva, then once again to become an adult crab.

ARTHROPOD CLASSIFICATION

Arthropods are split into four subphyla—Crustacea, Chelicerata, Hexapoda, and Myriapoda. Only marine species are included in the species numbers listed here.

CRUSTACEANS
Subphylum Crustacea

52,744 species
Crustaceans are the dominant marine arthropod group and include the familiar crabs, lobsters, shrimps, prawns, and barnacles as well as the smaller copepods, isopods, and krill. Most crustaceans have two pairs of antennae and three body segments: the head, the thorax, and the abdomen. The head and thorax are often fused together as the cephalothorax and may be protected by a shield or carapace. Paired appendages vary—some are sensory, while others are adapted for walking or swimming.

CHELICERATES
Subphylum Chelicerata

3,003 species
Spiders, scorpions, ticks, mites, horseshoe, crabs, and sea spiders belong to this group. A few species of spiders live in the intertidal zone, and some types of ticks and mites are either free-living or parasitic in marine habitats.
  Horseshoe crabs (class Merostomata) are completely marine. They have five pairs of legs; a body of two parts protected by a hinged carapace; and a characteristic long, tail-like telson. Sea spiders (class Pycnogonida) are all marine. Many species have a unique pair of appendages, called ovigers, overhanging the

head. The females use them for grooming, courtship, and also to transfer eggs to the ovigers of the male, where they remain until they hatch. Sea spiders are common in intertidal areas, but they are rarely seen due to their small size.

HEXAPODS
Subphylum Hexapoda

1,695 species
In total, there are over 1 million species of hexapods, the majority of which are insects—the largest of all animal groups. It includes beetles, flies, ants, and bees. Many insects live in coastal areas, but only a few live on the shore. Only one type of insect is truly marine—the marine skater, *Halobates*, a type of "true bug" (order Hemiptera). Of the 47 *Halobates* species, only five are able to spend their entire

life on the ocean. However, they require a solid object, such as a floating feather or lump of tar, on which to lay their eggs.

MYRIAPODS
Subphylum Myriapoda

78 species
The bodies of myriapods are divided into a head with one pair of antennae and a long body with many legs. The great majority of myriapods occur in terrestrial ecosystems; however, a few are found on the upper shore, especially around high-tide level, among rotting seaweed, in shingle, or under stones in mud. Most are centipedes or millipedes. Centipedes prey on small invertebrates, while millipedes feed on vegetable material or decaying matter—for example, in strand-line debris.

SUBPHYLUM CHELICERATA
Giant Sea Spider
Colossendeis australis

LENGTH 10 in (25 cm) (leg-span)	
WEIGHT Not recorded	
HABITAT Bottom dweller	

DISTRIBUTION Antarctic shelf and slope

Unlike most sea spiders, which have a leg-span of less than 1 in (2.5 cm), the giant sea spider—typical of other polar species—has a huge leg-span of about 10 in (25 cm). It has a large proboscis through which it sucks its food, but its body is so small that the sex organs and parts of its digestive system are situated in the tops of the legs. Sea spiders are somewhat unusual among arthropods in that some exhibit parental care, the males having a modified pair of legs to carry the eggs until they hatch.

SUBPHYLUM CRUSTACEA
Water Flea
Evadne nordmanni

LENGTH 1/32 in (1 mm)	
WEIGHT Not recorded	
HABITAT Open waters, to depths of 6,500 ft (2,000 m)	

DISTRIBUTION Temperate and cool waters worldwide

Most water fleas live in freshwater but this species and just a few others live as plankton in the ocean. It feeds on tiny bacteria, protists, and organic debris and is eaten by larger planktonic animals. It has a single conspicuous eye and feathery swimming appendages that are modified antennae. Females brood unfertilized eggs that hatch into more females. Sexual reproduction also occurs.

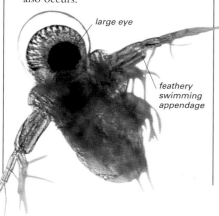

large eye

feathery swimming appendage

SUBPHYLUM CHELICERATA
American Horseshoe Crab
Limulus polyphemus

LENGTH Up to 24 in (60 cm)	
WEIGHT Up to 11 lb (5 kg)	
HABITAT Sandy or muddy bays to 100 ft (30 m)	

DISTRIBUTION Western Atlantic and Gulf Coast from southern Maine to the Yucatán Peninsula

Despite its name, the American horseshoe crab is more closely related to spiders than to crabs. It is mainly active at night and scavenges anything it can find, including small worms, bivalves, and algae. Adults have few predators—the horseshoe-shaped, greenish-brown shell, or carapace, provides effective protection. It has six pairs of thoracic appendages: the first pair (called chelicerae), is used for feeding; the other five are for walking and for grasping and tearing food. Five pairs of flattened abdominal appendages are used for swimming and scuttling along the bottom and also carry pagelike "book gills," through which oxygen is absorbed. Its long, rigid tail is used for steering and for righting itself. Hinged to the body, it can act as a lever.

The reproductive cycle is closely linked to spring high tides, when adults gather in large numbers on sandy beaches to breed. Females lay up to 80,000 eggs in a depression near the high-tide mark, providing a vital food source for birds and other marine creatures. The eggs are laid in batches mostly when the moon is in its full or new phases.

The eggs hatch into tailless "trilobite larvae," so called because they look a bit like fossil trilobites. Carried down the shore at high tide, the larvae swim around actively but also burrow into the sediment for safety. After a few days, they molt and become juveniles.

HUMAN IMPACT
MEDICAL RESEARCH

If the American horseshoe crab is injured, some of its blood cells form a clot, which kills harmful gram-negative bacteria. In order to exploit this property for human benefit, crabs are collected from shallow waters on the Atlantic coast of North America during the summer months. Researchers then remove about 20 percent of the blood from each crab. From this they extract a protein that is used to detect bacterial contamination in drugs and medical devices that will be in contact with blood. Taking blood from the crabs is sometimes fatal to them but most recover after they have been returned to the sea.

OCEAN LIFE

SUBPHYLUM CRUSTACEA

Cyclopoid Copepod

Oithona similis

LENGTH
¹⁄₅₄–³⁄₃₂ in (0.5–2.5 mm)

HABITAT
Surface waters to a
depth of 500 ft (150 m)

DISTRIBUTION Atlantic, Mediterranean, Southern Ocean, southern Indian and Pacific Oceans

Copepods make up a large percentage of zooplankton, and this is one of the most abundant, widespread species. As the name suggests, cyclopoid copepods have a single, central eye, which is light sensitive. They have a tiny, oval body that tapers to a thin tail. Most have six body segments and six pairs of swimming limbs. Tiny food particles are filtered from the water using specialized mouthparts. Females can be recognized when carrying egg sacs attached to their abdomens.

As part of the zooplankton, copepods of this genus are a vital element of oceanic food chains. They feed on marine algae and bacteria and in turn are an important source of protein for many ocean-dwelling animals. Every night cyclopoid copepods migrate from a depth of about 500 ft (150 m) to the surface layers of the ocean to feed. This daily journey, which is undertaken by many marine creatures, is one of the largest mass movements of animals on Earth.

SUBPHYLUM CRUSTACEA

Gooseneck Barnacle

Pollicipes polymerus

LENGTH
Up to 3 in (8 cm)

HABITAT
Intertidal zone of rocky
shores

DISTRIBUTION Eastern Pacific coast of North America, from Canada to Baja California, Mexico

So called because of its resemblance to a goose neck and head, the gooseneck barnacle forms dense colonies in crevices on rocky shores with strong waves. Barnacles anchor themselves to rocks by a tough, flexible stalk (peduncle), which also contains the gonads. This is actually their "head" end. Once the barnacle has attached itself to an object it secretes a series of pale plates at the end of its stalk, forming a shell around its featherlike legs, which comb through the water for food. The legs face away from the sea, enabling the barnacle to feed by filtering out particles of detritus from returning tidal water as it funnels past them through cracks in the rocks. These barnacles become sexually mature at about five years of age and may live for up to 20 years. The larval stage is free-living but depends on sea currents for its transport and survival. Colonies of gooseneck barnacle are susceptible to the damaging effects of oil pollution and they recover only slowly from disturbance.

SUBPHYLUM CRUSTACEA

Acorn Barnacle

Semibalanus balanoides

LENGTH
Up to ¹⁄₂ in (1.5 cm)
diameter

HABITAT
Intertidal zone of rocky
shores

DISTRIBUTION Northwest and northeast Atlantic, Pacific coast of North America

Like all adult barnacles, the adult acorn barnacle remains fixed in one place once it has anchored itself to a site. The free-swimming juveniles pass through several larval stages before molting into a form that can detect both other acorn barnacles and suitable anchoring sites. Once a larva fixes itself to a rock, using cement produced by glands in its antennae, it molts again. It then secretes six gray calcareous plates, forming a protective cone that looks rather like a miniature volcano. Four smaller, movable plates at the top of the cone open, allowing the acorn barnacle to feed. It does this when the tide is in by waving its modified legs, called cirri, in the water to filter out food. When the tide is out, the plates are closed to prevent the barnacle from drying out. Acorn barnacles are hermaphrodites that possess both male and female sexual organs, but they function as either a male or a female. They do not shed their eggs and sperm into the water; instead they use extendable penises, to transfer sperm to receptive neighbors.

CHARLES DARWIN

Before the British naturalist Charles Darwin (1809–1882) proposed his revolutionary theory of evolution in *The Origin of Species* (1859), he spent eight years studying barnacles. Realizing the impact his ideas on evolution would have on existing scientific and religious thinking, he delayed writing and instead produced four monographs on the classification and biology of barnacles. This work earned him the Royal Society's Royal Medal in 1853, validating his reputation as a biologist.

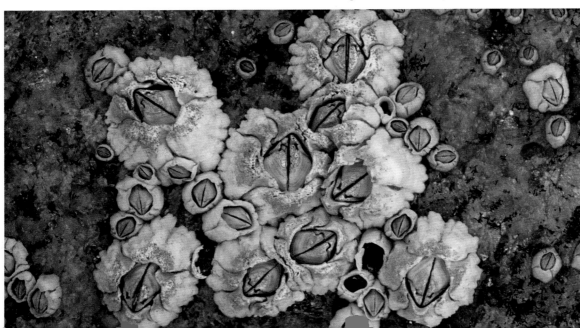

SUBPHYLUM CRUSTACEA

Giant Mussel Shrimp

Gigantocypris muelleri

LENGTH
½–¾ in (1.4–1.8 cm)

HABITAT
Planktonic, intermediate to deep sea

DISTRIBUTION Atlantic, Southern Ocean, western Indian Ocean

Like all ostracod crustaceans, the giant mussel shrimp has a carapace that encloses its body, so that its seven pairs of limbs are almost hidden from view. Its large, mirror eyes with parabolic-shaped reflectors focus light on to a flat plate in its center. It is planktonic, but lives at greater depth than many forms of plankton, usually below 650 ft (200 m), where it feeds on falling detritus. The picture shows a large female carrying embryos, which are clearly visible through the carapace.

SUBPHYLUM CRUSTACEA

Peacock Mantis Shrimp

Odontodactylus scyllarus

LENGTH
Up to 6 in (15 cm)

HABITAT
Warm water near reefs with sandy, gravelly, or shelly bottoms

DISTRIBUTION Indian and Pacific Oceans

The brightly colored peacock mantis shrimp is a stomatopod crustacean. Like all members of this group, it is a voracious predator. Its large, mobile, compound eyes have sophisticated stereoscopic and color vision that includes some ultraviolet shades. It uses sight when hunting, waiting quietly, like the praying mantis, for its unsuspecting prey to come within reach, then striking using its powerful, clublike second pair of legs with immense speed—about 75 mph (120 kph)—and force (up 100 times its own weight). Such power is created by a special, saddlelike hinge-joint in these legs, which acts like a spring. The peacock mantis shrimp can smash the shells of gastropods and crabs and tackles prey larger than itself. It excavates U-shaped burrows or lives in crevices in rocks or coral. After hatching, its larvae enter the plankton, where they develop over a few weeks before drifting down toward the sea floor to make their own homes.

SUBPHYLUM CRUSTACEA

Sea Slater

Ligia oceanica

LENGTH
Up to 1¼ in (3 cm)

HABITAT
Coasts with rocky substrata

DISTRIBUTION Atlantic coasts of northwestern Europe

Commonly found under stones and in rock crevices, the sea slater is a seashore-dwelling a member of the isopods (a group that also includes woodlice). It lives in the splash zone, but can survive periods of immersion in salt water. Its head, which has a pair of well-developed compound eyes and very long antennae, is not markedly separated from its body, which is flattened, about twice as long as it is broad, and ends in two forked projections called uropods.
As adults, sea slaters have six pairs of walking legs until their final molt, after which they have seven. The sea slater is not generally seen during the day unless it is disturbed, and it emerges from its hiding place only at night to feed on detritus and decaying seaweed. Sea slaters mature at about two years of age and usually breed only once before dying.

uropod

antenna

SUBPHYLUM CRUSTACEA

Sand Hopper

Orchestia gammarellus

LENGTH
1/16–3/8 in (2–10 mm)

HABITAT
Splash zone of sandy shores

DISTRIBUTION Atlantic coasts of northeastern Canada and northwestern Europe

Amphipod crustaceans, such as the sand hopper, live in large numbers in the splash zone of any shore where there is rotting seaweed. Their life cycle takes about 12 months and the female usually produces only one clutch of eggs, which she keeps in a brood pouch, where they hatch after one to three weeks. The young leave the pouch about a week later when their mother molts. Sand hoppers are also known as sand fleas because they jump about in a similar way and have similar laterally compressed bodies.

SUBPHYLUM CRUSTACEA

Antarctic Krill

Euphausia superba

LENGTH
Up to 2 in (5 cm)

HABITAT
Planktonic

DISTRIBUTION Southern Ocean

All oceans contain krill—small, shrimplike, planktonic crustaceans that live in open waters. Antarctic krill live in vast numbers in the subantarctic waters of the Southern Ocean, where they form a vital link in the food chain, being eaten in vast quantities by baleen whales, seals, and various fish. Krill rise to the surface at night to feed on phytoplankton, algae, and diatoms. For safety they sink to greater depths during the day. The feathery appearance of this species is due to its gills, which, unusually, are carried outside the carapace. Their filamentous structure increases the surface area available for gaseous exchange. Antarctic krill also have large light organs, called photophores. The light is thought to help them group together. They spawn in spring, during which females may release several broods of up to 8,000 eggs.

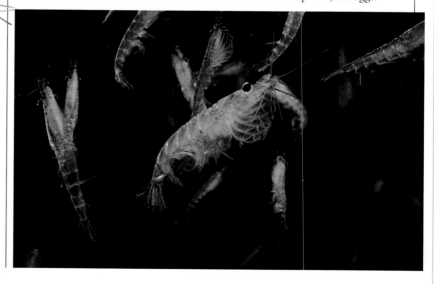

SUBPHYLUM CRUSTACEA

Deep Sea Red Prawn

Acanthephyra pelagica

LENGTH
Not recorded

HABITAT
Deep water

DISTRIBUTION Atlantic

In the low light levels of the deep ocean, red appears black, making the deep sea red prawn invisible to potential predators. Its hard outer casing, or exoskeleton, is thinner and more flexible than that of shallow-water crustaceans, which helps prevent them from sinking into the depths. The flesh of this prawn is also oily to aid further buoyancy. It uses its first three pairs of limbs to feed on small copepods. The remaining five pairs of limbs, the pereiopods, are used for locomotion. Gills attached to the tops of the legs are used for respiration.

SUBPHYLUM CRUSTACEA

Common Prawn

Palaemon serratus

LENGTH
Up to 4¼ in (11 cm)

HABITAT
Rock pools, shallow water, and lower parts of estuaries

DISTRIBUTION Eastern Atlantic from Denmark to Mauritania, Mediterranean, Black Sea

The common prawn has a semi-transparent body, making its internal organs visible, and is marked with darker bands and spots of brownish red. As with many other species of prawn, its shell extends forward between its stalked eyes to form a stiff, slightly upturned projection called a rostrum. This feature has a unique structure by which the common prawn can be distinguished from all other members of the same genus. The rostrum curves upward, splitting in two at the tip, where it has several toothlike projections on the lower and upper surfaces.

To either side of the rostrum there is a very long antenna that can sense any danger close by and is also used to detect food. Of the prawn's five pairs of legs, the rear three pairs are used for walking, while the front two pairs are pincered and used for eating. Attached to the abdomen is a series of smaller limbs called swimmerets that the

HUMAN IMPACT

PRAWN FARMING

Nearly all the world's farmed prawns come from developing countries such as Thailand, China, Brazil, Bangladesh, and Ecuador, which use intensive farming to meet demand. Cutting mangrove forests to construct prawn ponds is now being discouraged.

SHARED RESOURCES
Fishers in Honduras fish for wild prawns in a lagoon shared with prawn farmers.

prawn uses to swim. For a sudden, backward movement, the prawn flicks its tail. Females produce and look after about 4,000 eggs until they hatch into larvae. The larvae float among the plankton until they mature.

tail fan, or telson

pincered leg

SUBPHYLUM CRUSTACEA

Anemone Shrimp

Ancylocaris brevicarpalis

LENGTH
1 in (2.5 cm)

HABITAT
Shallow water reefs

DISTRIBUTION Indian Ocean, western Pacific

Nestling among the tentacles of an anemone, the anemone shrimp is safe from attack by predators. It rarely wanders far from its host, surviving by scavenging scraps that the anemone cannot eat. The shrimp may benefit the anemone by removing excess food particles, as well as any waste it produces. This type of relationship is called commensalism: one individual in the partnership profits from the liaison and the other comes to no harm. Removed from its host, this shrimp is defenseless. The anemone shrimp belongs to the same family (Palaeomonidae) as the common prawn (above) and so they have several features in common. These include a pair of long, sensory antennae used to sense danger and detect food and a rostrum (the elongated projection of the shell from between the eyes). The anemone shrimp is almost completely transparent, with a few purple and white spots.

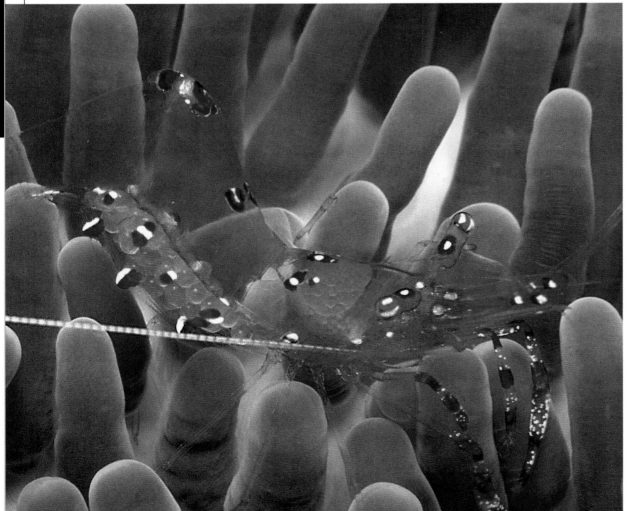

SUBPHYLUM CRUSTACEA

Spiny Lobster

Panulirus argus

LENGTH
24 in (60 cm)

HABITAT
Coral reefs and rocky areas

DISTRIBUTION Western Atlantic, Gulf of Mexico, Caribbean Sea

Being both nocturnal and migratory, the spiny lobster has excellent navigational skills. It can establish its position in relation to Earth's magnetic field and then follow a particular route as well as any homing pigeon. This lobster prefers warm

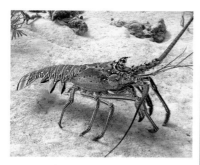

water and so remains in the shallows in summer before migrating in groups to deeper water in winter by walking in single file across the sea floor. It lacks the large claws of the European lobster (right) but is well protected from most predators by the sharp spines that cover its carapace.

SUBPHYLUM CRUSTACEA

European Lobster

Homarus gammarus

LENGTH
Up to 3 ft (1 m), typically 24 in (60 cm)

HABITAT
Rocky coasts

DISTRIBUTION Eastern Atlantic, North Sea, Mediterranean

In life the European lobster is brown or bluish—it only turns the familiar red when it is cooked. Individuals weigh up to 11 lb (5 kg). It has large, differently sized claws: the smaller one has sharper edges and is used for cutting prey, while the larger one is used for crushing. The European lobster lives in holes and crevices on the seabed. The European lobster is commercially important and in danger of overexploitation because it matures slowly, and is such a valuable commodity.

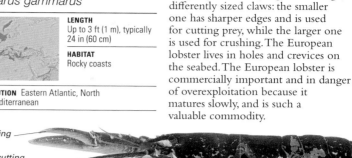

crushing claw

cutting claw

SUBPHYLUM CRUSTACEA

Reef Hermit Crab

Dardanus megistos

WIDTH (LEG-SPAN)
Up to 12 in (30 cm)

HABITAT
Near-shore tropical reefs

DISTRIBUTION Indian and Pacific Oceans

Like other hermit crabs, the reef hermit crab uses an empty gastropod mollusk shell to protect its soft abdomen. When it grows too big for its current shell, it simply looks for an unused larger one or evicts a weaker rival. It is while switching from one

shell to the next that the reef hermit crab is most vulnerable, as it risks exposing its soft, rather asymmetrical abdomen to predators. There are about 1,150 species of hermit crabs—the reef hermit crab lives in shallow-water tropical reef habitats, but some species live on coastal land. The reef hermit crab is a scavenger rather than a hunter, and drags itself over the seafloor looking for bits of animal matter and algae, tearing apart any carcasses that it finds with its dextrous mouthparts. Its close relative *Dardanus pedunculatus* attaches stinging anemones to its shell as protection from predators.

eye stalk

single large claw

SUBPHYLUM CRUSTACEA

Porcelain Crab

Petrolisthes lamarckii

WIDTH (SHELL)
Up to ¾ in (2 cm)

HABITAT
Pools on rocky beaches and shorelines

DISTRIBUTION Indian Ocean, Pacific coast of Australia, western Pacific

The flat, rounded body of the porcelain crab allows it to slip easily into small rock crevices to hide. However, if it becomes trapped by a predator or stuck beneath a rock, it can shed one of its claws in order to escape, and a new one will grow over time. This crab's abdomen is folded under its body, but it can be unfolded and moved like a paddle when swimming.

SUBPHYLUM CRUSTACEA

Robber Crab

Birgus latro

LENGTH
Up to 24 in (60 cm) across

HABITAT
Rock crevices and sandy burrows

DISTRIBUTION Tropical waters of Indian and Pacific Oceans

The robber crab, or coconut crab, is the largest terrestrial arthropod. When young, it lives inside a mollusk shell, like its hermit crab relatives, or even in a piece of broken coconut shell, but discards this as it grows bigger and tougher. The robber crab lives on oceanic islands and offshore inlets. It mainly eats fruit and nuts but will also scavenge for carrion. It can smash and eat coconuts but rarely does so. Adults live and mate on land, but females release their eggs into water.

SUBPHYLUM CRUSTACEA

Nodose Box Crab

Cyclozodion angustum

LENGTH
Not recorded

HABITAT
Offshore to depths of 50–650 ft (15–200 m)

DISTRIBUTION Western Atlantic, Gulf of Mexico, Caribbean Sea

Previously known as *Calappa angusta*, the nodose box crab is a true crab with a small abdomen that is tucked away underneath the body and four pairs of legs. This species may be recognized by the rows of nodules that radiate from behind its eyes across the upper surface of its yellowish shell, or carapace.

SUBPHYLUM CRUSTACEA

Japanese Spider Crab

Macrocheira kaempferi

WIDTH (SHELL) Up to 14½ in (37 cm)

HABITAT
Deep-water vents and holes to depths of 160–1,000 ft (50–300 m)

DISTRIBUTION Pacific Ocean near Japan

Not only is the giant Japanese spider crab the largest of all crabs, with a leg-span of up to 13 ft (4 m) and weighing 35–44 lb (16–20 kg), it may also be the longest living, estimated to live for up to 100 years. Living in the deep, cold waters around Japan, it moves slowly across the ocean floor on its spiderlike legs, scavenging for food.

PORCELAIN CRAB
The porcelain crab uses its flat body to crawl out of reach of predators. Here, the tentacles of an anemone provide a secure and permanent home for a porcelain crab in the Andaman Sea in the northern Indian Ocean. The mouthparts of the crab fan out and trap plankton, which it then brushes into its mouth.

Long-legged Spider Crab

Macropodia rostrata

LENGTH
Up to 1 in (2.5 cm)

HABITAT
Lower shore, usually not beyond 165 ft (50 m)

DISTRIBUTION Northeastern Atlantic from southern Norway to Morocco, Mediterranean

Also called the decorator crab because it camouflages itself using fragments of seaweed and sponges, the long-legged spider crab is covered in hook-shaped hairs that hold its disguise in place, enabling it to blend in with the seaweed among which it lives. This crab has a triangular-shaped carapace that extends forward between the eyes into an eight-toothed projection called a rostrum. Its spiderlike legs are at least twice as long as its body and can be used, somewhat ineffectively, for swimming. The long-legged spider crab feeds on small shellfish, algae, small worms, and detritus. Breeding occurs year-round on Atlantic coasts, but takes place between March and September in the Mediterranean. The male transfers sperm to the female using its first pair of abdominal legs. The female carries the eggs until they hatch into larvae that live in the plankton.

Spotted Reef Crab

Carpilius maculatus

LENGTH
About 3½ in (9 cm)

HABITAT
Shoreline to 33 ft (10 m), inshore reefs

DISTRIBUTION Indian Ocean and western Pacific

The conspicuous coloring of the spotted reef or coral crab is highly distinctive. Its smooth, light brown carapace has two large red spots behind each eye, three across the middle, and either two or four at the rear. Between the eyes the carapace has four small, rounded projections, which are also characteristic of the species. It is a nocturnal, slow-moving crab that uses its disproportionately large claws to feed on corals, snails, and other small marine creatures.

Edible Crab

Cancer pagurus

LENGTH
Up to 6 in (16 cm)

HABITAT
Intertidal zone to 330 ft (100 m), in rock pools and muddy sand offshore

DISTRIBUTION Northeastern Atlantic and North Sea; introduced to parts of the Mediterranean

The oval carapace of the edible crab has a characteristically "scalloped" or "piecrust" edge around the front and sides. Its huge pincers are distinctively black-tipped, while the body is purple-brown in small individuals and reddish brown in larger ones. Edible crabs mate at different times in different parts of their range, the females incubating their eggs for six to nine months. This crab is caught in large numbers and is highly valued as a luxury food.

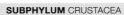

black eye on short eye stalk

mouth

Pea Crab

Pinnotheres pisum

LENGTH
Males ⅓ in (8 mm); females ½ in (14 mm)

HABITAT
Intertidal zone to 500 ft (150 m)

DISTRIBUTION Eastern Atlantic from northwestern Europe to West Africa, Mediterranean

Typically about the size of a pea, the tiny pea crab is usually found inside the shells of the common mussel. Protected from predators in the mantle cavity of its host, it feeds on plankton that becomes trapped on the mussel's gills as water passes over them. Whether the presence of this guest is harmful to the mussel is unclear. Female pea crabs are almost twice the size of males and have an almost translucent carapace through which their pink reproductive organs are visible. Males have harder, yellowish brown carapaces that protect them during the breeding season, which runs from April to October. During this time, males leave the safety of their host's shell and swim around looking for females with which to mate. In regions where shellfish are harvested commercially, the pea crab is considered a pest.

Common Shore Crab

Carcinus maenas

LENGTH
Up to 2½ in (6 cm)

HABITAT
Intertidal zone to 200 ft (60 m), all substrates; estuaries

DISTRIBUTION Northeastern Atlantic from Norway to West Africa; introduced elsewhere

The common shore crab tolerates a wide range of salt concentrations and temperatures and so can live in salt marshes and estuaries as well as along the shoreline. Its dark green carapace has five marked serrations on the edge behind the eyes. This opportunistic hunter preys voraciously on many types of animals, including bivalve mollusks, polychaetes, jellyfish, and small crustaceans. Where introduced, it may be detrimental to local marine life. On the west coast of the US, for example, it has had a considerable impact on the shellfish industry.

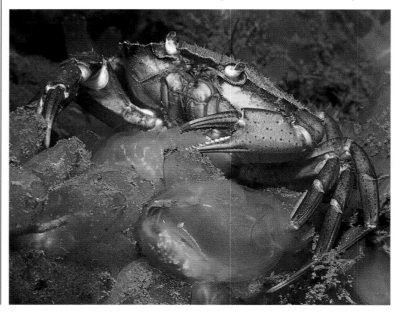

Blue Swimming Crab

Portunus pelagicus

LENGTH
Up to 2¾ in (7 cm)

HABITAT
Intertidal sandy or muddy seabeds to 180 ft (55 m)

DISTRIBUTION Coastal waters of the Indian and Pacific Oceans, eastern Mediterranean

SUBPHYLUM CRUSTACEA

Unlike most crabs, the blue swimming crab is an excellent swimmer and uses its fourth pair of flattened, paddlelike legs to propel itself through the water. Despite its name, only the males are blue—the females are a rather dingy greenish brown. Males also differ in having very long claws, more than twice as long as the width of their carapace. The claws are armed with sharp teeth that are used to snag small fish and other food items.

When a blue swimming crab feels threatened, it usually buries itself in the sand. If this measure fails to deter the threat, the crab adopts its own threat stance, extending its claws sideways in an attempt to look as large as possible. The natural range of this crab has been extended to a small part of the eastern Mediterranean by the opening of the Suez Canal. It is a popular food in Australia.

Ghost Crab

Ocypode saratan

SUBPHYLUM CRUSTACEA

LENGTH
About 1½ in (3.5 cm)

HABITAT
Sandy shores in deep burrows above the water line

DISTRIBUTION Coastal waters of western Indian Ocean, Red Sea

Resting in the cool of its burrow during the day, the fast-moving ghost crab emerges at twilight to hunt. It will eat anything it can find, including other crabs, and also scavenges whatever was brought in by the last tide. During the mating season, male ghost crabs defend their burrows, but they rarely fight, any disputes being settled by ritualistic displays. Their burrows can be over 330 ft (100 m) from the sea and over 3 ft (1 m) deep.

Orange Fiddler Crab

Gelasimus vocans

SUBPHYLUM CRUSTACEA

LENGTH
About 1 in (2.5 cm)

HABITAT
Near water on mud or sand

DISTRIBUTION Indian Ocean and western Pacific

Like its close relative the ghost crab (see above, right), the male orange fiddler crab also exhibits ritualistic displays to deter rivals. Males are easily recognized because one of their claws is greatly enlarged. In a mature adult this claw makes up more than half the crab's body weight and is used both to attract potential mates and to ward off rival males. Observing the distinctive "courtship wave" of fiddler crabs is helpful in identifying different species.

Orange fiddler crabs are active during the day. As well as digging a main burrow up to 12 in (30 cm) deep, they create a number of bolt holes into which they can retreat if danger threatens. At high tide, they seal themselves into their burrows with a small pocket of air. The presence of air is essential for their survival because fiddler crabs obtain oxygen from air, not water, despite having gills.

RITUAL DISPLAY

Each species of fiddler crab waves its claw in a slightly different way. If this ritual movement does not deter a rival male, then two crabs may "arm-wrestle" each other to resolve their dispute. The weaker individual usually retreats before any serious damage is done.

LEFT- OR RIGHT-HANDED?
Both crabs in this picture are right-handed, but in some males it is the left claw that is enlarged (see below).

RESTORING THE BALANCE
On arriving at the shore, the crabs head to the ocean, where they replace water and body salts lost during the arduous journey down from the forest plateau.

The remarkable annual migration of the red crab (*Gecarcoidea natalis*) from the forests of Christmas Island (southwest of Indonesia) down to the sea to spawn is one of the wonders of the natural world. Until recently, about 120 million of the crabs made the journey each year, spending the rest of the time on a forested plateau about 1,200 ft (360 m) above sea level. Red crabs are mainly herbivorous, feeding on fallen leaves, fruit, and flowers gathered from the forest floor, but they also eat other dead crabs and birds when the opportunity arises. They conserve water by restricting their activity to times of high humidity (over 70 percent), retreating to their burrows during drier periods. Like their marine counterparts, red crabs have gills for respiration, but the gill chamber of this species is lined with tissue that acts as a lung and maximizes gaseous exchange.

Red crabs on Christmas Island have undergone a noticeable decline in numbers, largely due to the accidental introduction of the yellow crazy ant in the 1930s. Since the mid-1990s, about 40 million red crabs are thought to have been killed by the ants, but the introduction in 2016 of a small wasp that disrupts the ants' food chain is having a positive effect. There is pressure to increase the number of phosphate mines on the island, which would involve deforestation, depriving the crab of its habitat in the affected areas.

The Sequence of Migration

The onset of the wet season on Christmas Island, usually in early November, signals the start of the red crabs' migration, which takes place over three lunar cycles. The males set off first, followed by females. It takes about a week for the crabs to reach the shore. After dipping in the ocean, the males compete fiercely for space to dig their burrows, where it is thought mating takes place. The males then return inland, leaving the females to brood their eggs in the burrows. They emerge after about two weeks to release their eggs into the sea at high tide at night during the last quarter of the moon.

STAGES OF MIGRATION

TRAVELING ACROSS LAND

MAN-MADE OBSTACLES *Although the distance to the shore is only about 1,600 ft (500 m), red crabs have to negotiate a number of obstacles, including roads and hot train tracks. In the past, it was not unusual for one million crabs to perish each year, yet this had little impact on the population. Today, various measures such as road closures and concrete underpasses offer some security to the crabs, but they still run the risk of dying from dehydration during their journey.*

RELEASING EGGS

EGG RELEASE *After incubating up to 100,000 eggs for 12–13 days, the females leave their burrows and gather on the shore to disperse them directly into the sea. They do so at night, as the high tide turns, by raising their claws and shaking their bodies vigorously to free the eggs from their pouches. Crabs on the cliffs may be 25 ft (8 m) above the water.*

MEGALOPAE LARVAE *Red crab eggs hatch as soon as they hit the water. The young remain in the ocean for up to 30 days, and pass through several larval stages before returning to the shallows as shrimplike megalop larvae. They metamorphose into tiny crabs after 3–5 days and leave the water to start life on land.*

FROM SEA TO LAND

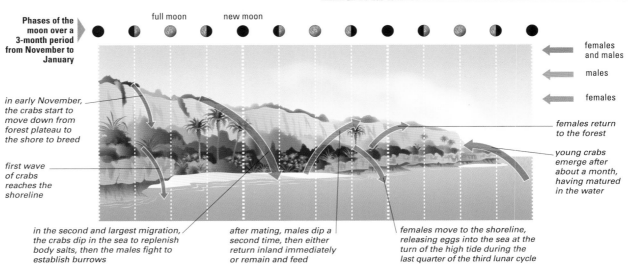

Phases of the moon over a 3-month period from November to January

full moon   new moon

females and males

males

females

in early November, the crabs start to move down from forest plateau to the shore to breed

first wave of crabs reaches the shoreline

in the second and largest migration, the crabs dip in the sea to replenish body salts, then the males fight to establish burrows

after mating, males dip a second time, then either return inland immediately or remain and feed

females move to the shoreline, releasing eggs into the sea at the turn of the high tide during the last quarter of the third lunar cycle

females return to the forest

young crabs emerge after about a month, having matured in the water

Shore Bristletail

Petrobius maritimus

LENGTH	²/₅ in (1 cm)
HABITAT	Rocky shores in the splash zone

DISTRIBUTION British Isles excluding Ireland

The shore bristletail, also known as the sea bristletail, derives its common name from the three long filaments extending from the tip of its abdomen. Its long body is well-camouflaged by drab-colored scales. It has long antennae and compound eyes that meet at the top of its head. The shore bristletail lives in rock crevices and feeds on detritus. It can move swiftly around the rocks using small spikes on its underside, called styles, to help it grip. When disturbed, it can leap small distances through the air by using its abdomen to catapult it away from the rock.

Rock Springtail

Anurida maritima

LENGTH	Up to ¹/₈ in (3 mm)
HABITAT	Upper intertidal zone of rocky shores

DISTRIBUTION Coasts of the British Isles

At low tide hundreds of rock springtails wander down the beach searching for food, returning to the shelter of their rock crevices an hour before the tide turns. Vast numbers of them squeeze together in the fissures to avoid being immersed at high tide. It is here that they molt and lay their eggs safe from submersion and many of their predators.

Rock springtails are blue-gray in color with segmented bodies that are wider at the posterior end. They have three pairs of appendages used for locomotion, which also allow them to swarm over the surface of calm rock pools without sinking—they cannot swim. Springtails are so named for their jumping organ, called the furcula, which acts like a spring, propelling the animal upward if threatened. Unlike other springtails, however, the rock springtail does not have this feature.

Dune Snail Bee

Osmia aurulenta

LENGTH	²/₅ in (1 cm)
HABITAT	Sand dune systems

DISTRIBUTION Coasts of northeastern Atlantic, North Sea, Baltic, and Mediterranean

Important in the pollination of sand-dune plants, the dune snail bee has a compact, brownish black body with a dense covering of golden red hairs that later fade to gray. Unlike the honey bee, which carries any pollen it collects in pouches on its legs, the dune snail bee carries its pollen in a brush of hairs under its abdomen.

Male bees of this species emerge between April and July, a little earlier in the year than the females, and seek out territories that contain a snail shell. They then leave scent marks (pheromones) on the stems of plants to attract passing females. Once a female has mated with her chosen partner, she adjusts the position of the shell so that the entrance is oriented in the most sheltered direction. Then she lays her eggs inside it.

Intertidal Rove Beetle

Bledius spectabilis

LENGTH	Up to ³/₄ in (2 cm)
HABITAT	Intertidal sandy and muddy shores

DISTRIBUTION Coasts of the British Isles and northern Europe

Unusual in that it lives in the intertidal zone after which it is named, this small arthropod has an elongated, smooth black body. Short reddish brown wing cases, or elytra, protect the wings but leave most of the flexible abdomen exposed. A mobile abdomen allows the intertidal rove beetle to squeeze into narrow crevices and also to push its wings up under the elytra.

Most rove beetles are active either by day (diurnal) or by night (nocturnal), but the life of the intertidal rove beetle is dictated by the tides. It builds a vertical, wine-bottle shaped burrow in the sand with a living chamber about ¹/₅ in (5 mm) diameter and retreats into it whenever the tide comes in. The burrow entrance is so narrow—about ¹/₁₀ in (2 mm) in diameter—that the air pressure within prevents any water from entering. The female lays her eggs in side chambers within the burrow and remains on guard, until her offspring have hatched and are mature enough to leave and construct their own burrows.

Marine Skater

Halobates sericeus

LENGTH	Females: ¹/₅ in (5 mm)
HABITAT	Ocean surface

DISTRIBUTION Pacific Ocean between 40° north and 5° north and south of the equator

This is a member of the only truly marine genus of insects. The marine skater spends its entire life on the surface of tropical and subtropical oceans where winter temperatures rarely fall below 68°F (20°C). Little is known about these insects due to the difficulty in studying them. Females are larger than males and after mating they lay 10–20 cream-colored, oval eggs on a piece of flotsam, such as a piece of floating wood. The eggs hatch into nymphs that molt through five stages before becoming adults.

Because this insect never dives below the surface, its diet is restricted to small organisms such as floating fish eggs, zooplankton, and dead jellyfish. It feeds by releasing enzymes onto the surface of its food and then drawing in the predigested material through its modified mouthparts.

Kelp Fly

Coelopa frigida

LENGTH	¹/₈–²/₅ in (3–10 mm)
HABITAT	Temperate shores with rotting seaweed

DISTRIBUTION North Atlantic and north Pacific shorelines

The most widely distributed of the seaweed flies, the kelp fly is found almost everywhere there is rotting seaweed along a strand line. They have flattened, lustrous black bodies, tinged with gray, and bristly, brownish yellow legs. Of the two pairs of wings, only the front pair is functional, the hind pair being modified to small club-shaped halteres that act as stabilizers when in flight. Kelp flies can crawl through vast layers of slimy seaweed without getting stuck, and if immersed in seawater they simply float up to the surface and fly off. Their larvae are equally waterproof. Strongly attracted to rotting seaweed by its smell, the female kelp flies seek out warm spots in which to lay their eggs. The larvae hatch and feed on the seaweed around them. After three molts they pupate; the adults emerge and complete the life cycle about 11 days after the eggs were laid. Kelp flies are an important food source for several coastal birds, including kelp gulls and sandpipers.

Bryozoans

DOMAIN	Eucarya
KINGDOM	Animalia
PHYLUM	Bryozoa
SPECIES	6,085

THESE COLONIAL ANIMALS live attached to the seabed and although numerous, they are often overlooked.

The individuals making up each colony are usually less than 1/32 in (1 mm) long, but the colonies may span over 3 ft (1 m). Bryozoans are also called ectoprocts or sea mats, the latter name referring to their tendency to encrust the surfaces of stones and seaweeds. Other colonial forms of bryozoan include coral-like growths, branched plantlike tufts, and fleshy lobes. Most species are marine, but a few live in fresh water.

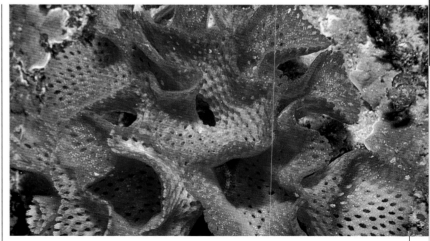

Anatomy

A bryozoan colony is made up of individuals called zooids, and may contain several or up to many millions. Each zooid is encased in a boxlike body wall of calcium carbonate or a gelatinous or horny material (chitin), and a small hole links it to other zooids. To feed, the animal pushes a circular or horseshoe-shaped structure out of an opening. Called a lophophore, this is crowned by tentacles covered in tiny, beating hairs that draw in planktonic food. In most species, fertilized eggs are stored in specialized zooids that form a brood chamber for developing larvae.

MAT OF ZOOIDS
This encrusting species of bryozoan has rectangular zooids joined in a single layer. The resulting mat spreads over seaweeds.

Habitats

With their great variety of body form, bryozoans can be found in almost any habitat from the seashore to the deep ocean, and from Arctic waters to tropical coral reefs. Colonies are most often found firmly attached to submerged rocks, seagrasses, seaweeds, mangrove roots, and dead shells, but some encrusting species even hitch a ride on the shells of living crustaceans and mollusks. A few unusual species do not need a surface for support and can live in the sand; these bryozoan colonies can move slowly over or through the sand by coordinated rowing movements of a long projection found on specialized zooids.

Bryozoan colonies originate from a single larva that settles on the seabed and becomes a zooid. More zooids are added to the colony by budding, a process in which a new zooid grows out from the side of the body wall. Most bryozoan larvae are short-lived and settle near the parent.

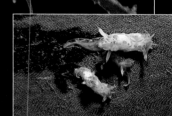

BRYOZOANS UNDER ATTACK
Sea slugs often make a meal of encrusting bryozoans, breaking into each zooid and eating the insides.

OCEAN LIFE

Gelatinous Bryozoan
Alcyonidium diaphanum

SIZE (HEIGHT)	Up to 12 in (30 cm)
DEPTH	0–656 ft (0–200 m)
HABITAT	Rocks and shelly sand

DISTRIBUTION Temperate waters of northeastern Atlantic

Colonies of this bryozoan have a firm, rubbery consistency and grow as irregular, lobed, or fingerlike growths that attach to the substratum with a small, encrusting base. This species may cause an allergic dermatitis when handled, and North Sea fishers are often affected when their trawl nets have gone through areas of dense bryozoan undergrowth.

Hornwrack
Flustra foliacea

SIZE (HEIGHT)	Up to 8 in (20 cm)
DEPTH	0–330 ft (0–100 m)
HABITAT	Stones, shells, rock

DISTRIBUTION Temperate and Arctic waters of northeastern Atlantic

This species is often mistaken for a brown seaweed. The colony grows up from a narrow base as thin, flat, fanlike lobes. These usually form dense clumps and cover the seabed like a crop of tiny brown lettuces. They litter the strandline on many shores in dried clumps and, by using a magnifying glass, an observer can easily see the individual, oblong colony members.

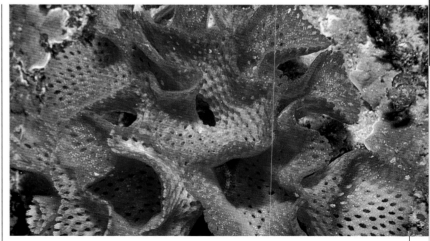

Pink Lace Bryozoan
Iodictyum phoeniceum

SIZE (WIDTH)	Up to 8 in (20 cm)
DEPTH	50–130 ft (15–40 m)
HABITAT	Rocky reefs

DISTRIBUTION Temperate and tropical waters around Australia

Pink lace bryozoan colonies feel hard and brittle to the touch because the walls of the individual zooids are reinforced with calcareous material. The colony is shaped like curly edged potato chips with a lacework of small holes. The beautiful dark pink to purple color remains even after the colony is dead and dried. This species prefers to live in areas with some current, and its holes may help reduce the force of the water against it. Similar species are found on coral reefs throughout the Indo-Pacific region.

Echinoderms

DOMAIN	Eucarya
KINGDOM	Animalia
PHYLUM	Echinodermata
CLASSES	5
SPECIES	About 7,483

THE NAME OF THIS PURELY MARINE group of invertebrates is derived from the Greek for "hedgehog skin." The group includes starfish, sea urchins, brittlestars, feather stars, and sea cucumbers. Echinoderms have radiating body parts, so most appear star-shaped, disk-shaped or spherical, and all have a skeleton of calcium carbonate plates under the skin. Inside is a unique system of water-filled canals, called the water-vascular system, that enables them to move, as well as to feed and breathe. Typically bottom-dwellers, they live on reefs, shores, and the seabed.

Anatomy

The echinoderm body is based on a five-rayed symmetry similar to the petals of a flower. This is apparent in starfish, brittlestars, and urchin shells (tests). Sea urchins are like starfish, with their arms joined to form a ball. Sea cucumbers resemble elongated urchins—their five-rayed symmetry can be seen end-on. The echinoderm skeleton is made of hard calcium carbonate plates, which are fused to form a rigid shell (as in urchins) or remain separate (as in starfish). Usually, it also features spiny or knoblike extensions that project from the body. Sea cucumbers have minimal skeletons reduced to a series of small, isolated plates. The water-vascular system consists of a ring canal with branches to each arm or body segment, as well as blind-ending tube feet that extend out through pores in the skin.

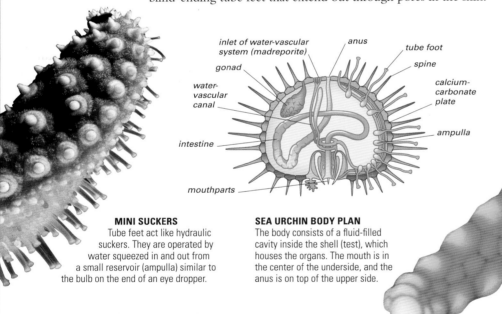

inlet of water-vascular system (madreporite)
anus
tube foot
gonad
spine
water-vascular canal
calcium-carbonate plate
ampulla
intestine
mouthparts

MINI SUCKERS
Tube feet act like hydraulic suckers. They are operated by water squeezed in and out from a small reservoir (ampulla) similar to the bulb on the end of an eye dropper.

SEA URCHIN BODY PLAN
The body consists of a fluid-filled cavity inside the shell (test), which houses the organs. The mouth is in the center of the underside, and the anus is on top of the upper side.

RADIAL SYMMETRY
This starfish has the typical five-rayed structure of echinoderms. Its arms are protected by hard plates, and its colors warn predators that it contains toxins.

ECHINODERM CLASSIFICATION

The echinoderms are split into five classes based on their shape, skeleton, and the position of their mouth, anus, and madreporite. A sixth class, "the Concentricycloidea," was at one time set up for the newly discovered sea daisies but these two species of disk-shaped animals are in fact a strange type of starfish and are now regarded as members of the class Asteroidea.

FEATHER STARS, SEA LILIES
Class Crinoidea

About 672 species
Also known as crinoids, these animals have a saucer-shaped body extending into five repeatedly branching, feathery arms used as filter-feeding appendages. Mouth and anus face upward. Sea lilies attach to the seabed by a jointed stalk, but feather stars break free when young to become swimming adults.

STARFISH OR SEA STARS
Class Asteroidea

About 1,921 species
The body of these mostly seabed scavengers is star-shaped, with five or more stout arms

merging into a central body disk. On the underside of the arms are rows of numerous tube feet and a groove, along which they pass food to the central mouth. The mouth is on the underside, and the anus and madreporite are on the upper surface. The skeleton is a layer of plates (ossicles) embedded in the body wall.

BRITTLESTARS, BASKET STARS
Class Ophiuroidea

About 2,111 species
These echinoderms have a disk-shaped body with five narrow, flexible arms. Basket stars' arms are branched and finely divided. The skeleton is a series of overlapping plates. The mouth, on the underside, doubles as an anus.

SEA URCHINS, SAND DOLLARS
Class Echinoidea

About 1,015 species
Body shape ranges from a disk (sand dollars) to a sphere (urchins) with five double rows of tube feet. The skeletal plates join to form a rigid shell (test) with movable spines.

SEA CUCUMBERS
Class Holothuroidea

About 1,764 species
These echinoderms have a sausage-shaped body with five double rows of tube feet, with those encircling the mouth modified into feeding tentacles. The skeleton comprises small, multishaped plates.

Reproduction

Most echinoderms have separate males and females, which reproduce by releasing sperm and eggs, respectively, into the water. Individuals often gather to spawn at the same time, thereby increasing their chance of success. This synchronized spawning is initiated by factors such as daylight length and water temperature. Each echinoderm group has its own type of larva with its own way of swimming, floating, and feeding. The larvae eventually transform into their adult form and settle on the seabed. However, some starfish keep their fertilized eggs and developing larvae in a pouch under their mouth, and nourishment comes in the form of yolk. In some brittlestars, the larvae are brooded in sacs inside the body, and the young are released after metamorphosis. In most echinoderm species, however, the fertilized eggs drift in the plankton and develop into free-floating larvae.

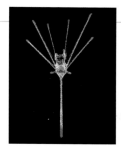

FLOATING AIDS
Long, paired arms help sea-urchin larvae, such as this one from a sea potato or heart urchin, to float in the plankton. Brittlestars have similar larvae.

RELEASING SPERM AND EGGS
By rearing up to spawn, sea cucumbers ensure that their eggs drift away to mix with sperm released by another individual.

Feeding

Echinoderms range from peaceful grazers and filter feeders to voracious predators. Carnivorous species of starfish extend their stomach over their prey and digest it externally. In contrast, most sea urchins are grazers, scraping rock surfaces using teeth that resemble the chuck of an electric drill. Combined with muscles and skeletal plates, they form a complex, powerful feeding apparatus called Aristotle's lantern. Sea cucumbers have an important role as seabed cleaners, vacuuming up organic debris and mud.

FILTER FEEDING BY TUBE FEET
Feather stars raise their arms to trap plankton. Specialized tube feet then pass the food particles along grooves in the arm branches, then along central arm grooves to the mouth.

Defense

If they can be broken open, sea urchins make a good meal for fish, sea birds, and sea otters. So, along with many other echinoderms, they protect themselves from predators with their long, sharp spines. These spines are mounted on ball-and-socket joints and can move in all directions, which turns them into fearsome weapons. If an echinoderm is attacked, spines may break off and embed themselves in the predator, creating a wound. Some spines are also venomous, such as those belonging to the crown-of-thorns starfish. Fire urchins and flower urchins also have enlarged and venomous, pincerlike pedicellariae (see below), which are strong enough to sting humans.

The cumbersome-looking sea cucumbers have no spines or protective plates, but they are far from defenseless. If attacked, many eviscerate their gut (and sometimes other internal organs) as a decoy and regrow them later. Some tropical sea cucumbers eject sticky white threads, called Cuverian tubules, which are strong enough to entangle and restrain an attacking crab. Brittlestars and starfish can break free from attack by discarding and regrowing arms.

PEDICELLARIAE
Slow-moving urchins and starfish can become overgrown by planktonic larvae looking for a place to settle. They defend themselves by using spines modified into tiny pincers, called pedicellariae, to catch and crush the larvae.

FIGHTING OVER OYSTERS

In European waters, the common starfish has a voracious appetite for oysters and mussels. So, fishers dredging the shellfish beds used to cut up the starfish and throw the pieces overboard. Unfortunately for the fishers, this tactic proved ineffective, because starfish can not only regenerate their limbs, but if a lost limb retains part of the central body disk, it is able to completely regenerate the body.

REGENERATING LIMBS
This common starfish is regrowing its two lost arms. Sometimes, regrowth produces one or more extra limbs.

CLASS ASTEROIDEA
Seven-arm Starfish
Luidia ciliaris

DIAMETER	Up to 24 in (60 cm)
DEPTH	0–1,300 ft (0–400 m)
HABITAT	Sediment, gravel, rock

DISTRIBUTION Temperate waters of northeastern Atlantic and Mediterranean

While the majority of starfish species have five arms, this large species has seven arms and very occasionally eight. Its body and arms have a velvety texture and are colored brick-red to orange-brown. Each arm is fringed with a conspicuous band of multidirectional, stiff white spines, which help the seven-arm starfish to bury

itself in sediment and delve after prey. Using its long tube feet, it can also move very quickly over rocks and gravel to latch onto its victims. This starfish is a voracious predator that feeds mainly on other echinoderms, including burrowing sea urchins, sea cucumbers, and brittlestars.

It breeds during the summer in southern Britain but much earlier, between November and January, in the Mediterranean.

multidirectional, stiff white spines on arm

velvety red or orange skin

CLASS ASTEROIDEA
Cushion Star
Culcita novaeguineae

DIAMETER	Up to 12 in (30 cm)
DEPTH	0–100 ft (0–30 m)
HABITAT	Coral reefs

DISTRIBUTION Adaman Sea and tropical waters of western Pacific

The cushion star looks more like a spineless sea urchin than a starfish; it gets its name from its plump, rounded body. Its arms are so short that they merge with its body and only their tips can be seen. Juveniles are much flatter than adults and have a clear pentagonal star shape, with obvious arms. They hide under rocks to escape predators, whereas the tougher adults are relatively safe in the open.

Cushion stars occur in a wide range of colors, from predominantly red to green and brown. The underside has five radiating grooves that represent the arms and are filled with tube feet. If the starfish is turned over, it can right itself by stretching out the tube feet on one side, anchoring them to the seabed, and pulling. It feeds mainly on detritus and fixed invertebrates, including live coral. Two other similar species are found in the Indo-Pacific tropics but this one is the most common and widespread.

thick, soft body

LIVE-IN GUESTS

The surface of the cushion star provides a home for a tiny shrimp, the sea-star shrimp *Periclimenes soror*. The shrimp does no harm to its host and is also found on other starfish. It often hides beneath the starfish and also matches its color to that of its host.

COMMENSAL SHRIMP
The sea-star shrimp seen here is on the underside of the cushion star but will venture out onto the top to feed.

CLASS ASTEROIDEA
Icon Star
Iconaster longimanus

DIAMETER	Up to 5 in (12 cm)
DEPTH	100–280 ft (30–85 m)
HABITAT	Deep reefs and slopes

DISTRIBUTION Tropical waters of Indian Ocean and western Pacific

This strikingly patterned species has long, thin arms and a flat disk. The arms and disk are edged by rows of skeletal plates that protect the starfish

and give it a rigid feel. These plates may be pale or dark and they form unique patterns on each individual, which enable researchers to recognize, track, and monitor individuals in the field. Data from such studies indicates that icon stars grow very slowly and suggests that the largest individuals may live for several decades. Although icon stars usually live in deeper, dark waters, they are common at depths of 15–65 ft (5–20 m) around Singapore, probably because the water there is turbid and light levels are low. Females produce large orange eggs that develop into tiny orange larvae; the eggs contain chemicals that deter fish predators.

CLASS ASTEROIDEA
Goosefoot Starfish
Anseropoda placenta

DIAMETER	Up to 8 in (20 cm)
DEPTH	30–1,600 ft (10–500 m)
HABITAT	Gravel, sand, mud

DISTRIBUTION Temperate and warm waters of northeastern Atlantic and Mediterranean

The goosefoot starfish gets its name from the appearance of the flattened, weblike disk that joins each of its five short arms together and produces an almost pentagonal shape. The central

portion of its disk is marked with a dark red patch, and conspicuous red lines radiate outward from this patch, along its arms. Its underside is colored yellow. This starfish glides slowly over the seabed searching for small crustaceans, mollusks, and other echinoderms to eat.

CLASS ASTEROIDEA
Mosaic Sea Star
Plectaster decanus

DIAMETER	Up to 6 in (16 cm)
DEPTH	30–600 ft (10–180 m)
HABITAT	Rocky reefs

DISTRIBUTION Temperate waters of South Australia

The incredibly bright colors of this starfish are a warning that it contains toxic chemicals. It should not be touched, as it can cause numbness. A mosaic of raised yellow ridges covers its red upper surface and it has a soft texture. The mosaic sea star feeds mostly on sponges, and these may be the source of its toxins. When the females spawn, the fertilized eggs are retained and brooded on the underside of the body.

CLASS ASTEROIDEA

Crown-of-thorns Starfish

Acanthaster planci

DIAMETER	Up to 20 in (50 cm)
DEPTH	3–65 ft (1–20 m)
HABITAT	Coral reefs

DISTRIBUTION Tropical waters of Indian and Pacific Oceans

Occasional plagues of this large and destructive starfish sometimes kill extensive areas of coral on the Great Barrier Reef of Australia on reefs in the Indo-Pacific. There has been much debate on whether such plagues are natural or are caused by overfishing of their few predators, such as the giant triton, *Charonia tritonis* (see p.285). Land runoff appears to stimulate plagues, and research on the Great Barrier Reef suggests that some outbreaks are caused by high input of nutrients from land. With up to 20 arms and a formidable covering of long spines, this species has few predators. The spines are mildly venomous and may inflict a painful wound if the starfish is picked up with bare hands. Crown-of-thorns starfish feed on corals by turning their stomach out through their mouth and digesting the coral's living tissue. Pure white coral skeletons indicate that this starfish has been feeding recently in the area. In the past decade, some outbreaks have been controlled using a single injection of bile salts, which has proved more effective than other chemicals. Research is ongoing into this and other control techniques, but tackling the reasons behind such outbreaks is also vital.

CLASS OPHIUROIDEA

Serpent Star

Astrobrachion adhaerens

DIAMETER	Up to 12 in (30 cm)
DEPTH	50–600 ft (15–180 m)
HABITAT	Antipatharian black corals

DISTRIBUTION Tropical and temperate waters of Australia and southwestern Pacific

arm coiled around black coral branch

Serpent stars are a type of brittlestar with long, flexible arms. During the day they wind their arms tightly round the branches of the deep-water black corals among which they live. At night they uncoil their arms and move around, feeding on the living polyps of their host. A black coral bush one or two yards high may be host for up to 40 or so serpent stars.

CLASS OPHIUROIDEA

Common Brittlestar

Ophiothrix fragilis

DIAMETER	Up to 5 in (12 cm)
DEPTH	0–500 ft (0–150 m)
HABITAT	Rocks, rough and gravely ground

DISTRIBUTION Temperate and warm waters of eastern Atlantic

This large brittlestar species gathers in dense groups that may cover several square miles of seabed in areas where there are strong tidal currents. They have been recorded at densities of 2,000 individuals per square yard. Each brittlestar holds up one or two arms into the current to feed on plankton, while linking its remaining arms with surrounding individuals to form a strong mat and prevent itself being swept away. They vary greatly in color, ranging from red, yellow, and orange to brown and gray, and they often have alternate light and dark bands on the arms. The fragile arms of this species are covered in long, untidy spines, while its small disk, which is only 1 in (2 cm) across, has a covering of shorter spines. In the intertidal zone, common brittlestars are not usually found in groups, but occur as individuals hiding in crevices and beneath stones.

OCEAN LIFE

CLASS ECHINOIDEA

Sand Dollar
Sculpsitechinus auritus

DIAMETER	Up to 4 in (11 cm)
DEPTH	0–165 ft (0–50 m)
HABITAT	Clean sand

DISTRIBUTION Tropical and warm waters of Indian Ocean, Red Sea, western Pacific

CLASS ECHINOIDEA

Long-spined Sea Urchin
Diadema savignyi

DIAMETER	Up to 9 in (23 cm)
DEPTH	0–230 ft (0–70 m)
HABITAT	Coral and rocky reefs

DISTRIBUTION Tropical waters of Indian and western Pacific Oceans

Sand dollars are sea urchins that have become extremely flattened as an adaptation for burrowing through sand. A mat of very fine spines covers the shell, or test, and the pattern of the animal's skeleton plates can often be seen through the skin. The mouth is on the underside. At the rear are two notches that open at the margins of the test, and water currents passing through these slits are thought to help to push the urchin down and prevent it being swept away.

Many divers on coral reefs have learned to avoid these sea urchins. They bristle with long, sharp spines that can easily wound, even through a wetsuit. The spines are mildly venomous and so brittle that they may break off in the wound. If a diver or predator comes near to it, this sea urchin waves its spines about vigorously. Only a few tough fish, such as the titan triggerfish, can successfully attack and eat such prickly prey. This species often has striped spines, while the other common Indo-Pacific long-spined species, *Diadema setosum*, has black spines.

CLASS ECHINOIDEA

Flower Urchin
Toxopneustes pileolus

DIAMETER	Up to 6 in (15 cm)
DEPTH	0–300 ft (0–90 m)
HABITAT	Sand, rubble, rocky reef

DISTRIBUTION Tropical waters of Indian Ocean, central and western Pacific Ocean

This species is extremely venomous and has caused rare fatalities. It has short, inconspicuous spines through which emerge an array of flowerlike appendages called pedicellariae. These help keep the urchin's surface clean but will sting animals that touch it. The pedicellariae also hold pieces of shell, rubble, and seaweed that shade the urchin from sunlight. Flower urchins may partially bury themselves, despite having few predators.

CLASS ECHINOIDEA

Edible Sea Urchin
Echinus esculentus

DIAMETER	Up to 6 in (16 cm)
DEPTH	0–160 ft (0–50 m)
HABITAT	Rocky areas

DISTRIBUTION Temperate waters of northeastern Atlantic

This large, spherical urchin is covered with uniform short spines that give it the appearance of a fat hedgehog. It is generally a pinkish color, with pairs of darker, radiating lines where its numerous tube feet emerge. These urchins are important grazers and can have much the same effect underwater as rabbits do on land, leaving the rocks covered only in hard pink encrusting algae. As their name suggests, the roe of this species can be eaten.

CLASS ECHINOIDEA

Purple Sea Urchin
Strongylocentrotus purpuratus

DIAMETER	Up to 4 in (10 cm)
DEPTH	0–130 ft (0–40 m)
HABITAT	Rocky reefs

DISTRIBUTION Temperate coastline of North America from Alaska to Mexico

This small sea urchin has been responsible for the demise of large areas of giant kelp forest off the North American coastline. Like most sea urchins, it feeds by scraping away at seaweeds and fixed animals and its favorite food is the giant kelp *Macrocystis pyrifera*. Unchecked by predators such as sea otters and California sheephead fish, numbers can reach densities of up to several hundred animals per square yard. Such large numbers can chew through kelp holdfasts, setting the plants adrift. In the past, when sea otters were hunted, urchin numbers increased explosively in some areas.

CLASS ECHINOIDEA

Sea Potato

Echninocardium cordatum

LENGTH	Up to 3¹/₂ in (9 cm)
DEPTH	0–650 ft (0–200 m)
HABITAT	Sand, muddy sand

DISTRIBUTION Temperate waters of northeastern Atlantic

Most sea urchins live in rocky areas, but the sea potato or heart urchin burrows in the sand. Unlike regular urchins it has a distinct front end and its spines are thin and flattened. Special spoon-shaped spines on the urchin's underside help it dig, while longer spines on its back allow water to funnel down into its burrow to be used for respiration. The dried shell, or test, of this urchin resembles a potato, hence the common name.

CLASS CRINOIDEA

Tropical Feather Star

Anneissia bennetti

DIAMETER	Up to 6 in (15 cm)
DEPTH	33–165 ft (10–50 m)
HABITAT	Coral reefs

DISTRIBUTION Tropical waters of western Pacific

All that can usually be seen of the tropical feather star is its numerous feathery arms held up into the water to trap food. This species has about a hundred arms, compared to the 10 that most temperate water feather stars have. The arms are attached to a small, disklike body and the mouth is on the upper side of the body, between the arms. The tropical feather star clings onto corals using numerous articulated, fingerlike appendages called cirri. It prefers elevated positions where it is exposed to food-bearing currents, and is active by both day and night. Like all feather stars, this species starts its early life by becoming attached to the seabed by a stalk—at this stage, it closely resembles a small sea lily. As it matures, the feather star breaks away and becomes free-living, leaving the stalk behind.

CLASS CRINOIDEA

Sea Lily

Neocrinus decorus

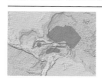

HEIGHT	Up to 24 in (60 cm)
DEPTH	500–4,000 ft (150–1,200 m)
HABITAT	Deep-sea sediments

DISTRIBUTION Tropical waters of western Atlantic Ocean

Sea lilies are stalked relatives of feather stars and usually remain fixed in the same place after developing from a settled planktonic larva. However, while *Neocrinus decorus* and other, similar sea lilies cannot swim in the way shallow-water feather stars do, they have been filmed dragging themselves over the seabed by their arms. To do this, they appear to break off the end of the stalk, then reattach to the substratum using flexible, fingerlike appendages on the stalk. In this way they can escape from predatory sea urchins. The stalk consists of a stack of disk-shaped skeleton pieces called ossicles, and looks like a vertebrate spinal column. Sea lilies feed by spreading out their numerous, feathery arms against the current and trapping plankton. Food particles are passed down the arms and into the mouth.

DISCOVERY

FOSSIL EVIDENCE

Today, there are relatively few sea lily species, all of which live in deep water, but this group once thrived in ancient seas. Entire fossil sea lilies such as this one are rare, but loose ossicles from their stalks are very common in some limestones.

CLASS CRINOIDEA

Passion Flower Feather Star

Ptilometra australis

DIAMETER	Up to 5 in (12 cm)
DEPTH	To at least 200 ft (60 m)
HABITAT	Rocky reefs, rubble

DISTRIBUTION Endemic to temperate waters of southern Australia

This stout feather star has 18–20 arms with long, stiff side branches called pinnules; the arms are different lengths, giving it a flowerlike appearance when viewed from above. They are called passion flowers by fishers because they are brought up in large numbers by commercial trawlers, embracing their nets with a tight grip. These feather stars are found in reefs and also in very shallow, sheltered bays and estuaries. Like most feather stars, the passion feather star is a filter feeder that grips onto the tops of rocks, sponges, and sea fans, where it spreads its arms wide to trap plankton and suspended detritus. It remains expanded both day and night but, like other feather stars, it can curl up its arms if disturbed or while resting. Its usual color is a burgundy red.

CLASS HOLOTHUROIDEA

Prickly Redfish

Thelenota ananas

LENGTH	Up to 28 in (70 cm)
DEPTH	16–100 ft (5–30 m)
HABITAT	Sandy areas of coral reefs

DISTRIBUTION Tropical waters of Indian Ocean and western Pacific

This massive sea cucumber looks like an animated rug as it crawls slowly over the seabed. The large, star-shaped papillae, called caruncles, that cover its body make it an unattractive proposition to potential predators. However, its appearance does not deter humans, and prickly redfish fetches high prices in parts of eastern Asia, where it is considered a delicacy. As a result, it is now endangered.

Large specimens reach up to 11 lb (5 kg) in weight and are traditionally collected by reef walking at low tide or by breath-hold diving. Other predators include various fish and crustaceans. The flat underside of the prickly redfish is covered with orange tube feet, which it uses to crawl over the seabed in search of detritus to eat. When spawning, prickly redfish gather together and rear up, then release eggs or sperm into the water from small pores near the head end of the body.

CLASS HOLOTHUROIDEA

Sea Cucumber

Pearsonothuria graeffei

LENGTH	Up to 12 in (30 cm)
DEPTH	15–165 ft (5–50 m)
HABITAT	Coral reefs

DISTRIBUTION Tropical waters of Indian Ocean, Red Sea, western Pacific

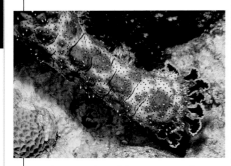

The juveniles of this sea cucumber look completely different from the adult (shown here). They are white with black lines and protruding yellow papillae, closely resembling sea slugs from the genus *Phyllidia*. These slugs are distasteful to fish, so this mimicry protects the young sea cucumbers. *Bohadschia graeffei* feeds by scooping sand and mud into its mouth using large black tentacles, which are modified tube feet. Organic material in the sediment is digested, while the remainder passes through the gut and is deposited outside, looking like a string of sausages. In areas where sea cucumbers are common, much of the surface sediment is vacuumed up and cleaned several times a year in this way.

CLASS HOLOTHUROIDEA

Edible Sea Cucumber

Holothuria edulis

LENGTH	Up to 14 in (35 cm)
DEPTH	15–100 ft (5–30 m)
HABITAT	Sand, rock, coral reefs

DISTRIBUTION Tropical waters of Indian and Pacific Oceans

As its name suggests, this is an edible species of sea cucumber, although it is not considered as good eating as others, such as the prickly redfish (above). It has a soft body, peppered with tiny warts, and is colored black on its back and pinkish red to beige underneath. Like many other large sea cucumbers, it is sometimes host to small pearlfish (family Carapidae) that live inside its body cavity.

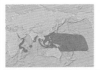

CLASS HOLOTHUROIDEA

Sea Apple

Pseudocolochirus violaceus

LENGTH	Up to 4 in (10 cm)
DEPTH	0–100 ft (0–30 m)
HABITAT	Rocks and reefs

DISTRIBUTION Tropical waters of Indo-Pacific region

The sea apple is one of the most colorful of all sea cucumbers and is widely collected for use in marine aquariums. It has a red and purple body and yellow tentacles. The species in the genus *Pseudocolochirus* are often hard to distinguish and for this reason the distribution map may include several species. The sea apple uses its tube feet to attach itself to rocks then extends its branched tentacles into the water to trap organic particles. From time to time, each tentacle is pushed into the mouth and the food it has trapped is wiped off.

CLASS HOLOTHUROIDEA

Deep-sea Cucumber

Laetmogone violacea

LENGTH	Not recorded
DEPTH	To at least 8,000 ft (2,500 m)
HABITAT	Soft sediments

DISTRIBUTION Deep, cold waters of Atlantic, Indian, and Pacific Oceans

In the deep ocean, sea cucumbers are one of the dominant sea-floor groups all over the world. *Laetmogone violacea* is one of a large number of species that crawl over the soft, muddy ocean bottom eating organic detritus. Its peglike "legs" may help to keep it from sinking too far into the mud. Ingested mud that the cucumber is unable to digest leaves its body as fecal casts, which may then be eaten again by other sea cucumbers. Like many deep-sea animals, this sea cucumber is almost colorless but glows all over with bioluminescent light; exactly how the animal uses this light is not yet known. Some deep-sea starfish are known to light up when approached by a predator, which may scare it away. It may be that the deep-sea cucumber uses its bioluminescence in the same way.

Small, Bottom-living Phyla

DOMAIN	Eucarya
KINGDOM	Animalia
PHYLA	More than 10
SPECIES	Many

MANY DISPARATE INVERTEBRATES play important parts in marine ecosystems but are seldom seen, because they are small or their habitats are difficult to study. Like all animals, they are grouped into phyla, each phylum representing an apparently distinct body plan. Most of these small, bottom-living phyla live in seabed sediment and are called "worms" due to their shape and burrowing lifestyle. However, the superabundant nematodes or roundworms of which there are at least 20,000 described species and possibly a million in total, live in a wide range of environments, including the seabed. A selection of these bottom-living phyla are represented below.

Sand-grain Animals

A community of tiny animals, referred to collectively as meiofauna, lives in the surface water film between the sand grains on beaches and in shallow water. They range in size from $\frac{1}{100}$ in to $\frac{1}{2}$ in (0.05 mm to 1 mm) and so can only be seen well with a microscope. Many have intricate and beautiful shapes. Almost every marine invertebrate phylum has representative species that live in this habitat, and some, such as gastrotrichs, occur virtually nowhere else. A diverse meiofauna is a good indication of a healthy environment, since these minuscule organisms are the basis for many marine food chains.

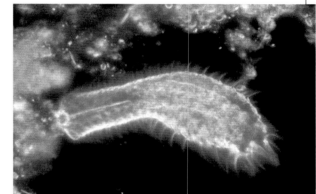

SAND COMMUNITY
Many invertebrates exist in the watery spaces between grains of coastal sand. The wormlike gastrotrich (phylum Gastrotricha) shown in this photo-micrograph is one of many species found in this habitat.

FEEDING APPARATUS
Female spoonworms (phylum Echiura) sweep up organic material and sediment with a scooplike proboscis, seen here extending from a burrow.

Mud-swallowers

The consistency and structure of seashore and seabed sediments depends largely on the many wormlike phyla that live there. Millions of these burrowing animals continually mix and rework the sediment, a process called bioturbation. Lugworms are famous for their ability to eat sand, depositing inedible material on the sand surface in the form of coiled heaps, or casts, but many of the less well-known groups, including peanut worms, acorn worms, and kinorhynchs, are as important. Organic material washed onshore with each tide or carried by currents is quickly incorporated into the sediment as the animals move around, or is processed as the surface mud is eaten.

New Phyla

Scientists have so far described only a fraction of the species that live in oceans. New species are being discovered all the time, mostly in groups such as sponges and soft corals that traditionally have been neglected. Occasionally a species is found that is fundamentally different from all other known organisms and so it is classified as belonging to a new phylum. Most of these exciting discoveries are from inaccessible areas such as deep-sea mud, and the animals are usually small. But when deep-sea hydrothermal vents (see pp.188–189) were visited by submersibles in the 1970s, gigantic tube worms like no others were found.

HORSESHOE WORMS
These sedentary worms (phylum Phoronida) live in small tubes buried in sand or mud or (as here) attached to seabed rocks. To feed, the worms extend a horseshoe shaped net of tentacles.

NEW SPECIES
Symbion americanus (above) lives on the mouthparts of the American lobster (left). It was first described in 2006, and is the second species in a new phylum of animals, the Cycliophora (see p.316).

OCEAN LIFE

Horseshoe Worm

Phoronis hippocrepia

LENGTH Up to 4 in (10 cm)

DEPTH 0–165 ft (0–50 m)

HABITAT Rocks and empty shells

DISTRIBUTION Shallow coastal waters of Atlantic Ocean, northeastern and western Pacific

Horseshoe worms are easily overlooked but they sometimes cover large areas of rock with their narrow, membranous tubes. The animal lives inside its tube, which encrusts rock surfaces or can bore into shells or limestone rock so that only the top part of the tube shows. The end of the wormlike body is thickened and anchors the animal in its tube. The feeding head with its horseshoe of delicate ciliated tentacles is extended to catch tiny planktonic animals while the body remains hidden in the tube. The feeding head is called a lophophore and is found in all members of the phylum. Horseshoe worms brood their egg masses within the lophophore, and larvae are continually released to drift and develop in the water.

PHYLUM SIPUNCULA

Peanut Worm

Golfingia vulgaris

LENGTH Up to 8 in (20 cm)

DEPTH 0–6,560 ft (0–2,000 m)

HABITAT Muddy sand and gravel

DISTRIBUTION Northeastern Atlantic and eastern Mediterranean; possibly Indo-Pacific, Southern Ocean

The peanut worm is shaped like a half-inflated sausage balloon. Its body is stout and has a long, thin region at the front called the introvert, which can be stretched right out or withdrawn completely inside the body. The animal has a crown of short tentacles around the mouth at the end of the introvert. It lives buried in sediment, which it eats as it burrows and digests any organic matter.

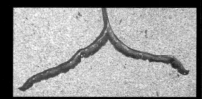

plump body · extended introvert

PHYLUM NEMATODA

Roundworm

Dolicholaimus marioni

LENGTH Up to ½ in (5 mm)

DEPTH Intertidal

HABITAT Among algae in rock pools

DISTRIBUTION Shores of the northeastern Atlantic

It is hard to see which end is which on a roundworm, as both ends of its thin body are pointed. The body is round in cross-section and has longitudinal muscles but no circular ones. This results in a characteristic way of moving in which the body is thrashed in a single plane forming C- or S-shapes in the process. This is a marine species, but roundworms also occur in vast numbers in the soil and fresh water.

PHYLUM ECHIURA

Spoonworm

Bonellia viridis

LENGTH Up to 6 in (15 cm)

DEPTH 3–330 ft (1–100 m)

HABITAT Muddy rocks

DISTRIBUTION Coastal temperate waters of northeastern Atlantic and Mediterranean

New genetic research has led some scientists to reclassify spoonworms as polychaetes (Phylum Annelida, see p.274). Female spoonworms have a proboscis that stretches out like an elastic band and can reach at least 3 ft (1 m) away in search of food. The worm's green, pear-shaped trunk remains hidden between rocks, safe from predators. In this species, the tip of the proboscis is forked, and usually this is all that can be seen of the worm. The proboscis collects food particles with the help of sticky mucus, and the food is moved along the proboscis and into the mouth by the whipping movements of hairlike cilia.

PHYLUM HEMICHORDATA

Purple Acorn Worm

Yoda purpurata

LENGTH 4½–7½ in (12–19 cm)

DEPTH About 8,200 ft (2,500 m)

HABITAT Deep-sea bed

DISTRIBUTION North Atlantic

Dressed in flamboyant pink, this small animal nevertheless lives an inconspicuous life browsing on the deep-sea floor. It was first discovered using a remotely operated submersible, more than 1.2 miles (2 km) below the surface. This species belongs to a group often referred to as acorn worms, which despite their wormlike appearance, provide clues to the evolutionary link between invertebrates and vertebrates. Unusually for a Hemichordate, it is a hermaphrodite with both ovaries and testes. The two earlike projections on either side of the head region may be responsible for its scientific name—Yoda is a fictional character in the *Star Wars* films, famous for his projecting ears.

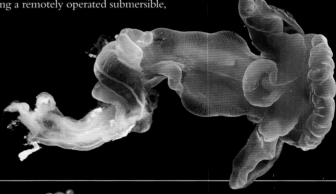

PHYLUM HEMICHORDATA

Pterobranch Worm

Rhabdopleura compacta

LENGTH Up to ½ in (5 mm)

DEPTH Not recorded

HABITAT Attached to sessile animals

DISTRIBUTION Cold waters of northern hemisphere

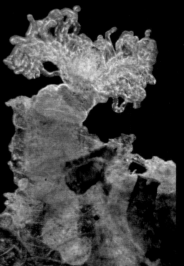

Like the acorn worms to which they are related (see above), pterobranch worms live in a thin tube and their bodies are divided into a proboscis, collar, and trunk. They also have a pair of arms covered in tentacles arising from the collar region. The tubes of many individuals are connected together with strands of soft tissue that join the trunks of the animals, enabling them to form a colony.

PHYLUM CEPHALORHYNCHA

Priapula Worm

Priapulus caudatus

LENGTH Up to 4 in (10 cm)

DEPTH 0–8,200 ft (0–2,500 m)

HABITAT Buried in sediment

DISTRIBUTION Cold waters of north Atlantic and Arctic Ocean

The stout, cylindrical body of this animal is divided into a short barrel-shaped proboscis at the head end, a longer trunk region, and a tail that consists of small bladders attached to a hollow stalk. The proboscis can be withdrawn into the trunk. The mouth on the end of the proboscis is edged with spines, which help the animal to seize other small marine worms for food.

PHYLUM CEPHALORHYNCHA

Mud Dragon
Echinoderes aquilonius

LENGTH	Less than 1mm
DEPTH	Shallow water
HABITAT	Muddy sediments
DISTRIBUTION	Northwestern Atlantic

Mud dragons resemble miniature insect pupae. The body appears segmented on the outside but this is superficial. It is covered with a thick, articulated cuticle and sharp spines on each body section. The tail end has a bunch of longer spines, and the head region has several rings of spines. The mouth is situated on the end of a cone-shaped structure. The animal can withdraw the entire head region into the rest of the body for protection, but it can also close the resulting hole with special plates, which are called placids. The head spines help the animal to push its way through the sediment, feeding on organic debris, bacteria, protists, and diatoms.
The 300 or so species of mud dragons are all marine. The sexes are separate but look similar. The eggs develop into free-living larvae that molt several times before attaining the adult form.

PHYLUM VESTIMENTIFERA

Lingulid Brachiopod
Glottidia albida

HEIGHT	1 in (2 cm)
DEPTH	0–1,476 ft (0–450 m)
HABITAT	Subtidal sediment
DISTRIBUTION	Coastal waters of northeastern Pacific, California to Mexico

Lingulid brachiopods resemble small clams with strange long tails. The tail is, in fact, a stalk, known as a pedicle, that emerges from between the brachiopod's shell valves.
Brachiopods have a two-part shell similar to a clam, but these two types of animals are not closely related. Many brachiopods use their pedicle to attach to rocks (see lamp shell below), but lingulid brachiopods live in burrows in sand and mud. The pedicle is two to three times the length of the shell and is used to make a burrow in the soft sediment in which it lives. When filtering plankton from the water it comes to the top of the burrow and opens its shell using special muscles, but the lingulid brachiopod can quickly disappear by

pulling itself down with its pedicle. Fossils of brachiopods with a similar shape appear in rocks that are 500 million years old.

PHYLUM BRACHIOPODA

Lamp Shell
Terebratulina septentrionalis

LENGTH	Up to 1¼ in (3 cm)
DEPTH	0–4,000 ft (0–1,200 m)
HABITAT	Rocks and stones
DISTRIBUTION	Temperate and cold waters of north Atlantic

It would be easy to mistake a lamp shell for a small bivalve mollusk, as both have a hinged shell in two parts and live attached to the sea floor. lamp shells, however, have a very thin, light shell and the two parts are different sizes, with the smaller one fitting into the larger. The shell valves cover the dorsal and ventral surfaces of the animal whereas in bivalve mollusks they are on the left and right side of the body. Most lamp shells attach their pear-shaped shell to hard surfaces by means of a fleshy stalk that emerges from a hole in the ventral shell valve. With the shell valves gaping open, the animal draws in a current of water that brings plankton with it. Taking up most of the space inside the shell is a feeding structure called the lophophore, which consists of two lateral lobes and a central coiled lobe covered in long ciliated tentacles. The beating of the cilia creates the water current. Lamp shells are found worldwide, but they are especially abundant in colder waters. In the northeastern Atlantic, *Terebratulina septentrionalis* is mostly found in deep water, while along the east coast of North America, it commonly occurs in shallow water. This species is very similar to *Terebratulina retusa*.

OCEAN LIFE

Pseudobiotus Water Bear

Pseudobiotus megalonyx

LENGTH Up to 1 mm

DEPTH Shallow water

HABITAT Muddy sediments

DISTRIBUTION Northwestern Atlantic

Although still tiny, this water bear is one of the largest and can be found by sampling tidal mud flats in the upper estuaries of rivers in northern Europe. The female in the photograph below has laid her eggs in her own molted cuticle, which she holds like a knapsack on her back. This is one of the few tardigrades that have been seen to mate. Males of this species grip the female and deposit sperm through the cloacal opening of her molted cuticle.

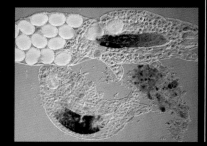

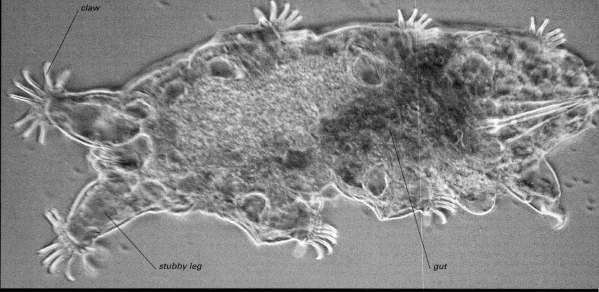

claw

stubby leg

gut

Echiniscoides Water Bear

Echiniscoides sigismundi

LENGTH Less than 1 mm

DEPTH Intertidal

HABITAT Marine sands

DISTRIBUTION Worldwide

This species of water bear lives in the spaces between sand grains in marine sediments, as do most of the other 200 or so marine species. Many other species live in fresh water, especially in the thin layer of water around damp-loving plants such as mosses, and in terrestrial environments. Water bears have a short, plump body without a well-defined head but with eyespots and sensory appendages at one end. There are four pairs of short stubby legs each ending in a bunch of tiny claws on which the animal lumbers slowly along. The relatively thick skin protects against abrasion from sand grains. The sexes are separate, but there are few males and the eggs can probably develop without being fertilized. The nearest relatives of these tiny animals are thought to be arthropods. Water bears are so tough that terrestrial species can withstand drying and freezing.

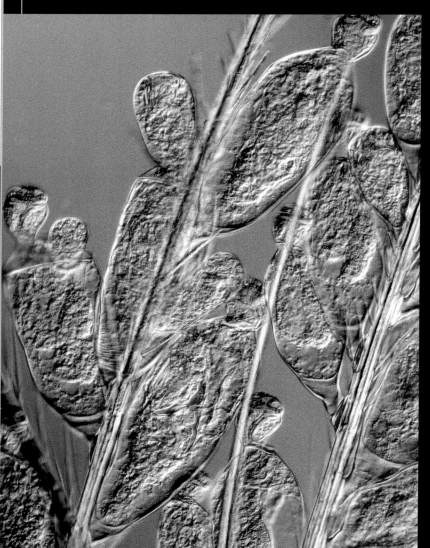

Cycliophoran

Symbion pandora

LENGTH 0.3 mm

DEPTH Not recorded

HABITAT Mouthparts of the Norway Lobster

DISTRIBUTION Northeast Atlantic

This species was first described in 1995 by two Danish biologists. It was found clinging to the mouthparts of a Norway lobster (*Nephrops norvegicus*) that was dredged up from the North Sea, and the biologists must have looked very closely to have seen it at all. Individuals have a rounded body and are attached to the substratum by a short stalk and adhesive disk. They feed by means of a mouth funnel surrounded by cilia and excrete via an anus next to the mouth. This feeding stage is neither male nor female, and the reproductive cycle is complex involving both sexually and asexually produced free-swimming larvae.

Symbion pandora was the first representative of the phylum Cycliophora. It has since been joined by a second species, *S. americanus,* discovered in 2006, which lives similarly on the American lobster (*Homarus americanus*). Molecular studies show that both species may be related to bryozoans (p.305) and tiny animals called entoprocts. It seems likely that, with a careful search, other species will soon be discovered.

Gastrotrich

Turbanella species

LENGTH Less than 1 mm

DEPTH Not recorded

HABITAT Well-oxygenated sediments

DISTRIBUTION Not recorded

Gastrotrichs are found in both fresh water and the sea, but *Turbanella* is a marine genus that lives in the spaces between sand grains in sea-floor sediments. It looks similar to a ciliated protist, but is a true multicellular animal with a mouth, gut, kidney cells, and other structures. It has several adhesive tubes, structures that secrete a sticky substance and help the animal attach to the substratum. By attaching and detaching the adhesive tubes at the front and rear of its body, it can loop around rather like a leech. Alternatively, it can glide using its cilia, searching for bacteria and protists to eat.

gut

cilia

Planktonic Phyla

DOMAIN	Eucarya
KINGDOM	Animalia
PHYLA	Ctenophora
	Chaetognatha
	Rotifera
SPECIES	About 2,337

OF THE MANY MAJOR GROUPS (phyla) of invertebrate animals, just a few are almost or entirely composed of planktonic animals. Three such phyla—comb jellies (Ctenophora), arrow worms (Chaetognatha), and rotifers (Rotifera)—are included here. In spite of their small size, many have a complex anatomy. All are ecologically important because the plankton community underpins all ocean food chains.

Predators

Carnivorous zooplankton have many methods of catching prey. Comb jellies are voracious predators—most trap their prey with a sticky secretion released from special "lasso" cells (colloblasts) lining their two trailing tentacles. Some draw in prey using negative pressure created by rapidly opening their mouths. Species of *Haeckelia* even recycle the stinging cells from their cnidarian prey. Arrow worms have vibration sensors to detect prey, which is caught and held by movable mouth hooks. Large prey is paralyzed by neurotoxins.

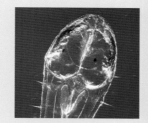

HOOKING PREY
This arrow worm has brown-colored hooks on either side of its circular mouth for holding prey.

Grazers

Some of the zooplankton are herbivorous, grazing on phytoplankton or filter feeding. Some planktonic rotifers feed on organic particles suspended in the water. The cilia on the crown that surrounds the oral cavity waft water into a food groove leading to the mouth. Here, food particles are sifted and returned to the pharynx where jawlike structures, called trophi, grind the food before it passes into the stomach. Trophi are unique to rotifers.

ROTIFER FEEDING METHOD
The rotifer's ciliated crown, used in locomotion and filter feeding, is visible on the left.

GLEAMING CILIA
Comb jellies, or sea gooseberries, swim by beating eight vertical rows of cilia combs, which shimmer with iridescent colors.

OCEAN LIFE

PHYLUM CTENOPHORA

Creeping Comb Jelly

Coeloplana astericola

DIAMETER	½ in (1 cm)
DEPTH	Not recorded
HABITAT	On the orange sea star

DISTRIBUTION Tropical waters of western Pacific

While most comb jellies live in the plankton, creeping comb jellies have taken up a bottom-living existence. Instead of the more usual rounded shape, they are flattened and look like a tiny squashed ball. The mouth is in the center of the underside with the statocyst, a balancing organ found in all comb jellies, opposite it on the upper side. Comb rows are absent as they have no need to swim, and they move by muscular undulations of the body rather like a small flatworm. This species lives on the orange sea star (*Echinaster luzonicus*), lying still and almost invisible during the day, its color and mottled pattern matching its echinoderm host. At night, it extends its two long feeding tentacles to ensnare planktonic prey.

creeping comb jelly

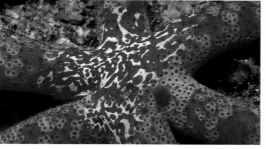

PHYLUM CTENOPHORA

Predatory Comb Jelly

Mnemiopsis leidyi

LENGTH	Up to 3 in (7 cm)
DEPTH	0–100 ft (0–30 m)
HABITAT	Open water

DISTRIBUTION Temperate and subtropical waters of western Atlantic, Mediterranean, and Black Sea

This comb jelly is a slightly flattened pear shape and has two rounded lobes on each side of the mouth that help it to surround and enclose larger prey. As well as two main feeding tentacles, there are smaller secondary tentacles in grooves surrounding the mouth. The long tentacles are armed with lasso cells that secrete a sticky material to ensnare prey.

The predatory comb jelly, which is native to the western Atlantic, was accidentally introduced to the Black Sea in the 1980s by the release of ship ballast water, and it has since spread to adjacent bodies of water, including parts of the eastern Mediterranean. In the Black Sea, it multiplied rapidly because of the ideal water conditions and the absence of its natural predators. This has had very serious effects on commercial fish catches because the predatory comb jelly is a planktonic predator and consumes fish larvae and fry.

PHYLUM CTENOPHORA

Venus's Girdle

Cestum veneris

LENGTH	Up to 6½ ft (2 m)
DEPTH	Near surface
HABITAT	Open water

DISTRIBUTION Tropical and subtropical waters of north Atlantic, Mediterranean, and western Pacific

The unusual name of this animal comes from the ribbon shape of its transparent, pale violet body. Eight rows of comb cilia are modified and run in two lines along one edge of the ribbon. The two main tentacles are short, and numerous other short tentacles occur along the lower edge of the body. As an escape response, Venus's girdle can swim rapidly by undulating its body. However, more usually it moves slowly by beating its comb cilia.

Tunicates and Lancelets

DOMAIN	Eucarya
KINGDOM	Animalia
PHYLUM	Chordata
SUBPHYLA	Tunicata
	Cephalochordata
CLASSES	5
SPECIES	About 3,056

TUNICATES HAVE A LONG, baglike body often attached to the sea floor; lancelets resemble small, stiff worms and live buried in sediment. Despite their simple appearance, these animals are included not with the world's other invertebrates but in the same phylum as backboned animals such as fish and mammals. This is because, uniquely among invertebrates, tunicates and lancelets possess an internal skeletal rod, or notochord. The best-known tunicates are sea squirts, some of which form colonies, whereas lancelets are all solitary.

Lifestyle

Sea squirts live attached to hard surfaces such as rocks, reefs, and shipwrecks. They spend their time filtering seawater, drawing in food-rich water through one siphon (inhalent) and releasing waste water through another (exhalent). Most sea squirts occur in shallow coastal waters where there is plenty of plankton, but there are also a few deep-water species. In sheltered sea lochs, they can cover hundreds of square yards of seabed. Some tunicates, including salps and pyrosomes, drift along on ocean currents with the plankton, often forming giant swarms. Lancelets are strong swimmers due to their flexible, muscular bodies, but they usually just burrow in sediment with only their head sticking out.

SWIMMING SQUIRT
Floating salps swim by jet propulsion, taking in water at one end and squirting it out of the other.

BOTTOM-LIVING SQUIRTS
Sea squirts sometimes grow together in clumps with cnidarians, and sea sponges, and they can be very colorful.

Anatomy

When tunicate larvae become adults, they lose the supporting notochord, but lancelets keep it during the adult stage. Tunicates are covered by a tough protective bag made out of cellulose called a tunic, which sticks to the sea floor by means of rootlike projections. Inside is a big sievelike structure, the pharynx, which connects the mouth and gut. This has a sticky mucus coating to trap plankton from the seawater passing through it. Lancelets also filter water through a pharynx, expelling it through an opening near the anus. A ring of stiff hairs (cirri) surrounding their mouth prevents sand getting in.

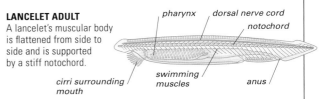

LANCELET ADULT
A lancelet's muscular body is flattened from side to side and is supported by a stiff notochord.

pharynx   *dorsal nerve cord*   *notochord*
cirri surrounding mouth   *swimming muscles*   *anus*

TUNICATE LARVA
The tadpole-shaped tunicate larva's nerve cord and notochord are reabsorbed when it changes into the adult form.

siphon   *nerve cord*   *notochord*
pharynx
attachment organ
heart   *inhalent siphon*

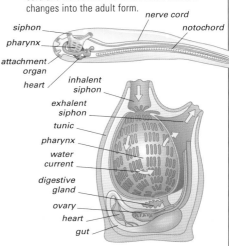

exhalent siphon
tunic
pharynx
water current
digestive gland
ovary
heart
gut

TUNICATE ADULT
Most of the space inside a tunicate is taken up by the huge pharynx, visible through the tunic of this translucent species.

Common Sea Squirt
Ciona intestinalis

HEIGHT	Up to 6 in (15 cm)
DEPTH	0–1,600 ft (0–500 m)
HABITAT	Any hard substrate

DISTRIBUTION Atlantic, Pacific, Indian, and Arctic Oceans; possibly Southern Ocean

The common sea squirt has no supporting structures in its adult form, so when it is seen out of water, it resembles a blob of jelly that may squirt out a jet of water when prodded. It is a typical solitary tunicate, whose internal structures are visible through its pale, greenish yellow, gelatinous outer covering, called a test or tunic, which is smooth and translucent. It has two yellow-edged

siphons and uses the larger of these, the inhalent siphon, to draw in water; the smaller, exhalent siphon is used to expel water, and its opening has six lobes, while that of the exhalent siphon has eight. The common

sea squirt lives up to its name and is found attached to a wide variety of rocks, reefs, seaweeds, and, in particular, man-made structures. The legs of oil platforms and jetties, for example, are often festooned with this sea squirt.

CLEANING UP

In sheltered sea lochs and harbors, the common sea squirt often covers large areas of rock or wall. In spite of its small size, it is able to filter several quarts of water per hour, filtering out plankton and other organic particles and leaving the water much clearer than it might otherwise be.

Colonial Sea Squirt

Atrolium robustum

HEIGHT	Up to 1¼ in (3 cm)
DEPTH	Shallow water
HABITAT	Coral reefs and rocks

DISTRIBUTION Widespread in tropical reef waters of Indian and western Pacific Oceans

Although it appears to be a solitary sea squirt, this species lives in urn-shaped colonies that share a single exhalent siphon (an opening through which water exits). The colony is dotted all over with the inhalent, or ingoing, siphons of the tiny zooids—the individuals that make up the colony. The colony's green color results from the presence of the symbiotic cyanobacterium *Prochloron*.

Star Sea Squirt

Botryllus schlosseri

WIDTH (CLUSTER)	Up to 6 in (15 cm)
DEPTH	Shore and shallows
HABITAT	Rock, stones, seaweeds

DISTRIBUTION Coastal Arctic and temperate waters of north Atlantic

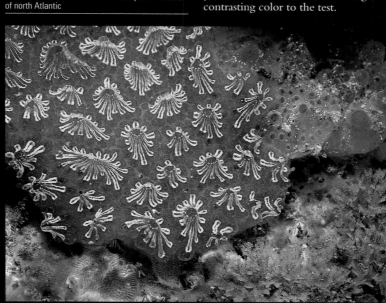

Individual star sea squirts are only about ³⁄₃₂ in (2 mm) long and cannot live on their own. Instead, they arrange themselves in star-shaped clusters, or colonies, embedded in a shared gelatinous casing, called a tunic or test. At the center of each star is a shared outgoing (exhalent) opening through which used water is voided. The colonies vary greatly in color and may be green, violet, brown, or yellow, with the individuals having a contrasting color to the test.

Sea Tulip

Pyura spinifera

HEIGHT	Up to 12 in (30 cm)
DEPTH	15–200 ft (5–60 m)
HABITAT	Rocky reefs

DISTRIBUTION Temperate waters of Australia

This giant sea squirt is held up into the water on the end of a long, thin stalk. This means that its large inhalent siphon is in a better position to pull in plankton-rich water. The sea tulip's body is covered in warty outgrowths and is naturally a bright yellow. However, the growth of an encrusting commensal sponge on many of these sea squirts gives them a pink appearance. During rough weather, sea tulips are often battered down onto the seabed, but they soon spring back on their flexible stalks.

Salp

Pegea confoederata

LENGTH	Up to 6 in (15 cm)
DEPTH	Near surface
HABITAT	Open water

DISTRIBUTION Warm waters worldwide

Salps are tunicates that resemble floating sea squirts. They swim by jet propulsion, taking in water through a siphon at one end of their bodies and expelling it at the other. Their transparent casing is loose and flabby and is encircled by four main muscles that form two distinct cross-bands. Individual salps are joined together in chains up to 12 in (30 cm) long, produced by the asexual reproduction (budding) of a young individual. The chains break up and disperse as they mature. Salps also reproduce sexually. Eggs are kept inside the body on the wall of the exhalent siphon, through which the developed larvae are expelled after being fertilized by sperm drawn in through the inhalent siphon.

Giant Pyrosome

Pyrostremma spinosum

LENGTH	Up to 33 ft (10 m) long
DEPTH	Near surface
HABITAT	Open water

DISTRIBUTION Warm waters between about 40° north and 40° south

The individuals that make up this giant, floating, colonial tunicate are only about 1 in (2 cm) long, but the colony, which resembles a gigantic hollow tube, can be large enough for a person to fit inside. Each individual lies embedded in the wall of the tube, with one end drawing in nutrient-laden water from outside and the other end expelling water and waste inside. The expelled water is used to propel the colony as a whole. A wave of bioluminescent light travels along the community if it is touched.

Appendicularian

Oikopleura labradoriensis

LENGTH	About ¼ in (5 mm)
DEPTH	Near surface
HABITAT	Open water

DISTRIBUTION Cold waters of north Atlantic, north Pacific, and Arctic

Appendicularians are shaped like tiny tadpoles and live inside flimsy mucus dwellings that they build to trap plankton. Water enters the dwelling via two inlets covered by protective grids, passes through fine nets that trap any plankton in mucus, and passes out through an aperture. The animal eats the plankton-loaded mucus and beats its tail to create water currents.

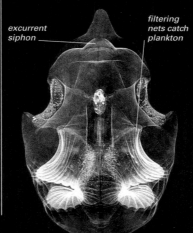

excurrent siphon

filtering nets catch plankton

Lancelet

Branchiostoma lanceolatum

LENGTH	Up to 2½ in (6 cm)
DEPTH	Shore and shallows
HABITAT	Coarse sand

DISTRIBUTION Coastal temperate waters of northeastern Atlantic, and Mediterranean

Looking like a thin, semitransparent, elongate leaf, the lancelet is difficult to spot in the coarse sediments in which it lives. It usually lies half-buried in the sand with its head end sticking out. Muscle blocks that run along both sides of its body show through the skin as a pattern of V-shaped stripes. At the head end, a delicate hood ringed by stiff tentacles overhangs the mouth. This feature filters out large sediment particles but allows smaller, organic particles to pass so that they can be ingested.

Jawless Fishes

DOMAIN	Eucarya
KINGDOM	Animalia
PHYLUM	Chordata
CLASSES	Myxini
	Cephalaspidomorphi
SPECIES	125

JAWLESS FISH FORM AN ANCIENT group of vertebrates encompassing a diverse range of extinct groups. Today, there are only two small groups: the lampreys and the hagfish. They are considered to be the most primitive living vertebrates, although scientific discussion continues over their exact relationships. Hagfish and lampreys look similar, with elongated bodies and jawless mouths, but the two groups evolved along separate lines. Lampreys live in fresh water and temperate coastal waters throughout the world and coastal species swim up rivers to breed. Hagfish are exclusively marine.

Anatomy

At first glance, lampreys and hagfish could easily be mistaken for eels due to their long, thin bodies and slimy, scaleless skin. However, they lack a bony skeleton, and have only a simple flexible rod called a notochord running along the length of the body. In lampreys, the mouth is in the center of a round oral disk armed with small, rasping teeth. Hagfish have a slitlike mouth surrounded by fleshy barbels on the outside and by tooth plates on the inside. The gills in both groups open to the outside through small, round pores behind the head, and there is a single nostril on top of the head.

LAMPREY MOUTH
The oral disk, or sucker, of lampreys is studded with horny teeth arranged in roughly concentric rows. Larger teeth surround the central mouth opening.

dorsal fin · gill openings

notochord · spinal cord · round, fleshy mouth

BODY SECTION
The bodies of lampreys (left) and hagfish are supported by a simple notochord flexed by a series of muscle blocks along the back. With no true bony vertebrae, this makes their bodies very flexible. They have a tail and dorsal fin, but lack paired fins.

HAGFISH
Hagfish find their way and detect carrion using fleshy barbels around the mouth. Their eyes are undeveloped and hidden beneath the skin, so they are nearly blind.

Reproduction

Coastal lampreys migrate from the sea into rivers and move upstream to spawn. Females excavate a nest in gravel and lay thousands of tiny white eggs, which hatch into wormlike larvae called ammocoetes. These small creatures have a horseshoe-shaped mouth without teeth. They live in muddy tunnels for several years, filter feeding on debris, then transform into adults and swim out to sea. Hagfish lay a few large eggs on the seabed, and these hatch into miniature adults.

HAGFISH EGGS
The eggs of hagfish are armed with tiny anchorlike hooks at both ends. When laid, they stick together like strings of sausages.

Feeding

While many freshwater species are parasitic or feed on small invertebrates, marine lampreys are all parasites, attacking both bony and cartilaginous fish. They attach to their living host using the teeth and lips of their oral disk to suck onto their victim. Teeth in the mouth are then used to rasp a hole in the fish, and its flesh, blood, and body fluids are all consumed. Sometimes, lampreys cause the death of their host through blood loss or tissue damage. In contrast, hagfish are mostly scavengers that feed on dead fish and whale carcasses, as well as live invertebrates. They can gain leverage to tear off chunks of flesh by literally tying themselves in a knot and using the knot to brace themselves against the carcass.

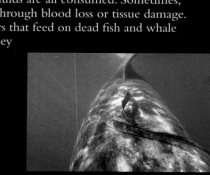

SHARK HOST
Large, slow-moving basking sharks are often parasitized by sea lampreys. The lampreys drop off when they have had their fill, leaving wounds that may get infected.

CLASS PETROMYZONTI

Sea Lamprey

Petromyzon marinus

LENGTH	Up to 4 ft (1.2 m)
WEIGHT	Up to 5½ lb (2.5 kg)
DEPTH	3–2,100 ft (1–650 m)

DISTRIBUTION Coastal temperate waters of, and rivers adjacent to, north Atlantic

With its long, cylindrical body, the sea lamprey might at first be mistaken for an eel, but closer inspection reveals differences. Unlike eels, the sea lamprey has no jaws. Its body is flattened toward the tail and it has two dorsal fins. Its circular mouth lies beneath the head, and is surrounded by a frill of tiny skin extensions. Inside the mouth, the teeth are arranged in numerous concentric arcs, which helps to distinguish it from the similar, but smaller, lampern (right). As a mature adult, the sea lamprey has dark mottling on its back. Adults live at sea and feed on dead or netted fish as well as attacking a wide variety of live ones. It uses a "sucker" to attach to its host, scrapes a hole through the skin, and sucks out flesh and fluids. It spawns in rivers and the larvae remain in fresh water for about five years before they mature and move out to sea. This species is now rare as a result of trapping, intentional poisoning, and the degradation of its river habitat.

CLASS PETROMYZONTII

Lampern

Lampetra fluviatilis

LENGTH	Up to 20 in (50 cm)
WEIGHT	Up to 5 oz (150 g)
DEPTH	0–30 ft (0–10 m)

DISTRIBUTION Coastal waters and rivers of northeastern Atlantic, northwestern Mediterranean

The lampern is also known as the river lamprey because the adults never stray far from the coast and often remain in estuaries. It may be distinguished from the sea lamprey (left) by its smaller size, uniform color, and the smaller number and different arrangement of its teeth. Larvae that hatch in rivers migrate to estuaries, where they spend a year or so feeding on herring, sprat, and flounder.

CLASS MYXINI

Hagfish

Myxine glutinosa

LENGTH	Up to 30 in (80 cm)
WEIGHT	Up to 1¾ lb (750 g)
DEPTH	130–4,000 ft (40–1,200 m)

DISTRIBUTION Coastal and shelf waters, below 55°F (13°C) in north Atlantic and western Mediterranean

This extraordinary fish can literally tie itself in knots and it does so regularly as a means of ridding itself of excess slime. Special slime-exuding pores run along both sides of the eel-like body, enabling it to produce sufficient slime to fill a bucket in a matter of minutes. The glutinous slime is usually more than adequate to deter most predators. Like all jawless fish, the hagfish has no bony skeleton but simply a supporting rod, called a notochord, allowing it great flexibility. Fleshy barbels surround its slitlike, jawless mouth, and it has only rudimentary eyes. There is a single pair of ventral gill openings about a third of the way along the body.

The hagfish spends most of its time buried in mud with only the tip of the head showing. It mainly eats crustaceans but will delve into whale carcasses to scavenge. Once the hagfish has latched onto a carcass with its mouth, it forms a knot near its tail, then slides the knot forward in order to provide itself with sufficient leverage to tear its mouth away along with a chunk of rotting flesh.

CLASS MYXINI

Pacific Hagfish

Eptatretus stoutii

LENGTH	Up to 20 in (50 cm)
WEIGHT	Up to 3 lb (1.4 kg)
DEPTH	65–2,100 ft (20–650 m)

DISTRIBUTION Coastal and shelf waters of northeastern Pacific

The Pacific hagfish is similar to the hagfish found in the Atlantic. It is usually a brownish red color and may have a blue or purple sheen. It has no true fins, only a dorsal finfold that continues around the tail but that has little function in swimming. The Pacific hagfish lives in soft mud and feeds mainly on carrion. It causes great damage to fish caught in static nets and will enter large fish through either the mouth or the anus and proceed to eat them from the inside out, consuming their guts and muscles.

dorsal finfold

CLASS MYXINI

Japanese Hagfish

Eptatretus burgeri

LENGTH	Up to 24 in (60 cm)
WEIGHT	Insufficient information
DEPTH	30–900 ft (10–270 m)

DISTRIBUTION Inshore temperate waters of northwestern Pacific

This species is also known as the inshore hagfish because it lives in relatively shallow water compared to other species of hagfish. It is similar in shape and size to the Pacific hagfish, with six gill apertures and a white line along its back. It lives buried in mud close to shore but migrates to deeper water to breed. Unlike other species of hagfish, it reproduces seasonally; this is thought to be a response to changing temperatures in the shallow waters in which it lives. Its tough skin is used to make leather.

OCEAN LIFE

Sharks, Rays, and Chimaeras

DOMAIN	Eucarya
KINGDOM	Animalia
PHYLUM	Chordata
CLASS	Elasmobranchii
SUBCLASSES	2
SPECIES	1,290

SHARKS, RAYS, AND CHIMAERAS are often informally grouped together as cartilaginous fish as they all have similar flexible skeletons. Within the group are some of Earth's most efficient predators, such as the white shark, as well as filter feeders, such as the manta way. Some have features unusual for fish, such as large brains, live birth, and warm blood. Fossils show that cartilaginous fishes have changed little in form in hundreds of millions of years. All have a skeleton of cartilage, teeth that are replaced by new ones when necessary, and toothlike scales covering their skin.

Anatomy

The internal skeleton of all the fishes in this group is made from flexible cartilage. In some species, parts of the skull and skeleton are strengthened by mineral deposits. The teeth are covered by very hard enamel and are formidable weapons. Sharks have several rows of teeth lying flat behind the active ones. These gradually move forward, and individual teeth may be replaced as often as every 8–15 days. Cartilaginous fishes have extremely tough skin. It is extra-thick in female sharks because males use their teeth to hold onto them when mating. A shark's skin is covered in tiny, backward-pointing, toothlike structures called dermal denticles, which feel like sandpaper. Rays have scattered denticles, some enlarged to form spines, while chimaeras mostly have no denticles. Unlike bony fishes (see pp.338–41), cartilaginous fishes do not have a gas-filled swim bladder. Sharks living in the open ocean, however, often have a very large, oil-filled liver, which aids buoyancy.

(see pp.338–41)

RAY BODY SHAPE
Skates and rays have flat bodies and large pectoral fins. The mouth is on the underside, so water for breathing is sucked in through a pair of holes, called spiracles, on the upper side, then passed over the gills.

Labels on ray diagram: snout, mouth, nostril, pectoral fin, gill slits, pelvic fin, cloaca, tail

Labels on shark diagram: first dorsal fin, second dorsal fin, gill slit, eye, anal fin, pelvic fin, pectoral fin, underslung mouth, heterocercal tail

SHARK BODY SHAPE
A typical shark has a sleek, streamlined body. Most sharks have a tail that is asymmetrical (heterocercal), and thethe paired pelvic fins are set far back. The mouth is underslung, and there are five gill slits on each side.

SANDTIGER SHARK
daggerlike point grips flesh

TIGER SHARK
serrated, bladelike edge cuts like a knife

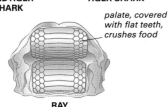

RAY
palate, covered with flat teeth, crushes food

TOOTH ADAPTATIONS
Sharks' teeth are shaped to suit their diet. Pointed ones are used for holding, while serrated teeth slice chunks from prey. Rays and chimaeras have teeth like grindstones to crush hard crustaceans and mollusks.

SHARK FISHING

Sharks are heavily fished all over the world for their meat, fins, liver oil, and skin. Most reproduce very slowly, producing only a few young every one to two years, and many species take a decade or more to reach full maturity. This slow rate of both reproduction and growth means that shark populations cannot sustain heavy fishing pressure. Many species are endangered, and unless adequate controls and protection are put in place, numbers will continue to decline.

SHARK FINS
Thousands of sharks are killed every year for their valuable fins, which are dried and then made into shark-fin soup. The body is often discarded while the shark is still alive.

Reproduction

In all cartilaginous fishes, the eggs are fertilized inside the female's body. Adult males have organs on the belly called claspers—rodlike appendages derived from the pelvic fins. During mating, one or both claspers are inserted into the female's cloaca (the shared opening of the digestive and reproductive tracts) to introduce the sperm. In sharks, mating can be a rough affair, although there may be some courtship.

Chimaeras, as well as some sharks and rays, lay eggs—that is, they are oviparous. The eggs are protected by individual leathery egg capsules, often known as mermaid's purses. The young then hatch out several months later. In contrast, most sharks and rays are viviparous: they give birth to live young after a long period of gestation. In some species, the eggs simply remain inside the mother until they hatch, sustained by yolk prior to birth (aplacental yolk sac viviparity). In about 10 percent of sharks, the young develop attached to a placentalike structure and are directly nourished by the female's body (placental viviparity). In all cases, the young are born fully formed, and they are able to hunt and feed. Immediately after birth, the female swims away and the young are left to fend for themselves.

CATSHARK EGGS
These egg capsules contain catshark embryos, which will hatch after about a year. The tendrils anchor the capsules to seaweeds.

LEMON SHARK BIRTH
Lemon sharks move to shallow, sheltered bays or lagoons to give birth. The young emerge tail first and swim away. The mothers then leave the nursery grounds.

HAMMERHEAD SHARKS
The broad head of these sharks provides space for extra electrical sense organs on the snout and gives a wide field of view, making them formidable hunters.

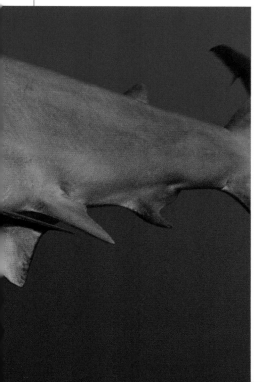

CARTILAGINOUS FISH CLASSIFICATION

Sharks and rays are classified together in one class (the Elasmobranchii), while chimaeras, which include ratfish and rabbitfish, are placed in their own class (Holocephali). Sharks comprise nine orders and rays at least four orders. Chimaeras have one order, the Chimaeriformes.

CHIMAERAS
Order Chimaeriformes

57 species
Chimaeras have a long, flabby body without scales, a large head with sensory canals, platelike teeth, and one gill opening. The first of two dorsal fins is erectile with a venomous spine. Reproduction is oviparous.

FRILL AND COW SHARKS
Order Hexanchiformes

6 species
These sharks have a long, thin body, six or seven pairs of gill slits, small spiracles, and a single dorsal fin near the tail. Frill sharks have three-pointed teeth; cow sharks' teeth are sawlike. Reproduction is yolk sac viviparity.

SLEEPER AND DOGFISH SHARKS
Order Squaliformes

145 species
Six families form this large, varied order: bramble, dogfish, rough, lantern, sleeper, gulper, and kitefin sharks. All have spiracles, five gill slits, and two dorsal fins, but no anal fin. They have yolk sac viviparity.

BRAMBLE SHARKS
Order Echinorhiniformes

2 species
Large, slow-moving deepwater sharks with thornlike skin denticles. They have yolk sac viviparity.

SAWSHARKS
Order Pristiophoriformes

10 species
Small, slender sharks, these species have a flattened head and sawlike snout with barbels. They have two spineless dorsal fins, no anal fin, and have yolk sac viviparity.

ANGELSHARKS
Order Squatiniformes

25 species
These flattened, raylike sharks have a rounded head with gill slits on the side, and spiracles. They have large pectoral and pelvic fins, two small dorsal fins, but no anal fin. Reproduction is yolk sac viviparity.

BULLHEAD AND HORN SHARKS
Order Heterodontiformes

9 species
Small, bottom-living sharks, these species have pointed front teeth and molarlike back teeth, a blunt, sloping head, nostrils connected to the mouth by a groove, paddlelike pectoral fins, an anal fin, and two spined dorsal fins. They are oviparous.

CARPETSHARKS
Order Orectolobiformes

46 species
These mainly bottom-living sharks include wobbegongs and nurse sharks. They have a broad, flattened head, barbels, and nostrils joined to the mouth by a deep groove. They have an anal fin and two spineless dorsal fins. Reproductive strategies vary.

MACKEREL SHARKS
Order Lamniformes

16 species
These large sharks, which include the white, basking, and megamouth sharks, have a cylindrical body, conical head, two dorsal fins, an anal fin, and a long upper tail lobe. Many can maintain a high body temperature. They have yolk sac viviparity.

GROUND SHARKS
Order Carcharhiniformes

At least 304 species
Body shapes vary in this the largest and most diverse shark group. All species have two spineless dorsal fins and an anal fin. Reproductive strategies vary.

RAYS AND SKATES
Orders Rajiformes, Myliobatiformes, Rhinopristiformes, Torpediniformes

740 species
These are mostly bottom-living fish with a flat, disk-shaped body, winglike pectoral fins joined to the head, and a long, thin tail. Reproduction is mostly viviparous with many live young, but some are oviparous.

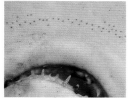

AMPULLAE OF LORENZINI
The black spots on a shark's snout are tiny electrical sense organs that help it find prey even in complete darkness.

Hunting Senses

Cartilaginous fishes have acute senses that help them to find prey, even if it is distant or buried in sediment. Predatory sharks smell or taste tiny amounts of blood as water passes over highly sensitive membranes in their nostrils, while catsharks also use smell to recognize each other. All cartilaginous fish have a system of pores called ampullae of Lorenzini that allows them to detect weak electrical signals given off by other animals. Most also have a lateral-line system, similar to that of bony fishes, which detects water movements. Cartilaginous fishes have eyes similar to those of mammals, and most have acute vision. They have no eyelids, but some sharks have a transparent "nictitating membrane," which protects their eyes when they are attacking prey.

BARBELS
Active at night, nurse sharks can find buried prey by touch and smell, using their sensory barbels.

ORDER CHIMAERIFORMES
Pacific Spookfish
Rhinochimaera pacifica

LENGTH	About 4¼ ft (1.3 m), plus tail filament
WEIGHT	Not recorded
DEPTH	1,100–4,900 ft (330–1,500 m)

DISTRIBUTION Parts of Pacific and eastern Indian Ocean

When scientists first hauled a Pacific spookfish up from the ocean depths, they were astonished by the sight of its enormously long, conical snout. The long, brownish body of this strange-looking fish tapers to a thin tail, so that the fish gives the impression of being pointed at both ends. The snout is whitish, flexible, and covered in sensory pores and canals.

Living in the dark depths of the ocean where its small eyes are of at best limited use, the spookfish uses its snout to find food and sense objects around it. The beak-shaped mouth under the base of the snout contains pairs of black, platelike teeth. Its tail has only a small lower lobe, while the upper lobe consists of a row of fleshy tubercles. Like other chimaeras, the spookfish relies mainly on its pectoral fins for propulsion rather than using its tail as most species of fish do. A very similar species of spookfish is found in the Atlantic Ocean.

ORDER CHIMAERIFORMES
Plownose Chimaera
Callorhinchus milii

LENGTH	Up to about 4¼ ft (1.3 m)
WEIGHT	Not recorded
DEPTH	At least 750 ft (230 m)

DISTRIBUTION Temperate waters in the southwest Pacific, off southern Australia and along the east coast of South Island, New Zealand

The plownose chimaera is also known as the elephant fish due to its most distinctive feature, a long, fleshy snout. The plownose uses this bizarre appendage to snuffle through the ocean floor mud in search of shellfish, which it crunches up using its platelike teeth. A network of prominent sensory canals crisscrosses its head. In spring, these fish come inshore into estuaries and bays to breed, and lay their eggs in horny, yellow-brown capsules. This chimaera is fished commercially for food.

REPRODUCTIVE EMBRACE

Mating underwater is a slippery business, so the male plownose chimaera has a retractable, clublike, spiny clasper on its head that helps it hang onto the female. The male transfers his sperm when he inserts his pelvic clasper into the female's cloaca.

retractable clasper

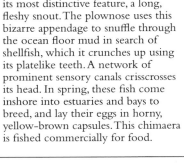

ORDER CHIMAERIFORMES
Spotted Ratfish
Hydrolagus colliei

LENGTH	Up to 3¼ ft (1 m)
WEIGHT	Not recorded
DEPTH	Close inshore to at least 2,950 ft (900 m)

DISTRIBUTION Northeastern Pacific

The scientific name of the spotted ratfish means "water rabbit," and it is also called the blunt-nosed chimaera. It belongs to the same family as the rabbit fish (see right) and is similar in shape, but unlike its relative, it does not have an anal fin on the underside next to the tail. Its pattern of white spots on a dark background may provide camouflage in the same way as the spots help to camouflage deer in a forest. The spotted ratfish uses its large pectoral fins to glide and flap its way over the seabed in search of its prey, which consists mainly of mollusks and crustaceans. Like other chimaeras, the female lays eggs, each one encased in a tadpole-shaped, protective capsule. The eggs are laid in the summer, two at a time, and are dropped onto the seabed.

The spotted ratfish is not fished commercially, since it is not very palatable, although it is sometimes unintentionally caught in nets along with other fish. It is not popular with fishers because it has the ability to inflict a nasty wound with its sharp dorsal spine and can also deliver a painful bite. The spotted ratfish is sometimes encountered at night by scuba divers, its large eyes glowing green by flashlight.

ORDER CHIMAERIFORMES
Rabbit Fish
Chimaera monstrosa

LENGTH	Up to 5 ft (1.5 m)
WEIGHT	Up to 5½ lb (2.5 kg)
DEPTH	Typically 1,000–1,300 ft (300–400 m)

DISTRIBUTION Eastern Atlantic and Mediterranean

Beautifully patterned with wavy brown and white lines, the rabbit fish belongs to a family called Chimaeridae, whose members have rounded snouts, long, tapering bodies, and tails ending in a long, thin filament, giving rise to their alternative name of ratfishes. The long, sharp spine in front of the first dorsal fin of the rabbit fish is venomous and can inflict a serious wound. Unlike sharks but in common with all other chimaeras, the rabbit fish can raise and lower this fin. The second dorsal fin is low and long and almost reaches the tail fin. Rabbit fish swim sluggishly in small groups and feed mainly on seabed invertebrates using their paired, rabbit-like teeth. These fish are often caught by accident in shrimp nets in the North Sea.

ORDER HEXANCHIFORMES

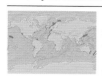

Frilled Shark

Chlamydoselachus anguineus

LENGTH	Up to 6½ ft (2 m)
WEIGHT	Not recorded
DEPTH	Mostly 66–4,921 ft (20–1,500 m)

DISTRIBUTION Worldwide but discontinuous

With its elongated, eel-like body and flattened head, the frilled shark bears little resemblance to other sharks. The most noticeable difference is that its mouth is at the front of its head instead of on the underside. In addition, while most modern sharks have five pairs of gill slits, the frilled shark has six, each with a frilled edge. Its small teeth are also unusual, each having three sharp points.

Frilled sharks have been observed swimming with their mouths open, displaying their conspicuous white teeth, leading to the suspicion that the teeth act as a lure for prey. This shark lives near the seabed in deep water but occasionally comes to the surface. It feeds on deep-water fish and squid. The male has two long claspers on the belly, which are used to transfer sperm to the female when mating. This species has yolk sac viviparity, meaning that the eggs hatch inside the mother, which then gives birth to live young. Up to 12 young are born as long as two years after fertilization.

Trawlers fishing for other deep-sea species often catch frilled sharks as by-catch. Because this species reproduces so infrequently, it is especially vulnerable, but it is not specifically targeted by deep-sea fishers.

ORDER HEXANCHIFORMES

Bluntnose Sixgill Shark

Hexanchus griseus

LENGTH	Up to 18 ft (5.5 m)
WEIGHT	Up to at least 1,300 lb (600 kg)
DEPTH	Up to 8,202 ft (2,500 m)

DISTRIBUTION Tropical and temperate waters worldwide

This enormous deep-water shark is sometimes spotted by divers in shallow water at night, but its more usual haunt is rocky seamounts and mid-ocean ridges. Its thick-set, powerful body has one dorsal fin, a large mouth lined with comblike teeth, and six gill slits. Its fins are soft and flexible, not rigid like those of most sharks. Fish, rays, squid, and bottom-living invertebrates are this shark's typical prey, although larger adults sometimes also hunt for seals and cetaceans.

ORDER HEXANCHIFORMES

Sharpnose Sevengill Shark

Heptranchias perlo

LENGTH	Up to 4½ ft (1.4 m)
WEIGHT	Not recorded
DEPTH	Up to 3,300 ft (1,000 m), typically 90–2,360 ft (27–720 m)

DISTRIBUTION Tropical and temperate waters worldwide, except northeastern Pacific

The sharpnose sevengill shark, as its name suggests, has a sharply pointed snout and is one of only two shark species that have seven gill slits—more than any other living shark species. It lives in deep water and hunts squid, crustaceans, and fish near the seabed. Like the sixgill sharks (see above), it has comb-shaped teeth. Young fish have black markings, which fade with age, on the tip of the single dorsal fin and on the upper part of the tail. Females have yolk sac viviparity and give birth to 6–20 young at one time. The sharpnose sevengill shark is rarely seen alive and little is known of its feeding and breeding behavior, but it is lively and aggressive on the rare occasions when it is captured. It is occasionally caught up as by-catch in trawl nets, and this may be contributing to a reduction in its numbers. It is listed as Near Threatened on the IUCN Red List of endangered species.

ORDER SQUALIFORMES

Piked Dogfish

Squalus acanthias

LENGTH	Up to 5 ft (1.5 m)
WEIGHT	Up to 20 lb (9 kg)
DEPTH	Up to 6,234 ft (1,900 m), typically 0–2,000 ft (0–600 m)

DISTRIBUTION Worldwide, except tropics, North Pacific, and polar waters

Sharks are not normally shoaling fish, but piked dogfish aggregate into huge groups numbering thousands of individuals, often all of one sex and size. Also known as the spurdog or spiny dogfish, this sleek, dark gray shark has two dorsal fins, each with a sharp spine in front of it. Irregular white spots decorate its sides, especially in young fish, and it has a pointed snout and large oval eyes. It was once very common and was possibly the most abundant species of shark, but it is now endangered globally as a result of overfishing. These sharks do not begin to breed until they are 10–20 years old and may live to be at least 40 or so years old. They grow very slowly, and the young take up to two years to develop inside the mother. Some populations migrate thousands of miles seasonally in order to avoid very cold water.

ORDER SQUALIFORMES

Velvet Belly Lanternshark

Etmopterus spinax

LENGTH	Up to 18 in (45 cm)
WEIGHT	Up to 2 lb (850 g)
DEPTH	Typically 650–1,650 ft (200–500 m), up to 8,200 ft (2,500 m)

DISTRIBUTION Eastern Atlantic and Mediterranean

As this small shark searches for fish and squid in the darkness, its black belly is illuminated by tiny, bright light organs called photophores. It has large eyes and two dorsal fins, each with a strong, grooved spine in front. It is one of about 30 similar species that include the smallest known sharks.

ORDER SQUALIFORMES

Greenland Shark

Somniosus microcephalus

LENGTH	Up to at least 21 ft (6.4 m)
WEIGHT	Up to at least 1,710 lb (775 kg)
DEPTH	0–8,694 ft (0–2,650 m)

DISTRIBUTION North Atlantic and Arctic waters

The sluggish Greenland shark has a heavy, cylindrical body that is usually brown or gray. It has a short, rounded snout and two equal-sized dorsal fins. As well as feeding on a variety of live prey, including fish, sea birds, and seals, this shark also scavenges dead cetaceans and drowned land animals, such as reindeer. It is often caught at ice holes while it hunts seals, but its uncooked flesh is poisonous. Carbon dating has now shown some may live for at least 260, possibly 500 years.

broad, interlocking teeth on lower jaw

ORDER SQUATINIFORMES

Pacific Angel Shark

Squatina californica

LENGTH	Up to 6 ft (1.8 m)
WEIGHT	Up to 60 lb (27 kg)
DEPTH	Typically 10–1,000 ft (3–300 m), up to 656 ft (200 m)

DISTRIBUTION Continental shelf of the eastern Pacific

Resembling something between a squashed shark and a ray, the Pacific angel shark spends most of its time lying quietly on the seabed. Its sandy or gray back, peppered with dark spots and scattered dark rings, provides good camouflage. Though superficially similar to a ray, this fish is marked out as a true shark by the gill slits on the side of its head, while rays have their

gills underneath. It draws water in through large, paired holes called spiracles behind its eyes and pumps it over the gills. Rearing up like a cobra, the Pacific angel shark ambushes passing fish including halibut, croakers, and other bottom-dwellers. It has also been known to snap at divers and fishers who have provoked it. At night, it swims for short distances above the seabed, sculling along with its tail. Females give birth to litters of six to ten pups after a gestation of nine to ten months. Young fish do not mature until they are at least ten years old and can live until they are 35 years old. This fish used to be abundant in the waters off California until intense fishing caused a population collapse in the 1990s. A gill net ban ended the fishery. This shark is categorized as Near Threatened on the IUCN Red List of endangered species.

ORDER SQUALIFORMES

Cookiecutter Shark

Isistius brasiliensis

LENGTH	Up to 20 in (50 cm)
WEIGHT	Not recorded
DEPTH	0–11,500 ft (0–3,500 m)

DISTRIBUTION Atlantic, Pacific, and southern Indian Ocean

Many cetaceans and large fish, including other sharks, suffer when cookiecutter sharks are around. This small, cigar-shaped shark is a parasite that bites chunks out of its prey. Using its unique thick, flexible lips to hold onto its victim by suction, it then twists itself around so that its razor-sharp lower teeth bite out a cookie-shaped piece of flesh. It is active at night, luring its victims with glowing green bioluminescent lights on its belly. It also preys on squid and crustaceans.

ORDER PRISTIOPHORIFORMES

Longnose Sawshark

Pristiophorus cirratus

LENGTH	Up to 5 ft (1.5 m)
WEIGHT	Not recorded
DEPTH	130–2,067 ft (40–630 m)

DISTRIBUTION Temperate and subtropical waters of southern Australia

Like all sawsharks, this species has a head that is flattened and extended to form a long, sawlike projection, or rostrum. This is edged with rows of large, sharp teeth. Two long sensory barbels hang down from the underside of the rostrum, which is studded with further sense organs, and the shark uses these to detect vibrations and electrical fields. It seeks out and kills prey, such as fish and crustaceans, by poking around on the seabed and slashing out sideways with its rostrum.

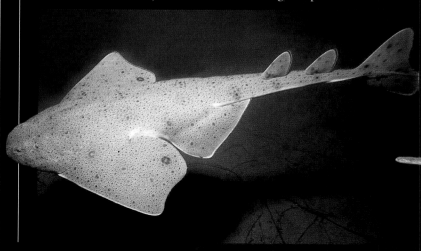

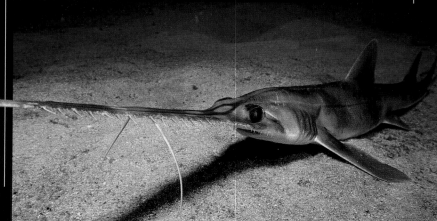

ORDER ORECTOLOBIFORMES

Tasseled Wobbegong
Eucrossorhinus dasypogon

LENGTH	At least 4¼ ft (1.3 m)
WEIGHT	Not recorded
DEPTH	At least 130 ft (40 m)

DISTRIBUTION Southwestern Pacific off northern Australia and Papua New Guinea

While it lies still, the tasseled wobbegong looks like a seaweed-covered rock, which is exactly its objective. It is one of a group of flattened, bottom-living sharks that are masters of camouflage. The squashed shape and broad, paired fins are further adaptations to an existence on the ocean floor. This species has a beautiful reticulated pattern of narrow, dark lines against a paler background. Around its mouth is a fringe of skin flaps that resemble weeds. During the day, it rests unseen under overhangs and ledges on coral reefs. At night, it emerges onto the reef to find a good vantage point from which to snap up passing fish. There is no escape from the gape of its huge jaws and its needlelike teeth for any fish straying near, as the tasseled wobbegong lunges up and grabs its prey. This species has been reported to bite divers who disturb it. Little is yet known of its biology, and reef destruction and overfishing have reduced its numbers.

MISLEADING SIMILARITY

The tasseled wobbegong looks remarkably similar to the angler (see p.351), which is an unrelated species of bony fish. Both of these predators, which specialize in ambushing their prey, are flattened, have broad heads, wide mouths disguised by skin flaps, and sharp, pointed teeth. Following a similar lifestyle, these two species have come up with similar answers, an example of convergent evolution.

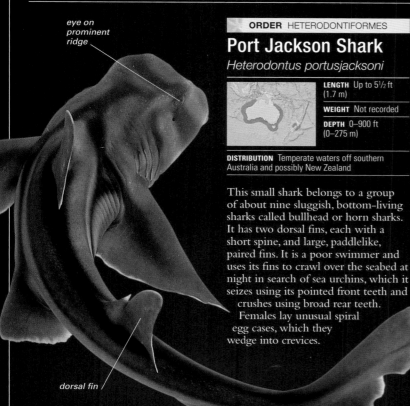

eye on prominent ridge

dorsal fin

ORDER HETERODONTIFORMES

Port Jackson Shark
Heterodontus portusjacksoni

LENGTH	Up to 5½ ft (1.7 m)
WEIGHT	Not recorded
DEPTH	0–900 ft (0–275 m)

DISTRIBUTION Temperate waters off southern Australia and possibly New Zealand

This small shark belongs to a group of about nine sluggish, bottom-living sharks called bullhead or horn sharks. It has two dorsal fins, each with a short spine, and large, paddlelike, paired fins. It is a poor swimmer and uses its fins to crawl over the seabed at night in search of sea urchins, which it seizes using its pointed front teeth and crushes using broad rear teeth. Females lay unusual spiral egg cases, which they wedge into crevices.

ORDER ORECTOLOBIFORMES

Zebra Shark
Stegostoma fasciatum

LENGTH	Up to 8 ft (2.4 m)
WEIGHT	Not recorded
DEPTH	0–210 ft (0–63 m)

DISTRIBUTION Indian Ocean and southwestern Pacific

The zebra shark is often seen by divers around coral reefs. Its long, ridged body and densely spotted skin make it unmistakable. Juveniles have stripes instead of spots and no ridges. This shark spends most of the day lying on the reef, usually facing into the current. At night, it squirms its flexible body into cracks and crevices on the reef, searching for mollusks, crustaceans, and small fish.

ORDER ORECTOLOBIFORMES

Tawny Nurse Shark
Nebrius ferrugineus

LENGTH	Up to 10½ ft (3.2 m)
WEIGHT	Not recorded
DEPTH	3–230 ft (1–70 m), typically 16–100 ft (5–30 m)

DISTRIBUTION Indian Ocean, western and southwestern Pacific

The docile, bottom-living tawny nurse shark is a favorite with underwater photographers because, although it may bite if harassed, it can be approached closely. During the day, it rests quietly in caves and channels in coral reefs, emerging at night to hunt for invertebrates. A pair of long sensory barbels on either side of the mouth helps the shark to find its prey, which it crushes using wide teeth.

ORDER ORECTOLOBIFORMES

Whale Shark

Rhincodon typus

LENGTH	40 ft or more (12 m)
WEIGHT	Over 13 tons (12 metric tons)
DEPTH	Surface, deep water in winter

DISTRIBUTION Tropical and warm temperate waters worldwide

The whale shark is a graceful, slow-moving giant and the largest fish in the world. At 5 ft (1.5 m) wide, its mouth is large enough to fit a human inside, but it is a harmless filter feeder that eats only plankton and small fish. To obtain the huge amount of food it needs, it sucks water into its mouth and pumps it out over its gills, where particles of food become trapped by bony projections called gill rakers and are later swallowed. This shark has the thickest skin of any animal, at up to 4 in (10 cm) thick. Prominent ridges run the length of its body, and it has a large, sickle-shaped tail. The pattern of white spots on its back is unique to each fish, enabling scientists, through analysis of photographs, to identify individuals. Their ocean travels are being increasingly tracked, and satellite tagging has shown that some whale sharks migrate across entire oceans. Whale shark eggs hatch inside the mother, and she gives birth to live young. Whale sharks are killed for their meat and fins, and although they are legally protected in some countries, they are endangered.

FEASTING ON PLANKTON

Every year, around April, many whale sharks migrate to Ningaloo Reef off northwestern Australia for a plankton feast. The plankton explosion results from a simultaneous mass spawning of the reef's corals, possibly triggered by the full moon.

ORDER LAMNIFORMES

Megamouth Shark

Megachasma pelagios

LENGTH	At least 18 ft (5.5 m)
WEIGHT	Not recorded
DEPTH	0–540 ft (0–165 m)

DISTRIBUTION Little known, but probably worldwide in the tropics

This gigantic shark was discovered in 1976, when one became entangled in the folds of a ship's sea anchor. Only about 100 have ever been sighted.

Like the whale shark, it is a filter feeder, engulfing huge mouthfuls of shrimp, which it may compress with its huge tongue. At night, it follows the shrimp toward the surface and may attract them with bioluminescent tissue inside its mouth.

ORDER LAMNIFORMES

Basking Shark

Cetorhinus maximus

LENGTH	20–36 ft (6–11 m)
WEIGHT	Up to 7.7 tons (7 metric tons)
DEPTH	3,937 ft (1,200 m)

DISTRIBUTION Cold- to warm-temperate coastal waters worldwide

The world's second-largest fish, the basking shark is protected in several countries. In summer, it swims open-mouthed at the surface, filtering out plankton. Every hour, the shark passes up to 395,000 gallons (1.5 million liters) of seawater through the huge gills that almost encircle its head. Its liver runs the length of the abdominal cavity and is filled with oil to aid buoyancy.

ORDER LAMNIFORMES

Sandtiger Shark

Carcharias taurus

LENGTH	Up to 10½ ft (3.2 m)
WEIGHT	Up to 350 lb (160 kg)
DEPTH	0–625 ft (0–190 m)

DISTRIBUTION Warm-temperate and tropical coastal waters, except eastern Pacific

Also known as the ragged-tooth shark and the gray nurse shark, the sandtiger shark is fearsome to look at. It is heavily built and its daggerlike, menacing teeth protrude, even when its mouth is closed. Many people will have seen these sharks in aquariums, and they are quite docile in captivity despite their chilling appearance. The sandtiger shark has a flattened, conical snout, is light brown in color, and its body is often speckled with darker spots. It lives in shallow coastal waters, especially on reefs and in rough, rocky areas with gullies and caves. Although it spends most of its time near the sea floor, it can hover in midwater by filling its stomach with air gulped in at the surface.

The mother gives birth to two live young at a time, one from each of a pair of uteri every other year. Within each uterus there are many other embryos, and the strongest embryo in each uterus kills and eats its siblings along with any unfertilized eggs before it is born. Sandtiger sharks are widely hunted for both sport and food.

ORDER LAMNIFORMES
White Shark
Carcharodon carcharias

LENGTH	Up to about 20 ft (6 m)
WEIGHT	Over 3.7 tons (3.4 metric tons)
DEPTH	0–4,300 ft (0–1,300 m)

DISTRIBUTION Wide range through most oceans except polar waters

The white shark, or great white, is one of the most powerful predators in the ocean and has a reputation as a killing machine. In fact, this shark is intelligent and capable of complex social interactions. It is, however, first and foremost a predator, feeding on prey that ranges from fish to marine mammals and birds.

Its powerful, tapered body and crescent-shaped tail are designed for sudden, swift attack, sometimes with such momentum that the shark leaves the water. It can sustain high speeds in cold waters because it can keep its swimming muscles, brain, and viscera warmer than the surrounding water due to adaptations in its circulatory system. This means that the shark's metabolism is more efficient than that of other sharks, allowing it to swim faster and with greater endurance.

Large numbers of these sharks are attracted to areas where there are sea mammal colonies, such as off South Africa. Satellite tags have shown that they migrate huge distances. Due to trophy hunting and fishing, it is critically endangered in the Mediterranean and classed as vulnerable worldwide.

FEARSOME TEETH
This shark's teeth can be up to 3 in (7.5 cm) long. They are extremely hard, with razor-sharp, serrated edges that can slice through the toughest flesh.

serrated edge

SHARK ATTACK

The white shark has made more fatal unprovoked attacks on humans than any other shark. Humans are not its natural prey, and many such attacks can be put down to the shark mistaking a diver for a seal or turtle. When stimulated by bait in the water, white sharks will bite anything, even a metal diving cage.

COUNTER-SHADED COLORATION
From above, the shark's dark back merges with the seabed; from below, its white belly blends with the downwelling light.

OCEAN LIFE

ORDER LAMNIFORMES
Goblin Shark
Mitsukurina owstoni

LENGTH	Up to 12¾ ft (3.9 m)
WEIGHT	Up to 460 lb (210 kg)
DEPTH	1,000–4,300 ft (300–1,300 m)

DISTRIBUTION Not fully known, but thought to be in temperate and tropical waters

One of the strangest-looking of all deep-water sharks, the goblin shark is pale pink, with a flabby body, tiny eyes, and a long, flattened, bill-like snout. This strange projection is covered in electroreceptors and is probably used to detect prey in the inky depths. Beneath the snout, the goblin shark has specialized jaws that can be shot forward to grab fish and octopuses using long, pointed teeth. Not very much else is known about

this shark except that it gives birth to live young and after death changes from pinkish to a dirty, brownish gray color. Adults are very rarely caught, and most catches have been of immature juveniles. Most data has come from sharks caught by boats fishing for deep-water fish using long lines. Fossils of sharks very similar to this species have been found in rocks over 100 million years old.

ORDER CARCHARHINIFORMES
Chain Catshark
Scyliorhinus retifer

LENGTH	2 ft (0.6 m)
WEIGHT	Not recorded
DEPTH	246–2,461 ft (75–750 m)

DISTRIBUTION North and western Atlantic, Caribbean

Most familiar catsharks belong to the Scyliorhinidae, which until recently included about 160 relatively small species. The family has now been split, with deep-water catsharks in the family Pentanchidae. With its chain-link pattern, this shark is unmistakable. Like other catsharks, its young develop inside horny capsules, each laid individually. Empty cases washed ashore are called mermaid's purses. Many deep-sea catsharks exhibit bright green fluorescence, which they may use to communicate.

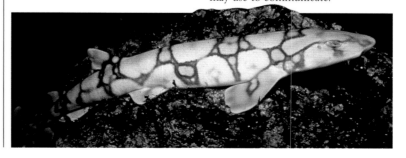

WHITE SHARK
One of the world's most formidable predators, the white shark catches its prey with a short, fast attack, typically from below. Sometimes its speed takes it and its prey clear of the water. The shark then often withdraws to wait for its victim to weaken before returning to finish it off.

ORDER CARCHARHINIFORMES

Blue Shark
Prionace glauca

LENGTH	Up to 13 ft (4 m)
WEIGHT	Up to 450 lb (200 kg)
DEPTH	0–1,150 ft (0–350 m)

DISTRIBUTION Temperate and tropical waters worldwide

A true ocean wanderer, the blue shark makes seasonal trans-ocean crossings from feeding to pupping areas. It is streamlined and elegant, with a long, pointed snout and white-rimmed black eyes. On long journeys, it rides ocean currents to help conserve energy. On the way, it makes frequent, deep dives, possiblyto help it get its magnetic bearings. When chasing fish, this shark may reach speeds of 43 mph (70 kph). It has been known to harass swimmers and has caused a few human fatalities. Once one of the commonest sharks, it is now critically endangered in the Mediterranean and near threatened globally.

ORDER CARCHARHINIFORMES

Tiger Shark
Galeocerdo cuvier

LENGTH	Up to at least 18 ft (5.5 m)
WEIGHT	Up to 1,750 lb (800 kg)
DEPTH	0–459 ft (0–140 m)

DISTRIBUTION Tropical and warm temperate waters worldwide

The tiger shark is the second most dangerous shark to humans, after the white shark (see p.331). It is huge and has a heavy head and a mouth filled with serrated teeth that have the characteristic shape of a cockscomb. One reason it is so dangerous is that it prefers coastal waters and is also found in river estuaries and harbors, and so it frequently comes into contact with humans. It is reputed to eat almost anything—as well as eating smaller sharks, including young tiger sharks, other fish, marine mammals, turtles, and birds, it is an inveterate scavenger, and a huge variety of garbage has been found in tiger shark stomachs. The young, born live after hatching from eggs inside the mother, begin life marked with blotches, which become "tiger stripes" in juveniles and fade by adulthood.

ORDER CARCHARHINIFORMES

Scalloped Hammerhead Shark
Sphyrna lewini

LENGTH	Up to 14 ft (4.3 m)
WEIGHT	Up to 330 lb (150 kg)
DEPTH	0–3,230 ft (0–1,000 m)

DISTRIBUTION Tropical and warm temperate waters worldwide

Along with the nine other known species of hammerhead sharks, the scalloped hammerhead has a strange, flattened, T-shaped head. In this species, the front of the head has three notches, which produces the scalloped shape from which it takes its name. The eyes are located at the sides of the head. Hunting near the seabed, the shark swings its head from side to side, looking for prey such as fish, other sharks, octopus, and crustaceans, and using sensory pits on its head to detect the electrical fields of buried prey such as rays. The head may also function as an airfoil, giving the shark lift and helping it to twist and turn as it chases its prey.

Scalloped hammerheads may be seen in large shoals of over a hundred individuals. They give birth to live young in shallow bays and estuaries, where the skin of the young darkens to give protection against sunlight.

ORDER CARCHARHINIFORMES

Whitetip Reef Shark
Triaenodon obesus

LENGTH	Up to 6½ ft (2 m)
WEIGHT	Up to 37 lb (18 kg)
DEPTH	Typically 25–130 ft (8–40 m), recorded at 1,080 ft (330 m)

DISTRIBUTION Tropical waters of the Indian Ocean and Pacific

One of the sharks most often seen by divers is the whitetip reef shark, which during the day may be found around coral reefs resting in caves and gullies, often in groups. The tip of its first dorsal fin and the upper tip of its tail are white, in contrast to its grayish brown back. At night, the whitetip comes out to hunt reef fish, octopus, lobsters, and crabs hidden among the coral. Packs sometimes hunt together, sniffing out the prey and bumping and banging the coral to get at them.

ORDER RAJIFORMES

Painted Ray

Raja undulata

LENGTH	Up to 4 ft (1.2 m)
WEIGHT	Up to 15 lb (7 kg)
DEPTH	150–650 ft (45–200 m)

DISTRIBUTION Eastern Atlantic and Mediterranean

Also known as the undulate ray, the painted ray is one of the most distinctive northern European rays. This species is patterned with long, wavy, dark lines edged with white spots that run parallel to the wing margins. Its ornate appearance makes it an attractive species for aquariums, but in its natural habitat, this pattern has the practical advantage of helping the ray to blend in with the gravel and sand on the seabed, where it feeds on flatfish, crabs, and other bottom-living invertebrates.

The biology of this beautiful ray has not been fully studied, but during the breeding season, males use paired claspers to transfer sperm during mating, and females are known to lay up to 30 eggs per year through spring and summer. Each egg is encased in a reddish-brown, oblong egg capsule up to 3½ in (9 cm) long, with a curved horn at each corner. Additional information on the distribution and status of this and other rays around Britain continues to be collected through the Shark Trust (UK) Great Eggcase Hunt. Members of the public are encouraged to collect empty egg cases that have been washed ashore, rehydrate them, and identify them. The number and location of egg cases found are collated each year.

ORDER RAJIFORMES

Blue Skate

Dipturus batis

LENGTH	5 ft (1.5 m)
WEIGHT	Unknown. *Dipturus intermedius* up to 220 lb (100 kg)
DEPTH	330–3,300 ft (100–1,000 m)

DISTRIBUTION Eastern Atlantic from northern Europe to southern Africa, Mediterranean

Once known as the common skate, this species was thought to be the largest and heaviest of the European rays. However, the species was recently divided into two species, and the flapper skate (*D. intermedius*) now has the record. Both species have long and pointed, and the front margin of the wings is strongly concave, giving them an overall angular shape. A blue skate's tail has a row of spines along its length, but unlike the large stinging spine of stingrays, these are not venomous.

This species is called blue skate because its underside is bluish gray. It can swim strongly and feeds on fish in midwater as well as hunting over the seabed for crabs, lobsters, bottom-dwelling fish, and other rays. Its oblong egg cases are up to 6 in (15 cm) long. Blue and flapper skates are both now critically endangered as a result of decades of overexploitation.

ORDER RHINOPRISTIFORMES

Atlantic Guitarfish

Pseudobatos lentiginosus

LENGTH	Up to 30 in (75 cm)
WEIGHT	Not recorded
DEPTH	0–100 ft (0–30 m)

DISTRIBUTION Coastal waters of Gulf of Mexico, Caribbean, and western Atlantic

Guitarfish are elongated rays with a triangular snout and narrow pectoral fins. Like sharks, these rays use their spineless tails for swimming, while other rays swim by flapping only their pectoral fins. The Atlantic guitarfish has two small dorsal fins set far back near the tip of its tail. Grayish brown with small white spots, it blends in with the sandy seabed, but this ray can be seen in shallow water searching for mollusks and crabs, probing the sand with its snout. Females have placental viviparity, giving birth to about six live young.

ORDER RHINOPRISTIFORMES

Smalltooth Sawfish

Pristis pectinata

LENGTH	Up to 25 ft (7.6 m)
WEIGHT	Up to 770 lb (350 kg)
DEPTH	0–33 ft (0–10 m)

DISTRIBUTION Subtropical waters in all oceans

Sawfish are elongated rays with a long, flat, sawlike snout, or rostrum, which they use to slash through shoals of fish and dig for shellfish and invertebrates. Like all rays, they have gill slits on the underside of the body rather than the sides. Females give birth to live young, which are about 2 ft (60 cm) long in the smalltooth species. The saws of the pups are sheathed and flexible at birth, in order to prevent injury to the mother. The smalltooth sawfish lives in coastal waters but also swims up river estuaries. Numbers of this species are severely depleted and it is listed as Critically Endangered on the IUCN Red List of endangered species.

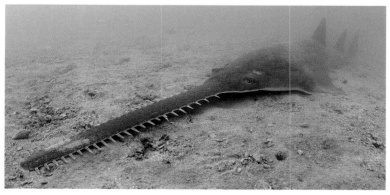

ORDER MYLIOBATIFORMES
Southern Stingray
Hypanus americanus

WIDTH (WINGSPAN)	6½ ft (2 m)
WEIGHT	Up to 300 lb (135 kg)
DEPTH	0–180 ft (0–55 m)

DISTRIBUTION Western Atlantic, Gulf of Mexico, Caribbean Sea

Stingrays are feared because their long tails are equipped with one or more daggerlike, venomous spines. The southern stingray has a single, serrated spine about midway along the tail and a flap of skin, also known as a finfold, on the underside of the tail. Its thick disk is dark gray on top and white underneath. This stingray spends most of the day lying buried in the sand at the bottom of shallow lagoons and off beaches. At night, the ray feeds by excavating holes in the sand and crunching up bivalve mollusks, crustaceans, and worms. Because its eyes are on top of its head, it cannot see its prey but uses smell and electro-receptors to detect it. While it is buried, its spiracles—through which it draws in water for breathing—are visible as a pair of holes in the sand. People are often stung when they inadvertently step on southern stingrays; the stinging spine is sharp enough to cause a serious wound and the venom causes severe pain.

ORDER MYLIOBATIFORMES
Round Stingray
Urobatis halleri

LENGTH	23 in (58 cm) including tail
WEIGHT	3 lb (1.4 kg)
DEPTH	0–300 ft (0–90 m)

DISTRIBUTION Eastern Pacific

As its name suggests, the round stingray has an almost circular disk. This species and its relatives the "stingarees" have shorter tails than other stingrays and the tail ends in a leaf-shaped fin. The round stingray varies in color from pale to dark brown and can be either plain or mottled with darker spots and reticulations. These rays are most often seen in summer, when they move inshore into inlets and bays to forage for invertebrates among seagrass and bask in the warm-water shallows. Females arrive in the shallows around June ready to breed, and the males, who are already there, swim along the shoreline looking for suitable mates. Sexually mature females are reported to give off an electrical field that the males can sense. Females give birth about three months after mating to about six live young. The juveniles stay inshore, where there are fewer predators, until they mature. When out foraging, they do not stray too far, remaining within an area of about 1 square mile (2.5 square km).

Predators of the round stingray include the northern elephant seal and the black sea bass. They are also likely to be hunted by large carnivorous fish such as sharks. The round stingray's sting is painful and can cause minor injuries.

ORDER RAJIFORMES
Giant Manta Ray
Mobula birostris

WIDTH (WINGSPAN)	Up to 26 ft (8 m)
WEIGHT	Up to 4,000 lb (1.8 metric tons)
DEPTH	0–80 ft (0–24 m); usually near surface

DISTRIBUTION Surface tropical waters worldwide, sometimes warm temperate areas

Divers often describe the experience of swimming beneath a manta ray as like being overtaken by a huge flying saucer, but the ones they mostly see are the smaller reef manta (*M. alfredi*). The giant manta ray is the biggest in the world, but like the biggest shark, the whale shark, it is a harmless consumer of plankton and small fish.

When feeding, it swims along with its cavernous mouth wide open, beating its huge triangular wings slowly up and down. On either side of the mouth, which is at the front of the head instead of on the underside as in other rays, are two long lobes, called cephalic horns, that funnel plankton into the mouth. The horns are the origin of mantas' other name: devil rays. A short, sticklike tail trails behind.

On coral reefs, reef manta rays tend to congregate over high points where currents bring plankton up to them. Reef mantas may be sociable with divers at some sites and have been known to "dance" with them. Small fish called remoras often travel attached to these giants. Despite their huge size, manta rays can "breach," leaping clear of the water. The IUCN classes giant mantas as endangered and reef mantas as vulnerable.

HUMAN IMPACT

STINGRAY CITY

Southern stingrays are not aggressive toward humans and only sting if stepped on or feel threatened. Their stings are used as defense against sharks, their natural predator. At a site in Grand Cayman in the Caribbean, called "Stingray City," they have become used to humans and will approach people. Visitors wade, swim, and dive among these graceful creatures.

ORDER MYLIOBATIFORMES

Blue-spotted Stingray

Taeniura lymma

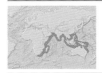

LENGTH	Up to 6½ ft (2 m) including tail
WEIGHT	Up to 65 lb (30 kg)
DEPTH	Shallow water to about 65 ft (20 m)

DISTRIBUTION Indian Ocean, western Pacific, Red Sea

Since it is active in the daytime, divers often see this beautifully colored ray on coral reefs. It is most often spotted lying on sandy patches under coral heads and rocks. Often, its blue-striped tail sticks out and gives away its hiding place. Large, bright blue spots cover the disk, which is greenish brown. Like all stingrays, it has a venomous spine on its tail. As the tide rises, these rays move in groups into shallow water to hunt for invertebrates such as mollusks, crabs, shrimp, and worms.

ORDER MYLIOBATIFORMES

Reticulate Whipray

Himantura uarnak

LENGTH	About 15 ft (4.5 m) including tail
WEIGHT	About 265 lb (120 kg)
DEPTH	65–165 ft (20–50 m)

DISTRIBUTION Coastal waters of Arabian Gulf, Red Sea, Indian Ocean, and western Pacific

This beautifully patterned stingray belongs to a group called whiprays, which have long, thin, flexible tails. Also known as the coach whipray, its upper surface is densely covered with wavy brown lines or reticulations. Its disk, or body, is about 5 ft (1.5 m) long and the tail can be nearly three times this length. Its stinger is a single, large spine located a short distance from the tail base; some individuals have two spines. The body is almost diamond-shaped and the snout is broadly triangular with a pointed tip. Found in warm waters, mainly near the coast, reticulate whiprays are sometimes seen by divers, lying quietly in sandy patches between rocks. There are other similar spotted species of Whipray, making management of this heavily fished, endangered species even more problematic.

ORDER MYLIOBATIFORMES

Spotted Eagle Ray

Aetobatus narinari

WIDTH (WINGSPAN)	Up to 10 ft (3 m)
WEIGHT	Up to 500 lb (230 kg)
DEPTH	3–260 ft (1–80 m)

DISTRIBUTION Tropical waters worldwide

Often solitary, spotted eagle rays also move around in huge shoals of at least a hundred individuals in open waters—a truly spectacular sight when silhouetted against a sunlit surface. Unlike most other rays, the spotted eagle ray is a very active swimmer.

Most of its swimming time is spent in open water, although it is also commonly seen inshore. It appears to "fly" through the water as it moves its pointed "wings"—enlarged pectoral fins—gracefully up and down. Besides the beautiful patterning of spots on its dorsal surface, another distinctive feature of the spotted eagle ray is its head, which ends in a flattened, slightly upturned snout that resembles a duck's bill. It has a long, thin whiplike tail with a venomous spine near the base.

These rays are very agile and can twist and turn to escape predatory sharks. Sometimes, small groups splash around at the surface, making spectacular leaps out of the water. Why they do this is not clear, but it may be to help dislodge parasites.

ORDER TORPEDINIFORMES

Atlantic Torpedo

Tetronarce nobiliana

LENGTH	Nearly 6½ ft (2 m) including tail
WEIGHT	Up to 200 lb (90 kg)
DEPTH	To 2,600 ft (800 m)

DISTRIBUTION Atlantic, Mediterranean

Electric rays use special organs to produce electricity, which they discharge to stun their prey or attack predators. The Atlantic torpedo is the largest electric ray and can produce a shock of up to 220 volts—enough to stun a person. It can easily be recognized by its circular, disklike body and short, thick tail ending in a large, paddle-shaped fin. It is a uniform dark brown or black on the back and white underneath. The electric organs are in the ray's wings, or pectoral fins, and like a battery, they can store electricity. When hunting, the Atlantic torpedo wraps its wings around its prey before stunning it.

Bony Fishes

DOMAIN	Eucarya
KINGDOM	Animalia
PHYLUM	Chordata
CLASSES	Actinopterygii
	Sarcopterygii
ORDERS	48
SPECIES	31,290

BONY FISHES EXCEED ALL OTHER VERTEBRATE groups both in number of living species and in their abundance. They have evolved into myriad shapes and sizes, suiting every aquatic lifestyle and habitat and range from the shore to the deepest depths and from polar seas to hot deep-sea vents. Bony fishes have an internal skeleton of bone, although that of a few primitive groups is part cartilage. The bony skeleton supports flexible fins that allow them to move with far greater precision than do the stiff fins of cartilaginous fishes. Just under half of bony fishes live only in fresh water, while the remainder lives in the oceans or migrates between the two.

Anatomy

Like other vertebrate animals, bony fishes have a skull, backbone, and ribs, but the skeleton also extends out into the fins as a series of flexible rays. Bony fishes, unlike sharks, can use their paired pectoral and pelvic fins for maneuvering, braking, and even swimming backward. Spiny-rayed fishes, a group that includes most bony fishes, also have sharp spines in the front portion of their dorsal, anal, and pelvic fins. A bony flap called the operculum covers the gills of bony fishes. It can be opened to regulate the flow of water in through the mouth and out over the gills. A covering of overlapping, flexible scales made of thin bone protects most bony fishes. Some primitive bony fishes, such as sturgeon, are armored with thick, inflexible scales or plates.

SWIMMING

The sideways force and backward force exerted when a fish moves its tail from side to side results in a thrust at an angle between the two. The resultant thrusts on left and right produce a net backward thrust and so the fish is propelled forward.

forward movement

movement of tail

sideways force

resultant thrust

backward force

BONY SKELETON

Flexible rays and hard spines support all the fins in bony fish, such as this cod. The fins connect to spines extending from the vertebrae. The fish can adjust the position of each fin precisely.

first dorsal fin
vertebrae linked into a flexible vertebral column
spine extends from vertebra to fin
second dorsal fin
third dorsal fin
skull
orbit (eye socket)
rib
pectoral fin
first anal fin
second anal fin
hinged jaw
bony gill covering (operculum)
pelvic fin
tail (caudal) fin

GILLS, VIEWED FROM ABOVE

As water passes over the gill filaments, gases are exchanged. Oxygen passes into the blood and carbon dioxide passes out into the water. Within the filaments, the blood flows in reverse relative to the water outside, so the concentration of gases in the fluids is opposed, which speeds the gases' transfer.

esophagus
gill filaments
oral valve
mouth
direction of water movement
gill arch, attachment point for filaments

SCALES

Bony fishes can be aged by their scales. Slow winter growth produces dark rings on the scales, so each dark ring indicates one year of life. The system works best for temperate-water fish such as cod.

surface concealed under adjacent scale
annual growth ring
exposed surface
exposed surfaces overlap to create smooth covering

HUMAN IMPACT

FISH FARMING

Many important bony fish stocks are fished at unsustainable levels. Relieving the pressure by fish farming is not easy as species such as cod are difficult to rear on a large scale. In contrast, almost all Atlantic salmon come from aquaculture. Many farmed fish eat feed prepared from other wild-caught fish.

SALMON FARM
Salmon farms, such as this one in Tasmania, are a common sight in temperate seas. However, there are problems with fish lice (see p.346) and with dilution of the wild gene pool by escapees.

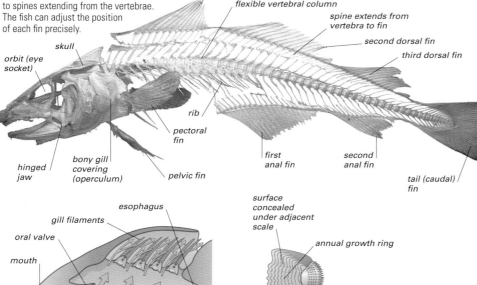

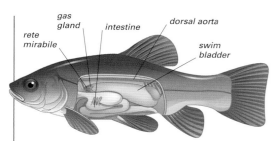

SWIM-BLADDER FUNCTION
A bony fish regulates its buoyancy by secreting gases, usually oxygen and carbon dioxide, from a gas gland into its swim bladder. The gland is supplied with blood (the source of the gas) by a network of capillaries called a rete mirabile.

FLEXIBLE APPENDAGES
A frogfish displays one of the many functions of bony fish fins. Since they are not required to give the fish lift, the warty frogfish's paired fins have evolved into flexible appendages with which it clambers over the seabed.

Buoyancy

Most bony fishes have a gas-filled swim bladder that allows them to adjust their buoyancy, enabling them to hover in midwater and not sink. This is especially useful to fish that spend their lives in midwater. Many bottom-living fish, such as flatfish, have a poorly developed swim bladder or none at all. To compensate for pressure changes as a fish swims toward or away from the surface, it regulates the amount of gas in the swim bladder, usually by secreting gas into and absorbing it back out through glands. In some primitive fish, such as the herring, the swim bladder is connected to the gut and is filled when the fish gulps air at the surface.

Many bony fishes can vibrate the swim bladder with special muscles to produce sounds. Cartilaginous fishes do not have a swim bladder. They can gain buoyancy through large, oil-filled livers and lightweight skeletons. However, cartilaginous fishes must also use their large pectoral fins and tail to give them lift. Bony fishes with a swim bladder have been freed from this necessity and, in many species, the fins have developed into versatile appendages used for courtship, feeding, attack, or defense.

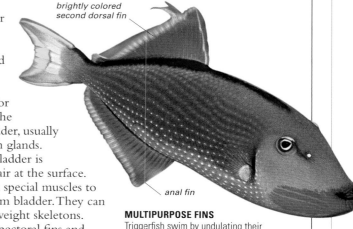

MULTIPURPOSE FINS
Triggerfish swim by undulating their second dorsal and anal fins, maintaining buoyancy with their swim bladder. Bright fins may also function as visual signals in communication, including courtship.

Senses

Bony fishes use vision, hearing, touch, taste, and smell. Vision is most important in well-lit habitats. Coral reef fish have good color vision, and they use colors and patterns for recognition, warning, deception, and courtship. Color receptors in the eyes do not operate well in dim light. Nocturnal fish and fish living in the twilight zone (see p.170) have large, sensitive eyes, but little sensitivity to different colors. Dark-zone fish often have only tiny eyes, but have a sharp sense of smell and use pheromones for long-distance communication. Sound also carries well underwater (see p.37) and some fish produce intense sounds with their swim bladder. Bony fishes move in unison in shoals with the help of their lateral-line sensory system, for which there is no equivalent in other vertebrates. Sense organs arranged in a canal along the head and sides of each fish pick up water movements created by the other fish. The wide field of view, due to having eyes set on the sides of the head, also helps precision shoaling.

SIGNALING COLORS
Color is effective in communication on well-lit coral reefs. The gaudiness of the mandarin fish may warn predators that it is unpalatable.

LATERAL LINE
The lateral-line system can be seen in many bony fishes, such as the pollack here, as a contrasting line along the sides of the fish. The shape of the line is a useful identification feature.

Reproduction

The majority of bony fish, when mature, simply shed their eggs and sperm directly into the sea, where fertilization takes place. The eggs develop and the larvae hatch while drifting on ocean currents. Death rates of eggs and larvae are high, but many eggs are laid—up to 100 million by the giant ocean sunfish. Once the larvae have grown to juvenile fish, they often congregate in nursery grounds in sheltered estuaries and bays. In contrast to most oceanic fish, many coastal bottom-living species are able to protect their offspring, so they lay fewer, larger eggs, often hiding them or caring for them until they hatch. Some have evolved elaborate forms of care, such as mouth brooding.

MOUTH BROODING
When a female jawfish has laid her eggs, the male collects them into his mouth to keep them safe. He will not feed until the eggs hatch and the fry disperse.

SEX CHANGE IN THE CUCKOO WRASSE

❶ Like most wrasse, cuckoo wrasse have a complex reproductive pattern featuring sex change. The majority of eggs develop first into pink females.

❷ Some older females develop the blue-and-orange pattern of males and change sex after about seven years of age. Others remain female.

❸ At the next spawning season, a sex-changed male acquires vibrant colors and courts all females in his territory, fertilizing their eggs (see p.361).

Hunting and Protection

All fish must eat and in doing so may expose themselves to the risk of being eaten if they come out into the open to forage. Their ultimate aim is to survive long enough to reproduce successfully and so pass on their genes to the next generation. Bony fish have evolved many ingenious methods for catching prey and defending themselves against predators. Camouflage is an effective strategy and can serve to hide a fish, both from its predators and from its prey. Color patterns can also deceive, and butterflyfish use false eye patterns to fool predators into lunging for their tail end. In the crowded environment of a coral reef, many small fish protect themselves with spines. Filefish erect a dorsal spine and lock it into position, thereby preventing larger fish from swallowing them. Out in the surface waters of the open ocean, there is nowhere to hide, and many small fish live in shoals for safety. Predators find it difficult to pick out a target as the shoal moves and swirls. Although the shoal is conspicuous, it is safer for each individual to join than to swim alone.

CAMOUFLAGE
Scorpionfish employ color, shape, and behavior in a combined camouflage strategy. Experts at keeping still, they can strike with lightning speed if a small fish strays within reach.

SHOALING
Even predatory fish need protection from larger predators, especially when young. Barracuda juveniles live in shoals during the day, while most adults hunt alone.

SHADOW-HUNTING
Trumpetfish often shadow predatory fish when they hunt, since the larger fish will often flush out suitable prey. This trumpetfish has chosen to swim with a Nassau grouper similar in color to itself.

BONY FISHES CLASSIFICATION

Bony fish comprise the ray-finned fishes (class Actinopterygii), and the lobe-finned fishes (class Sarcopterygii), including the lungfishes (freshwater) and the coelacanths, but also giving rise to tetrapods (see cladogram, p.206). Below are 30 marine orders of both classes.

COELACANTHS
Order Coelacanthiformes

2 species
One family—the only marine lobe-finned fish. Fins arise from fleshy, limblike lobes, vertebral column not fully formed.

STURGEONS AND PADDLEFISH
Order Acipenseriformes

28 species
Only sturgeon family is marine. Skeleton part bone, part cartilage. Sturgeons have asymmetrical tail and underslung mouth.

TARPONS AND TENPOUNDERS
Order Elopiformes

9 species
Two families, mostly marine. Spindle-shaped, silvery fish, one dorsal fin, forked tail. Unique bones in throat (gular plates). Swim bladder can be used as lung. Transparent larvae.

BONEFISH
Order Albuliformes

13 species
One family, mostly marine. Similar to tarpons, but smaller and very bony, with complex structural differences.

MORAY EEL, ORDER ANGUILLIFORMES

EELS
Order Anguilliformes

1,030 species
Marine and freshwater, 16 families. Body long and thin, no scales or pelvic fins, one long fin along back, tail, and belly.

SWALLOWERS AND GULPERS
Order Saccopharyngiformes

28 species
Four families of highly aberrant deep-sea, eel-like fish with huge, loose jaws; no tail fin, pelvic fins, scales, ribs, or swim bladder.

HERRINGS
Order Clupeiformes

435 species
Seven families, mostly marine. Silvery body with keeled belly and forked tail. Anchovies and herring comprise two biggest families.

MILKFISH
Order Gonorynchiformes

36 species
Four families, only milkfish and beaked salmon marine. Pelvic fins set far back.

CATFISH
Order Siluriformes

4,163 species
Only two marine families out of 33. Long body, up to four pairs of barbels around mouth. Sharp, sometimes venomous spine in front of dorsal and pectoral fins. Most with adipose fin.

SMELTS
Order Osmeriformes

46 species
Five families, mostly marine or anadromous. Small, slim relations of salmon.

SALMONS
Order Salmoniformes

242 species
One family with marine, anadromous, and freshwater members. Powerful, spindle-shaped fish, with large mouth and eyes. One fin plus adipose fin on back. Small, rounded scales. Pelvic fins abdominal.

LIGHTFISH, DRAGONFISH, AND HATCHETFISH
Order Stomiiformes

449 species
Four abundant, deep-ocean families. Mostly elongate predators with large teeth and photophores. Significant part of ocean's fishes.

GRINNERS
Order Aulopiformes

300 species
Eighteen families, all marine. Diverse, slim coastal and deep-sea fish. Large mouth with many small teeth. Pelvic fins abdominal, one fin plus adipose fin on back, no fin spines.

LANTERNFISH
Order Myctophiformes

278 species
Two deep-ocean, widely distributed, abundant families. Small, slim fish, large eyes and mouth. One fin plus adipose fin on back. Many photophores. Daily vertical migration.

VELIFERS, TUBE-EYES, AND RIBBONFISH
Order Lampriformes

27 species
Six families, all marine. Colorful, bright, often huge, open-water fish with crimson fins. Many have long rays from dorsal fin.

HERRING, ORDER CLUPEIFORMES

COD FISH
Order Gadiformes

652 species
Ten families, mostly marine and benthic. Most with two or three spineless dorsal fins and a chin barbel. Grenadiers have long, thin tails.

TOADFISH
Order Batrachoidiformes

85 species
One family, mostly marine, coastal, and benthic. Broad, flat head, wide mouth, eyes on top; one short, spiny and one long, soft dorsal fin.

CUSK EELS
Order Ophidiiformes

563 species
Five families, mostly marine. Eel-like fish with long dorsal and anal fin that may join with tail fin. Thin pelvic fins.

ANGLERFISH, ORDER LOPHIIFORMES

ANGLERFISH
Order Lophiiformes

399 species
Eighteen families, all marine. Large, flattened or rounded head with cavernous mouth and fishing lure on top. Shallow-water species benthic; deep-water species pelagic.

CLINGFISH
Order Gobiesociformes

186 species
One family, mostly marine. Small, shallow-water, benthic fish; pelvic fins forming sucker-disk. Eyes set high; single dorsal fin.

NEEDLEFISH
Order Beloniformes

283 species
Six families, marine and freshwater. Mostly long, thin fish with jaws extended as beaks. Flying fish have large pectoral and pelvic fins.

SILVERSIDES
Order Atheriniformes

393 species
Ten families, marine and freshwater. Small, slim, silvery fish; most with two dorsal fins, often in large shoals.

SQUIRRELFISH AND RELATIVES
Order Holocentriformes

92 species
One family, all marine. Deep-bodied, big eyes (except deep-water), dorsal fin spiny at front, forked tail, large scales. Most nocturnal.

DORIES AND ALLIES
Order Zeiformes

35 species
Six families, all marine. Deep-bodied but thin fish; large, spiny head and protrusile jaws. Long dorsal and anal fins with spines at front.

GOBIES
Order Gobiiformes

2,290 species
One major family and five to eight minor ones. Small, mostly bottom-living fish with two dorsal fins, pelvic fins often forming a sucker, and high-set eyes.

PIPEFISH AND SEAHORSES
Order Syngnathiformes

334 species
Five families, marine and freshwater. Long body encased in armor of bony plates. Small mouth at end of tubular snout.

PERCHLIKE FISH
Order Perciformes

Many thousand species, 50–100+ families, marine and freshwater. Largest and most diverse vertebrate order. Most have both spines and soft rays in dorsal and anal fins. Pelvic fins close to pectorals and with one spine. Perciform classification is currently undergoing considerable revision, hence uncertainty in numbers. Distinctive groups, such as gobies and blennies, have recently been assigned to separate orders. Others, such as scorpionfishes, are now included in the Perciformes. Ongoing work means many fish families remain to be assigned.

FLATFISH
Order Pleuronectiformes

825 species
Eleven families, mostly marine. Lie on seabed. Body flattened from side to side, both eyes on upper side. Start life as normal, symmetrical fish larvae in plankton.

PUFFERS AND FILEFISH
Order Tetraodontiformes

447 species
Ten families, marine and freshwater. Very diverse group ranges from triggerfish to ocean sunfish. Small mouth with few large teeth or tooth plates. Scales usually modified as plates, spines, or shields.

TURBOT, ORDER PLEURONECTIFORMES

OCEAN LIFE

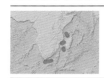

ORDER COELACANTHIFORMES

Coelacanth

Latimeria chalumnae

LENGTH	Up to 6 ft (2 m)
WEIGHT	Up to 210 lb (95 kg)
DEPTH	490–2,300 ft (150–700 m)

DISTRIBUTION Western Indian Ocean

When it was first discovered in 1938, the coelacanth was nicknamed "old four legs" because its pectoral fins had strange, fleshy, limblike bases. The only other primitive group to have a similar arrangement are the freshwater lungfish. It is from fish like these that the first four-legged land animals are thought to have developed. Coelacanths have a tail has an extra small lobe in the middle, and the body is covered in heavy scales, which are made up of four layers of bone and a hard mineral material. In life, these shimmer an iridescent blue with white flecks. Coelacanths live in deep water on steep, rocky reefs and, so far, have been found at only a few sites off the south and east coasts of Africa and the west coast of Madagascar. By using small submersibles to study these fish, scientists have discovered that they retreat into caves at night. When out searching for food, they drift along in ocean currents or scull slowly with their fins. Having located a fish or squid, the coelacanth then uses its powerful tail to propel itself forward so that it can seize its prey. The coelacanth is listed as Critically Endangered on the IUCN Red List of Endangered Species. International trade in this species is banned.

DISCOVERY

FOSSIL EVIDENCE

Coelacanths were thought to have become extinct about 65 million years ago. When a live coelacanth was caught in 1938, comparing it with fossil coelacanths enabled scientists to confirm its identity. A living specimen of a fossil group had been found.

FOSSIL COELACANTH
Fossil specimens of coelacanths have features almost identical to present-day coelacanths, including the unique three-lobed tail.

ORDER COELACANTHIFORMES

Indonesian Coelacanth

Latimeria menadoensis

LENGTH	Up to 4½ ft (1.4 m)
WEIGHT	Up to 200 lb (90 kg)
DEPTH	490–655 ft (150–200 m)

DISTRIBUTION Celebes Sea, north of Sulawesi, in the western Pacific

When the Indonesian coelacanth was discovered in 1998, it was at first thought to be the same species as the one found in African waters (see left). The two are indeed very similar, but molecular studies suggest that they are different species. An entire ocean separates them, and because coelacanths are slow swimmers, the populations are not thought to mix. The Indonesian coelacanth has the same white markings and distinctive gold flecks as the African species, but it is brown rather than bluish. As yet, little is known of its life history, but because it is so physically similar to the African species, it probably has the same behavior and could therefore be endangered by fishing.

ORDER ACIPENSERIFORMES

European Sturgeon

Acipenser sturio

LENGTH	11 ft (3.5 m)
WEIGHT	Up to 880 lb (400 kg)
DEPTH	13–295 ft (4–90 m)

DISTRIBUTION Coastal waters of northeastern Atlantic, Mediterranean, and Black Sea

Like most sturgeon, this species swims from the sea into large rivers to spawn in gravelly areas. These prehistoric-looking fish belong to a primitive group in which only the skull and some fin supports are made of bone. The rest of the skeleton consists mainly of cartilage. Instead of scales, five rows of distinctive bony plates, or scutes, run along the body. Two pairs of barbels hang down from the pointed snout and are used to search out bottom-living invertebrates. The European sturgeon is thought to live for at least 60 years.

HUMAN IMPACT

CAVIAR BAN

Once common, the European sturgeon is now extremely rare due to overfishing and poaching, and because locks and polluted estuaries have made many rivers unsuitable for spawning. Few active spawning sites remain. This sturgeon is critically endangered, and international trade in the fish itself and any products from it, including caviar (salted roe), has been banned.

ORDER ACIPENSERIFORMES

Beluga Sturgeon

Huso huso

LENGTH	16 ft (5 m)
WEIGHT	Up to 4,400 lb (2,000 kg)
DEPTH	230–590 ft (70–180 m)

DISTRIBUTION Northern Mediterranean, Black Sea, Caspian Sea, and associated rivers

The beluga is both the largest species of sturgeon and the largest European fish to enter fresh water. Stouter and heavier than the European sturgeon (see left), it has a more triangular snout with a very wide mouth. Four long barbels hang from the underside of the snout, reaching almost to the mouth. Like all sturgeon, the beluga has an asymmetrical, sharklike tail with the backbone extending into the large upper lobe. Reputed to be the most expensive fish in the world, it also produces the most prized caviar, with large fish containing 220–440 lb (100–200 kg). It is critically endangered due to overexploitation and loss of spawning habitat.

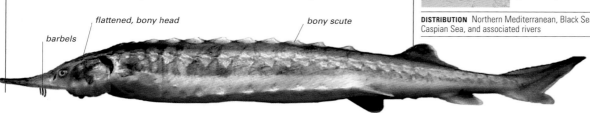

barbels flattened, bony head bony scute

ORDER ELOPIFORMES

Tarpon

Megalops atlanticus

LENGTH	Up to 8 ft (2.5 m)
WEIGHT	350 lb (160 kg)
DEPTH	0–100 ft (0–30 m)

DISTRIBUTION Coastal waters of western and eastern Atlantic

With its large scales and intensely silvery body, the tarpon resembles an oversized herring but, in fact, is closely related to the eels. It has an upturned mouth, and the base of the single dorsal fin is drawn out into a long filament, although this is not always easy to see. Living close inshore, this fish often enters estuaries, lagoons, and rivers. If it enters stagnant water, it surfaces and gulps air, which passes from the esophagus into its swim bladder; this then acts like a lung.

Tarpon spawn mostly in open water at sea. A large female can produce over 12 million eggs, but larval and juvenile mortality is high. The larvae, which are thin and transparent and very like eel larvae except that they have forked tails, drift inshore into estuarine nursery grounds. Tarpon larvae are also found in pools and lakes that become temporarily cut off from the sea. Fishers get to know the areas where tarpon shoals can regularly be seen from year to year, hunting for

other shoaling fish such as sardines, anchovies, and mullet. They will also eat some bottom-living invertebrates such as crabs. Considered an excellent game fish in US and Caribbean waters, tarpon make spectacular leaps when hooked. They are also fished commercially and, in spite of being rather bony, are considered delicious. Tarpon can live for 55 years and are often displayed in public aquariums. Their large scales are sometimes used in ornamental work.

ORDER ELOPIFORMES

Ladyfish

Elops saurus

LENGTH	Up to 3 ft (1 m)
WEIGHT	22 lb (10 kg)
DEPTH	0–165 ft (0–50 m)

DISTRIBUTION Coastal waters of western Atlantic and Caribbean Sea

The ladyfish has a single dorsal fin in the middle of its back and a tail that is deeply forked. Shoals of this slim, silvery blue fish can be found close to the shore and will skip along the surface if alarmed by a boat's engine noise. The adult fish move offshore to spawn in open water, and the young larvae, which resemble eel larvae, eventually drift back into sheltered bays and lagoons. Also known as the ten-pounder, the ladyfish is considered a good game fish and will leap out of the water when hooked. It is fished commercially, but it is not a very high-quality food fish and so is often used for bait.

ORDER ALBULIFORMES

Bonefish

Albula vulpes

LENGTH	Up to 3 ft (1 m)
WEIGHT	22 lb (10 kg)
DEPTH	0–280 ft (0–85 m)

DISTRIBUTION Tropical and subtropical coastal waters of western and eastern Atlantic

As its common name suggests, this fish is extremely bony. It is streamlined and silvery, with dark markings on its back, a single dorsal fin, and a blunt snout extending over the mouth. Bonefish have been found in tropical and subtropical waters of the Pacific Ocean, but it is not yet known if these populations are a different species from those found in the eastern and western Atlantic. Although bonefish do not make good eating, they are one of the world's most important game fish. Hunters enjoy stalking them through the shallows in bays and estuaries as they often swim at the surface with the dorsal fin showing.

ORDER ANGUILLIFORMES

European Eel

Anguilla anguilla

LENGTH	Up to 4¼ ft (1.3 m)
WEIGHT	14 lb (6.6 kg)
DEPTH	0–2,300 ft (0–700 m)

DISTRIBUTION Temperate waters of northeastern Atlantic, fresh water inland

Living most of its life in fresh water, the European eel swims thousands of miles down to the sea and across the Atlantic Ocean to the Sargasso Sea to spawn. After spawning in deep water, the eels die, leaving the eggs to hatch into transparent, leaflike (leptocephalus) larvae. Over the next year or so, the larvae drift back to the coasts of Europe. Nearing the coast, they change shape and become tiny transparent eels, or elvers, that swim and wriggle their way up rivers into fresh water. This once abundant fish is now critically endangered due to disease, habitat loss, illegal elver fishing, and river dams.

ORDER ANGUILLIFORMES

Chain Moray Eel

Echidna catenata

LENGTH	Up to 5½ ft (1.7 m)
WEIGHT	Not recorded
DEPTH	0–40 ft (0–12 m)

DISTRIBUTION Tropical reefs of western and central Atlantic

The chain moray eel is one of very few marine eels that can survive for some time out of water, and it will forage over wet rocks for up to 30 minutes at a time during low tide. As long as it remains wet, it can absorb some oxygen through its skin. The chain moray eel is easily recognized by its short, blunt snout and chainlike yellow markings. Some of its teeth are broad and molarlike and help it to cope with heavily armored prey such as crabs. It can swallow small crabs whole, but breaks up bigger ones first by twisting, tugging, and thrashing around. The chain moray eel is a member of a large family of moray eels (Muraenidae) that live on reefs throughout the tropics. Most species of moray eels are nocturnal, but the chain moray eel is usually active during the day.

VERSATILE HUNTER

Most species of moray eels spend the day in holes in a reef with just their heads sticking out, emerging at dusk to hunt. They rely on their excellent sense of smell to find fish resting between corals and rocks. Unusually, the chain moray eel also hunts over rocky shores and reefs at low tide during the day. It uses its sharp eyesight to search for fish and crustaceans in crevices and holes and, when it has located its prey, it strikes, rather like a snake. Other moray eels will sometimes strike at passing prey from their holes during the day.

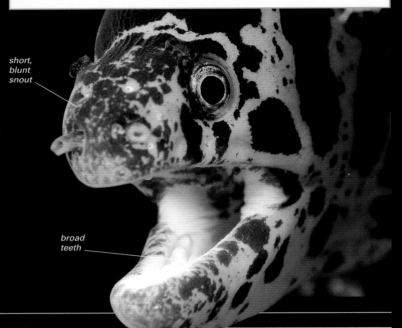

short, blunt snout

broad teeth

ORDER ANGUILLIFORMES

Ribbon Eel

Rhinomuraena quaesita

LENGTH	Up to 4¼ ft (1.3 m)
WEIGHT	Not recorded
DEPTH	3–200 ft (1–60 m)

DISTRIBUTION Tropical reefs of Indian and Pacific Oceans

Unlike most other eels, ribbon eels change color and sex during their life. Juveniles are nearly black with a yellow dorsal fin. As they mature, the black becomes bright blue and the snout and lower jaw turn yellow. This is the male color stage. When they reach a body length of about 4¼ ft (1.3 m), the males turn yellow and become fully functional females, which lay eggs. Ribbon eels live on coral reefs, mostly hiding in crevices. They have leaflike nostril flaps, which sense vibrations in the water.

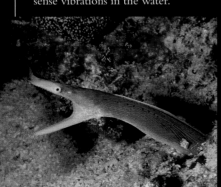

ORDER ANGUILLIFORMES

Conger Eel

Conger conger

LENGTH	Up to 10 ft (3 m)
WEIGHT	Up to 240 lb (110 kg)
DEPTH	1,600 ft (0–500 m)

DISTRIBUTION Temperate waters of northeastern Atlantic and Mediterranean

The large, gray head of a conger eel sticking out of a hole in a shipwreck is a familiar sight to many divers. Like their relatives the moray eels, conger eels hide in holes and crevices in rocky reefs during the day, only emerging at night to hunt for fish, crustaceans, and cuttlefish. This snakelike fish has a powerful body with smooth skin, no scales, and a pointed tail. A single dorsal fin runs along the back, starting a short distance behind the head, continuing around the tail, and ending halfway along the belly.

In the summer, adult conger eels migrate into deep water in the Mediterranean and Atlantic to spawn and then die. The female lays 3–8 million eggs, which hatch into long, thin larvae that slowly drift back inshore, where they grow into juvenile eels. They take 5–15 years to reach sexual maturity.

The conger eel is a good food fish and is caught in large numbers by anglers, but it sometimes manages to use its strength to escape with the bait.

ORDER ANGUILLIFORMES

Slender Snipe Eel

Nemichthys scolopaceus

LENGTH	Up to 4¼ ft (1.3 m)
WEIGHT	Not recorded
DEPTH	300–6,600 ft (90–2,000 m)

DISTRIBUTION Temperate and tropical seas worldwide

This long, slender, deep-sea eel has remarkable jaws, shaped like a bird's bill, with the ends turned out so that they can never fully close. It spends its life drifting in midwater, catching small crustaceans to eat. When the males mature and are ready to spawn, their jaws shorten, they lose all their teeth, and their front nostrils grow into large tubes. This probably enhances their sense of smell, helping them to find mature females. Little else is known of the slender snipe eel's lifestyle, as it is rarely caught.

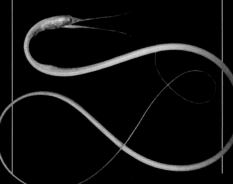

Spotted Garden Eel

Heteroconger hassi

LENGTH	Up to 16 in (40 cm)
WEIGHT	Not recorded
DEPTH	23–150 ft (7–45 m)

DISTRIBUTION Red Sea and tropical waters of Indian Ocean and western Pacific

These eels spend their lives swaying gracefully to and fro with their heads up in the water and their tails in their sandy burrows. Several hundred fish live together in a colony, or "garden," looking like evenly spaced plants blowing in the breeze. Garden eels are much slimmer than their close relatives, the conger eels. They are only about ½ in (14 mm) in diameter and have very small pectoral fins. The spotted garden eel usually has two large dark spots behind the head as well as many tiny ones all over the body. It has an upturned mouth that is designed to pick tiny planktonic animals from the water as the current flows by. Colonies of these eels occur only on sandy slopes that are exposed to currents but sheltered from waves. When danger threatens, the eels sink back down into their burrows, using their tails as an anchor until only their small heads and eyes are visible. They are very difficult to photograph underwater because they are able to detect the vibrations from a scuba diver's air bubbles and will disappear when they are approached.

Spotted garden eels stay in their burrows even when spawning. Neighboring males and females reach across and entwine their bodies before releasing eggs and sperm. Mixed colonies of spotted and whitespotted garden eels sometimes occur.

Banded Snake Eel

Myrichthys colubrinus

LENGTH	Up to 38 in (97 cm)
WEIGHT	Not recorded
DEPTH	Shallow water

DISTRIBUTION Tropical waters of Indian Ocean and western Pacific

Cleverly disguised to look like a venomous sea snake (yellow-lipped sea krait), the banded snake eel is avoided by most predators. This allows it to hunt safely over sand flats and seagrass beds near coral reefs for small fish and crustaceans. Most individuals of this species are banded with broad black and white bands, but in some areas these eels have dark blotches

distinctive banded markings

may eventually be identified as a different species. The banded snake eel has a pointed head with a pair of large tubular nostrils on the upper jaw that point downward. This gives the fish an excellent sense of smell that allows it to seek out prey hidden beneath the sand surface.

With only tiny pectoral and long dorsal fins, the banded snake eel swims by undulating its long body. When not hunting, it buries itself in the sand using the hard, pointed tip of its tail to burrow in tail-first. These fish are most active by night. They tend to remain in their burrows during the day and so are not often seen by divers.

The banded snake eel belongs to a large family (Ophichthidae) which includes around 250 snake and worm eels, most of which burrow into sand and mud. All the members of this family have flattened transparent,

hard, pointed tail tip

long, slender tail

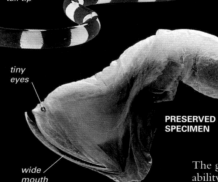

tiny eyes

wide mouth

PRESERVED SPECIMEN

Gulper Eel

Saccopharynx lavenbergi

LENGTH	Up to 5 ft (1.5 m)
WEIGHT	Not recorded
DEPTH	6,600–9,800 ft (2,000–3,000 m)

DISTRIBUTION Deep waters of eastern Pacific, from California to Peru

The gulper eel is best known for its ability to swallow prey as large as itself. This fish has a small head and tiny eyes but enormous jaws. Its mouth and throat can be hugely distended to engulf its prey and the teeth can be depressed backward. Its stomach can be similarly extended to accommodate its gargantuan meals. The body ends in a luminous organ on a long, whiplike tail. This feature may be used as a lure or a decoy, but this has yet to be confirmed, as this eel is rarely observed in the wild from submersibles. The gulper eel has planktonic eggs that develop into long, thin larvae, like those of its shallow-water relatives,

ORDER CLUPEIFORMES

Atlantic Herring

Clupea harengus

LENGTH	Up to 18 in (45 cm)
WEIGHT	Up to 2¼ lb (1 kg)
DEPTH	0–650 ft (0–200 m)

DISTRIBUTION North Atlantic, North Sea, and Baltic Sea

Until the middle of the 20th century, the Atlantic herring was the mainstay of many fishing communities bordering the North Sea and north Atlantic. Along the East Anglian coast of Great Britain, the fish were known as silver darlings. In the 20th century, excessive fishing using new techniques led to a steep decline in stocks. Today, the stocks are managed, but they are still under pressure.

The Atlantic herring feeds on plankton, coming to the surface at night after spending the day in deeper water. It lives in large shoals, and across its range the species is divided into distinct local races, which differ from each other in size and behavior. Each race has several traditional spawning grounds. The females produce up to 40,000 eggs each, which form a thick mat on the seabed.

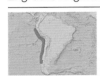

ORDER CLUPEIFORMES

Peruvian Anchoveta

Engraulis ringens

LENGTH	Up to 8 in (20 cm)
WEIGHT	Up to 1 oz (25 g)
DEPTH	10–260 ft (3–80 m)

DISTRIBUTION West coast of South America and southeastern Pacific

The distribution of this tiny, silvery relative of the herring depends on the yearly extent of the Peruvian Current. This cold, deep current comes to the surface along the west coast of South America, bringing rich supplies of nutrients with it. Enormous shoals of anchoveta feed on the plankton blooms triggered by the increase in nutrients. The fish shoal within about 50 miles (80 km) of the coast, and many local people depend on them, as do many birds, including pelicans.

ORDER GONORYNCHIFORMES

Milkfish

Chanos chanos

LENGTH	Up to 6 ft (1.8 m)
WEIGHT	Up to 31 lb (14 kg)
DEPTH	0–100 ft (0–30 m)

DISTRIBUTION Tropical and subtropical waters of Indian and Pacific Oceans

The milkfish is an elegant silvery fish with a streamlined body and a large, deeply forked tail. It is an important food fish in much of Southeast Asia and is extensively farmed. It feeds on plankton, soft algae, cyanobacteria, and small invertebrates. It is easy to keep in captivity as it is able to tolerate a wide range of salinity. Mature fish spawn in the sea, and the eggs and larvae drift inshore. Juveniles swim into estuaries and mangroves, where there are fewer predators, returning to the sea as they mature.

ORDER CLUPEIFORMES

South American Pilchard

Sardinops sagax

LENGTH	Up to 16 in (40 cm)
WEIGHT	Up to 17½ oz (485 g)
DEPTH	0–650 ft (0–200 m)

DISTRIBUTION West coast of South America and southeastern Pacific

Enormous shoals of these fish, made up of millions of individuals, were once found, but excessive fishing has reduced their numbers greatly.

Pilchards are an important food fish and are also used to produce oil and fish meal. The South American pilchard and California pilchard are thought to be the same species but different subspecies, and all pilchard species are similar. These silvery, medium-sized fish are blue-green on the back with a series of black marks along the sides.

ORDER CLUPEIFORMES

Allis Shad

Alosa alosa

LENGTH	Up to 33 in (83 cm)
WEIGHT	Up to 9 lb (4 kg)
DEPTH	0–16 ft (0–5 m)

DISTRIBUTION Temperate waters of northeastern Atlantic and Mediterranean Sea

The Allis shad is a silvery fish belonging to the herring family (Clupeidae) and is one of the few that enter fresh water. During April and May, mature adults migrate into rivers to spawn, swimming up to 500 miles (800 km) upstream. In some parts of its range, the species is known as the May fish. Its streamlined body is covered by large circular scales that form a keel under the belly, and it has a single dorsal fin. It is now very rare over much of its range.

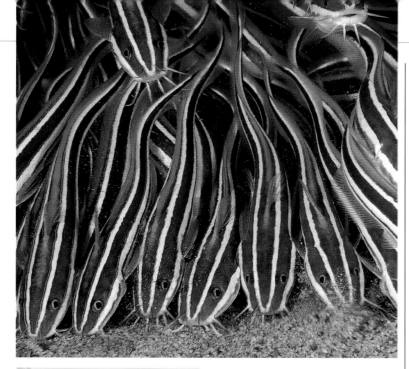

ORDER SILURIFORMES

Striped Catfish

Plotosus lineatus

LENGTH	Up to 13 in (32 cm)
WEIGHT	Not recorded
DEPTH	3–200 ft (1–60 m)

DISTRIBUTION Red Sea and tropical waters in Indian and Pacific Oceans

The juveniles of this distinctive black-and-white striped catfish of the family Plotosidae stay together in dense, ball-shaped shoals and are often seen by divers over coral reefs. Adults live on their own or in small groups, but are well protected by a venomous, serrated spine in front of the first dorsal fin and each of the pectoral fins. A sting from an adult striped catfish can be dangerous to humans and is very occasionally fatal. These fish hunt at night, using four pairs of sensory barbels around the mouth to find worms, crustaceans, and mollusks hidden in the sand. During the day, they hide among rocks. Plotosids are the only catfish found in coral reefs. This species also ventures along open coasts and into estuaries. It spawns in the summer months. Male striped catfish build nests in shallow, rocky areas and guard the eggs for about ten days. The larvae are planktonic.

ORDER OSMERIFORMES

European Smelt

Osmerus eperlanus

LENGTH	18 in (45 cm)
WEIGHT	Not recorded
DEPTH	To 160 ft (50 m)

DISTRIBUTION Temperate waters of northeastern Atlantic and Baltic Sea

Like salmon and trout, the European smelt has a dorsal fin and a small adipose (fleshy) fin on its back. The name of this fish derives from the fact that, when fresh, the European smelt has a strong smell that is reminiscent of cucumber. Adults swim in shoals in inshore waters, hunting small crustaceans and fish. They migrate up rivers to spawn, and the young fish are common in sheltered estuaries such as the Wash in southeast England (see p.128).

ORDER SILURIFORMES

Gafftopsail Sea Catfish

Bagre marinus

LENGTH	Up to 28 in (70 cm)
WEIGHT	Up to 10 lb (4.5 kg)
DEPTH	To 160 ft (50 m)

DISTRIBUTION Gulf of Mexico, Caribbean Sea, and subtropical waters of western Atlantic

The most conspicuous feature of this silvery catfish is the pair of very long mouth barbels that extend back almost to the end of the pectoral fins. It has another pair of short barbels under the chin. The first rays of the large dorsal fin and the pectoral fins are drawn out as long, flat filaments and these fins also have a venomous serrated spine. When threatened, this catfish erects its dorsal fin and spreads out its pectoral fins like the sails of a yacht.

deeply forked tail

sail-like dorsal fin

broad pectoral fin

ORDER ARGENTINIFORMES

Barrel-eye

Opisthoproctidae

LENGTH	Up to 4 in (10 cm)
WEIGHT	Not recorded
DEPTH	1,000–2,600 ft (300–800 m)

DISTRIBUTION Tropical and subtropical waters worldwide

Many fish that live in the twilight zone (see p.170), including barrel-eyes, have large eyes to make full use of what little light is available. As well as being large, the eyes of barrel-eyes are tubular and point upward. This arrangement probably helps the barrel-eye to stalk other fish from below. Looking up, it is likely that it can pick out the silhouette of its prey or spot fish with bioluminescent patches on their undersides. The barrel-eye *Macropinna microstoma* has a completely transparent top to its head. It can rotate its tubular eyes from pointing upward to forward to allow it to first pinpoint prey and then catch it.

ORDER OSMERIFORMES

Capelin

Mallotus villosus

LENGTH	Up to 10 in (25 cm)
WEIGHT	Up to 1⁴/₅ oz (52 g)
DEPTH	0–1,000 ft (0–300 m)

DISTRIBUTION North Pacific, north Atlantic, and Arctic Ocean

This small, silvery relative of salmon forms large shoals in cold and Arctic waters and is a vital food source for sea birds and marine mammals. The breeding success of some seabird colonies has been linked to the abundance of capelin, and this in turn depends on environmental factors and exploitation by fishing. It is a major food source for Inuit peoples. Capelin are slim fish, with an olive-green back fading into silvery white on the sides. Shoals of this fish swim along with their mouths open, straining out plankton, which is caught on their modified gills. While this is their main source of food, they also eat worms and small fish. In spring, the schools move inshore, the males arriving first and waiting for the females. The males develop a band of modified scales along their sides and use these to massage the female, stimulating her to lay her eggs in the sand.

TIDAL BREEDING

Capelin eggs make a good meal for many invertebrates and fish. To protect their eggs, large numbers of adult capelin swim into very shallow water at high tide and spawn on sandy beaches just below the tideline. Each female produces about 60,000 reddish, sticky eggs, which lie in the sand. When the eggs hatch after about 15 days, the larvae are washed out of the sand by the incoming tide and then swept out to sea on the outgoing tide.

Atlantic Salmon

Salmo salar

LENGTH	Up to 5 ft (1.5 m)
WEIGHT	Up to 100 lb (45 kg)
DEPTH	Mostly surface waters

DISTRIBUTION Temperate and cold waters of north Atlantic and adjacent rivers

While most marine fish would quickly die in fresh water, the Atlantic salmon can move easily between river and sea. Fish with this ability are called anadromous. Designed for long-distance swimming, this fish has a powerful, streamlined body and a large tail. During their spawning runs, Atlantic salmon swim against strong river currents and leap up waterfalls to reach their spawning grounds. Before spawning, the salmon roam the north Atlantic for several years feeding on other fish. The Atlantic salmon is highly prized as a game fish, but wild salmon are becoming increasingly rare.

HUMAN IMPACT

SALMON FARMING

Floating fish farms rearing Atlantic salmon are common in Scottish sea lochs and Norwegian fiords. Salmon fry from freshwater hatcheries are later moved to submerged nets in saltwater farms. Environmental concerns focus on fish lice, which spread from the farms and infect wild fish, and on chemicals used to treat farmed fish.

Coho Salmon

Oncorhynchus kisutch

LENGTH	Up to 3 ft (1 m)
WEIGHT	Up to 33 lb (15 kg)
DEPTH	0–820 ft (0–250 m)

DISTRIBUTION Temperate and cold waters of north Pacific and adjacent rivers

Like most other salmon, the coho salmon is a fast, streamlined predator with excellent eyesight for spotting its prey. This makes it a challenging game fish for anglers. When ready to breed, mature fish find their way from the ocean back to the same river in which they were born. While swimming upstream, they develop bright red sides and a green head and back. When they reach the shallow waters at the river's head, the females dig a nest, which is called a redd, in the gravel of the riverbed and lay their sticky eggs while the male fertilizes them. After spawning, the adults die and their bodies provide a feast for scavenging bears and other animals.

Arctic Char

Salvelinus alpinus

LENGTH	Up to 3 ft (1 m)
WEIGHT	Up to 33 lb (15 kg)
DEPTH	0–230 ft (0–70 m)

DISTRIBUTION Arctic Ocean and northern freshwater rivers and lakes

Arctic char are adapted for life in cold, oxygen-rich water and cannot tolerate warm or polluted water. There are two physiological races: a migratory form that lives in the sea but spawns in rivers, and a land-locked lake form. Migratory char grow to at least 3 ft (1 m) long and are regarded as excellent game fish. Shortly after the last ice age, they ranged much farther south but are now restricted to Arctic waters. The char that live in mountain lakes are relicts of this period.

Sloane's Viperfish

Chauliodus sloani

LENGTH	Up to 14 in (35 cm)
WEIGHT	Up to 1 oz (30 g)
DEPTH	1,600–6,000 ft (475–1,800 m)

DISTRIBUTION Tropical and temperate waters worldwide

Deep-water fish are some of the most bizarre of all fish, and Sloane's viperfish is no exception. At one end of its slender body it has a large head with huge, barbed teeth, while at the other it has a tiny forked tail. Rows of photophores run along the sides and belly and light the fish up like a night-flying airplane. During the day, it stays in deep water, but at night it migrates upward to feed where prey is more abundant. The single dorsal fin, just behind the head, has a very long first ray that can be arched over the head and may help entice prey within reach. Sloane's viperfish spawns throughout the year. It is one of nine species of viperfish, all deep-living.

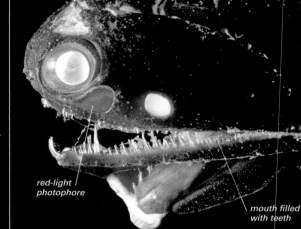

ORDER STOMIIFORMES

Pacific Blackdragon
Idiacanthus antrostomus

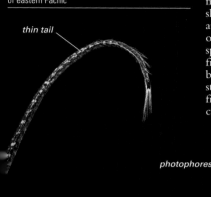

LENGTH Up to 15 in (38 cm)	
WEIGHT Up to 2 oz (55 g)	
DEPTH 650–3,300 ft (200–1,000 m)	

DISTRIBUTION Deep, tropical and temperate waters of eastern Pacific

The Pacific blackdragon haunts the depths of the ocean, its black snakelike body lit up by photophores along its belly. When it opens its mouth, it reveals a set of long, dagger-sharp teeth. Hanging off the lower jaw is a barbel tipped by a glowing lure that can be moved to entice prey to venture within reach.

The Pacific blackdragon is black on the inside as well as the outside, its black stomach preventing light from swallowed bioluminescent prey shining out. Male Pacific blackdragons are only about a quarter the size of the females. In the closely related species *Idiacanthus fasciola*, the young fish are similar in shape to the adults, but their eyes stick out on very long stalks. The stalks are absorbed as the fish grows and the eyes eventually come to lie in their sockets.

thin tail

photophores

snakelike body

barbel with lure

red-light photophore

mouth filled with teeth

ORDER STOMIIFORMES

Stoplight Loosejaw
Malacosteus niger

LENGTH Up to 9½ in (24 cm)	
WEIGHT Not recorded	
DEPTH 3,300–13,000 ft (1,000–4,000 m)	

DISTRIBUTION Deep tropical and temperate waters worldwide

Like many other deep-sea fish, the stoplight loosejaw is black, relatively small, and has a large mouth. However, it is unique in that it has no floor to its mouth, hence its name. Instead, a ribbon of muscle that joins the gill basket and the lower jaw contracts to shut the mouth. This arrangement may allow the fish a wider gape and a faster strike at prey. This fish is also a specialist in light production. It has two large photophores under each eye, one that produces normal blue-green bioluminescence and the other red. No natural red light reaches these depths, so most deep-sea creatures cannot see it. The red bioluminescence reflects well off a red animal, such as a shrimp, but the shrimp will be unaware that it has been spotlighted.

ORDER STOMIIFORMES

Lovely Hatchetfish
Argyropelecus aculeatus

LENGTH Up to 3 in (8 cm)	
WEIGHT Not recorded	
DEPTH 330–2,000 ft (100–600 m)	

DISTRIBUTION Tropical and temperate waters worldwide

An expert at hiding from predators, the hatchetfish's silvery coloration and use of bioluminescence conceals it against the downwelling light. They are also so thin that they are difficult to see head-on. This fish lives at medium depths and has large bulging eyes to make best use of what little light there is. At dusk, it rises up to 330–1,000 ft (100–300 m) to feed on small planktonic animals.

ORDER AULOPIFORMES

Tripodfish
Bathypterois grallator

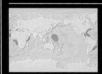

LENGTH Up to 15 in (37 cm)	
WEIGHT Not recorded	
DEPTH 2,900–11,500 ft (875–3,500 m)	

DISTRIBUTION Deep waters of Atlantic, Pacific, and Indian Oceans

ORDER AULOPIFORMES

Reef Lizardfish
Synodus variegatus

LENGTH 8–16 in (20–40 cm)	
WEIGHT Not recorded	
DEPTH 16–295 ft (5–90 m)	

DISTRIBUTION Tropical reefs in Red Sea, Indian Ocean, and western Pacific

The deep ocean floor where the tripodfish lives consists largely of soft mud. So, to prevent itself from sinking into the ooze while lying in wait for its prey, this fish perches on a tripod made from elongated rays of its pelvic and caudal fins. Facing into the current, it waits for small crustaceans to drift within reach, catching them in its mouth, which has a large gape. The tripodfish has very small eyes and is thought to detect its prey by feeling for tiny vibrations in the water.

The reef lizardfish habitually perches on the tops of rocks and corals, propped up on its long pelvic fins. From such vantage points, it keeps a lookout for passing shoals of fish, darting out and seizing one with its rows of sharp teeth. Its large mouth allows it to swallow quite big fish (as shown in the photograph). A variable blotchy brown and red coloration camouflages the reef lizardfish, hiding it from larger predators. It can also bury itself in patches of sand, leaving only its head and eyes showing. Confident of its disguise, this fish will remain completely still and allow divers to approach to within a few inches before darting away to a new perch. It is caught and eaten by reef fishers.

ORDER MYCTOPHIFORMES

Spotted Lanternfish
Myctophum punctatum

LENGTH Up to 4¼ in (11 cm)	
WEIGHT Not recorded	
DEPTH 0–3,300 ft (0–1,000 m)	

DISTRIBUTION Deep waters of north Atlantic and Mediterranean

The spotted lanternfish is one of over 250 species of lanternfish found in the world's oceans. Lanternfish are rather unprepossessing, small spindle-shaped fish with large eyes. However, in spite of their drab appearance they can put on an unrivaled display of light from an array of photophores along their sides and belly. In some species, males and females have different patterns of photophores, and this helps them to find each other in the dark depths. Photophore patterns also differ between species.

Large shoals of spotted lanternfish are common in the north Atlantic. Along with other lanternfish, it is an important food source for larger fish, sea birds, and marine mammals. During the day it stays in deep water, at 800–2,500 ft (250–750 m), but at night it swims up to within about 330 ft (100 m) or even right to the surface, where it feeds on planktonic crustaceans and fish fry.

OCEAN LIFE

OCEAN LIFE

ORDER LAMPRIFORMES

Opah

Lampris guttatus

LENGTH	Up to 6 ft (2 m)
WEIGHT	110–600 lb (50–275 kg)
DEPTH	330–1,300 ft (100–400 m)

DISTRIBUTION Tropical, subtropical, and temperate waters worldwide

Roaming the oceans worldwide, the opah leads a nomadic existence. Shaped like a gigantic oval dinner plate, this colorful fish is a steely blue and green with silvery spots and red fins. Although it is toothless, the opah is an efficient hunter, catching squid and small fish. Rather than using its tail to swim, like most fish, it flies through the water by beating its long, narrow pectoral fins like a pair of wings.

Opah regularly reach a weight of 110 lb (50 kg), although specimens as heavy as 600 lb (270 kg) have been reported. They spawn in the spring, laying eggs midwater, which hatch into larvae after 21 days. Also known as the moonfish, the opah is a valuable food fish in the Hawaiian Islands and on the west coast of mainland US. It is caught on long lines and with gill nets.

ORDER LAMPRIFORMES

Oarfish

Regalecus glesne

LENGTH	Up to 36 ft (11 m)
WEIGHT	Up to 600 lb (270 kg)
DEPTH	0–3,300 ft (0–1,000 m)

DISTRIBUTION Tropical, subtropical, and temperate waters worldwide

At up to 36 ft (11 m) in length, the oarfish is the longest bony fish known to science and is thought to be responsible for many sea serpent legends. Its bizarre appearance is enhanced by a crest of long red rays on its short, bluish head. These are followed by a bright red dorsal fin that runs the length of its silver body, which is marked with black streaks and spots. Its name comes from the pelvic fins, both of which extend as a single, long ray ending in an expanded tip, which looks like the blade of an oar. In the open ocean, the oarfish drifts in the currents, feeding on other fish and squid, its great length protecting it from most predators. Although it lives in tropical and temperate waters worldwide, the oarfish is rarely caught or seen alive, so little is known about its behavior.

ORDER GADIFORMES

Bib

Trisopterus luscus

LENGTH	Up to 18 in (46 cm)
WEIGHT	Up to 5½ lb (2.5 kg)
DEPTH	10–330 ft (3–100 m)

DISTRIBUTION Temperate waters of northeastern Atlantic and western Mediterranean

Divers often see shoals of striped bib around rocky reefs and shipwrecks. These are usually younger fish or adults that have moved inshore to spawn. Large old fish often lose their banded pattern and become very dark. The bib has a much deeper body than most of its relatives in the family Gadidae. A long chin barbel and long pelvic fins help it to find crustaceans, mollusks, and worms to eat.

ORDER GADIFORMES

Atlantic Cod

Gadus morhua

LENGTH	Up to 6 ft (2 m)
WEIGHT	Up to 200 lb (90 kg)
DEPTH	0–2,000 ft (0–600 m)

DISTRIBUTION Temperate and cold waters of north Atlantic

The Atlantic cod is a powerful, heavily built fish with a large head, an overhanging upper jaw, and a single long chin barbel. It has small, elongated scales. The coloration varies from reddish, especially in young fish, to a mottled brown with a conspicuous white lateral line. The Atlantic cod is a shoal-forming fish, living in water over the continental shelf, and usually feeding at 100–250 ft (30–80 m) above areas of flat mud or sand. Adults migrate to established breeding grounds to spawn, usually in the early spring, with each female releasing several million eggs into the water. Atlantic cod can live for at least 25 years and mature fish can reach a weight of over 200 lb (90 kg), but sophisticated modern fishing techniques means that most cod are caught long before they reach this age and weight. Fish over 33 lb (15 kg) are now rare. However, Atlantic cod and its close relative Pacific cod are still among the world's most important commercial species.

square-ended tail

FINS AND SCALES
The Atlantic cod has three dorsal fins and two anal fins. Its small scales have growth rings, which can be counted to give its age.

HUMAN IMPACT

OVERFISHING

Stocks of Atlantic cod were once thought to be inexhaustible, but numbers have declined drastically over most of its range. Cod exist as a number of discrete stocks that spawn in specific areas in water about 660 ft (200 m) deep. The collapse of stocks in Canada in the early 1990s led to fishing bans, but even today, numbers are nowhere near what they once were. Management of North Sea stocks prevented complete collapse, but the stock is still under pressure.

FORAGING FOR FOOD
Atlantic cod feed both in midwater and on the seabed. They eat shoaling fish, such as herring, and also crustaceans, worms, and mollusks.

ORDER GADIFORMES

Shore Rockling
Gaidropsarus mediterraneus

LENGTH
Up to 20 in (50 cm)

WEIGHT
Up to 2¼ lb (1 kg)

DEPTH
0–1,500 ft (0–450 m)

DISTRIBUTION Temperate waters of northeastern Atlantic and Mediterranean

Rockling are eel-like in appearance, with two dorsal fins. The first of these is a fringe of short rays that ripple constantly. The shore rockling can be found in rock pools, where it uses its mouth barbels to find food. Most are dark brown, but some are paler.

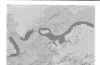

ORDER GADIFORMES

Torsk
Brosme brosme

LENGTH Up to 4 ft (1.2 m)

WEIGHT 26–66 lb (12–30 kg)

DEPTH 65–3,300 ft (20–1,000 m)

DISTRIBUTION Temperate and cold waters of north Atlantic

This heavily built member of the order Gadiformes lurks among rocks and pebbles in deep water offshore, where it searches for crustaceans and mollusks. It has thick lips, a long chin barbel, and a long dorsal and anal fin, each edged in white. In summer, two to three million eggs are laid, which float and develop near the surface. This species can live for 20 years. Torsk is fished commercially, especially off Norway, using trawls and lines, and it is also caught by anglers.

ORDER GADIFORMES

Pacific Grenadier
Coryphaenoides acrolepis

LENGTH
Up to 3 ft (1 m)

WEIGHT
Up to 6½ lb (3 kg)

DEPTH 1,000–12,000 ft (300–3,700 m)

DISTRIBUTION Deep, temperate waters of north Pacific

The Pacific grenadier is one of about 300 different species of grenadiers that are found just off continental shelves and are abundant in every ocean. Grenadiers are also known as rattails because they have a large, bulbous head with big eyes, a sharp snout, and a long, scaly tail. The Pacific grenadier is dark brown with a tall dorsal fin. Another low fin runs along the back and all the way around the tail.

This species spends most of its time near the seabed searching for food but it sometimes swims up into midwater, where it can catch squid, shrimp, and small fish.

ORDER OPHIDIIFORMES

Pearlfish
Carapus acus

LENGTH
Up to 8 in (21 cm)

WEIGHT
Not recorded

DEPTH
To 330 ft (100 m)

DISTRIBUTION Mediterranean; occasionally found in subtropical waters of eastern Atlantic

The adult pearlfish has a most unusual home—it lives inside the body cavity of sea cucumbers. To allow it to slip in and out of its host easily, it has an eel-like body, no pelvic fins, and no scales. It is a silvery-white color with reddish markings. At night, the pearlfish may swim out of the sea cucumber's rear end to go hunting for invertebrates to eat, returning to the body cavity tail first. However, the pearlfish may also eat the gonads and other organs of its host.

ORDER OPHIDIIFORMES

Spotted Cusk-eel
Chilara taylori

LENGTH Up to 15 in (37 cm)

WEIGHT Not recorded

DEPTH 0–900 ft (0–280 m)

DISTRIBUTION Temperate and subtropical waters of eastern Pacific

As it is a favorite food of sea lions, cormorants, and other diving birds, the spotted cusk-eel is most active at night or on gloomy, sunless days. If danger threatens, this eel-shaped fish can quickly slip between rocky rubble or bury itself tail-first in sand or mud, where it can remain in the safety of a mucus-lined burrow. Unlike true eels, it has scales and pelvic fins. The latter are reduced to one split ray set very far forward under the head. Its eggs are laid in open water and hatch into larvae that live close to the surface. These develop into juveniles that drift for an extended period before settling down to a seabed existence.

While there are many other species of cusk-eel, this is the only one in its genus.

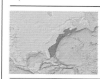

ORDER BATRACHOIDIFORMES

Oyster Toadfish
Opsanus tau

LENGTH Up to 17 in (43 cm)

WEIGHT Up to 4¼ lb (2.2 kg)

DEPTH 0–16 ft (0–5 m)

DISTRIBUTION Temperate and subtropical waters of northwestern Atlantic

Some people would consider the oyster toadfish an ugly animal, with its flat head, wide, toadlike mouth, and thick lips. It also has tassels around its chin, prominent eyes, and two dorsal fins, the first of which is spiny. Its shape and coloration provide camouflage in its home under rocks and debris. This hardy fish tolerates dirty and trash-strewn water and

COURTSHIP CALLS

People living in houseboats along the east coast of the US are sometimes kept awake at night during April to October by loud grunting noises. The culprits are male oyster toadfish calling to attract females to lay their eggs in nests dug under rocks. The male makes these noises by vibrating the walls of his swim bladder using special muscles. The swim-bladder wall acts like the skin over a drum. The male guards the eggs until they hatch after about four weeks.

is often found under jetties. It has been reared in captivity for use in experiments. It also does well in aquariums and is a popular game fish.

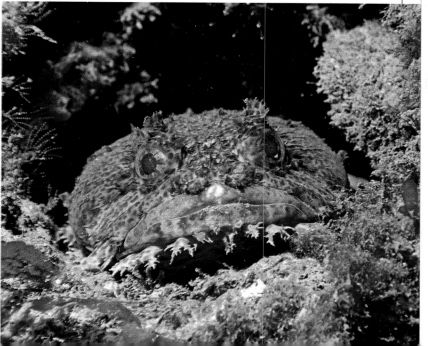

ORDER LOPHIIFORMES

Hairy Angler
Caulophryne jordani

LENGTH	Females up to 8 in (20 cm); males not recorded, but tiny
WEIGHT	Not recorded
DEPTH	330–5,000 ft (100–1,500 m)

DISTRIBUTION Deep water worldwide

Anglerfish include some of the most bizarrely shaped fish in the ocean and the hairy angler certainly fits into this category. It has a huge mouth, tiny eyes, and large dorsal and anal fins with very long projecting fin rays. It is also covered in sensory filaments that may act as a 3D array, detecting prey when touched. Like most anglerfish, it has a bioluminescent, movable lure on top of the head that is formed from the first spine of the dorsal fin. The biology of the hairy angler is poorly known because only a few specimens have ever been captured. However, in other deep-sea anglerfish, this lure is used to attract prey within reach. The fish then opens its mouth and creates a sudden, strong inward suction current. The prey is engulfed within a fraction of a second. Food is scarce in the deep sea and anglerfish living here usually have extra-large mouths and expandable stomachs that allow them to swallow prey as big or bigger than themselves. The hairy angler belongs to the family Caulophrynidae, also known as fanfins. The males of fish in this family are tiny and do not have lures. They live as parasites on the females when they are adults, and this was seen and filmed in 2018. Most caught specimens are damaged from contact with nets and from changes in pressure as they are brought to the surface.

large mouth

sensory hairs

ORDER LOPHIIFORMES

Polka-dot Batfish
Ogcocephalus radiatus

LENGTH	Up to 15 in (38 cm)
WEIGHT	Not recorded
DEPTH	0–230 ft (0–70 m)

DISTRIBUTION Subtropical waters of western Atlantic and Gulf of Mexico

Fish of the family Ogcocephalidae, to which the polka-dot batfish belongs, are among the most oddly shaped of the anglerfish. They prop themselves up on paired pectoral and pelvic fins that enable them to walk over the seabed in search of worms, crustaceans, and fish. Although the polka-dot batfish has a fishing lure, this is very short and evidence suggests it may secrete an odor that attracts potential prey. A hard, spiny skin protects these fish from predators, but they are so sluggish that divers can pick them up.

ORDER LOPHIIFORMES

Deep-sea Angler
Bufoceratias wedli

LENGTH	Females up to 10 in (25 cm); males not recorded
WEIGHT	Not recorded
DEPTH	1,000–5,700 ft (300–1,750 m)

DISTRIBUTION Gulf of Mexico, Caribbean Sea, and Atlantic

Living in the deep sea, this small, dark-colored anglerfish has a round body, delicate fins, and a luminescent lure at the end of a long rod called an illicium. A second, much smaller rod on the head is often hidden from view. It has a weak skeleton and small muscles that make it relatively light and able to float more easily. It has no need to swim much as it lures its prey within reach. Female deep-sea anglers have been caught undamaged by research submarines using a piece of equipment called a slurp gun that sucks animals into a container. The fish have then been photographed alive. Males have not yet been seen but are likely to be tiny and free-living.

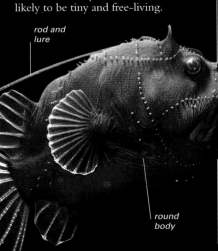

rod and lure

round body

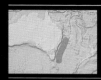

ORDER LOPHIIFORMES
Coffinfish
Chaunax endeavouri

LENGTH	Up to 9 in (22 cm)
WEIGHT	Not recorded
DEPTH	160–1,000 ft (50–300 m)

DISTRIBUTION Temperate waters of southwestern Pacific, off east coast of Australia

The coffinfish resembles a pink balloon covered in tiny spines and can make itself look bigger by inflating its body. It belongs to a family of

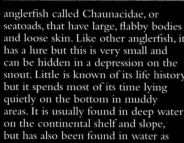

anglerfish called Chaunacidae, or seatoads, that have large, flabby bodies and loose skin. Like other anglerfish, it has a lure but this is very small and can be hidden in a depression on the snout. Little is known of its life history, but it spends most of its time lying quietly on the bottom in muddy areas. It is usually found in deep water on the continental shelf and slope, but has also been found in water as shallow as 165 ft (50 m).

ORDER LOPHIIFORMES
Sargassumfish
Histrio histrio

LENGTH	Up to 8 in (20 cm)
WEIGHT	Not recorded
DEPTH	About 0–36 ft (0–11 m)

DISTRIBUTION Tropical and subtropical seas worldwide; not recorded in eastern Pacific

This unusual frogfish (family Anternnariidae) lives in floating rafts of sargassum seaweed. It uses its prehensile, leglike pectoral fins to clasp clumps of weed and scramble around

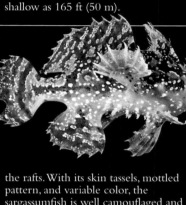

the rafts. With its skin tassels, mottled pattern, and variable color, the sargassumfish is well camouflaged and able to lure small fish and shrimps within striking range. If threatened, it can scramble onto the top of the seaweed raft. These fish are sometimes washed ashore with their rafts.

ORDER LOPHIIFORMES
Angler
Lophius piscatorius

LENGTH	Up to 6 ft (2 m)
WEIGHT	Up to 125 lb (57 kg)
DEPTH	65–3,300 ft (20–1,000 m)

DISTRIBUTION Northeastern Atlantic south to West Africa, Mediterranean, and Black Sea

The angler has a head like a flattened football fringed by a camouflage of seaweed-shaped flaps of skin, and a wide, flattened body that

tapers toward the tail. Its darkly marbled greenish-brown skin also helps the angler blend into the sediment of the sea floor. It lies patiently on the seabed, ready to suck in any fish that it can entice within range by flicks of the fleshy fishing lure on its dorsal fin. Large anglers have even been known to lunge up and catch diving birds. The species has well-developed pectoral fins, set on armlike bases, with sharp "elbows" that allow it to shuffle along over the sea floor. Anglerfish of the genus *Lophius* are also known as goosefishes or fishing frogs. This species is commercially exploited and sold as "monkfish."

ORDER LOPHIIFORMES
Common Blackdevil
Melanocetus johnsonii

LENGTH	Females 7 in (18 cm); males 1¼ in (3 cm)
WEIGHT	Not recorded
DEPTH	To 6,600 ft (2,000 m)

DISTRIBUTION Deep waters of Atlantic, Pacific, and Indian Oceans

This deep-sea anglerfish is also known as the humpback angler. The female common blackdevil has a huge head and large jaws with very long, daggerlike teeth,

which are used to catch prey that may be larger than herself. Her stretchy stomach and loosely attached skin help her accommodate these huge meals. Although the female is not completely blind, she has tiny eyes and probably cannot see prey until she has enticed it within range using her glowing lure. Compared to the female, the male is tiny and uses his acute sense of smell to find a mate. He has no teeth but hangs on to the female with special hooks on his snout. When she has laid her eggs and he has fertilized them, he swims away, but how much longer he lives is not known. Both male and female juveniles live near the surface, where they feed on small planktonic animals.

glowing lure

long, sharp teeth

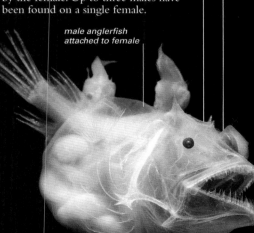

ORDER LOPHIIFORMES
Regan's Angler
Haplophryne mollis

LENGTH	Females 3 in (8 cm); males ¾ in (2 cm)
WEIGHT	Not recorded
DEPTH	650–6,600 ft (200–2,000 m)

DISTRIBUTION Tropical and subtropical deep waters worldwide

This unusual deep-sea anglerfish has unpigmented skin. The female of the species has an almost round body when mature, numerous very small teeth, spines above the eyes and behind the mouth, and a minimal fishing lure that consists of just a small flap on the snout. Like many other deep-sea anglerfish, the males of this species remain very small all their lives and their sole aim in life is to track down a female using their excellent sense of smell and latch onto her using special hooks. Finding a mate in

the depths of the ocean is difficult and by keeping the male attached, the female is assured that her eggs will be fertilized. The males eventually turn into parasites, biting into the female's skin. In time their blood supplies fuse and the male then becomes nourished by the female. Up to three males have been found on a single female.

male anglerfish attached to female

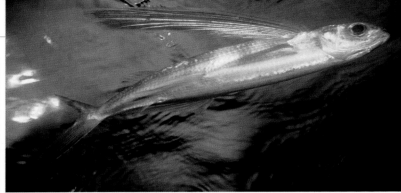

ORDER GOBIESOCIFORMES

Cornish Sucker

Lepadogaster purpurea

LENGTH	3 in (7 cm)
WEIGHT	Not recorded
DEPTH	0–6 ft (0–2 m)

DISTRIBUTION Temperate waters of northeastern Atlantic, Mediterranean, and Black Sea

Strong waves are no problem to this little fish—it can cling to rocks with a powerful sucker formed from its pelvic fins. It also has a low-profile body and a flattened, triangular head with a long snout that resembles a duck's bill. This shape allows the fish to slip easily between the rocks and, because it is only a few centimeters long, it may be difficult to spot, but it can be found by turning over rocks and seaweeds and searching in rock pools. The color of the Cornish sucker is variable, but it always has two blue spots outlined in brown, red, or black behind its head, and it has a small tentacle in front of each eye. In the spring or summer, females lay clusters of golden yellow eggs on the undersides of rocks on the shore. The eggs are guarded by the parent fish until they hatch.

ORDER BELONIFORMES

Atlantic Flyingfish

Cheilopogon heterurus

LENGTH	Up to 16 in (40 cm)
WEIGHT	Not recorded
DEPTH	Surface waters

DISTRIBUTION Tropical and warm temperate waters worldwide

Also known as the Mediterranean flyingfish, this species is distinguished by its very large, winglike pectoral and pelvic fins. If a predator, such as a tuna, attacks from below, the fish will beat its powerful forked tail rapidly, spread its "wings" at the last moment, and lift clear of the surface away from danger. The fish continues to beat its tail even in midflight and it can remain airborne for over 330 ft (100 m). The Atlantic flyingfish is edible, but is not commercially exploited.

ORDER BELONIFORMES

Hound Needlefish

Tylosurus crocodiles

LENGTH	Up to 5 ft (1.5 m)
WEIGHT	Up to 14 lb (6.5 kg)
DEPTH	0–43 ft (0–13 m)

DISTRIBUTION Tropical waters over coral reefs worldwide

Rendered almost invisible by its silvery color and needlelike shape, the hound needlefish swims along just beneath the surface, hunting for other fish that also live over coral reefs. Its long, thin snout is shaped like a spear, and it has been known to puncture small boats and cause injury to people by shooting up into the air when frightened. Although edible, the hound needlefish is not popular as a food fish because it has green-colored flesh.

ORDER BELONIFORMES

Atlantic Saury

Scomberesox saurus

LENGTH	Up to 20 in (50 cm)
WEIGHT	Not recorded
DEPTH	0–100 ft (0–30 m)

DISTRIBUTION North, northwestern, and eastern Atlantic and Mediterranean

Although not as thin as its needlefish relatives, the Atlantic saury has a similar narrow body and a long, beaklike snout lined with tiny teeth. The body is clear green above and bright silver on the sides. It has a single dorsal and anal fin, each followed by a series of small finlets. This fish lives in large schools that chase and capture smaller fish and shrimplike crustaceans while skimming along at the surface. It is fished commercially and is caught by attracting the fish to lights at night.

ORDER HOLOCENTRIFORMES

Whitetip Soldierfish

Myripristis vittata

LENGTH	Up to 10 in (25 cm)
WEIGHT	Not recorded
DEPTH	10–260 ft (3–80 m)

DISTRIBUTION Tropical waters of Indian and Pacific Oceans

Soldierfish are nocturnal coral-reef residents that hide in groups in caves and beneath overhangs on steep reefs during the daytime. The whitetip soldierfish is an orange red, as are most members of its family (Holocentridae). The leading edges of its median fins are white. At depth, where natural red light does not penetrate, the fish's red color appears black or gray, providing it with camouflage, especially on the deeper parts of the reef. Like many nocturnal fish, the whitetip soldierfish has large eyes, which help it pinpoint their prey in dim light. It has a short, blunt snout; large scales; and a deeply forked tail. Divers have observed that some individuals in a group of whitetip soldierfish often swim along upside down.

ORDER TRACHICHTHYIFORMES

Pineapplefish

Cleidopus gloriamaris

LENGTH	Up to 9 in (22 cm)
WEIGHT	Up to 16 oz (500 g)
DEPTH	10–650 ft (3–200 m)

DISTRIBUTION Temperate waters of eastern Indian Ocean and southwestern Pacific around Australia

The pineapplefish is completely encased in armor consisting of large, thick, modified scales studded with spines. Each yellow scale is outlined in black, resembling a segment of pineapple skin. This fish, which lives in dark caves and under ledges on rocky reefs, has a pair of light organs on its lower jaw containing bioluminescent bacteria. Normally hidden by the upper jaw, the lights only shine out when the fish opens its mouth. At night, the fish move out to hunt small fish and crustaceans that are attracted to the light.

ORDER TRACHICHTHYIFORMES

Eyelight Fish

Photoblepharon palpebratum

LENGTH	5 in (12 cm)
WEIGHT	Not recorded
DEPTH	3–164 ft (1–50 m)

DISTRIBUTION Tropical waters of western and central Pacific

The most characteristic feature of this small fish is the large light organ under each eye. The blue-green light can be turned on and off using a black membrane like an eyelid. These fish are active at night, often feeding in large groups, and use the light to signal to other individuals, startle predators, and find small planktonic animals to feed on. They also use bioluminescence to school in the dark. Eyelight fish are sometimes seen at night by divers on steep reef faces. Daytime sightings are rare, as these fish usually hide in caves during the day.

ORDER TRACHICHTHYIFORMES

Common Fangtooth

Anoplogaster cornuta

LENGTH	6–7 in (15–18 cm)
WEIGHT	Not recorded
DEPTH	1,600–16,000 ft (500–5,000 m)

DISTRIBUTION Deep waters in temperate and tropical waters worldwide

The huge, saberlike teeth of this deep-water predator are designed to make sure any fish it catches has no chance of escape. Suitable prey is rare in the deep ocean. The teeth are no good for cutting or chewing and so the common fangtooth swallows its prey whole, rather like a snake does. Adults are uniformly black or dark brown in color and are most common between 1,600 and 6,500 ft (500 and 2,000 m). They hunt by themselves or in small shoals, searching for other fish to eat. Juvenile common fangtooths look very different from the adults and were classified as a separate species until 1955. Paler in color and with long spines on the head, they live in water as shallow as 160 ft (50 m) and feed mainly on crustaceans.

Adult females shed their eggs directly into the sea, where they develop into planktonic larvae. The juveniles take on the adult shape when they are about 3 in (8 cm) long.

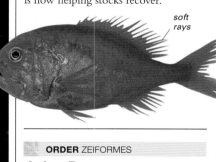

ORDER TRACHICHTHYIFORMES

Orange Roughy

Hoplostethus atlanticus

LENGTH	20–30 in (50–75 cm)
WEIGHT	Up to 15 lb (7 kg)
DEPTH	3,000–6,000 ft (900–1,800 m)

DISTRIBUTION North and south Atlantic, Indian Ocean, southwestern and eastern Pacific

This is one of the longest-lived fish species, with the age of some individuals determined as possibly 250 years old. Its bright, brick-red color appears black in the deep dark waters in which it lives, and this helps to hide it from predators.

This deep-bodied, spiny fish lives over rough ground and delicate, cold-water coral reefs (p.179). Deepsea fisheries target this fish, and because it grows slowly and reproduces late, it cannot sustain heavy exploitation. In Australasia, effective management is now helping stocks recover.

soft rays

ORDER ZEIFORMES

John Dory

Zeus faber

LENGTH	Up to 3 ft (90 cm)
WEIGHT	Up to 18 lb (8 kg)
DEPTH	15–1,300 ft (5–400 m)

DISTRIBUTION Eastern Atlantic, Mediterranean, Black Sea, Indian Ocean, western and southwestern Pacific.

The John Dory has one of the most distinctive appearances of all fish, with a rounded but very thin body, a heavy mouth, and tall fins. It is an expert hunter, stealthily approaching its prey head-on. In this attitude, its thin body is almost invisible and it can approach other fish closely. When it comes within striking range, it shoots out its protrusible jaws and engulfs its meal.

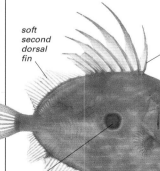

spiny first dorsal fin

soft second dorsal fin

dark mark like a thumbprint

DEEP-SEA FISHING
Trawling for fish in small boats is an arduous and often hazardous way of earning a living. Many fishers have perished at sea over the years.

Fishing

Exploitation of the sea's bounty provides humans with high-quality food and many useful by-products, and sustains coastal fishing communities. Fish have long been seen as a resource that could never run out. However, modern industrial-scale fishing methods are taking their toll. Many stocks have collapsed, and some are beyond recovery. The global recorded catch of marine fish and shellfish rose steadily from 18.4 million tons in 1950 to 96.7 million tons in 1996, with a few dips linked to poor anchovy catches in El Niño years (see p.68–69). Since then, maximum marine catches have both declined and risen, and in 2018 stood at 84.4 million tons.

The problems of ensuring a sustainable harvest from the sea are many. One fundamental difficulty is the "ownership" of stocks. There is little incentive for some to stop fishing in order to conserve fish if others continue, legally or illegally. It is difficult to police fisheries on the high seas, and illegal fishing is rife in some areas. It is notoriously problematic to accurately assess mobile fish stocks; and illegal fishing and trading and inaccurate reporting distort catch statistics.

Many large-scale fishing methods are indiscriminate. There is vast waste, as unwanted and over-quota fish and invertebrate species are discarded, and many turtles, cetaceans, and sea birds are inadvertently caught. Nets with escape hatches for turtles and marked long lines to prevent albatrosses from being hooked are two of a number of new methods to reduce by-catch. Sand eels and other small fish, often termed "whitebait," vital to sea birds such as puffins, are caught in industrial fisheries and turned into fishmeal for livestock. Yet not all fishing is unsustainable, and there is now increasing guidance for those consumers wishing to support well-managed fisheries and nondamaging fishing methods.

Traditional Fishing

Traditional fishing using small-scale fishing gear is rarely a threat to fish stocks. Fish is an important food source, particularly in countries in the developing world, where it provides up to 80 percent of total protein needs. Fishing is also a vital part of the economy in these countries. And yet, such localized, traditional fisheries take only about 10 percent of the global total catch.

STILT FISHING
This method of fishing is still practiced in parts of Sri Lanka and Thailand. The fishers cast their lines while perching on poles in shallow water.

FISHING AND THE ENVIRONMENT

DAMAGE AND WASTE

BOTTOM TRAWLING
Fishing gear dragged across the seabed damages the habitat and marine life and stirs up sediment, smothering and damaging nearby animals. Heavy metal scallop dredges are particularly harmful.

SHRIMP AND BYCATCH
In every catch of shrimp, up to 10 times their weight of other species is also caught in the net and subsequently discarded.

HAZARDS TO WILDLIFE

FISHING GEAR *Thousands of animals die needlessly each year entangled in fishing tackle. This Hawaiian monk seal is one of a total population of under 1,500.*

TURTLE
Drifting longlines for tuna, often dozens of miles long with thousands of hooks, also kill turtles, sharks, and marine birds.

FISH FARMING

SCALLOP FARMING *Farming scallops in hanging nets or on ropes avoids the adverse effects of trawling on the seabed and other species.*

PEN-RAISED TUNA *Fattening of wild-caught Atlantic blue-fin tuna in shallow-water enclosures may further deplete overfished stocks but is regulated by an international commission.*

OCEAN LIFE

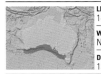

ORDER SYNGNATHIFORMES
Leafy Seadragon
Phycodurus eques

LENGTH	14 in (35 cm)
WEIGHT	Not recorded
DEPTH	13–100 ft (4–30 m)

DISTRIBUTION Eastern Indian Ocean, along the southern coast of Australia

It is hard to imagine anything less fishlike than the leafy seadragon. The bizarre tassels and frills that adorn its head and body form a spectacular camouflage that fools both predators and prey. Even its body and tail are bent and twisted to resemble seaweed stems. Closely related to seahorses, the leafy seadragon has a similar, but much longer, tubular snout. This is an effective feeding tool—the fish aims its snout at a small shrimp and then sucks hard, rather like a person would on a drinking straw. The leafy seadragon lives on rocky, seaweed-covered reefs and in seagrass beds. Unlike seahorses, it cannot coil its tail around an object. It moves very slowly and sways with the waves, mimicking the seaweed. Like seahorses and pipefish, the female deposits her eggs in a brood pouch under the male's tail and he carries them until they hatch.

ORDER SYNGNATHIFORMES
Trumpetfish
Aulostomus maculatus

LENGTH	Up to 3 ft (1 m)
WEIGHT	Not recorded
DEPTH	7–80 ft (2–25 m)

DISTRIBUTION Gulf of Mexico, Caribbean Sea, and subtropical waters of western Atlantic

The trumpetfish looks like a piece of drifting wood, hiding itself among sea fans and other corals. It has a long, slender, straight body, and when it flares open its mouth, its long snout resembles a thin trumpet. The trumpetfish is usually brown, but some individuals have a yellow body.

It hunts by lying in wait to ambush passing shoals of fish, but it is also known to follow predatory fish such as moray eels and steal some of the fish that they flush from their hiding places.

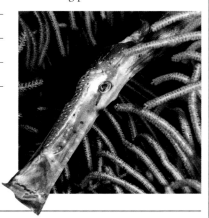

ORDER SYNGNATHIFORMES
Snake Pipefish
Entelurus aequoreus

LENGTH	Up to 24 in (60 cm)
WEIGHT	Not recorded
DEPTH	33–330 ft (10–100 m)

DISTRIBUTION Temperate waters of northeastern Atlantic

At first sight, the snake pipefish could easily be mistaken for a small sea snake. It has a long, elegant, rounded body tapering to a thin tail with a minute tail fin, and it is orange brown with pale-blue bands. Like all pipefish and seahorses, its head is drawn out into a distinctive tubular snout for sucking up small floating crustaceans and fish fry. Pipefish have no scales but, instead, the body is encased in segmented bony armor lying beneath the skin.

The female snake pipefish lays several hundred eggs into a shallow pouch along the male's belly during the summer. The young are released when they are about ¹/₂ in (1 cm) long. With sand eel numbers in decline, puffins may feed piperfish to their young. However, some young puffins choke or starve on this poor diet.

ORDER SYNGNATHIFORMES
Harlequin Ghost Pipefish
Solenostomus paradoxus

LENGTH	5 in (12 cm)
WEIGHT	Not recorded
DEPTH	Not recorded

DISTRIBUTION Tropical reefs in Indian Ocean and western and southwestern Pacific

The harlequin ghost pipefish looks as though it has wings attached to the sides of its long, thin body. In reality, these are greatly enlarged pelvic fins in which the female broods her eggs. The fins are modified to form a pouch, where the eggs remain until they hatch. This uncommon species occurs in a wide variety of bright colors and patterns that mimic the reef feather stars and black corals among which it lives. It also often swims head-down and so gains further camouflage by aligning its body with the branches among which it swims.

ORDER SYNGNATHIFORMES
Short-snouted Seahorse
Hippocampus hippocampus

LENGTH	6 in (15 cm)
WEIGHT	Not recorded
DEPTH	16–200 ft (5–60 m)

DISTRIBUTION Temperate and subtropical waters of northeastern Atlantic and Mediterranean

In the seahorse world, it is the males that give birth to the young. After an elaborate courtship dance, the female lays her eggs in a special pouch on the male's belly. The pouch seals over until the eggs hatch and the tiny baby seahorses emerge. This species is distinguished by its short snout, which is less than a third of the head length.

broad caudal fin

large eye

large pelvic fins

ORDER CALLIONYMIFORMES

Mandarinfish

Synchiropus splendidus

LENGTH
Up to 2½ in (6 cm)

WEIGHT
Not recorded

DEPTH
3–60 ft (1–18 m)

DISTRIBUTION Tropical waters of southwestern Pacific

With its yellow and orange body and distinctive green and blue markings, the mandarinfish is one of the most colorful of all reef fish. Its skin is covered with a distasteful slime and its bright colors warn predators not to touch it. Small groups live inshore on silt-covered seabeds among coral and rubble. Most members of the dragonet family (*Callionymidae*), to which it belongs, are colored to match their surroundings. It is a popular aquarium fish, but it is very difficult to maintain.

ORDER GOBIIFORMES

Yellow Shrimp Goby

Cryptocentrus cinctus

LENGTH
Up to 3 in (8 cm)

WEIGHT
Not recorded

DEPTH
3–50 ft (1–15 m)

DISTRIBUTION Tropical waters of northeastern Indian Ocean and southwestern Pacific

Shrimp gobies share their sandy burrows with snapping shrimps belonging to the genus *Alpheus*. The shrimps have strong claws and excavate and maintain the burrow, while the gobies have good eyesight and act as a lookout at its entrance. The yellow shrimp goby has bulging, high-set eyes; thick lips; and two dorsal fins. Although the usual coloration is yellow with faint, dusky bands, it can also be grayish white. This species lives in sandy areas of shallow lagoons and bays.

ORDER PERCIFORMES

Lionfish

Pterois volitans

LENGTH
Up to 15 in (38 cm)

WEIGHT
Not recorded

DEPTH
7–180 ft (2–55 m)

DISTRIBUTION Tropical waters of eastern Indian Ocean and western Pacific

Although the lionfish can inflict a painful sting, it is not usually dangerous to humans. Its flamboyant coloration of red stripes serves as a warning both to divers and to would-be predators. Also known as the turkeyfish, it hunts at night using its winglike pectoral fins to trap its prey of fish, shrimps, and crabs against the reef.

ORDER PERCIFORMES

Long-spined Bullhead

Taurulus bubalis

LENGTH
Up to 10 in (25 cm)

WEIGHT
Not recorded

DEPTH
0–330 ft (0–100 m)

DISTRIBUTION Temperate waters of northeastern Atlantic and western Mediterranean

Bullheads are small, cold-water relatives of scorpionfish and the stonefish (see above). Like them, they are stout, bottom-living fish with a broad head, large mouth, and spiny fins. The long-spined bullhead also has a long, sharp spine on each cheek. None of its spines is venomous. These small fish can be found in rock pools but are difficult to spot, as they match their color to their background. In the winter, the female lays clumps of eggs between rocks. These are then guarded by the male until they hatch between 5 and 12 weeks later.

ORDER PERCIFORMES

Three-spined Stickleback

Gasterosteus aculeatus

LENGTH
4 in (11 cm)

WEIGHT
Not recorded

DEPTH
0–330 ft (0–100 m)

DISTRIBUTION Temperate waters of north Atlantic and north Pacific

ORDER PERCIFORMES

Spotted Scorpionfish

Scorpaena plumieri

LENGTH
Up to 18 in (45 cm)

WEIGHT
Up to 3¼ lb (1.5 kg)

DEPTH
3–200 ft (1–60 m)

DISTRIBUTION Western Atlantic and eastern Atlantic around Ascension Island and St. Helena

Resting quietly on the seabed, the spotted scorpionfish is almost invisible thanks to its mottled color and weedlike skin flaps that cover its head. However, if this fish is disturbed, it can open its large pectoral fins to display dramatic black-and-white-spotted patches. Flashing these "false eyes" is often enough to frighten off a potential predator, but if this does not work, the spines on its dorsal fin can inflict a venomous sting.

The three-spined stickleback is equally at home in fresh water and shallow, coastal seawater. It has three sharp spines on its back and a series of bony plates along its sides. This species is known for its breeding behavior, which involves the male building a tunnel-like nest of plant material into which he entices one or more females to lay their eggs. He fans oxygenated water over the eggs as they develop.

ORDER PERCIFORMES

Stonefish

Synanceia verrucosa

LENGTH
Up to 16 in (40 cm)

WEIGHT
Up to 5½ lb (2.5 kg)

DEPTH
3–100 ft (1–30 m)

DISTRIBUTION Tropical waters of Indian Ocean and western Pacific

The stonefish is the world's most venomous fish and its sting is capable of killing a human. Each sharply tipped spine of the dorsal fin has a venom gland at the base from which a duct runs in a groove to the spine tip. Lying quietly on rocks or sediment in the shallows, the stonefish matches its color to its background and is easily trodden on. Its camouflage helps it ambush passing fish, which are sucked into its cavernous mouth with lightning speed.

OCEAN LIFE

ORDER PERCIFORMES

East Atlantic Red Gurnard

Chelidonichthys cuculus

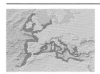

LENGTH	Up to 20 in (50 cm)
WEIGHT	Not recorded
DEPTH	50–1,300 ft (15–400 m)

DISTRIBUTION Temperate waters of northeastern Atlantic and Mediterranean

The East Atlantic red gurnard could be said to be a "walking-talking" fish. The first three rays of the pectoral fins are shaped as separate, thick, fingerlike feelers, which are covered with sensory organs. These feelers are used to "walk" over the seabed and probe for shrimps and crabs. The fish has a large head protected by hard, bony plates and spines and two separate dorsal fins.

These gurnards sometimes form shoals, and as the fish move around, they make short, sharp grunting noises by vibrating their swim bladder with special muscles and so stay in contact with other gurnards nearby. They spawn in spring and summer and the eggs and larvae float freely near the surface. Adults live for at least 20 years. Although caught commercially, this species is not a main target for fishing.

ORDER PERCIFORMES

Lumpsucker

Cyclopterus lumpus

LENGTH	Up to 24 in (60 cm)
WEIGHT	Up to 21 lb (9.5 kg)
DEPTH	7–1,300 ft (2–400 m)

DISTRIBUTION Temperate and cold waters of north Atlantic

The adult lumpsucker (or lumpfish) has a slightly grotesque appearance because the first dorsal fin becomes overgrown with thick, lumpy skin. Other bony lumps and bumps stick out in irregular rows along its large, rounded body. The pelvic fins form a strong sucker disk on its belly, which the lumpsucker uses to cling to wave-battered rocks near the shore where it spawns. The male guards the eggs from crabs and also fans them. Lumpsucker eggs are marketed as substitute caviar.

ORDER PERCIFORMES

Common Stargazer

Kathetostoma laeve

LENGTH	Up to 30 in (75 cm)
WEIGHT	Not recorded
DEPTH	200 ft (0–60 m), possibly 550 ft (150 m)

DISTRIBUTION Temperate waters of Indian Ocean around southern Australia

Looking like a cross between a bulldog and a seal, the common stargazer normally lies buried in shelly sand. It has its eyes set right on top of its large, square head and its mouth slants obliquely upward. This allows it to breathe and to see while remaining almost completely buried and is probably the reason behind its unusual name. Its large, white-edged pectoral fins help it lunge out of the sand and engulf passing fish and crustaceans. Common stargazers have also occasionally bitten divers who have inadvertently disturbed them while on night dives, when they are particularly difficult to spot. Anglers face a greater threat if they catch a common stargazer. Careless handling can result in a painful sting from a tough, venomous spine that lies behind each gill cover.

ORDER PERCIFORMES

Sergeant Major

Abudefduf saxatilis

LENGTH	Up to 9 in (23 cm)
WEIGHT	Up to 7 oz (200 g)
DEPTH	3–50 ft (1–15 m)

DISTRIBUTION Tropical and subtropical waters of Atlantic Ocean

This small fish is a familiar sight on most coral reefs in the Atlantic. It is one of the most common members of the damselfish family (*Pomacentridae*). It feeds on zooplankton in large groups, gathering above the reef to pick tiny animals and fish eggs from the water. In tourist areas, the fish are attracted to divers and boats and will eat almost anything that is offered. Male sergeant majors prepare a nesting area and guard the eggs laid by the females. A similar species, *Abudefduf vaigiensis*, is found on reefs in the Indo-Pacific region.

ORDER PERCIFORMES

Fairy Basslet

Pseudanthias squamipinnis

LENGTH	Up to 6 in (15 cm)
WEIGHT	Not recorded
DEPTH	0–180 ft (0–55 m)

DISTRIBUTION Red Sea and tropical waters of Indian Ocean and western Pacific

Fairy basslets live around coral outcrops and drop-offs. The larger, more colorful males have a long filament at the front of the dorsal fin, and they defend a harem of females. As they grow larger, the females change sex and turn into males.

ORDER PERCIFORMES

Potato Grouper

Epinephelus tukula

LENGTH	Up to 7 ft (2 m)
WEIGHT	Up to 240 lb (110 kg)
DISTRIBUTION	33–500 ft (10–150 m)

DISTRIBUTION Tropical waters of Red Sea, Gulf of Aden, and western Pacific

Groupers are large and important predators on coral reefs. They help maintain the health of a reef by picking off weak fish. Their large mouths and strong teeth also allow them to tackle crabs and lobsters. The potato grouper inhabits deep reef channels and seamounts. It has a large head and heavy body with a single long, spiny dorsal fin. Irregular dark blotches cover the body and dark streaks radiate from the eyes. These fish are territorial, and in some areas, individuals are hand-fed by divers. However this is not a practice that should be encouraged. Vulnerable to recreational fishing, this species is protected in some areas.

ORDER PERCIFORMES
False Clown Anemonefish
Amphiprion ocellaris

LENGTH	Up to 4 in (11 cm)
WEIGHT	Not recorded
DEPTH	3–50 ft (1–15 m)

DISTRIBUTION Tropical waters of eastern Indian Ocean and western Pacific

The most surprising thing about the false clown anemonefish is its home. It lives inside a giant stinging anemone. This small orange and white fish spends its whole life with its chosen anemone, which can be one of three species. At night, it sleeps among the bases of the tentacles on the anemone's disk. The fish is not stung and eaten because the anemone does not know it is there: a special slime covers the fish's body and prevents the anemone from recognizing it as food. Each anemone usually supports a large female, her smaller male partner, and several immature fish.

If the female dies, the male changes sex and becomes female and the largest immature fish takes on the male role. Both the false clown and the clown anemonefish are among the most popular aquarium fish, and numbers have been reduced in some areas by overcollecting.

ORDER PERCIFORMES
Red Bandfish
Cepola macrophthalma

LENGTH	Up to 30 in (80 cm)
WEIGHT	Not recorded
DEPTH	50–1,300 ft (15–400 m)

DISTRIBUTION Temperate and subtropical waters of northeastern Atlantic and Mediterranean

Very little was known about this strange fish until the 1970s, when divers discovered a population in shallow water around Lundy Island off the west coast of Britain. The red bandfish is shaped like an eel but flattened from side to side, with a long, golden-yellow fin running the length of the back and belly. In mature males, the fin has a bright blue edge. These fish live in deep mud burrows, emerging just far enough to feed on passing arrow worms and other plankton in the manner of tropical garden eels (see p.343). They also swim free of their burrows at times. In addition to single burrows, colonies of many thousands of individuals have been discovered. The burrows sometimes connect with those of burrowing crabs, and this may be a deliberate association.

ORDER PERCIFORMES
Ring-tailed Cardinalfish
Ostorhinchus aureus

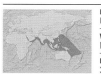

LENGTH	Up to 6 in (15 cm)
WEIGHT	Not recorded
DEPTH	3–130 ft (1–40 m)

DISTRIBTION Red Sea and tropical waters of Indian Ocean and western Pacific

Cardinalfish are small nocturnal reef fish. The ring-tailed cardinalfish hides under corals and in crevices during the day and emerges at night to feed on plankton. It has a distinctive black band around the tail base and two blue and white lines running from the snout through the eyes. Like all of the 350 or so species of cardinalfish, it has two separate dorsal fins. The male does not feed during the breeding season. Instead, after the female has laid her eggs, the male broods them in his mouth, protecting them until they hatch.

ORDER CARANGIFORMES
Bigeye Trevally
Caranx sexfasciatus

LENGTH	Up to 4 ft (1.2 m)
WEIGHT	Up to 40 lb (18 kg)
DEPTH	3–330 ft (1–100 m)

DISTRIBUTION Tropical waters of Indian Ocean and Pacific

During the day, shoals of bigeye trevally spiral lazily in coral reef channels and next to steep reef slopes, but at night, these fast-swimming predators split up and scour the reef for prey. Built for speed, these silvery fish have a narrow tail base, which is reinforced with bony plates called scutes, and a forked caudal fin. The first dorsal fin folds down into a groove to improve the streamlining of the fish and the pectoral fins are narrow and curved. There are many different species of trevally, which are difficult to tell apart. The bigeye trevally has a relatively large eye and the second dorsal fin usually has a white tip. These fish make good eating and are common in local markets in Southeast Asia. Juvenile bigeye trevally live close inshore and may enter estuaries and rivers.

ORDER PERCIFORMES
Harlequin Sweetlips
Plectorhinchus chaetodonoides

LENGTH	Up to 28 in (72 cm)
WEIGHT	Up to 15 lb (7 kg)
DEPTH	3–100 ft (1–30 m)

DISTRIBUTION Tropical waters of Indian Ocean and western Pacific

Small groups of harlequin sweetlips can often be seen gathered at dusk around large coral heads, waiting to be cleared of parasites by a cleaner wrasse (see p.360). These deep-bodied fish are patterned with small, brownish-black spots that break up their outline as they swim among the ever-changing shadows on the reef. Their name comes from their thickened lips, which they use to dig out invertebrates from sand.

JUVENILE COSTUME

Juvenile harlequin sweetlips have a different patterning than the adults. They have brown bodies and white spots edged in black. By swimming in a weaving, undulating fashion, the smallest juveniles mimic a toxic flatworm with a similar coloration and so escape predation. Their color may also warn that they themselves are unpalatable to predators.

ORDER PERCIFORMES

Cuckoo Wrasse

Labrus mixtus

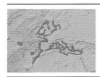

LENGTH	Up to 16 in (40 cm)
WEIGHT	Not recorded
DEPTH	7–650 ft (2–200 m)

DISTRIBUTION Temperate and subtropical waters of northeastern Atlantic and Mediterranean

The cuckoo wrasse is one of the most colorful fish in northern European waters. Large, mature males (shown here) are a beautiful blue and orange, while females are pink with alternate black and white patches along the back. When they are 7–13 years old, some females change color and sex

and become fully functional males. These males are known as secondary males and spawn in pairs with females. The male excavates a nest and attracts the female with an elaborate swimming display. To further complicate matters, it has been found that very few fish are born male but have the female coloring. These males are known as primary males and their role in reproduction is not yet fully understood.

ORDER PERCIFORMES

Cleaner Wrasse

Labroides dimidiatus

LENGTH	5½ in (14 cm)
WEIGHT	Not recorded
DEPTH	3–130 ft (1–40 m)

DISTRIBUTION Tropical reefs in Indian Ocean and southwestern Pacific

The cleaner wrasse spends its life grooming other fish, turtles, and occasionally even divers. This little fish is silvery blue with a black band running from snout to tail. The "client" recognizes it from its markings and does not try to eat it. Groups of cleaner wrasse usually consist of an adult male and a harem of females. If the male dies, the largest female changes sex and takes on the male role, becoming fully functional within a few days.

distinctive black band

small mouth with strong teeth

MUTUAL BENEFIT

Skin parasites are irritating and fish can be debilitated by a heavy infestation. On coral reefs, large fish line up at known "cleaning stations" such as a prominent coral head, spread their fins, and open their mouths. The resident cleaner wrasse picks off parasites and dead tissue and gets a good meal in return.

ORDER PERCIFORMES

Green Humphead Parrotfish

Bolbometopon muricatum

LENGTH	Up to 4¼ ft (1.3 m)
WEIGHT	Up to 100 lb (46 kg)
DEPTH	3–100 ft (1–30 m)

DISTRIBUTION Tropical reefs in Red Sea, Indian Ocean, and southwestern Pacific

Parrotfish are aptly named because not only are they brightly colored, but also their teeth are fused together to form a parrotlike beak. The green humphead parrotfish is much larger than most of its relatives. It has a huge crest-shaped hump on its head, a greenish body, large scales, and a single long dorsal fin. This destructive fish feeds by crunching up live coral, and it often breaks up the coral with its head. However, on the positive side, the coral sand it defecates after a meal helps consolidate the reef and build up patches of sand.

ORDER PERCIFORMES

Blackfin Icefish

Chaenocephalus aceratus

LENGTH	Up to 28 in (72 cm)
WEIGHT	Up to 7½ lb (3.5 kg)
DEPTH	16–2,500 ft (5–770 m)

DISTRIBUTION Polar waters of Southern Ocean around northern Antarctica

In the freezing waters around Antarctica, the temperature can fall to nearly 28°F (-2°C). This is below the temperature at which the blood of most fish would freeze. The blackfin icefish has a natural antifreeze in its blood that helps it survive in these conditions. It has no red blood cells and so appears a ghostly white. This makes its blood thinner so that it can flow freely in the cold temperatures. It is a sluggish hunter of small fish and krill and needs little oxygen.

ORDER PERCIFORMES

Wolf-fish

Anarhichas lupus

LENGTH	Up to 5 ft (1.5 m)
WEIGHT	Up to 53 lb (24 kg)
DEPTH	3–1,650 ft (1–500 m)

DISTRIBUTION North Atlantic and Arctic Ocean

This large and ferocious-looking fish is normally found on rocky reefs in deep water. However, north of the British Isles, divers regularly see them in shallow water. They are not aggressive to divers unless provoked.

The wolf-fish has a long body and a huge head with strong caninelike teeth at the front and molarlike teeth at the sides. These are used to break open hard-shelled invertebrates such as mussels, crabs, and sea urchins. Worn teeth are replaced each year. The skin is tough, leathery, and wrinkled and is usually grayish with darker vertical bands extending down the sides.

Spawning takes place during the winter. The female lays thousands of yellowish eggs in round clumps among rocks and seaweeds and the male guards them until they hatch. In spite of their unattractive appearance, wolf-fish are good to eat and are caught by anglers. They are also sometimes caught in trawl nets.

ORDER PERCIFORMES
Common Bluestripe Snapper
Lutjanus kasmira

LENGTH	Up to 16 in (40 cm)
WEIGHT	Not recorded
DEPTH	10–870 ft (3–265 m)

DISTRIBUTION Tropical reefs of Red Sea, Indian Ocean, and Pacific

Divers often see large shoals of common bluestripe snapper around coral and rock outcrops during the day. Their streamlined bodies mean that they can swim fast when they disperse at night to feed on smaller fish and bottom-dwelling crustaceans. They have a single long dorsal fin, which, like all their fins, is bright yellow. The common bluestripe snapper and many other similar species are important commercial fish. Their beautiful colors also make them popular specimens among aquarium-fish enthusiasts.

ORDER PERCIFORMES
Sand Eel
Ammodytes tobianus

LENGTH	Up to 8 in (20 cm)
WEIGHT	Not recorded
DEPTH	0–100 ft (0–30 m)

DISTRIBUTION Temperate waters of northeastern Atlantic and Baltic Sea

Shimmering shoals of sand eels are a familiar sight in many shallow sandy bays around northern Europe. These small, silvery fish have long, thin bodies with a pointed jaw and a single long dorsal fin. Large shoals of sand eels patrol the waters just above the seabed, feeding on planktonic crustaceans, tiny fish, and worms. If threatened, they dive into the sand. In winter, they spend most of the time buried. Various species of sand eel form an important part of the diet of larger fish, such as cod, herring, and mackerel, and of sea birds, especially Atlantic puffins. When the eels are scarce, local puffin colonies produce very few young. In some areas, overexploitation of sand eels for processing into fishmeal has been linked to sea-bird declines (see p.399).

ORDER PERCIFORMES
Atlantic Mackerel
Scomber scombrus

LENGTH	Up to 24 in (60 cm)
WEIGHT	Up to 7½ lb (3.5 kg)
DEPTH	0–650 ft (0–200 m)

DISTRIBUTION Temperate waters of north Atlantic, Mediterranean, and Black Sea

The Atlantic mackerel is designed for fast swimming. It has a torpedo-shaped, streamlined body; small dorsal fins; close-fitting gill covers; and small, smooth scales. In the summer, large shoals feed close inshore, voraciously preying on small fish and sieving plankton through their gills. From March to June, they lay their floating eggs in habitual spawning areas, the eggs hatching after a few days. In winter, the fish move into deeper water offshore and hardly feed. Several separate stocks exist within the north Atlantic, all of which are commercially exploited.

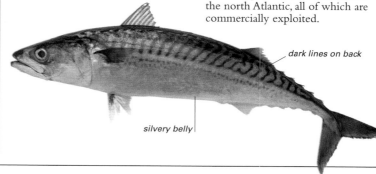

dark lines on back

silvery belly

ORDER PERCIFORMES
Greater Weever
Trachinus draco

LENGTH	Up to 20 in (50 cm)
WEIGHT	Up to 4½ lb (2 kg)
DEPTH	3–500 ft (1–150 m)

DISTRIBUTION Temperate waters of northeastern Atlantic and Mediterranean

Along with the lesser weever, this is one of very few venomous fish found in European waters. It has two dorsal fins, the first of which has venomous spines. During the day, the fish lies buried in the sand with just its eyes and fin-tip exposed. A painful wound can result from treading on the fish in shallow water.

ORDER PERCIFORMES
Northern Bluefin Tuna
Thunnus thynnus

LENGTH	Up to 15 ft (4.5 m)
WEIGHT	Up to 1,499 lb (680 kg)
DEPTH	0–9,900 ft (0–3,000 m)

DISTRIBUTION Northern and central Atlantic and Mediterranean

The northern bluefin tuna is one of the world's most valuable commercial fish and is heavily overexploited. Like mackerel, it is designed for high-speed swimming and is one of the fastest bony fish, attaining speeds of at least 43 mph (70 kph). The pectoral, pelvic, and first dorsal fins can be slotted into grooves to further streamline the torpedo-shaped body. To provide for long-distance, sustained swimming, the fish has large amounts of red muscle, which has a high fat content and can store oxygen. Other, similar species of bluefin tuna occur in the Pacific Ocean and southern parts of the Atlantic Ocean.

BUMPHEAD PARROTFISH
These bumphead parrotfish are patrolling a coral reef in the Info-Pacific in search of rich strands of coral. Schools of these huge fish crunch up live coral with their fused, beaklike teeth, often head-butting colonies to break them up. They void the indigestible skeleton as coral sand.

OCEAN LIFE

ORDER CARANGIFORMES

Pilotfish

Naucrates ductor

LENGTH	Up to 28 in (70 cm)
WEIGHT	Not recorded
DEPTH	0–100 ft (0–30 m)

DISTRIBUTION Tropical, subtropical, and temperate waters worldwide

Although a member of the predatory trevally family Carangidae, the pilotfish has taken up a scavenging, nomadic existence, traveling with large, ocean-dwelling bony fish, sharks, rays, and turtles. Its slim, silvery to pale bluish body is marked with six or seven bold black bands. These may help the host fish recognize it and so leave it alone. Darting in when its host has made a kill, the pilotfish eats any scraps it can find and also removes parasites. Young fish associate with jellyfish.

ORDER CARANGIFORMES

Sharksucker

Echeneis naucrates

LENGTH	Up to 3 ft (1 m)
WEIGHT	Up to 5 lb (2.3 kg)
DEPTH	65–165 ft (20–50 m)

DISTRIBUTION Tropical, subtropical, and temperate waters worldwide

The distinctive feature of the sharksucker is the powerful sucker disk on the top of its head, which enables it to attach itself securely to another fish. It feeds on its host's scraps and parasites, and also on small fish. Usually found attached to sharks or other large fish, cetaceans, and turtles, the sharksucker also swims freely over coral reefs. Its body is long and thin and ends in a fanlike tail.

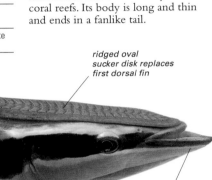

ridged oval sucker disk replaces first dorsal fin

lower jaw juts out beyond upper jaw

ORDER CARANGIFORMES

Dolphinfish

Coryphaena hippurus

LENGTH	Up to 7 ft (2.1 m)
WEIGHT	Up to 88 lb (40 kg)
DEPTH	0–280 ft (0–85 m)

DISTRIBUTION Tropical, subtropical, and temperate waters worldwide

With its shimmering colors, a dolphinfish leaping clear of the water is a spectacular sight. Metallic blues and greens cover its back and sides, grading into white and yellow on the underside. A fast ocean-dwelling fish, it is powered by a long, forked tail, with a single elongated dorsal fin providing stability. Also known as the dorado, it is a valuable market fish.

ORDER CARANGIFORMES

Atlantic Sailfish

Istiophorus albicans

LENGTH	Up to 10 ft (3.2 m)
WEIGHT	Up to 130 lb (60 kg)
DEPTH	0–650 ft (0–200 m)

DISTRIBUTION Temperate and tropical waters of Atlantic and Mediterranean

Like swordfish and marlin, the Atlantic sailfish has its upper jaw extended into a long spear. This is used to slash

through shoals of fish, stunning and maiming them. It has a huge sail-like dorsal fin, which is used for maneuvering and in displays but is folded away for fast swimming. A similar sailfish occurs in the Pacific and may be the same species.

ORDER CARANGIFORMES

Great Barracuda

Sphyraena barracuda

LENGTH	Up to 6½ ft (2 m)
WEIGHT	Up to 110 lb (50 kg)
DEPTH	0–330 ft (0–100 m)

DISTRIBUTION Tropical and subtropical waters worldwide

Barracuda are fast-moving predators with needle-sharp teeth and an undeserved reputation for ferocity. The great barracuda has a long, streamlined body with the second dorsal fin set far back near the tail. This fin arrangement, along with a large, powerful tail, allows it to stalk its prey and then accelerate at great speed. Large individuals in frequently dived sites will often allow divers to approach. Very occasionally, a lone fish may attack a diver if it mistakes a hand or shiny watch for a silvery fish. Eating even small amounts of barracuda can potentially result in ciguatera poisoning, caused by toxins accumulated from its food.

BARRACUDA SHOAL
While adults are normally solitary, juvenile great barracuda often swim together in large shoals in sheltered areas for protection.

GREAT BARRACUDA SKULL
Barracuda have flat-topped, elongated skulls with large, powerful jaws and knifelike teeth.

long front teeth

ORDER ACANTHURIFORMES

Bluecheek Butterflyfish

Chaetodon semilarvatus

LENGTH	Up to 9 in (23 cm)
WEIGHT	Not recorded
DEPTH	10–65 ft (3–20 m)

DISTRIBUTION Coral reefs in Red Sea and Gulf of Aden

Butterflyfish provide testimony to the health of a coral reef. A wide variety and plentiful numbers of these brightly colored, disk-shaped fish indicate that a reef is flourishing. Bluecheek butterflyfish are usually seen in pairs and often hide under table corals. The blue eye-patch hides the eye and confuses predators.

ORDER ACANTHURIFORMES

Bignose Unicornfish

Naso vlamingii

LENGTH	Up to 24 in (60 cm)
WEIGHT	Not recorded
DEPTH	3–165 ft (1–50 m)

DISTRIBUTION Tropical waters of Indian Ocean and southwestern Pacific

Unicornfish are so called because many have a hornlike projection on their forehead. However, the bignose unicornfish just has a rounded bulbous snout. At the base of the tail are two pairs of fixed, bony plates that stick out sideways like sharp knives, and the fish can inflict a serious wound on a potential predator. These blades are characteristic of surgeonfish (Acanthuridae), the family to which unicornfish belong.

Usually dark with blue streaks, the bignose unicornfish can pale instantly to a silvery gray. This often happens when the fish is being cleaned by a cleaner wrasse (see p.360). It favors steep reef slopes where it can feed on zooplankton in the open water.

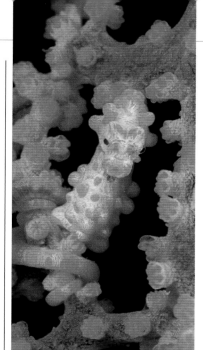

ORDER SYNGNATHIFORMES

Razorfish

Aeoliscus strigatus

LENGTH	6 in (15 cm)
WEIGHT	Not recorded
DEPTH	3–65 ft (1–20 m)

DISTRIBUTION Tropical reefs in Indian Ocean and western Pacific

While some reef fish habitually swim upside down, razorfish swim in synchronized groups in a vertical position, with their long, tubular snouts pointing down. These strange fish are encased in transparent bony plates that meet in a sharp ridge along the belly, like the edge of a razor, and also form a sharp point at the tail. A dark stripe along the body provides camouflage for the razorfish when hiding among sea urchins and branched corals.

ORDER ACANTHURIFORMES

Queen Angelfish

Holacanthus ciliaris

LENGTH	Up to 18 in (45 cm)
WEIGHT	Up to 3¼ lb (1.5 kg)
DEPTH	3–230 ft (1–70 m)

DISTRIBUTION Gulf of Mexico, Caribbean Sea, and subtropical waters of western Atlantic

ORDER SYNGNATHIFORMES

Pygmy Seahorse

Hippocampus bargibanti

LENGTH	1 in (2.5 cm)
WEIGHT	Not recorded
DEPTH	50–165 ft (15–50 m)

DISTRIBUTION Tropical waters of southwestern Pacific

This miniature seahorse lives on *Muricella* sea fans and was originally discovered when a sea fan was collected for an aquarium. It is very difficult to spot, as its body is covered in tubercles that exactly match the polyps of its host. Clinging on tightly with its prehensile tail, it reaches out into the water to suck in planktonic animals. Like other seahorses, it has a rigid body made up of bony plates and a head that is tucked in like a tightly reined carriage horse.

One of the most colorful Caribbean reef fish, the blue and yellow queen angelfish slips its slim body effortlessly between corals and sea fans. It uses its small mouth and brushlike teeth to nibble sponges, which are its main food. Like all angelfish, it has a sharp spine at the corner of the gill cover. Juveniles are brown and yellow with curved blue bars and feed on parasites that they pick from other fish.

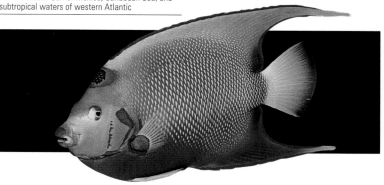

ORDER ACROPOMATIFORMES

Wreckfish

Polyprion americanus

LENGTH	Up to 7 ft (2 m)
WEIGHT	Up to 220 lb (100 kg)
DEPTH	130–2,000 ft (40–600 m)

DISTRIBUTION Atlantic, Mediterranean, Indian Ocean, and Pacific

The name of this fish comes from the juveniles' habit of accompanying drifting wreckage. This is a large, solid fish with a pointed head and a protruding lower jaw. It has a spiny dorsal fin and a bony ridge running across the gill cover. Adult wreckfish live close to the bottom of the sea floor and often lurk inside shipwrecks and caves. Juvenile wreckfish prefer surface waters and can often be approached by snorkelers, presumably because the fish consider the swimmers to be floating wreckage. As the young fish grow, they give up their nomadic life to live near the seabed. Adults are fished using lines and make good eating.

ORDER BLENNIIFORMES

Tompot Blenny

Parablennius gattorugine

LENGTH	Up to 12 in (30 cm)
WEIGHT	Not recorded
DEPTH	3–100 ft (1–30 m)

DISTRIBUTION Temperate and subtropical waters of northeastern Atlantic and Mediterranean

With its thick lips, bulging eyes, and a pair of tufted head tentacles, the tompot blenny is a comical-looking fish. Like all blennies, it has a long body, a single long dorsal fin, and peglike pelvic fins that it uses to prop itself up. Inquisitive by nature, the tompot blenny will peer out at approaching divers from the safety of a rock crevice.

ORDER PLEURONECTIFORMES

European Plaice

Pleuronectes platessa

LENGTH	Up to 3 ft (1 m)
WEIGHT	Up to 15 lb (7 kg)
DEPTH	0–655 ft (0–200 m)

DISTRIBUTION Arctic Ocean, northeastern Atlantic, Mediterranean, and Black Sea

This species is the most important commercial flatfish for European fisheries. Heavy fishing, however, has resulted in a progressive reduction in

the size and age of fish landed. It is a typical oval-shaped flatfish with long fins extending along both edges of its thin body. Flatfish have both eyes on one side of their body and are either "right-eyed" or "left-eyed." Plaice are right-eyed: they lie on the seabed with their left side down. Their upward-facing right side is brown with orange or red spots.

Plaice spend the day buried in the sand, emerging at night to feed on shellfish and crustaceans, which they crush using special teeth in the throat (pharyngeal teeth). Young plaice are also expert at nipping off the breathing siphons of shellfish that they spot sticking up out of the sand.

ORDER TETRADONTIFORMES

Scrawled Filefish

Aluterus scriptus

LENGTH	Up to 3½ ft (1.1 m)
WEIGHT	Up to 5½ lb (2.5 kg)
DEPTH	6–400 ft (2–120 m)

DISTRIBUTION Tropical and subtropical waters of Atlantic, Pacific, and Indian Oceans

Beautiful blue, irregular markings like a child's scribbles give this reef fish its name. Filefish are closely related to triggerfish (see below, left), but are thinner and, except for the scrawled filefish, usually smaller. This large species has one large and one tiny spine on its back over the eyes. The fish uses these spines to help wedge itself into crevices for safety.

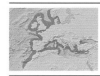

ORDER PLEURONECTIFORMES

Common Sole

Solea solea

LENGTH	Up to 28 in (70 cm)
WEIGHT	Up to 6½ lb (3 kg)
DEPTH	0–500 ft (0–150 m)

DISTRIBUTION Temperate waters of northeastern Atlantic, Baltic, Mediterranean, and Black seas

Common sole are masters of camouflage and can subtly alter their color to match the seabed on which they lie. The basic grayish brown color can be lightened or darkened and the pattern of darker splotches changed. Sole have a rounded snout and a semicircular mouth and their head is fringed with short filaments, giving them an unshaven appearance.

Like all flatfish, the common sole starts life as a tiny larval fish floating near the surface. As it grows, it gradually undergoes a radical metamorphosis. The eye on the left side moves around the head to join the eye on the right and the body starts to flatten. When it is about a

month old, it settles on the sea floor with its eyeless side facing down. The skin on the underside stays white but the upper side develops pigment. Although common sole can live to be nearly 30 years old, they are a valuable food fish and most are caught when only a few years old.

ORDER TETRADONTIFORMES

Spotted Boxfish

Ostracion meleagris

LENGTH	Up to 10 in (25 cm)
WEIGHT	Not recorded
DEPTH	3–100 ft (1–30 m)

DISTRIBUTION Tropical reefs in Indian Ocean and south Pacific, possibly extending to Mexico

ORDER TETRADONTIFORMES

Titan Triggerfish

Balistoides viridescens

LENGTH	Up to 30 in (75 cm)
WEIGHT	Not recorded
DEPTH	3–160 ft (1–50 m)

DISTRIBUTION Tropical reefs of Red Sea, Indian Ocean, and southwestern Pacific

The titan triggerfish is also known as the mustache triggerfish due to a dark line above its lips. It has large, strong front teeth and strong spines in its first dorsal fin. The first and longer spine can be locked in an upright position and released by depressing the second smaller "trigger" spine. This allows the fish to jam itself into a reef crevice, where it can rest safely, away from potential predators. The titan triggerfish preys on shellfish and crustaceans, which it crunches up

using its tough mouth and teeth. It can even make a meal of sea urchins by flipping them over and biting them on their vulnerable underside, where the spines are shorter.

PROTECTIVE PARENT

In the breeding season, titan triggerfish dig a nest in a sandy patch of coral rubble using their mouth as a water jet. The female lays her eggs in the nest and one or both of the parents remains nearby to guard it. Normally a wary fish, parent titan triggerfish will attack divers that come too close to the nest and can inflict severe bites that need medical attention.

Instead of a covering of scales, all boxfish are protected by a rigid box of fused bony plates under the skin. This means they cannot bend their body and must swim by beating their pectoral fins. A large tail gives some propulsion and is also used to help steer them like a rudder. Male spotted boxfish are more colorful than the females, which are brown with light spots. These fish secrete a poisonous slime from their skin that protects them from predators.

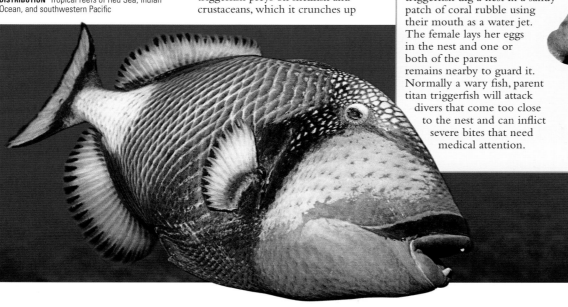

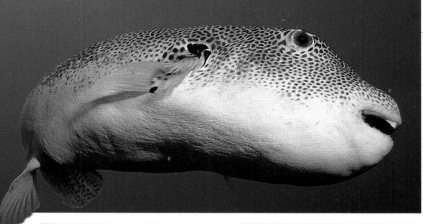

Ocean Sunfish
Mola mola

LENGTH	Up to 13 ft (4 m)
WEIGHT	Up to 5,000 lb (2,300 kg)
DEPTH	0–1,600 ft (0–480 m)

DISTRIBUTION Tropical, subtropical, and temperate waters worldwide

The ocean sunfish is the world's heaviest bony fish and has a distinctive disklike shape. Instead of a caudal fin, it has a rudderlike structure (clavus) formed by extensions of the dorsal and anal fin rays, and it swims by flapping its tall dorsal and anal fins from side to side. Its common name comes from the fish's habit of drifting in surface currents while lying on its side. It also swims upright with its dorsal fin sticking above the surface. The ocean sunfish has no scales, but its skin is very thick and stretchy. Like the porcupinefish (see below, left), to which it is related, the ocean sunfish has a single fused tooth-plate in each jaw, but it feeds mainly on soft-bodied jellyfish and other slow-moving invertebrates and fish. Females produce the most eggs of any bony fish, laying up to 100 million in the open ocean. Lone fish make grating noises with pharyngeal (throat) teeth, and this may help them to make contact with potential mates.

ORDER TETRADONTIFORMES

Star Pufferfish
Arothron stellatus

LENGTH	Up to 4 ft (1.2 m)
WEIGHT	Not recorded
DEPTH	10–200 ft (3–60 m)

DISTRIBUTION Tropical reefs in Indian Ocean and south Pacific

Compared with most other pufferfish, the star pufferfish is a relative giant. Its black-spotted skin is covered in small prickles and, if threatened, it will swallow water and swell up to an even larger size. At night, it searches out hard-shelled reef invertebrates and crushes them with powerful jaws that have fused, beaklike teeth.

HUMAN IMPACT

FUGU FISH

Pufferfish produce tetrodotoxin, a lethal poison that is stronger than cyanide and for which there is currently no antidote. In spite of this, these fish are eaten in Japan as a delicacy called "fugu." The poison is in the skin and some of the internal organs, and only licensed chefs, who have been specially trained, are permitted to prepare this dish. Pufferfish from the genus *Takifugu* are considered to be the best eating. A few people die every year from eating fugu, and the Emperor of Japan was historically banned from eating it for his own protection.

ORDER TETRADONTIFORMES

Porcupinefish
Diodon hystrix

LENGTH	Up to 35 in (90 cm)
WEIGHT	Up to 6½ lb (3 kg)
DEPTH	6–160 ft (2–50 m)

DISTRIBUTION Tropical and subtropical waters of Atlantic, Pacific, and Indian Oceans

When a porcupinefish is frightened, it pumps water into its body until it looks like a prickly soccer ball. Few predators are large enough or brave enough to swallow a fish in this state. Left to itself, the porcupinefish deflates and its long spines lie flat against its body. During the day, it hides in caves and reef crevices, emerging at night to feed on hard-shelled invertebrates such as gastropod mollusks.

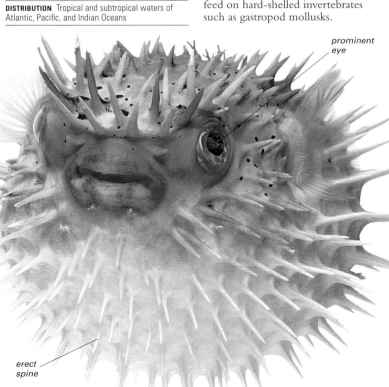

prominent eye

erect spine

Reptiles

DOMAIN	Eucarya
KINGDOM	Animalia
SUPERCLASS	Chordata
CLASS	Reptilia
ORDERS	4
SPECIES	11,570

DURING THE JURASSIC PERIOD, over 140 million years ago, reptiles were the largest animals in the oceans. Their place has since been taken by mammals, leaving few reptiles that are wholly marine. Of these, turtles are the most widespread, and sea snakes are the most diverse. Apart from the leatherback turtle, almost all are confined to warm-water regions, with the largest numbers around coasts and on coral reefs.

Anatomy

Marine reptiles have several adaptations for life in the sea. Turtles have a low, streamlined shell, or carapace, and broad, flattened forelimbs that beat up and down like wings. Marine lizards and crocodiles use their tails to provide most of the power when swimming, while most sea snakes have flattened tails that work like oars. Unlike land snakes, true sea snakes do not have enlarged belly scales, since they do not need good traction for crawling on land. All reptiles breathe air, and marine species have valves or flaps that prevent water entering their nostrils when they dive. Crocodiles also have a valve at the top of the throat, which enables them to open their mouths beneath the surface without flooding their lungs with water. Marine reptiles all need to expel excess salt. Sea snakes and crocodiles do this through salt glands in their mouths, while marine turtles lose salt in their tears. The marine iguana has salt glands located on its nose.

pointed scales (scutes)

streamlined shell (carapace)

head

short rear flippers

long front flippers

STREAMLINED SHELL
The hawksbill turtle has a tapering carapace with conspicuous scales, or scutes. Unlike most terrestrial tortoises, it cannot retract its head or legs inside its shell.

REPLACEMENT TEETH
A saltwater crocodile's teeth are constantly shed and replaced. During its lifetime, it may use over 40 sets.

Habitat

Most marine reptiles live close to the shore, or return to it to breed. The only fully pelagic species are true sea snakes—those in the subfamily Hydrophiinae. They remain in the open ocean for their entire lives. Sea snakes are also the deepest divers, feeding up to 330 ft (100 m) below the surface. Apart from the leatherback turtle, most marine reptiles depend on external warmth to remain active, which restricts them to tropical and subtropical waters. They also show striking variations in regional spread. This is particularly true of sea snakes: up to 25 species are found in some parts of the Indo-Pacific, but the Atlantic Ocean has none.

KEY
Number of sea snake species

- 12–25 species
- 2–12 species
- 1 species

PACIFIC OCEAN

ATLANTIC OCEAN

INDIAN OCEAN

SOUTHERN OCEAN

SEA SNAKES WORLDWIDE
Although diverse in the Indo-Pacific, sea snakes are absent from the Atlantic. Cold waters off southern Africa prevent them from spreading west.

Food and Feeding

Most marine reptiles are carnivorous. Sea snakes typically feed on fish, although a few are specialized predators of fish eggs. They use their venom mainly in feeding, rather than for defense, killing their prey by biting it, and then swallowing it whole. Green turtles feed on seagrass when they become adult, while other marine turtles are carnivorous throughout their lives. The marine iguana is the only marine reptile that is a fully herbivorous. When young, it feeds on algae close to the waterline, but as an adult, it grazes seaweed growing on submerged rocks. Reptiles are cold-blooded (ectothermic), so they use less energy than mammals or birds. This means that they need less food, and can go for long periods between meals. Sea snakes, for example, can survive on just one or two meals a month.

GRAZING ON ALGAE
Marine iguanas have blunter heads than most lizards, enabling them to tear seaweed from rocks. Sharp claws act as anchors.

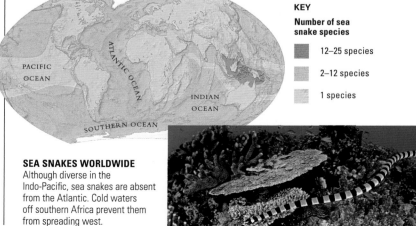

REEF SNAKE
A yellow-lipped sea krait searches for prey in a coral reef. Reefs are prime habitats for sea kraits, which generally live in shallow water.

Reproduction

True sea snakes are the only reptiles that reproduce at sea. They give birth to live young (they are viviparous) after a gestation period of up to 11 months—much longer than most terrestrial species. All other marine reptiles, including sea kraits and marine turtles, lay their eggs on land. Many of these animals breed on remote beaches and islands, and the adults sometimes arrive simultaneously and in large numbers. The eggs are incubated by ambient warmth, and in crocodiles and turtles, the nest temperature determines the sex ratio of the hatchlings. Once the eggs have hatched, growth is fast, but mortality can be high. Parental care is rare in marine reptiles; female crocodiles are an exception, guarding their nests and carrying their young to water after they have hatched.

NEST IN THE SAND
After excavating a nest, a female leatherback lays her eggs. Turtle eggs are almost spherical, and have soft, leathery shells, which tear open when they hatch.

REPTILE CLASSIFICATION

Three orders of living reptiles contain marine species. The fourth order includes only the tuataras, which are terrestrial. Snakes make up the vast majority of marine reptiles. Others, such as wart snakes and terrapins, live in fresh water, occasionally entering the sea.

TURTLES AND TORTOISES
Order Testudines

About 360 species
Seven turtle species are exclusively marine. Typical marine turtles (six species) have a hard carapace. The separately classified leatherback turtle has a rubbery carapace.

SNAKES AND LIZARDS
Order Squamata

10,980 species
About 70 species of snakes live in salt water. True sea snakes, belonging to the subfamily Hydrophiinae, spend their lives at sea, while sea kraits (members of the family Elapidae) breed on land. Seagoing lizards are all semiterrestrial; only one species, the marine iguana, gets all its food offshore.

CROCODILES AND ALLIGATORS
Order Crocodylia

26 species
Only the American crocodile and the saltwater crocodile live in both fresh water and the sea. Crocodilians usually feed at the surface, rarely diving more than a few yards when at sea.

MARINE ADAPTATIONS
Thanks to their low metabolic rate, marine reptiles can remain underwater for long periods. This young saltwater crocodile is lurking on the seabed off New Guinea.

ORDER CHELONIA

Green Turtle

Chelonia mydas

LENGTH	2½–3¼ ft (0.8–1 m)
WEIGHT	140–290 lb (65–130 kg)
HABITAT	Open sea, coral reefs, coasts

DISTRIBUTION Tropical and temperate waters worldwide

Elegantly marked and very effectively streamlined, this species is the most common turtle in subtropical and tropical waters, where it is often seen in eelgrass (seagrass) beds and on coral reefs. Its color varies from green to dark brown, but its scales and shell plates (scutes) are lighter where they meet, giving it a distinctive pattern. Like all marine turtles, it has front flippers that are long and broad and beat up and down like wings. They provide the power for swimming, while the much shorter rear flippers act as stabilizers. Young green turtles are carnivorous, eating mollusks and other small animals, but the adults feed mainly on eelgrass and algae—a diet that keeps them close to the coast.

Green turtles breed on isolated beaches, and they are remarkably faithful to their nesting sites. To reach them, some make journeys of more than 600 miles (1,000 km), navigating their way to remote islands that may be just a few miles across. They mate in the shallows, and the females then crawl ashore after dark to dig their nests and lay eggs. Green turtles lay up to 200 eggs, burying them about 30 in (75 cm) beneath the sand. The eggs take about 6–8 weeks to hatch. All the young emerge simultaneously and scuttle for the safety of the sea.

The green turtle has been hunted for centuries, mainly for food, and its numbers have declined significantly. Conservation measures include protection of the turtles' nest sites so that predators, including humans, do not dig up the nest and take the eggs.

EARLY LIFE

After hatching while buried in the sand, the young turtles use their front flippers to dig toward the surface. They then make a dash for the sea, trying to avoid becoming a meal for waiting predators, including birds, crabs, snakes, and ants. Very little is known about their early life, as young green turtles are rarely observed in the wild, but it is certain that they face many predators in the sea. Their growth rate is known to average more than 11 lb (5 kg) per year.

ORDER TESTUDINES

Hawksbill Turtle

Eretmochelys imbricata

LENGTH	2½–3¼ ft (0.8–1 m)
WEIGHT	100–165 lb (45–75 kg)
HABITAT	Coral reefs and coastal shallows

DISTRIBUTION Tropical and warm-temperate waters worldwide

Named after its conspicuous beaked snout, the hawksbill has a carapace with a raised, central keel and pointed shell plates (scutes) around its rear margin. It lives mainly in warm-water regions, rarely straying from shallows and coral reefs, where it feeds on sponges, mollusks, and other sedentary animals. It is less migratory than other marine turtles, breeding at low densities all over the tropics instead of gathering at certain beaches. On land, it has a distinctive gait, moving its flippers in diagonally opposite pairs—other marine turtles move their front flippers together—the same action they use when swimming.

The hawksbill is the chief source of tortoiseshell—detached, polished scutes. Despite being classified as Critically Endangered by the IUCN, hawksbills are often killed and stuffed when young to be sold as curios, particularly in Southeast Asia. Attempts at farming these turtles have not been successful.

ORDER TESTUDINES

Loggerhead Turtle

Caretta caretta

LENGTH	2¼–3¼ ft (0.7–1 m)
WEIGHT	165–350 lb (75–160 kg)
HABITAT	Open sea, coral reefs, coasts

DISTRIBUTION Tropical and warm temperate waters worldwide

After the leatherback (opposite), the loggerhead is the second-largest marine turtle. It has a blunt head, powerful jaws, and a steeply domed carapace. It hunts and eats hard-bodied animals, such as crabs, lobsters, and clams. This species takes about 30 years to mature and breeds every other year.

ORDER TESTUDINES

Leatherback Turtle

Dermochelys coriacea

LENGTH	4¼–6 ft (1.3–1.8 m)
WEIGHT	Up to 2,000 lb (900 kg)
HABITAT	Open sea

DISTRIBUTION Tropical, subtropical, and temperate waters worldwide

The leatherback is the world's largest marine turtle. Its carapace has a rubbery texture, having no hard plates, and has a tapering, pearlike shape. Its head is not retractable, and the leatherback is unique among turtles in having flippers without claws. It spends most of its life in the open sea, returning to the coast only when it breeds. It feeds on jellyfish and other planktonic animals, and while it gets most of its food near the surface, it can dive to depths of 3,300 ft (1,000 m).

Leatherbacks breed mainly in the tropics, on steeply sloping sandy beaches, laying up to nine clutches of eggs in each breeding season.

Unusually for a reptile, the leatherback turtle can keep its body warmer than its surroundings, thanks partly to the thick layer of insulating fat beneath its skin. This allows it to wander much more widely than other turtles, reaching as far north as Iceland and almost as far south as Cape Horn. Individuals may roam huge distances—one leatherback tagged off the coast of South America was later found on the other side of the Atlantic, 4,200 miles (6,800 km) away.

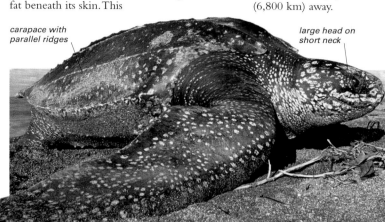

carapace with parallel ridges

large head on short neck

THROAT SPINES

The leatherback's throat contains dozens of backward-pointing spines that prevent jellyfish from escaping before they are completely swallowed. These endangered turtles often die after eating discarded plastic bags, which they mistake for jellyfish.

JELLYFISH TRAP
The leatherback's throat spines can be over ½ in (1 cm) long. They are regularly replaced during the animal's life.

ORDER TESTUDINES

Kemp's Ridley Turtle

Lepidochelys kempi

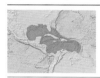

LENGTH	20–35 in (50–90 cm)
WEIGHT	55–90 lb (25–40 kg)
HABITAT	Coral reefs, coasts

DISTRIBUTION Caribbean, Gulf of Mexico, occasionally as far north as New England

Also known as the Atlantic Ridley turtle, this is the smallest marine turtle, and also the most threatened, largely as a result of its unusual breeding behavior. Unlike most marine turtles, Kemp's Ridleys lay their eggs by day, and the females crawl out of the sea simultaneously, during mass nestings called *arribadas* (Spanish for "arrivals"). At one time, these nestings took place throughout the turtle's range, but because the eggs were laid in such large concentrations in daylight, they were easy prey for human egg-harvesters and natural predators. Today, the vast majority of Kemp's Ridleys breed on a single beach in Mexico, where their nests are protected. These turtles were also often caught as bycatch in shrimp nets, but turtle excluding devices (TEDs) fitted to nets have helped to reduce this threat. Several weeks after an *arribada*, young Kemp's Ridleys emerge from their eggs in the thousands to make the dangerous journey down the beach and into the relative safety of the sea.

The adults are carnivorous bottom-feeders that mainly hunt crabs. They have an unusually broad carapace, and their small size makes them agile swimmers. The carapace changes color with age: yearlings are often almost black, while adults are light olive-gray. A closely related species, the olive Ridley turtle (*L. olivacea*), lives throughout the tropics. It is much less endangered than the Kemp's Ridley, thanks to its wider distribution.

ORDER TESTUDINES

Flatback Turtle

Natator depressa

LENGTH	3¼–4 ft (1–1.2 m)
WEIGHT	Up to 190 lb (85 kg)
HABITAT	Coasts, shallows

DISTRIBUTION North and northeastern Australia, New Guinea, Arafura Sea

Named after its carapace, which is only slightly domed, the flatback has the most restricted distribution of any marine turtle. It lives in shallow waters between northern Australia and New Guinea, reaching south along the Great Barrier Reef. When adult, it is largely carnivorous, feeding on fish and bottom-dwelling animals such as mollusks and sea squirts.

Despite their restricted range, adult flatbacks may swim over 600 miles (1,000 km) to reach nesting beaches. Females dig an average of three nests each time they breed and lay a total of about 150 eggs. The young feed at the surface on planktonic animals. Instead of dispersing into deep oceanic water, like the young of other turtle species, they remain in the shallows over the continental shelf.

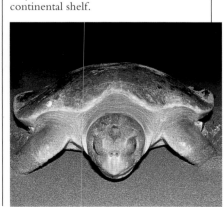

OCEAN LIFE

HAWKSBILL TURTLE
This turtle owes its common name to its sharp, powerful beak, shaped like that of a bird of prey. The specimen photographed here, on a reef in the southern Red Sea, is holding a piece of soft coral, but its jaws are strong enough to detach even hard corals. It has two claws on each flipper.

ORDER SQUAMATA

Yellow-lipped Sea Krait

Laticauda colubrina

LENGTH	3¼–10 ft (1–3 m)
WEIGHT	Up to 11 lb (5 kg)
HABITAT	Coral reefs, mangrove swamps, estuaries

DISTRIBUTION Eastern Indian Ocean and southwestern Pacific

This species is the most widespread of the sea kraits—a group of four closely related species that lay eggs on land, instead of giving birth at sea.

It has a pale blue body, marked with eye-catching dark blue rings, and distinctive yellow lips, which give it its common name. The yellow-lipped sea krait feeds on fish in shallow water, and although it has highly potent venom, it presents very little danger to humans because it is not aggressive and even when handled it rarely bites.

Unlike many other marine snakes, sea kraits have large ventral scales that give them good traction when they crawl, allowing them to move around comfortably on land. During the breeding season, they come ashore in large numbers to mate and lay clutches of up to 20 eggs. Once they have hatched, the young make their way to the shallows, before dispersing along coasts and out to sea.

ORDER SQUAMATA

Yellow-bellied Sea Snake

Hydrophis platurus

LENGTH	3¼–5 ft (1–1.5 m)
WEIGHT	Up to 3 lb (1.5 kg)
HABITAT	Open water

DISTRIBUTION Tropical and subtropical waters in Indian Ocean and Pacific

This boldly striped yellow-and-black snake has venom that is more toxic than that of a cobra. It is also the world's most wide-ranging snake and one of the very few that lives in the surface waters of the open ocean. Its distinctive colors warn that it is poisonous, protecting it from many predators. It feeds on small fish trying to shelter in its shade, swimming forward or backward with equal ease to grab them with its jaws. Although its fangs are tiny, its potent venom occasionally causes human fatalities.

At sea, these snakes may form vast flotillas hundreds of thousands strong, and after storms, they may be washed up on beaches that lie far outside their normal range. However, the species has never managed to colonize the Atlantic Ocean, because cold currents stand in its way. Yellow-bellied sea snakes give birth to up to six young each time they breed.

ORDER SQUAMATA

Beaked Sea Snake

Hydrophis schistosa

LENGTH	3¼–5 ft (1–1.5 m)
WEIGHT	Up to 4½ lb (2 kg)
HABITAT	Shallow inshore waters

DISTRIBUTION Indian Ocean and western Pacific, from Persian Gulf to northern Australia

Notoriously aggressive and readily provoked, this widespread species is responsible for nine out of every ten deaths from sea-snake bites. Light gray with indistinct blue-gray bands, it has a sharply pointed head, slender body, and paddlelike tail. Its fangs are less than ⅓ in (4 mm) long, but its jaws can gape widely to accommodate large prey. It feeds mainly on catfish and shrimp. swimming near the bottom in shallow, murky water, in coastal waters, mangrove swamps, estuaries, and rivers, locating its victims by smell and touch. Like all fish-eating snakes, it waits until its prey has stopped struggling, before turning it so that it can be consumed head-first.

Beaked sea snakes give birth to up to 30 young each time they breed, but their mortality is high, and only a small proportion of the young survive to become parents themselves. Despite their venom, these snakes are eaten by inshore predators, such as fish and estuarine crocodiles.

HUMAN IMPACT

DEADLY VENOM

The beaked sea snake's bite contains enough venom to kill 50 people—about twice as many as the most venomous terrestrial snakes, such as the king cobra or death adder. Most of the snake's human victims are bitten when wading or fishing in muddy water, although no reliable records exist of the numbers killed every year. Its deadly venom does not protect this snake from being caught in shrimp-trawling nets. This hazard affects many sea snakes, but the beaked sea snake is particularly susceptible because it lives in shallow water and eats shrimp that are targeted by trawlers.

ORDER SQUAMATA

Turtle-headed Sea Snake

Emydocephalus annulatus

LENGTH	2–4 ft (60–120 cm)
WEIGHT	Up to 3 lb (1.5 kg)
HABITAT	Coral reefs and coral sand banks

DISTRIBUTION Indian Ocean and Pacific, from northern Australia to Fiji

This Australasian sea snake is highly notable for its color variation, and also for its highly specialized lifestyle as a predator of fish eggs. The color it most commonly takes is a plain blue-gray, which is found throughout its range. A striking ringed form lives in some parts of the Great Barrier Reef, while a rarer, dark or melanistic form is found on isolated reefs farther east in the Coral Sea. The turtle-headed sea snake moves slowly among living corals, methodically searching for fish egg masses either glued to the coral's branches or laid directly on the coral sand. When it finds an egg mass, it scrapes the eggs off with an enlarged scale on its upper jaw, which works like a blade. In most cases, parent fish leave the eggs unguarded, so the snakes can feed unhindered, but some species—such as damselfish—guard their eggs aggressively and try to keep the snakes away.

Little is known about this snake's reproductive habits, apart from the fact that the females give birth to live young. In keeping with their lifestyle, turtle-headed sea snakes have tiny fangs (less than 1/32 in [1 mm] long) and they rarely try to bite. Their venom is one of the weakest of any sea snake, and instead of striking back at predators, they react to danger by disappearing into crevices in the reef.

ORDER SQUAMATA

Leaf-scaled Sea Snake

Aipysurus foliosquama

LENGTH	Up to 2 ft (60 cm)
WEIGHT	Up to 1 lb (0.5 kg)
HABITAT	Coral reefs and coral sand banks

DISTRIBUTION Timor Sea (Ashmore and Hibernia reefs)

This fish-eating snake has one of the most restricted ranges of any sea snake, being confined to a group of remote coral reefs about 185 miles (300 km) off the northwest coast of Australia. It is marked with contrasting bands or rings and gets its name from the characteristic shape of its dorsal scales. It lives in shallow water and rarely dives deeper than about 33 ft (10 m). Although venomous, it is rarely aggressive. Female leaf-scaled sea snakes are larger than the males and give birth to live young.

ORDER SQUAMATA

Olive Sea Snake

Aipysurus laevis

LENGTH	3–7 ft (1–2.2 m)
WEIGHT	Up to 6½ lb (3 kg)
HABITAT	Coral reefs, coastal shallows, estuaries

DISTRIBUTION Eastern Indian Ocean and western Pacific, from western Australia to New Caledonia

Plain brown or olive-brown above, with a paler underside, this common sea snake is one of six closely related species found in the reefs and shallow coastal waters of northern Australasia. Like its relatives, it has a cylindrical body, a flattened tail, and enlarged ventral scales—a feature normally found in snakes that spend some or all of their life on land. However, it is fully aquatic, hunting fish among the crevices and recesses of large corals. Instead of roaming throughout a reef, it often stays in the same small area of coral, rarely venturing into open water except after dark.

Olive sea snakes give birth to live young, producing up to five finger-sized offspring after a gestation period of nine months. Unlike the adults, the young are dark in color, with a boldly contrasting pattern of lighter bands. This is gradually lost as they become mature. Olive sea snakes are naturally inquisitive and often approach divers. They have short fangs and bite readily if provoked. Their venom is toxic and has been known to be fatal.

EQUIPPED TO GRIP
An adult marine iguana sprawls on the sand, displaying the broad feet and long claws it uses to grip submerged rocks while it tears off mouthfuls of food.

ORDER SQUAMATA

Marine Iguana

Amblyrhynchus cristatus

LENGTH	Up to 5 ft (1.5m), but often smaller
WEIGHT	Females 1⅛ lb (500 g); males 3⅓ lb (1.5 kg), sometimes larger
HABITAT	Rocky coasts

DISTRIBUTION Galápagos Islands

Restricted to the Galápagos Islands, this primeval-looking reptile is the only lizard that feeds exclusively at sea once it is an adult. The size and weight of this species varies between islands. It has a blunt head with powerful jaws and a distinctive spiky crest that runs down its head, neck, and back. This lizard's powerful claws help it clamber over rocks, while its tail propels its through water. It feeds on seaweeds and other algae. The young feed above the water, but adults dive up to 33 ft (10 m), and can hold their breath for over an hour. During the day, they spend their time feeding and sunbathing to raise their body temperature.

During the breeding season, male marine iguanas engage in lengthy headbutting contests as they compete for mates. Females lay up to six eggs in the sand, and the young emerge after an incubation period of up to three months. Marine iguanas have many natural predators, including sharks and birds of prey. Populations have also been severely affected by introduced animals, such as rats and dogs.

spiky crest

blunt snout

PROFILE OF A GRAZER
Unlike predatory lizards, the marine iguana has blunt but powerful jaws. It secretes the surplus salt derived from its diet from glands near its nose.

SURVIVING THE COLD

Although the Galápagos Islands are on the equator, they are bathed by the chilly Humboldt Current, which flows northward along the west coast of South America. Being a reptile, the marine iguana cannot generate its own body heat and needs special adaptations for feeding in these conditions. When it dives, its heart rate drops by about half, helping it conserve energy to keep its core temperature higher than the water around it. At night, the iguanas often huddle together to keep themselves warm.

BASKING IN THE SUN
When it returns to land, the marine iguana sprawls over sand or rocks to soak up warmth from the sun through its skin.

ORDER SQUAMATA

Water Monitor

Varanus salvator

LENGTH	Up to 9 ft (2.7 m)
WEIGHT	35–75 lb (15–35 kg)
HABITAT	Low-lying coasts, estuaries, rivers

DISTRIBUTION Indian Ocean, western Pacific, from Sri Lanka to the Philippines and Indonesia

An opportunistic predator with a wide-ranging diet, this is one of the largest lizards that regularly ventures into salt water. It has a long neck, strong legs, and a flattened tail, which lashes from side to side when it swims. It feeds on anything it can overpower, diving to catch prey in the shallows or running it down onshore. Like other monitors, it also feeds on carrion. In some places, it can often be seen on the outskirts of coastal villages, where it scavenges on discarded remains. Water monitors breed by laying eggs, which the female places at the end of a burrow.

conspicuous markings when young

ORDER SQUAMATA

Mangrove Monitor

Varanus indicus

LENGTH	Up to 4 ft (1.2 m)
WEIGHT	Up to 22 lb (10 kg)
HABITAT	Mangrove swamps, coastal forests, estuaries, rivers

DISTRIBUTION Western Pacific, from Micronesia to northern Australia

Similar in shape to the water monitor (left), this lizard has a comparable lifestyle, although it rarely swims far from the shore. Like all monitors, it has a long, supple neck and powerful clawed feet. Its tail is flattened laterally and is double the length of its body. They are very good swimmers and excellent climbers, hunting on the ground, in shallow water, and in trees. Fish make up a large part of their diet, although they eat a wide range of other food, including crabs, birds, other lizards, and even scavenged fishing bait.

The water monitor has had human help in expanding its range. In the past, it was introduced by humans throughout the western Pacific as a source of food, and more recently, the species has been introduced into some Pacific islands as a way of controlling rats. Water monitors lay up to a dozen eggs each time they breed and, like most lizards, their young hatch and develop without parental protection.

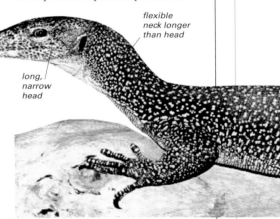

flexible neck longer than head

long, narrow head

OCEAN WANDERER
The saltwater crocodile is a strong swimmer. It has been seen 600 miles (1,000 km) from the nearest coast.

ORDER CROCODYLIA

American Crocodile

Crocodylus acutus

LENGTH	Up to 16½ ft (5 m)
WEIGHT	400–1,000 lb (180–450 kg)
HABITAT	Estuaries, open sea, coasts, lagoons

DISTRIBUTION Caribbean Sea, adjoining areas of Atlantic, Pacific coast of Central and South America

Of the four species of crocodile found in the Americas, this is the only one that—as an adult—is equally at home in both fresh water and the sea. When fully grown, it is olive brown, with a narrow-tipped snout, broad back, and a powerful, tapering tail. Its bony deposits (osteoderms) are smaller than those of other crocodiles. When young, American crocodiles feed on fish and small land animals, but adults often eat turtles, cracking them open in their jaws. Females bury their eggs in sand, laying about 40 every time they breed. Like all crocodiles, this species has been affected by being hunted for its skin, and by coastal development. Its stronghold is in Central America, but a few hundred individuals live in Florida, at the north of its range.

protruding eyes with vertical pupils

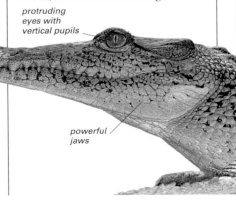

powerful jaws

ORDER CROCODYLIA

Saltwater Crocodile

Crocodylus porosus

LENGTH	Up to 23 ft (7 m)
WEIGHT	Up to 2,200 lb (1 metric ton)
HABITAT	Open sea, rivers, estuaries, coasts

DISTRIBUTION Indian Ocean, Pacific, from southern India to New Guinea and Australia

Also known as the estuarine or Indo-Pacific crocodile, this formidable predator is the world's largest reptile, and is also one of the few crocodilians that frequently swims far out to sea. Its power and ferocity are legendary, and it is thought to be responsible for more than 1,000 human deaths a year. The saltwater crocodile has powerful jaws housing teeth up to 5 in (13 cm) long. Its immensely tough skin is covered with thick scales. The scales on its back are armored with bony deposits called osteoderms, while its tail has a double row of upright bony plates (scutes). Its nostrils close when it dives, but it cannot exclude water from its mouth. Instead, it has a valve at the entrance to its throat, which opens only when it swallows food.

It controls its body temperature by cooling down in water and warming up in the sun. Like other large crocodiles, the saltwater crocodile hunts by stealth, lurking close to the shore, hiding beneath the water with little more than its eyes and nose visible. When an animal comes within range, it bursts out of the water with explosive force, grabs its victim, and then drags it under until it drowns. Crocodiles cannot chew their food—instead, they tear it to pieces, digesting scales, skin, and even bones. Their natural prey includes birds, fish, turtles, and a wide variety of mammals, such as wild boar, monkeys, horses, and water buffalo. Females lay up to 90 eggs in a waterside mound, carrying their young to the water when they hatch. Saltwater crocodiles are hunted in many parts of their range, making large specimens rarer than they once were.

tail with vertical scutes

pointed jaws

body slung between legs

ARMOR PLATING
The saltwater crocodile is protected by parallel rows of bony protruberances along its back.

Birds

DOMAIN	Eucarya
KINGDOM	Animalia
PHYLUM	Chordata
CLASS	Aves
ORDERS	29
SPECIES	9,500

BIRDS THAT HAVE ADAPTED to life at sea spend their lives in the air above the surface, in the upper layers of the open ocean, or along shorelines. Shore- (littoral) based birds rarely range far from land, and some visit the coast only at certain times of year. Others are pelagic, often remaining at sea for months on end and returning to land only to breed. Unlike land birds, many pelagic sea birds breed in large colonies on islands and cliffs, deserting them when the breeding season ends.

Anatomy

There is no such thing as a typical sea bird, although pelagic birds share many adaptations for life at sea. These include webbed feet, highly waterproof plumage, and glands that get rid of excess salt. Most terrestrial birds have hollow, air-filled bones (an adaptation that helps to save weight), but in diving species, such as penguins, the bones are denser and the air spaces reduced.

Some plunge-divers, including gannets and pelicans, have air sacs under their skin. These cushion the impact as they hit the water and help them to bob back to the surface with their prey. Compared to these marine species, shoreline birds show few specific adaptations for life in or near salt water but, like all birds, they have bills specialized for dealing with different kinds of food.

streamlined bill

narrow wings ideal for long-distance flight

FLYING DIVER
The northern gannet's streamlined shape is typical of a plunge diver. Its nostrils open inside its bill, enabling it to keep out water when it hits the surface.

webbed feet

BILL ADAPTATIONS
Apart from waterfowl, most birds of the sea and shore are carnivores, with bills that are adapted for different kinds of animal prey. A pelican's bill and pouch work like a scoop, while an albatross's hooked bill can grip slippery prey, such as jellyfish. Sea eagles catch their prey with their talons, but then use their bills to tear it into pieces. Curlews have long bills that can probe for animals buried in mud.

food pouch

PELICAN

tubular, external nostril excretes excess salt and aids sense of smell

hooked tip

ALBATROSS

long bill can probe deep into estuarine mud

hooked bill

SEA EAGLE

CURLEW

Habitats

Birds live throughout the world's oceans and shorelines, from the equator to the poles. Fewer than 200 species are truly pelagic, meaning that they ply the oceans. These oceanic birds include albatrosses, which have wingspans of up to 11 ft (3.5 m), and much smaller species, such as shearwaters and terns. Although they feed on sea animals, their true habitat is the air: the sooty tern, for example, hardly ever rests on the water and may spend its first five years entirely on the wing. Food is widespread but localized and hard to locate in open oceans, so most sea birds live closer to land.

Most diving sea birds feed in the shallow waters over continental shelves, while rocky coasts and mudflats are key habitats for waders and gulls. Estuaries are important habitats for coastal birds. Their muddy silt often harbors numerous worms and mollusks, accessible at low tide. In the tropics, mangrove swamps attract birds for the same reason; they also have the added bonus of trees, in which birds nest and roost.

MURRE DIVING
Using their wings as hydrofoils, common murres speed through icy water in search of fish. Members of the auk family, they are common in northern seas.

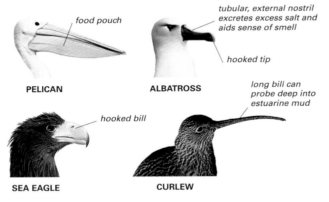

COASTAL WADERS
Eurasian oystercatchers feed in a variety of coastal habitats, from rocky shores to mudflats. These birds are waiting for the tide to turn so that they can start to feed.

OCEAN WANDERER
The black-browed albatross travels long distances in search of good feeding grounds. Its diet includes crustaceans, fish, squid, and carrion.

Feeding Methods

Marine birds have evolved several ways of hunting their food. Most spectacular are the plunge divers—birds such as gannets, boobies, and brown pelicans—which slam into shoals of fish from heights of up to 100 ft (30 m). Diving sea birds also include many that operate from the surface, such as cormorants and penguins. Emperor penguins typically dive to 490 ft (150 m) but have been recorded as deep as 1,755 ft (535 m)—the greatest depth for any bird.

Many oceanic birds, such as albatrosses and petrels, hunt on the wing, snatching animals or scraps from the surface. Kleptoparasitic birds, such as frigatebirds, which harass other birds into disgorging their catch, also hunt on the wing. Coastal birds often probe for food in the shallows or along the tideline, but skimmers slice through the water, holding their lower mandible underwater while in flight, a remarkable technique that works only if the surface is flat and calm.

brown pelican — 33 ft (10 m)
13 ft (4 m)
160 ft (50 m)
murre
cormorant
330 ft (100 m)
33 ft (10 m)
490 ft (150 m)
adelie penguin
660 ft (200 m)
emperor penguin
820 ft (250 m)

MAXIMUM FEEDING DEPTHS
Plunge divers (left), such as the brown pelican, rarely reach more than a few yards beneath the surface. Deeper divers, such as the penguins, use their wings or feet to propel themselves, often staying under for several minutes.

DAWN PATROL
Trailing its beak in the water, a skimmer searches for food in the calm waters of a lagoon.

HUMAN IMPACT

SEA BIRDS UNDER THREAT

The inexorable increase in fishing and shipping has had a significant impact on many coastal and marine birds. Sea birds are often harmed directly, becoming entangled in nets or caught in oil spills. They can also be harmed indirectly, when fishing reduces their food supply. Global warming poses yet another threat: changing sea temperatures can trigger major changes in the fish stocks on which birds feed.

COLLATERAL DAMAGE
Caught in a fishing net, this cormorant is one of thousands of birds that drown every day. Diving birds have difficulty seeing plastic netting underwater and often become trapped.

Dispersal and Migration

Marine birds can range over a huge distance in their lifetime. Some, such as the northern gannet, disperse over wide areas of ocean, returning to isolated colonies to breed. The dispersal instinct of northern gannets is strongest in young birds and slowly declines during the four years that it takes them to mature sexually. From then onward, adults congregate at their colonies in spring and summer, dispersing again when their chicks have left the nest.

Many other birds, such as the Red Phalarope, migrate between distinct summer and winter ranges. During their migrations, they can be seen "on passage" between their two homes. In the species profiles on the following pages, distribution maps show all the places where a species occurs—its summer and winter ranges, as well as those regions it migrates through.

NORTHERN GANNET
A typical dispersing species, this bird nests in colonies scattered around the north Atlantic. When not breeding, it wanders as far south as the tropics, usually over continental shelves.

■ summer distribution
□ winter distribution

RED PHALAROPE
This migrant nests in the high Arctic, and overwinters in the southeast Pacific and eastern Atlantic. An extensive network of migration routes means that it is seen in many parts of the world.

■ summer distribution
□ winter distribution

Breeding

Once they reach adulthood, all sea birds have to come to land to breed. Some species nest on their own, but many form large colonies— often because secure nesting sites are few and far between. Cliffs and islands are favorite locations, as they offer the best protection from predatory mammals. Petrels and shearwaters nest in burrows or fallen rocks, but most sea birds lay their eggs in the open, using little or no nesting material. Compared to terrestrial birds, they have small clutches.

Cormorants often lay three or four eggs, but many other marine birds, such as albatrosses and puffins, lay a single egg each year. These birds are often long-lived, but their low reproductive rate makes them vulnerable to environmental problems, such as oil spills or climate change.

TREE NESTER
Frigatebirds are unusual among marine birds in that they nest in shrubs and tress.

MIXED COLONY
Guanay cormorants, boobies, and brown pelicans nest in dense colonies on the desert islands off the coast of Peru—an area rich in fish.

Of the world's 40 bird orders, three are exclusively marine: the penguins, albatrosses and petrels, and tropicbirds. Eight other orders contain a mixture of terrestrial, coastal, and marine species.

WATERFOWL
Order Anseriformes

170 species
Most species of ducks, geese, and swans live on, or near, fresh water and often move to coasts for the winter. A few are totally marine, and live in inshore waters.

KING PENGUIN

PENGUINS
Order Sphenisciformes

18 species
These exclusively marine birds have lost the ability to fly. Most species are found in the Southern Ocean, but their range also extends northward in cold-current regions, reaching as far as the Galápagos Islands.

LOONS
Order Gaviiformes

5 species
These sleek, fish-eating birds dive from the water's surface, propelling themselves with their feet. Loons are found mainly in the far north. They breed inland by fresh water, but often overwinter at sea.

ALBATROSSES AND PETRELS
Order Procellariiformes

143 species
Totally marine birds occurring throughout all oceans, albatrosses and petrels return to land only to breed. Their external nostrils lend a good sense of smell. Most remain airborne for days, snatching food from the sea's surface.

GREBES
Order Podicipediformes

20 species
These fish-eating birds have lobed feet set far back along their bodies. Most grebes live in freshwater habitats, but some migrate to coastal waters after the breeding season.

TROPICBIRDS
Order Phaethontiformes

3 species
Marine birds that nest colonially on small islands, wander widely over tropical oceans, and dive for fish from the air. They have very long, flexible central tail feathers.

PELICANS AND RELATIVES
Order Pelecaniformes

111 species
Mostly long-legged birds that stalk prey in shallow water or in marshy habitats. Most live inland, but several are found on coasts and coral reefs and in mangrove swamps. Pelicans mostly inhabit lakes, but some feed at sea, scooping water and prey into their elastic bill pouch.

CORMORANTS AND RELATIVES
Order Suliformes

60 species
This large group of social sea birds includes cormorants, frigatebirds, and gannets. All have "totipalmate" feet, with all four toes joined by webs. They catch fish either by plunging into the water from the air or by diving from the surface. Found worldwide, they live on coasts. Some species, particularly cormorants, also frequent freshwater habitats.

BIRDS OF PREY
Order Accipitriformes

265 species
Predatory birds, these species have hooked bills and sharp talons for snatching their prey. As a group, birds of prey are largely terrestrial, but some species specialize in catching fish, and can often be seen on coasts. They rarely venture far out to sea.

WADERS, GULLS, AND AUKS
Order Charadriiformes

381 species
This diverse order contains coastal and oceanic species, including many long-distance migrants. Diets are varied and feeding methods range from plunge-diving to shoreline scavenging. Many species are gregarious, feeding and nesting in colonies.

KINGFISHERS AND RELATIVES
Order Coraciiformes

180 species
These are primarily birds of forests or fresh water, although some species feed along coasts and inshore waters. They dive on prey from the air, taking off again directly after catching it, although they can swim.

Brant Goose

Branta bernicla

LENGTH	22–26 in (55–66 cm)
WEIGHT	3–3½ lb (1.3–1.6 kg)
HABITAT	Estuaries, tundra, coastal grassland

DISTRIBUTION Arctic (breeding); North America, northwest Europe, China, Japan (nonbreeding)

A compact bird with a gray body, black head, and black neck, the brant goose breeds in the High Arctic but winters on coasts at temperate latitudes—a pattern followed by many other wildfowl. Its preferred food is eelgrass, a marine plant that grows in shallow water, but in its winter quarters it also grazes in coastal fields. Brant geese nest in colonies in low-lying coastal tundra, laying up to five eggs and raising a single brood each year. Like many birds in the High Arctic, their numbers undergo steep fluctuations. In mild summers, most of their goslings survive, but if conditions are unusually cold, very few young live long enough to migrate when summer comes to an end.

Common Eider

Somateria mollissima

LENGTH	20–28 in (50–71 cm)
WEIGHT	2¾–6¼ lb (1.2–2.8 kg)
HABITAT	Shallow coasts, estuaries

DISTRIBUTION Arctic Ocean, north Atlantic, north Pacific

This heavily built duck is a common sight on Arctic coasts, where it dives to catch mollusks and crabs, cracking them open with its powerful bill. The females are mottled brown, while the males (below) are mainly black and white, with a pink breast and greenish neck. Common eiders breed in groups, building their nests close to the sea. After the breeding season, they move to more temperate zones in the south of their range for the winter months.

HUMAN IMPACT

EIDERDOWN

To keep the eggs and young warm, female eiders line their nest with down feathers plucked from their breast. Eiderdown is a superb insulator and has long been used as a filling for clothes and bedding. It is still collected in Iceland, although demand has dwindled following the introduction of synthetic fibers.

DOWN OF THE COMMON EIDER

Flightless Steamer Duck

Tachyeres pteneres

LENGTH	24–30 in (61–76 cm)
WEIGHT	8¾–10 lb (4.0–4.5 kg)
HABITAT	Rocky coasts, inshore waters

DISTRIBUTION Southern South America

This heavily built duck is one of four closely related species, all from South America, that have lost the ability to fly. Like other steamer ducks, it has mottled gray plumage, yellow legs, and a robust, yellow-orange bill. It feeds on mussels, crabs and other small animals, diving among kelp beds to find its food. If threatened, it paddles noisily across the water with its wings, a behavior known as "steaming."

Red-breasted Merganser

Mergus serrator

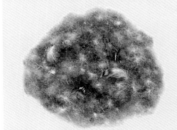

LENGTH	20–23 in (52–58 cm)
WEIGHT	2¼–2¾ lb (1–1.25 kg)
HABITAT	Coasts, estuaries, lakes, rivers

DISTRIBUTION Arctic and subarctic (breeding); temperate coasts (nonbreeding)

This is one the most widespread sawbill ducks—ducks that have narrow beaks with serrated edges, like the teeth of a saw. All these birds dive for fish, using their specially adapted bills to grip their slippery prey. Like other sawbills, the red-breasted merganser has an elongated body, a long neck, and orange-red legs. Males (right) have a metallic green head and shaggy crest, while the female's head is rust-colored, with a less flamboyant crest. These birds breed near fresh water, but spend the winter on coasts, where the water is less likely to freeze. Females build a nest in dense cover, or in a tree-hole, lining it with down. They lay up to 11 eggs, raising a single brood a year. Red-breasted mergansers are shot

Common Shelduck

Tadorna tadorna

LENGTH	23–26 in (58–67 cm)
WEIGHT	2–3¼ lb (0.85–1.45 kg)
HABITAT	Coasts, estuaries, salt lakes

DISTRIBUTION Europe, North Africa, Asia

With its brightly colored body and vivid red bill, the common shelduck is an eye-catching inhabitant of muddy shores. Normally seen in pairs, it feeds by dabbling in mud to collect small animals exposed by the falling tide. It nests in holes, and raises up to nine young each year. After breeding, common shelducks gather together to molt in flocks of up to 100,000 birds.

in some parts of their range in order to protect fish stocks, although there is little evidence that they actually do much harm.

ORDER SPHENISCIFORMES

King Penguin

Aptenodytes patagonicus

HEIGHT	33½–37½ in (85–95 cm)
WEIGHT	26–31 lb (12–14 kg)
HABITAT	Rocky coasts, open ocean

DISTRIBUTION Southern Ocean, subantarctic islands including Falkland Islands

This is the largest penguin found on shores outside Antarctica. Like its close relative the emperor penguin, it has a blue-black body with a white chest and conspicuous, yellow-orange markings on its head. Males and females look identical, and they share the task of incubating the single egg. Instead of building a nest, they cradle the egg on their broad webbed feet, where it is kept warm by a flap of skin. Their bodies are protected from the cold by short, densely packed feathers and a thick layer of blubber. King penguins feed on fish and squid, diving to depths of over 650 ft (200 m) to hunt their prey. At one time, these birds were exploited commercially for their blubber, oil, and feathers, but today they are fully protected.

BREEDING OUT OF STEP

King penguins have a breeding cycle found in no other sea bird. The cycle begins in November—the start of the southern summer—when the female lays her first egg. The chick takes 55 days to hatch, then stays with its parents for 11 months. Once the chick is independent, the female must complete her molt before laying again, this time in late fall. As a result, the king penguin's breeding cycle takes 18 months and moves in and out of phase with the calendar year.

KING PENGUIN CHICKS

OCEAN LIFE

382

ORDER SPHENISCIFORMES

Emperor Penguin

Aptenodytes forsteri

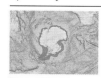

HEIGHT	43–45 in (110–115 cm)
WEIGHT	77–88 lb (35–40 kg)
HABITAT	Sea ice, rocky coasts, open ocean

DISTRIBUTION Southern Ocean, Antarctica

The emperor is the world's largest penguin and the only species that breeds in Antarctica during the southern winter. In shape and markings, it is very similar to the king penguin, but it can be over twice its weight. Rarely found outside Antarctic waters, it feeds among broken sea ice, diving to depths of up to 1,755 ft (535 m). It can remain underwater for as long as 20 minutes, and may travel up to 600 miles (1,000 km) in search of food. The emperor penguin breeds in scattered colonies on the ice itself. Adult females lay a single egg in early winter, and then transfer it to the male. During the winter darkness, while the females feed at sea, the males huddle together with their eggs balanced on their feet and protected within a fold of feathery skin. The incubation period lasts about two months. By the end of it, the males have lost about half their body weight. The females return when the chicks hatch, releasing the males, who head out to sea.

ORDER SPHENISCIFORMES

Chinstrap Penguin

Pygoscelis antarctica

HEIGHT	28–30 in (71–76 cm)
WEIGHT	6½–10 lb (3–4.5 kg)
HABITAT	Rocky coasts, open ocean

DISTRIBUTION Southern Ocean, Antarctic Peninsula, subantarctic islands

Easily identified by the black line around its chin, the chinstrap penguin is one of the most abundant penguin species. Males and females look identical, with blue-black bodies, white undersides, and straight black bills. They live at sea for most of the year, feeding in open water north of the polar ice. When swimming at high speed, they often leap clear of the water, or "porpoise," which allows them to breathe and coats their bodies with a layer of air bubbles, reducing friction with the water.

In November, chinstraps return to their breeding colonies on ice-free shores in Antarctica and on islands in the Southern Ocean. Here, they make their nests by scraping together small stones to form a shallow cup. Chinstraps tend to be more aggressive than other penguins, particularly when breeding. They steal stones from their neighbors and chase away any larger penguins that attempt to nest nearby. The female lays two eggs, and her chicks fledge and set off for the sea by February or March, when the southern fall begins. Chinstraps feed almost entirely on krill, and their current population growth, like that of Antarctica's krill-eating seals, may be linked to the decline of krill-eating baleen whales.

chinstrap marking

ORDER SPHENISCIFORMES

Macaroni Penguin

Eudyptes chrysolophus

HEIGHT	27½ in (70 cm)
WEIGHT	9¼ lb (4.2 kg)
HABITAT	Rocky coasts, open ocean

DISTRIBUTION Southern Ocean, Antarctic Peninsula, subantarctic islands, southern South America

Macaroni penguins are often seen together with a similar species of penguin, the rockhopper. However, macaronis are significantly larger and have distinctive flame-yellow crests that run above each eye and meet on the forehead. They are also found farther south, breeding on ice-free coasts on the Antarctic Peninsula. Their breeding colonies are extremely noisy, some containing over a million pairs spaced out just beyond pecking distance of each other. Macaroni Penguins lay two eggs a year, and both parents help with incubation, probably just of the larger second egg. Both share incubation at first before the female, then the male, each take a 12-day stint alone with the egg.

ORDER SPHENISCIFORMES

Little Penguin

Eudyptula minor

HEIGHT	16–18 in (40–45 cm)
WEIGHT	2¼ lb (1 kg)
HABITAT	Rocky and muddy coasts, open ocean

DISTRIBUTION Southern Australia, New Zealand, Tasman Sea and Southern Ocean

This is the smallest penguin, and it is also the only one that remains offshore during daylight, coming onto land after dark. It has a white underside, a gray-blue back and head, and no distinctive markings. During daylight, little penguins are often seen in small flotillas offshore, resting on the surface and periodically diving to catch fish.

SAFETY AFTER DARK

In some parts of their range— such as Phillip Island, near Melbourne—thousands of little penguins can be seen scrambling ashore as the light fades. This behavior protects them from most predators, although not from introduced mammals such as foxes and domestic dogs.

When feeding, they circle around small fish to concentrate them into a close-knit group, before swimming through the shoal and snapping them up. Unlike other penguins, they do not leave the water when they travel at speed. Little penguins usually nest in burrows or among fallen rocks, but may set up home in breakwaters and under houses and sheds. Each female lays a clutch of two eggs and raises up to two broods a year.

ORDER SPHENISCIFORMES

Magellanic Penguin

Spheniscus magellanicus

HEIGHT	28 in (71 cm)
WEIGHT	12 lb (5.5 kg)
HABITAT	Rocky coasts, open ocean

DISTRIBUTION Southern South America, Falkland Islands, south Atlantic and south Pacific

One of two species of black-and-white penguin from South America, the Magellanic penguin is identified by the two black bands across its breast (see below). It feeds in the cold waters that flow northward from the Southern Ocean, eating small, shoal-forming fish such as sardines. Like its close relative, the Humboldt penguin, it nests in burrows, raising up to two chicks each year.

ORDER SPHENISCIFORMES

Jackass Penguin

Spheniscus demersus

HEIGHT	24–28 in (60–70 cm)
WEIGHT	11 lb (5 kg)
HABITAT	Rocky coasts, open ocean

DISTRIBUTION Coastal waters of southern Africa, south Atlantic and southern Indian Ocean

Also known as the Cape penguin, the jackass is the only penguin that breeds in Africa. It is found from the Wild Coast east of Cape St. Francis westward, and northward to the Skeleton Coast near Walvis Bay in Namibia. Physically, it bears a strong resemblance to the Magellanic penguin (see left) from South America, although it has a single black breast band rather than two. It feeds on small fish such as pilchards, sardines, and anchovies, and gets its name from its braying call, which may be heard onshore when it breeds. Jackass penguins nest in burrows, and in the past, many of their nesting sites were destroyed by farmers collecting their droppings, or guano, for use as fertilizer. Today, depletion of food stocks due to overfishing and oil spills are two major threats that they face, along with competition from fur seals for breeding sites. Their numbers are in sharp decline.

EMPEROR PENGUINS
Emperor penguins are among the hardiest animals in the world, able to withstand blizzards on land and deep dives in the freezing waters of the Southern Ocean. Fast and agile swimmers, they can usually outrun and outmaneuver the leopard seals that hunt them, although they are vulnerable when entering and leaving the water.

ORDER GAVIIFORMES

Great Northern Diver

Gavia immer

LENGTH	28–35 in (70–90 cm)
WEIGHT	6½–10 lb (3–4.5 kg)
HABITAT	Freshwater lakes (breeding); coasts

DISTRIBUTION Northern North America, Greenland, Iceland, Europe, north Pacific, north Atlantic

This striking bird is best known for its haunting cry, which echoes across freshwater lakes during the summer breeding season. During the winter, the same bird is a common visitor in coastal waters, although at this time of year its black-and-white breeding plumage is replaced by less eye-catching shades of brownish black and gray. Like other divers, it has a streamlined body, small wings, and webbed feet set far back—a feature that makes it clumsy on land. On water, it is far more graceful. It floats with its bill held at a characteristic upward slant and can dive to depths of over 250 ft (75 m) to catch fish—its principal food.

During the summer months, Great Northern Divers usually live in pairs, carrying out spectacular courtship displays. When the breeding season comes to an end, they migrate to sheltered coasts, where there is less risk of icing. In North America, flocks of several hundred often gather on the Great Lakes, before heading south as far as the coast of Florida. In Europe, they winter on Atlantic coasts, dispersing as far south as Portugal.

ORDER PROCELLARIIFORMES

Black-footed Albatross

Phoebastria nigripes

LENGTH	27–29 in (68–74 cm)
WEIGHT	6½–7¾ lb (3–3.5 kg)
HABITAT	Open ocean, atolls, isolated islands

DISTRIBUTION North Pacific, Johnston Island and Marshall Islands

This dark-colored sea bird is often seen in summer off North America's west coast. One of three species of albatross found in the north Pacific, its dark underwings distinguish it from the other two. It is fond of scavenging, and often follows trawlers and shrimping boats to catch discarded offal. Black-footed Albatrosses breed in colonies on islands in the central and western Pacific. Like other albatrosses, they perform elaborate courtship displays. All albatrosses are monogamous, pairing up to breed with the same partner each fall.

HUMAN IMPACT

LONG-LINE FISHING

The Black-footed Albatross is a frequent victim of long-line fishing, which involves trailing lines that carry thousands of baited hooks. Albatrosses swallow the bait and become caught. Long-line fishing is estimated to kill at least 300,000 sea birds of all kinds each year.

VICTIM OF DROWNING
Albatrosses usually swallow their food whole, so long-line hooks become lodged in their stomachs and drag the birds underwater to drown.

ORDER PROCELLARIIFORMES

Short-tailed Albatross

Phoebastria albatrus

LENGTH	33–37 in (84–94 cm)
WEIGHT	6½–11 lb (3–5 kg)
HABITAT	Remote islands (breeding); open ocean

DISTRIBUTION North Pacific, Tori Shima Island and Senkaku Islands

This north Pacific albatross almost became extinct during the early 1900s due to demand for its feathers. By 1950, only about 20 birds were left. Thanks to conservation measures, the population now stands at more than 4,000, all breeding on remote islands in the far west of its range. Adults are mostly white, with black flight feathers, pink bills, and golden-yellow heads.

ORDER PROCELLARIIFORMES

Light-mantled Sooty Albatross

Phoebetria palpebrata

LENGTH	31–39 in (79–98 cm)
WEIGHT	5–10 lb (2.5–4.5 kg)
HABITAT	Remote islands (breeding); open ocean

DISTRIBUTION Southern Ocean, isolated islands in south Atlantic and southern Indian Ocean

Together with its close relative the Sooty Albatross, this is one of two southern albatrosses that have sooty brown plumage, as opposed to white and black. The Sooty Albatross is brown all over, but the Light-mantled species has a pale gray nape and back—the feature that gives it its name. A graceful glider, it feeds on fish, squid, and crustaceans. It is also highly inquisitive and often follows ships. After spending the winter at sea, it returns to its breeding sites by August, the start of the southern spring. Female birds lay a single egg in early summer, and the chicks become independent about four months after they hatch—a relatively rapid development compared with that of the larger albatrosses.

ORDER PROCELLARIIFORMES

Wandering Albatross
Diomedea exulans

LENGTH 3½–4½ ft
(1.1–1.35 m)

WEIGHT 18–25 lb
(8–11.5 kg)

HABITAT Remote islands
(breeding); open ocean

DISTRIBUTION Southern Ocean, south Atlantic,
southern Indian and Pacific oceans

This legendary sea bird has the largest recorded wingspan of any bird, at up to 11½ ft (3.5 m). It is restricted to the windswept southern oceans, where it feeds mainly on squid, snatching its food from the surface of the water. It is capable of remaining airborne for weeks at a time and frequently follows ships, soaring over the waves on its stiff, outstretched wings. The Wandering Albatross takes up to 11 years to mature, and during that time it gradually loses its juvenile plumage, becoming all white except for black markings on the tips and trailing edges of its wings. These birds nest on remote islands, typically breeding in alternate years.

LONG INCUBATION

Wandering Albatrosses build large, moundlike nests from mud, grass, and moss. Their single egg has one of the longest incubation periods of any egg, taking between 75 and 82 days to hatch. The solitary chick then remains in the nest for up to nine months, where it is fed by both its parents. During very severe weather, the chick may be left unattended for days at a time.

OCEAN LIFE

ORDER PROCELLARIIFORMES

Black-browed Albatross
Thalassarche melanophrys

LENGTH 33–37 in
(83–93 cm)

WEIGHT 6½–11 lb
(3–5 kg)

HABITAT Remote islands
(breeding); open ocean

DISTRIBUTION Southern Ocean, south Atlantic,
southern Indian and Pacific oceans

Also known as the Black-browed Mollymawk, this is the most numerous and widespread of the albatrosses. It is found from Antarctica to the edge of the tropics, and in places even farther north. Its wings, back, and tail are grayish black, and it has a distinctive black brow above each eye. It feeds on fish, squid, octopus, and crustaceans, and is also a frequent ship-follower, congregating in large numbers when waste is thrown overboard. Black-browed Albatrosses breed on remote islands and take at least five years to become mature. They are among the few southern albatrosses that regularly cross the Equator—isolated sightings have been recorded as far north as the British Isles.

ORDER PROCELLARIIFORMES

Southern Giant Petrel

Macronectes giganteus

LENGTH	34–39 in (86–99 cm)
WEIGHT	11 lb (5 kg)
HABITAT	Coasts, open sea; nests on ice-free coasts

DISTRIBUTION Southern hemisphere, from Antarctica as far north as the tropics

Part-scavenger and part-predator, this large petrel is often seen on the fringes of penguin colonies or near the carcasses of dead seals and whales. It uses its powerful bill to tear apart carrion and to kill young birds. Most adults have a pale head and a dark grayish brown back, but some are almost completely white with scattered black flecks.

tubular nostrils

ORDER PROCELLARIIFORMES

Northern Fulmar

Fulmarus glacialis

LENGTH	18–20 in (45–51 cm)
WEIGHT	1½–2 lb (700–900 g)
HABITAT	Rocky coasts, open sea

DISTRIBUTION North Pacific, north Atlantic, ice-free areas of Arctic Ocean

Often mistaken for a gull, this fulmar is actually a petrel and, like other petrels, has distinctive tubular nostrils. Common throughout northern waters, it is often seen flying over cliffs on its stiff, outstretched wings. Its weak feet make it clumsy on land, and its eyes are dark with a distinct brow ridge. Most northern fulmars in the Atlantic have white bodies and blue-gray upper wings, but in the Pacific many of the birds are much darker. Northern fulmars feed on small animals at or near the sea's surface, and they gather in large flocks to scavenge around fishing boats. They breed on exposed cliff ledges, with each female laying a single egg directly onto the rock. The incubation period is 52 days—almost twice as long as that in gulls of similar size. Despite its low reproductive rate, the northern fulmar has increased both in range and in numbers in recent years. It is exceptionally long-lived for its size, with ages of over 50 years recorded.

ORDER PROCELLARIIFORMES

Snow Petrel

Pagodroma nivea

LENGTH	12–14 in (30–35 cm)
WEIGHT	9–16 oz (250–450 g)
HABITAT	Rocky and ice-bound coasts

DISTRIBUTION Antarctica, subantarctic islands, Southern Ocean

Despite its dainty appearance, the snow petrel is one of the world's most southerly breeding birds. This entirely white petrel nests on ice-free cliffs in and near Antarctica, to within 680 miles (1,100 km) of the South Pole. It picks food from the surface of the sea, rarely straying far from the polar ice. Flocks of snow petrels are often seen sitting on icebergs.

ORDER PROCELLARIIFORMES

Bonin Petrel

Pterodroma hypoleuca

LENGTH	12 in (30 cm)
WEIGHT	8 oz (225 g)
HABITAT	Oceanic islands (breeding); open ocean

DISTRIBUTION Northwestern Pacific

There are over two dozen species of *Pterodroma* petrels, mostly in tropical and subtropical regions, and these are often difficult to distinguish at sea. The Bonin petrel is a typical example from the northwestern Pacific, where it nests on scattered islands westward from Hawaii. It has a small, short, slightly hooked bill and sharply pointed wings, and it is fast and agile as it speeds through the air just above the waves. It eats small planktonic animals, usually landing on the surface to feed. This petrel nests in burrows but has difficulty moving on land. To reduce the risk of attack from predators, the Bonin petrel generally returns to land at night, when it may deliver regurgitated food to its single chick. The parents share the task of egg incubation over about 49 days. On remote islands, petrel colonies can be decimated by introduced predators, such as rats and cats. This species is one that has been badly affected.

ORDER PROCELLARIIFORMES

Fairy Prion

Pachyptila turtur

LENGTH	10–11 in (25–28 cm)
WEIGHT	5–8 oz (150–225 g)
HABITAT	Islands (breeding); open ocean

DISTRIBUTION Southern Ocean and adjoining waters

A small, oceanic petrel, the fairy prion has a pale body and blue-gray upper wings with a distinct, M-shaped black band. It lives in flocks and feeds at night, using its bill to sieve planktonic animals from the water. It breeds on isolated coasts, laying a single egg either in a burrow or in a hollow deep among fallen rocks.

ORDER PROCELLARIIFORMES

Great Shearwater

Ardenna gravis

LENGTH	18–21 in (46–53 cm)
WEIGHT	1³/₄–2 lb (800–900 kg)
HABITAT	Oceanic islands (breeding); open ocean

DISTRIBUTION Atlantic Ocean, except off west coast of Africa south of Sierra Leone

This ocean-living sea bird is a wide-ranging migrant and is found across most of the Atlantic Ocean during the course of the year. It breeds in the far south, on some of the world's remotest islands. One of these, Nightingale Island in the Tristan da Cunha group, is home to about four million birds. Great shearwaters have pointed wings with dark brown upper surfaces. Their undersides are much paler, making the birds look alternately black then white during their tilting flight. Sometimes vocal at sea, they wail and scream noisily from their burrows when breeding. These calls are thought to help incoming birds to locate their mates after dark.

ORDER PROCELLARIIFORMES

Wilson's Storm Petrel

Oceanites oceanicus

LENGTH	6–7¹/₂ in (15–19 cm)
WEIGHT	1–1¹/₂ oz (30–40 g)
HABITAT	Coasts, islands (breeding); open ocean

DISTRIBUTION Worldwide except for north Pacific and extreme north Atlantic

Little bigger than a sparrow, Wilson's storm petrel is reputed to be the world's most numerous ocean-going sea bird. It breeds in widely scattered colonies, and its total population is unknown but may exceed 20 million. At sea this bird may be difficult to distinguish from its close relatives, but its plumage is uniformly sooty brown, apart from a band of white at the base of its tail. When feeding, storm petrels rarely settle on the water. Instead, they flutter their wings and patter the surface with their feet, pecking up planktonic animals. When food is abundant, they may suddenly appear in huge numbers

then disappear with equal abruptness. Wilson's storm petrel breeds as far south as Antarctica, digging a burrow with its bill and feet. It migrates northward when the southern summer comes to an end.

ORDER PROCELLARIIFORMES

Short-tailed Shearwater

Ardenna tenuirostris

LENGTH	16–17 in (41–43 cm)
WEIGHT	1–1¹/₂ lb (500–700 g)
HABITAT	Open ocean, offshore islands

DISTRIBUTION North Pacific, southwestern Pacific around southern coast of Australia

Awkward and ungainly on land, the short-tailed shearwater is a tireless flier, skirting around most of the north Pacific during its annual migration. Like other shearwaters, it travels just inches above the waves in fast-moving flocks, interrupting its flight whenever it spots food. However, it has a narrower bill than other shearwaters, and its overall color is a dark, smoky brown. It nests in vast island colonies, each pair producing

a single chick. Fed on a rich diet of oily food, the chicks weigh more than their parents by the time they leave the nest. For several centuries, shearwater chicks have been harvested for their oil and meat. The practice continues today, although the numbers killed are now strictly controlled.

MIGRATION

The short-tailed shearwater has a unique figure-eight migration route, 20,800 miles (33,500 km) long, that takes advantage of prevailing winds. After laying eggs in November and December, the birds head north in April and May, reaching the Bering Sea by August. They then move south along North America's west coast, before returning to their breeding colonies.

June–August

September

April–May

October

November–March

KEY

▬ breeding area

← migration route

← wind direction

ORDER PROCELLARIIFORMES

Leach's Storm Petrel

Oceanodroma leucorhoa

LENGTH	7¹/₂–8¹/₂ in (19–22 cm)
WEIGHT	1¹/₂–1³/₄ oz (40–50 g)
HABITAT	Coasts, islands (breeding); open ocean

DISTRIBUTION North Pacific, north Atlantic, coastal North America and Aleutian Islands

Leach's storm petrel is silent at sea, but it makes a high-pitched purring sound, interrupted by sharp whistles, in and near its nest. Unlike Wilson's storm petrel (see above), this species breeds in the Northern Hemisphere. It migrates southward in late summer, roaming throughout the north Pacific and much of the Atlantic. Small and brownish black, with a sharply forked tail, it flies rapidly, changing direction frequently as it scans the water's surface for food. It feeds on planktonic animals and small fish, pattering on the surface

with its feet and occasionally settling on the water to rest. Leach's storm petrels breed in colonies, laying a single egg and returning to their burrows at night with food for their hatched young. In the far north, some birds delay nesting until August to avoid the 24-hour daylight of the Arctic summer, during which they would be more vulnerable to predators.

ORDER PROCELLARIIFORMES

Common Diving Petrel

Pelecanoides urinatrix

LENGTH	8–10 in (20–25 cm)
WEIGHT	4–4¹/₂ oz (110–130 g)
HABITAT	Coasts, islands (breeding); open ocean

DISTRIBUTION Southern Ocean and adjoining waters and islands

This stubby bird with pointed wings and pale blue feet is the Southern Hemisphere's counterpart of the auklets (see p.399). Despite being unrelated, it shares the auklets' fast, low flight and their feeding technique. Instead of searching for food on the wing, like other petrels, it dives, using its wings to swim. It frequently flies straight through waves, emerging with rapidly whirring wings on the other side. This species nests in burrows, returning to its nests after dark. There are three other similar-looking species, all found in southern seas.

ORDER CICONIIFORMES

Gray Heron
Ardea cinerea

LENGTH	34–39 in (90–100 cm)
WEIGHT	3½–4½ lb (1.6–2 kg)
HABITAT	Estuaries, lagoons, coasts

DISTRIBUTION Europe, mainland Asia (except far north), Japan, Indonesia, Africa, Madagascar

Commonly seen in fresh water, the gray heron also frequently visits shores, especially in areas where lakes and ponds freeze in winter. Tall, gray-backed, and often immobile, it waits patiently for fish or other animals to come within range, then seizes them with a rapid jab of its daggerlike bill. On coasts, its feeding method restricts it to shallow water on rocky and low-lying shores, where it often follows the falling tide. Gray herons fly with slow wingbeats, their heads hunched into their shoulders and their legs trailing behind. They nest in trees, typically inland near water.

ORDER CICONIIFORMES

Little Egret
Egretta garzetta

LENGTH	22–27 in (56–65 cm)
WEIGHT	11–16 oz (300–450 g)
HABITAT	Muddy coasts, mangrove swamps

DISTRIBUTION Southern Europe, Africa, southern Asia, Southeast Asia, Australasia

Pure white with black legs, a black bill, and bright yellow feet, the little egret is usually seen on its own or in scattered groups, wading quietly through shallow water on coasts. This bird feeds on fish and other shoreline animals that are disturbed by its approach. During the breeding season, both males and females grow long, lacy feathers on their heads and backs. They nest in trees, building flimsy nests out of sticks.

ORDER PELECANIFORMES

Red-billed Tropicbird
Phaethon aethereus

LENGTH	Up to 19½ in (50 cm) excluding tail
WEIGHT	1¼–1¾ lb (600–800 g)
HABITAT	Coasts, islands (breeding); open ocean

DISTRIBUTION Eastern Pacific, Caribbean, tropical Atlantic, northeast Indian Ocean

The largest of the three species of tropicbird, this elegant sea bird spends most of its life flying over the open ocean, often hundreds of miles from land. From a distance the red-billed tropicbird resembles a dove, but for two highly distinctive tail streamers that flutter behind it as it flies. It feeds by plunge-diving, hovering to locate its prey before diving with half-folded wings into the sea. Despite being very buoyant, it seldom swims. Like other tropicbirds, it nests on remote coasts and oceanic islands and is rarely seen outside tropical waters.

ORDER CICONIIFORMES

Pacific Reef Egret
Egretta sacra

LENGTH	24–27 in (60–70 cm)
WEIGHT	14–26 oz (400–750 g)
HABITAT	Coastal and freshwater wetlands

DISTRIBUTION Australasia, Pacific islands, western Pacific coast from Southeast Asia to Japan

This compact shoreline egret has two contrasting color forms, so different that they look like separate species. One form (or morph) is completely white, with a pale yellow bill and yellow-gray legs. The other form has a similarly colored bill and legs, but its plumage is dark gray. The balance between the two forms varies. In some islands in the tropical Pacific the white form predominates, but in New Zealand, the overwhelming majority are gray. Pacific reef egrets forage alone or in small groups, feeding on small fish, crabs, and mollusks. When hunting, they hold their heads and bodies almost horizontally and often shade the water with their half-spread wings. Unlike most egrets, they frequently nest on the ground, among fallen rocks or in coastal caves, as well as in low-growing trees.

ORDER PELECANIFORMES

Great Frigatebird
Fregata minor

LENGTH	34–39 in (86–100 cm)
WEIGHT	3–4 lb (1.4–1.8 kg)
HABITAT	Coasts, islands (breeding); open ocean

DISTRIBUTION Tropical regions in Indian Ocean and Pacific, sporadic in tropical Atlantic

With their extraordinarily long wings and slender bodies, frigatebirds are unrivaled experts at gliding flight. The five species all have glossy black plumage, strong, hooked bills, and small, webbed feet. The males also have a bright red throat pouch, which they inflate during courtship displays. Despite weighing less than a large gull, the great frigatebird has a wingspan of up to 7½ ft (2.3 m), allowing it to glide for hours while making only the merest flick of its wings. As it flies, it observes other sea birds as they feed, then pursues them to steal their catch. Frigatebirds also hunt their own food, such as flying fish, snapping it up from the sea's surface. They nest in coastal bushes, where they make flimsy nests out of twigs.

ORDER PELECANIFORMES
Brown Pelican
Pelecanus occidentalis

LENGTH	4–5¼ ft (1.2–1.6 m)
WEIGHT	7¾–10 lb (3.5–4.5 kg)
HABITAT	Coastal waters, estuaries, islands

DISTRIBUTION Pacific and Atlantic coasts of North and South America, Galápagos Islands

Commonly seen inshore and in harbors, the brown pelican is the heaviest sea bird that fishes by plunge-diving. Groups of birds often fish together, skimming over the waves before rising into the air, folding back their wings, and hitting the water with a spectacular splash. The pelican's throat pouch balloons outward underwater, scooping up prey, which it then swallows at the surface.

ORDER PELECANIFORMES
Blue-footed Booby
Sula nebouxii

LENGTH	30–33 in (76–84 cm)
WEIGHT	2¼–4½ lb (1.5–2 kg)
HABITAT	Inshore waters, rocky coasts, islands

DISTRIBUTION Pacific coast of Central America, Galápagos Islands

This is one of six species of boobies—a group of plunge-diving birds, closely related to gannets, that often have brightly colored feet. The blue-footed booby is brown with white undersides. Its feet are grayish brown in juveniles but brilliant turquoise-blue in adults. Blue-footed boobies often feed in flocks, hitting the water almost simultaneously when they locate a shoal of fish. Smaller than gannets, they are able to fish closer inshore, sometimes diving into water less than 3 ft (1 m) deep. They nest in small colonies on offshore islands, laying their eggs on the ground.

pale, streaked head plumage

distinctive blue webbed feet

ORDER PELECANIFORMES
Northern Gannet
Morus bassanus

LENGTH	34–39 in (87–100 cm)
WEIGHT	6¼–7 lb (2.8–3.2 kg)
HABITAT	Open sea, rocky coasts, offshore islands

DISTRIBUTION Eastern and western coasts of north Atlantic

This highly streamlined bird with its gleaming white body and black-tipped wings is the most striking plunge-diver in the north Atlantic. Northern gannets roam the seas with a distinctive pattern of flapping and gliding flight, attacking shoals of fish by diving from heights of up to 100 ft (30 m). They breed in crowded colonies on rocky islands and clifftops, laying a single egg each year. Juveniles take five years to mature, gradually losing their brown plumage. During that time, they roam far over the ocean before returning to their native colony to breed.

ORDER PELECANIFORMES
Brown Booby
Sula leucogaster

LENGTH	25–30 in (64–76 cm)
WEIGHT	1½–3¼ lb (0.7–1.5 kg)
HABITAT	Inshore waters, rocky coasts, islands

DISTRIBUTION Tropical oceans worldwide, except southeastern Pacific

This booby is a superb diver. It is the most widespread booby and has distinct color variations. Most brown boobies are brown all over, apart from a white underside. However, birds from the eastern Pacific have white heads and their bills are gray rather than the typical bright yellow. They all live in the same way, diving for fish and squid from heights of up to 100 ft (30 m). They also skim low over the surface, looking for flying fish, which they catch in midair. They often fly in front of ships, watching for fish caught up in the bow-waves, and they like to fish close to land, roosting on buoys or coastal trees. Despite their agility in the air, they are clumsy at takeoff and landing.

PLUNGE-DIVING

Gannets and boobies all show adaptations for a plunge-diving lifestyle: forward-facing eyes, streamlined heads and bills; and nostrils with no external openings. Their wings fold back along the body just before the moment of impact, and the force is absorbed by air sacs under their skin.

AERIAL ATTACK
This sequence of photos shows how the wings fold during a dive.

ORDER PELECANIFORMES

Guanay Cormorant

Phalacrocorax bougainvillii

LENGTH 29–31 in (74–78 cm)	
WEIGHT 4–5 lb (1.75–2.25 kg)	
HABITAT Desert coasts, islands, inshore waters	

DISTRIBUTION Pacific coast of Peru and northern Chile

Boldly marked in black and white, with a conspicuous red patch around each eye, the Guanay cormorant nests in huge colonies along the coast of the Atacama Desert, the most arid region on Earth. It feeds on anchovetas—small fish that abound in the cold waters of the Humboldt Current. Like other cormorants, it pursues fish underwater, holding its wings against its body and propelling itself with its legs. It floats low down in the water, periodically dipping its head beneath the surface to check for food. Guanay cormorants have nested on the same offshore islands for millennia, depositing deep layers of desiccated droppings known as guano. During El Niño years, when the ocean temperature rises, shortage of food forces these cormorants to forage far afield, often as far north as Panama.

ORDER PELECANIFORMES

Great Cormorant

Phalacrocorax carbo

LENGTH	32–40 in (80–101 cm)
WEIGHT	4¼–5½ lb (2–2.5 kg)
HABITAT	Coasts, inshore waters, rivers, lakes

DISTRIBUTION Northeast North America, Europe, Africa, Asia, Australasia

Equally at home in fresh water and at sea, the great cormorant can be found across a vast swath of the world, from Greenland to Australasia. From a distance, its plumage looks jet black, but close up it has a greenish metallic sheen, with white patches that vary between local races. Like its many relatives, it fishes by pursuit diving and its feathers are only partly waterproof. After feeding, it rests with its wings spread apart to dry. Great cormorants have a strong, direct flight, with steady flapping interspersed with short glides. They can often be seen in small groups, skimming just above the surface of the sea or following rivers inland. They nest on rocky ledges and in trees, making a platform out of seaweed, flotsam, or twigs, and the females lay three or four greenish-white eggs. Great cormorants are sometimes persecuted by anglers, particularly in trout-fishing regions, but they remain highly successful.

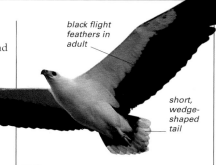

black flight feathers in adult

short, wedge-shaped tail

ORDER ACCIPITRIFORMES

White-bellied Sea Eagle

Haliaeetus leucogaster

LENGTH	28–35 in (70–90 cm)
WEIGHT	5½–9¼ lb (2.5–4.2 kg)
HABITAT	Inshore waters, rivers, lakes, reservoirs

DISTRIBUTION South and Southeast Asia, New Guinea, Australia

This black-and-white eagle makes an impressive sight as it soars over water with its wings, up to 6½ ft (2 m) wide, held in a shallow V shape. Its wide-ranging diet includes fish, water birds, turtles, and sea snakes, which it snatches from the surface, rarely entering the water. It also scavenges and forces smaller sea birds to drop their catch. It breeds close to water, building a large nest in a high tree.

ORDER ACCIPITRIFORMES

Brahminy Kite

Haliastur indus

LENGTH	17–20 in (43–51 cm)
WEIGHT	14–25 oz (400–700 g)
HABITAT	Beaches, estuaries, rivers

DISTRIBUTION South and Southeast Asia, northern Australia, islands of western Pacific

A common scavenger in parts of its range, the brahminy kite is also an effective hunter, crisscrossing the water from a height of a few yards, dropping to the surface to catch fish, or to pick up scraps of waste. It also feeds on beaches and mudflats, and is seen in the outskirts of coastal towns. Adults have deep chestnut plumage, and a distinctive white chest and head. Their breeding season varies according to location, but they often nest in mangroves, making a platform-shaped nest from seaweed and sticks. Both parents help to raise the one to two young.

ORDER ACCIPITRIFORMES

Osprey

Pandion haliaetus

LENGTH	20–26 in (50–65 cm)
WEIGHT	2¾–4½ lb (1.2–2 kg)
HABITAT	Coasts, reefs, lagoons, rivers, lakes

DISTRIBUTION Worldwide except polar regions, southern South America and New Zealand

This fish-eating raptor has one of the widest distributions of any bird of prey, breeding mainly in the Northern Hemisphere and migrating south for the winter. The osprey is easy to distinguish from other birds of prey on coasts, thanks to its light build, its conspicuous, dark eye-stripe, and its narrow, slightly kinked wings. It feeds entirely on fish, plunging from heights of up to 165 ft (50 m) and entering the water feet-first. Its wings are strong, its legs are heavily muscled, and its toes have long, hooked talons and spiny soles—an adaptation that gives it a firm grip on its slippery prey. These birds have been known to take prey that approaches their own weight. They nest in the tops of high trees and hatch a single brood of two to three chicks each year. During the 20th century, ospreys suffered severely as a result of pesticide pollution, particularly from DDT. Their population has now recovered, and in some regions—for example, northern Britain—they have resumed breeding after a gap of many years.

AIRBORNE ATTACK

The osprey cruises high above water looking for food. Once it spots a fish, it hovers for a few seconds before half-folding its wings and going into a steep dive. It hits the water at high speed, sometimes partly submerging, before gripping its prey with one foot and climbing laboriously back into the air. Once airborne, it shakes the water off its plumage, before heading to a perching post or to its nest.

ORDER CHARADRIIFORMES

Snowy Sheathbill

Chionis alba

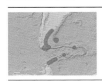

LENGTH	13½–16 in (34–41 cm)
WEIGHT	1–1¾ lb (450–775 g)
HABITAT	Rocky coasts, inshore waters, sea ice

DISTRIBUTION Antarctic Peninsula, subantarctic islands, southern South America, Falkland Islands

Sheathbills are the only birds with nonwebbed feet that breed on the shores of Antarctica. Stocky and short-legged, they bear a superficial resemblance to chickens, particularly when they escape from danger by running away. Almost wholly carnivorous, they scavenge carrion along the shoreline, and also loiter around penguin colonies to steal eggs and food from adult birds.

ORDER CHARADRIIFORMES

Eurasian Oystercatcher

Haematopus ostralegus

LENGTH	15½–19 in (40–48 cm)
WEIGHT	14–28 oz (400–800 g)
HABITAT	Rocky shores, damp inland habitats

DISTRIBUTION Iceland, Europe, N. and E. Asia, (breeding); S. Europe, Africa, S. Asia (nonbreeding)

With its bright orange bill and loud piping call, this is one of the most conspicuous waders on European shores. Often seen in small parties, it feeds on mussels, limpets, and other mollusks, using its bill to smash or pry apart their shells. To locate good feeding sites, it often flies along the tideline, calling loudly to other oystercatchers. On coasts, it nests on shingle and gravel, laying two to four camouflaged eggs. The Eurasian oystercatcher is one of 11 species of oystercatchers (family Haematopodidae). All have the same overall shape and brightly colored bills, but in some species, the plumage is totally black.

brightly colored bill

ORDER CHARADRIIFORMES

Black-necked Stilt

Himantopus mexicanus

LENGTH	14–16 in (35–40 cm)
WEIGHT	5–7 oz (150–200 g)
HABITAT	Shallow coasts, salt marshes, wetlands

DISTRIBUTION Worldwide except far north and northeast Asia; summer visitor only in north of range

The black-necked stilt's immensely long legs trail far behind its tail when it flies. It breeds in a broad range of wetland habitats. It feeds in calm fresh or salt water, striding through the shallows, scything its bill through the water to catch small animals or picking them from the surface.

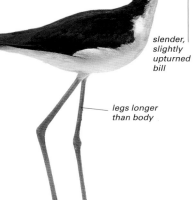

slender, slightly upturned bill

legs longer than body

ORDER CHARADRIIFORMES

Pied Avocet

Recurvirostra avosetta

LENGTH	16½–18 in (42–45 cm)
WEIGHT	8–14 oz (225–400 g)
HABITAT	Shallow coasts, salt marshes, wetlands

DISTRIBUTION Europe, temperate Asia (breeding); W. Europe, Africa, S. and S.E. Asia (nonbreeding)

Instantly recognizable by their long upturned bills, avocets are elegant waders that feed in shallow water, both on coasts and inland. There are four species, all similar in shape and size. Of these, the pied avocet is by far the most widespread and is the only species that is found in Europe and Africa, as well as Asia. Pied avocets feed by dipping their bill in water, and then sweeping it from side to side. The tip of the bill is highly sensitive to touch, so the bird can catch food even in the turbid water of estuaries and lagoons. Pied avocets swim well and sometimes upend to find food in the same way as dabbling ducks. They nest in groups, making cup-shaped hollows on mudflats, where they lay a clutch of four eggs. Despite their dainty appearance, they can be aggressive if their nests are threatened. Parents charge at intruders with their heads lowered, and they are able to chase away much bulkier birds, such as geese and ducks.

ORDER CHARADRIIFORMES

Gray Plover

Pluvialis squatarola

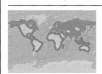

LENGTH	10–11 in (26–28 cm)
WEIGHT	6–8 oz (170–240 g)
HABITAT	Arctic tundra, coasts, estuaries

DISTRIBUTION Arctic (breeding); temperate and tropical coasts worldwide (nonbreeding)

This long-distance migrant, one of the most widespread waders, is found on coasts in every continent except Antarctica. In their breeding plumage, seen only in the Arctic tundra, males have a black underside and face, but by the time they head south to winter on coasts, both sexes are a speckled gray. Gray plovers feed on insects in summer and on marine worms and crustaceans in winter.

ORDER CHARADRIIFORMES

Ruddy Turnstone

Arenaria interpres

LENGTH	8½–10 in (21–25 cm)
WEIGHT	3–4 oz (80–110 g)
HABITAT	Rocky/sandy coasts, coastal lowlands

DISTRIBUTION Arctic coasts (breeding); temperate and tropical coasts worldwide (nonbreeding)

Found on coasts all over the world, the ruddy turnstone feeds in a distinctive way, scuttling along the tideline, flicking stones aside with a deft movement of its bill. This often reveals sandhoppers and other small animals, which it snaps up or chases. Ruddy turnstones, like many waders, nest in the far north, but their feeding habits restrict them to coastal areas. After breeding, their southward migration takes them to coasts on every continent except Antarctica.

ORDER CHARADRIIFORMES

Whimbrel

Numenius phaeopus

LENGTH	15½–18 in (40–46 cm)
WEIGHT	10–16 oz (270–450 g)
HABITAT	Arctic tundra, coasts, reefs, wetlands

DISTRIBUTION N. Europe, Arctic (breeding); temperate and tropical coasts worldwide (nonbreeding)

Using its long, downcurved bill, the whimbrel feeds by probing into wet mud or by extracting animals from rocky crevices. It is one of eight similar species, collectively known as curlews, that have mottled brown plumage, sharply pointed wings, and bills up to 8 in (20 cm) long. The whimbrel's bill is only half this length, but it is a precision instrument, with sensitive nerve-endings at its tip that enable the bird to feel for buried food. The whimbrel is strongly migratory, nesting inland across much of the far north, in marshy open country. At this time of the year, the male sings from high in the air, gradually descending on widely spread wings. After breeding, whimbrels head south along coastlines, reaching as far south as the tip of South America and New Zealand.

ORDER CHARADRIIFORMES

Gray Phalarope

Phalaropus fulicarius

LENGTH	8–9 in (20–22 cm)
WEIGHT	2–3 oz (50–75 g)
HABITAT	Marshy coastal tundra, plankton-rich open ocean

DISTRIBUTION Arctic coasts (breeding); South Atlantic and eastern South Pacific (nonbreeding)

Also known as the red phalarope, this short-billed wader shows a remarkable reversal of roles when it breeds. Unlike most birds, the female—shown here—has a much brighter breeding plumage than the male. Once she has mated and laid her eggs, she takes no part in incubation or raising the young. By comparison with other waders, gray phalaropes are highly aquatic birds and spend much of their time afloat. They breed close to coasts, and once they have migrated south, they often overwinter far out at sea.

ORDER CHARADRIIFORMES

Dunlin

Calidris alpinus

LENGTH	6½–8½ in (16–22 cm)
WEIGHT	1½–1¾ oz (40–50 g)
HABITAT	Coasts, marshes, tundra

DISTRIBUTION Arctic, subarctic (breeding); temperate and tropical coasts in N. hemisphere (nonbreeding)

In winter, flocks of dunlins create a breathtaking spectacle, as they wheel in the thousands over coastal feeding grounds. Up close, the dunlin is a typical calidrid wader, one of over two dozen similar species that feed on coasts worldwide. It has a compact body, narrow wings, a tapering tail, and a black, finely pointed bill. Its plumage is variable, but breeding males usually have a black patch on the underside, which fades when they molt. Dunlins mainly eat small crustaceans and mollusks that live just beneath the surface of the shore. When feeding, they usually stay close to the water's edge, alternately pecking into the mud or sand, and then running forward at high speed. Dunlins breed in the Arctic and subarctic, where they nest in a range of habitats from moorland to tundra, often some distance inland. Both parents help to incubate the eggs and raise the young. After breeding, they gather in flocks to migrate to warmer coasts, but rarely travel into the Southern Hemisphere. Other members of this genus include many other flock-forming species, such as the red knot and sanderling, most of which travel as far north as the Arctic Ocean to breed.

WINTER FLOCKS

Wintering waders form some of the largest bird flocks to be found on coasts. Flocking makes it harder for predators to approach unseen and helps young birds to locate good feeding sites by following adults. Some waders, such as the purple sandpiper and ruddy turnstone, frequently form mixed flocks.

AERIAL MANEUVERS
Flocks of mixed shorebirds show extraordinary coordination, with thousands of birds changing direction almost simultaneously.

OCEAN LIFE

Swallow-tailed Gull

Creagrus furcatus

LENGTH	21½–23½ in (55–60 cm)
WEIGHT	21–32 oz (600–900 g)
HABITAT	Coasts, inshore waters, open sea

DISTRIBUTION Galápagos Islands and Malpelo Island (breeding); Pacific coast of South America

Distinguished by its sharply forked tail, this South American gull is atypical in feeding at night. It eats squid and fish, spotting them with its large eyes, which are surrounded by distinctive red rings and angled forward to give a wide field of binocular vision. Swallow-tailed gulls nest on islands and disperse far out to sea during the rest of the year.

Great Black-backed Gull

Larus marinus

LENGTH	28–31 in (71–79 cm)
WEIGHT	2¾–4¾ lb (1.2–2.1 kg)
HABITAT	Rocky coasts, islands, inland in winter

DISTRIBUTION North Atlantic, breeding north to Svalbard

With a wingspan of up to 5½ ft (1.7 m), the great black-backed is one of the world's largest gulls. Heavily built, with black upperwings and a powerful bill, it scavenges food, but it is also a highly predatory bird. It frequently preys on other sea birds and their young, and will attack mammals as large as rabbits. It breeds alone or in colonies, nesting on cliff ledges or on open ground.

European Herring Gull

Larus argentatus

LENGTH	22–26 in (56–66 cm)
WEIGHT	1¾–2¾ lb (750 g–1.25 kg)
HABITAT	Coasts, reservoirs, urban areas

DISTRIBUTION Northwest Europe

Noisy, assertive, and always on the lookout for a meal, this is an abundant gull in northwest Europe. It has gray upperparts and black wingtips, and a large yellow bill with a conspicuous red spot near the tip. Young herring gulls are mottled brown, and it takes them three years to develop the full adult plumage. Often seen in flocks, herring gulls are highly adaptable birds, feeding on anything edible that they can find. They rarely venture far out to sea, but their range extends a long way inland, where they are often associated with humans—following tractors to eat earthworms turned up by the plow, or wheeling noisily over garbage dumps. Herring gulls nest on the ground and on rooftops, typically laying three eggs. They can be highly aggressive if their nests are disturbed. It is replaced by very similar species in North America, south and east Europe, and Asia.

SCAVENGING

Scavenged food forms a large part of the herring gull's diet, both on land and at sea. This gull has benefited from urban expansion and the growth in fishing, both of which generate a large supply of edible waste. Herring gulls may cause problems at inland garbage dumps by picking up waste and carrying it away.

FOOD OVERBOARD
Large numbers of herring gulls follow fishing ships operating close to coasts. Unlike pelagic birds, they usually return to land at night.

ORDER CHARADRIIFORMES

Black-legged Kittiwake

Rissa tridactyla

LENGTH 15½–18 in (39–46 cm)

WEIGHT 11–18 oz (300–500 g)

HABITAT Rocky coasts, inshore waters, open sea

DISTRIBUTION Northern hemisphere; breeds north to Svalbard and Greenland

Kittiwakes get their name from their call—a loud, three-syllable shriek that echoes around their nesting colonies on northern coasts. A medium-sized, gray-backed gull, the black-legged kittiwake breeds on narrow cliff ledges but spends the rest of the year wandering far out to sea. It feeds mainly on small fish, and often follows fishing vessels. Unlike most gulls, however, it rarely shows any interest in scavenging food on land. Black-legged kittiwakes have evolved several adaptations for breeding on bare rock. Their feet

have longer claws than those of most other gulls, and they build cup-shaped nests out of seaweed and mud, which help to keep their eggs secure. Both parents help to incubate the eggs and feed the young, and the adults' recognition calls can make a deafening noise when several hundred pairs nest close together. After breeding, these birds disperse away from the coast, traveling as far south as tropics off West Africa. They are monogamous, with pairs meeting up again at the same nesting site after spending up to eight months apart.

ORDER CHARADRIIFORMES

Laughing Gull

Leucophaeus atricilla

LENGTH 15–17 in (38–43 cm)

WEIGHT 11–18 oz (300–500 g)

HABITAT Coasts, inshore waters

DISTRIBUTION North America, Caribbean, Central America (breeding); N. South America (nonbreeding)

A widespread summer visitor to North American coasts, the laughing gull rarely wanders far inland. It feeds mainly by scavenging and often follows ferries and fishing boats. Bold and self-confident, it is a familiar sight to picnickers on beaches, where it pushes larger gulls aside in the competition to get at food. It nests in large colonies on coasts. Like many dark-headed gulls, it loses its black cap during the nonbreeding season, when its head turns a dull white.

ORDER CHARADRIIFORMES

Ivory Gull

Pagophila eburnea

LENGTH 16–18 in (40–46 cm)

WEIGHT 1–1¼ lb (450–600 g)

HABITAT Coasts, open sea, sea ice

DISTRIBUTION Arctic Ocean, north Atlantic, wintering in south of range

Completely white, apart from its yellow-tipped bill, black eyes, and black feet, the ivory gull is the world's most northerly breeding bird. With its buoyant flight and pigeonlike walk, it ranges across open water and sea ice, and it can be found almost anywhere over the Arctic Ocean. It feeds largely by scavenging and is quickly attracted to the carcasses of dead seals and whales. The ivory gull is currently undergoing a steep decline. The reasons for this are unclear.

ORDER CHARADRIIFORMES

Brown Noddy

Anous stolidus

LENGTH 16–18 in (40–45 cm)

WEIGHT 7–9 oz (200–250 g)

HABITAT Open sea, inshore, oceanic islands

DISTRIBUTION Worldwide in tropical waters; present on some islands year-round

Noddies are dark, tropical terns that often feed far out to sea. There are three species of noddies and the brown noddy is the largest and most widespread. Brownish black all over, apart from a paler crown, it has slender wings, a long, sharp bill, and small, jet-black legs. Brown noddies feed mainly on fish and squid, hovering and then plunging in the same way as terns. They nest on islands throughout the tropics, making nests from twigs and seaweed in trees or on the ground.

ORDER CHARADRIIFORMES

Caspian Tern

Hydroprogne caspia

LENGTH 19–23 in (48–59 cm)

WEIGHT 1¼–1¾ lb (550–750 g)

HABITAT Coasts, lakes, reservoirs, gravel pits

DISTRIBUTION North America, Eurasia, Africa, Australia (breeding); northern South America, Southeast Asia (nonbreeding)

Despite its name, this large, black-crested tern has a global distribution. Gray-backed, with a large, dark red bill, it has a black cap that is darkest when it breeds. It plunge-dives for food in shallow water, and nests in colonies, laying its eggs directly on gravel or mud.

ORDER CHARADRIIFORMES

White Tern

Gygis alba

LENGTH 11–13 in (28–33 cm)

WEIGHT 3½–4½ oz (100–125 g)

HABITAT Open sea, inshore, oceanic islands

DISTRIBUTION Tropical waters worldwide

Also known as the fairy tern, this delicate and graceful bird wanders far out over tropical oceans, where it is known for its habit of fluttering close to boats. Slim and lightly built, with

black eyes and a straight black bill, it is the only tern whose plumage is entirely white. It spends most of its time flying a few yards above the surface, periodically dropping down in order to catch small fish and squid. Unlike most terns, it is a solitary breeder, nesting on widely scattered islands. It lays its single egg on a rocky ledge, or in a slight hollow in a sloping branch. The parents take turns cradling the egg throughout its five-week incubation period—an unusually long time for an egg of its size. The chick emerges with strong feet and claws for clinging to its nesting site.

ORDER CHARADRIIFORMES

Inca Tern

Larosterna inca

LENGTH 16–17 in (40–42 cm)

WEIGHT 6–8 oz (175–225 g)

HABITAT Coasts and inshore waters

DISTRIBUTION Pacific coast of South America from Ecuador to central Chile

With its curling white "mustache" plumes, this South American tern is easy to identify. It feeds in the cold, nutrient-rich waters of the Humboldt Current, dipping down to the surface to catch small fish. Inca terns often follow sea lions and whales, preying on shoals of fish as they try to escape the larger predators. They nest among rocks or in abandoned burrows.

ORDER CHARADRIIFORMES

Black Skimmer
Rynchops niger

LENGTH	16–20 in (40–50 cm)
WEIGHT	9–14 oz (250–400 g)
HABITAT	Estuaries, lagoons, lakes, coasts

DISTRIBUTION Pacific and Atlantic coasts of North, Central, and South America, north to Massachusetts

Similar to terns in overall shape, skimmers have remarkable and highly distinctive bills. The lower part, or mandible, of the bill is at least a third longer than the upper part and is laterally compressed, giving it a shape like a scissor blade. When feeding, a skimmer flies low over calm water with its lower mandible slicing through the surface. If the mandible touches food, the skimmer snaps its bill shut, flicking its catch into its mouth. The black skimmer is one of three species of skimmers, all of which are dark above, with white underparts. Like its relatives, it often feeds at dawn and dusk, and it will also feed during the night if the moonlight is bright enough. It lives in small flocks and nests on beaches and sand spits, laying its eggs in an unlined hollow on the ground. It is migratory in the far north and south of its range.

ORDER CHARADRIIFORMES

Arctic Skua
Stercorarius parasiticus

LENGTH	18–26 in (46–65 cm)
WEIGHT	14–21 oz (400–600 g)
HABITAT	Coasts, tundra, moorland, open sea

DISTRIBUTION Northern waters (breeding); throughout Southern Hemisphere (nonbreeding)

This slender-winged sea bird, also called the parasitic or Arctic jaeger, is exceptionally fast and maneuverable in the air—a skill that is central to the way it feeds. There are several color forms, which differ in their proportion of brown and white, but all Arctic skuas have streamers that give their tails a sharp central point. This species catches fish, but it is better known as a kleptoparasite, which steals food from other birds. It swoops down on gulls and terns as they return from the sea, chasing them and often gripping their tail feathers with its bill. Its victims react by disgorging food, which the skua deftly intercepts in midair. Arctic skuas also hunt small land animals and steal eggs and chicks from nests. They nest on the ground and winter at sea.

ORDER CHARADRIIFORMES

Great Skua
Stercorarius skua

LENGTH	20–26 in (51–66 cm)
WEIGHT	2³⁄₄–3¹⁄₂ lb (1.2–1.6 kg)
HABITAT	Coasts, inshore waters, open sea

DISTRIBUTION North Atlantic (breeding), dispersing south to equator (nonbreeding)

Powerfully built, with short, broad wings, the great skua is shaped like an unusually thickset gull, but it has mottled, dark brown plumage that changes only slightly as it matures. It is a rapacious predator, eating fish, small mammals, and also other birds, as well as raiding nests for eggs and chicks. Normally slow and ponderous in the air, it becomes swift and agile when it hunts, and chases birds as large as gannets to force them to regurgitate their food, which it then eats. The great skua nests on the ground and spends the rest of the year at sea.

ORDER CHARADRIIFORMES

Common Murre
Uria aalge

LENGTH	15¹⁄₂–16¹⁄₂ in (39–42 cm)
WEIGHT	1³⁄₄–2¹⁄₂ lb (850 g–1.1 kg)
HABITAT	Inshore waters, rocky coasts, open sea

DISTRIBUTION North Atlantic, north Pacific

Conspicuously marked in brownish black and gleaming white, the common murre spends most of the year at sea. It dives for fish from the surface, swimming underwater using its wings. In spring, common murres crowd together on narrow cliff ledges, where each female lays a single egg directly on to the rock. When the chick is fully grown, the male parent escorts it into the sea.

ADAPTED EGGS

Guillemot eggs are pointed at one end, adding stability and allowing them to be incubated with less likelihood that they will move around on a flat or sloping bare ledge. Their color varies greatly, and their irregular surface markings of dark blotches and intricate scribbling may aid identification by the parents.

markings unique to each egg

Atlantic Puffin

Fratercula arctica

LENGTH 11–12 in (28–30 cm)	
WEIGHT 14 oz (400 g)	
HABITAT Inshore waters, rocky coasts, open sea	

DISTRIBUTION North Atlantic, breeding north to Greenland and Svalbard

With its vividly marked bill, bright red feet, and red-and-black eye patches, this is the most colorful sea bird in the north Atlantic. Like other members of the auk family, it feeds by pursuing fish underwater, using its strong, stubby wings to swim. In the air, it flies rapidly on fast-beating wings, skimming over the waves as it returns to its nest with food. Atlantic puffins breed in large clifftop colonies, digging burrows in coastal turf. The parents take turns incubating the single egg, and they both help to feed the developing nestling. Instead of regurgitating food, as most sea birds do, they return with small fish held in their bills, carrying about six fish simultaneously, arranged alternately head to tail. Each nestling is fed continuously for about six weeks, after which the parents abandon it and head out to sea. After going without food for several days, the young bird crawls out of the burrow and flutters down to the sea after dark. Puffins disperse out to sea in fall, when they lose the bright bill colors that make them so conspicuous during the summer months.

HUMAN IMPACT

COMPETING FOR FOOD

The puffin population has fallen sharply of late, especially in the eastern Atlantic. This may be due to the growing fishery for sand eels, a fish that puffins rely on, especially in breeding season. Sand eels are used in fertilizers, animal foods, and as a source of edible oil.

UNFAIR SHARES
A catch of sand eels is brought aboard a boat. These finger-shaped fish, unrelated to true eels, are an important food for some fish and sea birds.

Least Auklet

Aethia pusilla

LENGTH 6 in (15 cm)	
WEIGHT 3 oz (85 g)	
HABITAT Inshore waters, rocky coasts, open sea	

DISTRIBUTION North Pacific, breeding mainly in the Aleutian Islands and islands in Bering Sea

Crested Auklet

Aethia cristatella

LENGTH 9½–10½ in (24–27 cm)	
WEIGHT 9 oz (250 g)	
HABITAT Inshore waters, rocky coasts, open sea	

DISTRIBUTION North Pacific, breeding mainly in the Aleutian Islands and islands in Bering Sea

The north Pacific is home to more species of auks than anywhere else. The crested auklet is a typical example, with a compact body, sooty-gray plumage, and a feathery crest that curves forward from its forehead over its orange-red bill. Like other auks, it flies low on rapidly whirring wings and feeds in flocks so dense that they resemble swarms of insects wheeling over the water. Crested auklets breed among fallen rocks on island coasts, in colonies containing thousands of birds. Their courtship displays are energetic and noisy, as they throw back their heads and make loud grunts and trumpeting sounds. When the breeding season is over, they disperse out to sea and spend the winter as far south as Japan.

This tiny bird is probably the most abundant species of auk, a family that also includes murres and puffins. Short, plump, and gray-backed, with a stubby red-tipped bill, it nests in vast colonies off the Alaskan coast, some of which contain more than a million birds. Least auklets also feed together, floating on the surface in large gatherings known as "rafts." They are pursuit divers that eat mainly zooplankton.

Pied Kingfisher

Ceryle rudis

LENGTH 10 in (25 cm)	
WEIGHT 3¼ oz (90 g)	
HABITAT Coasts, lagoons, estuaries, rivers, marshes	

DISTRIBUTION Africa, Middle East, south Asia

This boldly patterned, black-and-white bird is the only kingfisher that regularly fishes offshore. Instead of watching for prey from a perch, as many other kingfisher species do, it flies rapidly above the surface with its head facing down as it scans the water below. If it spots food, it hovers on the spot, and then dives down to make a catch. It can also eat while in flight, another unique adaptation. Male and female pied kingfishers look similar, although the female has a double breast band compared to the male's single band. Pairs nest in burrows in sandy banks and are often helped by the previous year's young to collect food for the nestlings. The adults have a loud, high-pitched call, which may be heard as they speed past.

Collared Kingfisher

Todirhamphus chloris

LENGTH 11 in (28 cm)	
WEIGHT 4½ oz (120 g)	
HABITAT Forests, coasts, beaches, mangrove swamps, estuaries	

DISTRIBUTION Red Sea, Persian Gulf, Southeast Asia, Australasia

Also known as the mangrove kingfisher, this bird lives in a variety of habitats, although in Australia it is restricted to the coast. Greenish blue above, with a white belly and collar, it has a black eye-stripe and a sharply pointed bill. On coasts, it hunts crabs as well as fish and, like all kingfishers except the pied (see above), beats its prey against a perch before swallowing it. It often nests in hollows in mangrove trees, and lays three or four eggs. In the far south of its range, this bird is a summer visitor only.

Mammals

DOMAIN	Eucarya
KINGDOM	Animalia
PHYLUM	Chordata
CLASS	Mammalia
ORDERS	27
SPECIES	About 5,500

ONLY A SMALL MINORITY OF THE WORLD'S mammals live in seawater, but taken together, they show an extraordinary range of shapes, sizes, and lifestyles. They include cetaceans (whales and dolphins), sirenians (manatees and dugongs), and carnivores, particularly the pinniped carnivores (seals, sea lions, and walruses). All marine mammals breathe air, like their terrestrial counterparts, and they give birth to live young, either in the sea or onshore. Many species are migratory, with a sophisticated navigational sense.

Anatomy and Physiology

Marine mammals have many adaptations for life at sea, not only in their anatomy, but also in their physiology, regulating how their bodies work. Cetaceans and sirenians have lost all visible traces of hind limbs (not including the pelvic girdle); instead, they propel themselves with their tail flippers or flukes, which beat up and down. Fur seals and sea lions swim with their front flippers, while true seals use their rear flippers, bringing them together like a pair of hands. Despite needing to breathe air, many marine mammals are superb divers. Some, such as the elephant seal, can reach depths in excess of 6,600 ft (2,000 m) and stay underwater for up to two hours. When they dive, their heart rate drops, and blood flow is modified so that vital organs receive enough oxygen until they resurface. Instead of breathing in before they dive, the deepest divers often exhale. This helps them to avoid decompression sickness, or the "bends."

DIVING MAMMAL
When a Harbor Seal dives, its heart rate falls below 10 beats a minute. Blood diverted from its muscles and digestive system flows to its heart and brain.

heart beats rapidly after surfacing
heart rate drops as seal dives
rate remains low throughout dive

SHARED PATTERNS
A sea lion's front flipper has the same arrangement of bones as a human arm. The "arm" bones are short and sturdy, helping to bear the animal's bulk on land. Long finger bones make up the flipper's blade.

humerus
phalange
radius
metacarpal
ulna
scapula

FLIPPERS AND FLUKES
A humpback whale's flippers contain bones, and beat like a pair of wings. Its flukes, or tail fins, are made of rubbery tissue, and contain no bones at all.

INSULATING BLUBBER
Compared to air, seawater drains much more heat from mammals' bodies. To keep warm, many polar species, such as this walrus, have a thick layer of insulating fat, called blubber, under the skin.

blowhole
sonic lips (source of sound)
outgoing clicks (to prey)
melon
incoming (reflected) clicks
ear drum
sound channel in jaw

USING ECHOLOCATION
Dolphins and toothed whales use pulses of high-pitched sound to locate prey. The forehead contains an oil-filled organ called the melon, which is thought to function as an "acoustic lens" to focus outgoing sound.

VARIED DIET
Penguins are just one item on the leopard seal's menu. Despite its reputation for ferocity, at least half of its diet consists of krill, which it filters with its cheek teeth.

Feeding

Apart from plant-eating manatees and dugongs, most marine mammals are exclusively carnivorous. In open water, many pursue individual prey, tracking it by sight or by echolocation. Some seals have a twin strategy. They catch prey individually, but they can also filter out planktonic animals in bulk, using complex cheek teeth that interlock to form a sieve. This efficient feeding method reaches extremes in the baleen whales, which cruise through shoals of fish or krill, often swallowing over 220 lb (100 kg) of food at a time. Not all marine mammals catch moving prey. Sea otters dive to collect clams, mussels, and sea urchins, while walruses and gray whales suck mollusks out of seabed sediment.

Breeding

Marine mammals typically produce a single young each time they breed. Cetaceans and sirenians give birth in water, as do sea otters, but all other marine mammals have to return to land. In species with a harem system, such as fur seals and elephant seals, fighting between rival males for control of mates can be ferocious. After mating, the females of most marine mammals raise their young on their own. For their size, true seals develop fastest, some being weaned in as little as four days. At the other end of the spectrum, a dolphin calf may suckle for over 20 months—the start of a mother–calf bond that can last for six years.

SEA OTTER PUP
A young sea otter rides on its mother's chest, while she floats in calm water. The pup depends on her for at least five months.

HUMAN IMPACT

THREATS AND CONSERVATION

Historically, marine mammals have been heavily exploited for food, oil, and fur, bringing some species close to extinction. Whales and seals are the primary targets. In 1986, the International Whaling Commission agreed on a moratorium on all commercial whaling. Despite dissent, this ban remains in force. Seals continue to be hunted, or culled, to control populations, but the rarest species are protected by international agreements.

ENGRAVED WHALE TOOTH
The art of scrimshaw, or engraving on whale teeth and walrus tusks, was popular among whalers during the 17th and 18th centuries. Whalebone carving still takes place in areas where small-scale native whaling is permitted.

COLONY BREEDING
Many seals and sea lions, such as these South American sea lions, are highly sociable in the breeding season, forming large colonies on beaches to mate and have their pups.

MARINE MAMMAL CLASSIFICATION

Two orders of mammals—the cetaceans and sirenians—are wholly marine. Seals and sea lions are also aquatic, but like other members of the carnivore order, they give birth on land. Several other carnivore species feed at sea, but of these only the sea otter is entirely marine.

CARNIVORES
Order Carnivora

302 species
Most carnivores are terrestrial, but a few spend some of their lives in the sea. The polar bear is equally at home on dry land, on sea ice, and in salt water. Seven species of otter often enter salt water, but the sea otter is the only one to spend all of its time offshore. The most fully aquatic carnivores are the 34 species of pinnipeds, until recently classified in their own order, the Pinnipedia. They are split into three families. One family comprises the sea lions and fur seals, which have external ears, use their forelimbs for propulsion, and use all four flippers to move on land. The second family is composed of the true seals, which lack external ears, use hind limbs for propulsion and are less mobile on land. The final family contains only the walrus, which has very wrinkled skin and long tusks.

CETACEANS
Order Cetartiodactyla

85 species
Cetaceans are divided into two suborders. The 13 baleen whales lack teeth, and filter food from the water using a fibrous material called baleen. The 72 toothed whales are predators that hunt individual prey. Cetaceans give birth at sea, and are helpless if stranded on land.

SIRENIANS
Order Sirenia

4 species
Living mainly in the tropics, sirenians, or sea cows, are barrel-shaped vegetarians that live in salt and fresh water. They include the dugong and three species of manatees, such as the Caribbean manatee, below. Slow-moving and thick-skinned, sirenians have broad muzzles, paddlelike front flippers, and a broad, horizontally flattened tail.

thick fur over
layer of
blubber for
insulation

large paws,
furred on
both sides

ORDER CARNIVORA

Polar Bear

Ursus maritimus

LENGTH	Up to 9 ft (2.8 m)
WEIGHT	Females up to 550 lb (250 kg); males up to 1,750 lb (800 kg)
HABITAT	Arctic tundra, pack ice, open sea

DISTRIBUTION Circumpolar in the Arctic, southward as far as Newfoundland and the Pribilof Islands

Icon of the Arctic, the polar bear is the largest mammalian carnivore and has incomparable stamina, resilience, and power. Its body is streamlined, the head grading almost imperceptibly into a long, powerful neck. Its huge paws may be over 12 in (30 cm) wide and are furred on their undersides, providing grip while retaining body heat. Its hearing and sense of smell are acute: it can hear prey that is under 3 ft (1 m) or more of ice and can smell carrion 3 miles (5 km) away. Polar bears spend most of the year at sea, roaming the drifting pack ice and swimming across open areas. Naturally buoyant, they can swim for hours, although they hunt mainly on the ice.

The main prey of polar bears is seals, often caught at breathing holes. They also eat sea birds and fish, and the corpses of beached whales are a favorite food. During the summer, many of them live on land and eat a wider range of food, from reindeer to berries. Females give birth in winter, suckling their cubs in a den dug in the snow. For centuries, the polar bear has been hunted by native peoples of the Arctic, without its numbers declining. However, climate change is reducing sea ice and, as a result, bear numbers are falling in some areas.

ORDER CARNIVORA

Sea Otter

Enhydra lutris

LENGTH	2¼–5¼ ft (0.7–1.6 m) including tail
WEIGHT	33–100 lb (15–45 kg)
HABITAT	Inshore waters along rocky coasts

DISTRIBUTION North Pacific from Japan to Alaska and California

Unlike other otters, the sea otter is able to spend its whole life in the ocean. It has a blunt head, a stocky body, webbed rear feet, and small front paws with sharp claws. It uses these to gather food and pick up large stones. At the surface, it floats on its back, using a stone that rests on its chest as an anvil to smash open its prey. Sea otters feed on mollusks, sea urchins, and crabs. While they can dive to 320 ft (97 m), they rarely venture more than ½ mile (1 km) from the shore.

SLEEPING SECURELY

Sea otters often sleep in beds of giant kelp, using the seaweed to keep from drifting away. Their fur is the densest of any mammal. The hairs are packed so tightly that they prevent water penetration, ensuring that the otter's skin never gets wet. This is a vital adaptation, because sea otters live in cold water and do not have insulating fat.

ORDER CARNIVORA

Marine Otter

Lontra felina

LENGTH	Up to 4 ft (1.2 m) including tail
WEIGHT	7–13 lb (3–6 kg)
HABITAT	Exposed rocky shores

DISTRIBUTION Pacific coast of South America from Peru to Cape Horn

This lithe predator lives on some of the world's stormiest coastlines, particularly in the remote southern part of its range. The marine otter's closest relatives live mainly in fresh water, but it spends almost all its time in the sea. Like typical river otters, this coast-dwelling otter has short brownish yellow fur, webbed toes on all four feet, and sensitive whiskers that help it to find prey. It fishes along rocky coasts in the rich, cool waters of the Humboldt Current, and instead of making burrows, it shelters in sea caves just above the level of the highest tides.

Marine otters have long been hunted for their pelts, and current estimates of the population are less than 1,000 animals. The species is now protected, but preservation of its habitat may be equally important in guaranteeing its long-term survival.

ORDER CARNIVORA

European Otter

Lutra lutra

LENGTH	2¾–4¼ ft (83–132 cm) including tail
WEIGHT	11–31 lb (5–14 kg)
HABITAT	Rivers, lakes, estuaries, rocky coasts

DISTRIBUTION Temperate and tropical Eurasia, south to Indonesia

Once widespread throughout Europe and Asia, the European otter has been badly affected by pollution and habitat change and by being hunted for its fur. It has a streamlined body, short but dense coat, and webbing on all four paws, and is extraordinarily agile underwater, twisting and turning to catch fish. Inland, European otters are largely nocturnal, spending the daytime in their dens, or holts. Those that live on the coast, however, can often be seen during the day.

ORDER CARNIVORA

Antarctic Fur Seal
Arctocephalus gazella

LENGTH	4–6½ ft (1.2–2 m)
WEIGHT	90–420 lb (40–190 kg)
HABITAT	Rocky coasts, open sea in polar waters

DISTRIBUTION Southern Ocean

Ranging further south than any other fur seal, this polar species feeds on fish, squid, and krill in the icy waters off Antarctica. In spring it comes ashore after spending winter at sea. Males are up to three times heavier than females, with an imposing mane and thickened neck that gives them a front-heavy appearance. This species breeds on islands, such as South Georgia and Kerguelen, and is rising in number. This may be a side effect of the whaling industry, which has reduced competition for krill.

ORDER CARNIVORA

Northern Fur Seal
Callorhinus ursinus

LENGTH	4½–7 ft (1.4–2.1 m)
WEIGHT	90–600 lb (40–270 kg)
HABITAT	Coasts and sea in cold-water regions

DISTRIBUTION North Pacific, Bering Sea

Of the nine species of fur seals, most live in the Southern Hemisphere; this is the only northern species that exists in significant numbers. Like other fur seals, it has a thick, dark coat, external ears, and long front flippers that it uses for swimming and for moving around on land. Males can be five times heavier than females, but both sexes have short muzzles, giving them a characteristic snub-nosed look. Their large eyes allow them to see at night, which is when they do most of their feeding, as their prey is closer to the surface. They feed mainly on fish, but also on squid, and migrate far out into the Pacific after they breed. Most northern fur seals breed on islands in the Bering Sea. Decimated by commercial hunters from the mid-1700s onward, they are now protected by hunting controls.

ORDER CARNIVORA

South American Fur Seal
Arctocephalus australis

LENGTH	4½–6¼ ft (1.4–1.9 m)
WEIGHT	90–440 lb (40–200 kg)
HABITAT	Coasts and sea in cold-water regions

DISTRIBUTION Pacific and Atlantic coasts of southern South America, Falkland Islands

Once found along the entire length of South America's southern coasts, this fur seal now breeds on offshore islands, where it faces less disturbance from humans. It is blackish gray, with paler undersides in females, and is agile on land, using its flippers to climb steep rocks. Males may be about three times the weight of females. This species feeds mainly at night, hunting fish, squid, lobsters, and mollusks, and is itself hunted by sharks and killer whales.

ORDER CARNIVORA

California Sea Lion
Zalophus californianus

LENGTH	6½–7¾ ft (2–2.4 m)
WEIGHT	240–880 lb (110–400 kg)
HABITAT	Rocky coasts and open sea

DISTRIBUTION Pacific coast of the US, Galapagos Islands

Famed for its acrobatic antics in marine aquariums, the California sea lion is just as agile in the wild. Its sleek body is covered with short fur, which ranges in color from brownish black in males to light brown in females and young; mature males may be more than three times as heavy as females and have a distinctive bony hump on their heads. They feed on fish and squid.

ORDER CARNIVORA

Walrus
Odobenus rosmarus

LENGTH	8¼–11½ ft (2.5–3.5 m)
WEIGHT	1,750–4,000 lb (800–1,800 kg)
HABITAT	Coasts and shallow open water

DISTRIBUTION Arctic Ocean, Bering Sea, Hudson Bay

Instantly recognizable by its tusks, the walrus is the second-largest pinniped after the elephant seals. Its skin is unlike any other mammal's, with deep creases and wrinkles, but very little hair. Its skin color varies from dark gray brown to pink, depending on the amount of blood flowing into it. Beneath the skin is a thick layer of fat, or blubber, which keeps their bodies warm. Walruses feed on marine invertebrates, which they find in the seabed sediment at depths of up to 165 ft (50 m). They locate their food mainly by touch, using stiff whiskers that resemble a mustache. It was once thought that they used their tusks to dredge up food, but it is now known that they uncover it by squirting water with their mouths. Once their prey has been uncovered, they separate the soft parts from the shells. It is unclear how they do this, but their technique involves more suction than crushing, because intact shells are often found around their breathing holes.

Females give birth to a single calf after a 15-month gestation (including delayed implantation), and they may not breed again for another three years. Walruses are highly gregarious, making them easy prey for hunters. They have been hunted by indigenous peoples for at least 15,000 years, both for food and for their hides.

AUSTRALIAN SEA LIONS
Foraging in the cold waters of the Great Australian Bight, Australian sea lions feed on squid and cuttlefish, as well as fish. Like other sea lions, they use their front flippers to swim. On land, their rear flippers swivel forward, allowing them to move on all fours as fast as a human can run.

OCEAN LIFE

ORDER CARNIVORA

Common Seal

Phoca vitulina

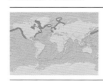

LENGTH	4–6¼ ft (1.2–1.9 m)
WEIGHT	175–310 lb (80–140 kg)
HABITAT	Inshore waters, estuaries, rivers

DISTRIBUTION North Pacific and north Atlantic, reaching as far south as Baja California

Also known as the harbor seal, this species has the widest distribution of any seal and the widest variety of markings. Its background color ranges from pale gray to brown, with dark spots and rings and sometimes a dark stripe along the back. It has a smoothly domed head and a doglike muzzle. It feeds primarily on fish, often catching them in shallow water

clawed front flipper

Harp Seal

ORDER CARNIVORA

Pagophilus groenlandicus

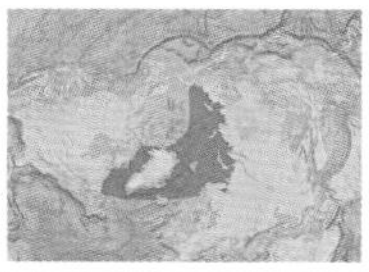

LENGTH	5¼–5½ ft (1.6–1.7 m)
WEIGHT	265–300 lb (120–135 kg)
HABITAT	Polar waters

DISTRIBUTION North Atlantic and adjoining regions of the Arctic Ocean, extending eastward to Siberia

One of the most common seals in the far north, the harp seal is born with an exceptionally luxurious coat of long white fur, which camouflages the pups as they lie on sea ice. Adult harp seals are silvery-gray with a mottled pattern

close to the shore. It dives for up to five minutes, but rarely to any great depth. The common seal spends much of its time on rock flats and sandbanks, and it is here that the females give birth. Most pups shed their soft natal coat before they are born, starting life with a dark version of the adult coat, unlike the pups of some other seals. Although they can swim almost immediately, they often use their front flippers to ride on their mother's back. They are weaned at about four weeks. True to their name, common seals are still abundant, but in the North Sea they have been adversely affected by pollution, and also by a highly infectious viral disease that broke out in the late 1980s.

of dark patches, which become more prominent as they age. They feed mainly on fish and shrimp, living on the southern edge of the Arctic pack ice, and resting on it when they molt. In early spring, adult females give birth to a single pup each, which they wean after just 12 days. At this point, the pup gradually sheds its white coat and takes up life in the sea. For many decades, the pups have been the subject of a controversial hunt, which supplies their pelts to the fur trade. Despite campaigns by conservationists, over 250,000 pups are still culled every year. Harp seals are also hunted by sharks, polar bears, and killer whales.

ORDER CARNIVORA

Ringed Seal

Pusa hispida

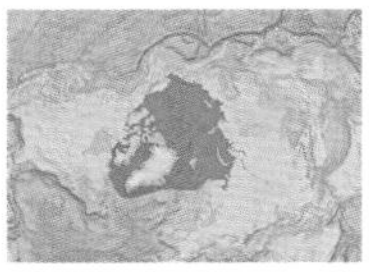

LENGTH	3½–5¼ ft (1.1–1.6 m)
WEIGHT	100–210 lb (45–95 kg)
HABITAT	Polar waters around sea ice

DISTRIBUTION Arctic Ocean, north Pacific, north Atlantic, Baltic Sea, Sea of Okhotsk

Named after its conspicuous circular markings, the ringed seal is found throughout the Arctic, in open water near sea ice and also under the ice

ORDER CARNIVORA

Gray Seal

Halichoerus grypus

LENGTH	5½–7½ ft (1.7–2.3 m)
WEIGHT	220–685 lb (100–310 kg)
HABITAT	Rocky coasts, offshore islands

DISTRIBUTION Discontinuous populations in northwest Atlantic, Iceland, British Isles, Baltic Sea

The gray seal has a distinctive convex muzzle, which gives it a "Roman-nosed" appearance. Adults vary in color: males are usually gray overall, with pale patches on their undersides, while females often have a marbled pattern of dark patches over a much lighter background. Males may be two or three times heavier than females—a difference exceeded by few other true seals. When not hunting for their usual diet of fish, gray seals spend

ORDER CARNIVORA

Mediterranean Monk Seal

Monachus monachus

LENGTH	8¼–9 ft (2.5–2.7 m)
WEIGHT	550–660 lb (250–300 kg)
HABITAT	Rocky coasts in warm-water regions

DISTRIBUTION Coasts of Greece, west Turkey, northeast Morocco, and northwest Algeria

itself, where it digs breathing holes. It can dive for up to 45 minutes, feeding on fish and zooplankton. Female ringed seals breed on the ice, where they dig dens in the snow. These seals are a favorite prey of polar bears, which hunt them in their dens and when they surface to breathe.

their time either resting on rocks or "bottling"—sleeping in the water with their bodies vertical and their nostrils just above the surface. They breed onshore, hauling themselves out onto beaches or grass farther inland. Their pups have a white natal coat, and they stay onshore for two to three months before venturing into the sea.

Both of the two species of *Monachus* seals are endangered. Their coats vary from dark brown to light tan. Males are slightly larger than females, and the pups, unusually for seals, are born with black fur and a small but variably sized white area on their abdomens. This seal was once common, but centuries of hunting and disturbance have reduced its population to a few hundred. Most exist in the Mediterranean, but the largest colony is on the Atlantic coast of Morocco. Its closest living relative is the rare Hawaiian monk seal.

ORDER CARNIVORA

Northern Elephant Seal

Mirounga angustirostris

LENGTH	7¼–13 ft (2.2–4 m)
WEIGHT	660–5,500 lb (300–2,500 kg)
HABITAT	Islands in deep water off rocky coasts

DISTRIBUTION Pacific coast of North America, from Bering Sea to Baja California

Male elephant seals are the largest of all pinnipeds, and the colossal males dwarf the females. There are two species, one in each hemisphere. They are very similar in appearance and have similar life histories. The northern elephant seal is gray or brown, with no obvious markings. The male has a huge, muscular neck, powerful jaws, and an inflatable proboscis resembling a shortened trunk. Both sexes have a layer of insulating blubber and a short, stiff coat, without any soft underfur. They are superb divers: the northern species has been tracked to depths of over 1 mile (1.6 km). They eat squid and deep-water fish, although it is still not clear exactly how they find their prey.

ORDER CARNIVORA

Weddell Seal

Leptonychotes weddellii

LENGTH	9¼–10¾ ft (2.8–3.3 m)
WEIGHT	880–1,300 lb (400–600 kg)
HABITAT	Polar waters around sea ice

DISTRIBUTION Southern Ocean, extending northward to South Georgia

The Weddell seal is found in the waters around the entire coast of Antarctica. It is the world's most southerly ranging marine mammal and lives permanently on the continent. Its head looks small in proportion to its body, and it has a short, dense coat of bluish black fur, with light streaks on the sides. It feeds mainly on fish, diving to depths of over 2,000 ft (600 m), and is able to stay underwater for over an hour. Weddell seals are so well adapted to life in cold water that they bask on ice in preference to bare ground. They breed on ice, and their winter survival depends on keeping open their breathing holes. They gouge these out with their canine teeth, starting when the ice is thin, and maintaining them as the ice thickens, to depths of up to 6½ ft (2 m).

ORDER CARNIVORA

Crabeater Seal

Lobodon carcophaga

LENGTH	6½–7¾ ft (2–2.4 m)
WEIGHT	440–660 lb (200–300 kg)
HABITAT	Polar waters around sea ice

DISTRIBUTION Southern Ocean and adjoining regions north of the Antarctic Convergence

Despite its name, this seal feeds only on krill and other planktonic animals. It filters water using its strange molar teeth, which have elongated cusps that look like a set of stubby fingers. When its jaws close, the cusps act like a sieve, letting water out but keeping food in. Crabeater seals have slender bodies, with fur that may be light or dark brown and darker flippers. They live close to pack ice and breed on it, and they are extremely nimble on land. Their mummified remains have been found over 30 miles (50 km) inland in Antarctica's Dry Valleys. Their total population is thought to be 10–20 million, making them more numerous than all other seal species combined.

ORDER CARNIVORA

Leopard Seal

Hydrurga leptonyx

LENGTH	7¾–11 ft (2.4–3.4 m)
WEIGHT	440–1,300 lb (200–590 kg)
HABITAT	Polar waters, rocky coasts

DISTRIBUTION Southern Ocean and adjoining regions north of the Antarctic Convergence

With its long muzzle and sharply constricted neck, this solitary predator looks very different from other seal species found off Antarctica. Unlike most true seals, it propels itself forward through the water with its front flippers rather than its rear ones— a characteristic that it shares with fur seals and sea lions. Its body is black or dark gray with a silvery underside, marked with darker flecks and spots. Its jaws are exceptionally powerful, with an unusually wide gape, and they are armed with long incisors and canine teeth, as well as elaborate cheek teeth that can strain food from the water. About half of the leopard seal's diet consists of krill, but the remainder is made up of much larger animals that it hunts individually. For example, leopard seals are adept at catching penguins as they enter the water, throwing them into the air to rip the skin and feathers from their bodies. They also prey on squid, fish, and other seals. Females give birth to a single pup each year, weaning it at the age of four weeks.

OCEAN LIFE

ORDER CETACEA

North Atlantic Right Whale

Eubalaena glacialis

LENGTH	50–54 ft (15–16.5 m)
WEIGHT	Up to 77 tons (70 metric tons)
HABITAT	Temperate and subpolar waters

DISTRIBUTION Northwestern Atlantic and vestigial populations in northeastern Atlantic

The North Atlantic right whale was one of the first whales to be hunted commercially and is now one of the most critically endangered species, with a total population of 300–500 individuals. A deep bluish black, apart from white markings on its belly, it has a deeply arched mouth, with a lower jaw shaped like a gigantic scoop. Its head its covered with distinctive areas of hard pale skin, known as callosities, which scientists use to identify individuals. Like all baleen whales, it feeds by filtering food from seawater, using brushlike strips of baleen that hang from its upper jaw. North Atlantic and North Pacific right whales both feed at high latitudes but migrate to warmer waters to breed. An almost identical species, the southern right whale, is found in the Southern Hemisphere. Unlike its northern counterparts, its population is increasing and is estimated to be about 10,000.

HUMAN IMPACT

WHALING

Commercial whaling has exploited many species. The North Atlantic right whale was the first to be seriously affected. This whale was decimated by Basque whalers who then expanded operations to Canada in the 1500s. The sperm whale was the quarry of American whalers in the Pacific from the 1780s. Modern whaling, targeting species such as the blue whale, expanded rapidly in the 20th century, using factory ships and explosive harpoons.

WHALING STATION
Hauled onto a ship in the Southern Ocean, a whale is flensed, or stripped of its blubber and flesh.

ORDER CETACEA

Bowhead Whale

Balaena mysticetus

LENGTH	45–66 ft (14–20 m)
WEIGHT	Over 110 tons (100 metric tons)
HABITAT	Polar and subpolar waters

DISTRIBUTION Arctic Ocean, Bering Sea, adjoining regions of north Atlantic and north Pacific

Named after its arching lower jaw, the bowhead has the longest baleen plates of any whale at up to 13 ft (4 m). Grayish black with a paler chin, it has a huge head in proportion to its body and remarkably thick blubber, which insulates it in near-freezing water. Bowheads can break upward through ice at least 8 in (20 cm) thick, allowing them to maintain open water holes throughout the Arctic winter.

ORDER CETACEA

Gray Whale

Eschrichtius robustus

LENGTH	43–46 ft (13–14 m)
WEIGHT	15–39 tons (14–35 metric tons)
HABITAT	Temperate and subpolar coastal waters

DISTRIBUTION North Pacific, Bering Sea, Arctic Ocean

Unlike other baleen whales, the gray whale feeds on the sea floor, filtering animals out of the sediment. Its body is gray with white mottling, and it has a narrowish head, with yellowish baleen plates up to 16 in (40 cm) long. Its entire body is often heavily encrusted with barnacles and whale lice. Although gray whales stay close to the coast, they carry out record-breaking migrations. On the west coast of North America, large numbers migrate between the Bering Sea and Baja California in Mexico, a round trip of up to 12,400 miles (20,000 km). Unfortunately, their coast-hugging habits make them easy prey for whalers. By the mid-1900s, they had been almost wiped out, but legal protection has allowed their numbers to recover.

ORDER CETACEA

Humpback Whale

Megaptera novaeangliae

LENGTH	49–56 ft (15–17 m)
WEIGHT	33–37 tons (30–34 metric tons)
HABITAT	Open oceans, from subpolar to tropical

DISTRIBUTION Worldwide, except extreme north and south

The humpback's lively behavior makes it a favorite with whale-watchers. This whale has a blue-black body, deeply notched tail fins (flukes) and extremely long, winglike flippers. Its flukes and flippers are often splashed with white markings—the pattern, unique as a fingerprint, is used to identify individuals. Unlike most baleen whales, humpbacks often trap their prey by lunging upward from below. To concentrate shoals of fish or krill, they often spiral around them while exhaling air. This "bubble-netting" may be carried out by several individuals working as a team. Humpbacks spend the summer in cold, food-rich waters, moving to lower latitudes to give birth in winter. They often feed near coasts. Although protected, current humpback populations are about a fifth of those of prewhaling days.

ORDER CETACEA
Common Minke Whale
Balaenoptera acutorostrata

LENGTH 23–30 ft (7–9 m)

WEIGHT 2.2–3 tons (2–2.7 metric tons)

HABITAT Open ocean and coastal waters

DISTRIBUTION Worldwide, except extreme north and south

This is the smallest of the rorquals— a name given to baleen whales that have expandable, pleated throats. It is also the most numerous, with a global population as high as 1 million. Like its much larger relative, the blue whale, it has a torpedo-shaped body with a single dorsal fin set far back, toward its tail. It is gray or brown above, with a paler underside, and short, pointed flippers that may have a white band.

Minke whales live alone or in small groups. They are naturally inquisitive and regularly approach boats. They eat small fish and planktonic animals and, like other rorquals, they feed mainly in cold-water regions, eating much less during the breeding season, when they migrate toward the tropics. The common minke is the only rorqual still hunted commercially, despite a moratorium observed by most member countries of the International Whaling Commission (IWC).

WHALE SONG

Like all whales, mature male humpbacks use sound to communicate. They produce the longest, most complex sound sequences of any animal, with each "song" lasting up to 30 minutes. The song is heard miles away by other humpbacks. Each regional population has its own song, sung only in the breeding season. It is thought that humpbacks use folds in their larynx, which act like vocal cords, to produce sound and "sing."

HUMPBACK WHALE
In common with all baleen whales, the humpback whale has large jaws and a long head in relation to the rest of its body. It has widely spaced throat grooves and knoblike projections on the upper and lower jaws. Despite its great size, it is an energetic swimmer and often breaches spectacularly.

ORDER CETACEA

Blue Whale

Balaenoptera musculus

LENGTH 100–110 ft (31–33 m)	
WEIGHT 125–165 tons (113–150 metric tons)	
HABITAT Open ocean	

DISTRIBUTION Tropical, temperate, subpolar waters worldwide, except in regions with permanent sea ice

The blue whale, one of the rorqual whales, is probably the largest animal that has ever lived. Its heart is the size of a small car and its call, at about 180 decibels, is louder than the sound of a jet aircraft taking off. This animal's future hangs in the balance after decades of whaling. Although it is no longer hunted, it remains seriously endangered.

The blue whale has a flattened head, a pointed snout, and a pleated, expandable throat. The rest of the body tapers to a pair of enormous tail fins (flukes). Blue whales are a mottled blue mixed with gray on their backs, but their undersides vary from white to yellow. They feed by filtering small animals, mostly krill and other small crustaceans, from the water. Their baleen plates can collect over 6,600 lb (3,000 kg) of food a day. Females give birth to a single calf about every 3 years.

BALEEN

Instead of teeth, baleen whales have flexible strips of baleen, or whalebone, which hang from the upper jaw. To feed, the whale takes in a mouthful of water, then sieves it through its baleen. The water is expelled, leaving small animals trapped, which the whale then swallows.

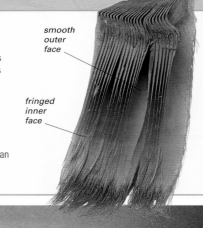

smooth outer face

fringed inner face

BALEEN STRIPS
Baleen strips are made of keratin, like human fingernails. The inner face of each strip is divided into hundreds of parallel fibers.

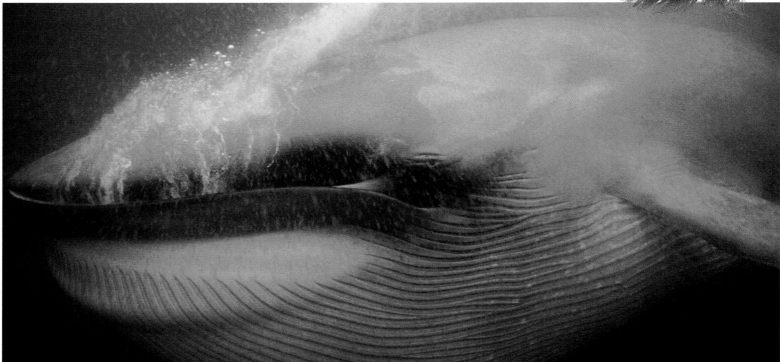

ORDER CETACEA

Sperm Whale

Physeter macrocephalus

LENGTH Up to 65 ft (20 m)	
WEIGHT On average 55 tons (50 metric tons)	
HABITAT Deep water, especially close to edges of continental shelves	

DISTRIBUTION Worldwide, except extreme north and south

The largest toothed whale, the sperm whale is also the largest predator that hunts individual prey. Even in poor light, it is unmistakable, with a huge, square-ended head. Adult males are typically 13 ft (4 m) longer than females and twice as heavy. This species has wrinkled skin and a row of knobby projections between its dorsal fin and its tail. It dives to 6,560 ft (2,000 m) to hunt giant squid. Its head contains a waxy oil, which was thought to be a buoyancy regulator, but research now suggests it is used to produce and channel beams of high-pitched sound, which are used to detect prey.

ORDER CETACEA

Cuvier's Beaked Whale

Ziphius cavirostris

LENGTH	18–23 ft (5.5–7 m)
WEIGHT	Up to 3.9 tons (3.5 metric tons)
HABITAT	Deep water

DISTRIBUTION Tropical, subtropical, and temperate waters worldwide, except in far north and south

There are at least 20 species of beaked whales, but little is known about most of them. Cuvier's beaked whale is probably one of the most widespread, because stranded specimens have been found in many parts of the world.

Like its relatives, it has an almost cylindrical body, a small dorsal fin placed far back, and relatively short flippers for its size. Its jaws are short and beaklike, with an upturned mouthline. Females are toothless, but in males, the lower jaw has two peglike teeth at its tip, which project when the mouth is closed. Adults are generally dark gray or brown with lighter-colored heads. Linear battle scars on males are also light in color.

Cuvier's beaked whale lives in deep water and can dive for more than half an hour. Its feeding behavior is poorly known, apart from the fact that it preys primarily on squid. It has never been hunted commercially, but it is occasionally an accidental bycatch in fishing nets, an occurrence that has become more common with the spread of deep-water trawling.

ORDER CETACEA

Northern Bottlenose Whale

Hyperoodon ampullatus

LENGTH	19½–33 ft (6–10 m)
WEIGHT	Up to 11 tons (10 metric tons)
HABITAT	Deep water

DISTRIBUTION Arctic Ocean, temperate and subpolar waters of the north Atlantic

One of the largest beaked whales, this species has a gray body and a bulbous forehead, which sometimes overhangs its jaws. Males have two to four nonfunctional teeth at the tip of the lower jaw; females rarely have any. Its tail fins (flukes) are large and powerful, but the front flippers are small and set far forward, just behind the head. These whales are exceptionally good divers, capable of staying underwater for over two hours. Unlike other beaked whales, this species was commercially hunted for many years, but it is still locally abundant.

ORDER CETACEA

Narwhal

Monodon monoceros

LENGTH	13–16½ ft (4–5 m)
WEIGHT	Up to 1.8 tons (2 metric tons)
HABITAT	Polar waters, open leads in sea ice

DISTRIBUTION Arctic Ocean, north as far as Svalbard and Franz Josef Land

The male narwhal is instantly recognizable by its unicornlike tusk, which is up to 10 ft (3 m) long. A highly modified upper tooth, the tusk emerges through the animal's upper lip, developing spiral grooves as it grows. Apart from this outstanding feature, males and females are similar, with a long cylindrical body, a bulbous head, and very short, beaklike jaws. They are dappled gray above and pale or white beneath. The function of the males' tusk is much debated, but recent work suggests it is used for ritual combat between males, with females mating more frequently with larger-tusked males.

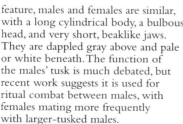

ORDER CETACEA

Beluga Whale

Delphinapterus leucas

LENGTH	9½–14¾ ft (3–4.5 m)
WEIGHT	Up to 1.8 tons (1.6 metric tons)
HABITAT	Coastal waters, sometimes rivers

DISTRIBUTION Arctic Ocean, Bering Sea, Sea of Okhotsk, Hudson Bay, Gulf of St. Lawrence

With its distinctive, yellowish white coloration, the beluga or white whale is easy to identify. In overall shape it is similar to its close relative the narwhal (see left), although it has no tusk. Its color changes with age: newborn belugas are dark gray, and it can take them up to ten years to assume the adult color, which comes with sexual maturity. Belugas are slow swimmers and feed on a wide variety of fish and other animals. They often live close inshore during summer months and may enter the lower reaches of large rivers. They are remarkably sociable and vocal, making a range of different sounds, including trills, clicks, and chirps. In the days of wooden sailing-ships, these sounds were easily audible through hulls—earning belugas the nickname "canary of the seas."

Formerly abundant throughout the Arctic, belugas have been reduced to localized populations by centuries of hunting. They are still hunted today, although on a reduced scale, but they face growing threats from pollution and shipping traffic.

MASSED RANKS

During the breeding season, belugas gather in herds that may be thousands strong. Within each herd, the whales are grouped according to age and sex, with pregnant and nursing mothers staying close together with their young. Belugas communicate using sound, but they can also make facial expressions—a unique attribute among whales. They also often hunt in groups.

ORDER CETACEA

Indo-Pacific Humpback Dolphin

Sousa chinensis

LENGTH	6½–9¼ ft (2–2.8 m)
WEIGHT	Up to 620 lb (280 kg)
HABITAT	Coastal waters, lagoons, estuaries

DISTRIBUTION Red Sea, Persian Gulf, Indian Ocean, and southwestern Pacific

This warm-water dolphin gets its name from the hump beneath its dorsal fin. Generally seen alone or in groups of less than 10, it shows a wide variation in color—some specimens are bluish gray, while others are almost white, particularly when they age. It feeds on fish, octopus, and squid, and it rarely strays far from the shore. A similar species exists off the Atlantic coast of Africa, and both species roll their bodies as they surface to breathe, keeping most of their body submerged.

ORDER CETACEA

Common Bottlenose Dolphin

Tursiops truncatus

LENGTH	6½–13 ft (2–4 m)
WEIGHT	Up to 1,400 lb (635 kg)
HABITAT	Coastal waters, open oceans

DISTRIBUTION Temperate and tropical regions worldwide

A familiar sight worldwide in marine aquariums, the common bottlenose dolphin is a playful and inquisitive mammal, with a habit of interacting with humans in the wild. Its color varies from slate blue to light gray, with a paler underside. It has a pronounced beak, a slightly hooked dorsal fin, and up to 27 pairs of peglike teeth in each jaw.

These dolphins are highly sociable and often travel in groups of 2–15. Like other dolphins, they find their prey by echolocation, but they also use sound to communicate, using a complex repertoire of whistles, clicks, and squeaks. They frequently ride the bow-waves of ships, and they also play with human swimmers. Females give birth to a single calf every 3–6 years.

ORDER CETACEA

Long-snouted Spinner Dolphin

Stenella longirostris

LENGTH	4½–12½ ft (1.4–3.8 m)
WEIGHT	Up to 175 lb (80 kg)
HABITAT	Open oceans

DISTRIBUTION Tropical and subtropical waters worldwide

Graceful, energetic, and highly acrobatic, this dolphin gets its name from its habit of leaping out of the water and then spinning around up to seven times before splashing back into the sea. Smaller than many other oceanic dolphins, it is dark gray with white on its underside—the white varies from a small patch to a wide zone extending from its head almost to its tail. It has up to 64 pairs of teeth in each jaw, and it feeds on fish, often far out to sea. Females give birth to a single calf, suckling it for up to two years.

Spinner dolphins are sociable, swimming in groups that range in size from less than 50 to several thousand and often traveling with other species. These dolphins and their close relatives often swim in large groups above shoals of yellowfin tuna, and thousands are drowned every year in purse-seine nets, which are intended to catch tuna but trap other marine life indiscriminately.

ORDER CETACEA

Short-beaked Common Dolphin

Delphinus delphis

LENGTH	5½–8 ft (1.7–2.4 m)
WEIGHT	Up to to 440 lb (200 kg)
HABITAT	Coastal waters, open oceans

DISTRIBUTION Temperate, subtropical, and tropical waters worldwide

The short-beaked common dolphin is beautifully marked with a complex pattern of colored bands and has inspired artists since classical times.

Its markings are extremely variable, and it is only in recent years that it has been separated from the similar Long-beaked common dolphin, which has a more restricted distribution. Often seen in very groups, this dolphin is highly active and acrobatic, and is among the fastest swimmers of all cetaceans, with a top speed of about 29 mph (47 kph).

Short-beaked common dolphins usually feed far out to sea, where they prey on fish and cephalopods. Adult females give birth every 1–3 years. This dolphin is one of the most common cetaceans and has a global population estimated at several million. However, like other oceanic dolphins, it is threatened by both the expansion of fishing and deliberate hunting.

ORDER CETACEA

Risso's Dolphin

Grampus griseus

LENGTH	12½–13½ ft (3.8–4.1 m)
WEIGHT	Up to 1,100 lb (500 kg)
HABITAT	Deep water

DISTRIBUTION Tropical and warm-temperate waters worldwide

Also known as the gray grampus, this large dolphin is typically blackish blue with a square head that is quite different from the pointed heads of beaked dolphins. Close up, this dolphin's skin often appears scarred, especially in older individuals. Scarring is mainly due to fights between rivals, but some of it is due to encounters with squid, which make up a large proportion of its prey. When feeding, it can dive for up to half an hour. Risso's dolphin is less sociable than many other dolphins, but it often swims alongside ships.

Killer Whale

Orcinus orca

LENGTH 18–33 ft (18–10 m)

WEIGHT Up to 7.3 tons (6.6 metric tons)

HABITAT Open waters, areas of broken sea ice

DISTRIBUTION Tropical, temperate, and polar waters worldwide

With its conspicuous black-and-white markings, the killer whale, or orca, is—despite its name—the largest and most striking member of the dolphin family (Delphinidae). Apart from its bold patterning, its most eye-catching feature is its huge dorsal fin, which is up to 6 ft (1.8 m) high in older males. It has large, paddle-shaped flippers and a massive, barrel-shaped body that tapers toward streamlined jaws, which are armed with interlocking teeth up to 4 in (10 cm) long. Killer whales are the largest hunters of warm-blooded prey. Their diet includes fish, squid, birds, seals, and other whales. Their hunting strategy is remarkably varied: they deliberately upend ice floes to tip seals into the sea, and they even lunge onto beaches to catch seals lying near the waterline. Intelligent, vocal, and highly sociable, they live in stable groups (pods), which develops their own cultural characteristics. Despite their ferocity toward prey animals, killer whales are easily tamed in captivity and have never been known to attack humans in the wild.

PODS AND CLANS

An average killer whale pod contains 20 animals, which stay together for life, often sharing care of the young. Pods within the same geographical range make up a clan—a regional group that is thought to have a distinctive "dialect" that is passed on from adults to their young.

ON THE MOVE
Female humpback whales typically breed every 2–3 years. The mother is very protective of her calf, which she suckles for 6–10 months.

Many land mammals make long treks in search of existing feeding grounds, but the distances they cover are dwarfed by the vast annual journeys made by some whale species. Humpback whales, for example, spend the summer in rich feeding grounds in cold waters (see below), gorging themselves on krill, zooplankton, and fish. With the onset of winter, their food supply dwindles, and they migrate toward the equator. In these waters, there is little for the whales to feed on, and they fast for several months. However, these warm and sheltered waters provide a suitable environment in which to give birth and begin to rear their calves.

Whale migration is a complex subject, and it is only in recent years—with the advent of satellite radio-tracking—that is has become possible to chart the path of individual animals, uncovering their migration routes. Some whales, such as bowheads and narwhals, migrate only a limited distance, staying in Arctic waters but moving in step with seasonal changes in the sea ice. In other species, migration routes are much longer, and they vary among different populations: humpback whales are a good example of this. Killer whales show a mixed pattern of migration. "Resident" pods move very little during the course of the year, but others can migrate thousands of miles.

It is not known precisely how whales navigate on these long journeys. Although the mechanism is unclear, magnetite (an oxide of iron) has been found in tissues around the brains of some cetaceans, including humpbacks. It is thought that this helps the whales to sense gradients in the Earth's eomagnetic field, which in turn acts as a guide to navigation.

Humpback Whale Migration

Humpback whales from the Northern Hemisphere spend the summer in feeding grounds in the northern Pacific and Atlantic Oceans. In winter, these whales migrate south to warmer waters to breed. Humpback whales from the Southern Hemisphere feed in waters off Antarctica and breed in warmer waters off Australia, the Pacific islands, southern Africa, or South America. The northern Indian Ocean population is resident all year round.

→ main migration
major feeding areas (summer)
major breeding areas (winter)

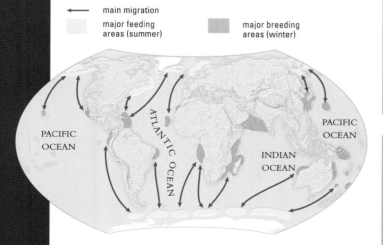

PACIFIC OCEAN
ATLANTIC OCEAN
PACIFIC OCEAN
INDIAN OCEAN

WHALE BEHAVIOR

FEEDING

ALASKAN FEEDING GROUNDS
Humpback whales feed by scooping water into their huge mouths, from which they mostly filter krill and other large zooplankton. In Alaska, they gather close inshore in summer to feed on schools of small fish.

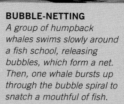

BUBBLE-NETTING
A group of humpback whales swims slowly around a fish school, releasing bubbles, which form a net. Then, one whale bursts up through the bubble spiral to snatch a mouthful of fish.

MIGRATING

SOLITARY BLUE WHALE
Although blue whales are known to make major migrations, they swim farther offshore than other migratory whales and do not have defined breeding areas, so their migration is less well understood.

SPY-HOPPING *Many whales, such as this gray whale, "spy-hop," rising vertically in the water with the head well above the surface. They may be checking for landmarks while migrating.*

BREEDING

WARM-WATER DISPLAY *From June to December each year, thousands of southern right whales gather close to shore in bays east of the Cape Peninsula, South Africa. The females give birth to a single calf in these waters, then mate immediately afterward. The area offers the best land-based whale-watching in the world, providing spectacular views, such as this whale breaching.*

WHALE-WATCHING

CLOSE ENCOUNTER *Whale-watching is a growth industry that is worth over $2.3 billion a year, with over 13 million people taking part. These tourists are observing a gray whale in its summer breeding grounds in the waters off Baja California, Mexico.*

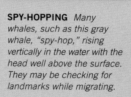

ORDER CETACEA

Long-finned Pilot Whale

Globicephala melas

LENGTH	20–23 ft (6–7 m)
WEIGHT	Up to 2.5 tons (2.3 metric tons)
HABITAT	Cold coastal waters, open oceans

DISTRIBUTION Temperate and subpolar waters worldwide, except north Pacific

There are two species of pilot whales, distinguished primarily by the length of their flippers—a feature that is difficult to observe at sea. The long-finned pilot whale lives mainly in cold-water regions. It has glossy, jet black coloration, with an anchor-shaped pale patch on the throat and chest. This species has a bulbous head and short jaws. Its long dorsal fin has a hooked shape in males. Its flippers have a sharp backward bend, or "elbow," and are up to a fifth of its body length. Long-finned pilot whales feed mainly on deep-water squid and octopus. They are highly gregarious, living in groups that can be hundreds strong, and often associate with other cetaceans. They easily become disoriented in shallow coastal waters, often becoming stranded in large numbers. This tendency to herd together has been exploited for centuries by whale hunters, who were able to drive them into shallow water for slaughter. In some locations—such as the Faroe Islands—pilot whales are still hunted today.

STRANDING

Pilot whales often become stranded on beaches. If one whale strands, others frequently follow, leading to a mass stranding. Theories to explain stranding involve factors that disrupt the whales' navigational systems, such as temporary anomalies in Earth's magnetic field, ships' sonar, sickness, and storms.

ORDER CETACEA

Harbor Porpoise

Phocoena phocoena

LENGTH	4¼–6½ ft (1.3–2 m)
WEIGHT	Up to 165 lb (75 kg)
HABITAT	Coastal waters, tidal regions of rivers

DISTRIBUTION Cold-temperate and subpolar waters in Northern Hemisphere

One of the most common cetaceans in the Northern Hemisphere, the harbor porpoise, as its name suggests, rarely strays into deep water. It prefers shallow, coastal waters and sometimes swims into rivers. It has a short, barrel-like body, with small flippers and a blunt dorsal fin. Its overall color is dark gray, while its underside is paler. Unlike most dolphins, this porpoise has a blunt snout, which houses 21–28 pairs of spade-shaped teeth in each jaw. Harbor porpoises often live alone, or sometimes in pairs or small groups; they feed on fish and cephalopods.

Females give birth after a gestation period of up to 11 months, and the single calf is tiny by cetacean standards, weighing as little as 13 lb (6 kg). In the past, harbor porpoises were often hunted for meat and as a source of oil. Today, a greater threat is posed by fishing nets—being small, it is easy for them to become accidentally trapped.

ORDER SIRENIA

West African Manatee

Trichechus senegalensis

LENGTH	Up to 11 ft (3.5 m)
WEIGHT	Up to 1,100 lb (500 kg)
HABITAT	Mangrove swamps, lagoons, inland waterways, estuaries

DISTRIBUTION West Africa, from Senegal to Angola

One of three species of manatees, this docile vegetarian lives mainly in fresh water but also feeds in the mangrove swamps on Africa's west coast. It has a barrel-shaped body covered in coarse gray skin and front flippers with tiny nails. Like all sirenians, it has no hind limbs and swims with its spoon-shaped tail, which slowly beats up and down as it cruises through the shallows.

Using its fleshy lips, it feeds on plants above and below the water line, taking in various invertebrates, such as small crabs, while doing so. Manatees lack the complex chambered stomachs of terrestrial plant-eaters such as cattle and antelopes. Most digestion occurs in their intestines, which may be 150 ft (45 m) long. West African manatees are usually solitary; however, groups may be seen where food is plentiful or during the mating season. At birth, West African manatees are about 3 ft (1 m) long. Their slow reproductive rate makes them vulnerable to environmental change, and to hunters who target them for meat and skin.

ORDER SIRENIA

West Indian Manatee

Trichechus manatus

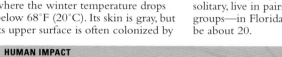

LENGTH	8¼–12¾ ft (2.5–3.9 m)
WEIGHT	Up to 3,570 lb (1,620 kg)
HABITAT	Coastal waters, inland waterways

DISTRIBUTION Western Atlantic from southeast US to northeast South America, Caribbean Sea

This is the largest species of manatee, and also the best studied—something explained partly by its distribution, which extends northward as far as Florida. It often ventures into coastal waters, although it avoids regions where the winter temperature drops below 68°F (20°C). Its skin is gray, but its upper surface is often colonized by algae, which gives it a greenish tinge. Its vision and hearing, provided by small eyes and ears, are not very acute, but its mobile lips are covered with sensitive bristles, which it uses to find underwater plants, such as seagrass, in depths of up to about 13 ft (4 m). It needs to consume approximately one-quarter of its body weight in food each day. Although its diet is mainly vegetarian, it sometimes eats fish to obtain protein.

Manatees and dugongs (see below) owe their blimplike shapes partly to the large amounts of gas generated as they digest their food. To compensate for this, they have unusually dense bones, which help them to maintain neutral buoyancy. Manatees can be solitary, live in pairs, or form larger groups—in Florida, group size can be about 20.

COLLISION RISK

In the past, West Indian manatees were hunted for their meat, skin, and oil, which was sometimes used in lamps. Today, the main threats facing them are pollution and collisions with boats. In Florida, where boat traffic is heavy, many manatees bear the scars of their encounters with boats.

PROPELLER INJURY
These parallel scars on a manatee's back were caused by a propeller. Fortunately, this type of injury isn't necessarily fatal.

ORDER SIRENIA

Dugong

Dugong dugon

LENGTH	6½–10¾ ft (2–3.3 m)
WEIGHT	at least 1,260 lb (570 kg)
HABITAT	Coastal shallows, lagoons, estuaries

DISTRIBUTION Indian Ocean and western Pacific, from East Africa to South Pacific islands

Unlike manatees, the dugong is essentially a marine animal, grazing in seagrass beds in warm, shallow waters. Its body is blimp-shaped, like that of manatees, but it has a crescent-shaped tail and a broad head with a large, U-shaped upper lip. Part of its diet consists of buried stems or rhizomes, which it collects by nuzzling its way into the sediment, while steadying itself with its front flippers. Dugongs feed in scattered herds, which may contain more than a hundred animals.

Their main predators are sharks, but they are more threatened by hunting in many places. The species is already extinct in the Mediterranean, where it may have existed until classical times, and it is under threat in many parts of the Indian Ocean. However, it appears to be thriving around the coastline of Australia, which is home to over half the world's dugongs.

STELLER'S SEA COW

A close relative of the dugong, Steller's sea cow lived in the icy waters of the Bering Sea, feeding on kelp and other seaweeds. It was hunted to extinction in 1768, 27 years after it was first recorded by the German naturalist Georg Steller (1709–1746).

ARTIST'S IMPRESSION
Steller's sea cow weighed up to 12 tons (11 metric tons) and was probably the largest marine mammal of its time, after whales.

ATLAS OF THE OCEANS

Oceans of The World

OCEANS COVER 71 PERCENT of Earth's surface and contain 97 percent of its water. The geography of the ocean basins tells us much about Earth's past and the geological forces that continue to shape the world.

There are five oceans separating the world's major landmasses and numerous marginal seas, gulfs, and connecting straits. The continents are surrounded by shallow shelves, which extend a variable distance from the shore before descending into the deep ocean basins. The ocean basins contain the flattest parts of Earth's surface—the abyssal plains—but also the greatest extremes of elevation, from deep ocean trenches to the peaks of the world's largest volcanoes. The longest mountain chains on Earth are the mid-ocean ridges, which circle the planet along the boundaries of the major tectonic plates.

The maps in this chapter draw on the most detailed knowledge of the topography of the global sea floor yet assembled. They combine the latest measurements from satellite altimeters with more than 100 years of ship-borne hydrographic surveys to give the clearest possible portrayal of the shape of the seabed.

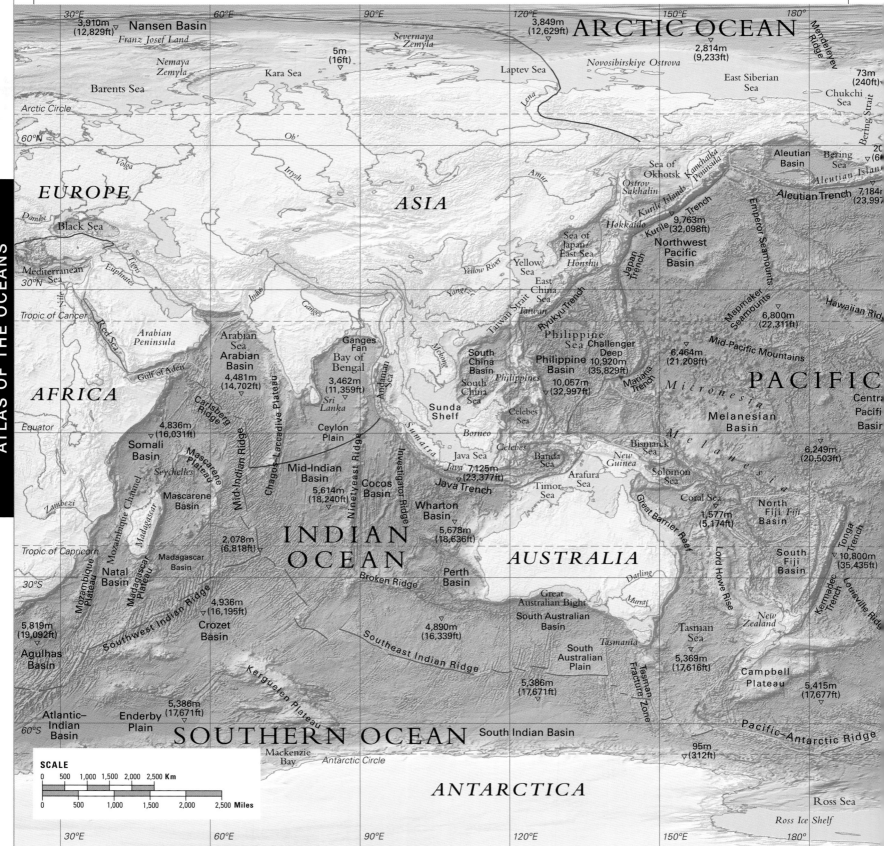

SCALE

0 500 1,000 1,500 2,000 2,500 Km

0 500 1,000 1,500 2,000 2,500 Miles

TECTONIC PLATES

Earth's surface is split into 7 major plates and at least 15 minor ones. Their boundaries are most clearly expressed in the topography of the sea floor, where mid-ocean ridges, ocean trenches, and fracture zones represent divergent, convergent, and transform boundaries, respectively. Although some plates have continental and oceanic parts, the largest, the Pacific Plate, is only oceanic crust.

Tectonic Plates map labels: EURASIAN PLATE, Arabian Plate, Philippine Plate, Indian Plate, AUSTRALIAN PLATE, Fiji Plate, ANTARCTIC PLATE, Okhotsk Plate, Juan de Fuca Plate, Rivera Plate, PACIFIC PLATE, Bismarck Plate, Caroline Plate, Easter Plate, Cocos Plate, Nazca Plate, Juan Fernandez Plate, NORTH AMERICAN PLATE, Caribbean Plate, SOUTH AMERICAN PLATE, Scotia Plate, Shetland Plate, Sandwich Plate, EURASIAN PLATE, AFRICAN PLATE

KEY

sea level	□ land
800 ft (250 m)	△ seamount
1,600 ft (500 m)	▽ sea depth
3,300 ft (1,000 m)	▼ maximum depth on map
6,500 ft (2,000 m)	— convergent boundary
9,800 ft (3,000 m)	— divergent boundary
16,400 ft (5,000 m)	— transform boundary
	— uncertain boundary

ATLAS OF THE OCEANS

Map labels

Canada Basin — 3,718m (12,199ft)
Beaufort Sea, Banks Island, Victoria Island, Queen Elizabeth Islands, Ellesmere Island, Baffin Island, Baffin Bay, Baffin Basin, Greenland, Greenland Sea, Spitsbergen, Norwegian Sea, Arctic Circle

Yukon River, Gulf of Alaska, Vancouver Island, Hudson Bay, Great Lakes, NORTH AMERICA, Rio Grande, Gulf of Mexico, Greater Antilles, Caribbean Sea

Mendocino Fracture Zone — 5,999m (19,683ft)
Murray Fracture Zone, Molokai Fracture Zone, Hawaiian Islands, Clarion Fracture Zone, Clipperton Fracture Zone

Labrador Sea, Labrador Basin, Newfoundland, Northwest Atlantic Mid-Ocean Channel — 13m (43ft), Newfoundland Basin, Oceanographers Fracture Zone, Sohm Plain, Mid-Atlantic Ridge — 3,780m (12,402ft), Reykjanes Basin, Reykjanes Ridge, Denmark Strait, Iceland, Iceland Basin, Charlie Gibbs Fracture Zone, Norwegian Basin, North Sea, Ireland, Britain, Scandinavia, EUROPE, 4,139m (13,580ft), Iberian Plain, Azores, Madeira Plain, Canary Islands, Mediterranean Sea

5,464m (17,927ft), Hatteras Plain, Sargasso Sea, Nares Plain, Cape Verde Basin, Cape Verde Islands, Barracuda Fracture Zone, 8,952m (29,404ft), ATLANTIC OCEAN, AFRICA, Tropic of Cancer, Sierra Leone Basin, Guinea Basin, Niger, Equator

OCEAN, Middle America Trench, Guatemala Basin — 3,806m (12,487ft), Demerara Plain, Orinoco, Amazon, SOUTH AMERICA, São Francisco, Ascension Fracture Zone, Brazil Basin — 5,706m (18,721ft), Angola Basin — 5,042m (16,543ft), Congo

5,451m (17,885ft), Galapagos Fracture Zone — 4,567m (14,984ft), Marquesas Islands, Marquesas Fracture Zone, Tiki Basin, Polynesia, Galapagos Islands, Bauer Basin, Peru Basin, Peru-Chile Trench, Nazca Ridge, Mendaña Fracture Zone, Yupanqui Basin, Easter Fracture Zone, 8,069m (26,474ft), Chile Basin, Roggeveen Basin, Challenger Fracture Zone — 1,426m (4,679ft), Chile Rise, East Pacific Rise, Mornington Abyssal Plain — 6,034m (19,798ft), Menard Fracture Zone

Rio Grande Rise, Mid-Atlantic Ridge, Rio Grande Fracture Zone — 1,739m (5,706ft), Walvis Ridge, Cape Basin, Cape of Good Hope, Tropic of Capricorn, Gough Fracture Zone, Atlantic-India Ridge

Southwest Pacific Basin, Easter Fracture Zone, Eltanin Fracture Zone, Agassiz Fracture Zone, Udintsev Fracture Zone, 4,283m (14,058ft), Amundsen Plain, Amundsen Sea, Southeast Pacific Basin, Bellingshausen Plain, Bellingshausen Sea, Drake Passage, Cape Horn, Antarctic Peninsula, Argentine Basin, Falkland Islands, South Georgia, Scotia Sea, 7,152m (23,466ft), America-Antarctica Ridge, Weddell Plain, Weddell Sea, SOUTHERN OCEAN, Ronne Ice Shelf, ANTARCTICA, Antarctic Circle

The Arctic Ocean

THE SMALLEST OF THE OCEANS, the Arctic Ocean is nearly enclosed by Asia, Europe, Greenland, and North America. In winter it is almost entirely covered by pack ice, which halves in area during summer. Exploration of the Arctic in the 18th and 19th centuries was driven by the search for trade routes between the Atlantic and Pacific Oceans. The North Pole was first reached in 1909 by an American expedition using dogs and sleds, led by Robert Peary.

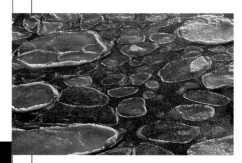

ARCTIC SEA ICE
Sea ice covering the Arctic expands from less than 3 to about 6 million square miles (4.5 to 15 million square km) from summer to winter.

Ocean Circulation

The Arctic receives a huge influx of fresh water from the great Siberian rivers—the Ob', Yenisey, and Lena. Together with the freezing and melting of sea ice, this produces a layer of relatively fresh surface water. A clockwise gyre is established over the Canada Basin, while the Transpolar Current flows from the Chukchi Sea to the Greenland Sea. Warm, salty water enters the Arctic from the Atlantic at moderate depth, while very cold, very salty "bottom water" flows out into the Atlantic. Eighty percent of the Arctic's water exchange is with the North Atlantic and 20 percent is with the Pacific. About two percent of the water leaving the Arctic is in the form of icebergs calved from the Greenland Ice Sheet. Arctic sea ice has declined in area at about 13 percent per decade since 1979, and hit a record low of 1.4 million square miles (3.63 million square km) in September 2012. At the present rate, climate change could cause the ocean's sea ice to disappear by 2050.

ICEBREAKER
Ships with strengthened bows and powerful engines—icebreakers—are needed to penetrate Arctic sea ice.

Ocean Floor

The floor of the Arctic Ocean consists of two main basins separated by the sharp Lomonsov Ridge. On the North American side lie the Canada and the Makarov basins, separated by the Alpha Cordillera. On the Eurasian side the Fram and Nansen basins are split by the Gakkel Ridge—an extension of the Mid-Atlantic Ridge. The young Arctic Basin started to open about 36 million years ago, completing the separation of North America from Europe, and connecting the Arctic to the Atlantic. There is an unusually broad continental shelf on the Asian side of the ocean, with shallow seas extending more than 1,000 miles (1,600 km) from the coast in places, compared with the more typical 30–75 miles (50–125 km) on the North American side.

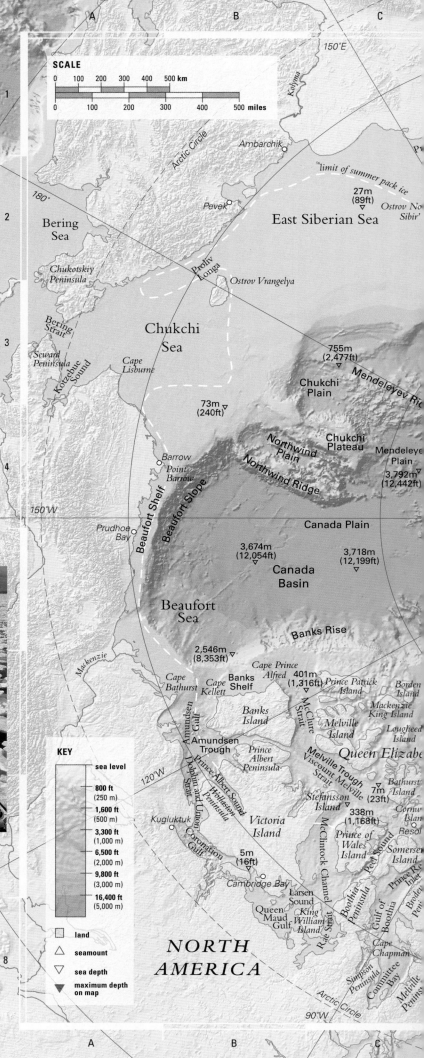

KEY

| sea level |
| 800 ft (250 m) |
| 1,600 ft (500 m) |
| 3,300 ft (1,000 m) |
| 6,500 ft (2,000 m) |
| 9,800 ft (3,000 m) |
| 16,400 ft (5,000 m) |

☐ land
△ seamount
▽ sea depth
▼ maximum depth on map

Map labels

ASIA

Lena
Tiksi
6m (20ft)
Bjorkhaya Guba
nitraya Lapteva
Ostrov Bol'shoy Lyakhovskiy
P'nyy
Novosibirskiye Ostrova

Laptev Sea

Proliv Vil'kitskogo
Ostrov Bol'shevik
Severnaya Zemlya
Poluostrov Taymyr
Dikson
Gydanskiy Poluostrov
Obskaya Guba
Yenisey
Arctic Circle
Poluostrov Yamal
Baydaratskaya Guba

3,849m (12,629ft)
Ostrov Oktyabr'skoy Revolyutsii
Ostrov Komsomolets

ARCTIC OCEAN

2,814m (9,233ft)
Wrangel Plain

Makarov Basin
Pole Plain
Lomonosov Ridge
Gakkel Ridge
Nansen Basin

4,484m (14,712ft)
North Pole

3,910m (12,829ft)

Alpha Cordillera
1,250m (4,101ft)
2,590m (8,498ft)

Voronin Trough
Central Kara Plateau
Syvataya Anna Trough
Franz Josef Land

Kara Sea
5m (16ft)
Ostrov Belyy
60°E
Kara Strait
328m (1,076ft)
East Novaya Zemlya Trough
Novaya Zemlya

170m (558ft)
109m (358ft)

Barents Sea

102m (335ft)
Kvitøya
limit of summer pack ice
limit of winter pack ice
Thor Iversen Bank

Barents Plain
Stor Bank

30°E

EUROPE

Yermak Plateau
Litke Trough
Nordaustlandet
Edgeøya
Barentsøya
Spitsbergen
Longyearbyen
Storfjordrenna
Bjornöya Bank
Bjørnøya
Barents Trough

257m (843ft)
North Cape
Hammerfest

5

Spitsbergen Fracture Zone
Lena Trough
5,601m (18,377ft)
Knipovich Ridge
Kap Morris Jesup
Wandel Sea
Nord
Hovgaard Fracture Zone
Boreas Plain
Fugløya Bank
Tromsø

Lincoln Sea
Cape Columbia
Alert
Greenland Fracture Zone
2,580m (8,465ft)
Bodø
Vestfjorden

Alex Heiberg Island
15m (49ft)
3,900m (12,796ft)
Belgica Bank
Greenland Plain
Greenland Sea
Mohns Ridge
Norwegian Sea
222m (738ft)
Røst Bank
Halten Bank

6

Ellesmere Island
Nares Strait
80°N
Dumshaf Plain
1,280m (4,200ft)
Voring Plateau

Grise Fiord
Qaanaaq
Daneborg
210m (689ft)
Jan Mayen Fracture Zone
Jan Mayen
Norwegian Basin
Norwegian Trench

77m (253ft)
Kolbeinsey Ridge
Jan Mayen Ridge

7

evon
Sound
Cape Sherard
Devon Shelf 732m (2,402ft)
Devon Slope
aster Trough
Bylot Island

Baffin Basin
Kullorsuaq
Ittoqqortoormiit

Greenland
Iceland Plateau
Denmark Strait

68m (223ft)
Faeroe-Shetland Trough
Shetland Islands

Faeroe Islands

0°

2,377m (7,799ft)
Upernavik

Baffin Bay
Uummannaq
Qeqertarssuaq
70°N
60°W
30°W
Arctic Circle
Iceland
Reykjavik

Britain

8

Inset maps

SURFACE CURRENTS
Transpolar Current
Beaufort Gyre
West Greenland Current
East Greenland Current
Norwegian Current

SURFACE WINDS
Polar Easterlies
Polar Easterlies
Polar Easterlies
Westerlies

Beaufort Scale	Speed
0–3	0–10 mph (0–16 kph)
3–5.5	10–25 mph (16–40 kph)
over 5.5	over 25 mph (over 40 kph)

Northwest Passage

THE ATLANTIC AND PACIFIC OCEANS are linked by the
Northwest Passage through the Arctic Ocean. A tough route to
navigate, it includes many narrow straits between islands, and its
surface is often frozen, even in summer. Shrinking sea-ice has made
it more navigable, with the passage open most years since 2007.
The first cruise ship, with 1,500 passengers, passed through in 2016.

ARCTIC OCEAN E2

Baffin Bay

AREA 266,000 square miles (689,000 square km)

MAXIMUM DEPTH 6,900 ft (2,100 m)

INFLOWS Arctic Basin, Labrador Sea, glaciers of
West Greenland

Baffin Bay lies between Greenland
and Baffin Island, and is really the
extreme northwest arm of the Atlantic.
The surface ices over each winter, but
warmer water from the Labrador Sea
flows up its eastern shore, keeping
parts of the adjacent Greenland coast
ice-free. Water returns south along
the western shore as the cold Labrador

GLACIER MEETING SEA

Current, often carrying icebergs into
the North Atlantic. Seals have long
been hunted in the area. Large numbers
were killed each year as recently as the
1980s, but commercial hunting of
marine mammals is now controlled.

DISCOVERY

EXPLORING THE PASSAGE

Much effort was expended in
the 17th century in search of
the Northwest Passage from the
Atlantic to East Asia, but no
viable route was found due to
the year-round presence of sea ice
and the numerous islands. Interest
revived in the 19th century when
expeditions were undertaken
by the Royal Navy. In 1820 an
expedition from Baffin Bay got as
far as Melville Island before being
blocked by ice. John Franklin's
expedition of 129 men was lost
in 1848. British explorer Robert
McClure crossed from
the Beaufort Sea to Baffin Bay
in 1854, but he had to walk part
of the way. It was the Norwegian
Roald Amundsen who finally
sailed via Lancaster Sound,
south of Victoria Island, and out
through the Bering Strait in 1906.

ARCTIC OCEAN A1

Beaufort Sea

AREA 184,000 square miles (476,000 square km)

MAXIMUM DEPTH 15,350 ft (4,680 m)

INFLOWS Chukchi Sea, Arctic Basin, Rivers
Mackenzie, Colville

The deep Beaufort Sea lies to the west
of the Canadian Arctic Archipelago. Oil
was discovered off the Alaskan shore
in 1968, and is also extracted off the
MacKenzie delta. Artificial islands have
been built to protect some production
wells from drifting sea ice. Gas has also
been found near Melville Island.

OIL EXPLORATION

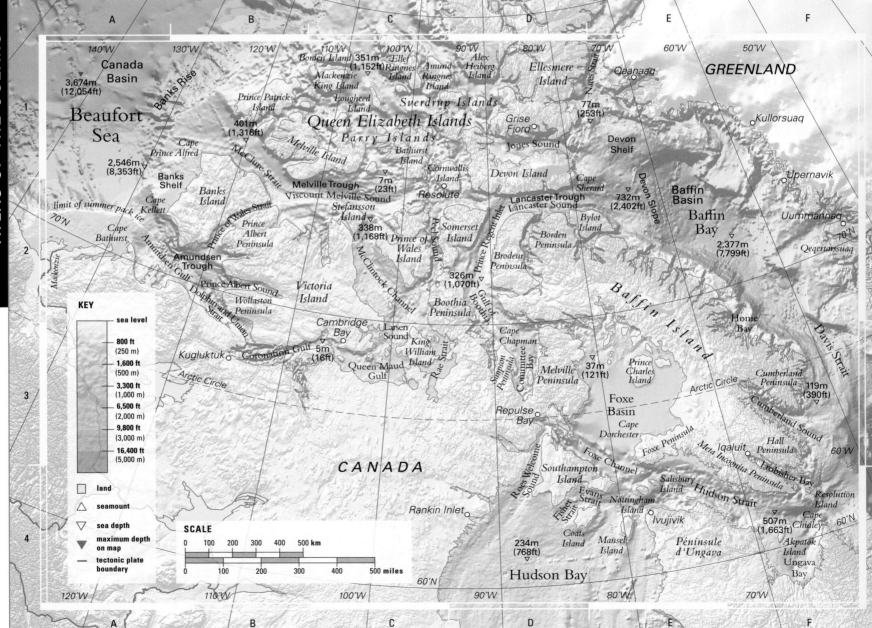

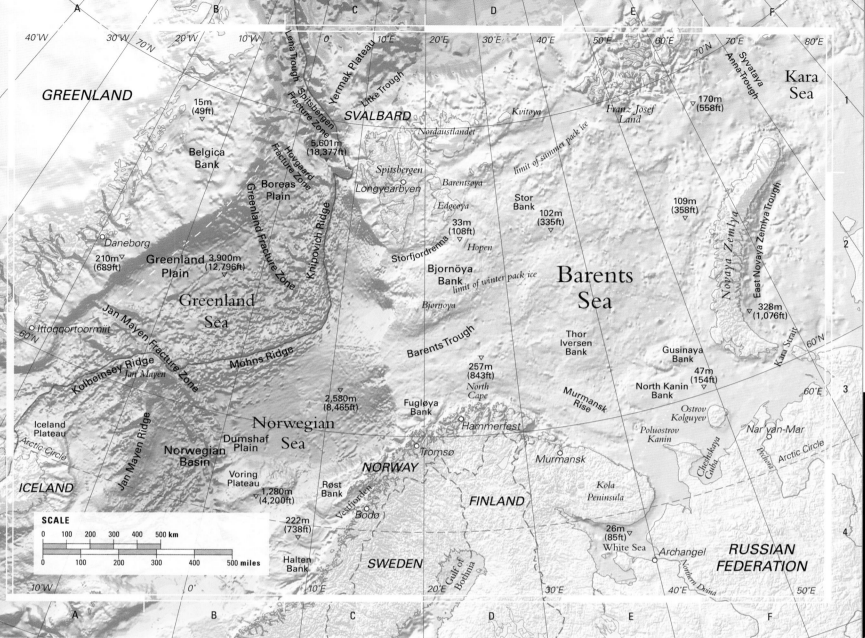

The Barents Sea

THE BARENTS AND GREENLAND SEAS MARK the boundary between the Arctic Ocean and the Atlantic. Most of the Arctic Ocean's water exchange is with the north Atlantic. This exchange occurs at the Fram Strait, north of the Greenland Sea.

ARCTIC OCEAN E2

Barents Sea

AREA 540,000 square miles (1.4 million square km)

MAXIMUM DEPTH 2,000 ft (600 m)

INFLOWS Norwegian Sea, Arctic Basin

The Barents Sea is relatively shallow, lying north of Europe and south of the islands of Svalbard and Franz Joseph Land. To the east, Novaya Zemlya is an extension of the Ural Mountains, which mark the geographical boundary between Europe and Asia. Large areas along the mainland and around the islands are continental shelf of less than 660 ft (200 m) deep. Warm water from the North Atlantic Drift flows in from the southwest, keeping most of the sea ice-free in summer. The Russian port of Murmansk

remains free of ice even in the winter. The warm, salty Atlantic water meets cold, less saline Arctic water, and warm, moderately salty coastal water, producing an area of high biological productivity. The spring bloom of phytoplankton starts near the ice edge, where fresh water from melting ice produces a stable surface layer. The phytoplankton form the basis of a food chain that supports a rich fishery, and cod is the most important catch. During the Cold War, Russia maintained a large northern fleet of warships and submarines. Many of these vessels now lie deteriorating in naval ports along the Kola Peninsula, raising fears of possible damage to the marine environment. Particular concerns have been raised about contamination from the nuclear reactors of abandoned submarines.

ARCTIC OCEAN B2

Greenland Sea

AREA 463,000 square miles (1.2 million square km)

MAXIMUM DEPTH 16,000 ft (4,800 m)

INFLOWS Arctic Basin, Norwegian Sea

The Greenland Sea, which stretches between Greenland, Svarlbad, and Jan Mayen Island, is a major area of sea ice formation in the Arctic Ocean. The East Greenland Current carries surface water and ice south along the coast of Greenland, but the Jan Mayen Current takes some surface water to the east. This divergence leaves an area of open water where new sea ice is

continually formed in the winter. An ice tongue, known as the Odden, develops eastward from the main ice edge, and dissolved salt is left behind in a layer of cold, briny water beneath the new ice. Being more dense, this very salty water sinks to the seafloor, where it pools before spilling over the ridges between Greenland and Jan Mayen to the south. This downwelling plays a major role in the global thermohaline circulation (see pp.60–61) of the oceans.

ICY PANCAKES

Pancake ice is formed in rough water as cakes of icy slush bump into each other, producing a raised rim.

The Atlantic Ocean

THE ATLANTIC SEPARATES the "old world" of Europe and North Africa from the "new world" of the Americas. The Portuguese pioneered the ocean's exploration, with Bartolomeu Dias reaching Africa's southern tip in 1488, but a Spanish expedition led by Christopher Columbus was the first to cross the Atlantic to reach the West Indies in 1492.

ATLANTIC SURF BATTERS BERMUDA

Ocean Circulation

A clockwise gyre controls surface currents in the north Atlantic. The strong Gulf Stream current brings in warm, salty water from the Caribbean. It then continues northwestward as the North Atlantic Drift, giving western Europe a milder climate than its latitude alone allows. Some water moves north toward the Arctic Ocean, but most returns south as the Canaries Current. Warm water returns to the west with the North Equatorial and Guiana currents. The South Atlantic Gyre turns counterclockwise, with the Antarctic Circumpolar Current forming its southern boundary. The cold Benguela Current flows north up the coast of Africa. In the west, the Brazil Current flows rather weakly south from the equator.

TIDAL POWER
The Atlantic's high tidal range makes many locations suitable for harnessing tidal power, such as this installation off Devon, England.

Ocean Floor

The Atlantic Ocean floor is dominated by the Mid-Atlantic Ridge, marking where the continents to the east and west are splitting apart. The ridge breaks through the sea surface at some points, most notably as the island of Iceland. Transverse faults scar the flanks of the ridge for many hundreds of miles east and west, until they are buried under the marine sediments of the abyssal plains. These flat areas are quite narrow in the Atlantic, occupying the edge of the deep ocean between the broad ridge and the continental rise. The sedimentary margins of the Atlantic bear some rich mineral deposits, including oil and gas in the North Sea, the Gulf of Mexico, off Venezuela, and West Africa. The Atlantic has only two deep ocean trenches: the Puerto Rico Trench and the South Sandwich Trench.

ICEBERG SCOURING
This image shows a trench scoured out by the keel of a drifting iceberg—a fairly common occurrence in shallow waters around Greenland.

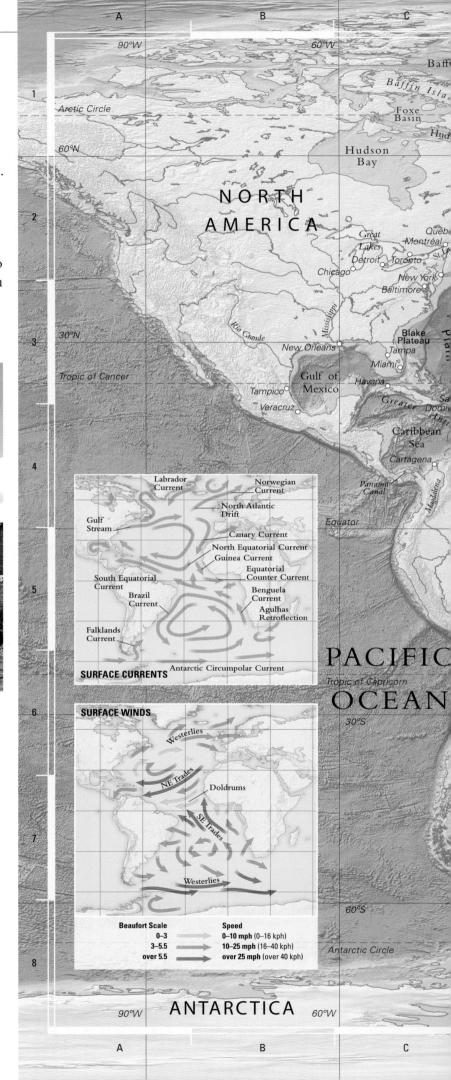

SURFACE CURRENTS

Labrador Current · Norwegian Current · North Atlantic Drift · Gulf Stream · Canary Current · North Equatorial Current · Guinea Current · Equatorial Counter Current · South Equatorial Current · Brazil Current · Benguela Current · Agulhas Retroflection · Falklands Current · Antarctic Circumpolar Current

SURFACE WINDS

Westerlies · NE Trades · Doldrums · SE Trades · Westerlies

Beaufort Scale	Speed
0–3	0–10 mph (0–16 kph)
3–5.5	10–25 mph (16–40 kph)
over 5.5	over 25 mph (over 40 kph)

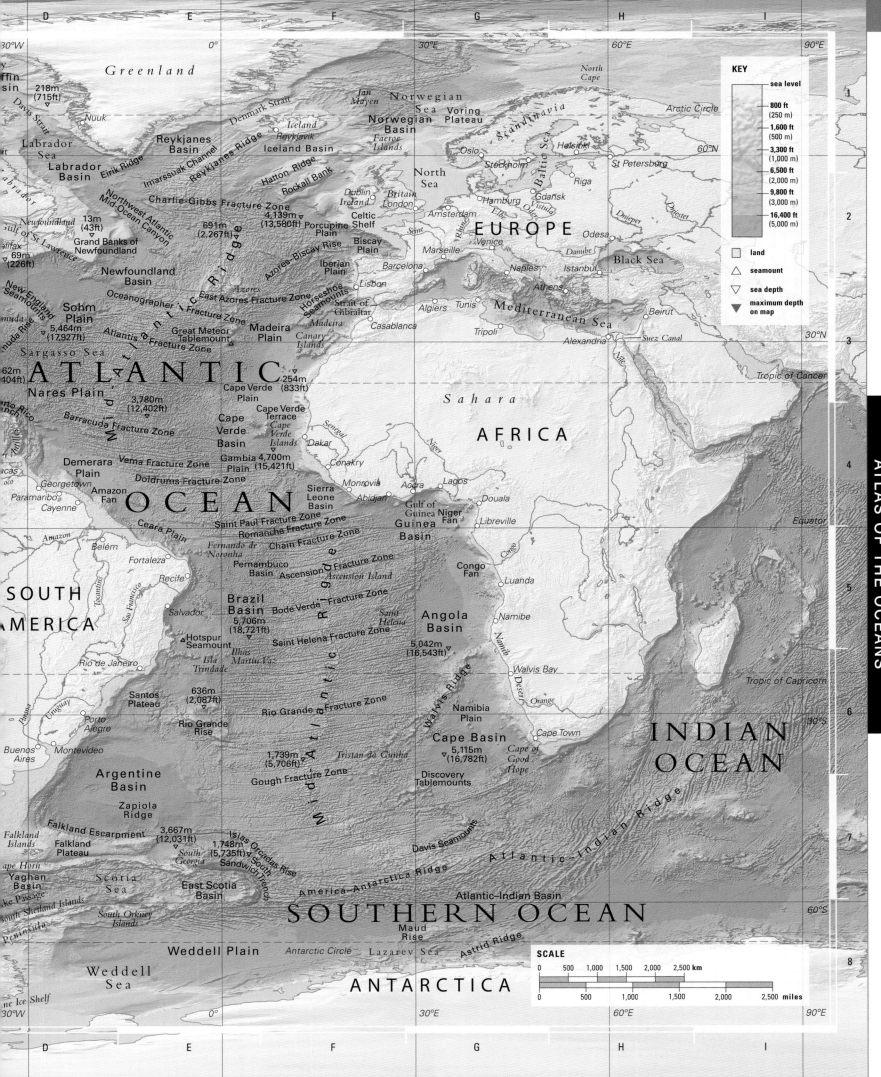

KEY

sea level

800 ft (250 m)

1,600 ft (500 m)

3,300 ft (1,000 m)

6,500 ft (2,000 m)

9,800 ft (3,000 m)

16,400 ft (5,000 m)

□ land

△ seamount

▽ sea depth

▼ maximum depth on map

Greenland

218m (715ft)

Nuuk

Davis Strait

Labrador Sea

Labrador Basin

Eirik Ridge

Reykjanes Basin

Iceland

Reykjavik

Reykjanes Ridge

Iceland Basin

Denmark Strait

Imarssuak Channel

Jan Mayen

Norwegian Sea

Voring Plateau

North Cape

Norwegian Basin

Scandinavia

Arctic Circle

Oslo

Stockholm

Helsinki

St Petersburg

60°N

Faeroe Islands

Hatton Ridge

Rockall Bank

Charlie-Gibbs Fracture Zone

Northwest Atlantic Mid-Ocean Canyon

13m (43ft)

Gulf of St Lawrence

Halifax

69m (226ft)

Grand Banks of Newfoundland

Newfoundland Basin

691m (2,267ft)

4,139m (13,580ft)

Porcupine Plain

Celtic Shelf

Dublin Ireland

Britain London

North Sea

Hamburg

Amsterdam

Seine

Oder

Elbe

Vistula

Gdańsk

Riga

Baltic Sea

Dnieper

Dniester

EUROPE

2

New England Seamounts

Sohm Plain

5,464m (17,927ft)

Bermuda Rise

Oceanographer Fracture Zone

Atlantis Fracture Zone

Biscay Plain

Iberian Plain

Azores

Azores-Biscay Rise

East Azores Fracture Zone

Horseshoe Seamounts

Lisbon

Marseille

Venice

Barcelona

Naples

Danube

Odesa

Istanbul

Black Sea

Athens

Beirut

3

Mid-Atlantic Ridge

Great Meteor Tablemount

Madeira Plain

Madeira

Canary Islands

Strait of Gibraltar

Algiers

Tunis

Mediterranean Sea

Casablanca

Tripoli

Alexandria

Suez Canal

30°N

Sargasso Sea

62m (404ft)

Nares Plain

3,780m (12,402ft)

254m (833ft)

Cape Verde Plain

Sahara

Tropic of Cancer

ATLANTIC

Barracuda Fracture Zone

Puerto Rico

Cape Verde Basin

Cape Verde Terrace

Cape Verde Islands

Senegal

Dakar

AFRICA

4,700m (15,421ft)

Gambia Plain

Niger

Nile

OCEAN

Demerara Plain

Georgetown

Paramaribo

Cayenne

Amazon Fan

Vema Fracture Zone

Doldrums Fracture Zone

Sierra Leone Basin

Conakry

Monrovia

Abidjan

Accra

Lagos

Douala

Libreville

Gulf of Guinea

Niger Fan

Guinea Basin

Equator

Amazon

Belém

Fortaleza

Recife

Ceara Plain

Saint Paul Fracture Zone

Romanche Fracture Zone

Fernando de Noronha

Chain Fracture Zone

Pernambuco Basin

Ascension Fracture Zone

Ascension Island

Congo

Congo Fan

Luanda

5

SOUTH AMERICA

Tocantins

San Francisco

Salvador

Brazil Basin

5,706m (18,721ft)

Bode Verde Fracture Zone

Mid-Atlantic Ridge

Saint Helena

Saint Helena Fracture Zone

Angola Basin

5,042m (16,543ft)

Namibe

Walvis Bay

Namib Desert

INDIAN OCEAN

Hotspur Seamount

Isla Trindade

Ilhas Martin Vas

636m (2,087ft)

Rio de Janeiro

Santos Plateau

Rio Grande Rise

Rio Grande Fracture Zone

Walvis Ridge

Namibia Plain

Tropic of Capricorn

Parana

Uruguay

Porto Alegre

1,739m (5,706ft)

Tristan da Cunha

Cape Basin

5,115m (16,782ft)

Cape Town

Orange

Cape of Good Hope

30°S

6

Buenos Aires

Montevideo

Argentine Basin

Gough Fracture Zone

Discovery Tablemounts

Zapiola Ridge

Falkland Escarpment

3,667m (12,031ft)

Islas Orcadas Rise

Atlantic-Indian Ridge

Davis Seamounts

7

Falkland Islands

Falkland Plateau

1,748m (5,735ft)

South Georgia

South Sandwich Trench

America-Antarctica Ridge

Cape Horn

Yaghan Basin

Drake Passage

Scotia Sea

East Scotia Basin

South Shetland Islands

South Orkney Islands

SOUTHERN OCEAN

Atlantic-Indian Basin

60°S

Antarctic Peninsula

ne Ice Shelf

Weddell Plain

Antarctic Circle

Lazarev Sea

Astrid Ridge

Maud Rise

8

Weddell Sea

ANTARCTICA

SCALE

0 500 1,000 1,500 2,000 2,500 km

0 500 1,000 1,500 2,000 2,500 miles

30°W 0° 30°E 60°E 90°E

KEY

	sea level
	800 ft (250 m)
	1,600 ft (500 m)
	3,300 ft (1,000 m)
	6,500 ft (2,000 m)
	9,800 ft (3,000 m)
	16,400 ft (5,000 m)

☐ land
△ seamount
▽ sea depth
▼ maximum depth on map
— tectonic plate boundary

SCALE

0 50 100 150 200 250 km

0 50 150 200 250 300 miles

Iceland

THE ISLAND OF ICELAND straddles the Mid-Atlantic Ridge and is one of the few places on Earth where it is possible to walk on newly created oceanic crust. It is the site of sea floor spreading that was responsible for linking the Atlantic to the Arctic Ocean around 36 million years ago. The surrounding seas are areas of water and heat exchange between the two oceans.

ATLANTIC OCEAN B1

Denmark Strait

LENGTH 300 miles (480 km)

MINIMUM WIDTH 180 miles (290 km)

Most of the water leaving the Arctic Ocean flows into the north Atlantic through the Denmark Strait, propelled by the East Greenland Current. Icebergs from the eastern side of the Greenland Ice Sheet are carried south by this cold current, while the warm North Atlantic Drift flows northeast on the eastern side of the island, between Iceland and the Faeroe Islands. At depth, cold, dense Arctic bottom water pools to the northeast

of Iceland until it overflows the Greenland–Iceland Rise and cascades 6,500 ft (2,000 m) down into the main Atlantic basin.

This is the start of a global journey as the dense water circulates around the deepest parts of the world's oceans—the deep-water leg of the "great ocean conveyor belt" (see p.61).

In winter, sea ice builds up along the Greenland coast. Sometimes, cold winds blow east off the Greenland Ice Sheet, pushing sea ice offshore. More sea ice is created as the wind cools the exposed surface water, and a tongue of sea ice can extend south from the Greenland Sea through the Denmark Strait.

ATLANTIC OCEAN B3

Reykjanes Ridge

LENGTH 930 miles (1,500 km)

HEIGHT ABOVE SEA FLOOR 6,500 ft (2,000 m)

RATE OF SPREAD ¾ in (1.8 cm) per year

The Reykjanes Ridge is the part of the Mid-Atlantic Ridge that rises up to the ocean surface to the southwest of Iceland. The ridge clearly displays the parallel ridges and valleys that are left behind on either side of the central rift as the sea floor spreads at a divergent plate boundary. Here, the North American and Eurasian plates are moving apart at ½–1 in (1–2 cm)

per year. The parallel features become less distinct away from the ridge, as the older crust is draped in sediment in both the Reykjanes Basin and the Iceland Basin.

Before this rifting started, Greenland and Britain were almost adjacent, connected by a land bridge. The Hebrides and Faeroe Islands, Rockall, and the other banks on the eastern side of the Iceland Basin are the result of basalt floods associated with the early stages of the rift.

SURTSEY

Surtsey was born in 1963 when a volcano on the western flank of the Mid-Atlantic Ridge breached the surface of the sea off Iceland.

The Western North Atlantic

IN THE WESTERN NORTH ATLANTIC, the warm, fast-flowing Gulf Stream pulls away from the North American coast and runs northeastward as the North Atlantic Drift. The Labrador Current brings cold water south along the coast as far as the Gulf of Maine.

ATLANTIC OCEAN C2

Gulf of St. Lawrence

AREA 60,000 square miles (155,000 square km)

MAXIMUM DEPTH 7,550 ft (2,300 m)

INFLOWS Atlantic Ocean, St. Lawrence River

The gulf lies between the mouth of the St. Lawrence River and the islands of Newfoundland and Cape Breton. The Laurentian Trough, between the two islands, was scoured out by the Laurentide Ice Sheet during the last ice age. It channels sediment from the river over the edge of the continental shelf and onto the Laurentian Fan. The St. Lawrence River is the largest freshwater input to the Atlantic from the North American east coast, and its mouth is the largest estuary of its type in the world. The St. Lawrence Seaway, which was opened in 1959, gives vital shipping access to the Great Lakes.

SEA ICE IN THE GULF OF ST. LAWRENCE

ATLANTIC OCEAN B3

Gulf of Maine

AREA 35,000 square miles (90,700 square km)

MAXIMUM DEPTH 1,240 ft (377 m)

INFLOWS Atlantic Ocean; St. John, Penobscott Rivers

Like much of the continental shelf off the east coast of North America, the Gulf of Maine was above sea level during the last ice age. Georges Bank stands 330 ft (100 m) above the floor of the Gulf, and was an island until 6,000 years ago. Cape Cod and the islands of Nantucket and Martha's Vineyard are the highest standing of a series of moraines left behind as the glaciers retreated and the sea level rose.

Occasionally, the Gulf Stream lies not far offshore and the temperature of the sea off Nantucket beaches can be several degrees higher than it is off nearby Cape Cod.

North of the Gulf of Maine, the Bay of Fundy extends more than 120 miles (200 km) inland. The bay acts like a funnel, producing a tidal range of 43 ft (13 m) at its northern end, which is the highest in the world.

ATLANTIC OCEAN E2

Grand Banks

AREA 108,000 square miles (280,000 square km)

AVERAGE DEPTH 330 ft (100 m)

The Grand Banks is a large area of continental shelf, extending up to 310 miles (500 km) off Newfoundland. The area is renowned for dense sea fogs, which arise when warm, moist air from the south is chilled by the cold Labrador Current, causing condensation. The Labrador Current presents another shipping hazard by bringing icebergs to the area—the *Titanic* famously sank south of the Grand Banks in 1912.

Although turbidity currents have never been directly observed, their power was felt in 1929 when an earthquake triggered a huge sediment flow (submarine landslide) down the continental slope off the Grand Banks. Submarine telegraph and telephone cables were broken over a distance of 500 miles (800 km)—from the timing of the breaks, the speed of the flow was estimated at 25–34 mph (40–55 kph).

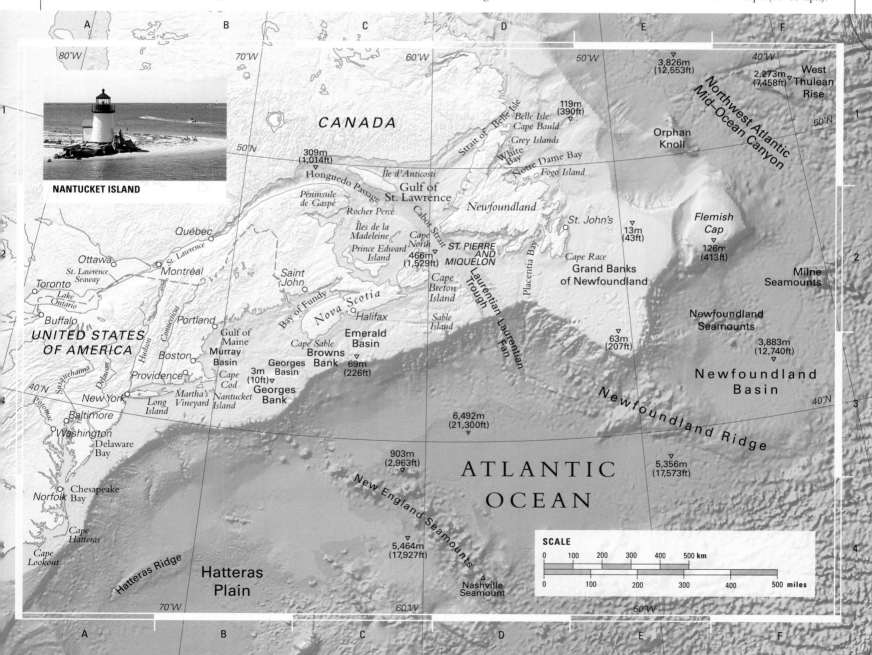

NANTUCKET ISLAND

CANADA

UNITED STATES OF AMERICA

Toronto
Lake Ontario
Buffalo
Ottawa
St. Lawrence Seaway
Montréal
Québec
St. Lawrence
Saint John
Péninsule de Gaspé
Rocher Percé
Îles de la Madeleine
Prince Edward Island
Honguedo Passage
Île d'Anticosti
Gulf of St. Lawrence
Cape North
Cabot Strait
466m (1,529ft)
ST. PIERRE AND MIQUELON
Cape Breton Island
Laurentian Trough
Strait of Belle Isle
Belle Isle
Cape Bauld
Grey Islands
White Bay
Notre Dame Bay
Fogo Island
Newfoundland
St. John's
119m (390ft)
13m (43ft)
Placentia Bay
Cape Race
Grand Banks of Newfoundland
63m (207ft)
Laurentian Fan
Newfoundland Ridge
Orphan Knoll
Flemish Cap
126m (413ft)
Milne Seamounts
Newfoundland Seamounts
3,883m (12,740ft)
Newfoundland Basin
3,826m (12,553ft)
2,273m (7,458ft)
West Thulean Rise
Northwest Atlantic Mid-Ocean Canyon

309m (1,014ft)
New York
Baltimore
Washington
Delaware Bay
Norfolk
Chesapeake Bay
Cape Hatteras
Cape Lookout
Hatteras Ridge
Hatteras Plain
New England Seamounts
Nashville Seamount
5,464m (17,927ft)
903m (2,963ft)
6,492m (21,300ft)
5,356m (17,573ft)
ATLANTIC OCEAN

Portland
Boston
Providence
Long Island
Martha's Vineyard
Nantucket Island
Cape Cod
Gulf of Maine
Murray Basin
Georges Basin
Georges Bank
Browns Bank
Cape Sable
Nova Scotia
Bay of Fundy
Halifax
Emerald Basin
Sable Island
3m (10ft)
69m (226ft)

Connecticut
Hudson
Delaware
Susquehanna
Potomac

SCALE

| 0 | 100 | 200 | 300 | 400 | 500 km |

| 0 | 100 | 200 | 300 | 400 | 500 miles |

The North Sea and Baltic Sea

A NUMBER OF SHALLOW SEAS COVER the continental shelf off the northwest coast of Europe, the largest of which is the North Sea. The North Atlantic Drift brings warmer water into the region, producing the mild climate enjoyed by adjacent coastal areas.

ATLANTIC OCEAN D1

Norwegian Sea

AREA 534,000 square miles (1.4 million square km)

MAXIMUM DEPTH 13,020 ft (3,970 m)

INFLOWS Central North Atlantic, numerous Norwegian fjords

The Norwegian Sea lies between Norway and Iceland, separated from the main part of the north Atlantic by the submarine Faeroe–Iceland Ridge. Although situated at high latitude, this sea is kept free of ice by the warm, salty North Atlantic Drift, which flows from the southwest between Scotland and Iceland and continues into the Barents

PULPIT ROCK AT LYSEFJORD
Fjords cut into the Atlantic coast of Norway, showing where ice-age glaciers scoured deep valleys below today's sea level.

Sea (see p.427) as the Norwegian Atlantic Current. This relatively warm water is the reason for the Norwegian port of Bergen's reputation as the wettest place in Europe, with rain expected at least 275 days a year.

ATLANTIC OCEAN G5

Baltic Sea

AREA 149,000 square miles (386,000 square km)

MAXIMUM DEPTH 1,473 ft (449 m)

INFLOWS Vistula, Oder, Western Divina Rivers

The Baltic is a shallow, virtually enclosed inland sea with little tide. It does not benefit from the warmth of the North Atlantic Drift, and its northern branches, the Gulf of Bothnia and the Gulf of Finland, ice over in the

winter. A large influx of river water gives the Baltic a low salinity—it is the largest area of brackish water in the world. Its only outflow is to the North Sea via the Danish Straits (three channels linking the Baltic to the Kattegat), Kattegat Bay, and the Skagerrak Strait. There is a weak influx of dense salt water at depth that isolates the basin floor from the surface waters, producing an oxygen-depleted dead zone. Without significant out-flows, the Baltic Sea is vulnerable to pollution carried in by rivers and from large population centers on its coasts. Although there is a sea route to the North Sea, there is also a shorter, more sheltered route through the Kiel Canal.

ACROSS THE ORESUND STRAIT
The 10-mile (16-km) long Oresund Bridge links the Danish capital Copenhagen with Malmo in Sweden.

ATLANTIC OCEAN D6

North Sea

AREA 220,000 square miles (570,000 square km)

MAXIMUM DEPTH 2,300 ft (700 m)

INFLOWS North Atlantic; Elbe, Weser, Ems, Rhine, Scheldt, Thames, Humber Rivers

Water from the Atlantic enters the North Sea between the Shetland and Orkney Islands, flowing south down the Scottish and English coasts. Warmer Atlantic water also enters from the English Channel and flows east along the Dutch coast, resulting in a counterclockwise circulation. The largest sand banks on the North Sea floor, including the Dogger, Jutland,

and Fisher banks, are terminal moraines marking the southern edge of the ice sheet during the last ice age, when the bottom of the sea was exposed by lower sea levels. A trough located to the west of the Norwegian Trench is buried under thick sediments that contain oil and gas deposits.

OIL RIG IN THE NORTH SEA

D · E · F · G · H · I

Norwegian Basin

Voring Plateau
1,280m
(4,200ft)

Norwegian Sea

Røst Bank

Vestfjorden

Bodo

Traena Deep

Arctic Circle

SCALE
0 50 100 150 200 250 km
0 50 100 150 200 250 miles

KEY
sea level
800 ft (250 m)
1,600 ft (500 m)
3,300 ft (1,000 m)
6,500 ft (2,000 m)
9,800 ft (3,000 m)
16,400 ft (5,000 m)

□ land
△ seamount
▽ sea depth
▼ maximum depth on map
— tectonic plate boundary

Traena Bank
222m (738ft)

Sklinna Bank

Halten Bank

SWEDEN

Luleå

Oulu

85m (279ft)

1,320m (4,331ft)

Trondheimsfjorden
Frøya Bank
94m (308ft)

Hitra

Trondheim

Umeå

Gulf of Bothnia

FINLAND

Vaasa

Shetland Islands

Lerwick

Norwegian Trench

Sognefjorden

NORWAY

Bergen

60°N

Turku (Åbo)

Helsinki

St Petersburg
RUSSIAN FEDERATION

Gulf of Finland

4

ir Isle

Hardangerfjorden

Walker Bank

Boknafjorden

Stavanger

Kristiansand

Lindesnes

700m (2,279ft)

Oslo

Gävle

Stockholm

Åland

Tallinn

ESTONIA

Lake Peipus

Vänern

Norrköping

175m (574ft)

Hiiumaa

Lake Pskov

uth nk

Gotland Basin

Saaremaa

Gulf of Riga

LATVIA

75m (246ft)

Devil's Hole

67m (200ft)

Skagerrak

Kattegat

Vättern

Gothenburg

Halmstad

Kalmar

Öland

Gotland

Ventspils

Riga

Western Dvina

5

North Sea

Great Fisher Bank

Jutland Bank

13m (43ft)

DENMARK

Aalborg

Laeso

Århus

Jylland

Karlskrona

Helsingborg

9m (30ft)

Baltic Sea

Courland Lagoon

Liepaja

LITHUANIA

Klaipeda

6

Barmade Bank

Dogger Bank

Weiss Bank

15m (49ft)

Odense

Fyn

Copenhagen

Sjaelland

Malmo

Bornholm

Gulf of Danzig

RUSSIAN FEDERATION

Kaliningrad

ingston on Hull

The Wash

Well Bank

Vlieland Bank

Friesian Islands

Kiel

Kiel Bay

Kiel Canal

Helgoland Bay

Lolland Falster Bay

Mecklenburg Bay

Rugen

Rugen

Pomeranian Bay

Gdansk

POLAND

Szczecin

BELARUS

NETHERLANDS

Amsterdam

Rotterdam

Hamburg

Bremen

Ems

Weser

Elbe

Mittelandkanal

Oder

Elbe

7

ondon

Dover

Strait of Dover

Calais

BELGIUM

Antwerp

Maas

Meuse

GERMANY

Rhine

Main

LUXEMBOURG

CZECH REPUBLIC

50°N

8

e Havre

FRANCE

Seine

Mosel

10°E

20°E

STRAIT OF DOVER
The Strait of Dover is one of the world's busiest shipping lanes, linking the English Channel to the North Sea and giving shipping access to Europe's ports. It also provides ferry links between Britain and the European mainland, although much of this traffic now uses the Channel Tunnel.

MIDDELGRUNDEN WIND FARM
Arranged in an elegant curve 1.2 miles
(2 km) east of Copenhagen, Denmark,
each of this wind farm's 20 turbines can
produce 2 megawatts of electricity.

Declining oil stocks, threats to fuel supplies, and the risks of climate change are increasingly focusing attention on alternative sources of energy that do not generate greenhouse gases. After hydroelectricity, wind-farming is the most advanced source of renewable energy. It is easiest to build wind farms on land, but many people do not want turbines built near their homes. Wind farms built at sea are less controversial, although costlier to build and maintain. Since it is usually windier at sea and there are no hills or trees to cause turbulence, offshore turbines are more efficient than land-based ones.

The world's first commercial offshore wind farm was built in the Baltic (see below). Other large wind farms are in place or under construction in the Kattegat sea area; in British, German, Danish, Dutch, and Belgian areas of the North Sea; and off the coast of China. There are also wind farms in the Mediterranean and off Ireland, Japan, Norway, and Spain, and one is planned in South Korea. The first US offshore wind farm was completed in 2016 off the East Coast. Further North American farms are proposed for the West Coast, Great Lakes, and Canada.

Because they burn no fossil fuels, wind farms can help reduce greenhouse gases. However, the overall benefits may take some time to appear. It can take 1,000 tons of concrete just to build the foundations of an offshore turbine, and concrete production is one of the biggest sources of greenhouse gases.

Baltic and Kattegat Wind Farms

The countries around the Baltic Sea and the Kattegat sea area have played a central role in the development of offshore wind energy. The world's first commercial offshore wind farm was commissioned in 1991 near the Danish fishing port of Vindeby (now decommissioned). The Baltic and Kattegat are ideal for wind-farming because of their low average depths. The first wind farms in this region were sited in water less than 33 ft (10 m) deep.

● wind farms   ● wind farms under construction or with consent authorized

NORWAY

North Sea

SWEDEN

Kattegat

DENMARK

Middelgrunden

Vindeby   *Baltic Sea*

Rødsand ll

POLAND

GERMANY

OFFSHORE WIND FARMS

UNDER CONSTRUCTION

MASSIVE BLADES
Wind-turbine blades are assembled on the dockside, then carried to the tower on a barge. The rotors reach 558 ft (170 m) in diameter.

FOUNDATIONS
The concrete foundations of a turbine are cast onshore in a dry dock. They are then floated out to sea and sunk in shallow water on site.

POWER DISTRIBUTION

SUBSTATION   The electricity generated by offshore turbines has to be transmitted to land. This substation collects power from 72 turbines at the Nysted Offshore Wind Farm off southern Denmark, and transforms it from 33,000 to 132,000 volts for transmission ashore via 30 miles (48 km) of submarine cable.

LARGE WIND

ENBW BALTIC 2 Although many Baltic wind farms have now been surpassed in both number of turbines and peak capacity, the Baltic Sea remains a good location for renewable energy. EnBW Baltic 2, pictured here, is spread over 10 square miles (27 square km). It has 80 turbines that can generate 288 megawatts of power, serving around 340,000 households.

CASUALTIES

WHITE-TAILED EAGLE Wind energy brings huge benefits, but turbines can harm birds. Each year, between six and nine white-tailed eagles are found dead at one Norwegian wind farm. Simple steps, like painting the turbine blades, can reduce casualties by 70 percent.

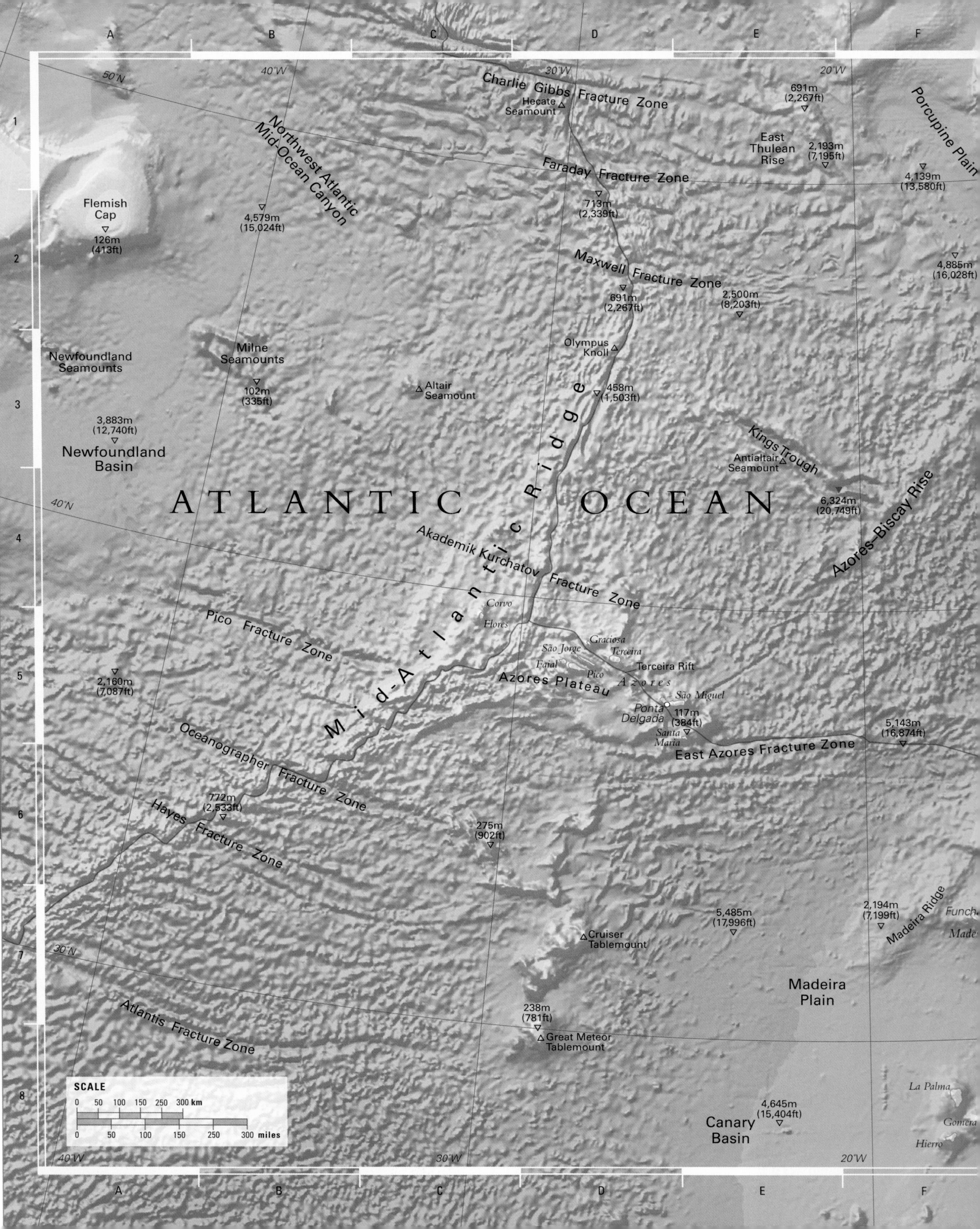

A B C D E F

1

50°N

40°W 30°W 20°W

Charlie Gibbs Fracture Zone

Hecate
Seamount

691m
(2,267ft)

Northwest Atlantic
Mid-Ocean Canyon

Faraday Fracture Zone

East
Thulean
Rise

2,193m
(7,195ft)

Porcupine Plain

4,139m
(13,580ft)

Flemish
Cap

126m
(413ft)

4,579m
(15,024ft)

713m
(2,339ft)

Maxwell Fracture Zone

2

4,885m
(16,028ft)

Newfoundland
Seamounts

Milne
Seamounts

691m
(2,267ft)

2,500m
(8,203ft)

Olympus
Knoll

3

102m
(335ft)

Altair
Seamount

458m
(1,503ft)

Kings Trough

3,883m
(12,740ft)

Antialtair
Seamount

Newfoundland
Basin

ATLANTIC   OCEAN

40°N

6,324m
(20,749ft)

Azores–Biscay Rise

4

Akademik Kurchatov Fracture Zone

Pico Fracture Zone

Corvo

Flores

Graciosa
Terceira

São Jorge

Terceira Rift

Mid-Atlantic Ridge

Faial
Pico

Azores

5

2,160m
(7,087ft)

Oceanographer Fracture Zone

Azores Plateau

São Miguel

Ponta
Delgada

117m
(384ft)

5,143m
(16,874ft)

772m
(2,533ft)

Santa
Maria

Hayes Fracture Zone

6

275m
(902ft)

East Azores Fracture Zone

5,485m
(17,996ft)

2,194m
(7,199ft)

Madeira Ridge

Funch

7

30°N

Cruiser
Tablemount

Madeira
Plain

Mad

Atlantis Fracture Zone

238m
(781ft)

Great Meteor
Tablemount

La Palma

8

SCALE

0 50 100 150 250 300 km

0 50 100 150 250 300 miles

4,645m
(15,404ft)

Canary
Basin

Gomera

Hierro

40°W 30°W 20°W

A B C D E F

Map labels (left page):

G | H | I

10°W IRELAND UNITED KINGDOM 0° 50°N
Porcupine Bank
Cork Bristol Channel
Celtic Sea Plymouth Isle of Wight
Porcupine Seabight
38m (125ft) Isles of Scilly Land's End English Channel Seine
Goban Spur Channel Islands Cherbourg 1
Celtic Shelf Golfe de St-Malo
99m (325ft) Brest FRANCE
Pointe du Raz Loire
Belle Île Nantes 2
Plateau de Rochebonne
Île de Ré
Île d'Oléron
Biscay Plain Bay of Biscay Bordeaux
4,870m (15,978ft)
Charcot Seamounts Cabo de Ajo Biarritz 3
Theta Gap Cabo Ortegal Costa Verde Santander Donostia–San Sebastián
Gijón
La Coruña
492m (1,614ft) Galicia Bank Cabo Fisterra Vigo Douro
Iberian Plain SPAIN 40°N 4
Oporto PORTUGAL
Cabo Mondego Tagus
5,536m (18,164ft) Guadiana
Lisbon
Cabo de Roca Cabo Espichel Cabo de Sines
Tagus Plain Algarve Guadalquivir 5
Horseshoe Seamounts Gorringe Ridge Cabo de São Vicente Gulf of Cadiz Gibraltar Mediterranean Sea
Gettysburg Seamount Strait of Gibraltar Ceuta
Tangiers
Ampere Seamount Rabat 6
20m (66ft) Seine Plain Casablanca KEY
Seine Seamount 4,265m (13,993ft)
Safi
MOROCCO 7
Dacia Seamount Agadir Canyon Cap Rhir Agadir
Conception Bank
Canary Islands Lanzarote
Santa Cruz Fuerteventura 8
Tenerife Cap Juby
Las Palmas
WESTERN SAHARA 10°W
G H I

KEY
— sea level
— 800 ft (250 m)
— 1,600 ft (500 m)
— 3,300 ft (1,000 m)
— 6,500 ft (2,000 m)
— 9,800 ft (3,000 m)
— 16,400 ft (5,000 m)

☐ land
△ seamount
▽ sea depth
▼ maximum depth on map
— tectonic plate boundary

The East Atlantic

THE EASTERN NORTH ATLANTIC is renowned for its winter storms, which batter the western coasts of Europe. The energy for these storms is provided by the Gulf Stream feeding warm water into the North Atlantic Drift, which flows to the northeast. The North Atlantic Gyre pulls some of this water south along the African coast as the Canaries Current.

ATLANTIC OCEAN I3

Bay of Biscay

AREA 86,000 square miles (223,000 square km)

MAXIMUM DEPTH 15,535 ft (4,735 m)

INFLOWS Loire, Dordogne, Garonne, Adour Rivers

The Bay of Biscay lies between Brest, on the Brittany Peninsula, and the north coast of Spain. The northern half of the bay is quite shallow, overlying the continental shelf, but this steeply drops away to the Biscay Plain, which is a small, partially opened ocean basin. Ships crossing the bay experience heavy seas, as

LIGHTHOUSE ON THE BAY
The Old Lighthouse at La Raz Cap is one of several that mark treacherous rocks off the Brittany Peninsula in the Bay of Biscay.

full-size Atlantic rollers are amplified by the sudden shallow depth. There is a weak counterclockwise surface current within the bay. The Charcot Seamounts, Azores–Biscay Rise, and Kings Trough mark an inactive crustal fracture where the sea floor was once splitting apart.

ATLANTIC OCEAN D5

Azores

TYPE Volcanic islands

AREA 890 square miles (2,300 square km)

NUMBER OF ISLANDS 9

The Mid-Atlantic Ridge is the dominant seafloor feature in the eastern Atlantic region, with a central trough and numerous transform fracture zones. Just east of the ridge, the Azores island group straddles the triple junction between the Eurasian, African, and North American plates.

The islands rise from the extensive Azores Plateau, an area of thickened ocean crust. Although volcanic in origin, the oldest islands also include substantial accumulations of limestone and clay sediments. A mantle hotspot (see p.51) underlies the plateau and seems to be slowly spreading it apart at the Terceira Rift, a fracture that links the East Azores Fracture Zone to the Mid-Atlantic Ridge. The last volcanic eruption in the Azores was in 1957, when the Capelinhos volcano produced a cinder island (an island composed of lava fractures called cinders) off Faial's coast.

ATLANTIC OCEAN G8

Canary Islands

TYPE Volcanic islands

AREA 2,900 square miles (7,400 square km)

NUMBER OF ISLANDS 7

The name of these islands derives not from the yellow bird of the same name, but from the Latin word for dogs, *Canaria*. The islands are volcanic, overlying a mantle hotspot. Pico del Teide on Tenerife is the third largest volcano on Earth, rising more than 12,000 ft (3,700 m) above sea level, or almost 23,000 ft (7,000 m) from

the sea floor. It last erupted in 1909. Teide's slopes are unstable, and there is evidence that huge landslides have occurred in the past. There is also a risk that volcanic activity or earth tremors could cause part of La Palma island to slip into the sea, resulting in an enormous tsunami. Such an event would threaten the coasts of the north Atlantic, including heavily populated parts of North America, with inundation. La Palma's volcano, Cumbre Vieja, erupted in 1949, 1971, and again in 2021, causing the evacuation of thousands of people and destroying hundreds of homes.

TENERIFE ISLAND

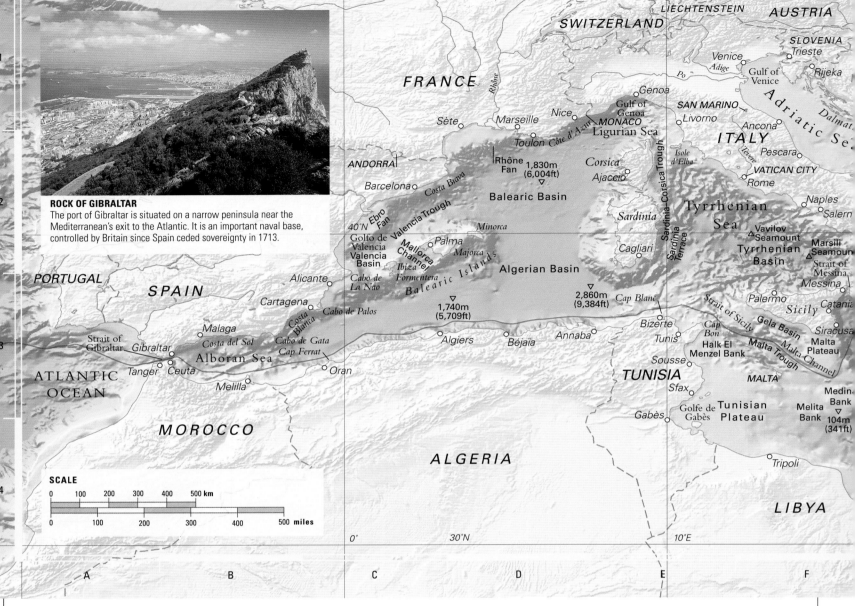

ROCK OF GIBRALTAR
The port of Gibraltar is situated on a narrow peninsula near the Mediterranean's exit to the Atlantic. It is an important naval base, controlled by Britain since Spain ceded sovereignty in 1713.

SCALE

0   100   200   300   400   500 km

0   100   200   300   400   500 miles

The Mediterranean Sea and Black Sea

THE MEDITERRANEAN IS AN ALMOST enclosed sea, with high evaporation and salinity, a very small tidal range, and a complex floor. The adjacent Black Sea is the last remnants of the Tethys Ocean, which closed as Africa converged with Eurasia.

ATLANTIC OCEAN D2

Western Mediterranean

AREA 328,000 square miles (850,000 square km)

MAXIMUM DEPTH 11,800 ft (3,600 m)

INFLOWS Atlantic Ocean; Ebro, Rhône Rivers

The entire Mediterranean loses three times more water by evaporation than it gains from rainfall and rivers combined. This loss is balanced by a surface inflow from the Atlantic through the Strait of Gibraltar. The inflow continues as an eastward current along the north African coast, giving rise to a counterclockwise circulation

in the western Mediterranean. At depth there is a strong undercurrent of outflowing salty water. The flat floors of the Algerian and Balearic basins are underlain by deep sediments. In contrast, the Tyrrhenian Sea contains many seamounts and ridges. A chain of active volcanoes (including Etna, Stromboli, and Vesuvius) is found on the sea's eastern margin, where the African Plate is subducting beneath the Eurasian Plate. The eastward flow of surface water continues through the Strait of Sicily into the eastern Mediterranean. The narrower Strait of Messina, between Italy and Sicily, is notorious for its whirlpools, the inspiration for the Greek mythological sea monsters Scylla and Charybdis.

ATLANTIC OCEAN H3

Eastern Mediterranean

AREA 637,000 square miles (1.65 million square km)

MAXIMUM DEPTH 16,720 ft (5,095 m)

INFLOWS Black Sea; Adige, Nile, Po Rivers

The eastern and western parts of the Mediterranean are separated by Sicily and the submerged Malta and Tunisian plateaus. The eastward flow from the western Mediterranean continues along the African coast, and a counterclockwise circulation prevails in the eastern Mediterranean, and in the Ionian, Aegean, and Adriatic seas. Surface water becomes more saline through evaporation as it travels east,

and starts to sink after cooling by winter winds. It then returns westward, exiting through the Strait of Gibraltar about 150 years after entering. The sea floor is dominated by the Mediterranean Ridge, a result of compression between the convergent African and Eurasian plates. These sediments are older—70 million years compared with 25 million years in the western Mediterranean. The Adriatic Sea is a shallow branch of the eastern Mediterranean. Rising sea levels at the end of the last ice age flooded valleys parallel to its eastern shore, giving rise to the islands of the Dalmatian coastline.

VENICE LAGOON

Venice was built in the shallow waters of a lagoon in the Adriatic. Its merchants grew rich by controlling access to the Silk Route.

KEY

	sea level
	800 ft (250 m)
	1,600 ft (500 m)
	3,300 ft (1,000 m)
	6,500 ft (2,000 m)
	9,800 ft (3,000 m)
	16,400 ft (5,000 m)

☐ land

△ seamount

▽ sea depth

▼ maximum depth on map

— tectonic plate boundary

VOLCANIC ISLANDS
The islands of Santorini in the Aegean Sea are the remains of an explosive volcanic eruption about 3,500 years ago.

ATLANTIC OCEAN H2

Aegean Sea

AREA 83,000 square miles (214,000 square km)

MAXIMUM DEPTH 10,800 ft (3,294 m)

INFLOWS Black Sea, Mediterranean Sea

The Aegean Sea contains more than 1,000 islands and is the source of most of the Mediterranean's cold, saline deep water. Before 1990 this source was in the Adriatic, but climate changes have led to increased winter cooling in the Aegean. It is a geologically complex area, as the Aegean microplate and the Anatolian Plate to the east are caught between the converging African and Eurasian plates. The Aegean crust is of continental thickness, but has been stretched and thinned, notably in the area of the Cretan Trough, so that much of it is now below sea level. The Hellenic Trough and Pliny Trench mark where the African Plate is subducting beneath

the Aegean microplate. The Aegean Volcanic Arc stretches from Greece to Turkey through the southern Cyclades. These volcanoes are dormant or extinct, but earthquakes still occur at a depth of 95–105 miles (150–170 km). The islands of Santorini, in the southern Cyclades, are the remains of an explosive volcanic eruption around 1640 BCE.

This was the largest volcanic event of the last 10,000 years and may have caused the downfall of Crete's Minoan civilization. Behind the volcanic arc, the main Cyclades sit on top of a subsided plateau. At the northern end of the Aegean, a transform fault marks the contact with the Eurasian Plate, an area prone to strong, shallow earthquakes.

SKIATHOS ISLAND
Aegean islands consist mostly of hard metamorphic and volcanic rocks, so their coasts often show steep cliffs, headlands, and wave-cut features.

ATLANTIC OCEAN J2

Black Sea

AREA 163,000 square miles (422,000 square km)

MAXIMUM DEPTH 7,200 ft (2,200 m)

INFLOWS Mediterranean Sea, Sea of Azov; Danube, Dniester, Dnieper, Kizil Irmak Rivers

The Black Sea is an enclosed inland sea, connected to the Mediterranean Sea via the Dardanelles, the Sea of Marmara, and the Bosporus. There is negligible exchange of water with the Mediterranean, and the surface waters of the Black Sea are about half as saline as the eastern Mediterranean. A previous small outflow through the Bosporus to the Aegean appears to have been reversed due to reduced inflow after the damming of some of the rivers feeding the Black Sea.

Although the surface waters are relatively fresh, below about 330–490 ft (100–150 m) lies a highly saline water body with very slow turnover. Decaying organic matter consumes all the oxygen in this water, making the Black Sea the world's largest oxygen-free marine system—the deep water is essentially dead. The basin is an isolated

BLACK SEA SHIPPING
The Bosporus, the narrowest strait open to international navigation, connects the Black Sea with the Sea of Marmara.

remnant of the north shore of the ancient Tethys Ocean. The southern part of the Black Sea is deep, but it is not as deep as the Mediterranean, and the underlying crust is thicker than most ocean crust.

The northern parts—the Sea of Azov and the Gulf of Odessa—overlie a shallow continental shelf. The delta of the Danube, Europe's longest river, extends from the western shore, and Danube waters have carried sediment across the edge of the shelf to build up a thick cone of sediment.

The Gulf of Mexico and Caribbean Sea

THE CARIBBEAN SEA AND GULF OF MEXICO form a semi-enclosed extension of the north Atlantic. A low input of fresh water and high evaporation rates make the surface waters highly saline. The area is subject to violent storms and some volcanic activity.

ATLANTIC OCEAN B3

The Gulf of Mexico

AREA 618,000 square miles (1.6 million square km)

MAXIMUM DEPTH 17,070 ft (5,203 m)

INFLOWS Caribbean Sea; Mississippi, Rio Grande, Apalachicola Rivers

The Gulf of Mexico is almost enclosed by parts of North America. The broad continental shelves to the north and south of the deep central basin are rich in oil deposits.

ATLANTIC OCEAN E6

The Caribbean Sea

AREA 1.06 million square miles (2.75 million square km)

MAXIMUM DEPTH 25,215 ft (7,685 m)

INFLOWS Atlantic Ocean, Magdalena River

The Caribbean Sea is a tropical body of water bounded to its south and west by South and Central America and to its north and east by the Greater and Lesser Antilles. Most of the Antilles, and some parts of the mainland coast, are fringed with coral reefs and small, low-lying islets called cays (or keys).

The underlying Caribbean Plate was once part of the Pacific Ocean floor, and it is still moving slowly eastward between the North and South American plates. A subduction zone separates the Caribbean Plate from the Atlantic Plate

Circulation is weak, and the water becomes more salty as it is heated up. Inflows from rivers and the Caribbean are balanced by an outflow of warm, salty water—the beginnings of the Gulf Stream—via the Straits of Florida to the east. This channel runs between two limestone plateaus—the Florida peninsula, above sea level to the north, and the Bahamas, a submerged plateau topped by low-lying islands to the south. The coasts of the Gulf are affected by powerful hurricanes in the late summer and fall.

to the east, giving rise to the volcanic island arc of the Lesser Antilles.

An east-to-west surface current permeates the whole of the Caribbean, with water from the Guiana Current flowing in via gaps between the small islands in the east, and flowing out in to the Gulf of Mexico via the Yucatán Channel in the northwest. in the northwest.

SOUFRIERE VOLCANO ON MONTSERRAT

HUMAN IMPACT

PANAMA CANAL

Opened in 1914, the Panama Canal links the Atlantic Ocean with the Pacific, allowing ships to avoid the long journey around Cape Horn. Its construction was one of the most difficult engineering projects ever attempted, taking 10 years and costing many lives. Each year, 14,000 ships use the canal.

PANAMA GATES IN 1913

ATLANTIC OCEAN H2

The Sargasso Sea

AREA 2 million square miles (5.2 million square km)

MAXIMUM DEPTH 23,000 ft (7,000 m)

INFLOWS None

The Sargasso Sea is a large area of the north Atlantic southeast of Bermuda. It is bounded by ocean currents: the Gulf Stream to its west and north, the Canary Current far to the east, and the North Equatorial Current to its south. The area between these currents rotates slowly in a clockwise direction and is often quite calm.

Large mats of yellow-brown sargassum seaweed float on its surface, providing shelter and food for communities of small crustaceans and fish, including freshwater eels. Adult eels migrate to the Sargasso every year to mate and spawn, and their young are carried back to the rivers of North America and Europe by the Gulf Stream. Deep water in this part of the Atlantic flows from north to south.

FLORIDA KEYS
The Florida Keys are small, low-lying islands composed mainly of ancient coral reefs that are underlain by limestone.

HURRICANE LILI
Hurricanes occur frequently in the Gulf of Mexico and Caribbean Sea, often causing much damage to coastal regions.

D   E   F   G   H   I

80°W   70°W   60°W

Norfolk
Cape Hatteras

F AMERICA

Nashville
Seamount

Myrtle Beach   Onslow Bay
Long Bay
Charleston
Savannah   Richardson Hills   Hatteras Ridge
Hoyt Hills   Slope

Sohm Plain
5,464m
(17,927ft)

BERMUDA

Escarpment
Florida-Hatteras Slope

Bermuda Rise

30°N

Jacksonville

5,255m
(17,242ft)

Sargasso Sea

Blake Spur
Blake-Bahama Ridge

Blake Plateau
Blake Abyssal Plain
Blake Basin

Hatteras Plain

Eastward Knoll
Researcher Seamount

Cape Canaveral

West Florida Shelf

Tampa

Little Bahama Bank
Great Abaco Canyon
Great Abaco Island
Great Bahama Island

ATLANTIC OCEAN

Fort Lauderdale
Miami

Great Abaco Canyon
BAHAMAS   Andros Island
Tongue of the Ocean
Nassau
Eleuthera Island
Bahama Escarpment
Bahama Basin

Tropic of Cancer

Slope
Florida Keys
Pourtales Escarpment
Straits of Florida
Cay Sal Bank
Great Bahama Bank
Santaren Channel

Exuma Sound
Exuma Valley
Long Island

Vema Gap

Nares Plain

6,081m
(19,952ft)

20°N

Havana
Golfo de Batabano
Isla de la Juventud
Archipiélago de Sabana
Archipiélago de Camagüey
Cienfuegos

Acklins Island
TURKS & CAICOS ISLANDS
Great Inagua

5,868m
(19,253ft)

CUBA
Greater

Mouchoir Passage

Yucatan Basin

Golfo de Guacanayabo
Santiago de Cuba
Guantanamo Bay

Hispaniola Basin
8,962m
(29,404ft)
Puerto Rico Trench

Leeward Islands

6,691m
(21,953ft)

CAYMAN ISLANDS
Cabo Cruz
Cayman Ridge
Cayman Trench   7,680m
5,830m   (25,198ft)
(19,128ft)

Windward Passage
Cap-Haïtien
Golfe de la Gonave
HAITI
Jérémie
Port-au-Prince
Hispaniola
DOMINICAN REPUBLIC

Mona Canyon
647m
(2,123ft)
Mona Passage
PUERTO RICO
San Juan

BRITISH VIRGIN ISLANDS
Virgin Passage
Anegada Gap
Anegada Passage
Tintamarre Spur
ANGUILLA

Barracuda Ridge

Montego Bay
NAVASSA ISLAND
Jamaica Ridge

Santo Domingo
Ponce
VIRGIN ISLANDS
ST KITTS & NEVIS
St John's
ANTIGUA & BARBUDA

JAMAICA
Kingston Channel
Antilles
5,550m
(18,210ft)
Muertos Trough
Gibbs Seamount
MONTSERRAT
Basse-Terre
GUADELOUPE
Guadeloupe Passage

Pedro Bank
960m
(3,150ft)
4,536m
(14,883ft)

Dominica Passage
DOMINICA

Nicaraguan Rise
Rosalind Bank
Sue Ridge

Beata Ridge

Venezuelan Basin

Martinique Passage
Fort-de-France
MARTINIQUE

Caribbean Sea

4,557m
(14,952ft)
4,828m
(15,841ft)

St Lucia Channel
ST LUCIA

Aves Ridge
Grenada Basin
2,997m
(9,833ft)
St Vincent Channel
BARBADOS
Bridgetown
Barbados Trough

Nicaraguan Rise
Mosquito Bank
Pedro Escarpment
Aruba Gap

ST VINCENT
Tobago Basin
GRENADA
St George's
2,319m
(7,609ft)

San Andrés Trough
Calarca Bank
Saury Seamount

Colombian Basin

Punta Gallinas
ARUBA
Oranjestad
Curaçao
Bonaire
Willemstad
Los Roques Basin
Islas los Roques
Isla de Margarita
TRINIDAD & TOBAGO

10°N

Ibara Trough
Mono Rise
Clark Basin
3,531m
(11,585ft)
3,493m
(11,460ft)

Ríohacha
NETHERLANDS ANTILLES
Bonaire
1,390m
(4,561ft)
Bonaire Basin
Coro
Isla de Tortuga
Cumaná
Gulf of Paria
Port-of-Spain
Trinidad

Barranquilla
Santa Marta
Cartagena
Gulf of Venezuela
Lake Maracaibo
Maracaibo
Caracas
Barcelona

nón
Volcán Bank
Colón
Panama Canal
Panama City
Gulf of Darien
Magdalena

VENEZUELA

PANAMA
Orinoco

80°W   70°W   60°W

COLOMBIA

SCALE
0 100 200 300 400 500 km
0 100 200 300 400 500 miles

KEY
sea level
800 ft (250 m)
1,600 ft (500 m)
3,300 ft (1,000 m)
6,500 ft (2,000 m)
9,800 ft (3,000 m)
16,400 ft (5,000 m)
land
seamount
sea depth
maximum depth on map
tectonic plate boundary

The Central Atlantic

THE MID-ATLANTIC RIDGE, THE WORLD'S LONGEST mountain chain, is the main sea-floor feature in the central Atlantic. Either side of the ridge are two flat abyssal plains, the Angola and Brazil basins. The dominant Atlantic gyres meet in the central Atlantic. Both are westward-flowing near the Equator, but are separated by the Equatorial Countercurrent, a strong eastward surface flow, and the Equatorial Undercurrent, an even stronger flow 330 ft (100 m) deep. The Canaries Current flows south along the North African coast, becoming the North Equatorial Current. In the south, the cold Benguela Current flows up the African coast, then away from the coast as the South Equatorial Current. This current splits where it reaches South America, becoming the Guiana and rather weak Brazil currents.

PILLOW LAVA
Pillows of lava form at the Mid-Atlantic Ridge when extruded lava rapidly cools upon contact with cold water.

ATLANTIC OCEAN G3

Mid-Atlantic Ridge

LENGTH 6,200 miles (10,000 km)

AVERAGE HEIGHT ABOVE SEA FLOOR 9,800 ft (3,000 m)

RATE OF SPREAD 1–2 in (2–5 cm) per year

The Mid-Atlantic Ridge dissects the entire length of the Atlantic in a series of rifts and fractures. In the central Atlantic, with relatively narrow continental shelves, it is easy to see where the coasts on either side of the ocean were once joined. The ridge mostly lies 4,900–9,800 ft (1,500–3,000 m), below sea level, although the volcanoes of Ascension Island and Saint Helena breach the surface. The

ASCENSION ISLAND
Rising just to the west of the Mid-Atlantic Ridge, Ascension Island has 44 distinct volcanic craters.

sea floor gets deeper and older away from the center, and smoother as its features are covered in sediments. The featureless abyssal plains of the Angola and Brazil basins lie to the east and west respectively. The ridge is displaced east–west at numerous points by transform faults, where the African and South American plates are moving past each other. These fracture zones extend some distance from the ridge, sometimes as active faults where parts of the same plate are moving at different speeds.

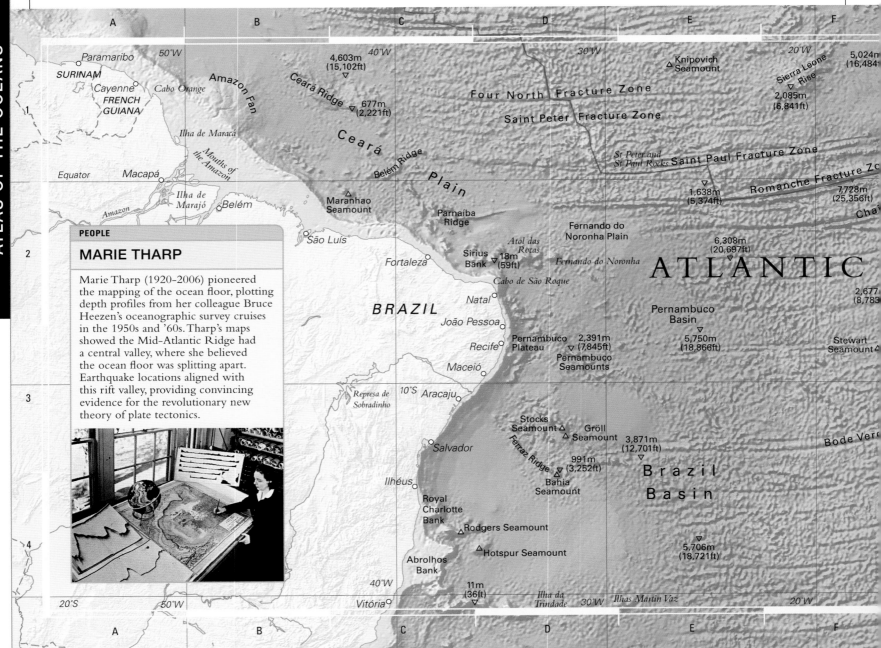

PEOPLE

MARIE THARP

Marie Tharp (1920–2006) pioneered the mapping of the ocean floor, plotting depth profiles from her colleague Bruce Heezen's oceanographic survey cruises in the 1950s and '60s. Tharp's maps showed the Mid-Atlantic Ridge had a central valley, where she believed the ocean floor was splitting apart. Earthquake locations aligned with this rift valley, providing convincing evidence for the revolutionary new theory of plate tectonics.

ATLANTIC OCEAN I1

Gulf of Guinea

AREA 500,000 square miles (1.4 million square km)

MAXIMUM DEPTH 17,070 ft (5,204 m)

INFLOWS Atlantic Ocean, Niger, Volta Rivers

Part of the north Atlantic's Canaries Current continues along the African coast and into the Gulf of Guinea as the eastward-flowing Guinea Current. The main freshwater input to the gulf is provided by the River Niger, which has an extensive depositional fan, up to 2.5 miles (4 km) thick. An even greater source of fresh water for the south Atlantic is from the Congo River to the south. Large oil and gas reserves have accumulated in the sediments of the Niger Delta and Fan, and Nigeria is Africa's biggest oil producer. Smaller deposits lie in the Congo Fan and in the continental shelf off Gabon, and deeper water in the Gulf of Guinea is now being explored for oil. When the Atlantic Ocean basin started to open 180 million years ago, three rifts opened up in the crust, forming a tectonic triple-junction. Two of the rifts continued opening to the south and the west, forming today's south Atlantic Ocean. Activity in the third rift, to the northeast, ceased rather quickly. The site of this stalled spreading center is marked by a chain of extinct volcanoes, including the islands of Annabon, São Tomé, Principe, and Bioco in the Gulf of Guinea, and Mount Cameroon inland. São Tomé rises 6,640 ft (2,020 m) above sea level.

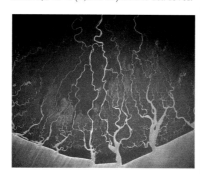

NIGER DELTA
This image, taken from a space shuttle, shows the delta coastline of the Niger River, and sediments being carried offshore.

ATLANTIC OCEAN K4

Skeleton Coast

TYPE Secondary coast

LENGTH 870 miles (1,400 km)

The cold Benguela Current hugs the west coast of southern Africa and dominates its climate. Although prevailing winds are from the sea, the air above the cold water carries little moisture, and the adjacent coast is a desert. When warm air from the land meets the cold sea air, dense fogs often form. This can be a hazard to navigation, as testified to by the numerous ship hulks along the notorious Skeleton Coast. Even without the fog, any vessel disabled by engine trouble is driven toward the shore by wind and current, and the nearest ports are quite distant. Many sailors who have survived being shipwrecked here have had little choice but to attempt the arduous journey out of the Namib Desert on foot. Namibia's coastal waters are dredged for diamonds, as the Benguela Current carries sediments from the Orange River north along the coast. These sediments include large quantities of gem-quality diamonds washed down from the South African interior. A rich fishery is another by-product of the Benguela Current, which causes upwelling of nutrient-rich waters.

SHIP'S SKELETON
Although there are doubtless human skeletons on the Skeleton Coast, the ones most likely to be seen are those of rusting ships.

ATLAS OF THE OCEANS

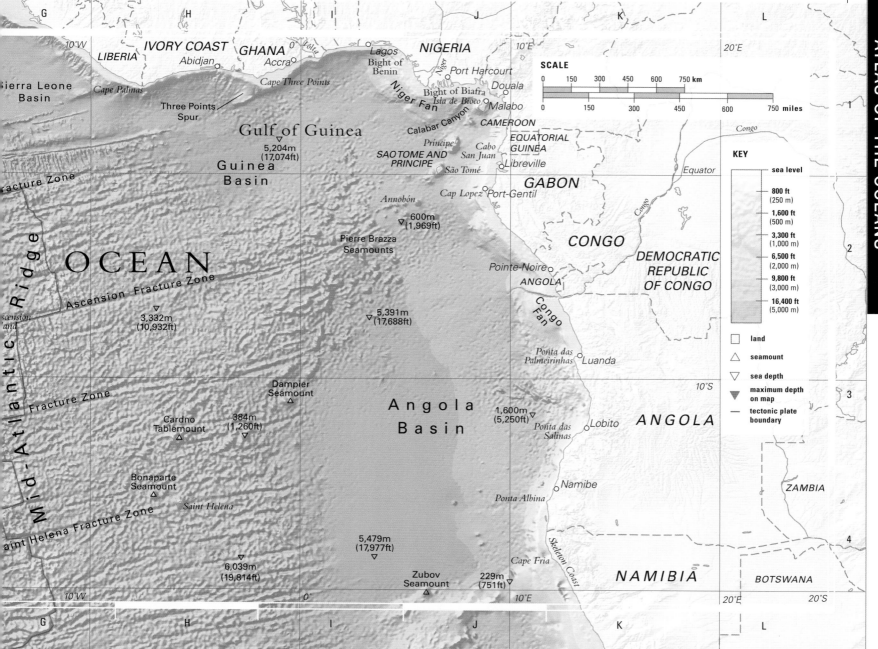

Map features:
- Sierra Leone Basin
- LIBERIA, IVORY COAST, Abidjan, GHANA, Accra, Cape Palmas, Cape Three Points, Three Points Spur
- NIGERIA, Lagos, Bight of Benin, Port Harcourt, Niger Fan, Bight of Biafra, Isla de Bioco, Malabo, Douala, CAMEROON, Calabar Canyon
- Gulf of Guinea 5,204m (17,074ft)
- Guinea Basin
- Principe, SAO TOME AND PRINCIPE, Cabo San Juan, São Tomé, EQUATORIAL GUINEA, Libreville, Equator
- Annobón 600m (1,969ft)
- Pierre Brazza Seamounts
- GABON, Cap Lopez, Port-Gentil
- OCEAN
- Ascension Fracture Zone
- Mid-Atlantic Ridge
- Ascension Island 3,332m (10,932ft)
- 5,391m (17,688ft)
- CONGO, Pointe-Noire, ANGOLA, Congo Fan
- DEMOCRATIC REPUBLIC OF CONGO
- Fracture Zone
- Dampier Seamount
- Cardno Tablemount 384m (1,260ft)
- Angola Basin
- Ponta das Palmeirinhas, Luanda
- 1,600m (5,250ft), Ponta das Salinas, Lobito, ANGOLA
- Bonaparte Seamount, Saint Helena
- Saint Helena Fracture Zone
- Namibe, Ponta Albina, ZAMBIA
- 5,479m (17,977ft)
- 6,039m (19,814ft), Zubov Seamount
- 229m (751ft), Cape Fria, Skeleton Coast, NAMIBIA, BOTSWANA

SCALE
km: 0 150 300 450 600 750 km
miles: 0 150 300 450 600 750 miles

KEY
- sea level
- 800 ft (250 m)
- 1,600 ft (500 m)
- 3,300 ft (1,000 m)
- 6,500 ft (2,000 m)
- 9,800 ft (3,000 m)
- 16,400 ft (5,000 m)
- □ land
- △ seamount
- ▽ sea depth
- ▼ maximum depth on map
- — tectonic plate boundary

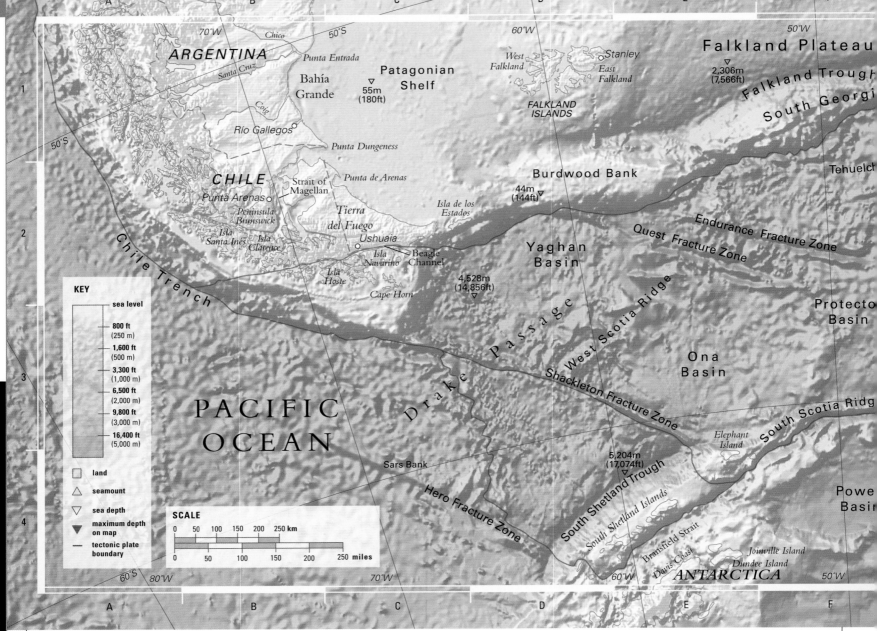

KEY

sea level

800 ft (250 m)
1,600 ft (500 m)
3,300 ft (1,000 m)
6,500 ft (2,000 m)
9,800 ft (3,000 m)
16,400 ft (5,000 m)

☐ land
△ seamount
▽ sea depth
▼ maximum depth on map
— tectonic plate boundary

SCALE

0 50 100 150 200 250 km

0 50 100 150 200 250 miles

(Map labels: ARGENTINA, Chico, Punta Entrada, Patagonian Shelf, West Falkland, Stanley, East Falkland, Falkland Plateau, 2,306m (7,566ft), Falkland Trough, South Georgi..., Santa Cruz, Bahía Grande, 55m (180m), FALKLAND ISLANDS, Coig, Río Gallegos, Punta Dungeness, Tehuelch..., CHILE, Strait of Magellan, Punta de Arenas, Burdwood Bank, 44m (144ft), Punta Arenas, Peninsula Brunswick, Tierra del Fuego, Isla de los Estados, Endurance Fracture Zone, Isla Santa Inés, Isla Clarence, Ushuaia, Yaghan Basin, Quest Fracture Zone, Chile Trench, Isla Navarino, Beagle Channel, Isla Hoste, 4,528m (14,856ft), West Scotia Ridge, Protecto... Basin, Cape Horn, Drake Passage, Shackleton Fracture Zone, Ona Basin, PACIFIC OCEAN, South Scotia Ridge, Elephant Island, 5,204m (17,074ft), Sars Bank, South Shetland Trough, South Shetland Islands, Powe... Basin, Hero Fracture Zone, Bransfield Strait, Davis Coast, Joinville Island, Dundee Island, ANTARCTICA, 70°W, 50°S, 60°W, 50°W, 60°S, 80°W, 70°W, 60°W, 50°W)

The Scotia Sea

THE COLD SCOTIA SEA AND THE SUBPOLAR waters that adjoin it lie between the south Atlantic and the Southern Ocean. Sea ice is present around the region's shorelines in winter, and icebergs calved from the Antarctic ice sheets can be found year-round.

ATLANTIC OCEAN G2
Scotia Sea

AREA 350,000 square miles (900,000 square km)

MAXIMUM DEPTH 18,300 ft (5,576 m)

INFLOWS Southern Ocean

The Scotia Sea is bounded by Tierra del Fuego and South Georgia to the north, the South Shetland and South Orkney islands to the south, and the South Sandwich Islands to the east.

It is swept by the Antarctic Circumpolar Current, which flows from the Pacific into the Atlantic through the Drake Passage. Part of this flow turns north along the eastern shore of South America as the cold Falklands Current. Where it meets the warm waters of the Brazil Current north of the Falkland Islands,

upwelling of nutrients supports a rich fishery. The Scotia Plate is moving eastward relative to the South American and Antarctic plates. The separation of South America and Antarctica began around 100 million years ago, opening up a route for Pacific Ocean currents to flow into the young south Atlantic and Indian Ocean basins—the first step in the thermal isolation of Antarctica.

ROCKHOPPER PENGUINS

ATLANTIC OCEAN B2 AND C2
Strait of Magellan

LENGTH 330 miles (530 km)

MINIMUM WIDTH 2½ miles (4 km)

The first European known to have sailed from the Atlantic into the Pacific was Portuguese explorer Ferdinand Magellan, and the strait he used between the South American mainland and Tierra del Fuego is named after him. The route is sheltered from the full might of the Southern Ocean, although it has some narrow passages that can be hazardous to navigate. It was the preferred route for Atlantic–Pacific sea trade until the confirmation of an open ocean route around Cape Horn in 1616.

Another sheltered route through the Tierra del Fuego archipelago is the Beagle Channel, named after the survey ship that carried British naturalist Charles Darwin on his scientific voyage of 1831–1836. Cape Horn is the southernmost point of South America, situated on Hoorn Island, one of the Hermite Islands to the south of Tierra del Fuego. The most

southerly passage between the major oceans, Cape Horn was discovered and named Kaap Hoorn in 1616 by a merchant navigator, in honor of his sponsors in the Dutch town of Hoorn. However, most commercial traffic between the Atlantic and the Pacific now travels via the Panama Canal.

CAPE HORN
Cape Horn is notorious for its atrocious weather conditions. Sailing around it is the peak of many sailors' ambitions.

G   H   I   J   K   L

Ridge 110m (361ft)

Fracture Zone

4,314m (14,154ft)

1,139m (3,737ft)

5,576m (18,295ft)

187m (614ft)

Scotia Sea

South Orkney Islands

Orkney Deep

Endurance Ridge

SOUTHERN OCEAN

Northwest Georgia Rise

1,608m (5,276ft)

South Georgia Rise

South Georgia

East Scotia Basin

Guevara Seamounts

Bruce Ridge

210m (689ft)

East Scotia Ridge

1,077m (3,534ft)

South Sandwich Islands

3,140m (10,302ft)

South Sandwich Trench

8,325m (27,314ft)

Ligeti Ridge

South Sandwich Fracture Zone

7,152m (23,466ft)

Islas Orcadas Rise

1,748m (5,735ft)

ATLANTIC OCEAN

3,099m (10,168ft)

5,404m (17,731ft)

1,780m (5,840ft)

SOUTH SANDWICH VOLCANO
Mount Belinda, on Montague Island, entered an eruptive phase in 2001, and was still active when this satellite image was taken in 2005.

40°W   30°W   20°W   10°W

50°S   60°S

ATLANTIC OCEAN J2

South Sandwich Trench

LENGTH 600 miles (965 km)

MAXIMUM DEPTH 27,300 ft (8,325 m)

RATE OF CLOSURE 2¾ in (7 cm) per year

Although discovered by James Cook in 1775, the South Sandwich Islands were not visited until 1818, when seal hunters landed. They were never permanently settled and remain uninhabited. With volcanic peaks rising up to 3,300 ft (1,000 m) above sea level, the islands are mostly composed of basaltic lava and covered by glaciers. North of the islands is the Protector Shoal—an undersea volcano that rises to within 100 ft (30 m) of the surface. The South Sandwich Islands mark the eastern boundary of the Scotia Sea, and the

South Sandwich Trench lies a little farther to the east. Both features are caused by tectonic processes occurring where the Scotia and South Atlantic plates meet. The Scotia Plate is split and spreading at the East Scotia Ridge, forming a new plate at its eastern end—the South Sandwich microplate. This plate is geologically young, at about 8 million years old, and buoyant. Moving eastward at about 2¾ in (7 cm) per year, it is converging with the South Atlantic Plate, resulting in the older South Atlantic Plate sinking beneath the South Sandwich Plate at a subduction zone. This zone is marked by the South Sandwich Trench and the volcanic island arc of the South Sandwich Islands (or the Scotia Arc).

SEA ICE
Sea ice clings to the shore of Bellinghausen Island, of the South Shetland group, named after the Russian explorer who discovered it in the 19th century.

ATLANTIC OCEAN F3

South Georgia Ridge

LENGTH 1,600 miles (2,500 km)

HEIGHT ABOVE SEA FLOOR 9,800 ft (3,000 m)

RATE OF RELATIVE MOTION ¼ in (0.7 cm) per year

The South Georgia Ridge marks the northern edge of the Scotia Plate, a boundary that continues east through the Tierra del Fuego archipelago. This is a transform boundary (see p.48) with the South Atlantic Plate to the north. There is a similar transform boundary marked by the South Scotia Ridge, with the Antarctic Plate to the south. Fragments of continental crust, such as Burdwood Bank and South Georgia, seem to have been left behind as South America moved west. The island of South Georgia was named by James Cook in 1775, but may have been sighted as early as 1675. It was a base for seal hunters in the 19th century, and in

the 20th century, seven whaling stations were established on the more sheltered northern shore. The last of these closed in 1965. North of the South Georgia Ridge lies the Falkland Plateau, an area of thickened ocean crust of moderate depth, and the broad continental shelf off the east coast of South America—the Patagonian Shelf. The Falkland Islands are a continental fragment left over from the breakup of Gondwana (see p.44) and the subsequent opening of the south Atlantic.

ABANDONED WHALING STATION
Old, rusting whaling ships lie in the harbor at Grytviken, a whaling station from 1904–1965, on South Georgia.

The Indian Ocean

THE INDIAN OCEAN IS THE THIRD-LARGEST ocean on Earth, lying between Africa and Australia. Sea routes across the northern Indian Ocean were opened up by traders from the Persian Gulf and by the Chinese Admiral Zheng between 1405 and 1433. The Portuguese explorer Vasco de Gama was the first European to circumnavigate Africa, reaching India in 1498.

SUNRISE IN THE STRAIT OF MALACCA

Ocean Circulation

The southern Indian Ocean is dominated by the counterclockwise South Indian Gyre. This drives the South Equatorial Current, which in turn feeds the Agulhas Current. The circulation north of the Equator is complicated by the Indian subcontinent and the annual wind reversal that characterizes the monsoon climate. High pressure over India from November to April pushes surface water in the Arabian Sea away from India, generating the North Equatorial Current and Equatorial Countercurrent. In the summer, low pressure over India gives rise to southwesterly winds, the Southwest Monsoon Current replaces the North Equatorial Current, and the Somali Current flows strongly northeast along the East African coast.

STILT FISHING
Poles are used as perches by some fishers in Sri Lanka so as not to scare away the fish.

Ocean Floor

The Indian Ocean floor is dominated by three mid-ocean ridges—the Southwest Indian Ridge, the Mid-Indian Ridge, and the Southeast Indian Ridge—which meet at a triple junction. The Indian Ocean started opening when Africa separated from Antarctica and Australia, achieving its present form when India collided with Asia 36 million years ago. Two long, linear features record India's rapid movement northward: Ninety East Ridge and the Chagos–Lacadive Plateau. The Indian Ocean has just one large oceanic trench—the Java–Sunda Trench, where the Australian and Indian plates are subducting beneath the Eurasian Plate. The Indian and Australian plates now appear to be moving independently, but the location of their boundary is uncertain.

INDIAN WATERS
Clear water off the Kenyan coast reveals rock shoals, coral growth, and sand bars.

KEY

- sea level
- 800 ft (250 m)
- 1,600 ft (500 m)
- 3,300 ft (1,000 m)
- 6,500 ft (2,000 m)
- 9,800 ft (3,000 m)
- 16,400 ft (5,000 m)
- □ land
- △ seamount
- ▽ sea depth
- ▼ maximum depth on map

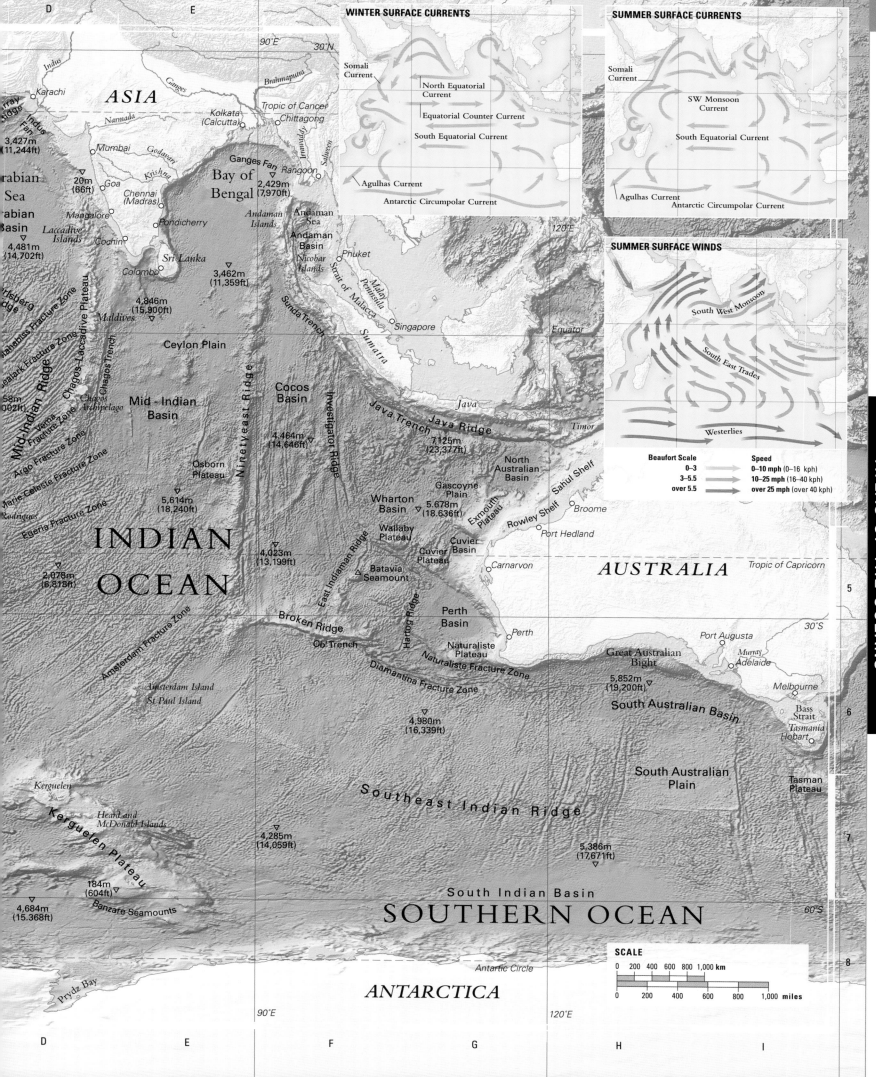

WINTER SURFACE CURRENTS

Somali
Current

North Equatorial
Current

Equatorial Counter Current

South Equatorial Current

Agulhas Current

Antarctic Circumpolar Current

SUMMER SURFACE CURRENTS

Somali
Current

SW Monsoon
Current

South Equatorial Current

Agulhas Current

Antarctic Circumpolar Current

SUMMER SURFACE WINDS

South West Monsoon

South East Trades

Westerlies

Beaufort Scale	Speed
0–3	0–10 mph (0–16 kph)
3–5.5	10–25 mph (16–40 kph)
over 5.5	over 25 mph (over 40 kph)

D   E   90°E   30°N

Indus

Karachi

ASIA

Ganges   *Brahmaputra*

*Murray
Ridge*

*Indus
Fan*
3,427m
(11,244ft)

Narmada   *Kolkata
(Calcutta)*   *Chittagong*
Tropic of Cancer

*Arabian
Sea*

20m
(66ft)   *Mumbai*   *Goa*   *Godavari*   *Krishna*

*Arabian
Basin*
4,481m
(14,702ft)   *Mangalore*   *Chennai
(Madras)*

*Laccadive
Islands*   *Cochin*   *Pondicherry*

Sri Lanka

Colombo   ▽
3,462m
(11,359ft)

Ganges Fan   ▽
Rangoon
2,429m
(7,970ft)

Bay of
Bengal

*Andaman
Islands*   *Andaman
Sea*

*Andaman
Basin*

*Nicobar
Islands*   *Phuket*

120°E

*Carlsberg
Ridge*

Rodriguez Fracture Zone

*58m
(002ft)*

Mid – Indian
Ridge

*Vema
Fracture Zone*

Chagos-Laccadive Plateau

*Chagos
Archipelago*   *Chagos Trench*

Ceylon Plain

4,846m
(15,900ft)   ▽
Maldives

Sunda Trench

Sumatra

Strait of Malacca

*Malay
Peninsula*

Singapore

Equator

Java   *Timor*

Argo Fracture Zone

Marie-Celeste Fracture Zone

Rodriguez

2,078m
(6,818ft)   ▽

Mid – Indian
Basin

*Osborn
Plateau*

5,614m
(18,240ft)

Ninetyeast Ridge

Cocos
Basin

4,464m
(14,646ft)   ▽

Investigator Ridge

Java Trench   Java Ridge
7,125m
(23,377ft)

INDIAN

OCEAN

Egeria Fracture Zone

Amsterdam Fracture Zone

4,023m
(13,199ft)   ▽

East Indianman Ridge

Wharton
Basin
5,678m
(18,636ft)   ▽

*Batavia
Seamount*

Broken Ridge

Ob' Trench

Hartog Ridge

*Amsterdam Island
St Paul Island*

Naturaliste
Plateau

Naturaliste Fracture Zone

Diamantina Fracture Zone

4,980m
(16,339ft)   ▽

Kerguelen

*Heard and
McDonald Islands*

Kerguelen Plateau

184m
(604ft)   ▽

4,684m
(15,368ft)   ▽

Banzare Seamounts

Southeast Indian Ridge

4,285m
(14,059ft)   ▽

South Indian Basin

SOUTHERN OCEAN

Prydz Bay

ANTARCTICA

90°E   120°E

Gascoyne
Plain

North
Australian
Basin

Sahul Shelf

Wallaby
Plateau

Cuvier
Plateau   Cuvier
Basin

Exmouth
Plateau

Rowley Shelf

Broome

Port Hedland

Carnarvon

AUSTRALIA   Tropic of Capricorn   5

Perth
Basin

Perth

30°S

Great Australian
Bight

5,852m
(19,200ft)   ▽

Port Augusta

Murray
Adelaide

Melbourne

South Australian Basin   *Bass
Strait*
Tasmania
Hobart

6

South Australian
Plain

*Tasman
Plateau*

5,386m
(17,671ft)   ▽

7

Antarctic Circle

8

SCALE				
0	200 400 600 800	1,000 km		
0	200	400	600	800 1,000 miles

D   E   F   G   H   I

The Red Sea and Arabian Sea

CIRCULATION IN THE NORTHEAST Indian Ocean uniquely reverses twice a year due to the monsoon winds. For thousands of years, navigators used this to run trade routes in the region. Today, oil and the Suez Canal make the area strategically important.

INDIAN OCEAN B4
Red Sea

AREA 175,000 square miles (450,000 square km)

MAXIMUM DEPTH 9,975 ft (3,040 m)

INFLOWS Arabian Sea

The Red Sea is an embryonic ocean, and it has been opening over the last 25 million years, ever since the Arabian Plate began its gradual rift away from Africa. A central trough is flanked by relatively shallow shelves, and its warm waters contain many fringing coral reefs. Since 1869, the Red Sea has been linked with the Mediterranean Sea via the 100-mile (160-km) long Suez Canal.

THE SUEZ CANAL

INDIAN OCEAN G5
Arabian Sea

AREA 1.5 million square miles (3.9 million square km)

MAXIMUM DEPTH 19,038 ft (4,481 m)

INFLOWS Indus, Namada Rivers

The Arabian Sea lies between the Arabian Peninsula and India. It is underlain by the abyssal plain of the Arabian Basin. This oceanic part of the Indian Plate is bounded to the west by the Owen Fracture Zone

INDIAN OCEAN I8
Maldives

TYPE Coral atoll islands

AREA: 115 square miles (298 square km)

NUMBER OF ISLANDS 1,192

The Maldives lie midway along the Chagos–Laccadive Plateau. The Laccadive Islands and a number of submerged banks mark the northern end of the ridge. There are more than 1,000 Maldive islands, grouped into 27 atolls, composed of coral and sandbars. The highest island is less than 10 ft (3 m) above sea level. With a warm climate, shallow lagoons, and refreshing sea breezes, the Maldives are an idyllic vacation destination.

MALDIVE ISLAND

Although tourism plays an increasingly important role in the economy of the islands, fishing remains the main occupation of the islanders.

INDIAN OCEAN D2
Persian Gulf

AREA 93,000 square miles (241,000 square km)

MAXIMUM DEPTH 360 ft (110 m)

INFLOWS Tigris, Euphrates, Karun Rivers

The Persian Gulf (also known as the Gulf) is a warm, semienclosed sea, mostly less than 330 ft (100 m) deep. It is connected to the Arabian Sea via the Strait of Hormuz and the Gulf of Oman. The shallow waters are well mixed and more productive than the Red Sea, owing to the nutrient runoff from the land to the north and east. Corals have adapted to the very warm water temperature, which can reach 91°F (33°C).

The Arabian Plate, spreading from the Red Sea rift, is moving northeast and sliding under the Eurasian Plate, so the northeastern side of the Persian Gulf is deeper. This tectonic activity has folded and uplifted sediments up to 280 million years old and produced structural traps for oil, which has accumulated in large reservoirs beneath the Gulf and surrounding land. Oil now dominates the region's economy.

and to the south by the Carlsberg Ridge, where India and Africa are diverging.

To the west lies the Gulf of Aden, a precursor to the Red Sea rift, with a well-established spreading ridge. On the northern shore, earthquakes within the Makran subduction zone sometimes trigger the eruption of mud volcanoes. One of these appeared suddenly as a new island off the Pakistani town of Gwadar in 2013. Such features usually subside within a few months.

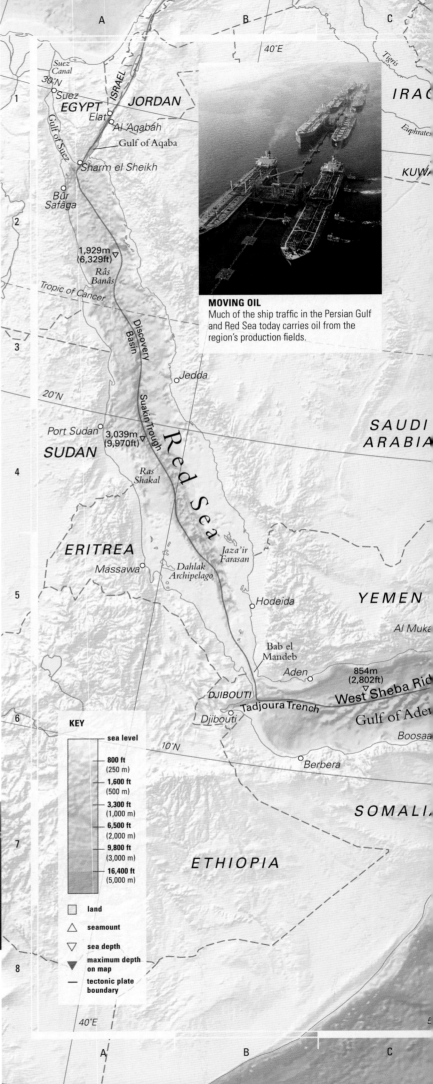

MOVING OIL
Much of the ship traffic in the Persian Gulf and Red Sea today carries oil from the region's production fields.

KEY

sea level

- 800 ft (250 m)
- 1,600 ft (500 m)
- 3,300 ft (1,000 m)
- 6,500 ft (2,000 m)
- 9,800 ft (3,000 m)
- 16,400 ft (5,000 m)

□ land

△ seamount

▽ sea depth

▼ maximum depth on map

— tectonic plate boundary

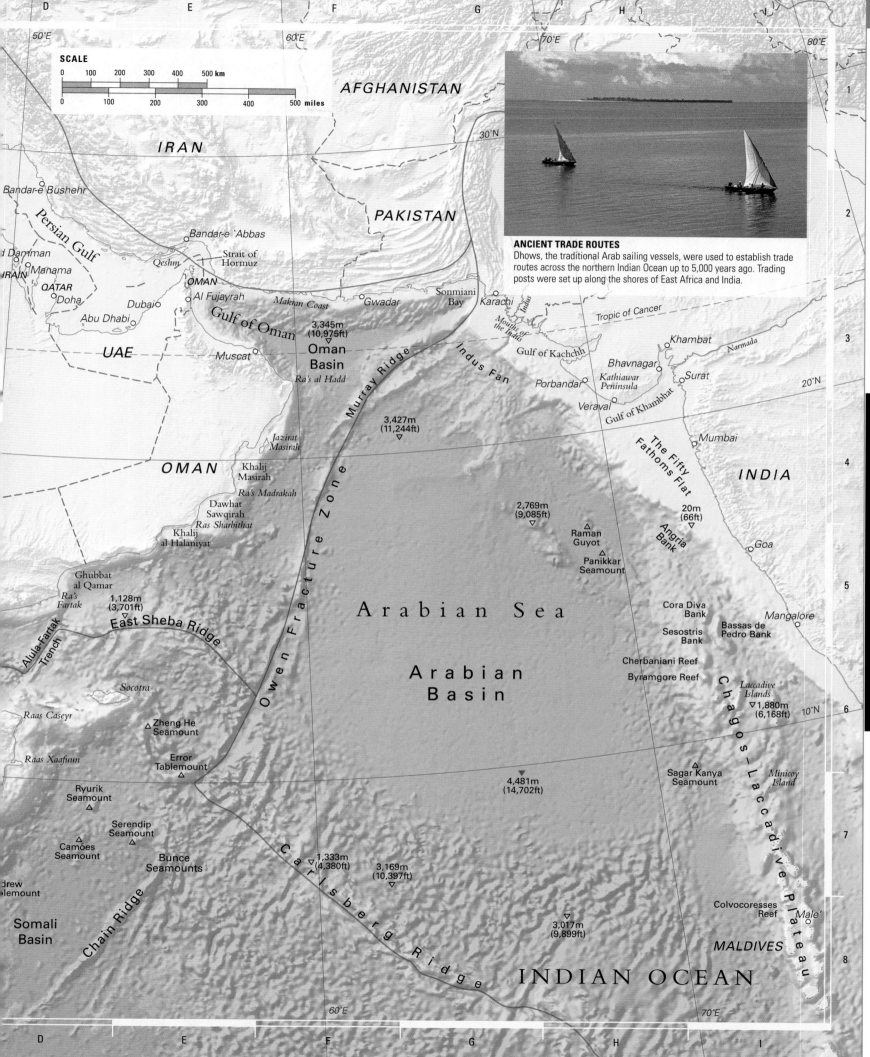

SCALE

0 100 200 300 400 500 km

0 100 200 300 400 500 miles

AFGHANISTAN

IRAN

PAKISTAN

Bandar-e Bushehr

Persian Gulf

l Dammam

Manama

QATAR

Doha

BAHRAIN

Bandar-e ʿAbbas

Qeshm

Strait of Hormuz

OMAN

Dubai

Abu Dhabi

UAE

Al Fujayrah

Gulf of Oman

Makran Coast

Gwadar

Sonmiani Bay

Karachi

Indus

Mouths of the Indus

Tropic of Cancer

3,345m (10,975ft)

Oman Basin

Muscat

Raʾs al Hadd

Murray Ridge

Indus Fan

Gulf of Kachchh

Khambat

Bhavnagar

Kathiawar Peninsula

Surat

Narmada

20°N

3,427m (11,244ft)

Porbandar

Veraval

Gulf of Khambhat

Mumbai

Jazirat Masirah

OMAN

Khalij Masirah

Raʾs Madrakah

Dawhat Sawqirah

Ras Sharbithat

Khalij al Halaniyat

The Fifty Fathoms Flat

INDIA

2,769m (9,085ft)

20m (66ft)

Raman Guyot

Angria Bank

Goa

Panikkar Seamount

Ghubbat al Qamar

Raʾs Fartak

1,128m (3,701ft)

East Sheba Ridge

Owen Fracture Zone

Arabian Sea

Cora Diva Bank

Sesostris Bank

Mangalore

Bassas de Pedro Bank

Alula-Fartak Trench

Socotra

Arabian Basin

Cherbaniani Reef

Byramgore Reef

Chagos–Laccadive Plateau

Laccadive Islands

1,880m (6,168ft)

10°N

Raas Caseyr

Zheng He Seamount

Raas Xaafuun

Error Tablemount

4,481m (14,702ft)

Sagar Kanya Seamount

Minicoy Island

Ryurik Seamount

Serendip Seamount

Camões Seamount

Bunce Seamounts

drew lemount

Chain Ridge

Carlsberg Ridge

1,333m (4,380ft)

3,169m (10,397ft)

Colvocoresses Reef

Male'

Somali Basin

3,017m (9,899ft)

MALDIVES

INDIAN OCEAN

ANCIENT TRADE ROUTES

Dhows, the traditional Arab sailing vessels, were used to establish trade routes across the northern Indian Ocean up to 5,000 years ago. Trading posts were set up along the shores of East Africa and India.

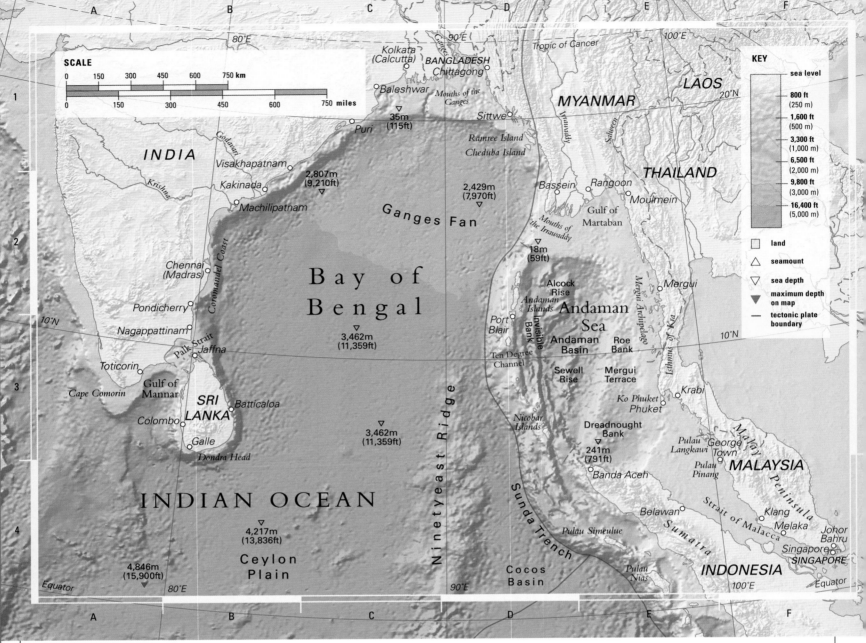

SCALE

| 0 | 150 | 300 | 450 | 600 | 750 km |
| 0 | 150 | 300 | 450 | 600 | 750 miles |

KEY

sea level

800 ft (250 m)
1,600 ft (500 m)
3,300 ft (1,000 m)
6,500 ft (2,000 m)
9,800 ft (3,000 m)
16,400 ft (5,000 m)

□ land
△ seamount
▽ sea depth
▼ maximum depth on map
— tectonic plate boundary

The Bay of Bengal

THE NORTHEAST CORNER of the Indian Ocean is enclosed on three sides by land. This area of tropical sea is subject to a monsoon climate, and vulnerable to cyclones between the months of June and November.

INDIAN OCEAN C2

Bay of Bengal

AREA 1.1 million square miles (2.9 million square km)

MAXIMUM DEPTH 15,400 ft (4,695 m)

INFLOWS Ganges, Brahmapurta, Mahanadi, Godavari, Krishna, Kaveri, Irrawaddy Rivers

Circulation in the Bay of Bengal is clockwise during the northwest monsoon, with a westward flow in the main ocean. This flow reverses during the southwest monsoon. The northern half of the bay is underlain by the Ganges Fan, a thick cone of sediment extending from the continental rise across the abyssal plain. This is the fastest-accumulating sediment in the world, originating high in the Himalayas, and supplied by the Brahmaputra River and the Ganges.

INDIAN OCEAN F4

Strait of Malacca

LENGTH 600 miles (963 km)

MINIMUM WIDTH 9 miles (15 km)

The Strait of Malacca links the Indian Ocean with the Pacific Ocean, via the South China Sea. About 140 ships pass through this busy waterway daily. The cargo includes about one quarter of the world's oil, headed from the Persian Gulf to markets including Japan and China. Such heavy traffic has always attracted pirates. After a rise in piracy during the 1990s, attacks dropped away with the advent of naval patrols from 2005. The world saw a peak of 450 pirate attacks in 2010, but by that time most were on the other side of the Indian Ocean, off the coast of Somalia.

INDIAN OCEAN E3

Andaman Sea

AREA 308,000 square miles (798,000 square km)

MAXIMUM DEPTH 12,400 ft (3,777 m)

INFLOWS Bay of Bengal, Strait of Malacca; Irriwaddy, Salween Rivers

The Andaman Sea lies between the Andaman Islands, Sumatra, and the Malay Peninsula. There is a broad continental shelf in the east and the north, where the sediment is dredged for cassiterite, an ore of tin. Alcock Rise and Sewell Rise are separated by an area of deep ocean floor, where a spreading center has been pushing the Burma and Sunda microplates apart for the last 3–4 million years. This divergence created the Andaman Sea. The eastern half of the sea lies over the Sunda Plate, which includes most of Sumatra and the Malay Peninsula. The western half includes the Andaman and Nicobar Islands and sits on the Burma Plate, which forms a junction with the Indian Plate at the Sunda Trench. At this subduction zone, the Indian Plate is being overridden by the younger Burma Plate. The southern part of this zone was the source of the 2004 Indian Ocean tsunami.

HAVELOCK ISLAND
Mangroves line the eastern shore of Havelock Island, part of the Andaman Islands group. These volcanic islands are also fringed by coral reefs.

The Java Trench

IN THE EASTERN INDIAN OCEAN, the South Equatorial Current carries water from east to west during the southwest monsoon, but shifts south during the northeast monsoon to be replaced by the eastward-flowing Equatorial Counter Current. The area includes the deepest part of the Indian Ocean, the Java Trench, where the Australian Plate meets the Eurasian Plate.

INDIAN OCEAN D2

Java Trench

LENGTH 1,600 miles (2,600 km)

MAXIMUM DEPTH 23,377 ft (7,125 m)

RATE OF CLOSURE 2½ in (6 cm) per year

The Java Trench is a continuation of the Sunda Trench, where the oceanic part of the Australian Plate is being subducted beneath the continental Eurasian Plate. A string of volcanoes has resulted behind the trench, strung out across Sumatra, Java, and the Lesser Sunda Islands. The Australian Plate is moving north at a rate of 2½ in (6 cm) per year. The Indian Ocean floor south and west of the trench shows tectonic features aligned in this northerly direction. Between Investigator Ridge and Ninety East Ridge lie a series of north–south fractures formed to accommodate different rates of motion in the Australian and Indian plates. The Ninety East Ridge itself is the longest underwater mountain chain, at 3,100 miles (5,000 km) in length. It followed in the wake of India's rapid motion north as the Indian Ocean opened up. The ridge represents piles of extruded volcanic material formed above the Kerguelen Hotspot that were carried north as the sea floor spread between India and Antarctica.

INDIAN OCEAN F2

Timor Sea

AREA 235,000 square miles (610,000 square km)

MAXIMUM DEPTH 10,800 ft (3,300 m)

INFLOWS Indian Ocean, Arafura Sea

The Timor Sea marks the eastward boundary of the central Indian Ocean. Pacific water flows in from the Arafura Sea (see p.473) during the southwest monsoon, feeding the South Equatorial Current. This flow is reversed during the northeast monsoon. Australia's aboriginal people probably arrived from southeast Asia by island-hopping across the Timor Sea. It is mainly shallow, but with the deep Timor Trough lying along its northern edge. Significant reserves of oil and gas are thought to lie in the continental shelf sediments beneath the sea, and exploitation rights are disputed between Australia and East Timor. The warm shallow tropical waters make the Timor Sea a breeding ground for tropical storms and cyclones from January to March. Such storms proceed southwestward into the Indian Ocean, sometimes turning inland to hit the coast of Western Australia. There are fishing grounds, including a shrimp fishery, in coastal waters on the Australian side of the Timor Sea.

TIMOR FISHING
Traditional, shore-based fishing is practiced by Timorese fishers, seen here hauling in a net at Areia Branca Beach near Dili.

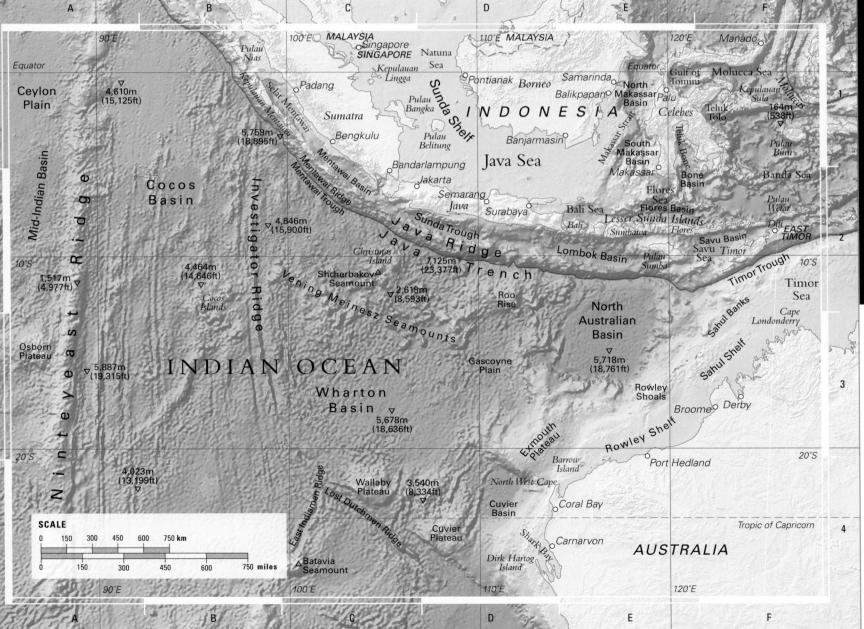

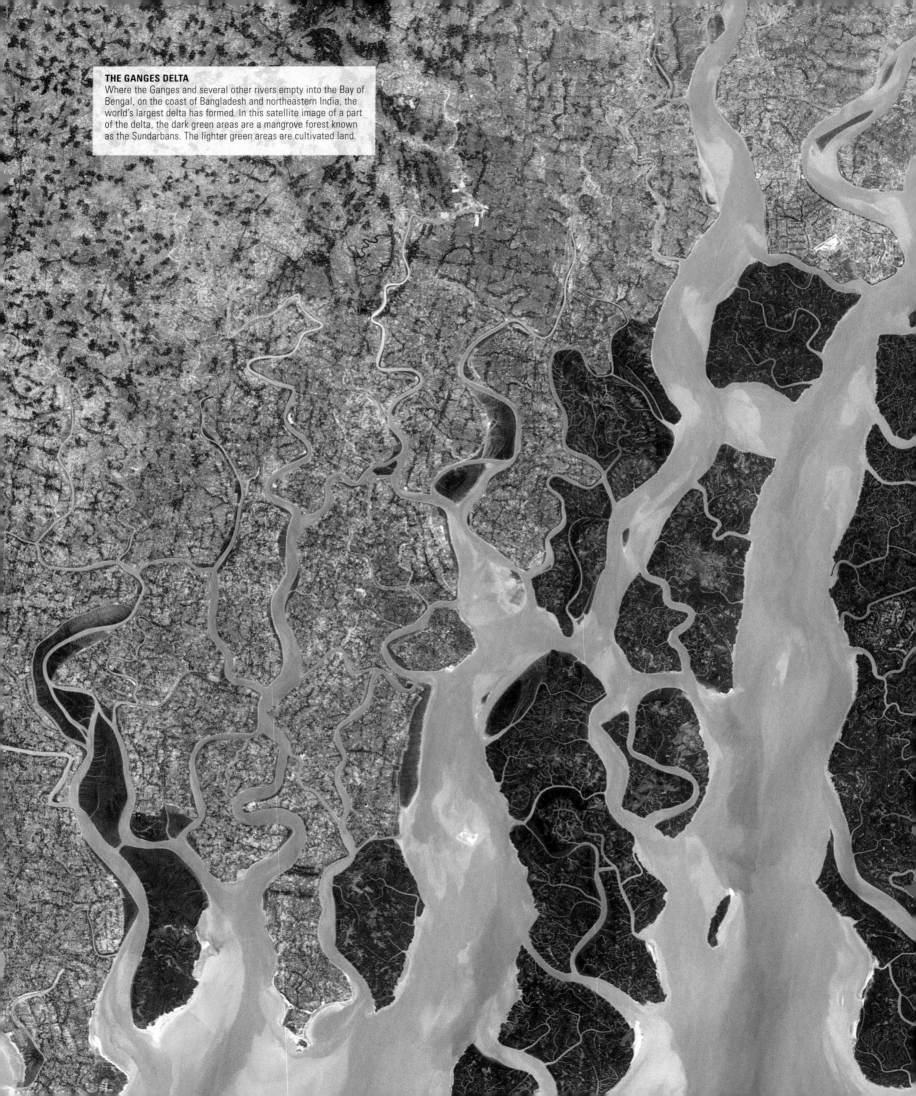

THE GANGES DELTA
Where the Ganges and several other rivers empty into the Bay of
Bengal, on the coast of Bangladesh and northeastern India, the
world's largest delta has formed. In this satellite image of a part
of the delta, the dark green areas are a mangrove forest known
as the Sundarbans. The lighter green areas are cultivated land.

KEY

sea level

- 800 ft (250 m)
- 1,600 ft (500 m)
- 3,300 ft (1,000 m)
- 6,500 ft (2,000 m)
- 9,800 ft (3,000 m)
- 16,400 ft (5,000 m)

☐ land

△ seamount

▽ sea depth

▼ maximum depth on map

— tectonic plate boundary

Equator

40°E

50°E

60°E

SOMALIA

Kismaayo

KENYA

Tana

191m (623ft)

Coco-de-Mer Seamounts

Madingley Rise

North Kenya Bank

Pate Island

Mombasa

4,886m (16,031ft)

Somali Basin

Amirante Islands

Seychelles Bank

Victoria Inner Islands

Mascarene Plateau

Tanga

Pemba

Amirante Ridge

Amirante Basin

Zanzibar

Zanzibar

Fred Seamount

Dar es Salaam

SEYCHELLES

Rufiji

Mafia

Amirante Trench

Fortune Bank

Wilkes Rise

Anton Bruun Ridge

Ritchie Bank

TANZANIA

914m (3,087ft)

Aldabra Group

Cosmoledo Group

Providence Reef

Bulldog Bank

Farquhar Group

Agalega Islands

10°S

Rufiji

Giraud Seamount

Ruvuma

Cabo Delgado

Grande Comore

COMOROS

Nosy Glorieuses

Hydra Seamount

4,801m (15,752ft)

Mascarene Basin

Moroni

Anjouan

Geyser Reef

Leven Bank

Tanjona Bobaomby

Bardin Seamount

Wormley Seamount

Mohéli

MAYOTTE

Antsiranana

Pemba

Nosy Be

INDIAN

Comoro Basin

3,301m (10,831ft)

Tanjona Vilanandra

Mahajanga

Tanjona Masoala

Tromelin

Cargado Carajos Bank

MOZAMBIQUE

MALAWI

Lake Nyasa

Davie Ridge

Nosy Sainte Marie

Mascarene Plain

Wilshaw Ridge

Soudan Bank

Quelimane

Zambezi

Mozambique Channel

Betsiboka

Toamasina

La Pérouse Seamount

Port Louis

MAURITIUS

PontaTimbue

Zambezi Canyon

MADAGASCAR

Tsiribihina

St-Denis

Mascarene Islands

Beira

20°S Baia de Sofala

Mangoky

4,976m (16,326ft)

RÉUNION

Mauritius Trench

Ilha do Bazaruto

Bassas da India

Tanjona Ankabna

Hall Tablemount

Jaguar Seamount

Île Europa

Toliara

Tropic of Capricorn

Inhambane

69m (226ft)

Tanjona Vohimena

Mozambique Plateau

Natal Basin

Madagascar Basin

7,023m (23,042ft)

Madagascar Plateau

1,984m (6,510ft)

4,769m (15,647ft)

40°E

50°E

60°E

Map labels (left page)

70°E
MALDIVES
Equator

Mabahiss Fracture Zone

Sealark Fracture Zone

Chagos–Laccadive Plateau

4,301m
(14,112ft)

Mid-Indian Ridge

Chagos Bank

Chagos Archipelago

Diego Garcia

Chagos Trench

3,658m
(12,002ft)

6,102m
(20,021ft)

Vema Fracture Zone

10°S

ya de
alha Bank

Argo Fracture Zone

OCEAN

Mid-Indian Basin

569m
(1,869ft)

4

Marie Celeste Fracture Zone

5

lix
Seamount

odrigues Ridge
Rodrigues

Egeria Fracture Zone

Mid-Indian Ridge

20°S

6

Tropic of Capricorn   7

2,078m
(6,818ft)

Southwest Indian Ridge

8

SCALE

0 100 200 300 400 500 km

0   100   200   300   400   500 miles

70°E

The Seychelles and Madagascar

THE EASTERN INDIAN OCEAN FLOOR is littered with scars documenting the breakup of Gondwana over the last 150 million years. The warm waters of the Indian Ocean have also proved to be an ideal environment for diverse marine life.

INDIAN OCEAN D3

Seychelles

VOLCANIC Continental islands

AREA 176 square miles (455 square km)

NUMBER OF ISLANDS 115

The main islands of the Seychelles—the Inner Islands—are made of granite, rising over 3,000 ft (900 m) above sea level on top of the Seychelles Bank. The other islands to the southwest—the Outer Islands—are coral islands (atolls) on top of seamounts. The Seychelles Bank is the most northerly part of the submarine Mascarene Plateau, which extends as far as the island of Réunion in the

GRANITE BOULDERS IN THE SEYCHELLES

south. This continental fragment broke off from India around 65 million years ago as the current Mid-Indian Ridge started spreading.

INDIAN OCEAN B5

Mozambique Channel

AREA 386,000 square miles (1 million square km)

MAXIMUM DEPTH 370 ft (110 m)

INFLOWS Zambezi, Rio Lúrio Rivers

The Mozambique Channel separates Madagascar from the mainland of Africa. The area is home to the ancient coelacanth, found on both sides of the channel and off the Comoros. A counterclockwise gyre is found around the Comoros, and counterclockwise eddies dominate the flow in the main part of the channel. The warm Agulhas Current arises over the Natal Basin, fed by the South Equatorial Current.

INDIAN OCEAN H3

Mid-Indian Ridge

LENGTH 2,100 miles (3,400 km)

AVERAGE HEIGHT ABOVE SEA FLOOR 5,000 ft (1,500 m)

RATE OF SPREAD 1¼ in (3 cm) per year

The Indian and African plates are moving apart due to spreading at the Mid-Indian Ridge, which is marked by a series of transform fracture zones. Rifting was triggered 65 million years ago when the Réunion Hotspot erupted a vast amount of basalt through the Indian continental plate, forming a plateau called the Deccan Traps. An older spreading ridge, which first separated India from Africa, lies subsided between the Mascarene Basin and the Mascarene Plain.

INDIAN OCEAN F6

Mauritius and Réunion

TYPE Volcanic islands

AREA 1,800 square miles (4,550 square km)

NUMBER OF ISLANDS 2

The Mascarene Islands, Mauritius and Réunion, are the largest and youngest islands associated with the Mascarene Plateau, rising 21,300 ft (6,500 m) above the sea floor. Like the older banks of the plateau to the northeast and the Rodrigues Ridge to

the east, they are volcanic in origin, having formed above a deep mantle hotspot. After the Deccan Traps eruption (see above), the Réunion Hotspot continued to punch through the crust as India moved north, leaving a trail of volcanic structures across the ocean floor, including the Laccadive and Maldive islands and the Chagos Bank on the other side of the Mid-Indian Ridge. Réunion's main peak, Piton de la Fournaise, is one of the most active volcanoes in the world.

FRINGING REEF
A reef fringes the lagoon on the north coast of the volcanic island of Mauritius.

The Pacific Ocean

THE PACIFIC IS THE LARGEST OCEAN. It is twice the size of the Atlantic and covers more than a third of the planet's surface. Many Pacific islands were colonized by Micronesians and Polynesians before Europeans arrived in the 16th century. The Portuguese explorer Ferdinand Magellan died after crossing the Pacific in 1521, leaving his crew to complete the first circumnavigation of the world.

BORA BORA ISLAND IN THE SOUTH PACIFIC

Ocean Circulation

The Pacific is cut off from the Arctic Ocean but exchanges water with the Southern Ocean. The North Equatorial Current is the world's longest westward-flowing current, carrying water 9,000 miles (14,500 km) across the ocean. The warm Kuroshio Current flows north as the North Pacific's western boundary current, and the Kuroshio Extension returns warm water to the western Pacific. The counterclockwise gyre in the South Pacific is formed by the South Equatorial Current, the warm East Australia Current, the Antarctic Circumpolar Current, and the Humboldt Current. A strong upwelling occurs where the cold Humboldt Current diverges from the coast, but this routinely fails as part of the El Niño Southern Oscillation (see pp.68–69).

ROUGH SEAS
The north Pacific is a breeding ground for storms. Here, the bow of a ship plows through violent storm waves in the Bering Sea.

Ocean Basin

The Pacific Basin has been shrinking since the opening of the Atlantic and Indian Oceans. It has more subduction zones, where oceanic crust is consumed, than any other ocean. Violent volcanic eruptions are associated with these zones, producing the Ring of Fire around the Pacific's shores (see p.184). The western Pacific is studded with chains of volcanic islands and marked by deep ocean trenches where the Pacific Plate meets the continental Eurasian Plate and smaller oceanic plates. The floor of the eastern Pacific is fairly smooth in comparison, sloping gently away from the coast of North America and the East Pacific Rise. Mid-ocean island chains and seamounts have arisen from intermittent eruptions above mantle hot spots.

EAST PACIFIC RISE
Pillow lavas are extruded at all mid-ocean ridges and form the top layer of the crust throughout the oceans.

KEY

- sea level
- 800 ft (250 m)
- 1,600 ft (500 m)
- 3,300 ft (1,000 m)
- 6,500 ft (2,000 m)
- 9,800 ft (3,000 m)
- 16,400 ft (5,000 m)

□ land
△ seamount
▽ sea depth
▼ maximum depth on map

SCALE

| 0 | 500 | 1,000 | 1,500 | 2,000 | 2,500 km |

| 0 | 500 | 1,000 | 1,500 | 2,000 | 2,500 miles |

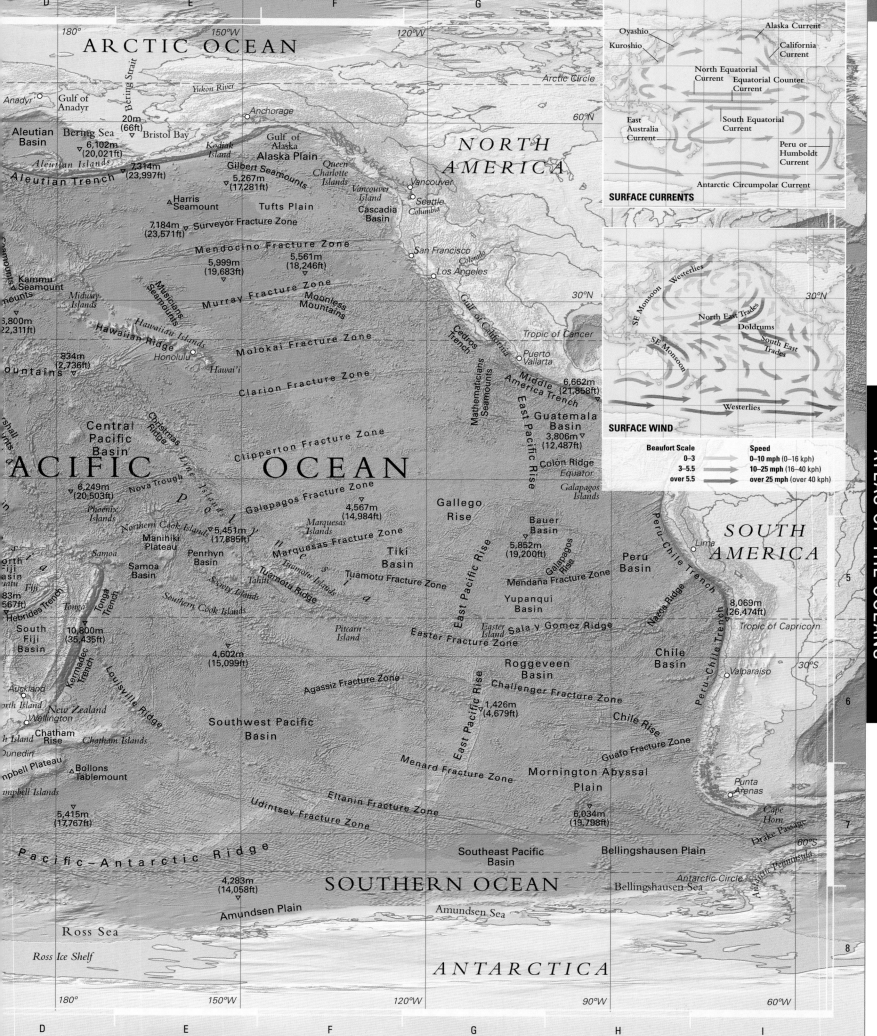

ARCTIC OCEAN

Anadyr°   Gulf of Anadyr

Bering Strait

Yukon River

Arctic Circle

60°N

NORTH AMERICA

Anchorage

20m (66ft)   Bristol Bay

Aleutian Basin

Bering Sea   6,102m (20,021ft)

Kodiak Island

Gulf of Alaska

Alaska Plain

Queen Charlotte Islands

Vancouver Island

Vancouver

Seattle
Columbia

Cascadia Basin

Aleutian Islands   7,314m (23,997ft)

Gilbert Seamounts   5,267m (17,281ft)

Aleutian Trench

Harris Seamount

Tufts Plain

7,184m (23,571ft)   Surveyor Fracture Zone

San Francisco

Mendocino Fracture Zone

5,999m (19,683ft)

5,561m (18,246ft)

Kammu Seamount

Los Angeles

30°N

mounts

Midway Islands

Musicians Seamounts

Murray Fracture Zone

Moonless Mountains

Colorado

5,800m (22,311ft)

Hawaiian Islands

Hawaiian Ridge

Cedros Trench

Gulf of California

Tropic of Cancer

834m (2,736ft)   Honolulu

Molokai Fracture Zone

Hawai'i

Puerto Vallarta

ountains

Clarion Fracture Zone

Mathematicians Seamounts

Middle America Trench

6,662m (21,858ft)

Central Pacific Basin

Christmas Ridge

Line Islands

Clipperton Fracture Zone

East Pacific Rise

Guatemala Basin   3,806m (12,487ft)

PACIFIC   OCEAN

Colón Ridge

Equator

Galápagos Islands

Nova Trough

6,249m (20,503ft)

Galapagos Fracture Zone

4,567m (14,984ft)

Gallego Rise

SOUTH AMERICA

Phoenix Islands

Northern Cook Islands   5,451m (17,885ft)

Marquesas Islands

Bauer Basin

Lima

Peru-Chile Trench

Manihiki Plateau

Penrhyn Basin

Marquesas Fracture Zone

Tiki Basin

5,852m (19,200ft)

Galápagos Rise

Peru Basin

orth Fiji asin

Samoa

Samoa Basin

Society Islands

Tuamotu Islands

Tuamotu Ridge

Tahiti

Tuamotu Fracture Zone

Mendaña Fracture Zone

Yupanqui Basin

Nazca Ridge

anu 83m 567ft)

Fiji

Tonga

Tonga Trench

Southern Cook Islands

Pitcairn Island

Easter Island

Sala y Gomez Ridge

8,069m (26,474ft)

Tropic of Capricorn

Hebrides Trench

Easter Fracture Zone

Chile Basin

South Fiji Basin

10,800m (35,435ft)

4,602m (15,099ft)

East Pacific Rise

Auckland

orth Island

New Zealand
Wellington

Kermadec Trench

Louisville Ridge

Agassiz Fracture Zone

Roggeveen Basin

Challenger Fracture Zone

1,426m (4,679ft)

Chile Rise

Valparaiso

30°S

h Island
Dunedin

Chatham Rise

Chatham Islands

Southwest Pacific Basin

Peru-Chile Trench

npbell Plateau

Bollons Tablemount

Menard Fracture Zone

Guafo Fracture Zone

mpbell Islands

5,415m (17,767ft)

Eltanin Fracture Zone

Udintsev Fracture Zone

Mornington Abyssal Plain

6,034m (19,798ft)

Punta Arenas

Cape Horn

Drake Passage

60°S

Pacific–Antarctic Ridge

Southeast Pacific Basin

Bellingshausen Plain

Antarctic Peninsula

4,283m (14,058ft)

SOUTHERN OCEAN

Bellingshausen Sea

Antarctic Circle

Amundsen Plain

Amundsen Sea

Ross Sea

ANTARCTICA

Ross Ice Shelf

180°   150°W   120°W   90°W   60°W

D   E   F   G   H   I

SURFACE CURRENTS

Oyashio

Kuroshio

Alaska Current

California Current

North Equatorial Current

Equatorial Counter Current

East Australia Current

South Equatorial Current

Peru or Humboldt Current

Antarctic Circumpolar Current

SURFACE WIND

Westerlies

SE Monsoon

North East Trades

Doldrums

South East Trades

SE Monsoon

Westerlies

Beaufort Scale	Speed
0–3	0–10 mph (0–16 kph)
3–5.5	10–25 mph (16–40 kph)
over 5.5	over 25 mph (over 40 kph)

The Bering Sea And Gulf of Alaska

THE COLD, STORMY SUBPOLAR SEAS of the North Pacific are highly productive, supporting a rich fishery. Geologically, the area is dominated by a subduction zone, and the area's volcanoes and earthquakes pose an ever-present danger.

PACIFIC OCEAN

Aleutian Trench

LENGTH 2,000 miles (3,200 km)

MAXIMUM DEPTH 25,663 ft (7,822 m)

RATE OF CLOSURE 3 in (8 cm) per year

The Bering Sea is bounded to the south by the Aleutian Islands. On the Pacific side of the islands lies the Aleutian Trench, marking where the Pacific Plate is plunging beneath the North American Plate. It is this subduction zone that gives rise to the volcanic arc of islands, the most northerly link in the Pacific Ring of Fire. The trench continues to the east,

SEALS IN THE ALEUTIAN ISLANDS

where the contact is between ocean crust and continental crust. The largest volcanic event of the 20th century was the eruption of Mount Katmai on the Alaskan Peninsula in 1912. This boundary can also produce powerful earthquakes, such as the event that destroyed part of Anchorage in 1964.

PACIFIC OCEAN D3

Bering Sea

AREA 890,000 square miles (2.3 million square km)

MAXIMUM DEPTH 20,021 ft (6,102 m)

INFLOWS Pacific Ocean; Yukon, Anadyr' Rivers

The Bering Sea is named after a Danish navigator in the Russian Navy, who explored the area in 1741. It lies between mainland Asia and North America and is bounded by the Aleutian Islands to the south and linked to the Arctic Ocean in the north by the narrow Bering Strait. There is a flow of cold Arctic water south through this strait, feeding a counterclockwise circulation. The main freshwater input is the Yukon River, which has deposited an extensive delta at its mouth. The Bering Sea is one of the world's richest fisheries, helping Alaska account

THE BERING STRAIT
This satellite image shows ice from the Chukchi Sea streaming south through the Bering Strait.

for about half of the total US fish and shellfish catch. Harbor seals and gray whales also take advantage of these productive waters. In contrast to the deep ocean basin beneath the southwestern half of the sea, the broad continental shelf in the northwest is very shallow. Much of this area formed a land bridge during the last ice age, when sea levels were up to 390 ft (120 m) lower than they are today. This route was ice-free for extended periods, allowing several species, including humans, to migrate from Asia to North America on foot for the first time.

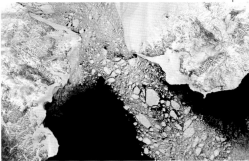

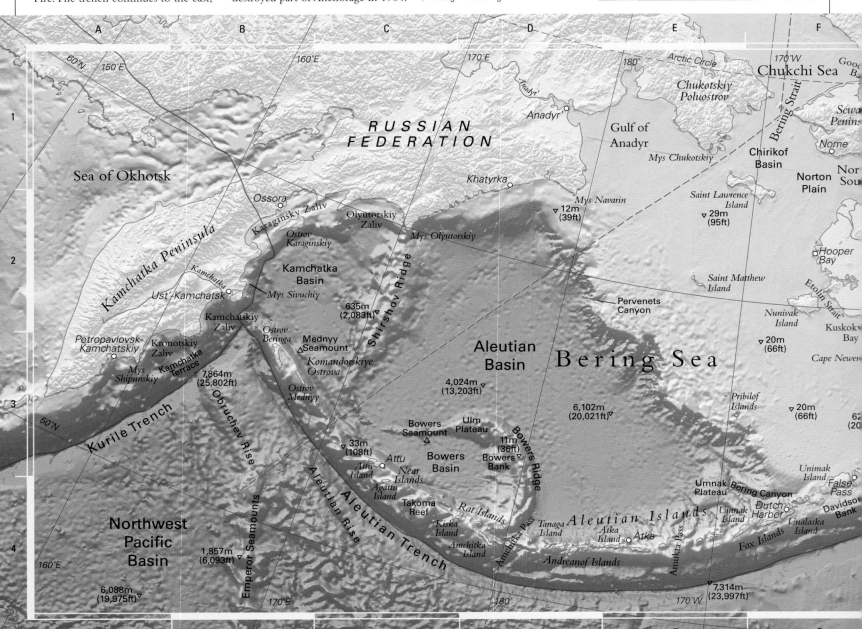

PACIFIC OCEAN I3

Gulf of Alaska

AREA 600,000 square miles (1.5 million square km)

MAXIMUM DEPTH 16,400 ft (5,000 m)

INFLOWS Susitna, Copper Rivers; icebergs from numerous glaciers

A counterclockwise subpolar gyre extends across the north Pacific and into the Gulf of Alaska, fed by the warm waters of the northern Kuroshio Extension, the extension of the Kuroshio Current. The surface waters are cooled and become less saline due to precipitation as they cross the ocean. Many of the storms that lash the west coast of Canada originate in the Gulf of Alaska. The circulation is completed as the Alaska Current and the Aleutian Current return west along the Alaskan coast and south of the Aleutian Islands. The gulf's waters are very productive, providing feeding grounds for many species of fish. Pacific salmon spend up to five years at sea, much of it in the gulf and adjacent seas, before returning to spawn in the Asian and North American rivers where they

ALASKAN FJORD
The valleys and fjords of the Alexander Archipelago testify to extensive erosion by glaciers during the last ice age.

were born. The floor of the Gulf of Alaska is peppered with seamounts. There are two main chains: the Patton and Gilbert seamounts, and the Kodiak Seamounts, both running away from the Alaska Peninsula. Their origin is the Cobb Hotspot, situated beneath the spreading center of the Juan de

Fuca Plate west of Vancouver Island. The seamounts were created above the hotspot over the last 30 million years, then carried northwest by seafloor spreading. Since 1977, oil has been shipped through ports on the south coast of Alaska. In 1989, Prince William Sound was the site of one of the worst maritime environmental disasters, when the tanker *Exxon Valdez* ran aground, releasing about 30 million gallons (114 million liters) of crude oil.

PACIFIC OCEAN L4

Cascadia Basin

AREA 66,000 square miles (170,000 square km)

MAXIMUM DEPTH 9,600 ft (2,930 m)

INFLOWS Pacific Ocean; Columbia, Fraser Rivers

The Cascadia Basin is the last remnant of the original eastern Pacific oceanic plate, the Farallon Plate, which has been almost entirely subducted beneath North America. The Cascade Range of volcanoes in Oregon and Washington State, including Mount St. Helens, are a product of this subduction. Mount St. Helens erupted in a catastrophic explosion in 1982, killing 57 people, and still shows signs of activity. Earthquakes and associated tsunamis are also a risk in the area, although the last major earthquake is thought to have been in 1700. The underlying ocean crust appears to be split into three small plates. The largest is the Juan de Fuca Plate, named after a Greek sea captain who explored the area for Spain in 1592. The Explorer Plate lies to the north and the Gorda Plate to the south.

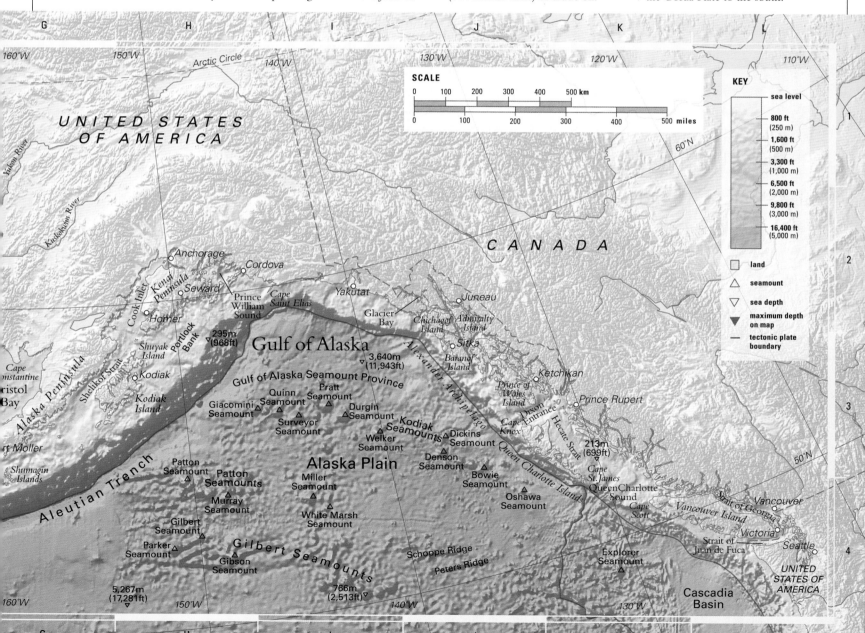

SCALE

0 100 200 300 400 500 km

0 100 200 300 400 500 miles

KEY

- sea level
- 800 ft (250 m)
- 1,600 ft (500 m)
- 3,300 ft (1,000 m)
- 6,500 ft (2,000 m)
- 9,800 ft (3,000 m)
- 16,400 ft (5,000 m)
- ☐ land
- △ seamount
- ▽ sea depth
- ▼ maximum depth on map
- — tectonic plate boundary

UNITED STATES OF AMERICA

CANADA

Yukon River

Kuskokwim River

Anchorage

Cordova

Seward

Kenai Peninsula

Prince William Sound

Cape Saint Elias

Yakutat

Juneau

Homer

Cook Inlet

Portlock Bank

295m (968ft)

Glacier Bay

Chichagof Island

Admiralty Island

Gulf of Alaska

3,640m (11,943ft)

Sitka

Baranof Island

Shuyak Island

Shelikof Strait

Kodiak

Kodiak Island

Alaska Peninsula

Cape Constantine

Bristol Bay

Gulf of Alaska Seamount Province

Quinn Seamount

Pratt Seamount

Durgin Seamount

Kodiak Seamounts

Dickins Seamount

Alexander Archipelago

Prince of Wales Island

Ketchikan

Prince Rupert

Giacomini Seamount

Surveyor Seamount

Welker Seamount

Denson Seamount

Cape Knox

Dixon Entrance

Hecate Strait

213m (699ft)

Fort Moller

Shumagin Islands

Aleutian Trench

Patton Seamount

Patton Seamounts

Miller Seamount

Alaska Plain

Bowie Seamount

Queen Charlotte Islands

Cape St. James

Queen Charlotte Sound

Murray Seamount

White Marsh Seamount

Oshawa Seamount

Cape Scott

Vancouver Island

Strait of Georgia

Vancouver

Gilbert Seamount

Parker Seamount

Gibson Seamount

Gilbert Seamounts

Schoppe Ridge

Peters Ridge

Explorer Seamount

Victoria

Strait of Juan de Fuca

Seattle

UNITED STATES OF AMERICA

5,267m (17,281ft)

766m (2,513ft)

Cascadia Basin

Arctic Circle

60°N

50°N

The Northwestern Pacific

LIKE THE BERING SEA TO THE NORTHEAST, this part of the Pacific is shaped by subduction at the edge of the Pacific Plate. Volcanoes, earthquakes, and tsunamis present a risk to human life, particularly in the densely populated islands of Japan.

PACIFIC OCEAN E4

Sea of Okhotsk

AREA 600,000 square miles (1.6 million square km)

MAXIMUM DEPTH 11,063 ft (3,372 m)

INFLOWS Sea of Japan/East Sea; Amur, Uda, Okhota, Penzhina Rivers

A subarctic shelf sea, the Sea of Okhotsk is a branch of the northwestern Pacific. It is enclosed to the north by the Asian landmass, bounded to the east by the Kurile

DANGEROUS SEA ICE
Ice can pose a hazard to shipping from November to June, with its location dependent on winds and ocean currents.

Islands, and linked in the south to the Sea of Japan/East Sea by two narrow straits. Navigation of the sea is restricted by sea ice in the winter, when ice formation on ship hulls also presents a danger to shipping. The southern part of the sea is notorious for its sea fogs throughout the year. The Sea of Okhotsk is very productive, accounting for nearly 70 percent of Russia's East Asian fish catch. It is home to several endangered species of marine life, including Kurile harbor seals and gray whales. The Okhotsk Plate includes the continental crust of the Kamchatka Peninsula, with its string of volcanoes, and the islands of Sakhalin and Hokkaido. In much of the area, the sea floor is quite shallow, but it is deeper in the Kurile Basin, where the ocean crust has stretched and thinned.

HUMAN IMPACT

OIL EXPLORATION

A rush to exploit rich oil and gas deposits on the island of Sakhalin Island started in 1996. The area became the largest recipient of foreign investment in Russia, and oil production is expanding offshore into the Sea of Okhotsk. The large amount of construction in such a short time has raised concerns about the impact on this wilderness environment, including disturbance of the marine life of the Sea of Okhotsk.

SUPPLY LINE
Pipelines link Sakhalin's offshore oil and gas fields with Russia's main Pacific port, Vladivostok, 1,120 miles (1,800 km) to the south.

PACIFIC OCEAN F5 AND E8

Kurile and Japan Trenches

LENGTH 2,390 miles (3,850 km)

MAXIMUM DEPTH 26,575 ft (8,100 m)

RATE OF CLOSURE 3 in (8 cm) per year

Subduction at the Kurile Trench has produced volcanoes along the Kamchatka Peninsula and the volcanic island arc of the Kuriles. The Kuriles form an almost complete submarine ridge between Hokkaido and Kamchatka, with only two deep water channels from the Pacific into the Sea of Okhotsk. The highest Kurile island

is Atlasov. The islands of Honshu and Hokkaido represent a more mature island arc, where crustal thickening has resulted from prolonged subduction at the Japan Trench and the joining of multiple island arcs. With so much tectonic activity, earthquakes are rampant. A magnitude-8 quake in 1923 killed more than 100,000 people in Tokyo and Yokohama. In 2011, the magnitude-9 Tohoku event, 43 miles (70 km) offshore and 19 miles (30 km) beneath the Japan Trench, thrust the seabed up by 19–26 ft (6–8 m), triggering a tsunami up to 131 ft (40 m) high (see pp.462–463).

KURILE ISLAND
Large craters, such as the flooded center of this island, form when a volcano's magma chamber collapses or explodes.

PACIFIC OCEAN C7

Sea of Japan/ East Sea

AREA 378,000 square miles (978,000 square km)

MAXIMUM DEPTH 12,276 ft (3,743 m)

INFLOWS East China Sea; Tumen, Ishikari, Shinano, Agano, Mogami, Teshio Rivers

Circulation within the Sea of Japan (also known as the East Sea) is counterclockwise, with warm water entering from the East China Sea through the Korea Strait. There are rich fishing grounds here and in the north. Squid are among the species sought by Japanese and Korean fishermen, who attract the animals

to the surface at night using powerful lights. The continental shelf is slightly wider on the eastern side than on the western side, and particularly narrow off the coast of Korea. There are three main basins: the Yamato Basin in the east, the Japan Basin in the north, and the Tsushima Basin in the southwest. Between these basins lies the Yamato Ridge, possibly a remnant of the spreading center that opened up the sea. The Sea of Japan/East Sea is a geologically complex basin bisected by the junction between the Okhotsk Plate and the Eurasian Plate. In 1983, a magnitude-7.7 earthquake on the sea floor off northern Honshu triggered a destructive tidal wave that reached a height of 46 ft (14 m) at the coast, killing 107 people in Japan and Korea.

PACIFIC OCEAN G6

Northwest Pacific Basin

AREA 2.4 million square miles (6.3 million square km)

MAXIMUM DEPTH 21,800 ft (6,650 m)

INFLOWS Bering Sea, Philippine Sea

The Oyashio Current (from the Japanese for "mother stream") brings cold water south from the Bering Sea, along the western edge of the Northwest Pacific Basin, forming the western arm of the subarctic gyre. Where the Oyashio Current meets the warm waters of the Kuroshio Current ("black stream") off Japan, there is a productive fishery.

KEY

sea level

800 ft (250 m)
1,600 ft (500 m)
3,300 ft (1,000 m)
6,500 ft (2,000 m)
9,800 ft (3,000 m)
16,400 ft (5,000 m)

land

seamount

sea depth

maximum depth on map

tectonic plate boundary

CHINA

NORTH KOREA
Kimch'aek
Hungnam
Wonsan

SOUTH KOREA
Tonghae
P'ohang
Pusan
Cheju-do

Korean Plateau
Korea Strait
Tsushima
Kitakyus
Fukuoka
Kyushu

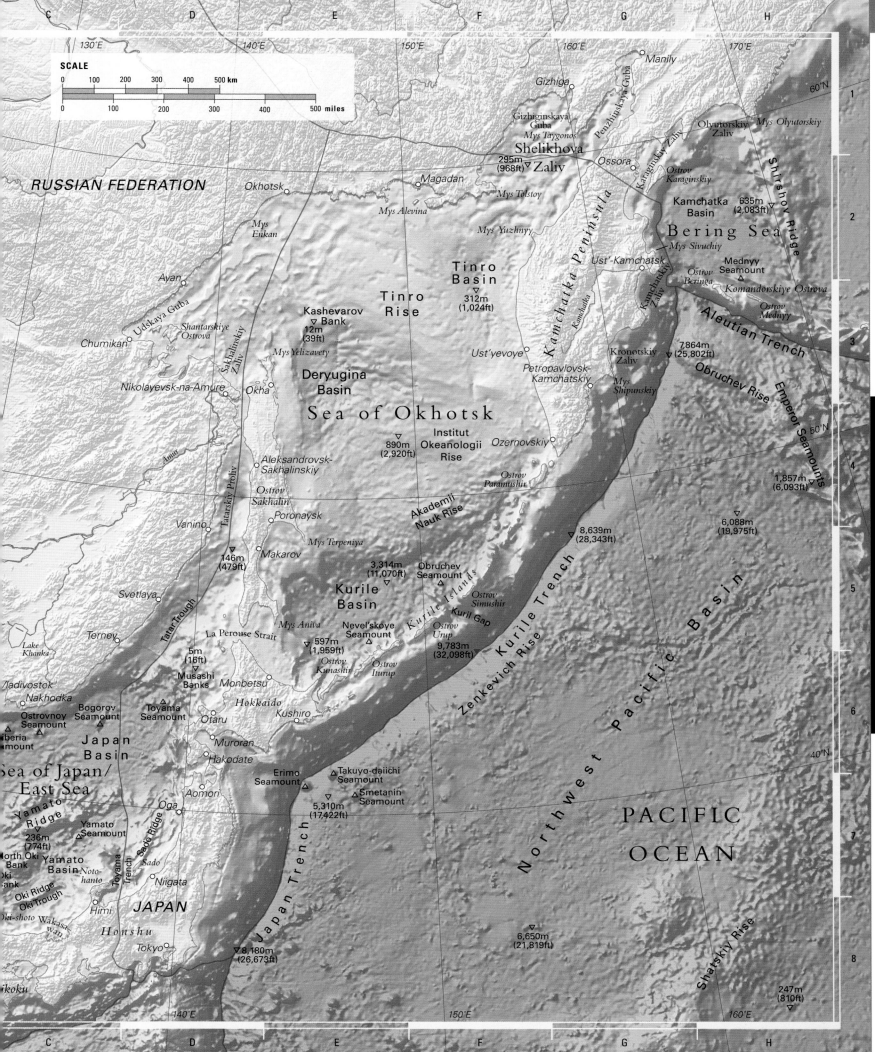

SCALE

| 0 | 100 | 200 | 300 | 400 | 500 km |

| 0 | 100 | 200 | 300 | 400 | 500 miles |

130°E 140°E 150°E 160°E 170°E

RUSSIAN FEDERATION

Manily

Gizhiga

Gizhiginskaya
Guba
Mys Taygonos

Shelikhova
295m
(968ft) ▽ **Zaliv**

Ossora

Olyutorskiy
Zaliv

Mys Olyutorskiy

60°N

Okhotsk

Magadan

Mys Tolstoy

Penzhinskaya Guba

Karaginskiy Zaliv

Ostrov
Karaginskiy

**Kamchatka
Basin**

635m
(2,083ft)

Shirshov Ridge

Mys Alevina

Mys Yuzhnyy

Kamchatka Peninsula

Mys Sivuchiy

Bering Sea

Ayan

**Tinro
Basin**

Ust'-Kamchatsk

Mednyy
Seamount

Ostrov
Beringa

Komandorskiye Ostrova

Udskaya Guba

Shantarskiye
Ostrova

Kashevarov
▽ **Bank**
12m
(39ft)

**Tinro
Rise**

312m
▽
(1,024ft)

Kamchatka

Ostrov
Mednyy

Aleutian Trench

Chumikan

Mys Yelizavety

Ust'yevoye

Kronotskiy
Zaliv

7,864m
(25,802ft)
▽

Obruchev Rise

Nikolayevsk-na-Amure

Sakhalinskiy Zaliv

Okha

**Deryugina
Basin**

Petropavlovsk-
Kamchatskiy

Mys
Shipunskiy

Emperor Seamounts

Amur

Sea of Okhotsk

Institut
Okeanologii
Rise

Ozernovskiy

50°N

Aleksandrovsk-
Sakhalinskiy

890m
(2,920ft)
▽

1,857m
(6,093ft)
▽

Ostrov
Sakhalin

Poronaysk

Ostrov
Paramushir

Vanino

Tatarskiy Proliv

Mys Terpeniya

**Akademii
Nauk Rise**

6,088m
(19,975ft)
▽

146m
(479ft)
▽

Makarov

3,314m
(11,070ft)
▽

Obruchev
Seamount

8,639m
(28,343ft)
▽

Svetlaya

Tatar Trough

**Kurile
Basin**

Kurile Islands

Ostrov
Simushir

Kurile Trench

Terney

Mys Aniva

Nevel'skoye
Seamount

Kuril Gap

Lake
Khanka

La Perouse Strait

5m
(16ft)
▽

Ostrov
Urup

9,783m
(32,098ft)
▽

Zenkevich Rise

Northwest Pacific Basin

Vladivostok

Musashi
Banks

Ostrov
Kunashir

Ostrov
Iturup

Nakhodka

Monbetsu

Bogorov
Seamount

Ostrovnoy
△ Seamount

Toyama
Seamount

Hokkaido

Kushiro

Iberia
mount

**Japan
Basin**

Muroran

**Sea of Japan/
East Sea**

Hakodate

40°N

Erimo
Seamount

Takuyo-daiichi
△ Seamount

Yamato
Ridge

Oga

Aomori

Smetanin
△ Seamount

PACIFIC

236m
(774ft)
▽

Yamato
△ Seamount

5,310m
(17,422ft)
▽

North Oki
Bank

**Yamato
Basin**

Noto-
hanto

Sado Ridge

Sado

Niigata

OCEAN

Oki
Bank

Toyama
Trench

Oki Ridge

Oki Trough

Himi

6,650m
(21,819ft)
▽

Oki-shoto

Wakasa
wan

JAPAN

Honshu

Japan Trench

Tokyo

8,180m
(26,673ft)
▽

Shatskiy Rise

247m
(810ft)
▽

ikoku

140°E 150°E 160°E

TSUNAMI STRIKES JAPAN
A tsunami wave surges ahead and begins to engulf homes on the coast of Natori in Miyagi Prefecture, Honshu, on March 11, 2011.

The undersea earthquake and consequent tsunami that hit Japan's Tohoku region (the northeast part of Honshu, Japan's largest island) on March 11, 2011, is one of the greatest disasters in human terms ever to hit Japan, and financially the costliest natural disaster ever. It was reported in 2021 that the catastrophe killed 19,747 people, with 2,556 missing, and has had an estimated economic cost of $360 billion. In addition, it has had serious environmental results, due to breakdowns, explosions, and leaks caused at the Fukushima nuclear energy plant.

The tsunami resulted from a sudden fracture in Earth's crust along a fault under the sea floor approximately 43 miles (70 km) off Honshu. This rupture triggered a massive earthquake. As sections of the sea floor suddenly sprang upward by around 20-27 ft (6-8 m), powerful tsunami waves were generated. On reaching the coast, these swept through towns and across fields, roads, and airports, smashing dwellings, vehicles, and boats. When the water later receded, a colossal jumble of debris was dumped on the landscape.

Within days, it was apparent that tens of thousands of buildings had been destroyed and hundreds of thousands of people displaced. At the Fukushima plant, three reactors suffered overheating and gas explosions after damage to backup power and containment systems. Subsequently, there were leaks of harmful radioactive materials from the plant into the atmosphere, ocean, and ground around the plant.

Wave height map

The map below shows maximum wave heights recorded along the affected coasts of Japan. At one location, the water rose for a short time to an estimated 127 ft (38.9 m) above sea level, but in most of the worst affected areas the rise was 10-40 ft (3-12 m). The waves washed over anti-tsunami seawalls, and the water then surged inland for distances of up to 6 miles (10 km). Over an hour or so, an area of nearly 216 square miles (560 square km) was inundated. The tsunami also traveled across the Pacific, causing significant damage in locations thousands of miles away.

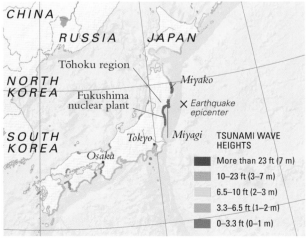

CHINA

RUSSIA   JAPAN

Tōhoku region

NORTH
KOREA   Fukushima
nuclear plant

Miyako

✕ Earthquake
epicenter

SOUTH
KOREA   Tokyo   Miyagi

Osaka

TSUNAMI WAVE HEIGHTS
- More than 23 ft (7 m)
- 10–23 ft (3–7 m)
- 6.5–10 ft (2–3 m)
- 3.3–6.5 ft (1–2 m)
- 0–3.3 ft (0–1 m)

STORY OF THE TSUNAMI

PRELUDE

EARTHQUAKE HITS HONSHU
The magnitude 9.0 earthquake, at around 2:45 pm local time, caused severe damage to roads and buildings in Tohoku and some areas surrounding Tokyo. It also set two oil refineries on fire.

TSUNAMI WAVES APPEAR
Within a few minutes, media reports and film footage were showing massive tsunami waves sweeping relentlessly toward the coast of the Tohoku region.

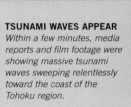

WARNINGS GO OUT
In Japan's coastal areas, sirens warn of any approaching tsunami, and signs indicate to where people should evacuate. But this time, the waves were so big that the warnings had a limited impact.

津波避難場所
Safety Zone for Tsunami
緑町町内会館
Midoricho Chonaikaikan
80m先右折
80m Right Turn Ahead

WAVES OVERWHELM COAST

TSUNAMI SURGE Within 10–50 minutes of the earthquake, colossal quantities of water were surging through harbors and across fields.

FLOTSAM AND JETSAM
Buildings and their contents were smashed or lifted up and aggregated into floating islands. Many fires broke out.

RELIEF OPERATION

EVACUATION SHELTERS
As relief operations began, evacuees were housed in shelters improvised from gymnasia. But with difficulties getting food, water, and medicines to survivors, the Japanese government was faced with huge challenges.

AFTERMATH

DAMAGED NUCLEAR PLANT At the Fukushima nuclear plant, the tsunami stopped pumps that ran vital cooling systems. This caused explosions and leakages of radioactive material that were still being dealt with in 2021.

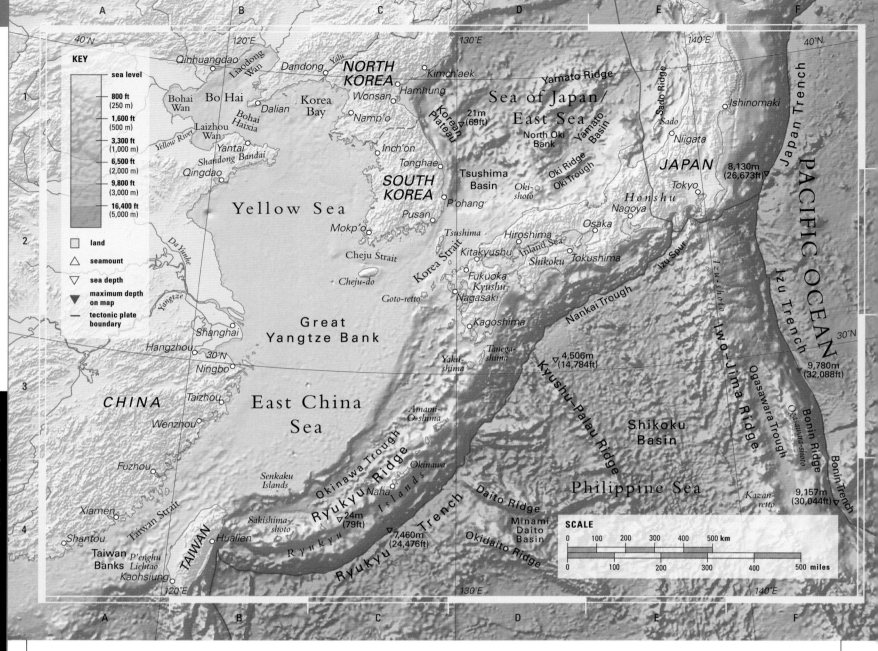

KEY

sea level

800 ft
(250 m)
1,600 ft
(500 m)
3,300 ft
(1,000 m)
6,500 ft
(2,000 m)
9,800 ft
(3,000 m)
16,400 ft
(5,000 m)

☐ land
△ seamount
▽ sea depth
▼ maximum depth
on map
— tectonic plate
boundary

SCALE

0 100 200 300 400 500 km

0 100 200 300 400 500 miles

The East China Sea

SHALLOW SEAS OVERLIE THE BROAD continental shelf off the coast of northern China. Warm water flows through the area from the south, feeding the Pacific's western boundary current, the Kuroshio Current, which allows the survival of the world's most northerly colonies of coral in the coastal waters of Japan. The area is vulnerable to cyclones moving in from the southwest.

PACIFIC OCEAN B3

East China Sea

AREA 290,000 square miles (751,000 square km)

MAXIMUM DEPTH 8,912 ft (2,717 m)

INFLOWS South China Sea, Philippine Sea, Yangtze River

The East China Sea is a warm, shallow, productive shelf sea that lies between the Chinese mainland and the Ryukyu Islands. It is linked to the South China Sea through the Taiwan Strait, and to the Sea of Japan through the Korea Strait. In spring and summer the warm Tsushima Current flows north through the Korea Strait, but this is suppressed by northerly winds during winter. The region is

also occasionally hit by typhoons (hurricanes) during the summer (see pp.70–71). The continental shelf beneath the South China Sea extends a long way from shore, partly due to sediments deposited by the Yangtze (Chang Jiang), Asia's longest river. The Yangtze is navigable by ocean-going ships up to 1,000 miles (1,600 km) inland, and China's main port, Shanghai, lies at its mouth. Fishing is an important source of income for the region, and the East China Sea is a shipping route between the South China Sea, Japan, and the north Pacific. There are also deposits of natural gas beneath the sea floor of the East China Sea, which China started developing in 2003.

PACIFIC OCEAN B2

Yellow Sea

AREA 205,000 square miles (530,000 square km)

MAXIMUM DEPTH 338 ft (103 m)

INFLOWS Yellow, Yangtze, Liao He, Luan He, Yalu, Han Rivers

Enclosed to the north, the Yellow Sea is an extension of the East China Sea, lying between the Chinese coast and the Korean Peninsula. It gets its name from the sand carried in suspension by the waters of the Yellow River (Huang He), the largest inflowing river. The sea is very shallow, and tidal ranges along the Korean side are some of the largest in the world. These strong tides also contribute to the color of the sea by stirring up sediment that has settled on the sea floor. The area around the northernmost bay of Bo Hai is one of the most industrialized in China; Dalian is China's third-largest port.

RYUKYU ISLANDS
The climate of the Ryukyu Islands is subtropical, with many of the islands fringed by coral reefs.

PACIFIC OCEAN C4

Ryukyu Trench

LENGTH 868 miles (1,398 km)

MAXIMUM DEPTH 24,476 ft (7,460 m)

RATE OF CLOSURE 2½–3 in (6–8 cm) per year

The Philippine oceanic plate is in contact with the Eurasian Plate to the south of Japan, resulting in a subduction zone marked by the Ryukyu Trench and the Nankai Trough. Volcanic island arcs have resulted to the northwest of the trenches. The Ryukyu Islands are a relatively young island arc compared with the mature arc of the Japanese islands of Honshu and Hokkaido, which has grown to a considerable land mass. The Ryukyu Islands include Okinawa, home to a large American naval base.

The South China Sea

THE SOUTH CHINA SEA IS THE LARGEST body of water after the five oceans, and its productive waters account for more than eight percent of the world's fish catch. This tropical sea is fed with water from the Java Sea through the Sunda Strait, which flows weakly north, and out through the Taiwan Strait. The northern part of the area can be battered by typhoons in the late summer.

PACIFIC OCEAN C2

South China Sea

AREA 1.4 million square miles (3.7 million square km)

MAXIMUM DEPTH 16,457 ft (5,016 m)

INFLOWS Xi Jiang, Red, Tha Chin, Mekong Rivers

The deep central basin of the South China Sea is surrounded by broad continental shelves for the most part. There are oil and natural gas fields in the Gulf of Tongking and the Vereker Banks, and off Borneo and across the Sunda Shelf. The South China Sea is dotted with shallow banks and reefs containing more than 200 tiny islands. Ownership of these and any associated subsea oil or minerals is disputed between China, Vietnam, Taiwan, Malaysia, Brunei, and the Philippines. In recent years, China has been asserting its claim by enlarging some islands and establishing military bases on them. This dispute is motivated by the presence of minerals, and it is estimated that vast oil reserves lie underneath the islands. Typhoons blow in from the Pacific in late summer. Typhoon Haiyan, the strongest to reach land, hit the Philippines in November 2013, with wind speeds over 186 miles (300 km) per hour producing a 16 ft (5 m) storm surge (see pp.72–73). On reaching Vietnam a few days later, it had weakened into a tropical storm.

PACIFIC OCEAN A2

Gulf of Thailand

AREA 124,000 square miles (320,000 square km)

MAXIMUM DEPTH 262 ft (80 m)

INFLOWS South China Sea, Chao Phraya River

A semienclosed extension of the South China Sea, the Gulf of Thailand lies between the Malay Peninsula and Indochina. The gulf contains offshore natural gas, and some oil deposits. Its shallow waters are largely fed by fresh river water inflow, principally from the Chao Phraya River. This river input gives the surface waters a relatively low salinity, while salt water from the main part of the South China Sea only enters deep down, pooling in areas deeper than 160 ft (50 m). Coral reefs thrive in the warm water, and there is a strong tourist industry based around good dive sites, such as the island of Ko Samui. To the south of the island, the Sunda Shelf was a land bridge during the last ice age, connecting Borneo, Sumatra, and Java to the Asian mainland.

PACIFIC OCEAN D3

Sulu Sea

AREA 100,300 square miles (260,000 square km)

MAXIMUM DEPTH 18,400 ft (5,600 m)

INFLOWS Celebes Sea

The Sulu Sea is a deep, tropical sea surrounded by islands. Currents follow the monsoon winds, coming from the south in the summer and the north in the winter. Small-scale fishing is the main economic activity in both the Sula Sea and the neighboring Celebes Sea, with shrimp, turtles, and pearl oysters among the catch.

PEARL-COLLECTING IN THE SULU SEA

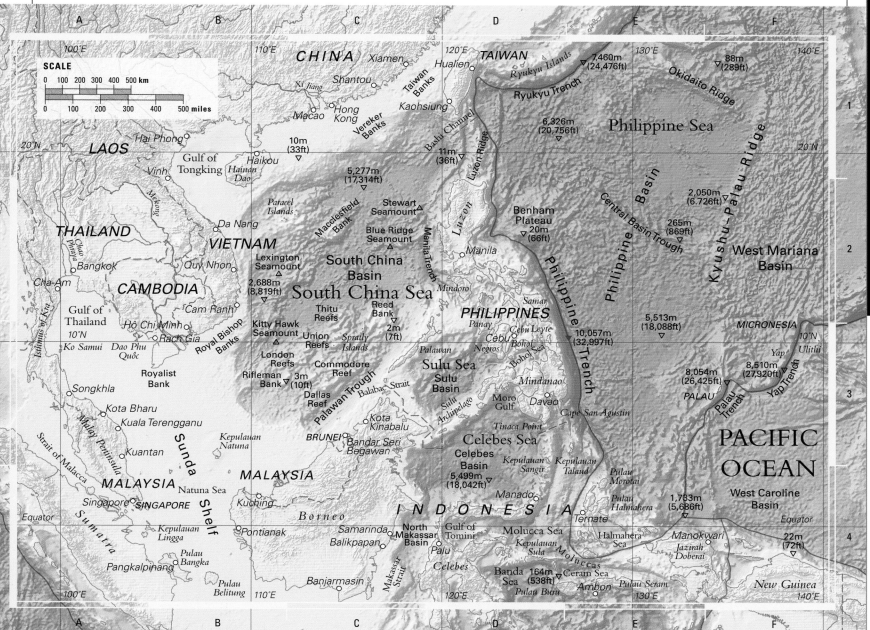

A B C D E F

Yellow Sea

SOUTH KOREA
P'ohang
Pusan

Sea of Japan/ East Sea

JAPAN
Honshu
Tokyo
Nagoya

8,130m
(26,673ft)

6,650m
(21,819ft)

Hiroshima
Kitakyushu
Osaka

Inland Sea
Shikoku
Tokushima

Northwest Pacific Basin

Cheju Strait
Cheju-do

Korea Strait
Fukuoka
Kyushu
Nagasaki

1,339m
(4,393ft)

Goto-retto

Kagoshima

9,780m
(32,088ft)

Shanghai

Great Yangtze Bank

Tanega-shima
Yaku-shima

Nankai Trough

4,506m
(14,784ft)

Shikoku Basin

Isakov Seamount

30°N

CHINA

East China Sea

Amami-O-shima

Makarov Seamount

Senkaku Islands

Okinawa Trough
Ryukyu Islands
Okinawa

Daito Ridge

Kazan-retto

Iwo-Jima Ridge

Ogasawara Trough

Bonin Ridge
Ogasawara-shoto

Choyo Seamount

5,661m
(18,574ft)

Sakishima-shoto

Naha

24m
(79ft)

Ryukyu Ridge

88m
(289ft)

Bonin Trench

Michelson Ridge

TAIWAN
Hualien

7,460m
(24,476ft)

Ryukyu Trench

Okidaito Ridge

9,157m
(30,044ft)

Marcus Island

1,082m
(3,550ft)

Basin Channel

6,326m
(20,756ft)

Kyushu-Palau Ridge

PACIFIC

20°N

Luzon Ridge

11m
(36ft)

Philippine Basin

West Mariana Ridge

Mariana Trench

6,464m
(21,208ft)

Luzon

Philippine Sea

2,050m
(6,726ft)

Magellan Seamounts

Benham Plateau

Central Basin Trough

265m
(869ft)

NORTHERN MARIANA ISLANDS

Mariana Ridge

20m
(66ft)

West Mariana Basin

Saipan
Tinian
9,888m
(32,442ft)

East Mariana Basin

Manila

Philippine Trench

Mindoro

GUAM

Mi

PHILIPPINES
Panay
Cebu
Cebu
Leyte

Samar

10,057m
(32,997ft)

Challenger Deep
10,920m
(35,829ft)

10°N

Negros
Bohol
Bohol Sea

5,513m
(18,088ft)

Ulithi

ic

Sulu Sea

Yap
Yap Trench

Caroline Islands

8,054m
(26,425ft)

8,510m
(27,920ft)

Caroline Ridge

Chuuk Islands

Mindanao
Moro Gulf
Davao

PALAU

West Caroline Rise

Sorol Trough

Palau Trench

Cape San Agustin

West Caroline Trough

MICRONESIA

Tinaca Point

Eauripik Rise

Tinaca Point
Kepulauan Talaud

4,859m
(15,942ft)

1,554m
(5,099ft)

Lyra Basin

1,402m
(4,560ft)

Celebes Sea
Celebes Basin

Kepulauan Sangi

West Caroline Basin

East Caroline Basin

Mussau Trench

5,499m
(18,042ft)

Manado

Pulau Morotai

4,262m
(13,984ft)

7,249m
(23,784ft)

Ontong J

Equator

Ternate

Pulau Halmahera

New Guinea Trench

Manus Trench

Molucca Sea

1,733m
(5,686ft)

Bismarck Archipelago

Lyra Reef

Kepulauan Sula

Halmahera Sea

Manokwari
Jazirah Doberai

22m
(72ft)

Manus Island

Bismarck Sea

New Ireland

164m
(538ft)

Ceram Sea

INDONESIA

PAPUA NEW GUINEA

Celebes
Pulau Buru
Pulau Seram
Ambon

Jayapura

New Guinea

Wewak
Sepik

Rabaul

Banda Sea

130°E

140°E

150°E

A B C D E F

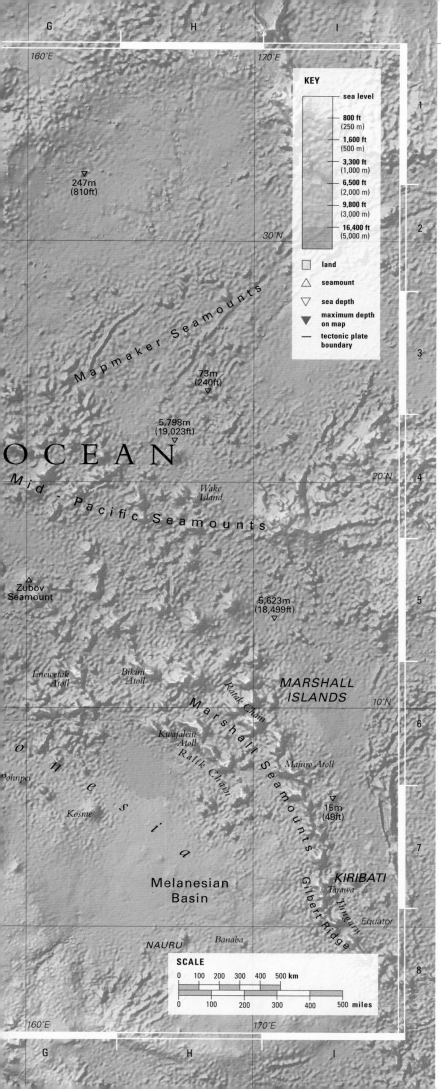

Micronesia

THE NAME MICRONESIA APPLIES TO AN AREA in the western Pacific, north of the equator. Its stretches to the Caroline and Mariana islands in the west, and Nauru, the Marshall Islands, and Kiribati (or Gilbert) Islands to the east.

PACIFIC OCEAN B4

Philippine Sea

AREA	1.9 million square miles (5 million square km)
MAXIMUM DEPTH	35,580 ft (10,540 m)
INFLOWS	Pacific Ocean, South China Sea

The Philippine Sea stretches east to west between the Marianas Islands and the Philippines, and from north to south between Japan and Palau. This warm sea is swept by the North Equatorial Current, which turns north to form the Kuroshio Current. The water becomes very warm in the summer, and the area is a breeding ground for typhoons. The Philippine Sea is underlain by the Philippine Plate, an oceanic plate that is subducting at the Philippine and Ryukyu trenches. The plate is split into two main basins. The westernmost

TYPHOON DESTRUCTION

Philippine Basin is the deepest and oldest, separated from the West Mariana Basin by the Kyushu–Palau Ridge. This ridge, and the Iwo-Jima and West Mariana ridges, are the remnants of island arcs associated with ancient subduction zones.

PACIFIC OCEAN E5

Mariana Trench

LENGTH	1,580 miles (2,542 km)
MAXIMUM DEPTH	35,827 ft (10,920 m)
RATE OF CLOSURE	1½ in (4 cm) per year

At the eastern edge of the Philippine Plate lies the volcanic island arc of the Northern Mariana Islands. To the east lies the Mariana Trench, where the Pacific Plate is subducting beneath the Philippine Plate. The Mariana Trench includes Challenger Deep—at 35,827 ft (10,920 m), this is the deepest

MARIANA ISLANDS
The southern members of the Mariana Islands are limestone platforms with fringing coral reefs.

known part of the ocean. Named after a British survey ship that measured its depth in 1951, it was explored for the first time in 1960 by the deep-sea submersible Trieste. This was followed by remotely operated submarines in 1995–1998 and 2009; *Deepsea Challenger* in 2012; and American and Chinese scientists in 2019, 2020, and 2021.

PACIFIC OCEAN I6

Marshall Islands

TYPE	Coral atoll islands
AREA	70 square miles (180 square km)
NUMBER OF ISLANDS	34

Most of the seamounts scattered across the floor of the western Pacific are far from any plate boundary. The seamounts are found in groups, often strung out in lines running southeast–northwest—the direction of motion of the Pacific Plate. They are caused by hotspots in Earth's mantle, which periodically punch through the ocean crust to form volcanoes. Some may reach the surface as islands and in the Marshall Islands, coral atolls were formed as the plate moved away from the hotspot and the volcanic islands subsided. The atolls of Enewetak and

Bikini were chosen for their remote location as the site of American nuclear bomb tests in the 1940s and 1950s. Today, the low-lying islands of Kiribati are threatened by sea-level rise due to the warming climate.

GARDEN EELS
These garden eels are among the sea life to be found at the bottom of Rongelap Atoll in the Marshall Islands.

Midway and Hawaii

OCEAN CIRCULATION IS CLOCKWISE in the central North Pacific, with the Kuroshio Extension flowing eastward in the north, turning south in the east, and returning as the North Equatorial Current south of the Hawaiian Islands. The seafloor is dominated by seamount chains, mostly running from northwest to southeast, and fracture zones, oriented roughly east–west.

PACIFIC OCEAN B1

Emperor Seamounts

TYPE Volcanic seamount chain

LENGTH 1,200 miles (2,000 km)

NUMBER OF SEAMOUNTS 17

The Emperor Seamounts stretch over 1,200 miles (2,000 km) from the northwestern Hawaiian Islands in the south to the Aleutian Trench in the north. They are the oldest part of the Emperor–Hawaii seamount chain—more than 40 million years old—with age and depth increasing to the north. The individual seamounts are named after emperors of Japan.

PACIFIC OCEAN C3

Midway Islands

TYPE Coral atoll islands

AREA 2½ square miles (6.2 square km)

NUMBER OF ISLANDS 4

These four islands form a coral atoll that lies roughly halfway between North America and Asia. Its position has made it useful for a variety of purposes over the years, first as a telegraph cable and radio station, then as a flying boat stopover, and as an important naval and air station in wartime. Today, it is administered by the US government as a wildlife refuge.

PACIFIC OCEAN E3

Hawaiian Islands

TYPE Volcanic islands

AREA 6,400 square miles (16,600 square km)

NUMBER OF ISLANDS 19

The Hawaiian Islands are a volcanic chain, with large, active volcanic islands at the eastern end, and older, subsided seamounts and atolls at the western end of the chain. They are a continuation of the Emperor seamount chain, created by the same hot spot beneath the seabed as the Pacific Plate first moved north, and then west-northwest. This change in direction of plate motion about 40 million years ago accounts for the different alignments of the Emperor and Hawaii chains. Hawaii's Mauna Loa is a broad shield volcano, the largest on Earth. From its base on the sea floor it rises about 55,800 ft (17,000 m)—much taller than Mount Everest. Its weight has caused the underlying ocean crust to sag, producing the Hawaiian Trough to the north and east of the main island. The seas around Hawaii are tropical, and warm enough for coral reefs to grow. The prevailing winds in this region are the trade winds from the northeast, making the northeast coasts of the islands considerably wetter and greener.

HAWAIIAN LAVA
Fluid, basaltic lava from the Kilauea volcano reaches the Pacific Ocean on the Island of Hawaii.

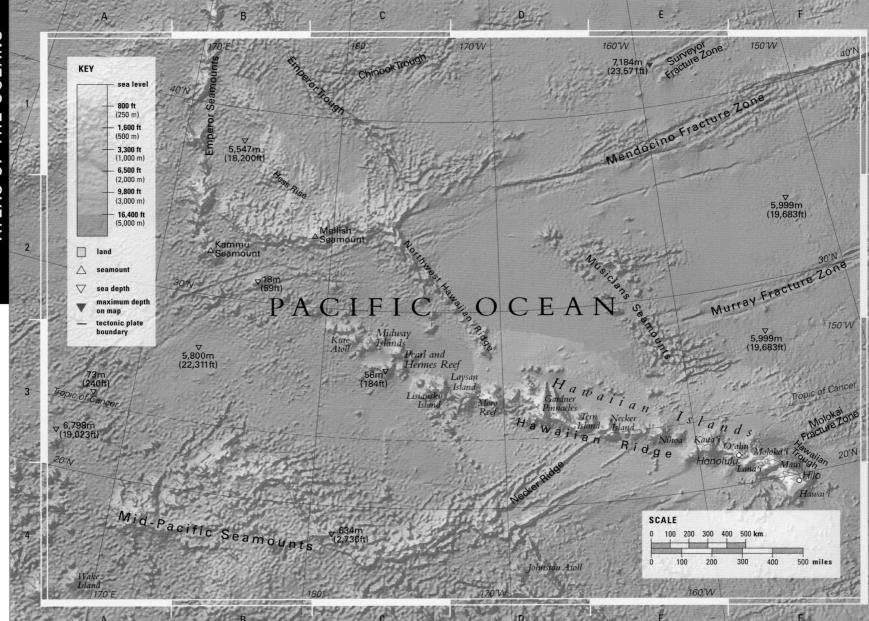

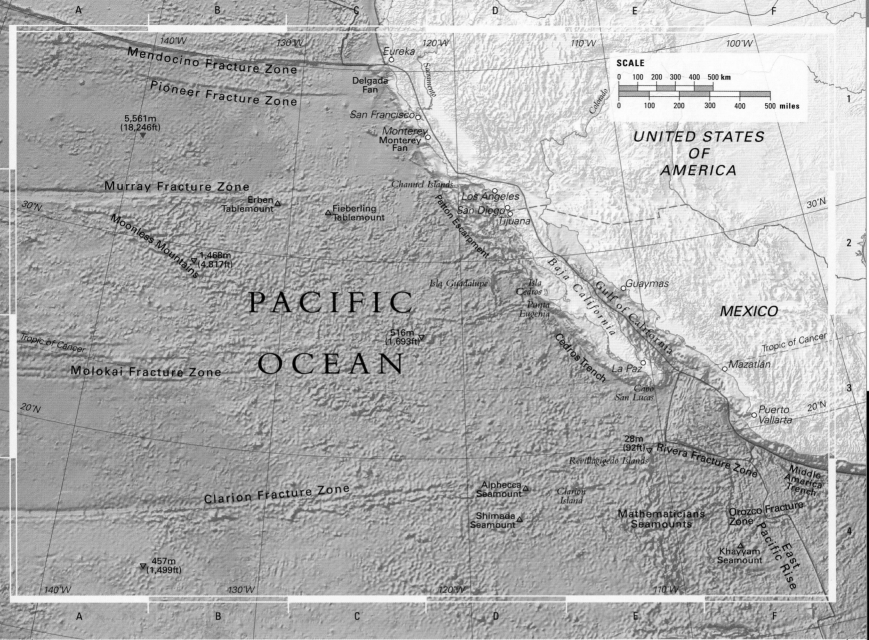

Gulf of California

THE CALIFORNIA CURRENT FLOWS south in the western North Pacific, forming the western arm of the North Pacific Gyre. South of the Tropic of Cancer it sweeps west to become the North Equatorial Current. Against the California coast, a cold coastal current runs north. The floor of the western Pacific slopes gently away from North America, scarred by long fracture zones.

PACIFIC OCEAN C1
San Francisco Bay

AREA 1,600 square miles (4,160 square km)

MAXIMUM DEPTH 1,970 ft (600 m)

INFLOWS San Joaquin, Sacramento Rivers

San Francisco Bay and its associated water bodies form the world's largest natural harbor. In addition to San Francisco itself, there are major ports in Oakland, Richmond, Stockton, Sacramento, and San Pablo Bay. The bay is an estuary, containing large areas of salt marsh that support a number of endangered species. It is famous for its sea fogs, which arise when sea breezes blow over cold inshore waters.

THE GOLDEN GATE BRIDGE

PACIFIC OCEAN C1
Monterey Bay

AREA 620 square miles (1,600 square km)

MAXIMUM DEPTH 10,700 ft (3,250 m)

INFLOWS Pacific Ocean

Monterey Bay is one of the most diverse marine ecosystems in the world, and is home to a variety of fish, invertebrate, seabird, and mammal species, including sea otters, harbor and elephant seals, bottlenose dolphins, and turtles. It is also famous for its fish-processing industry, which grew up in the early part of the 20th century around Cannery Row. The bay is the site of one of the longest undersea canyons in the world, Monterey Canyon, which reaches a depth of more than 9,800 ft (3,000 m) and runs 60 miles (100 km) offshore. A mud volcano north of the canyon is the source of cold hydrocarbon seeps (see pp.188–189), which support an ecosystem based on metabolizing sulfide compounds, rather than using sunlight for energy. Similar deep-sea communities are found at hydrothermal vents on the mid-ocean ridges.

PACIFIC OCEAN E2
Gulf of California

AREA 62,000 square miles (160,000 square km)

MAXIMUM DEPTH 10,000 ft (3,050 m)

INFLOWS Fuerte, Sonora, Yaqui, Colorado Rivers

The Gulf of California was originally named the Sea of Cortez by Spanish explorers and is still known locally by that name. It marks the boundary between the North American and Pacific plates. The peninsula of Baja California lies on the Pacific Plate and is moving northwest, away from Mexico, at about 2 in (5 cm) each year. To the north, large earthquakes are quite frequent along the San Andreas Fault. The waters of the gulf support a rich ecosystem and a healthy commercial fishery. In addition to the native species, migratory visitors include humpback whales, manta rays, and leatherback turtles. The California gray whale completes the longest migration of any mammal, spending a few weeks each year in breeding grounds off Baja California, before returning to the Bering Sea, 5,000 miles (8,000 km) away.

GREEN SAND BEACH
Papakolea or Mahana Beach, on the volcanic Big Island of Hawaii, is one of only a handful of beaches in the world composed of predominantly green-colored sand. The distinctive hue comes from the mineral olivine, eroded from rocks in the volcanic cinder cone that partly encircles the beach.

Melanesia

THE NAME MELANESIA COMES FROM THE GREEK for "black islands." The area includes the islands north of Australia, from Celebes and New Guinea in the west, to Fiji and Samoa in the east. The surrounding seas are tropical, their warm waters fed by the westward flow of the South Equatorial Current. The region is geologically complex, with some parts volcanically active.

PACIFIC OCEAN F2

Solomon Sea

AREA 278,000 square miles (720,000 square km)

MAXIMUM DEPTH 29,300 ft (8,940 m)

INFLOWS Pacific Ocean, Coral Sea

The Solomon Sea lies between the Solomon Islands and the island of New Guinea, with the island of New Britain to the north and the Louisiade Archipelago to the south. The area is geologically complex, forming the remains of a closing ocean basin caught between the Australian Plate moving north and the Pacific Plate moving west. The floor of the Solomon Sea appears to consist of one or two very small tectonic plates (microplates) of oceanic origin. The Solomon Sea Microplate is spreading from the area of the Pockington Trough and rotating clockwise, subducting to the north and possibly to the southwest. Volcanic activity is particularly intense off the New Georgia Islands, where the spreading ridge is being subducted: the submarine volcano Kavachi breached the surface explosively in 2002. On the other side of the sea, the tectonic upheavals have resulted in uplift of New Guinea's Huon Peninsula, where raised coral terraces are found some distance inland.

RABAUL VOLCANO
When Tavurvur erupted in 2014, it was far less damaging than the 1994 eruption that destroyed much of nearby Rabaul after the city and surrounding villages were evacuated.

PACIFIC OCEAN F1

Bismarck Sea

AREA 124,000 square miles (320,000 square km)

MAXIMUM DEPTH 9,200 ft (2,800 m)

INFLOWS Pacific Ocean, Solomon Sea

The Bismarck Sea lies off the north coast of New Guinea. It is surrounded by volcanic islands, the largest being New Britain. The underlying Bismarck Microplate is caught between the Australian Plate, moving north, and the Pacific and Caroline plates, moving west. The northern islands of the archipelago arise from the subduction of the Caroline and Pacific plates, marked by the Manus Trench to the north. Volcanoes on the south side of the sea are currently more active, arising from the subduction of the Solomon Sea Microplate at the New Britain Trench. To the east of the Bismarck Sea lies the Ontong Java Rise. This submarine plateau is one of the world's largest expanses of igneous rock. It is composed of flood basalts, some of which date back to 120 million years ago.

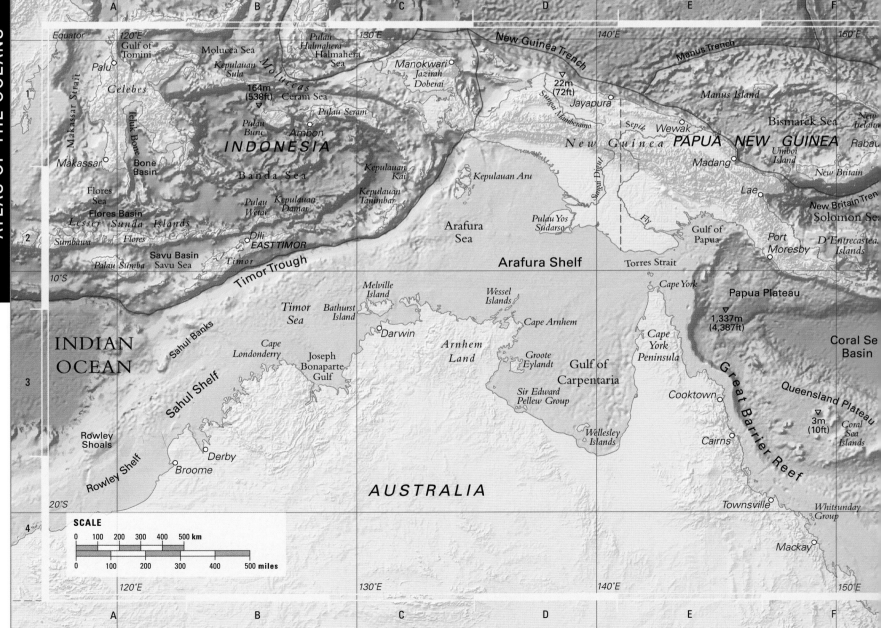

Arafura Sea

AREA 251,000 square miles (650,000 square km)

MAXIMUM DEPTH 260 ft (80 m)

INFLOWS Coral Sea, Timor Sea, Banda Sea

Lying over a shallow continental shelf, the Arafura Sea marks the boundary between the Pacific and Indian Oceans. During the Indian summer monsoon, water flows westward into the Indian Ocean, pulled by the South Equatorial Current, but the flow is reversed when the Equatorial Counter Current is active during the Indian winter.

SUNSET OVER THE ARAFURA SEA

Coral Sea

AREA 1.8 million square miles (4.8 million square km)

MAXIMUM DEPTH 30,070 ft (9,165 m)

INFLOWS West Central Pacific Ocean; Fly, Purari, Kikori Rivers

This tropical sea earns its name from the presence of coral reefs along most of its coasts. The Great Barrier Reef (see p.161) grows out to the edge of the Australian continental shelf, on the western side of the sea. Warm water enters from the Pacific, circulating weakly before leaving through the Torres Strait to the west, or to the

GREAT BARRIER REEF

The largest coral reef structure in the world, the Great Barrier Reef stretches for more than 1,200 miles (2,000 km).

south as the East Australia Current. The eastern and northern sides of the Coral Sea are marked by deep trenches, where the oceanic part of the Australian Plate is subducting. Volcanism has resulted in the Solomon Islands and the Vanuatu chain. Explosive eruptions in these islands can create pumice, a volcanic rock that floats due to gas bubbles trapped inside. Lumps of pumice are sometimes washed up on the western shore of the Coral Sea.

Fiji Plateau

AREA 30,900 square miles (80,000 square km)

AVERAGE DEPTH Less than 820 ft (250 m)

The Fiji Plateau is the thickest part of the Fiji Plate. The islands of Fiji were originally part of a continuous volcanic island arc alongside Vanuatu and the Solomons. They were moved east when the Pacific Plate changed its direction of motion, triggering the creation of new ocean crust in the North Fiji Basin. The Hunter Ridge to the south marks the transform fracture that allowed this eastward motion, while the Vityaz Trench to the north marks the subduction zone that created the islands, though the trench is now inactive. Over time a substantial platform of limestone accumulated around the original volcanic islands, as they were uplifted, faulted, and folded by the opening of the young, buoyant North Fiji Basin. Growth of the platform continues today, thanks to the coral reefs that fringe Fiji's hundreds of islands.

KEY

- sea level
- 800 ft (250 m)
- 1,600 ft (500 m)
- 3,300 ft (1,000 m)
- 6,500 ft (2,000 m)
- 9,800 ft (3,000 m)
- 16,400 ft (5,000 m)
- □ land
- △ seamount
- ▽ sea depth
- ▼ maximum depth on map
- — tectonic plate boundary

CLOSE ENCOUNTER
A hawksbill turtle swims past a group of scuba divers in the Ras Mohammed protected area near the popular Red Sea diving resort of Sharm el-Sheikh, Egypt.

The pioneering work in 1943 of French divers Jacques Cousteau and Emile Gagnan in creating scuba (self-contained underwater breathing apparatus) moved diving away from being the domain of the military, scientists, and a privileged few and opened up the wonders of the underwater world to a wider public. Although there is an element of risk, a vast array of training agencies exist to assist the potential diver in taking up the sport safely. In line with the increased accessibility of diving, dive operations have sprung up in every location where there is a combination of a tourist market and easy access to a body of water suitable for scuba diving. The COVID pandemic that began in 2020 brought diving tourism almost to a halt, although by mid-2021, there were signs of recovery. Anecdotal evidence suggests that as tourism resumes, divers will find that the health of marine ecosystems has improved somewhat from being left relatively undisturbed for many months.

The modern diver is faced with a bewildering array of choices when it comes to the dive experience itself. The simplest form of diving is a course that is based within a resort, which supplies all of the equipment and training, and guides the diver around local sites. More specialized is the dedicated dive vacation, where the qualified diver seeks out a location specifically to explore local attractions. Some diving enthusiasts choose to base themselves on dive vessels that explore some of the more remote locations of the undersea world. For the most committed divers, it is possible to join any of a variety of diving expeditions or projects engaged in conservation or surveying work.

The Divers' Code

The vast majority of modern dive operations and tourist authorities insist on a strict set of guidelines to minimize the effect of too many divers on reefs. The Divers' Code includes the following points: no contact with coral; maintain buoyancy control so that accidental contact with the reef is avoided; no collection of shells, coral, or other mementos; no touching, harassing, or feeding of any marine animals; and an exclusion zone is to be maintained around large marine animals, such as whales.

SAFETY FIRST
A dive instructor teaches novice divers how to respond if they lose their mask. Hand signals are used to communicate underwater.

DIVING ACTIVITIES

DIVE SITES

WRECKS *It is not just the living world that attracts divers. In some parts of the ocean, such as here in the Caribbean, the sea floor is littered with wrecked ships. Exploring the hulks can be like traveling back in time.*

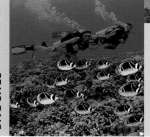

REEFS *Coral reefs provide some of the most spectacular sights on Earth. A healthy reef positively teems with a wide variety of wildlife, including many brightly colored fish species, such as these raccoon butterflyfish on a reef in Hawaii.*

BENEFITS AND DRAWBACKS

ECOTOURISM *Some conservation-minded divers spend their vacations assisting scientists in their research of coral reefs.*

LOCAL ECONOMIES *Residents of prime dive locations can benefit by offering diving services. But if a location becomes too popular, the sheer number of visitors can cause severe damage to the reefs.*

INTERACTING WITH WILDLIFE

FEEDING SHARKS *The feeding of top predators, such as this Caribbean reef shark, is a complex issue. Although the behavior of the sharks is altered by such activities, such diving trips are becoming increasingly popular in many locations, resulting indirectly in the protection of the sharks by local operators.*

BAD PRACTICE *A boy diving in the Red Sea touches a spiny pufferfish, causing it to inflate. The handling of marine wildlife can harm both the animal and the diver and is discouraged by the Divers' Code.*

10°N

5m
(16ft)

Central
Pacific
Basin

Magellan Rise

5,582m
(18,345ft)

Christmas Ridge

Kingman Reef
Palmyra Atoll
Teraina

Tabuaeran

Clipperton Fractu

Kiritimati

Howland Island

Equator

Baker Island

KIRIBATI

Nova Trough

Line

P A C I F I C

27m
(89ft)

6,249m
(20,503ft)

Phoenix Islands
Kanton

Otona

Enderbury Island

Jarvis Island

Malden Island

Galapagos Fractu

P

Starbuck Island

TUVALU

Funafuti

Atafu Atoll

Nukunonu Atoll

TOKELAU

Fakaofo Atoll

o

5,451m
(17,885ft)

l

Penrhyn

y

Vostok Island

n

4,370m
(14,338ft)

Millennium
Island

10°S

Robbie Ridge

Rakahanga

Pukapuka

Nassau

Manihiki

Manihiki
Plateau

e

Flint Island

s

Îles Wallis

SAMOA

Savai'i

Upolu

Northern Cook Islands

Suwarrow

Penrhyn
Basin

i

a

Ile Futuna

WALLIS
AND
FUTUNA
Zephyr
Bank

Apia

Tutuila

AMERICAN
SAMOA

COOK
ISLANDS

Society Ridge

Bora-Bora

Moorea

Tuam

Vanua
Levu

FIJI

Suva Koro
Sea
Viti Levu

Lau Group

Vava'u
Group

TONGA

Samoa Basin

Palmerston

Aitutaki

Atiu

Papeete

Tahiti

Society Islands

FRENCH
POLYNESIA

Tuamot

20°S

4m
(13ft)

Lau
Basin

10,587m
(34,736ft)

NIUE

Tongatapu
Group

Lau Ridge

Tonga Ridge

Tonga Trench

10,800m
(35,435ft)

5,901m
(19,361ft)

Rarotonga

Mangaia

Southern Cook Islands

3,780m
(12,402ft)

South Fiji Basin

Ozbourn
Seamount

1,039m
(3,409ft)

Tomaszeski
Seamount

Southwest
Pacific Basin

Îles Australes

President Thiers
Seamount

Kermadec
Islands

Kermadec Trench

Louisville Ridge

4,602m
(15,099ft)

30°S

10,047m
(32,964ft)

180°

170°W

5,655m
(18,554ft)

160°W

150°W

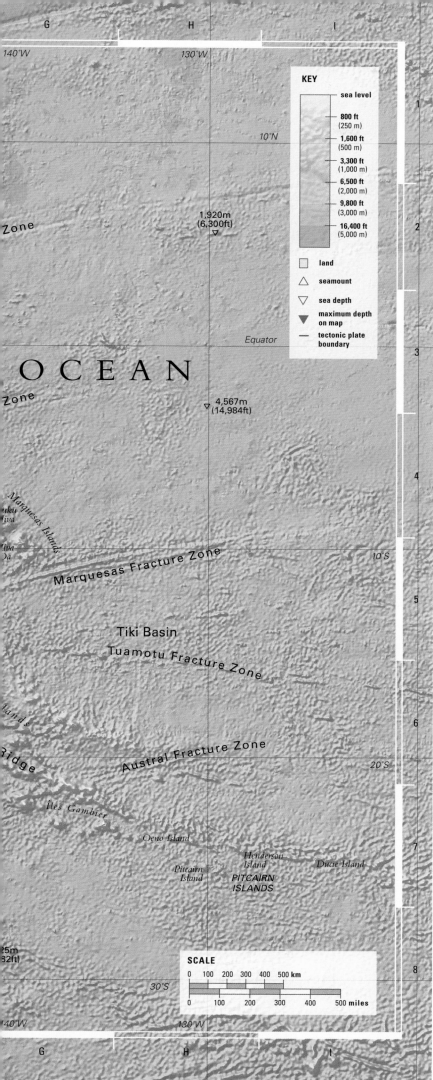

Polynesia

THE FLOOR OF THE SOUTHWEST PACIFIC is dotted with chains of islands and seamounts, and cut by major fracture zones. Its deepest part lies in the Tonga Trench, where the Pacific Plate is subducting beneath the Australian Plate. The counterclockwise South Pacific Gyre controls the ocean currents, with eastward Equatorial Countercurrents north and south of the equator.

PACIFIC OCEAN G4

Marquesas Islands

TYPE	Volcanic islands
AREA	490 square miles (1,270 square km)
NUMBER OF ISLANDS	15

The Marquesas Islands are volcanic in origin, overlying a mantle hotspot, with high mountain peaks and ridges. They lie in the path of the strong South Equatorial Current, flowing from the east, producing eroded coastlines, with steep cliffs dotted with sea caves. Coral reefs are limited to a few sheltered bays. The Marquesas lie farther from a continental landmass than any other island group on Earth.

Despite this isolation, the islands suffered the effects of an earthquake in Alaska in 1946 when they were hit by the resultant tsunami, which rose up to 40 ft (12 m) high in some places.

ERODED VOLCANIC COASTLINE

PACIFIC OCEAN D5

Cook Islands

TYPE	Volcanic islands
AREA	93 square miles (240 square km)
NUMBER OF ISLANDS	15

The Cook Islands are split into two groups: the low-lying coral atolls of the northern group, most of which rise from the Manihiki Plateau; and the mainly high volcanic islands in the south. The largest island, Rarotonga, rises 2,140 ft (652 m) above sea level and 14,800 ft (4,500 m) above the ocean floor. The islands were settled in about 300 BCE by people originating from eastern Melanesia, who migrated via Fiji, Tonga, and Samoa. Polynesians were expert ocean explorers and perfected the use of many navigation aids at a time when European mariners relied on keeping the land in sight. Their techniques included using the stars; knowledge of currents, winds, and wave patterns; and the flight of birds. Captain James Cook was the first European to sight the Cook Islands in 1773. Russian sailors named the islands in his honor in the 19th century.

RAROTONGA
Like most of the Southern Cook Islands, Rarotonga has a fringing coral reef several hundred yards offshore, which protects a shallow lagoon and coastal plain.

PACIFIC OCEAN F6

Tuamotu Islands

TYPE	Coral atoll islands
AREA	340 square miles (885 square km)
NUMBER OF ISLANDS	78

The Tuamotu Islands are the longest chain of coral atolls in the world, stretching over 1,200 miles (2,000 km). The Tuamotus were settled by Polynesians by 700 CE. Ferdinand Magellan was the first European to chart the group in 1521. The islands became known for their rare black pearls, and these still form a major part of the economy. The Tuamotu Islands are part of French Polynesia and the islands of Mururoa and Fantgataufa were used as sites for about 200 French nuclear weapon tests between 1966 and 1996. The atolls were formed as volcanic islands subsided, leaving behind their fringing coral reefs. The islands rise from the Tuamotu Ridge, a plateau of volcanic material formed about 63–40 million years ago. The Gambier Islands at the southeast end of the ridge are younger, with their volcanic peaks still standing above sea level.

Map labels:

140°W · 130°W · 10°N · Equator · 10°S · 20°S · 30°S

KEY

sea level
800 ft (250 m)
1,600 ft (500 m)
3,300 ft (1,000 m)
6,500 ft (2,000 m)
9,800 ft (3,000 m)
16,400 ft (5,000 m)

□ land
△ seamount
▽ sea depth
▼ maximum depth on map
— tectonic plate boundary

1,920m (6,300ft) ▽

4,567m (14,984ft) ▽

OCEAN

Zone

Marquesas Islands
nuku Hiva
Hiva Oa

Marquesas Fracture Zone

Tiki Basin
Tuamotu Fracture Zone

Austral Fracture Zone

Ridge

Îles Gambier

Oeno Island
Pitcairn Island
Henderson Island
Ducie Island
PITCAIRN ISLANDS

25m 32ft)

SCALE
0 100 200 300 400 500 km
0 100 200 300 400 500 miles

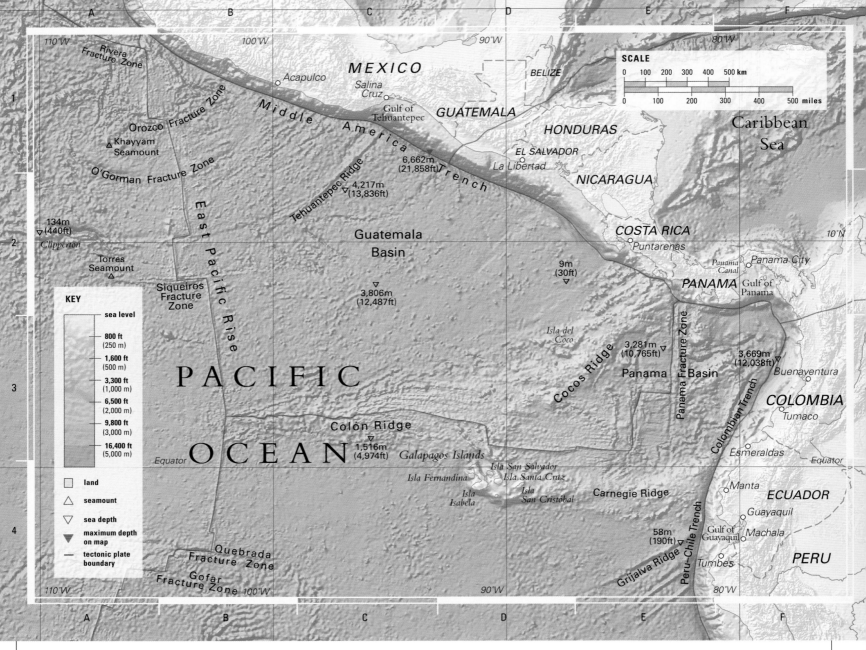

KEY

	sea level
▬	800 ft (250 m)
▬	1,600 ft (500 m)
▬	3,300 ft (1,000 m)
▬	6,500 ft (2,000 m)
▬	9,800 ft (3,000 m)
▬	16,400 ft (5,000 m)

- ☐ land
- △ seamount
- ▽ sea depth
- ▼ maximum depth on map
- — tectonic plate boundary

Galápagos Islands

THE EQUATORIAL COUNTERCURRENT flows into the eastern equatorial Pacific from the west, feeding a counterclockwise gyre over the Guatemala Basin. From the south, the Humboldt Current feeds into the South Equatorial Current, which runs westward across the Pacific. The area is underlain by two plates, the Cocos and Nazca, which are remnants of the original eastern Pacific plate.

PACIFIC OCEAN B2

East Pacific Rise

LENGTH 5,600 miles (9,000 km)

HEIGHT ABOVE SEA FLOOR 3,280 ft (1,000 m)

RATE OF SPREAD 4½–6 in (11–15 cm) per year

The East Pacific Rise is the fastest-spreading mid-ocean ridge in the world, producing a broad, gently sloping ridge with few transform offsets. It was here that the first submarine hydrothermal vents, or black smokers (see pp.188–189), were discovered. These vents give rise to oases of life on the deep-ocean floor, supporting complex communities of tube worms, clams, shrimp, and crabs, fueled by nutrients in the vent fluids.

PACIFIC OCEAN C1

Middle America Trench

LENGTH 1,700 miles (2,750 km)

MAXIMUM DEPTH 21,858 ft (6,662 m)

RATE OF CLOSURE 3½ in (9 cm) per year

The Cocos Plate is subducting beneath the North American and Caribbean plates at the Middle America Trench. A chain of volcanoes has arisen along Central America's western coast, with volcanism most active in the southern part of the subduction zone behind the trench. Earthquakes in the area are triggered by plate movement.

PACIFIC OCEAN D3

Galápagos Islands

TYPE Volcanic islands

AREA 3,030 square miles (7,850 square km)

NUMBER OF ISLANDS 19

The Galápagos Islands first appeared on maps drawn by Flemish cartographers Abraham Ortelius and Gerardus Mercator in 1570. The Galápagos take their name from the old Spanish word for tortoise, as early visitors found giant tortoises roaming the islands. There are many other species that have made unique adaptations to the local environment, including the marine iguana. It is the only iguana to feed in the sea, diving up to 50 ft (15 m) to forage for marine algae. The cold waters of the Humboldt Current allow Galápagos penguins to survive at the equator. The islands are the result of volcanic eruptions above a mantle hotspot. The same hotspot is responsible for driving the Cocos and Nazca plates apart at the Colon Ridge. The Nazca Plate on which the islands sit is moving eastward, so the oldest of the islands are found in the east. They have been volcanically extinct for several million years, but some of the younger islands are still active volcanoes. Farther east, the submarine Carnegie Ridge is also built from Galápagos Hotspot material.

GALÁPAGOS ISLAND IGUANA
Study of the islands' unique animals, including marine iguanas, helped British naturalist Charles Darwin to formulate his theory of evolution.

Easter Island

IN THE EASTERN SOUTH PACIFIC, the cold Humboldt Current flows north up the coast of South America, forming the eastern arm of the South Pacific Gyre. It then turns west in the tropics, feeding the South Equatorial Current. In some years, this current is weakened and warm water pools in the east, disrupting weather patterns over a wide area of the Pacific Ocean.

PACIFIC OCEAN F2

Peru-Chile Trench

LENGTH	3,650 miles (5,900 km)
MAXIMUM DEPTH	26,474 ft (8,069 m)
RATE OF CLOSURE	3 in (7.8 cm) per year

The Peru–Chile Trench (also called the Atacama Trench) is the longest ocean trench, marking the point at which the Nazca Plate meets the South American Plate. The Nazca Plate is primarily dense ocean crust and so is being subducted beneath the more buoyant South American continental plate. The South American crust has been deformed and thickened by the convergence, creating the Andes Mountains. Melting of the rocks around the subducting slab has led to volcanism and many of the Andes' tallest peaks are volcanoes. Earthquakes along the trench produced nine large tsunamis during the 20th century, resulting in more than 2,000 deaths. The trade winds drive surface waters offshore throughout most years, leading to upwelling of nutrient-rich deep water off the coast of Peru. This upwelling makes the water very productive and yields large fish catches, predominantly anchovies and sardines. Under El Niño (see pp.68–69) conditions, however, the wind direction reverses and the fish catch plummets.

PACIFIC OCEAN B3

Easter Island

TYPE	Volcanic island
AREA	63 square miles (164 square km)
NUMBER OF ISLANDS	1

Easter Island lies near the East Pacific Rise, which separates the Pacific Plate to the west from the Nazca Plate to the east. The island is the highest point of the Easter Fracture Zone, a series of ridges and trenches marking a transform fault running 3,650 miles (5,900 km) across the floor of the South Pacific, from the Peru–Chile Trench in the east to the Tuamotu Archipelago (see p.477) in the west. Easter Island was named in 1722 by Dutch sailors, who came across it on Easter Sunday. It had been settled up to 1,000 years earlier by Polynesians, who today call the island Rapa Nui. The island is famous for its giant stone statues, which are known as moai, found in groups along the coast. About half of the 900 statues remain unfinished in the quarry—it seems statue-carving stopped abruptly

about a century before the first European explorers arrived. It is thought that the island's population collapsed in the face of environmental degradation, triggering a spiritual transformation away from the ancestor cult represented by the statues.

MOAI
Easter Island's enigmatic moai statues were carved from soft volcanic rock taken from Ranu Raraku, one of the island's many volcanic craters.

ATLAS OF THE OCEANS

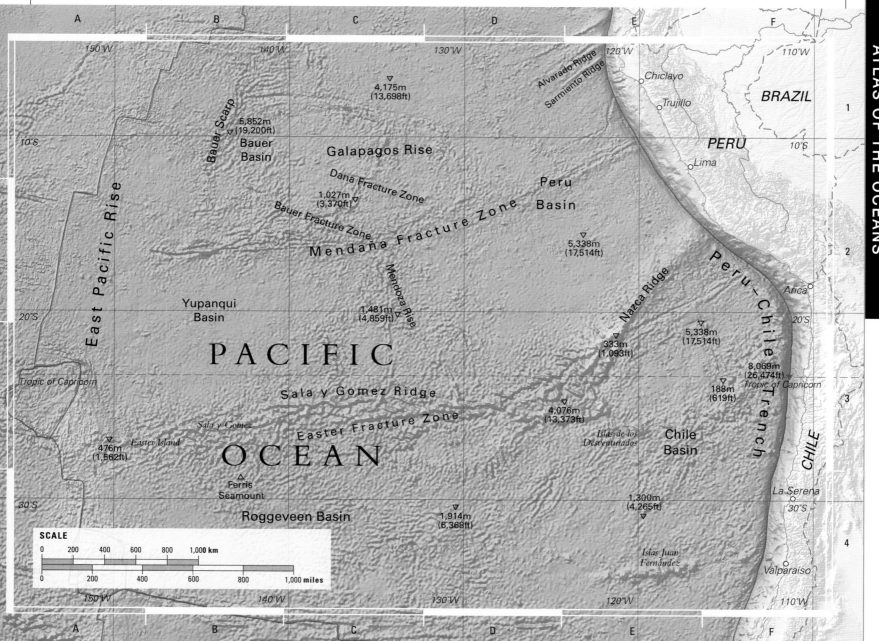

Easter Island · Peru-Chile Trench · Bauer Scarp 5,852m (19,200ft) · Bauer Basin · Galapagos Rise · 4,175m (13,698ft) · Alvarado Ridge · Sarmiento Ridge · Chiclayo · Trujillo · BRAZIL · PERU · Lima · Dana Fracture Zone · 1,027m (3,370ft) · Bauer Fracture Zone · Mendaña Fracture Zone · Peru Basin · 5,338m (17,514ft) · Mendoza Rise · Yupanqui Basin · 1,481m (4,859ft) · PACIFIC · Nazca Ridge · 5,338m (17,514ft) · Arica · 333m (1,093ft) · 8,069m (26,474ft) · 188m (619ft) · Peru–Chile Trench · Sala y Gomez Ridge · 4,076m (13,373ft) · Easter Fracture Zone · Islas de los Desventurados · Chile Basin · Sala y Gomez · CHILE · Easter Island · 476m (1,562ft) · OCEAN · Ferris Seamount · Roggeveen Basin · 1,914m (6,368ft) · 1,300m (4,265ft) · La Serena · Islas Juan Fernández · Valparaíso

East Pacific Rise · Tropic of Capricorn

SCALE
| 0 | 200 | 400 | 600 | 800 | 1,000 km |
| 0 | 200 | 400 | 600 | 800 | 1,000 miles |

Southeast Australia and New Zealand

TO THE NORTHEAST OF AUSTRALIA, ocean currents flow from the east, feeding the warm East Australia Current, which sweeps south along the Australian coast, before turning east to flow north of New Zealand. South of New Zealand, the Antarctic Circumpolar Current flows from west to east. New Zealand straddles a major tectonic boundary between the Pacific and Australian plates.

PACIFIC OCEAN G4

Kermadec–Tonga Trench

LENGTH 1,550 miles (2,500 km)

DEPTH 35,430 ft (10,800 m)

RATE OF CLOSURE 6–9 in (15–24 cm) per year

The Kermadec–Tonga Trench runs between the North Island of New Zealand and the island of Tonga. The Pacific Plate converges with the Australian Plate in this subduction zone. At its northern end, closure rates of 9 in (24 cm) per year have been measured—the fastest recorded plate motion yet. The older oceanic crust of the Pacific Plate is sinking below the more buoyant crust of the young oceanic Australian Plate. The Tonga Ridge and the older Lau Ridge formed as arcs of volcanoes. The rapid motion has caused extension of the Australian Plate and the opening of a back-arc basin between the two ridges, in the Lau Basin. The latest addition to this island chain connected the islands of Hunga-Ha'apai and Hunga Tonga after undersea volcanic eruptions in 2009 and 2014. Together with Fiji and Samoa, the 36 islands of Tonga are the cradle of the Polynesian seafaring culture, which had stretched across the South Pacific by the 12th century.

PACIFIC OCEAN B6

Bass Strait

LENGTH 250 miles (400 km)

MINIMUM WIDTH 62 miles (100 km)

The Bass Strait separates Tasmania from Australia, overlying a shallow shelf around 160 ft (50 m) deep. Strong winds and currents from the Southern Ocean combine with the shallow depth to make its waters notoriously rough. Hundreds of ships were wrecked on its shores during the 19th century, before the erection of lighthouses made navigation safer. Natural gas fields were discovered beneath the eastern Bass Strait in the 1960s and 1990s.

PHILLIP ISLAND

PACIFIC OCEAN C5

Tasman Sea

AREA 890,000 square miles (2.3 million square km)

MAXIMUM DEPTH 19,500 ft (5,945 m)

INFLOWS Southern Ocean, Coral Sea

The Tasman Sea has a subtropical climate in the north, but the influence of cold sub-Antarctic water makes it temperate in the south, where it lies in the path of the westerly winds known as the "roaring forties." The sea is named after Dutch explorer Abel Tasman, who sailed across it in 1642, becoming the first European to reach the islands of Tasmania, New Zealand, Tonga, and Fiji. On a later voyage in 1644, he succeeded in reaching Terra Australis (the Southern Land), which we now know as Australia.

PACIFIC OCEAN H3

Southwest Pacific Basin

AREA 8.9 million square miles (23 million square km)

MAXIMUM DEPTH 18,500 ft (5,655 m)

INFLOWS Pacific Ocean, Southern Ocean

The Southwest Pacific Basin lies east of New Zealand and the Kermadec–Tonga Trench. It is bounded in the east by the East Pacific Rise (see p.478), in the south by the Pacific–Antarctic Rise, and in the north by the Polynesian island chains. The Louisville Ridge is the only significant chain of seamounts and much of the basin floor is an abyssal plain. There are extensive deposits of manganese in the northern and southern parts of the basin.

LORD HOWE ISLAND
The warm waters of the East Australia Current allow Lord Howe Island to host the world's most southerly coral reef.

PACIFIC OCEAN E5

New Zealand

TYPE Micro-continental island group

AREA 103,700 square miles (268,680 square km)

NUMBER OF ISLANDS 2 main islands (700 smaller islands)

New Zealand separated from Australia and Antarctica 80 million years ago, and is now positioned at the boundary between the Pacific and Australian plates. The largely transverse Alpine Fault runs 435 miles (700 km) across the South Island. Crustal compression and distortion across a 155-mile (250-km) wide zone has raised the Southern Alps over 13,000 ft (4,000 m) above sea level. The plate boundary continues north as the Hikurangi Trench, a classic subduction zone producing volcanism on North Island, and south as the Macquarie Ridge, where shallow subduction has uplifted the Australian Plate. A swarm of earthquakes hit the South Island between 2010 and 2012, the largest of which killed 185 people and destroyed the historic cathedral in Christchurch in 2011. The main islands of New Zealand are the highest points of an extensive area of continental crust that includes the Campbell Plateau, Challenger Plateau, and Chatham Rise. To the southeast, Campbell Plateau is the world's largest area of submerged continental crust.

CAMPBELL ISLAND
The southernmost of New Zealand's subantarctic islands, Campbell Island is primarily volcanic in origin.

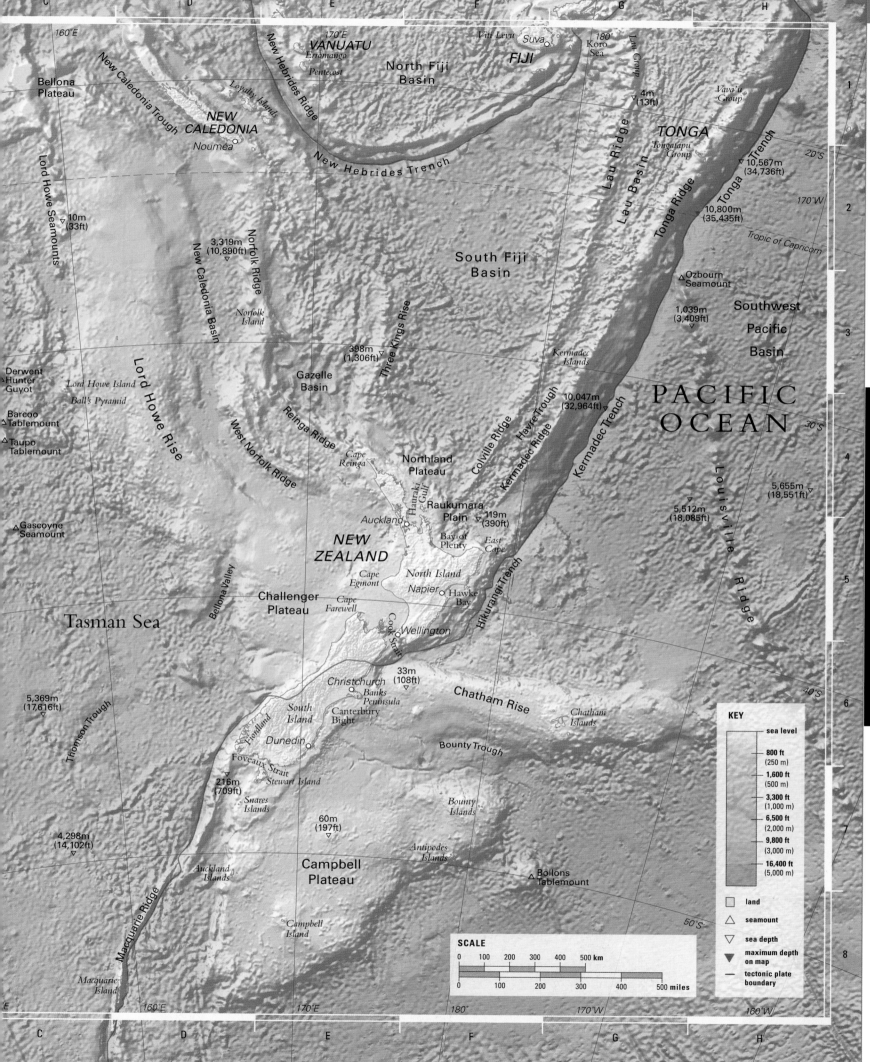

C   D   E   F   G   H

160°E

170°E
Viti Levu   Suva
VANUATU   FIJI   180°   Koro
Erömango   Sea
Pentecost   Vava'u
Bellona   New Hebrides Ridge   North Fiji   Group
Plateau   Basin   4m
New Caledonia Trough   ▽ (13ft)
Loyalty Islands   Lau Ridge
NEW   TONGA
CALEDONIA   Tongatapu
Nouméa   Group
New Hebrides Trench   10,567m
▽ (34,736ft)   20°S
Lord Howe Seamounts   Tonga Trench
10m   Lau Basin
▽ (33ft)   10,800m   170°W
▽ (35,435ft)   2
3,319m   Tropic of Capricorn
(10,890ft)   South Fiji
Norfolk Ridge   ▽   Basin   Ozbourn
New Caledonia Basin   △ Seamount
Southwest
Norfolk   1,039m   Pacific
Derwent   Island   (3,409ft)   Basin   3
△ Hunter   ▽
Guyot   Lord Howe Island   398m
Ball's Pyramid   (1,306ft)   Kermadec
△ Barcoo   ▽   Islands   PACIFIC
Tablemount   Lord Howe Rise   Three Kings Rise
△ Taupo   Gazelle   10,047m   OCEAN   30°S
Tablemount   Basin   Reinga Ridge   (32,964ft)
West Norfolk Ridge   Cape   Northland   ▽
Reinga   Plateau   Colville Ridge
△ Gascoyne   Hauraki   Raukumara   119m   5,655m
Seamount   Gulf   Plain   (390ft)   5,512m   (18,551ft)
Auckland ▽   ▽   (18,085ft)   ▽
NEW   Bay of   East
ZEALAND   Cape   Plenty   Cape   Louisville Ridge
Egmont   North Island
Napier
Cape   Hawke
Challenger   Farewell   Bay
Plateau   Cook   Wellington   Hikurangi Trench
Tasman Sea   Strait
Bellona Valley
5,369m   33m
(17,616ft)   Christchurch   (108ft)   Chatham Rise
▽   Banks   ▽
Thomson Trough   Peninsula   Chatham
South   Canterbury   Islands   40°S
Island   Bight
Fiordland   Dunedin   Bounty Trough
4,298m   216m
(14,102ft)   Foveaux   (709ft)
▽   Strait   ▽   Bounty
Stewart Island   Islands
Snares
Islands   60m
(197ft)
▽
Antipodes
Auckland   Islands   Boilons
Islands   Campbell   Tablemount
Macquarie Ridge   Plateau

KEY
sea level
800 ft
(250 m)
1,600 ft
(500 m)
3,300 ft
(1,000 m)
6,500 ft
(2,000 m)
9,800 ft
(3,000 m)
16,400 ft
(5,000 m)

☐ land
△ seamount
▽ sea depth
▼ maximum depth
on map
— tectonic plate
boundary

Campbell
Island

Macquarie
Island

SCALE
0   100   200   300   400   500 km
0   100   200   300   400   500 miles

160°E   170°E   180°   170°W   160°W

C   D   E   F   G   H

The Southern Ocean

THE SOUTHERN OCEAN COMPLETELY surrounds
Antarctica and links the Indian, Atlantic, and Pacific
Oceans. Antarctica's coast was not sighted until 1820,
and its shores were not fully explored until the 20th
century. The Southern Ocean is
generally described as being south
of 60° latitude but is physically
better defined by the extent of the
Antarctic Circumpolar Current.

WEDDELL SEAL DIVING BENEATH THE ICE

Ocean Circulation

The eastward-flowing Antarctic Circumpolar Current is the strongest in
the world. It flows up to 9,800 ft (3,000 m) deep and carries 4.8 million
cubic ft (135,000 cubic meters) of water per second through Drake Passage.
It diverts heat flowing from the equator, isolating Antarctica and causing
the buildup of the thick Antarctic ice cap. Cold currents branch off up the
eastern sides of the Indian, Atlantic, and Pacific Oceans. The Circumpolar

Current (also known as the West
Wind Drift) is driven by the
prevailing westerly winds, which
blow uninterrupted by any
landmass. Wind speeds in the
Southern Ocean are the highest
in the world: the "roaring
forties" give way to the "furious
fifties" and the "screaming
sixties" as one sails south.

SEA ICE AND ICEBERGS
Tabular icebergs and sea ice can be
found drifting in much of the Southern
Ocean throughout the year.

Ocean Floor

The Southern Ocean is unusual in not having a well-defined basin
bounded by land masses: it surrounds the South Pole and extends for
360° of longitude. There are a series of deep basins lying between the
continental shelf and the ridges at the edge of the Antarctic Plate. The
continental shelf is narrow and deeper than that of other continents due
to the depression of the crust under the weight of the 1½-mile (2.5-km)
thick Antarctic Ice Sheet. The surrounding spreading ridges drove the
break up of the supercontinent of Gondwana, starting when Africa
began to move north 165
million years ago. An area
of thick ocean crust, the
Kerguelen Plateau, lies
between the Southern
and Indian Oceans. This
is one of the largest
submarine plateaus, and
it is composed of flood
basalts that erupted about
97 million years ago.

FROZEN WATERFALLS
Fresh glacial water refreezes on
contact with cold ocean water,
which is kept liquid by its salt.

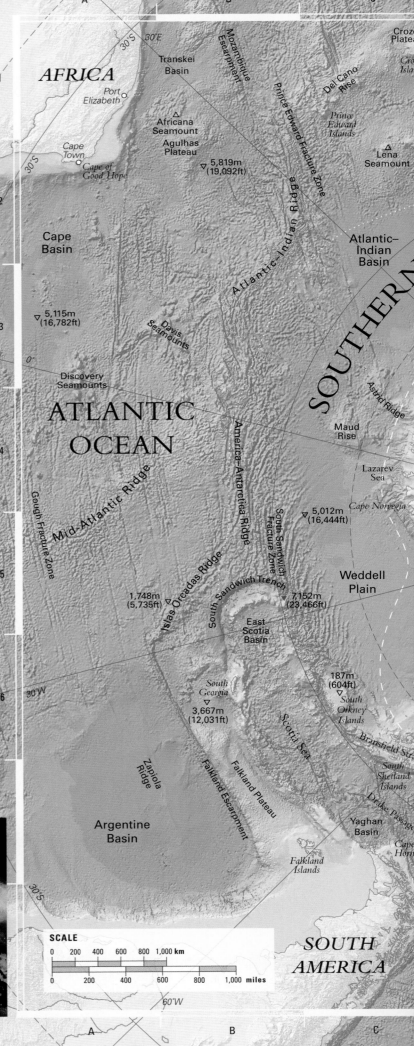

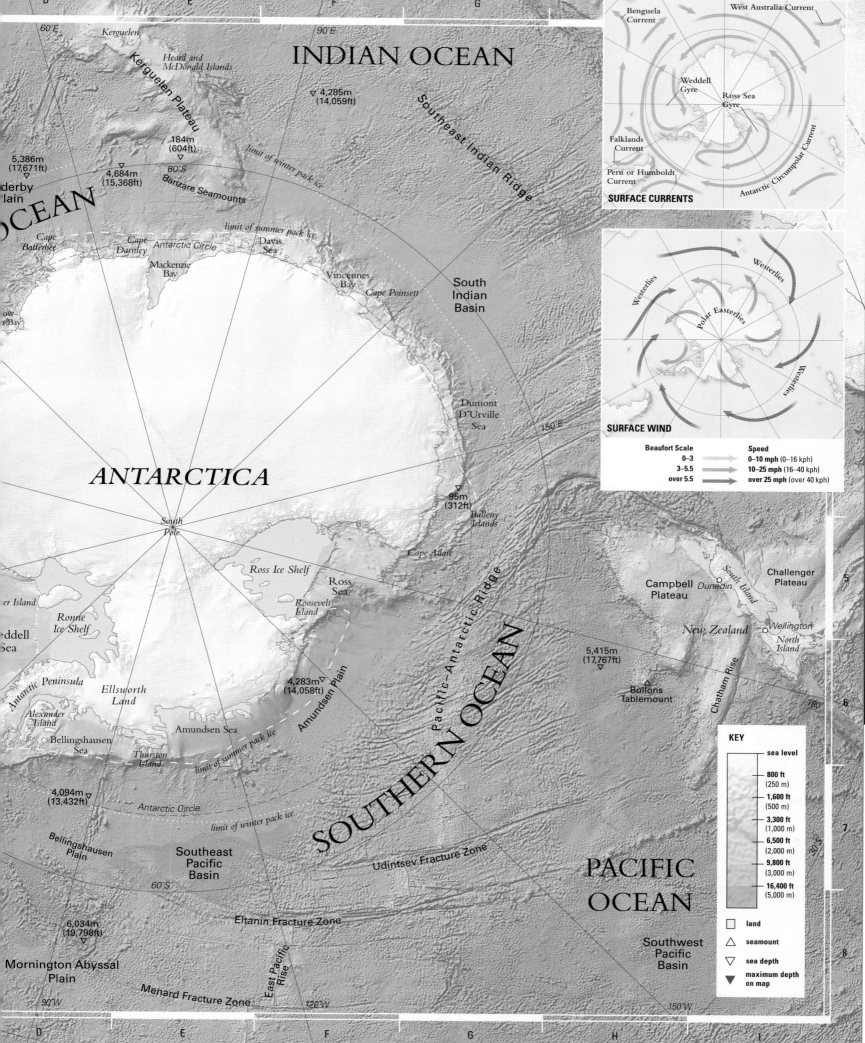

The Antarctic Peninsula

THE ANTARCTIC PENINSULA EXTENDS north from Antarctica toward South America, separating the Weddell Sea from the Bellingshausen Sea. At its northern end, a chain of islands marks the edge of the Antarctic Plate.

SOUTHERN OCEAN D2

Weddell Sea

AREA 1.1 million square miles (2.8 million square km)

MAXIMUM DEPTH 16,440 ft (5,012 m)

INFLOWS Southern Ocean, Filchner-Ronne Ice Shelf, Larsen Ice Shelf

The Weddell Sea is named after British seal hunter James Weddell, who reached a latitude of 74°34' South in 1832, the most southerly point that would be reached for the next 80 years. It is mostly covered with pack ice, even in summer, and is the source for 70 percent of the cold Antarctic bottom water. A clockwise gyre carries

YOUNG WEDDELL SEAL

ice west, then north up the Antarctic Peninsula, before turning east with the Circumpolar Current. The Antarctic continental shelf is at its widest here, with shallow banks extending out from under the Filchner–Ronne Ice Shelf.

SOUTHERN OCEAN C3 AND D4

Filchner–Ronne Ice Shelf

AREA 166,000 square miles (430,000 square km)

MAXIMUM THICKNESS 3,000 ft (900 m)

INFLOWS Arctic Ice Sheet

The Filchner Ice Shelf was first sighted by the German explorer Wilhelm Filchner in 1912. It lies to the east of Berkner Island. To the west of Berkner Island lies the Ronne Ice Shelf. It was charted from the air by American naval commander Finn Ronne in 1947. Together these two shelves make up the second-largest floating ice shelf by area, and the largest by volume. The bedrock of the ice-covered Berkner Island in fact lies below sea level. Although fed by glaciers from the continental ice cap, and grounded on its landward side, most of the area of the Filchner–Ronne Ice Shelf is floating in the Weddell Sea. The ice shelf itself is up to 3,000 ft (900 m) thick, and the underlying seafloor is up to 4,600 ft (1,400 m) deep.

SOUTHERN OCEAN A1

Bellingshausen Sea

AREA 232,000 square miles (600,000 square km)

MAXIMUM DEPTH 13,400 ft (4,094 m)

INFLOWS Southern Ocean, George VI Ice Shelf

The Bellingshausen Sea is named after Fabian von Bellingshausen, an officer in the Imperial Russian Navy, who was the first to sight the coast of Antarctica, in 1820. It is one of several parts of the Southern Ocean that is rich in krill, the basis of a productive marine food chain.

ROTHERA RESEARCH STATION.

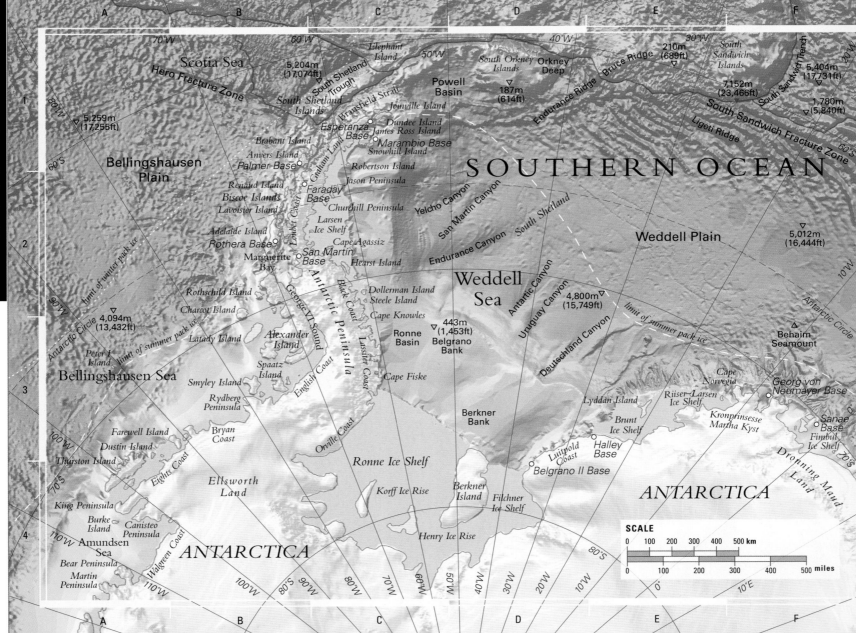

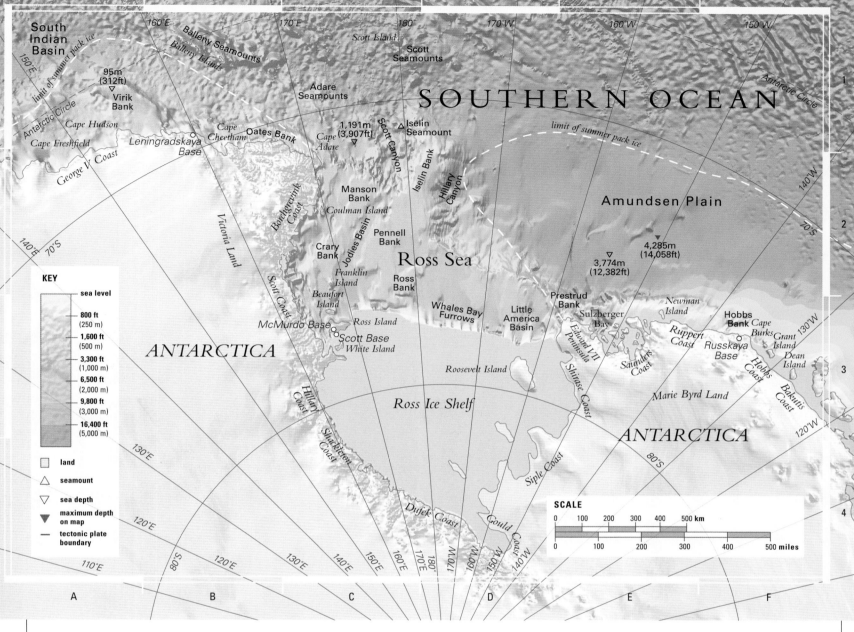

SOUTHERN OCEAN

The Ross Sea

THE ROSS SEA IS A LARGE BAY situated between Marie Byrd Land and Victoria Land. Water circulates in a clockwise gyre, with a westward coastal current being turned north along the coast of Victoria Land to join the eastward Circumpolar Current.

SOUTHERN OCEAN D2

Ross Sea

AREA 370,000 square miles (960,000 square km)

MAXIMUM DEPTH 8,300 ft (2,500 m)

INFLOWS Southern Ocean, Ross Ice Shelf

The Ross Sea is named after the British naval officer James Clark Ross, who charted this part of the Antarctic coast in 1841. The volcanoes Mount Erebus and Mount Terror on Ross Island are named after his two ships during this expedition. Ross Island is home to the largest scientific base on Antarctica, the American research

station at McMurdo Sound. In the past, iceberg calving and melting was balanced by the accumulation of new ice, but Antarctica is now losing about 220 billion tons (200 billion tonnes) of ice to the ocean each year.

ICEBREAKER AT SEA
A Russian icebreaker is shown here among drifting ice in the Ross Sea, with the Transantarctic Mountains in the background.

SOUTHERN OCEAN C3

Ross Ice Shelf

AREA 188,000 square miles (487,000 square km)

MAXIMUM THICKNESS 2,600 ft (800 m)

INFLOWS Antarctic Ice Shelf

The southern half of the Ross Sea is overlain by the world's largest floating ice shelf, the Ross Ice Shelf, which extends up to 280 miles (450 km) from the shore of Antarctica. The Norwegian explorer Roald Amundsen started his successful expedition to the South Pole in 1911 by crossing this ice shelf. It ranges in thickness from about 820 ft (250 m) at the ice front to 2,600 ft (800 m) inland. Ice floats with most of its volume underwater, making the height of the ice front above sea level about 65–100 ft (20–30 m). The shelf flows seaward at about 3,000 ft (900 m) per year, propelled by the accumulating weight of compacted snow that falls on the high plateau of the Antarctic Ice Sheet. Accumulation on the ice cap is thought to be balanced by iceberg calving at the front of Antarctica's ice shelves and melting on their underside.

DISCOVERY

ICE SHELF SEAS

Ice-core drilling, seismic sounding, and ice-penetrating radar have all been used to measure the thickness of floating ice shelves from the surface. Now, autonomous underwater vehicles (AUVs) are being deployed to explore the "cave seas" beneath the ice. They are able to gather more detailed information about sea floor depth, ice thickness, ocean temperature, pressure, and salinity.

SEA FLOOR BENEATH THE ICE

ICE-SHELF REMNANT
Although most of the Larsen B ice shelf broke up in 2002, a part of it, visible on the left, remains. This contains deep crevasses, making it prone to further disintegration.

Ice shelves are extensions of land ice sheets that are floating over the sea. Continually pushed away from the land by the weight of accumulating snow, an ice shelf gradually advances over the ocean until its front breaks off to form a tabular iceberg. This advance and retreat is part of a natural cycle, but in the Antarctic Peninsula, some small ice shelves have suffered catastrophic collapses as a result of regional warming of 5.4°F (3°C) over the last 50 years. In 2008, for example, the Wilkins Ice Shelf lost around 200 square miles (500 square km).

Although the loss of floating ice does not affect global sea levels, it seems that the adjacent continental ice sheet may become unstable if it loses the "buffer zone" provided by an ice shelf. After the Larsen B Ice Shelf collapsed in 2002, scientists measured nearby glaciers flowing between two and eight times faster than they had before. Small ice shelves and coastal glaciers are also in retreat along the Amundsen Sea coast of West Antarctica. If the regional warming continues and the West Antarctic Ice Sheet collapses as a result, global sea levels could rise more than 16 ft (5 m), threatening densely populated coastal areas worldwide.

Larsen Ice Shelf Collapse

The Larsen Ice Shelf occupies the eastern shore of the Antarctic Peninsula. In 1995, the northern part of the ice shelf, Larsen A, broke into tiny fragments during a storm. In 2002, most of the central part, Larsen B, disintegrated in a similar manner over a few weeks. At the moment, the largest part of the shelf, Larsen C to the south, seems to be stable, although it, too, lost 12 percent of its area when the large iceberg A–68 broke away in 2017.

Weddell Sea

Larsen Ice Shelf
Ronne Ice Shelf
West Antarctica
Transantarctic Mountains
East Antarctica
Ross Ice Shelf
Ross Sea

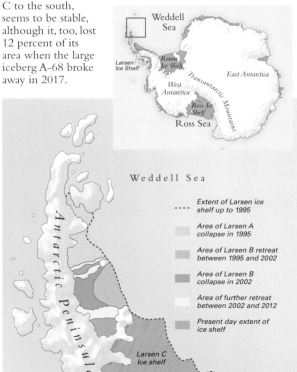

Weddell Sea

Antarctic Peninsula

Larsen C Ice shelf

- - - - Extent of Larsen ice shelf up to 1995

Area of Larsen A collapse in 1995

Area of Larsen B retreat between 1995 and 2002

Area of Larsen B collapse in 2002

Area of further retreat between 2002 and 2012

Present day extent of ice shelf

ICE SHELVES IN RETREAT

CRACKING UP

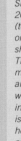

MELTWATER POOL *In the summer, meltwater collects in low-lying parts of the surface of an ice shelf, including crevasses and depressions. Melting also occurs on the underside.*

CRACK IN LARSEN A *The additional weight of meltwater may increase the pressure at the base of a crevasse, causing it to penetrate deeper into the ice shelf and to widen.*

DISAPPEARING ICE SHELF

LARSEN B COLLAPSE
Extensive meltwater pools are visible in a satellite image of the Larsen B Ice Shelf taken on January 31, 2002, before it broke up (top). The collapse itself, on March 7, 2002, is shown in the lower image. The ice shelf broke into a multitude of small fragments, and a few larger ones, which quickly dispersed into the Weddell Sea. It is possible that meltwater helped push surface crevasses through the entire 720 ft (220 m) thickness of the Larsen B Ice Shelf.

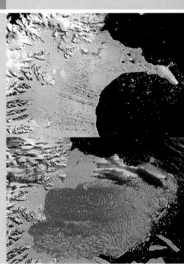

ICEBERG B-15 *One of the largest icebergs ever seen, at 185 miles (300 km) long, 25 miles (40 km) wide, and 200 ft (60 m) high, B-15 broke off from the Ross Ice Shelf in March 2000. It drifted around the Ross Sea for several years, disrupting navigation and penguin migration. The iceberg eventually broke into smaller pieces, some of which were seen not far from New Zealand in November 2006.*

GIANT ICEBERGS

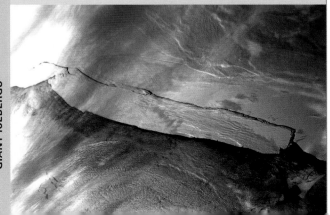

Glossary

A

abyssal Relating to oceanic depths greater than about 6,500 ft (2,000 m). The **abyssal plain** is the flattish plain at 13,000–20,000 ft (4,000–6,000 m) that forms the bed of most ocean basins. The **abyssal zone** is the region of both seabed and open water between 6,500 ft (2,000 m) and the abyssal plain. See also *bathyal, hadal*.

air mass A body of air with relatively uniform temperature and pressure, forming above a given region of Earth's surface. Its characteristics derive from this surface region and are distinct from surrounding air masses. "Tropical maritime" and "polar continental" are examples.

albedo The extent to which incoming radiation is reflected from a surface. Ice has a high albedo, reflecting most of the Sun's radiation reaching it.

algae Simple plants and plantlike protists that can photosynthesize, ranging from seaweeds (**macroalgae**) to microscopic plankton (**microalgae**). Some types of algae, such as **green microalgae** and green seaweeds, are often classified as plants. Red and brown seaweeds are also algae, but are often classified separately. Singular **alga**. See also *cyanobacteria, photosynthesis, protists, seaweed*.

amphipods A group of small, common, shrimplike crustaceans, flattened from side to side, which live mainly on the sea floor and feed on detritus (dead material).

anadromous Of fish: living most of their lives at sea but entering rivers to breed, for example, the salmon. See also *catadromous*.

anaerobic Relating to processes occurring without oxygen, or to organisms that are able to live in the absence of oxygen. See also *anoxic*.

anemones Solitary cnidarians that live attached to surfaces and grab passing prey with their stinging tentacles. See also *cnidarians, coral, polyp*.

annelids see *segmented worms*.

anoxic Of a habitat: without available oxygen for living creatures.

Antarctic Circle Line of latitude in the Southern Hemisphere south of which there is at least one day of 24-hour sunshine and at least one day of 24-hour darkness per year.

anticyclone A pattern of circulating air in the atmosphere with high pressure in the center; usually associated with settled weather. See also *cyclone*.

archaea A group of tiny, single-celled organisms. Like bacteria, they have no cell nucleus, but are classified separately. They often live in extreme environments, such as deep-sea vents. See also *bacteria*.

Arctic Circle Line of latitude in the Northern Hemisphere north of which there is at least one day of 24-hour sunshine and at least one day of 24-hour darkness per year.

arthropods A major group (phylum) of invertebrate animals with jointed legs and a hard outer skeleton. It includes crustaceans (crabs, shrimp, and relatives), insects, spiders. See also *crustaceans, exoskeleton*.

asexual reproduction Reproduction that does not involve combining the genes from two individuals (sex). It can consist of splitting or fragmenting the body, the budding of new individuals, or specialized structures forming, such as spores.

atoll A low, ring-shaped island, or series of arc-shaped islands, forming the rim of a shallow lagoon. The structure results from an accumulation of coral on top of a sunken volcano. See also *lagoon*.

authigenic Of sediments: formed locally in the ocean (especially via chemical processes), not transported from elsewhere. See also *sediment*.

autotroph An organism, such as a plant, that can make its own food, rather than eating or absorbing food produced by other organisms. See also *photosynthesis, primary producer*.

B

backshore Part of the shore above the average high-water mark, affected by the sea only during the highest tides and storms. See also *foreshore*.

backwash The flow of water back to the sea after a wave has broken on a beach.

bacteria Microscopic single-celled organisms abundant in all ecosystems. Their cells are much smaller than those of animals and plants and have no nucleus. See also *archaea, cyanobacteria, protists*.

baleen Horny plates in the mouths of some whales that are used to strain food, such as krill, from the water.

bank A shallow region of sea surrounded by deeper water. Often the site of productive fisheries.

bar A long, narrow, offshore deposit of sediment lying parallel to a coastline. Bars may be permanently submerged, or exposed at low tide. A bar that is always exposed is a barrier island. A bar across the mouth of a bay and attached to the coast is a **baymouth bar**. See also *barrier island, spit*.

barbel Sensitive fleshy projections often found in pairs around the mouths of some fish.

barnacles Specialized crustaceans whose adults live attached to rocks and other surfaces. They are protected by hard shell-like plates and filter-feed using highly modified limbs. See also *crustaceans, filter feeding*.

barrage Human construction built across an estuary or inlet to protect against flooding by heavy seas.

barrier island A permanently exposed bar of sand or pebbles lying parallel to a coastline. A **barrier beach** is a similar structure, but can be attached to the mainland at one or both ends.

barrier reef A coral reef parallel to, but some distance from, a shoreline.

basalt A common volcanic rock; originally solidified lava. The rock of the ocean floor is mainly basalt that has spread from mid-ocean ridges.

basin see *ocean basin*.

bathyal Relating to ocean depths between about 660 and 6,500 ft (200 and 2,000 m). The **bathyal zone** is the region of seabed and water column between these depths. See also *abyssal*.

beach face The steeply sloping part of a beach, below a berm. See also *berm*.

benthos Living organisms that live on or in the seabed (**benthic organisms**).

berm A ridge of sediment high on a beach, left behind by a high tide. Also called a **beach ridge**.

biodiversity The diversity or variety of living organisms; determined by, for instance, the number of species, or the variation within species.

biogenic Formed by the action of living organisms.

bioluminescence The production of light by living organisms.

biomass The total mass or weight of living organisms in a given area.

biome Any large-scale association of plants and animals, especially one dependent on particular climatic conditions. **Mangrove swamp** and the **abyssal plain** are marine biomes.

bioturbation Disturbance and mixing of seafloor sediments, usually by burrowing animals.

bivalves Mollusks, such as clams, mussels, and oysters, that have a shell made up of two hinged halves. Most bivalves move slowly or not at all, and are filter feeders. See also *filter feeding, mollusks*.

black smoker A hydrothermal vent in which the emerging hot water is colored black with dark minerals.

bloom A rapid growth of plankton, often turning the water cloudy and greenish; usually a response to an increase in the availability of nutrients in the water. See also *phytoplankton*.

blue-green algae see *cyanobacteria*.

bony fishes The large group that includes all fish species except jawless fishes, sharks, and other cartilaginous fishes. See also *cartilaginous fishes*.

bore see *tidal bore*.

brackish Saltier than fresh water, but less salty than typical ocean water.

breaker zone The zone of a beach, or other shoreline, where waves break.

breakwater An artificial barrier built in the sea, usually near a harbor, to protect against waves and heavy seas.

brittlestars Echinoderms with narrow, jointed, flexible arms; related to starfish. See also see *echinoderms, starfish*

bryozoans Filter-feeding colonial animals that live attached to surfaces, such as seaweed fronds, either as flat sheets or as tufty, plantlike growths. Sometimes called "moss animals."

bycatch In fishing, the portion of a catch made up of non-target species.

C

calcareous Consisting of or containing calcium carbonate.

calcium carbonate The chemical $CaCO_3$. It is the main constituent of coral skeletons and mollusk shells, limestone, and chalk.

calve To shed icebergs into the sea. See also *icebergs*.

carapace The upper shell of a turtle; the protective outer covering of some other animals, such as crabs.

carbon cycle The cycling of carbon through the environment. During the cycle, carbon exists in the bodies of living things, in carbon dioxide in the atmosphere and oceans, in fossil fuels, and in rocks such as limestone.

cartilaginous fishes Fish, such as sharks, rays, skates, and chimaeras, whose skeleton is of cartilage, not bone. See also *bony fishes*.

catadromous Of fish: living most of their lives in fresh water but migrating to the sea to breed. Eels are an example. See also *anadromous*.

cephalopods A group of swimming mollusks that includes squid, cuttlefish, octopuses, and nautiluses. They have large brains and demonstrate complex behavior. See also *mollusks*.

chemosynthesis A process in which some organisms make their own food using the energy from naturally occurring chemicals such as hydrogen sulfide. See also *photosynthesis, autotroph, primary producer*.

chlorophyll The green pigment of plants and seaweeds that allows them to make their own food by using the Sun's energy. See *photosynthesis*.

chromatophore A skin cell in which the distribution of colored pigment can be altered, allowing an animal to change color. Color change may be fast, as in cephalopods, or slower, as in crustaceans and some fish.

cilia Tiny beating hairlike structures on the surfaces of some cells. Used to aid movement in small organisms, or to create water currents. Singular **cilium**.

cloaca The combined opening of the digestive, urinary, and reproductive systems of many vertebrates (e.g., fish, birds) and some invertebrates.

cnidarians A major group (phylum) of invertebrate animals with simple bodies bearing tentacles that surround a single opening (mouth). Cnidarians include corals, anemones, and jellyfish, and are often colonial. Their two typical body forms are the polyp and the medusa. In some cnidarians, both forms occur during the life cycle. See also *colonial, coral, medusa, nematocyst, polyp*.

coast See *concordant coast, depositional coast, discordant coast, drowned coast, emergent coast, erosional coast, primary coast, secondary coast*.

cold seep A natural seepage of oil or other energy-containing chemicals on the sea floor, often supporting dense concentrations of marine life.

colonial Of an animal: living in colonies. A colony can consist of separate individuals, as in the case of sponge shrimp, or animals joined by strands of living tissue, as in the case of many marine invertebrates, such as corals and bryozoans. Individuals may be specialized for different roles, such as feeding, reproduction, and defense, in which case the colony may behave like a single animal. See also *bryozoans, cnidarians, zooid*.

comb jellies see *ctenophores*.

commensal Living in close association with an organism of another species, for example, by sharing its burrow, without either helping or damaging it. See also *mutualism, symbiosis*.

concordant coast Coast on which hills and valleys are roughly parallel to the shore, resulting either in a straight coastline or one with rocky islands running parallel to the shoreline. See also *discordant coast*.

continental crust The material in Earth's crust that forms the continents, including the continental margins. It is lighter and thicker than oceanic crust.

continental margin A continent's edge below sea level, including the continental shelf and continental slope.

continental rise The gently sloping seabed around the edge of ocean basins that adjoins the bottom of the continental slope.

continental shelf The gently sloping seabed around the edges of most continents, formed of continental crust and averaging around 425 ft (130 m) deep.

continental slope Sloping seabed at the seaward edge of the continental shelf. It descends relatively steeply to the continental rise.

convection Circulating currents in a fluid—for example air, water, or hot rock—that result from heated portions rising because they are less dense, and sinking later as they cool.

copepods Small, swimming crustaceans, usually less than $\frac{1}{16}$ in (2 mm) long, that make up a large part of the zooplankton. There are also many parasitic and burrowing species. See also *zooplankton*.

coral Any of various cnidarians that live fixed to the ocean bottom, secrete skeletons for support, and are usually colonial. The true corals lay down hard skeletons of calcium carbonate outside their bodies that eventually form coral reefs. Other coral groups include the sea fans. See also *cnidarians, sea fans, zooxanthellae*.

coral bleaching Phenomenon in which coral animals lose their tiny symbiotic algae (zooxanthellae), usually in response to a stress in the environment. Bleached corals may later die. See also *zooxanthellae*.

coral reef A rocklike, often ridge-shaped structure of calcium carbonate built in shallow tropical seas by generations of coral animals. See also *barrier reef, fringing reef*.

coralline Resembling coral; mainly applied to red seaweeds that form hard, calcareous crusts on rocks or in coral reefs.

Coriolis effect Phenomenon resulting from the rotation of Earth, in which winds and currents traveling toward or away from the equator are deflected to the right in the Northern Hemisphere and to the left in the Southern Hemisphere. The effect helps to explain the direction of prevailing winds and the existence of gyres.

crabs see *crustaceans*.

crinoids Stalked echinoderms, also called sea lilies, that filter-feed using their branching arms. Some species have no stalks and are known as **feather stars**. See also *echinoderms*.

crustaceans The most diverse and abundant group of arthropods in the oceans. It includes crabs, lobsters, shrimps, barnacles, krill, copepods, isopods, and amphipods. Their jointed appendages are variously modified as claws, legs, swimming organs, or filter-feeding devices, depending on the species. See also *arthropod*.

ctenophores Transparent jellyfish-like animals that hunt in the plankton. They swim using beating hairlike structures arranged in rows called comb plates. Also called **comb jellies**.

current Any sustained horizontal flow of water. See also *drift, surface current, thermohaline circulation, turbidity current, western boundary current*.

cusp Any shape formed by two concave lines meeting at a point. Cusp-shaped ridges of sand are often created on beaches by wave action.

cyanobacteria A group of minute, single-celled organisms, which can photosynthesize like plants. They are classified as bacteria, because they have a similar structure. Also called **blue-green algae**, although they are not closely related to other algae. See also *photosynthesis*.

cyclone (1) Also called a **depression**, a pattern of circulating air in the atmosphere with low pressure at the center. Cyclones normally form over oceans outside the tropics and are associated with wet and windy weather. **(2)** See *tropical cyclone*.

D

dark zone Vertical zone of the seabed and water column at around 3,300–13,000 ft (1,000–4,000 m), between the twilight zone and abyssal zone. Virtually no light penetrates this deep. See also *abyssal, twilight zone*.

delta An often fan-shaped structure of sediment built by the deposition of material by a river at its mouth.

demersal Of a fish: living mainly near the sea floor.

deposit feeding Feeding by extracting food particles from mud or other deposits. See also *filter feeding*.

depositional coast A coast that is growing seaward due to deposition of sand and other sediment supplied by rivers or ocean currents. See also *emergent coast, erosional coast*.

detritus Fragments of dead organisms and organic waste material, often mixed with sediment or suspended in ocean currents. A **detritivore** is an animal that feeds on detritus.

diatoms A group of plantlike protists that are part of the algae and major primary producers in the plankton. They are single-celled but often grow as chains or colonies. Diatoms secrete intricate cases of silica around themselves. See also *algae, primary producer, protists*.

dimorphism see *sexual dimorphism*.

dinoflagellates A group of protists that bear two flagella. They are common in ocean plankton. Some are animal-like (eating other organisms), while others are plantlike (photosynthesizing) and are therefore part of the alga. See also *algae, flagellum, protists*.

discordant coast Coast on which hills and valleys are roughly at right angles to the shore, resulting in an indented coastline of headlands and bays. See also *concordant coast*.

doldrums The region of very light winds close to the equator.

dorsal Relating to the back or upper surface of an animal. See also *ventral*.

drift A broad, slow-moving flow of surface water; for example, the North Atlantic Drift.

drowned coast A coast where the land has sunk or the sea level has risen compared with the previous level. It may show features such as rias or fjords. See also *emergent coast, fjord, ria*.

dune A hill or ridge-shaped structure of sand formed by wind action along some coasts and in deserts. Coastal dunes are usually formed on low-lying land behind beaches.

E

echinoderms A major group (phylum) of marine invertebrates that includes starfish, brittle stars, sea urchins, sea lilies, sea cucumbers, and sea daisies. Echinoderms have bodies arranged in parts rather like the spokes of a wheel (so-called "radial symmetry"). They have chalky protective plates under their skin, and use a unique system of hydraulic "tube feet" for moving, or for capturing prey, or both.

echolocation Method of locating and characterizing nearby objects, used by dolphins, bats, and some other animals, by emitting high-pitched sounds and interpreting their echoes.

echo-sounding The use of sound equipment to measure the depth of objects or the ocean floor; also used as a synonym for echolocation. See also *sonar*.

eddy A circular motion of any size and speed in a fluid. **Mesoscale eddies** of more than 60 miles (100 km) across are important features of ocean circulation. In tidal currents and whirlpools, an eddy is a circular motion slower than a whirlpool. See also *gyre, vortex, whirlpool*.

Ekman effect Tendency for a wind or current to cause air or water above or below it to move, but in a different direction to the original wind or current. The effect results from the rotation of Earth. At the ocean surface, the net result is usually that a prevailing wind creates a water current at 90° to the wind direction. See also *Coriolis force*.

El Niño Phenomenon by which the waters of the eastern Pacific off South America become warmer than usual every 4–7 years. The opposite phenomenon, in which eastern Pacific waters are unusually cold, is called **La Niña**. The term El Niño is also used as shorthand for the larger phenomenon called the El Niño–Southern Oscillation. See *ENSO*.

emergent coast A coast where the land has risen or sea level has fallen compared with a former level. See also *drowned coast, isostasy*.

ENSO Used as an abbreviation for the El Niño–Southern Oscillation.

A worldwide variation in Earth's climate pattern and ocean circulation, including the El Niño phenomenon, associated with a change in the position of warm surface waters in the eastern Pacific.

erosional coast A coast that is being eroded by the action of the sea. Rocky coasts are typically erosional, but so are some low-lying, sandy coasts. See also *depositional coast*.

estuary The mouth of a large river. Used more broadly, the term includes any bay or inlet where sea water becomes diluted with fresh water.

eustatic Of sea-level changes: occurring worldwide simultaneously, for example, as a result of melting ice sheets. See also *isostasy*.

eutrophication The altering of an aquatic ecosystem by the addition of plant nutrients, such as nitrate and phosphate. Often caused by humans, it can greatly change the character of an ecosystem by, for example, causing algal blooms. See also *bloom*.

exoskeleton A skeleton on the outside of an animal's body, often also acting as a protective barrier. Arthropods, such as crustaceans and insects, have an exoskeleton. See also *arthropods*.

F

fast ice Sea ice forming a continuous sheet. See also *sea ice, pack ice*.

fathom The traditional unit of depth measurement at sea, equivalent to 6 ft (1.83 m).

fault A fracture in Earth's crust where rocks have moved relative to one another either vertically or horizontally.

feather stars see *crinoids*.

Ferrel cell A large-scale circulation of air in temperate regions, involving air rising at around 60°N and S, flowing southward at a high altitude, descending at around 30°N or S, and returning north as the westerlies (westerly winds). See also *Hadley cell*.

fertilization The union of a male and female sex cell (such as a sperm and an egg cell in animals) as the first step in the production of a new organism by sexual reproduction. Some marine animals release eggs and sperm into the sea to meet by chance (**external fertilization**), while in others, the male transfers sperm directly into the female's body (**internal fertilization**).

fetch The distance of open ocean across which a wind is able to blow, and across which waves generated by the wind are traveling. A longer fetch tends to result in larger swell waves. See also *swell wave*.

filter feeding Feeding by collecting and separating food particles from the environment. When the food particles are suspended in water it is also called **suspension feeding**. See also *deposit feeding*.

fjord A narrow, steep-sided, deep inlet of the sea, once occupied by a glacier. Fjords have a shallower **sill** where they meet the open sea. See also *ria*.

flagellum A flexible, microscopic, hairlike structure used for propulsion by some single-celled organisms and for creating a water current by sponges. It is longer than a cilium. Plural **flagella**. See also *cilia, sponges*.

flatworms A major group (phylum) of invertebrates with simple, usually flattened bodies. Free-living forms are carnivorous; there are also many parasitic species, including tapeworms.

fluke Either of the lobes forming a whale's, dolphin's, or dugong's tail.

foraminiferans A group of protists whose empty, chalky skeletons are a major part of some deep-sea sediments. They are animal-like (they feed on other organisms) and include both planktonic and bottom-living types. See also *protists*.

forced wave A water wave created by storm winds at sea. Forced waves are taller and have a shorter wavelength than swell waves. See also *swell wave*.

foreshore The part of a shoreline that lies between the average high- and low-water marks. See also *tides*.

frazil ice Ice in the form of tiny crystals floating on or near the sea surface. It is the first stage in the formation of sea ice. See *sea ice*.

fringing reef A coral reef just offshore, without an intervening lagoon or stretch of water. See also *barrier reef*.

front A vertical or oblique region at the boundary of two masses of air or water with different characteristics.

G

gabion A wire cage filled with stones. Gabions are used to protect coastlines artificially against erosion.

gastropods The group of mollusks that includes snails, slugs, and pteropods (sea-butterflies). See also *mollusks*.

gill rakers Projections on the insides of the gill supports of some fish that sieve particles entering their mouths.

glacier An elongated mass of compressed ice that flows slowly downhill. Glaciers that reach the sea give rise to icebergs.

grease ice Stage of formation of sea ice in which frazil ice crystals congeal to form a soupy texture. See *frazil ice, sea ice*.

greenhouse gas A gas, such as water vapor, carbon dioxide, or methane, that prevents heat from radiating from Earth, causing Earth's surface to warm (the **greenhouse effect**). Some greenhouse gas emissions are natural; others are caused by human activities.

groyne An artificial barrier built down a beach and into the sea to hinder

transport of materials by longshore drift. See also *longshore drift*.

guyot A flat-topped submarine mountain, also called a **tablemount**. See also *seamount*.

gyre A large-scale circulation of surface ocean currents, typically spanning a whole ocean. See also *eddy*.

H

hadal Relating to the deepest oceanic regions below 20,000 ft (6,000 m), within ocean trenches; deeper than the abyssal zone. See also *abyssal*.

Hadley cell A large-scale circulation of air in warmer regions, caused by warmed air rising near the equator, traveling to mid-latitudes, cooling and descending, and returning to the equator as the *trade winds*.

halocline A boundary between waters of different salinities, across which salinity changes rapidly. See also *pycnocline, thermocline*.

headland A promontory on a shoreline, usually high and rocky and under strong forces of coastal erosion. See also *erosional coast*.

heat capacity The amount of heat energy that a given substance can absorb for a given rise in temperature. Water has a high heat capacity and so can act as a store of heat.

hermaphrodite An animal that is both male and female. Animals that are both sexes at once are called **simultaneous hermaphrodites**. Others start as males then become females, or vice versa. Some species change sex repeatedly.

holdfast A rootlike structure that anchors a seaweed to rocks but does not absorb nutrients like a true root.

holoplankton Planktonic organisms that spend all of their life as plankton. See also *meroplankton, plankton*.

holothurians Soft-bodied, sausage-shaped echinoderms, also called **sea cucumbers**, that feed mainly by swallowing mud and detritus. Their radial symmetry is not obvious at first glance. See also *echinoderms*.

hotspot A localized region of Earth that experiences large-scale upwelling of magma. As oceanic crust moves over a hotspot, a line of volcanic islands, such as the Hawaiian islands, may form over millions of years.

hurricane (1) A name for a tropical cyclone, especially one occurring in the Atlantic. See *tropical cyclone*. **(2)** A wind speed greater than 72 mph (116 km/h).

hydrocarbon Any chemical compound made only of carbon and hydrogen atoms.

hydroids Cnidarians that grow as small, branching colonies of polyps attached to rocks or seaweed. Each polyp is specialized either for feeding,

reproduction, or sometimes for defense. See also *cnidarians, polyp*.

hydrothermal vent A fissure in a volcanically active region of the ocean floor from which superheated, chemical-laden water emerges. The energy in the chemicals fuels rich biological communities via the activities of chemosynthetic bacteria and archaea. See also *chemosynthesis*.

I

ice age Any episode in which Earth's temperatures were much lower than today and ice cover more extensive. The **Ice Age** (with capitals) refers to a series of such episodes within the last 2 million years, the last ending around 10,000 years ago.

iceberg A large fragment of ice calved from the end of a glacier or ice sheet that is in contact with the sea. See also *calve*.

ice cap A mass of permanent ice similar to an ice sheet but smaller in extent.

ice lead A channel of open water among sea ice.

ice rafting Transport of rocky debris out to sea, frozen into icebergs. When the icebergs melt, the material is deposited as sediment.

ice sheet A very large mass of permanent ice covering land, such as the Antarctic Ice Sheet.

ice shelf An extension of an ice sheet into the ocean. Ice shelves are anchored to the sea floor at their landward end, but farther from the coast, they float on water.

igneous rock Any rock that originates from the cooling of magma, such as basalt or granite.

intermediate coast A coast whose features are intermediate between a primary and secondary coast. See also *primary coast, secondary coast*.

internal wave A wave occurring at the boundary of two different layers of the same fluid rather than at the surface—for example, at the boundary between two layers of ocean water.

intertropical convergence zone The region of air close to the equator where the north and south trade winds converge.

invertebrate Any animal without a backbone, ranging from flatworms to spiders. Of a total of 30 major groups (phyla) of animals, 29 are composed of invertebrates.

irradiance The amount of radiation falling on a given area.

island arc Chain of islands, usually including active volcanoes, created by the collision of the oceanic crust of two tectonic plates. One of the plates is subducted beneath the other, creating a trench on one side of the arc. See also *subduction, ocean trench*.

isopods A group of crustaceans that usually have flattened bodies. The group is mainly marine but also includes the land-living woodlice.

isostasy A state of equilibrium; applied especially to the relatively light rocks of the continental crust, which can be thought of as floating like icebergs among the heavier rocks of the ocean floor and mantle. **Isostatic rebound** is the tendency of land that was formerly ice-covered to rise slowly to its equilibrium level, often creating emergent coasts. See also *continental crust, emergent coast.*

IUCN The initials still used to designate the World Conservation Union (formerly the International Union for the Conservation of Nature). This organization carries out conservation-related activities, including gathering and publishing information on the current status of endangered species.

J

jawless fishes Two groups of primitive fish called lampreys and hagfish, which branched off the line of fish evolution before jaws had evolved.

jellyfish Cnidarians that typically drift among the plankton and catch prey using stinging tentacles. The body form of true jellyfish is a medusa. Some apparently similar forms such as the Portuguese man-of-war are not true jellyfish, but siphonophores. See also *cnidarians, medusa, siphonophores.*

K

katabatic wind A wind that blows downward from an ice sheet, glacier, or cold valley, usually at night.

krill Swimming, shrimplike crustaceans typically growing to ¾–2⅓ in (2–6 cm) long, which form a large part of the zooplankton and an important link in the Southern Ocean's food chain.

L

La Niña see *El Niño.*

lagoon A stretch of coastal water almost cut off from the sea by a spit or other barrier; also, the shallow water within the ring of an atoll.

larva A young stage of an animal, especially when completely different in structure from the adult. The larvae of many marine animals, such as starfish, live as part of the plankton. See also *metamorphosis.*

latent heat The heat absorbed or released when a substance changes its state—from gas to liquid, for example. The heat released when water vapor condenses is the main source of energy for hurricanes.

latitude A position on Earth expressed in terms of its angle north or south of the plane of the equator. **Low latitudes** are those close to the

equator, while **high latitudes** are nearer the poles.

levee A natural raised bank around some rivers, or an artificial bank built around a river or estuary.

littoral Relating to the area of shore between high- and low-water marks.

longitude A position on Earth expressed in terms of its angle east or west of an agreed line called the prime meridian circling Earth from pole to pole and passing through Greenwich, London, UK.

longshore drift Process by which sediment is transported along a coast as a result of waves breaking at an oblique angle to the shoreline.

M

magma Molten rock rising from deep inside Earth.

mangrove Any of various trees growing on muddy shores in the tropics and adapted to live with their roots and lower trunks immersed in salt water.

mangrove swamp Forestlike ecosystem formed by mangroves growing in muddy tidal areas and river mouths. Mangrove swamps only occur in the tropics and subtropics.

mantle All the rock lying between Earth's crust and its core. The mantle extends to a depth of about 1,800 miles (2,900 km).

medusa One of the two main body forms of cnidarians. Medusae are wide and saucer-shaped, as well as usually free-floating and able to swim. A jellyfish is an example of a medusa. See also *cnidarians, polyp.*

meroplankton Planktonic animals that are the larvae of animals that are not planktonic as adults, such as crabs.

metamorphosis The process of transforming body form from that of the young (larval) form to a radically different adult form. It is common in marine invertebrates such as starfish, whose larvae live in the plankton but whose adults live on the sea floor.

mid-ocean ridge A submerged range of mountains running along any part of the deep-ocean floor, marking the place where seafloor spreading is taking place. Also called a **spreading ridge**. See also *seafloor spreading.*

mimicry Phenomenon in which one species of animal has evolved to look similar to another, unrelated animal.

mixed layer The upper layer of the ocean that is kept mixed by winds and currents, so that its temperature and chemical characteristics are roughly uniform throughout.

mollusks A major group (phylum) of invertebrate animals that includes the gastropods (snails and slugs), bivalves (clams and relatives), and cephalopods (octopuses, squid, cuttlefish, and nautiluses). Mollusks are soft-bodied and typically have hard shells, though

some subgroups have lost the shell during their evolution.

mucus A sticky or slimy substance secreted by animals for protection, trapping prey, helping with movement, or other purposes.

mutualism A close relationship between two different species in which both benefit.

N

nanoplankton Planktonic organisms of 0.002–0.2 mm in diameter. Not as small as picoplankton. See also *picoplankton, plankton.*

neap tide The tide with the smallest range within an approximately two-week cycle, caused by the gravity of the Sun partly canceling out the effect of the Moon. See also *spring tide, tides.*

nearshore The part of the shore affected by waves and tides under normal conditions. It includes the foreshore plus an area beyond whose bed is shallow enough to be stirred up by wave action. See also *foreshore.*

nekton Animals of the open ocean that can swim strongly enough not to be at the mercy of ocean currents. Nekton include squid, adult fish, and marine mammals. See also *plankton.*

nematocyst The coiled structure within the stinging cell of a jellyfish or other cnidarian that shoots out and injects toxin via a dartlike tip. See also *cnidarians.*

nudibranchs see *sea slugs.*

O

ocean basin A region of low-lying oceanic crust within which a deep ocean (or part of one) is contained, and usually surrounded by land or shallower seas.

oceanic crust The type of Earth's crust that forms the deep ocean bed. Made mainly of basalt, it is thinner, denser, and heavier than continental crust.

ocean trench Elongated low-lying region of the ocean floor. Trenches are the deepest parts of the ocean. See also *subduction.*

ooze Sediment on the deep ocean floor containing a large proportion of the remains of the skeletons of planktonic organisms, such as foraminiferans or radiolarians.

overfall A stretch of rough water produced when a tidal current flows in the opposite direction to the wind.

ovoviviparous Producing live young by retaining eggs so that they hatch while still in the female's body.

P

pack ice A mosaic of floating ice formed when continuous sea ice

is broken up by storms or waves. See also *fast ice, sea ice.*

pancake ice Stage of formation of sea ice consisting of small flat areas of ice, curled at the edges where they bump into each other.

pectoral fin Either of the front pair of fins in most fish and marine mammals, mainly used for steering but sometimes for propulsion. See also *pelvic fin.*

pelagic Relating to or living in the waters of the open ocean, without immediate contact with the shore or the sea bottom. See also *demersal.*

pelvic fin Either of the pair of fins located further back than the pectoral fins in most fish. See also *pectoral fin.*

perennial Of plants: living for three or more years.

pheromone An odor produced by an animal to communicate with others of the same species, to attract the opposite sex, for example.

photic zone see *sunlit zone.*

photophore A light-producing organ.

photosynthesis Process in green plants, algae, and cyanobacteria whereby the Sun's energy is used to build energy-containing food molecules from carbon dioxide and water. See also *chemosynthesis, chlorophyll.*

phylum The highest-level grouping in the classification of the animal kingdom. Each phylum has a unique basic body plan. Mollusks, arthropods, and echinoderms are examples.

phytoplankton Planktonic organisms, such as microscopic algae and cyanobacteria, which produce their own food by photosynthesis.

picoplankton The smallest planktonic organisms, typically bacteria, of 0.0002–0.002 mm in diameter. See also *nanoplankton.*

plankton Marine or freshwater organisms, living in open water, that cannot swim strongly and so drift with the currents. Although small life forms dominate, larger creatures, such as jellyfish, are also planktonic. See also *nanoplankton, nekton, phytoplankton, zooplankton.*

plate boundary A border between two tectonic plates. The plates may be converging (**destructive boundary**), diverging (**constructive boundary**), or sliding past (**conservative or strike-slip boundary**). See also *transform fault.*

plate, tectonic see *tectonic plate.*

plate tectonics Phenomena linked to the relative movement of Earth's tectonic plates, including continental drift, seafloor spreading, earthquakes, and mountain-building; also, the theory explaining these occurrences.

polychaetes A large subgroup of segmented worms common in the oceans, often with bristles down the sides of the body. (Polychaete means "many bristles"). Some species can move around, while others anchor themselves within tubes or burrows and filter-feed. See also *segmented worm, tube worm*.

polynya An area of open water in an otherwise ice-covered sea, especially in the Arctic.

polyp One of the two main body-forms of cnidarians. An anemone or coral animal is a polyp. Polyps are typically tubular and attached to a surface at their base. See also *cnidarians, medusa*.

prevailing wind A wind that tends to blow from a particular direction. See *trade winds, westerlies*.

primary coast A coast whose features have not been significantly altered by marine erosion, the activity of animals such as corals, or human intervention. See also *secondary coast*.

primary producer Often called simply a producer, an organism that makes food, using energy either from the Sun or from naturally occurring inorganic chemicals. See also *autotroph, chemosynthesis, photosynthesis*.

productivity Rate at which living material is produced by organisms by growth and reproduction. See also *primary producer*.

prokaryotes Organisms such as bacteria and archaea, whose cells are smaller and simpler in structure than the cells of animals, plants, and protists. Cells of prokaryotes have no nucleus. See also *archaea, bacteria*.

protein A large molecule built by organisms from smaller molecules called amino acids. Proteins range from the enzymes that promote chemical reactions in body cells, to structural materials such as keratin—the tough protein that makes up hair, horn, and nails.

protists A wide grouping of often unrelated, microscopic organisms, traditionally classified as a single kingdom. It includes mostly single-celled forms, either animal-like (formerly called protozoa) or plant-like (many of which are termed algae). Some experts also include larger algae (seaweeds). Protist cells contain nuclei, like the cells of animals and plants, but unlike those of bacteria.

pteropods Swimming, planktonic gastropod mollusks, also called **sea butterflies**. The crawling foot of their snail-like ancestors has evolved into muscular "wings" that propel them along. See also *gastropod, plankton*.

pycnocline A boundary region in ocean waters within which density changes rapidly. It typically results from a combination of temperature and salinity levels, both of which affect density. See also *thermocline*.

R

radiation The emission of high-energy particles or waves. **Electromagnetic radiation** consists of electromagnetic waves: listed from long-wave to short-wave forms, these are radio waves, microwaves, infrared (heat) rays, visible light, ultraviolet light, X-rays, and gamma rays. Short-wavelength electromagnetic radiation has the highest energy.

radiolarians Single-celled predatory organisms mainly living as plankton, often with a delicate, perforated, spherical skeleton. Radiolarian remains of are an important part of some oceanic sediments.

reclamation The artificial conversion of a former coastal sea or wetland area into dry land.

reef see *coral reef*.

refraction The change of direction of a wave when it passes into a different medium—for example, light waves passing from air into water. Ocean waves are also refracted when they reach shallow water.

respiration **(1)** Breathing. **(2)** Also called **cellular respiration**, the biochemical processes within cells that break down food molecules, usually by combining them with oxygen, to provide energy for an organism. See also *anaerobic*.

revetment A sloping structure of spaced wooden or concrete beams, constructed to protect a beach or low cliff against erosion.

ria A winding inlet of the sea, a drowned former river valley. Most present-day rias were created when sea levels rose at the end of the last ice age. Unlike a fjord, a ria was never occupied by a glacier.

ribbon worms A major group (phylum) of narrow-bodied, unsegmented marine worms, also called proboscis worms, some of which can reach 160 ft (50 m) in length.

rip current A current flowing away from a shoreline, carrying water that has been pushed shoreward by waves. See also *tide rip*.

rip-rap Boulders piled deliberately on a shoreline to prevent erosion.

S

salinity Degree of saltiness.

salps Barrel-shaped, delicate-bodied tunicates that live as filter-feeders in the plankton. See also *tunicates*.

salt marsh An ecosystem developing on sheltered, flat, muddy coastlines, where tidal flats are colonized by salt-tolerant land plants. See also *tidal flat*.

sand dune see *dune*.

scute Any of the horny plates that form the outer covering of the shells of turtles; also used to described a similar protective structure on some fish and other animals.

sea arch A natural arch on a rocky shoreline, usually created by two sea caves on either side of a headland eroding into each other.

sea butterflies see *pteropods*.

sea cave A cave created at the foot of a cliff by wave action.

sea cucumbers see *holothurians*.

sea fans Fan-shaped corals belonging to the gorgonian or horny coral group. Though often growing on coral reefs, they are not reef formers themselves. See also *coral*.

sea pens A group of soft-bodied, colonial cnidarians. Each colony resembles a single individual, with one large, burrowing polyp anchoring the colony in seafloor mud, and smaller polyps feeding and reproducing. See also *cnidarians, polyps*.

sea slugs Shell-less marine gastropods, often with bright colors and tufty gills (**ctenidia**) on their backs. Sea slugs are carnivores and are not closely related to land slugs. Also called **nudibranchs**. See also *gastropods*.

sea stack An isolated pillar of rock left standing offshore on a rocky coastline after all the surrounding land has been eroded away.

sea urchins A group of echinoderms, usually with a rigid case called a **test**, a globular body, long spines, and a downward-facing mouth. Most graze algae from hard surfaces, though the heart urchins and sand dollars are burrowers. See also *echinoderms*.

seafloor spreading The creation of new oceanic crust by the upwelling of magma at mid-ocean ridges and consequent spreading of the sea floor on either side. See also *plate tectonics*.

seagrasses Any of various plants able to grow and root in shallow, sandy seabed along coastlines, especially in warmer seas. Although not actually grasses, they are true flowering plants, unlike seaweeds, which are algae.

sea ice Ice that forms on the surface of the sea, as distinct from ice shelves and icebergs, which originate on land. Some sea ice forms only in winter, while other sea ice is semi-permanent. Sea ice forms and evolves in several stages. See *frazil ice, grease ice, pack ice, pancake ice*.

seamount A submarine mountain, usually an extinct volcano.

sea spiders A group of eight-legged predatory marine arthropods. It is not agreed whether sea spiders are closely related to land spiders or not.

sea squirts see *tunicates*.

seaweed A member of any of three main groups of large-bodied algae. Seaweeds can make their own food by photosynthesis, but they lack roots. Their classification is not agreed, but green seaweeds seem to be related to plants, while red and brown seaweeds may represent two unrelated lines of evolution. See also *algae*.

secondary coast A coast with features significantly altered by marine erosion, the activity of animals such as corals, human intervention, or all three. See also *primary coast*.

sedentary Of animals such as worms: habitually staying in one position. See also *sessile*.

sediment An accumulation of solid particles that have settled out from water; also used for deposits left by other agencies such as the wind.

sedimentary rock Any rock originating from sediment that has later become compacted and hardened, such as sandstone.

segmented worms A major group (phylum) of worms, also called **annelids**, whose body is built from repeating units (segments) each bearing copies of organs, such as kidneys. The phylum includes earthworms, plus many marine species, mostly within a subgroup called the polychaetes. See also *phylum, polychaetes, worm*.

sessile Of an animal: attached permanently to a surface, especially without a stalk, and not able to move around. See also *sedentary*.

sexual dimorphism Situation in which the males and females of a species differ in appearance, for example, in color, shape, or size.

shrimp Any of various small, usually swimming crustaceans. True shrimps are relatives of crabs and lobsters.

siphon In mollusks: a fleshy tubular extension of the body that aids the flow of oxygenated seawater to the gills or sometimes transports food particles for filtering. Cephalopods use their siphons for jet propulsion. See also *cephalopods*.

siphonophores Floating, predatory, colonial cnidarians, such as the Portuguese man-of-war. The colony members have specialized functions but act together so that the colony functions like a single animal. See also *cnidarians, colonial, polyp, zooid*.

sonar A method of echo-sounding; often used more broadly as a synonym for echolocation. See also *echolocation, echo-sounding*.

Southern Oscillation see *ENSO*.

spit A peninsula of sand or shingle or both created by longshore drift, usually at a point where the shoreline changes direction. See also *bar, barrier island, longshore drift, tombolo*.

sponges A large group (phylum) of marine animals with a very simple structure that feed by creating currents through their bodies and filtering small particles from the water. They have no muscles or nerve cells, and sometimes no symmetry.

spore **(1)** A tiny structure produced (usually in large quantities) by non-flowering plants, fungi, and some protists, from which a new individual can grow. Spores are much smaller than seeds and usually produced asexually, sometimes forming part of a complex life history. **(2)** The inactive, resistant form of some bacteria that helps them survive unfavorable conditions. See also *asexual reproduction*.

spreading ridge see *mid-ocean ridge*.

spring tide The highest high tide and lowest low tide within an approximately two-week cycle, caused by the Sun and the Moon being in positions in which their gravitational effects add together most strongly. See also *tides, neap tide*.

squid see *cephalopod*.

stack see *sea stack*.

standing wave A wave that stays in the same position rather than moving along, found in particular situations such as tidal races.

starfish A group of echinoderms, also called **sea stars**, having five or more "arms" (extensions to the body) and both mouth and anus on the underside. They swallow their prey whole, which can be very large for their size. See also *echinoderms*.

storm beach The topmost ridge of sediment on a beach, usually formed by the highest spring tides in combination with storm conditions. See also *berm, spring tide*.

storm surge A rapid rise in sea level caused by storm winds driving water toward a shoreline. It can cause disastrous coastal flooding, especially if occurring at the same time as a high spring tide.

subantarctic Relating to latitudes immediately north of the Antarctic Circle.

subarctic Relating to latitudes immediately south of the Arctic Circle.

subduction The forcing down of oceanic crust belonging to one tectonic plate beneath another plate when two plates are colliding. Ocean trenches are the location of such **subduction zones**.

sublittoral Relating to the coastal marine environment below the low-water mark.

submersible A vessel built to operate underwater. Some submersibles are designed to be able to withstand great pressures in order to explore the ocean depths.

sunlit zone The topmost layer of ocean water, where enough light penetrates for photosynthesis to occur. Also called the **photic zone**, it extends from the surface to up to 660 ft (200 m). See also *dark zone, twilight zone*.

surf zone The zone on a shore where waves break and create foaming, turbulent water.

surface current Any current flowing at the surface of the ocean—for example, the Gulf Stream. Surface currents are mainly caused by friction from prevailing winds. See also *current, thermohaline circulation*.

surface tension The attraction between water molecules at a water surface, which creates a thin film with the strength to resist small deflections, allowing some insects, for example, to walk on the water surface.

suspension feeding see *filter feeding*.

swash The movement of turbulent water up a shore after a wave breaks. The **swash zone** is the zone of a shore where swash typically occurs.

swell waves Regular, smoothly traveling waves on the open ocean, especially when at a distance from the winds or storms that originally caused them. See also *fetch*.

swim bladder A gas-filled organ in many fish, used to control buoyancy, and sometimes for other purposes such as sound production.

symbiosis A close living relationship between two species, especially one in which both benefit. See also *mutualism, commensalism*.

T

tablemount see *guyot*.

tabular Of an iceberg: very wide and flat-topped.

tectonic plate Any of the large rigid sections into which Earth's crust and uppermost mantle are divided, whose relative movement is the subject of plate tectonics. The African Plate and the Pacific Plate are examples. See *plate tectonics*.

terrigenous Of marine sediments: originating on the land (for example, carried to the sea by rivers).

thermocline A region at a particular depth in the ocean or height in the air where average temperature changes rapidly. See also *pycnocline*.

thermohaline circulation The part of the ocean's water circulation powered by differences in the salinity and temperature of different water masses, rather than by the wind. Thermohaline circulation is the cause of most deep-water and some surface currents. See also *surface current*.

tidal bore A single large wave created when an incoming tide moves up a narrowing channel, such as an estuary.

tidal bulge or trough see *tides*.

tidal current see *tides*.

tidal flat A flat, muddy area covered at high tide; characteristic of sheltered areas such as estuaries.

tidal race A strong current created when a tide-generated water flow moves through a narrow channel.

tide rip A stretch of turbulent water where different tidal currents meet.

tides Fluctuation in sea level resulting from the gravitational attraction of the Sun and the Moon on Earth's oceans, combined with Earth's own rotation. In the open oceans, each tidal cycle of just over 12 hours generates a small but measurable vertical rise (**tidal bulge**) and fall (**tidal trough**) in the water. Tidal effects are much more obvious near the coast, and lead to horizontal water movements (**tidal currents**) as well as vertical movements.

tombolo A spit linking an island to the mainland or another island. See *spit*.

trade winds Prevailing winds blowing from the east toward the equator in subtropical and tropical latitudes.

transform fault A fault in which the rocks on either side are displaced horizontally. Numerous transform faults occur at right angles to mid-ocean ridges. See also *plate tectonics*.

trench see *ocean trench*.

tropical Relating to the warm regions of Earth that lie between the equator and the tropics of Cancer and Capricorn, at latitudes of 23.5° north and south, respectively. The term is sometimes used loosely for phenomena typical of these regions, even when occurring north or south of the two tropics.

tropical cyclone A large-scale, circulating weather system in warmer latitudes, called by different names, such as hurricane and typhoon, in different parts of the world. It generates intense winds and torrential rain. Its energy comes from the water vapor rising from warm seas and then condensing. A less powerful version of the phenomenon is called a **tropical storm**. See also *cyclone, hurricane, latent heat, typhoon*.

tsunami A sometimes huge water wave usually generated by displacement of water by an earthquake and capable of devastating shorelines thousands of miles from its origin. Sometimes inaccurately called a "tidal wave."

tube worms Worms that live anchored and protected in tubes, which are either secreted or built of material such as sand grains. Tube worms include the giant worms living around some hydrothermal vents, as well as many segmented worms. See also *polychaetes*.

tunicates A group of mainly filter-feeding marine invertebrates closely related to backboned animals (vertebrates). There are both solitary and colonial species. They include non-moving attached forms (**sea-squirts**) and others that drift in the plankton. See also *salps*.

turbidity current A phenomenon similar to an underwater avalanche or landslide, involving water laden with sediments slipping down a slope.

twilight zone The vertical zone of the water column and seabed lying between approximately 660 and 3,300 ft (200 and 1,000 m) deep, into which some light penetrates, but not enough to support photosynthesis.

typhoon see *tropical cyclone*.

U

upwelling The upward motion of deep-ocean water toward the surface. Some upwelling increases ocean fertility by recirculating nutrients from deeper layers.

V

ventral Relating to the lower surface or belly of an animal. See also *dorsal*.

vertical migration Behavior of many zooplankton, fish, and squid of the open ocean, in which they rise nearer the surface by night and sink deeper by day, probably to escape predators.

vertical transport Any large-scale vertical flow of ocean water.

vortex A fast-rotating eddy in a fluid; sometimes used as a synonym for whirlpool. See also *eddy, whirlpool*.

W

water column The volume of water between the ocean surface and the bottom of the ocean.

wave A motion or disturbance that transfers energy. The water in a wave crossing the open ocean does not move significantly except up and down as the wave passes. The high point of a wave is its **crest** and the low point its **trough**. Water motion becomes more complex and turbulent in waves breaking on shores (**breakers**).

westerlies Prevailing winds that blow from the west. Westerlies are the most common winds in temperate regions.

western boundary current A relatively narrow, fast-moving surface current formed at the western boundary of an ocean basin, usually as part of a gyre. The Gulf Stream is an example. Deep-water western boundary currents also exist. See also *gyre*.

whirlpool A powerful eddy or vortex formed at the sea's surface, often caused when two separate tidal currents meet. See also *eddy, vortex*.

white smoker A deep-ocean hydrothermal vent in which the emerging hot water appears white because of light-colored mineral particles suspended in it.

worm Any of a variety of usually non-swimming invertebrate animals that are long, slender and flexible, and lack legs and shells. See *flatworms, ribbon worms, segmented worms, tube worms*.

Z

zoea The planktonic larval stage of certain crustaceans, including crabs. They are different in structure from their adult forms, having long spines.

zooid An individual in a colony of interconnected animals, such as bryozoans. The term is not applied to colonial coral animals, which are termed polyps. See also *polyp*.

zooplankton Any animals or animal-like protists that are part of the plankton. See also *plankton*.

zooxanthellae Symbiotic, microscopic algae living in the tissues of many corals. See also *symbiosis*.

INDEX

Acknowledgments

Dorling Kindersley would like to thank several people for their help in the preparation of this book. At the American Museum of Natural History, Udayan Chattopadhyay was unfailingly helpful and John Sparks provided many valuable comments on the text and images. Georgina Garner and Erin Richards worked on early versions of the contents list. Frances Dipper and Robert Dinwiddie drew up the original lists of species and physical features described in the catalog sections. Additional design work was done by Janis Utton and Pankaj Sharma. Tamlyn Calitz and Amy Walters provided editorial assistance, and Klara Kayser contributed design assistance. Neil Fletcher did additional picture research for the Birds section.

PICTURE CREDITS

Dorling Kindersley would like to thank the following for their help in supplying images: Romaine Werblow in the DK Picture Library; All at SeaPics.com; All at FLPA; Jonathan Hamston at OSF; Teresa Riley at Getty Images; All at Alamy Images.

KEY:
(a-above; b-below/bottom; c-center; f-far; l-left; r-right; t-top)

SIDEBAR IMAGES
Corbis: David Keaton (*Atlas of the Oceans*); Jeffrey L. Rotman (*Ocean Environments*). **Getty Images:** National Geographic/Raul Touzon (*Ocean Life*); Photonica/Anna Grossman (*Introduction*).

1 Getty Images: Taxi/Peter Scoones. **2–3 Getty Images:** Stone/Warren Bolster. **4 Corbis:** (tc); Lawson Wood (bc). **4–5 Getty Images:** National Geographic/Brian Skerry. **5 Getty Images:** Image Bank/Mike Kelly (tc). **NASA:** Jacques Descloitres, MODIS Rapid Response Team, NASA/GSFC (cra). **6–7 FLPA:** Minden Pictures/Norbert Wu (*Background*). **8–9 FLPA:** Minden Pictures/Chris Newbert, **Carrie Vonderhaar. 10–11 naturepl.com:** Brandon Cole. **112 Alamy Stock Photo:** Nature Picture Library. **13 naturepl.com:** Visuals Unlimited. **14–15 Alamy Stock Photo:** Nature Picture Library. **16 Getty Images:** UCG / Contributor. **16–17 David Hall (www.seaphotos.com)**. **18–19 naturepl.com:** Andy Murch. **20 Alamy Stock Photo:** David Fleetham. **21 FLPA:** Minden Pictures/Frans Lanting. **22–23 Alamy Stock Photo:** Nature Picture Library. **24–25 Corbis:** NASA. **26–27 Shutterstock.com:** AlyoshinE. **28–29 Getty Images:** Taxi/Jason Childs. **30 Alamy Images:** Pictor International/ImageState (bc). **DK Images:** Frank Greenaway (bl). **30–31 Alamy Images:** Hawkeye (c). **31 Alamy Images:** Bryan & Cherry Alexander Photography (br). **DK Images:** (fbl); Brian Cosgrove (bc); Zena Holloway (bl). **NASA:** GSFC/MODIS Rapid Response Team, Jacques Descloitres (tl); Liam Gumley, MODIS Atmosphere Team, University of Wisconsin-Madison Cooperative Institute for Meteorological Satellite Studies (ca). **32 Alamy Images:** David Wall (br). **Science & Society Picture Library:** Science Museum, London (bl). **33 Alamy Images:** PHOTOTAKE Inc./Carolina Biological Supply Company (bc); Stephen Frink Collection/James D. Watt (ca); Visual&Written SL/Kike Calvo (cra). **NASA:** Provided by the SeaWiFS Project, Goddard Space Flight Center, and ORBIMAGE (br). **34 NASA:** MODIS Instrument Team, NASA Goddard Space Flight Center, (c); **NOAA:** (tr). **35 Alamy Images:** Roger Cracknell (cl); Chris A Crumley (tr). **DK Images:** Frank Greenaway/Courtesy of the University Marine Biological Station, Millport, Scotland (crb). **SeaPics.com:** Bob Cranston (ca). **36 Alamy Images:** Reinhard Dirscherl (cb). **Dive Gallery/Jeffrey Jeffords (www.divegallery.com):** (bc). **Robert Dinwiddie:** (cl). **Image Quest Marine:** Y. Kito (bl). **36–37 Alamy Images:** Visual&Written SL/Takaji Ochi (c). **37 AguaSonic Acoustics:** Mark Fischer (crb). **Alamy Images:** Sue Cunningham Photographic (cra); James Davis Photography (cr); Dinodia Images/Ashvin Mehta (tr). **Science Photo Library:** (cb). **38–39 Corbis:** Brenda Tharp. **42 Alamy Images:** Danita Delimont (c). **Corbis:** Raymond Gehman (bc). **DK Images:** Harry Taylor (cra). **43 NASA:** JPL (br). **Science Photo Library:** Bill Bachman (tr). **45 Corbis:** Bettmann (tc); David Lawrence (cra). **46 Alamy Images:** Norman Price (tr). **Planetary Visions** (bl). **46–47 Alamy Images:** Nordicphotos/Sigurgeir Sigurjonsson (b). **47 Planetary Visions** (tr). **48 DK Images:** Colin Keates/Courtesy of the Natural History Museum, London (tr). **49 Alamy Images:** Douglas Peebles Photography (br). **Woods Hole Oceanographic Instititution:** Jayne Doucette (tr). **50–51 Getty Images:** AFP/Lothar Slabon. **52–53 Corbis:** Image by Digital image © 1996 CORBIS; Original image courtesy of NASA. **NASA:** Jesse Allen, Earth Observatory. Image interpretation provided by Dave Santek and Jeff Key, University of Wisconsin-Madison. **54 ESA:** Eumetsat (cr). **55 Alamy Images:** Kos Picture Source (t). **DK Images:** Peter Wilson (bl). **56 Action Images:** Reuters/Carlo Borlenghi. **57 Alamy Images:** Kos Picture Source (br). **Corbis:** Emmanuelle Thiercelin (cr). **Getty Images:** Brian Carlin/Team Vestas Wind (crb); Clive Mason (cra); Joseph Prezioso (tr). **Clipper Round the World Yacht Race:** Clipper Ventures Plc (cr). **58–59 NASA:** Image courtesy the SeaWiFS Project, NASA/Goddard Space Flight Center, and ORBIMAGE (c). **59 Alamy Images:** Chris Linder (br). **Corbis:** Bettmann (tr). **NASA:** (cr); Image processed by Robert Simmon based on data from the SeaWiFS project and the Goddard DAAC (c). **60 SeaPics.com:** Doug Perrine (bl). **61 S.M.R.U.:** Simon Moss (bl). Courtesy of **Andreas M. Thurnherr:** (br). **62 Alamy Stock Photo:** Pally. **63 Alamy Stock Photo:** Brandon Cole Marine Photography (ca). **Getty Images:** Moment/s0ulsurfing—Jason Swain (cb); Westend61 (tr). **NASA:** NASA's Earth Observatory, NASA Goddard Space Flight Center/Michon Scott (cr). **The Ocean Cleanup:** Olika 3D (br). **64 Alamy Images:** Danita Delimont (bl). **65 Corbis:** Bettmann (cla); Sygma/Gyori Antoine (bl). **Getty Images:** Photographer's Choice/Kerrick James (tr). **NASA:** Image courtesy Jacques Descloitres, MODIS Land Rapid Response Team at NASA GSFC (clb). **NOAA:** Michael Van Woert, NOAA NESDIS, ORA (br). **NASA:** LANCE/EOSDIS MODIS Rapid Response Team at NASA GSFC (clb). TopFoto.co.uk: RIA Novosti / Sergey Mamontov (bl). **66 Alamy Images:** Aflo Foto Agency (tr); Boating Images Photo Library/Keith Pritchard (cra); Michael J. Kronmal (br); Tribaleye Images/J Marshall (bl). **Corbis:** Image by Digital image © 1996 CORBIS; Original image courtesy of NASA (clb). **67 Alamy Images:** Mark Lewis (b); PHOTOTAKE Inc./Dennis Kunkel (tc). **Getty Images:** Photographer's Choice/Malcolm Fife (tr). Courtesy of **US Navy:** Photo courtesy of Ian R. MacDonald, Texas A&M Univ. Corpus Christi (cla). **68 Alamy Images:** Bill Brooks (cb); Images&Stories (br). Courtesy of **Chris Baisan, University of Arizona:** (bl). **NASA:** JPL/Caltech (cl). **69 Corbis:** Jonathan Blair (ca); EPA/Josue Fernandez (b). **NOAA:** Lieutenant Mark Boland, NOAA Corps (cla). **Corbis:** National Geographic Society / Robb Kendrick (c). **Reuters:** Reuters Photographer (b). **70 NASA:** Jeff Schmaltz, MODIS Rapid Response Team. Caption by Rebecca Lindsey, NASA Earth Observatory. (cl, c, cr). **NOAA:** Aircraft Operations Center (br). **71 Corbis:** EPA/Alejandro Ernesto (bl). **OSF/photolibrary:** Warren Faidley (t). **SeaPics.com:** Doug Perrine (cr). **stevebloom.com:** (br). **72–73 Corbis:** Reuters/Erik De Castro. **73 Corbis:** Bryan Denton (crb); Demotix/Herman Lumanog (cra/Typhoon); EPA/Dennis M. Sabangan (cra/Typhoon Haiyan aftermath, cr, br); Reuters/ Erik De Castro (fcrb, crb/Food Drop). **NASA:** NASA Goddard MODIS Rapid Response Team (cb/Haiyan 8th Nov, cb/Haiyan 11th Nov). **Planetary Visions Limited:** (cra/Eye of the Storm). **74–75 Getty Images:** Lonely Planet Images/Karl Lehmann. **76 Alamy Images:** David Gregs (cr); ImagePix (bc). **iStockphoto.com:** Dan Brandenburg (cra). **NOAA:** Captain Andy Chase (bl). **OSF/photolibrary:** Pacific Stock (crb). **77 Alamy Images:** Michael Diggin (clb). **Getty Images:** Taxi/Helena Vallis (t). **iStockphoto.com:** Paul Topp (crb). **78 Alamy Images:** Mooch Images (clb). **78–79 Getty Images:** Robert Harding World Imagery/Lee Frost (b). **80 Alamy Images:** Mary Evans Picture Library (br); Ian Simpson (bl). **Don Dunbar (www.easternmaineimages.com):** (tl) (c). **81 Alamy Images:** Malcolm Fife (cb); Peter L. Hardy (bc); Colin Palmer Photography (tr). **www.uwphoto.no:** Erling Svensen (cl). **82 Alamy Images:** Shaughn F. Clements (b); phototramp.com/Maciej Tomczak (cr). **83 Corbis:** Dave Bartruff (l); Christie's Images (br). **84–85 Corbis:** Lawson Wood. **86–87 Getty Images:** Photonica/Photolibrary.com. **88 Corbis:** Yann Arthus-Bertrand (br). **NASA:** (clb). **89 Corbis:** Michael Busselle (bc); Lloyd Cluff (t); Ecoscene/John Wilkinson (clb). **DK Images:** Colin Keates/Courtesy of the Natural History Museum, London (crb). **90 Alamy Stock Photo:** Travelscapes. **91 Alamy Images:** Louise Murray (br); robertharding (tr). **Corbis:** Matthieu Paley (c); Sygma/Kapoor Baldev (crb). **Getty Images:** Ashley Cooper (cr). **92 Alamy Images:** Michael Howell (b). **Corbis:** Yann Arthus-Bertrand (cra); Jack Fields (c); Frans Lanting (tr). **93 Alamy Images:** FLPA (clb); geogphotos (fcla). **iStockphoto.com:** Andrew Dorey (ca); Gregor Erdmann (cla). **NASA:** GSFC/METI/ERSDAC/JAROS, and U.S./Japan ASTER Science Team. **94 Alamy Images:** Jack Stephens (cl). **Rob Havemeyer Acadia National Park ME:** (tr). **94–95 Steven Russell (www.pbase.com/nodfather):** (b). **95 Alamy Images:** Eric Nathan (cr). **Corbis:** Kevin Fleming (tc). **DK Images:** Jon Spaull (tr). **96 Alamy Images:** Atmosphere Picture Library/Bob Croxford (tr). **Corbis:** Ric Ergenbright (bl). **DK Images:** Rough Guides/Ian Aitken (crb). **www.undiscoveredscotland.co.uk:** (cla). **97 Alamy Images:** Sean Burke (t); CuboImages srl/Marco Casiraghi (bl). **NASA:** Johnson Space Center - Earth Sciences and Image Analysis (br). **98 Corbis:** Peter Johnson (c); Richard T. Nowitz (br). **Wombat Pitts:** (cra). **99 Alamy Images:** Ross Jardine (cl); Simon Reddy (cr). **Dreamstime.com:** Kawee Wateesatogkij (b). **NASA:** Image courtesy Jacques Descloitres, MODIS Land Rapid Response Team at NASA GSFC (tc). **100–101 Corbis:** Digital image © 1996 CORBIS; Original image courtesy of NASA. **Andy Biggs. 102 Alamy Images:** Danita Delimont (bc). **Dreamstime.com:** Riccardocalli (ca); Dan Hershman: (cr). **Dr Sandy Tudhope, Institute of Geology and Geophysics, Edinburgh University:** (bl). **103 Alamy Images:** Danita Delimont (b). **Getty Images:** Stone/James Randklev (cl). **Marco Nero:** (cr). **104 WaterLand Neeltje Jans:** RWS MD afd. Multimedia. **105 Alamy Images:** Florida Images (cr); geogphotos (cra); Rodger Tamblyn (tc). **Getty Images:** Finnbarr Webster (bl). **Natural Visions:** Heather Angel (c). **NOAA:** NOAA Restoration Center, Chris Doley (cb). **Sky Pictures luchtfotografie (www.skypictures.nl):** (br). **106 Alamy Images:** Patrick Mallette (cl). **Getty Images:** Altrendo/altrendo nature (tr); Lonely Planet Images/Bethune Carmichael (clb). **106–107 Corbis:** Martin Harvey (b). **107 Alamy Images:** Guillen Photography (cra); Wildscape (c). **DK Images:** Shaen Adey (tr); James Stevenson (tl). **108 Alamy Images:** Danita Delimont (tc); Peter Lewis (b). Paul Yung: (cr). **109 Alamy Images:** Atmosphere Picture Library/Bob Croxford (bl); imagebroker/Harald Theissen (cla). **Corbis:** Jason Hawkes (r). **DK Images:** Geoff Dann (br). **110 Alamy Images:** Mark Boulton (tl). **Corbis:** Tony Arruza (bl); Yann Arthus-Bertrand (cr). **111 Alamy Images:** Simon Reddy (tc); Laurie Wilson (b). **Corbis:** (cr). **112 Alamy Images:** Ian Dagnall (tr); Danita Delimont (bc). **Corbis:** Douglas Peebles (br). **DK Images:** Lloyd Park (cl). **113 Alamy Images:** Jon Arnold Images (crb); Clint Farlinger (ca). **Corbis:** Neil Rabinowitz (bl). **Getty Images:** National Geographic/Skip Brown (t). **iStockphoto.com:** Judi Ashlock (tr). **114 Corbis:** Post-Houserstock/Dave G. Houser (cl). **114–115 Alamy Images:** Jon Arnold Images/Doug Pearson (c). **115 Alamy Images:** Tim Graham (ca). **Corbis:** (bc). **DK Images:** Dave King (tr). **OSF/photolibrary:** Richard Herrmann (cra). **116 Getty Images:** Stone/Paul Souders (bl); Stone/Tom Bean (cr). **US Geological Survey:** (bc). **116–117 Getty Images:** Moment/Aurélien Pottier (t). **117 Corbis:** Dale C. Spartas (bl). **NASA:** Jacques Descloitres, MODIS Rapid Response Team, NASA/GSFC (br). **118 Alamy Stock Photo:** Stock Connection Blue (tr). **Corbis:** Reuters/Sergio Moraes (cr). **NASA:** Jacques Descloitres, MODIS Rapid Response Team, NASA/GSFC (b). **119 Alamy Stock Photo:** Doug Pearson (tr). **Corbis:** Sygma/Annebicque Bernard (bc). **Getty Images/iStock:** Kutredrig (cl). Courtesy of **Clive Griffin (www.pbase.com/clivegriffin):** (clb). **120 Alamy Images:** Jack Sullivan (cr). Pierre-Yves Lagrée, LMM CNRS Université Paris VI: (t). **NASA:** Earth Sciences and Image Analysis Laboratory at Johnson Space Center (b). **121 Alamy Images:** Karsten Wrobel (tr). **Corbis:** (clb); Yann Arthus-Bertrand (cla). **DK Images:** Rob Reichenfeld (crb). **122 Alamy Images:** Eddie Gerald (b). **NASA:** Provided by the SeaWiFS Project, NASA/Goddard Space Flight Center, and ORBIMAGE (tl); Imaginechina (bl). **Shutterstock.com:** HelloRF Zcool (cr). **123 Alamy Images:** Tibor Bognar (cl). **Corbis:** Peter Guttman (br); Galen Rowell (b); Michael S. Yamashita (tr). **124 Alamy Images:** Florida Images (crb); Renee Morris (bc). **FLPA:** Skylight (cl). **125 Corbis:** James L. Amos (bc); Carol Havens (t). **DK Images:** Mike Linley (cr). **Getty Images:** Photographer's Choice/Cameron Davidson (bl). **Natural Visions:** Heather Angel (br). **126 Alamy Images:** Jon Sparks (t). **Corbis:** Rob Howard (bl). **Dreamstime.com:** Ondrej Prosický (cr). **127 Alamy Images:** David Poole (br); J. Schwanke (tr). **Corbis:** Annie Griffiths Belt (cla); Reuters/Darren Staples (crb). **128 Alamy Images:** Mark Boulton (cb); Rod Edwards (ca); Robert Harding Picture Library Ltd (br). **129 Corbis:** Natalie Fobes (b); Steve Kaufman (br). **ESA:** Landsat 8 OLI & Landsat 4 TM (tc). **Nial Moores/Birds Korea (www.birdskorea.org):** Mr. Jeon Shi-Jin (cla). **130 Alamy Images:** David Hosking (br). **Getty Images:** National Geographic/Tim Laman (tr). **SeaPics.com:** Jeremy Stafford-Deitsch (bc). **130–131 Oceanwide Images:** Bob Halstead (c). **131 Alamy Images:** Danita Delimont (cra); Reinhard Dirscherl (tr). **Corbis:** Michael S. Yamashita (br). **SeaPics.com:** D.R. Schrichte (ca). **132 DK Images:** Rough Guides/Demetrio Carrasco (br); Peter Wilson (ca). **132–133 SeaPics.com:** Masa Ushioda (t). **133 Alamy Images:** Mireille Vautier (clb). **Corbis:** Stephen Frink (br). **US Geological Survey:** (cra). **134 Alamy Images:** Tim Graham (bc). **Corbis:** (crb); Yann Arthus-Bertrand (ca). **135 Alamy Stock Photo:** Panama Landscapes by Oyvind Martinsen (br). **Corbis:** Arne Hodalic (cl). **Getty Images:** National Geographic/Timothy Laman (tr). **SeaPics.com:** Jeremy Stafford-Dietsch (bc). **136–137 naturepl.com:** Jurgen Freund. **138–139 naturepl.com:** Aflo. **Alamy Images:** WaterFrame. **140 Alamy Images:** Aqua Image (cla). **DK Images:** Frank Greenaway/Courtesy of the Natural History Museum, London (crb). **iStockphoto.com:** Ingvald Kaldhussæter (cra). **Sue Scott:** (bl). **SeaPics.com:** Mark Conlin (c). **141 British Marine Aggregate Producers Association (www.bmapa.org):** (br). **Getty Images:** Image Bank/Astromujoff (t). **OSF/photolibrary:** Michael Brooke (bc). **142 DK Images:** Frank Greenaway/Courtesy of the Weymouth Sea Life Centre (bc); Jerry Young (cr). **Sue Scott:** (cl). **SeaPics.com:** Mark Conlin (bl). **142–143 David Hall (www.seaphotos.com):** (b). **143 Image Quest Marine:** Jim Greenfield (crb). **Marine Wildlife:** Paul Kay (tr). **NOAA:** Dr. James P. McVey, NOAA Sea Grant Program (bc). **Sue Scott:** (c). **144 DK Images:** Frank Greenaway (ca). **Sue Scott:** (cl) (bl) (br). **144–145 Getty Images:** National Geographic/Bill Curtsinger (t). **145 Alamy Images:** Guillen Photography (bl). **Sue Scott:** (crb) (cra). **146 Alamy Images:** Fabrice Bettex (bc); Gavin Parsons (cr). **DK Images:** Tim Ridley (tl). **Marine Wildlife:** Paul Kay (cla). **Sue Scott:** (clb). **147 Corbis:** Ralph A. Clevenger (b). **OSF/photolibrary:** Tobias Bernhard (tc). **Sue Scott:** (cla) (cra). **148 Corbis:** (tl). **Sue Scott:** (cra) (bl). **SeaPics.com:** Doug Perrine (crb). **149 Alamy Images:** Mark Lewis (b); PNR Photography (crb). **Dr. Alberto V. Borges/Chemical Oceanography Unit from the University of Liège, Belgium:** (tr). **150 Alamy Images:** Ross Armstrong (tr); Joel Day (clb); Andre Seale (cla). **Sue Scott:** (bc). **US Fish and Wildlife Service National Image Library:** Chris Dau (br). **151 SeaPics.com:** Phillip Colla. **152 Alamy Images:** Danita Delimont (c); Nick Hanna (cl). **Corbis:** Yann Arthus-Bertrand (cr). **152–153 OSF/photolibrary:** Pacific Stock (c). **153 Dive Gallery/Jeffrey Jeffords (www.divegallery.com):** (tc). **JM Roberts, Scottish Association for Marine Science:** (crb). **SeaPics.com:** Clay Bryce (br); James D. Watt (cra). **154 Alamy Images:** Michael Patrick O'Neill (bl); Sylvia Cordaiy Photo Library Ltd (tc). **DK Images:** Jerry Young (clb). **155 Alamy Images:** Stephen Frink Collection (cra); Karen & Ian Stewart (c). **Dive Gallery/Jeffrey Jeffords (www.divegallery.com):** (br). **SeaPics.com:** Andrew J. Martinez (tl); James D. Watt (bl). **156 Alamy Images:** Stephen Frink Collection (bl). **Corbis:** Stephen Frink (c); Lawson Wood (tc). **157 Corbis:** Bob Krist (b). **SeaPics.com:** Rodger Klein (tr). **158 Alamy Images:** Nick Hanna (tr); Martin Harvey (bl); Zute Lightfoot (br). **159 Alamy Images:** Steve Allen Travel Photography (tr); Slick Shoots (cr). **Corbis:** Cordaiy Photo Library Ltd/John Parker (cla). **SeaPics.com:** Marc Bernardi (b). **160 Alamy Images:** Aqua Image (cla). **Getty Images:** Photodisc/Giordano Cipriani (cr). **SeaPics.com:** James D. Watt (b). **161 Alamy Images:** Robert Harding Picture Library Ltd (crb); Andre Seale (clb). **Corbis:** Reuters/Handout (cra). Brian McMorrow: (tr). **SeaPics.com:** James D. Watt (br). **162–163 Oceanwide Images:** Gary Bell. **164 DK Images:** Frank Greenaway (clb). **NASA:** Image and animations provided by the SeaWiFS Project and the NASA GSFC Scientific Visualization Studio (cra). **Sue Scott:** (tl) (bc) (br). **165 Alamy Images:** Jeremy Inglis (br); Andre Seale (bl). **Image Quest Marine:** Scott Tuason (t). **166–167 Getty Images:** Stone/Kim Westerskov. **168 Science Photo Library:** Alexis Rosenfeld (bc). **169 DeepSeaPhotography.Com:** Kim Westerskov (tc). **Image Quest Marine:** Peter Parks (br). **Science Photo Library:** Susumu Nishinaga (cb). **SeaPics.com:** Ingrid Visser (c). **170 OSF/photolibrary:**

(cla); Howard Hall (b). **SeaPics.com:** Peter Parks/iq3-d (tc). **171 ExploreTheAbyss.Com:** Peter Batson (cr) (clb) (bl) (br). **Image Quest Marine:** Peter Herring (tr). **NOAA:** Archival Photography by Steve Nicklas, NOS, NGS (c). **OSF/photolibrary:** Pacific Stock (tc). **172 Getty Images:** National Geographic/Paul Nicklen. **173 Enrique Alvarez:** (br). **DeepFlight:** (cra/Deepflight Super Falcon). **NOAA:** (cr, crb). **www.uboatworx.com:** David Pearlman (fcra). **174 Science Photo Library:** (tr). **175 DeepSeaPhotography.Com:** Kim Westerskov (br). **NOAA:** Image Courtesy of the Deep Atlantic Stepping Stones Science Party, IFE , URI-IAO, and NOAA (bc); Office of Ocean Exploration (cb). **Science Photo Library:** Dr Ken MacDonald (t). **SeaPics.com:** Mark Conlin (bl). **176 NOAA:** Commander John Bortniak, NOAA Corps (bl); Fisheries Collection (br). **Dr. P. J. Ramsay/African Coelacanth EcoSystem Programme:** (cl). **177 Alamy Images:** Ron Scott (br). **ExploreTheAbyss.Com:** Peter Batson (bc). **NASA:** Image provided by the USGS EROS Data Center Satellite Systems Branch (tr). **NOAA:** OAR/National Undersea Research Program (NURP); University of Connecticut (ca); Ocean Explorer (cb). **178 www.uwphoto.no:** Erling Svensen. **179 ExploreTheAbyss.Com:** Peter Batson (cra). **FLPA:** D. P. Wilson (c). **Jason Hall-Spencer/Marine Conservation Society:** (crb). **JM Roberts, Scottish Association for Marine Science:** (bc); AWI & Ifremer 2003 (tc) (br). **180 Alamy Images:** Travelpix (clb). **NASA:** Jacques Descloitres, MODIS Rapid Response Team, NASA/GSFC (br). **NOAA:** National Geophysical Data Center (cla). **SeaPics.com:** David Wrobel (bl). **181 Alamy Images:** Phototake Inc./Dennis Kunkel (bc). **Image Quest Marine:** Peter Parks (t). **Science Photo Library:** Steve Gschmeissner (cb). **SeaPics.com:** D.R. Schrichte (br). **182 Alamy Images:** Blickwinkel (clb). **ExploreTheAbyss.Com:** Peter Batson (cla). **SeaPics.com:** Doug Perrine (bl). **Craig Smith & Mike Degruy:** (br). **182–183 NOAA:** OAR/National Undersea Research Program (NURP) (c). **183 naturepl.com:** David Shale (bl). **Naval Historical Foundation, Washington, D.C.:** (br). **Science Photo Library:** US Geological Survey (tr). **184 Alamy Images:** Fabrice Bettex (tr). **Getty Images:** Phil Walter (bl). **184–185 Getty Images:** Moment/by wildestanimal (b). **185 Alamy Images:** David Tipling (c). **Corbis:** Cordaiy Photo Library Ltd/John Farmar (ca); Ralph White (tl). **Planetary Visions:** Lamont-Doherty Earth Observatory (br). **186 Planetary Visions. 187 NASA:** Canadian Space Agency/National Snow and Ice Data Centre (crb); GSFC (cra); JPL (c); JPL-Caltech (cb). **NASA:** (clb). **Planetary Visions:** (tr, ca). **Science Photo Library:** David Vaughan (br). **University College London:** (cb). **188** Image courtesy of **Karen L. Von Damm. Image obtained from the DSV Alvin, with funding provided by the U.S. National Science Foundation:** (tr). **Science Photo Library:** Southampton Oceanography Centre/B. Murton (bl). **188–189 Woods Hole Oceanographic Institution:** (c). **189 ExploreTheAbyss.Com:** Peter Batson (tc) (c) (br). **Richard T. Lutz:** (cr). **NOAA:** Ocean Explorer (bl). **190–191 FLPA:** Minden Pictures/Norbert Wu. **192 Bridgeman Art Library:** Royal Geographical Society, London, UK (tr). **Corbis:** Ecoscene/Graham Neden (cb). **193 Alamy Images:** Rosemary Calvert (b); John Digby (tl). **SeaPics.com:** Franco Banfi (tr). **194 Alamy Images:** Blickwinkel (br); Eric Ghost (fbr); K-Photos (tc). **M.A. Felton:** (bc). **NOAA:** Michael Van Woert, NOAA NESDIS, ORA (crb). **194–195 Getty Images:** Photographer's Choice/Siegfried Layda (c). **195 Alamy Images:** Giles Angel (crb); Nordicphotos/Kristjan Fridriksson (tr). **196 Corbis:** Sygma. **Science Photo Library:** NOAA. **197 Corbis:** Bettmann (ca) (cra); Hulton-Deutsch Collection (tr); Ralph White (crb) (cb) (br). **Henning Pfeifer:** (clb). **Rex Features:** ITV (ITV/TPC) (c). **Science Photo Library:** NOAA. **198 Alamy Images:** Bryan & Cherry Alexander Photography (fbl) (br). **naturepl.com:** Bryan and Cherry Alexander (r). **NOAA:** Michael Van Woert, NOAA NESDIS, ORA (bl). Courtesy of **Don Perovich:** (fbr). **199 Corbis:** Bettmann (cra). **DK Images:** Harry Taylor (crb). **NOAA:** Michael Van Woert, NOAA NESDIS, ORA (bc) (br). **SeaPics.com:** John KB Ford/Ursus (c). **200 NASA:** Jacques Descloitres, MODIS Land Rapid Response Team, NASA/GSFC (cl). **200–201 Alamy Images:** Brandon Cole Marine Photography (b). **201 Alamy Images:** Popperfoto (br). **Corbis:** Paul A. Souders (tc). **SeaPics.com:** Bryan & Cherry Alexander (c); iq3-d/Peter Parks (cra). **202–203 Dreamstime.com:** Irochka. **204–205 naturepl.com:** Alex Mustard. **206 Alamy Images:** Bruce Coleman/Tom Brakefield (cl/*Kingdom*); Norma Jospeh (c); Visual&Written SL/Kike Calvo (cl/*Genus*). **Corbis:** Brandon D. Cole (b/*Hagfish*). **DK Images:** (cl/*Domain*); Martin Camm (cl/*Phyllum*) (cl/*Order*) (cl/*Family*) (cl/*Species*); Geoff Dann (b/*Lamprey*) (b/*Ray-Finned Fish*); Frank Greenaway (b/*Cartilaginous Fish*); David Peart (cl/*Class*). **SeaPics.com:** Mark V. Erdmann (b/*Lobe-Finned Fish*). **207 DK Images:** (*Fungi*); Neil Fletcher (*Plants*); Dave King (*Red Seaweeds*); Jane Miller (*Animals*); Karl Shone (*Brown Seaweeds*). **SeaPics.com:** iq3-d/Peter Parks (*Protists*). **208 DK Images:** (*Echinoderms*); Frank Greenaway (*Molluscs*); Dave King (*Arthropods*). **209 DK Images:** Jerry Young (*Chordates*). **210 FLPA:** Minden Pictures/Chris Newbert. **211 Conservation International:** Robert Thacker (ca); Jeffrey T. Williams/Smithsonian Institution (c). **FLPA:** Minden Pictures/Norbert Wu (crb). **Dr J. Frederick Grassle, Rutgers University:** (br). **Sue Scott:** (tr) (cra). **SeaPics.com:** Phillip Colla (cb). **212 naturepl.com:** Steven Kazlowski (br) **213 Getty Images:** Moment/Brent Durand (tr). **iStockphoto.com:** Dan Schmitt (cla). **SeaPics.com:** Doug Perrine (b). **214 DeepSea-Photography.Com:** Kim Westerskov (clb). **FLPA:** D. P. Wilson (cla). **OSF/photolibrary:** Mark Jones (br). **Sue Scott:** (cra) (cr) (bc). **215 M. Boyer/edge-of-reef.com:** (bc). **OSF/photolibrary:** Richard Herrmann (t). **SeaPics.com:** Masa Ushioda (br). **216 iStockphoto.com:** Dan Schmitt (bl). **Sue Scott:** (tr) (cl) (ca). **216–217 SeaPics.com:** Espen Rekdal (b). **217 Dive Gallery/Jeffrey Jeffords (www.divegallery.com):** (br). **Marine Wildlife:** Paul Kay (ca) (cr). **Sue Scott:** (tr). **218 Dreamstime.com:** Donyanedomam (bl). **FLPA:** Linda Lewis (cra). **Getty Images/iStock:** Alberto Carrera (br). **218–219 FLPA:** Minden Pictures/Norbert Wu (c). **219 Alamy Images:** Danita Delimont (bl). **220 Alamy Images:** SCPhotos/Tom & Pat Leeson (bl); Bruce Coleman/Patrice Ricard (fbr). **iStockphoto.com:** Steffen Foerster (cl). **naturepl.com:** Pete Oxford (cr). **SeaPics.com:** Mark Conlin (clb); Chris Huss (fbl) (br). **221 Getty Images:** National Geographic/Brian J. Skerry (t). **OSF/photolibrary:** Doug Allan (bl). **222 Alamy Images:** Brandon Cole Marine Photography (bc). **FLPA:** Minden Pictures/Norbert Wu (c). **223 ExploreTheAbyss.Com:** Peter Batson (ca) (bl). **Charles G. Messing/Nova Southeastern University, Florida:** (br). **NOAA:** OAR/National Undersea Research Program (NURP); Univ. of Hawaii (cra). **OSF/photolibrary:** Norbert Wu (cla). **224 DK Images:** (bl). **Image Quest Marine:** (c); Y. Kito (br). **224–225 ExploreTheAbyss.Com:** Peter Batson (t). **225 ExploreTheAbyss.Com:** Peter Batson (cl). **OSF/photolibrary:** (br). **SeaPics.com:** iq3d/Peter Parks (cr). **226 DK Images:** Colin Keates (bc) (bl). **Science Photo Library:** Ria Novosti (cra); Sinclair Stammers (cl). **226–227 FLPA:** Minden Pictures/Fred Bavendam (b). **227 The Academy of Natural Sciences:** Ted Daeschler (cl). **DK Images:** Colin Keates (cla) (tr) (cra); Harry Taylor/Courtesy of the Royal Museum of Scotland, Edinburgh (c). **SeaPics.com:** Doug Perrine (tl). **228 Alamy Images:** Natural Visions/Heather Angel (tc). **Bridgeman Art Library:** Private Collection (cra). **DK Images:** Harry Taylor/Courtesy of the Hunterian Museum (University of Glasgow) (c); Harry Taylor/Courtesy of the Natural History Museum, London (bc). **OSF/photolibrary:** Karen Gowlett-Holmes (br). **Science Photo Library:** David Parker (clb). **229 Alamy Images:** David Fleetham (tc); Stephen Frink Collection/James D. Watt (cra). **DK Images:** (c/*Terrestrial Mammal*) (c/*Jawless Fish*); Bedrock Studios (c/*Armoured Fish*) (c/*Turtle*); Robin Carter (c/*Placodont*); Neil Fletcher (c/*Penguin*); Giuliano Fornari (c/*Ichthyosaurus*) (c/*Plesiosaur*); Jon Hughes (c/*Whale*); Colin Keates (bl) (c/*Cambrian*) (c/*Ediacaran*) (c/*Ammonite*); Harry Taylor/Courtesy of the Natural History Museum, London (c/*Shark*); Harry Taylor/Courtesy of the Royal Museum of Scotland, Edinburgh (c/*Lobe-Finned*). **Getty Images:** Science Faction/G. Brad Lewis (br). **230–231 Getty Images:** Taxi/Gary Bell. **232 MicroScope/Woods Hole:** D. J. Patterson (cb). **NOAA:** OAR/National Undersea Research Program (NURP); Lousiana Univ. Marine Consortium (cr). **Oceanwide Images:** Gary Bell (b). **OSF/photolibrary:** Phototake Inc/Dennis Kunkel (ca). **University of Illinois at Urbana-Champaign:** (cra). **233 DK Images:** M.I. Walker (tc). **Image Quest Marine:** Peter Parks (cl). **Oceanwide Images:** Rudie Kuiter (cr). **Laura K. Sycuro, Fred Hutchinson Cancer Research Center, Seattle:** (br). **234 Alamy Stock Photo:** Scenics & Science (clb). **Science Photo Library:** Wim Van Egmond (bl). **234–235 Science Photo Library:** Mint Images/Frans Lanting (c). **235 Science Photo Library:** Wim Van Egmond (bc). **238 Algaebase.org:** M.D. Guiry (cr) (br); John Huisman (cl). **Sue Scott:** (bl). **239 Alamy Images:** Bob Gibbons (t). **Algaebase.org:** M.D. Guiry (bl). **Natural Visions:** Heather Angel (br). **Sue Scott:** (cr). **240–241 Alamy Stock Photo:** Blue Planet Archive. **242 Sue Scott:** (cra). **242–243 Image Quest Marine:** Roger Steene (c). **243 Rob Houston:** (bc). **Sue Scott:** (tc) (cr). **244 naturepl.com:** Sue Daly (cr). **245 Alamy Stock Photo:** Premaphotos (crb). **246 Alamy Stock Photo:** Minden Pictures (cr). **Algaebase.org:** Coastal Imageworks/Colin Bates (bc); M.D. Guiry (tr). **Sue Scott:** (cl) (cr) (br). **247 Alamy Images:** Olivier Digoit (cra); Sami Sarkis (br); Kevin Schafer (tl). **Algaebase.org:** John Huisman (clb). **SeaPics.com:** Andrew J. Martinez (crb). **248 Natural Visions:** Heather Angel (cra) (bc). **Charles J. OKelly:** (tc). **Science Photo Library:** Alexis Rosenfeld (c). **249 Natural Visions:** Heather Angel (cl). **Jonathan Sleath:** (bl) (crb); Dr David Holyoak (cra). **250 Alamy Images:** Andrew Woodley (cl). **Alamy Stock Photo:** Nature Photographers Ltd (cr). **Sue Scott:** (bl) (br). **SeaPics.com:** Jeremy Stafford-Deitsch (ca). **251 Alamy Images:** Nature Picture Library/Jose B. Ruiz (cl); Wildscape/Jason Smalley (cr). **Sue Scott:** (cla) (b). **252 Alamy Images:** Roger Eritja (tr); Marilyn Shenton (tl). **OSF/photolibrary:** Kathie Atkinson (br). **Sue Scott:** (bl). **US Geological Survey:** Forest & Kim Starr (cl). **253 Corbis:** FLPA/Peter Reynolds (tc). **DK Images:** Richard Watson (bl). **FLPA:** Minden Pictures/Tui De Roy (tl). **OSF/photolibrary:** (br). **254 Getty Images:** Lonely Planet Images/Grant Dixon (r). **Natural Visions:** Heather Angel (bc). **OSF/photolibrary:** Phototake Inc. (cl). **255 MicroScope/Woods Hole:** David Patterson & Aimlee Laderman (br). **Einar Timdal/University of Oslo:** (cl) (tr) (cb) (bl). **256 Alamy Images:** Brandon Cole Marine Photography (ca); Robert Fried (cla). **FLPA:** Minden Pictures/Fred Bavendam (bc). **Andy Murch/Elasmodiver.com:** (br). **OSF/photolibrary:** (clb). **SeaPics.com:** Mark Conlin (crb); Doug Perrine (tr). **257 Alamy Images:** Dave and Sigrun Tollerton (cla). **FLPA:** Minden Pictures/Birgitte Wilms (b). **SeaPics.com:** Phillip Colla (tc). **258 Dr. Frances Dipper:** (fcl) (cl) (br). **Natural Resources Canda:** The Sponge Reef Project (clb). **NOAA:** (cr). **259 Alamy Images:** Wolfgang Pölzer (bl). **Dr. Frances Dipper:** (tl) (tr) (ca) (cb). **Dreamstime.com:** Nanisub (c). **University of Florida:** (br). **260 Alamy Images:** Tribaleye Images/J. Marshall (ca). **OSF/photolibrary:** Pacific Stock/David Fleetham (bl). **SeaPics.com:** Mark Conlin (c); Doug Perrine (cla). **260–261 Getty Images:** National Geographic/Paul Nicklen (b). **261 Dr. Frances Dipper:** (bc). **DK Images:** Frank Greenaway (cla). **OSF/photolibrary:** (cr). **SeaPics.com:** David Wrobel (ca). **262 Alamy Images:** Reinhard Dirscherl (crb). **Dr. Frances Dipper:** (cb). **SeaPics.com:** iq3-d/Chris Parks (t); David Wrobel (bc). **Shutterstock.com:** RLS Photo (bl). **263 Richard L. Lord. 264 Alamy Stock Photo:** Visual&Written SL (cra). **Marine Wildlife:** Paul Kay (br). **SeaPics.com:** Jeremy Stafford-Deitsch (cl); Steven Wolper (bc). **265 Alamy Images:** Andre Seale (clb). **Dr. Frances Dipper:** (br). **NOAA:** Mr. Mohammed Al Momany, Aqaba, Jordan (t). **SeaPics.com:** Doug Perrine (bc). **266 Alamy Images:** Michael Patrick O'Neill (cr) (bl); Wolfgang Pölzer (tc). **Dr. Frances Dipper:** (crb). **Sue Scott:** (c). **267 Sue Scott:** (cl) (tc) (c) (br). **SeaPics.com:** Franco Banfi (cr). **268 Alamy Images:** Aqua Image (cl). **OSF/photolibrary:** Tobias Bernhard (b). **SeaPics.com:** Masa Ushioda (cra). **Dr. Charlie Veron/Australian Institute of Marine Science:** Photo by Mary Stafford-Smith (c). **269 Alamy Images:** Mark Morgan (cl). **Dr. Frances Dipper:** (cra). **naturepl.com:** Scotland: The Big Picture (crb); Solvin Zankl / GEOMAR (br). **SeaPics.com:** Doug Perrine (tr). **270 Alamy Images:** Nature Picture Library/Jose B. Ruiz (b). **Dr. Frances Dipper:** (ca). **Sue Scott:** (t). **SeaPics.com:** Doug Perrine (cr). **271 Alamy Images:** FLPA (c). **M. Boyer/edge-of-reef.com:** (cr) (bl). **OSF/photolibrary:** Tobias Bernhard (tr). **Sue Scott:** (br). **272 M. Boyer/edge-of-reef.com:** (cl) (cla) (bl). Courtesy of **John J. Holleman:** (crb). **Michael D. Miller:** (tr) (cr). **273 Image Quest Marine:** Peter Parks (cr); Roger Steene (cl). **Marine Wildlife:** Paul Kay (bl). **Kåre Telnes/Seawater.no:** (br). **274 Corbis:** Lawson Wood (tr). **Keith Hiscock:** (cb). **Image Quest Marine:** Roger Steene (ca). **Marine Wildlife:** Paul Kay (crb). **SeaPics.com:** Larry Madrigal (cra). **275 Dr. Frances Dipper:** (cr). **Dive Gallery/Jeffrey Jeffords (www.divegallery.com):** (c). **DK Images:** Steve Gorton (cla). **ExploreTheAbyss.Com:** Peter Batson (br). **Dr. Dieter Fiege:** (tc). **Image Quest Marine:** Jim Greenfield (bl). **276 DK Images:** Matthew Ward (c) (clb). **SeaPics.com:** Clay Bryce (ca). **276–277 Getty Images:** Image Bank/Mike Severns (c). **277 Alamy Images:** Robert Harding Picture Library Ltd/Sylvain Grandadam (tr). **Dive Gallery/Jeffrey Jeffords (www.divegallery.com):** (cr) (br). **DK Images:** Andreas Von Einsiedel (cra). **OSF/photolibrary:** Karen Gowlett-Holmes (cla) (ca). **SeaPics.com:** Doc White (c). **278 Alamy Images:** Daniel L. Geiger/SNAP (tc); Wildscape/Jason Smalley (tr). **SeaPics.com:** Mark Strickland (b). **279 Alamy Images:** Nature Picture Library/Jose B. Ruiz (ca). **Keith Hiscock:** (tr). **Image Quest Marine:** Peter Parks (cra); Scott Tuason (cr). **SeaPics.com:** John C. Lewis (tl). **Alamy Images:** age fotostock (cla). **280 Marine Wildlife:** Paul Kay (t). **SeaPics.com:** Marilyn & Maris Kazmers (bc); Espen Rekdal (cl). **281 DK Images:** Andreas von Einsiedel (ca). **FLPA:** Minden Pictures/AUSCA/D. Parer & E. Parer-Cook (crb). **Jon Moore/Coastal Assessment Liaison & Monitoring, Pembroke:** (tl). **SeaPics.com:** D. R. Schrichte (b). **282–283 Getty Images:** Taxi/Pete Atkinson. **284 Alamy Images:** Natural Visions/Heather Angel (cl). **DK Images:** Dave King (tc); Frank Greenaway/Courtesy of the Natural History Museum, London (bc). **FLPA:** Minden Pictures/Norbert Wu (cra). **Image Quest Marine:** Roger Steene (crb) (br). **Alamy Images:** Juniors Bildarchiv GmbH (bc). **285 Alamy Images:** Liquid-Light Underwater Photography (br). **Oceanwide Images:** Gary Bell (t). **SeaPics.com:** James D. Watt (clb). **286 Alamy Images:** Carol Buchanan (tl). **Dive Gallery/Jeffrey Jeffords (www.divegallery.com):** (c). **Marine Wildlife:** Lucy Kay (cra). **NOAA:** National Estuarine Research Reserve Collection (br). **287 Alamy Images:** Andre Seale (br). **FLPA:** Minden Pictures/Norbert Wu (t). **288 Corbis:** Jeffrey L. Rotman (b). **Image Quest Marine:** Peter Batson (ca). **Oceanwide Images:** Gary Bell (cra). **SeaPics.com:** Doug Perrine (tl); Jeff Rotman (crb). **289 Alamy Images:** f1 online (bl). **Image Quest Marine:** Peter Batson (bc). **Oceanwide Images:** Gary Bell (t). **SeaPics.com:** Marc Chamberlain (br). **290 Alamy Images:** Daniel L. Geiger/SNAP (cla). **DK Images:** Colin Keates/Courtesy of the Natural History Museum, London (cra). **NHPA:** Ken Griffiths (bl). **OSF/photolibrary:** Barrie Watts (clb). **SeaPics.com:** Espen Rekdal (tr). **291 Dive Gallery/Jeffrey Jeffords (www.divegallery.com):** (b). **Image Quest Marine:** Jez Tryner (cla) (ca) (cra). **NOAA:** Jamie Hall (r). **292 DK Images:** Frank Greenaway/Courtesy of the Natural History Museum, London (cra). **Image Quest Marine:** Roger Steene (cl). **naturepl.com:** Fred Bavendam (tr); Christophe Courteau (cl). **NOAA:** Dr. Bradley Stevens (crb). **293 FLPA:** Minden Pictures/Fred Bavendam (r). **NOAA:** Hopcroft (bl). **SeaPics.com:** Franco Banfi (cl). **294 Alamy Stock Photo:** Jeff Rotman (bl). **iStockphoto.com:** Ian Campbell (br). **SeaPics.com:** Marli Wakeling (tr). **295 Alamy Stock Photo:** Minden Pictures (br). **Dive Gallery/Jeffrey Jeffords (www.divegallery.com):** (cr). **Ifremer (www.ifremer.fr):** A. Le Magueresse (clb). **Natural Visions:** Heather Angel (cl). **296 Dive Gallery/Jeffrey Jeffords (www.divegallery.com):** (b). **DK Images:** Frank Greenaway (cr). **OSF/photolibrary:** (cl). **Still Pictures:** Lynda Richardson (tr). **297 DK Images:** Jane Burton (c); Andreas von Einsiedel (bc); Dave King (cra). **Image Quest Marine:** Masa Ushioda (crb). **David Kusner:** (tc). **OSF/photolibrary:** Green Cape Pty Ltd (bl). **298–299 Steve Smithson. 300 DK Images:** Jane Burton (bl). **Marine Wildlife:** Paul Kay (cl) (br). **SeaPics.com:** Masa Ushioda (cr). **301 Alamy Images:** Maximilian Weinzierl (tr). **DK Images:** Kim Taylor & Jane Burton (cr). **Oceanwide Images:** Gary Bell (tl). **SeaPics.com:** David B. Fleetham (b). **302 naturepl.com:** Jurgen Freund. **303 Corbis:** Roger Garwood & Trish Ainslie (tr). **SeaPics.com:** Ralf Kiefner (c) (crb). **304 Laurent Dabouineau/University U.C.O. Bretagne Nord, France:** (cl). **FLPA:** Foto Natura/Jef Meul (ca). **Nature Portfolio (www.natureportfolio.co.uk):** Bob Ford (br). **305 Alamy Images:** SNAP/Daniel L. Geiger (cl). **Karen Gowlett-Holmes:** (crb). **Marine Wildlife:** Lucy Kay (cra). **Sue Scott:** (bl) (bc). **Kåre Telnes/Seawater.no:** (tr). **306 SeaPics.com:** Phillip Colla (cl). **306–307 Alamy Stock Photo:** Brandon Cole Marine Photography (c). **307 Alamy Images:** David Fleetham (tr). **Corbis:** FLPA/Douglas P. Wilson (tc). **Sue Scott:** (br). **SeaPics.com:** Marli Wakeling (bl). **Still Pictures:** P. Danna (cr). **308 Alamy Images:** David Fleetham (clb). **M. Boyer/edge-of-reef.com:** (tr). **Marine Wildlife:** Paul Kay (crb). **Oceanwide Images:** Gary Bell (br). **Sue Scott:** (cla). **SeaPics.com:** D.R. Schrichte (bl). **309 Dive Gallery/Jeffrey Jeffords (www.divegallery.com):** (t). **OSF/photolibrary:** Tobias Bernhard (bl). **Sue Scott:** (br). **310 M. Boyer/edge-of-reef.com:** (tr) (cl). **Marine Wildlife:** Paul Kay (cr). **SeaPics.com:** David Wrobel (bl). **311 Alamy Stock Photo:** Minden Pictures (tl). **Dr. Frances Dipper:** (tr). **DK Images:** Colin Keates (br). **NOAA:** (bl). **312 Alamy Images:** F. Jack Jackson (clb). **Australian Institute of Marine Science:** (t). **Dr. Jacob Dafni:** Dr. A. Diamant (bc). **Image Quest Marine:** Peter Herring (br); Roger Steene (cra). **Corbis:** Visuals Unlimited / David Wrobel (br). **313 Alamy Images:** Lawrence Stepanowicz (cra). **ExploreTheAbyss.Com:** Peter Batson (br). **OSF/photolibrary:** Stephen Foote (bl). **SeaPics.com:** Andrew J. Martinez (cl). **Peter Funch, University of Aarhus:** (crb). **Science Photo Library:** Andrew J. Martinez (br). **314 Corbis:** Lawson Wood (clb). **FLPA:** D. P. Wilson (c). **naturepl.com:** David Shale (cra). **www.uwphoto.no:** Erling Svensen (tl) (cl) (br). **315 ExploreTheAbyss.Com:** Peter Batson (ca). **NOAA:** OAR/National Undersea Research Program (NURP); College of William & Mary (tr). **SeaPics.com:** Scott Leslie (b). **Getty Images:** Visuals Unlimited / Richard Herrmann (tr). **316 FLPA:** Foto Natura/Jan Van Arkel (t). **Peter Funch, University of Aarhus:** (bl). **M. Antonio Todaro, University of Modena e Reggio Emilia:** (br). **FLPA:** Minden/FN/Jan Van Arkel (cla). **317 Alamy Images:** David Fleetham (br). **Alamy Stock Photo:** Science History Images (cl). **M. Boyer/edge-of-reef.com:** (c). **ExploreTheAbyss.Com:** Peter Batson (tr). **Image Quest Marine:** Peter Parks (tc). **Lyubomir Klissurov:** (bl). **318 M. Boyer/edge-of-reef.com:** (c). **ExploreTheAbyss.Com:** Peter Batson (cl). **Sue Scott:** (bc). **SeaPics.com:** Espen Rekdal (br). **319 M. Boyer/edge-of-reef.com:** (cla). **Dr. Frances Dipper:** (cra). **Photo by Per R. Flood © Bathybiologica.no:** (bc). **Natural Visions:** Heather Angel (br). **Sue Scott:** (ca).

SeaPics.com: David Wrobel (clb). **320 Alamy Stock Photo:** Buiten-Beeld (tr); Pally (bl). **Corbis:** Brandon D. Cole (crb). **SeaPics.com:** Jonathan Bird (br). **321 Corbis:** Brandon D. Cole (cr). **OSF/photolibrary:** (tr); Zig Leszczynski (cla). **www.uwphoto.no:** Erling Svensen (bl). **322 Alamy Images:** Brandon Cole Marine Photography (cra). **DK Images:** Frank Greenaway (cl); James Stevenson (ca). **322–323 Getty Images:** Dmitry Miroshnikov (bc). **323 DK Images:** Frank Greenaway (br); Dave King (tc). **SeaPics.com:** Doug Perrine (cl) (bc). **324 Janna Nichols:** (clb). **SeaPics.com:** Doug Perrine (tl) (ca). **www.uwphoto.no:** Erling Svensen (br). **325 Alamy Stock Photo:** Kelvin Aitken/VWPics (crb). **imagequestmarine.com:** Kelvin Aitken/V&W (cl). **Andy Murch/Elasmodiver.com:** (cra). **OSF/photolibrary:** Paul Kay (bl). **326 Alamy Images:** Stephen Frink Collection/Marty Snyderman (br). **OSF/photolibrary:** Gerard Soury (bl). **SeaPics.com:** Saul Gonor (cra); Espen Rekdal (cla). **327 Alan Chow:** (bc). **Dr. Frances Dipper:** (br). **DK Images:** Frank Greenaway (bl). **Andy Murch/Elasmodiver.com:** (t). **328 Andy Murch/Elasmodiver.com:** (t) (br) (bc). **naturepl.com:** Bruce Rasner/Jeff Rotman (cl). **329 Alamy Images:** Jeff Rotman (cla). **DK Images:** Harry Taylor/Courtesy of the Natural History Museum, London (cl). **John A. Scarlett:** (tr). **SeaPics.com:** Scott Michael (br); David Shen (clb). **330–331 Steve Bloom Images. OceanwideImages.com:** C & M Fallows. **332 OSF/photolibrary:** Pacific Stock (br). **Powder River Photography/Todd Mintz:** (cl). **SeaPics.com:** Richard Herrmann (tl). **333 Alamy Images:** Michael Patrick O'Neill (bl). **Alamy Stock Photo:** Juniors Bildarchiv GmbH (tl). **Ardea:** Paulo di Oliviera (cr). **SeaPics.com:** Doug Perrine (br). **334 Andy Murch/Elasmodiver.com:** (br). **SeaPics.com:** Randy Morse (cl). **334–335 Marine Wildlife:** Alexander Mustard (t). **335 Alamy Images:** M. Timothy O'Keefe (c). **D Images:** Frank Greenaway (tl). **Image Quest Marine:** Carlos Villoch (br). **SeaPics.com:** Doug Perrine (clb); Tim Rock (br). **336 Alamy Images:** WorldFoto (cra). **DK Images:** Colin Keates/Courtesy of the Natural History Museum, London (clb). **336–337 Alamy Images:** Reinhard Dirscherl (c). **337 Alamy Images:** Mark Boulton (br). **Dive Gallery/Jeffrey Jeffords (www.divegallery.com):** (cb). **OSF/photolibrary:** David Fleetham (tr). **338 Alamy Images:** Reinhard Dirscherl (bl); Stephen Frink Collection (cl); Images&Stories (tr). **DK Images:** Frank Greenaway (cla). **Naoko Kouchi:** (cla/*Background*). **SeaPics.com:** Doug Perrine (br). **339 Alamy Images:** Blickwinkel (br). **FLPA:** Minden Pictures/Fred Bavendam (cl). **OSF/photolibrary:** Dr. F. Ehrenstrom & L. Beyer (ca) (br). **340 DK Images:** Steve Gorton (clb); Colin Keates/Courtesy of the Natural History Museum, London (tc). **Getty Images:** Taxi/Peter Scoones (cl). **Image Quest Marine:** Masa Ushioda (crb). **SeaPics.com:** Mark V. Erdmann (cr). **Andreas Svensson/Norwegian University of Science and Technology:** (bl). **341 Marine Wildlife:** Alexander Mustard (t). **Robert A. Patzner, University of Salzburg, Austria:** (br). **342 Ardea:** Pat Morris (cb). **Rick J. Coleman:** (bl). **DK Images:** (cla). **Marine Wildlife:** Alexander Mustard (tr). **Dr. Volker Neumann:** (br). **343 DK Images:** (clb). **FLPA:** Minden Pictures/Norbert Wu (crb). **Marine Wildlife:** Alexander Mustard (tl). **344 Corbis:** Paul A. Souders (bl). **naturepl.com:** Dan Burton (tl). **SeaPics.com:** Mark Conlin (c); Jeff Jaskolski (br). **345 Alamy Images:** Andre Seale (tl). **Ardea:** Paul Souders / Danita Delimont (bl). **BluePlanetArchive.com:** David Wrobel (crb). **SeaPics.com:** Shedd Aquar/Ceisel (tr). **346 Alamy Images:** Wolfgang Pölzer (b). **Getty Images:** National Geographic/Paul Nicklen (t). **347 FLPA:** Minden Pictures/Norbert Wu (cla). **Getty Images:** National Geographic/Wolcott Henry (bc). **Image Quest Marine:** Peter Herring (tr). **OSF/photolibrary:** Paulo De Oliveira (bl). **348 Ardea:** Paulo de Oliveira (ca). **DK Images:** Frank Greenaway (cr). **Keith Hiscock:** (cra). **OSF/photolibrary:** Doug Allan (b). **349 Peter Ajtai:** (clb). **SeaPics.com:** Marilyn & Maris Kazmers (br). **www.uwphoto.no:** Erling Svensen (cla) (tc). **350 Alamy Stock Photo:** Nature Picture Library (tr). **Image Quest Marine:** Justin Marshall (br). **SeaPics.com:** James D. Watt (clb). **351 marinethemes.com:** Kelvin Aitken (tc). **Natural Visions:** Peter David (clb). **OSF/photolibrary:** Neil Bromhall (br); Rodger Jackman (ca). **Science Photo Library:** Dante Fenolio (cr). **FLPA:** Biosphoto/Bruno Guenard (clb). **352 Dr. Frances Dipper:** (clb). **FLPA:** D. P. Wilson (cla). **Oceanwide Images:** Gary Bell (bc); Rudie Kuiter (crb). **OSF/photolibrary:** Richard Herrmann (tr). **Alamy Images:** WaterFrame (ca). **353 Alamy Stock Photo:** Pally (br). **DK Images:** Dave King (br). **New Zealand Seafood Industry Council Ltd:** (cr). **354 Magnum Photos:** Harry Gruyaert. **355 Alamy Images:** Charles Bowman (bl); Jeff Rotman (crb). **FLPA:** Minden Pictures/Norbert Wu (ca). **naturepl.com:** Michael Pitts (cr). **OSF/photolibrary:** Sue Scott (cb). **Sue Scott:** (cra). **SeaPics.com:** Richard Herrmann (br). **Brian Skerry:** (tc). **356 Alamy Stock Photo:** Buiten-Beeld (br). **NHPA:** A.N.T. Photo Library (t). **Robert A. Patzner, University of Salzburg, Austria:** (cb). **357 Alamy Images:** Papilio/Steve Jones (br). **Dive Gallery/Jeffrey Jeffords (www.divegallery.com):** (clb). **DK Images:** Jane Burton (tl). **SeaPics.com:** Doug Perrine (cra). **358 Alamy Images:** Blickwinkel (bl); Reinhard Dirscherl (br). **Dr. Frances Dipper:** (cra) **Marine Wildlife:** Paul Kay (tc). **SeaPics.com:** Doug Perrine (ca). **359 Dr. Frances Dipper:** (tr). **Dive Gallery/Jeffrey Jeffords:** (www.divegallery.com) (tl). **DK Images:** Jerry Young (br). **Robert A. Patzner, University of Salzburg, Austria:** (clb). **SeaPics.com:** V&W/Hal Beral (crb). **360 DK Images:** Jerry Young (c). **Marine Wildlife:** Paul Kay (tc). **OSF/photo library:** Doug Allan (br). **SeaPics.com:** Jonathan Bird (cl); Jeremy Stafford-Deitsch (cra). **361 OSF/photolibrary:** David Fleetham (cla). **DK Images:** Geoff Dann (cra). **Alamy Images:** Wolfgang Pölzer (bl). **OSF/photolibrary:** Richard Herrmann (br). **362–363 Alamy Stock Photo:** Steve Bloom Images. **364 DK Images:** Frank Greenaway (cla). **SeaPics.com:** Masa Ushioda (tr). **Alamy Images:** Reinhard Dirscherl (bl). **DK Images:** Colin Keates/Courtesy of the Natural History Museum, London (br). **OSF/photolibrary:** Pacific Stock (cra). **365 Dive Gallery/Jeffrey Jeffords:** (www.divegallery.com) (c). **Alamy Images:** Blickwinkel (bl); Wolfgang Pölzer (tc) (br). **OSF/photolibrary:** Pacific Stock (bc). **Dr. Frances Dipper:** (cla). **366 Alamy Images:** Reinhard Dirscherl (tc); Sami Sarkis (c). **Dive Gallery/Jeffrey Jeffords (www.divegallery.com):** (crb). **Marine Wildlife:** Alexander Mustard (cra). **SeaPics.com:** Doug Perrine (bl). **367 DK Images:** Dave King (bl). **OSF/photolibrary:** Richard Herrmann (br). **SeaPics.com:** Jez Tryner (tl). **368 Corbis:** Staffan Widstrand (cra). **DK Images:** Dave King (ca). **FLPA:** Minden Pictures/J. H. Editorial/Cyril Ruoso (cl); Minden Pictures/Tui De Roy (crb). **Oceanwide Images:** Gary Bell (bc). **369 Getty Images:** Image Bank/Tobias Bernhard (b). **OSF/photolibrary:** Olivier Grunewald (cla). **370 Getty Images:** Image Bank/Pete Atkinson (t). Brook Mathews, Sydney: (br). **Ilan Ben Tov, Israel:** (bl). **371 FLPA:** Peter Reynolds (cra); S. A. Team/Foto Natura (tc). **Getty Images:** National Geographic/Bill Curtsinger (bl). **SeaPics.com:** Doug Perrine (br). **372–373 Harald Slauschek/UnderwaterVisions.net. 374 Dick Bartlett:** (cra). **naturepl.com:** Constantinos Petrinos (l). **Queensland Museum, Australia (www.Qmuseum.qld.gov.au):** (br). **375 Getty Images:** Taxi/Gary Bell (r). **SeaPics.com:** Gary Bell (cla); Steve Drogin (clb). **376 Alamy Images:** Pep Roig (t). **Dreamstime.com:** Kseniya Ragozina (br). **Kraig Haver Photography:** (bl). **377 FLPA:** Minden Pictures/Mike Parry (c). **Adam Slavický:** (bl). **Scott Solar/Amazon Reptile Center:** (tl). **Dr. Adam P. Summers:** (br). **Frank Bambang Yuwono:** (tr). **378 Alamy Stock Photo:** Jim Zuckerman (bc). **DK Images:** Ken Findlay (clb/*Albatross*); Chris Gomersall (clb/*Curlew*); Frank Greenaway/Courtesy of The National Birds of Prey Centre, Gloucestershire (clb/*Sea Eagle*); Rob Reichenfeld (clb/*Pelican*). **iStockphoto.com:** Hans F. Meier (cla). **OSF/photolibrary:** Survival Anglia (crb). **379 Alamy Images:** PhotoStockFile/Paul Wayne Wilson (bc). **OSF/photolibrary:** Doug Allan (t). **SeaPics.com:** Richard Herrmann (br). **380 Alamy Images:** Petr Svarc (cla); WorldFoto (cra). **DK Images:** (crb); Frank Greenaway (bl). **381 Alamy Images:** Malcolm Schuyl (br); David Tipling (bl). **DK Images:** Steve Gorton (clb). **Neil Fletcher:** Tomi Muukonen (cla). **Getty Images:** Elmar Weiss / 500px (cr). **Dr. Paul Hofmann:** (c). **382 Alamy Images:** Bryan & Cherry Alexander Photography (l). **DK Images:** Neil Fletcher (br). **383 Alamy Images:** Kim Westerskov (bl). **Neil Fletcher:** Barry Hughes (cra). **OSF/photolibrary:** Konrad Wothe (clb). **SeaPics.com:** Hiroya Minakuchi (ca); Kevin Schafer (bc). **384–385 FLPA:** Fritz Polking. **Getty Images:** National Geographic/Paul Nicklen. **386 Alamy Images:** INFOCUS Photos/Malie Rich-Griffith (bc). **naturepl.com:** Peter Reese (cr). **OSF/photolibrary:** Daniel Cox (tr) (tl). **387 Alamy Images:** ImageState/Pete Oxford (bl). **Neil Fletcher:** Hanne & Jens Eriksen (tr). **388 Alamy Images:** George McCallum Photography (ca); INFOCUS Photos/Malie Rich-Griffith (cb). **Neil Fletcher:** Hanne & Jens Eriksen (bl); Jonathan Grey (cla); Just Birds (br). **389 Alamy Images:** Nature Photographers Ltd/Paul Sterry (crb). **Neil Fletcher:** George Reszeter (t). **SeaPics.com:** Doug Perrine (crb). **390 Alamy Images:** Barry Bland (tr); Chris Mercer (cl). **DK Images:** Kim Taylor (fcl). **Neil Fletcher:** Just Birds (clb). **SeaPics.com:** Robert Shallenberger (br). **391 Alamy Images:** Robert E. Barber (cr); f1 online/Pölzer (bc). **Neil Fletcher:** Joe Fuhrman (tl); Mike Read (tr). **SeaPics.com:** Phillip Colla (br). **392 Alamy Images:** Blickwinkel (b). **FLPA:** Minden Pictures/Tui De Roy (ca) (tr). **393 Alamy Images:** WoodyStock/Ingo Schulz (cra). **Dreamstime.com:** Brian Kushner (br). **Neil Fletcher:** Ian Montgomery (tc); Mike Read (cla). **394 Alamy Images:** Mike Lane (bl); PhotoStockFile/Paul Wayne Wilson (br). **DK Images:** Cyril Laubscher (cr); Frank Greenaway/Courtesy of the Natural History Museum, London (ca). **Neil Fletcher:** Barry Hughes (cla). **395 Alamy Images:** Bryan & Cherry Alexander Photography (cr); R. & M. Thomas (b). **Alamy Stock Photo:** BIOSPHOTO (cb). **Getty Images/iStock:** drferry (tr). **SeaPics.com:** Richard Herrmann (tl). **396 Alamy Images:** George McCallum Photography (br); The Photolibrary Wales (bl). **Neil Fletcher:** Just Birds (cl). **SeaPics.com:** Scott Leslie (tr). **397 Alamy Images:** Robert E. Barber (cb); Scott Camazine (tl). **DK Images:** Cyril Laubscher (br). **Neil Fletcher:** Joe Fuhrman (bl); George Reszeter (cra). **398 Alamy Images:** Blickwinkel (br). **DK Images:** Harry Taylor/Courtesy of the Natural History Museum, London (bc). **Neil Fletcher:** Dudley Edmonson (cra); Barry Hughes (tc). **Getty Images:** Image Bank/Roine Magnusson (c). **399 Alamy Images:** Kevin Schafer (bl). **DK Images:** Irv Beckman (crb). **OSF/photolibrary:** David Tipling (tc). **Still Pictures:** Mark Edwards (cra). **400 Alamy Images:** Brandon Cole Marine Photography (cl). **DK Images:** Philip Dowell (cr). **Marine Wildlife:** Doug Allan (bl). **Still Pictures:** Steven Kazlowski (clb). **400–401 OSF/photolibrary:** Mark Jones (c). **401 Alamy Images:** Steven J. Kazlowski (ca). **Alamy Stock Photo:** Jan Carroll (cr). **OSF/photolibrary:** Pacific Stock (br). **402 DK Images:** Jerry Young (tl). **FLPA:** Foto Natura/Wil Meinderts (br). **Getty Images/iStock:** Kara Capaldo (cb). **Howard Hall Productions:** (bl). **403 FLPA:** Minden Pictures/Tui De Roy (bc). **Getty Images:** Image Bank/Joseph Van Os (tr). **Brian Lockett (www.air-and-space.com):** (bl). **Marine Wildlife:** Paul Kay (cla). **NOAA:** Captain Budd Christman, NOAA Corps (crb). **404–405 Getty Images:** National Geographic / David Doubilet. **406 Alamy Stock Photo:** AGAMI Photo Agency (tr); Nature Picture Library (bl). **406–407 Steve Smithson. 408 Ardea:** Francois Gohier (tr). **Corbis:** The Mariners' Museum (cl). **FLPA:** Minden Pictures/Flip Nicklin (bl). **SeaPics.com:** Howard Hall (bc). **409 Alamy Images:** Brandon Cole Marine Photography (t); Stephen Frink Collection/James D. Watt (b). **410–411 naturepl.com:** Tony Wu. **412 DK Images:** Frank Greenaway (tr). **naturepl.com:** Doc White (ca). **SeaPics.com:** Doug Perrine (b). **413 Alamy Images:** Andre Seale (cla). **FLPA:** Minden Pictures/Flip Nicklin (crb). **Getty Images:** National Geographic/Brian Skerry (tr). **SeaPics.com:** John K. B. Ford/Ursus (bl). **414 FLPA:** Minden Pictures/Flip Nicklin (tr). **SeaPics.com:** Thomas Jefferson (cl); Robert L. Pitman (br). **Getty Images:** Gerard Soury (bc). **415 Alamy Images:** Stock Connection Blue/Tom Brakefield. **416 Marine Wildlife:** Sue Flood. **417 FLPA:** Minden Pictures/Michio Hoshino (ca); Minden Pictures/Flip Nicklin (cra); Minden Pictures/Norbert Wu (bc). **naturepl.com:** Todd Pusser (crb). **Mike Scott:** (cb). **SeaPics.com:** Phillip Colla (c). **418 Image Quest Marine:** Masa Ushioda (br). **naturepl.com:** Doug Perrine (tr). **SeaPics.com:** Florian Graner (cl). **Alamy Stock Photo:** imageBROKER (cl). **419 DK Images:** Peter Visscher (crb). **FLPA:** Minden Pictures/Chris Newbert (tr). **Getty Images:** Photographer's Choice/Pete Atkinson (b). **420–421 NASA:** Jacques Descloitres, MODIS Rapid Response Team, NASA/GSFC. **424 Alamy Images:** Bryan & Cherry Alexander Photography (cla); LOOK Die Bildagentur der Fotografen GmbH (cb). **426 Alamy Images:** Jack Stephens (cla). **Corbis:** Lowell Georgia (cra). **427 Alamy Images:** Nordicphotos/Kristjan Fridriksson (br). **428 Alamy Images:** Greenshoots Communications (c); David Sanger Photography (cla). **Corbis:** Ralph White (bl). **430 Alamy Images:** FLPA (br). **431 DK Images:** David Lyons (cl). **NASA:** Jacques Descloitres, MODIS Land Rapid Response Team, NASA/GSFC (ca). **432 Alamy Images:** Ace Stock Ltd (bc); allOver photography (clb). **DK Images:** Linda Whitwam (ca). **433 Alamy Images:** Nick Hanna (br). **434 Mads Eskesen. 435 Alamy Images:** Mike Lane (br). **EnBW Energie Baden-Württemberg AG:** (crb). **Mads Eskesen:** (tr) (cra). **Horns Rev Havmøllepark (www.hornsrev.dk):** Medvind Fotografi/Bent Sørensen (bc). **Nysted Havmøllepark (www.nystedhavmoellepark.dk):** (crb). **437 Alamy Images:** Wild Places Photography/Chris Howes (br). **Corbis:** Sygma/Bernard Annebicque (cr). **438 DK Images:** Christopher & Sally Gable (tl); John Heseltine (br). **439 Alamy Images:** Vehbi Koca (crb); Rob Rayworth (bl). **NASA:** Image courtesy NASA/GSFC/MITI/ERSDAC/JAROS, and U.S./Japan ASTER Science Team (c). **440 Corbis:** (tr) (bl); Sygma/Harford Chloe (cb). **NASA:** Jacques Descloitres, MODIS Land Rapid Response Team, NASA/GSFC (br). **442 Alamy Stock Photo:** Granger Historical Picture Archive (bl). **Corbis:** Cordaiy Photo Library Ltd/John Farmar (tr). **Science Photo Library:** Southampton Oceanography Centre/B. Murton (cla). **443 Alamy Images:** Wild Places Photography/Chris Howes (cra). **Corbis:** (ca). **444 FLPA:** Colin Monteath (br). **iStockphoto.com:** Patrick Roherty (bc). **445 Alamy Images:** Bryan & Cherry Alexander Photography (bl). **NASA:** Jesse Allen, NASA Earth Observatory and the HIGP Thermal Alerts Team (cr). **NOAA:** Lieutenant Philip Hall, NOAA Corps (br). **446 Alamy Stock Photo:** J. Pie (cl). **Corbis:** Yann Arthus-Bertrand (bl); Reuters/Supri (cla). **448 Alamy Images:** Blickwinkel (bc); Tor Eigeland (tr). **Corbis:** Jonathan Blair (cl). **449 Alamy Images:** Images of Africa Photobank/Peter Williams (tr). **450 Alamy Images:** Neil McAllister (br). **451 Alamy Images:** Julio Etchart (cra). **452–453 Photoshot:** Planet Observe. **455 iStockphoto.com:** Wesley Drake (cra). **OSF/photolibrary:** Michael Brooke (br). **456 Alamy Images:** Danita Delimont (cla). **Corbis:** Ralph White (bc). **NOAA:** Commander Richard Behn, NOAA Corps (cl). **458 NASA:** George Riggs, NASA GSFC (cra). **US Fish and Wildlife Service National Image Library:** Alaska Maritime National Wildlife Refuge/Kevin Bell (cla). **459 Corbis:** Neil Rabinowitz (tc). **460 Alamy Images:** FocusRussia (cra); Iain Masterton (cl). **Corbis:** Michael S. Yamashita (crb). **462–463 Corbis:** Nippon News/Aflo/Newspaper/Mainichi/ Mainichi. **463 Kevin Jaako:** www.jaako.com (cra/Signboard). Reuters: Yomiuri Yomiuri (cra). **464 Alamy Images:** Chris Willson (br). **465 Getty Images:** ullstein bild (cra). **467 Alamy Images:** Andre Seale (br). **Corbis:** Douglas Faulkner (cr); Reuters/Alex De La Rosa (tr). **468 Dreamstime.com:** Martinmark (cra). **469 Alamy Images:** Dennis Hallinan (bc). **470–471 Getty Images:** Tom Benedict. **472 Getty Images:** Ness Kerton/AFP (tc). **473 iStockphoto.com:** Angela Bell (cla). **SeaPics.com:** Gary Bell (tc). **474 Getty Images:** AFP/Tarik Tinazay. **475 Alamy Images:** Nick Hanna (bl); Images&Stories (cra). **Corbis:** Stephen Frink (br); Jeffrey L. Rotman (bc); Lawson Wood (crb). Getty Images: Image Bank/Zac Macaulay (tr). **OSF/photolibrary:** David Fleetham (ca). **477 Alamy Images:** Danita Delimont (cra); INTERFOTO Pressebildagentur (cb). **478 OSF/photolibrary:** Mike Hill (br). **479 iStockphoto.com:** Michal Wozniak (tr). **480 Alamy Images:** LOOK Die Bildagentur der Fotografen GmbH (cl); Bruce Percy (br). **Corbis:** Paul A. Souders (cra). **482 Alamy Images:** Bryan & Cherry Alexander (cl). **Alamy Stock Photo:** Minden Pictures (bc). **naturepl.com:** Norbert Wu (tl). **484 Corbis:** Eye Ubiquitous/C. M. Leask (cra). **NOAA:** Commander John Bortniak, NOAA Corps (cla). **485 Alamy Images:** Blickwinkel (bl); Kim Westerskov (br). **486 Ardea:** Edwin Mickleburgh. **Dreamstime.com:** Staphy. Back Endpapers: **Getty Images:** David Doubilet **487 Alamy Images:** Graphic Science (tr); Steve Morgan (ca). **NASA:** MODIS Land Science Team (cra); Jacques Descloitres, MODIS Land Science Team (br); NASA/GSFC/LaRC/JPL, MISR Team (crb). JACKET IMAGES: *Front:* **Dreamstime.com:** Digitalbalance; *Back:* **Dreamstime.com:** Digitalbalance; *Spine:* **Dreamstime.com:** Digitalbalance.

Data for the bathymetric maps in the Atlas of the Oceans chapter provided by Planetary Visions based on ETOPO2 global relief data, SRTM30 land elevation data, and the Generalised Bathymetric Chart of the Ocean. ETOPO2 published by the U.S. Department of Commerce, National Oceanic and Atmospheric Administration, National Geophysical Data Center, 2001. SRTM30 published by NASA and the National Geospatial Intelligence Agency, 2005, distributed by the U.S. Geological Survey. GEBCO One Minute Grid reproduced from the GEBCO Digital Atlas published by the British Oceanographic Data Centre on behalf of the Intergovernmental Oceanographic Commission of UNESCO and the International Hydrographic Organisation, 2003.

All other images © Dorling Kindersley
For further information see: www.dkimages.com

AMERICAN MUSEUM OF NATURAL HISTORY
www.amnh.org

The American Museum of Natural History in New York City is one of the largest and more respected museums in the world. Since its founding in 1869, the Museum has grown to encompass more than 40 permanent exhibition halls, including those in the Rose Center for Earth and Space and the Hayden Planetarium, as well as galleries for temporary exhibitions. The Museum's scientists draw on a world-class permanent collection of more than 34 million specimens and artifacts, some of which are billions of years old, and on one of the largest natural history libraries in the world. Through its Richard Gilder Graduate School, the Museum grants the Ph.D. degree in Comparative Biology and the Master of Arts in Teaching (MAT) degree, the only such freestanding, degree-granting program at any museum in the United States. Plan a trip to the Museum, home of the world's largest collection of dinosaur fossils, or visit online at www.amnh.org.